Lecture Notes in Networks and Systems

Volume 447

AF869683

The series "Lecture Notes in Networks and Systems" publishes the latest developments in Networks and Systems—quickly, informally and with high quality. Original research reported in proceedings and post-proceedings represents the core of LNNS.

Volumes published in LNNS embrace all aspects and subfields of, as well as new challenges in, Networks and Systems.

The series contains proceedings and edited volumes in systems and networks, spanning the areas of Cyber-Physical Systems, Autonomous Systems, Sensor Networks, Control Systems, Energy Systems, Automotive Systems, Biological Systems, Vehicular Networking and Connected Vehicles, Aerospace Systems, Automation, Manufacturing, Smart Grids, Nonlinear Systems, Power Systems, Robotics, Social Systems, Economic Systems and other. Of particular value to both the contributors and the readership are the short publication timeframe and the world-wide distribution and exposure which enable both a wide and rapid dissemination of research output.

The series covers the theory, applications, and perspectives on the state of the art and future developments relevant to systems and networks, decision making, control, complex processes and related areas, as embedded in the fields of interdisciplinary and applied sciences, engineering, computer science, physics, economics, social, and life sciences, as well as the paradigms and methodologies behind them.

Indexed by SCOPUS, INSPEC, WTI Frankfurt eG, zbMATH, SCImago.

All books published in the series are submitted for consideration in Web of Science.

For proposals from Asia please contact Aninda Bose (aninda.bose@springer.com).

More information about this series at https://link.springer.com/bookseries/15179

Xin-She Yang · Simon Sherratt · Nilanjan Dey ·
Amit Joshi
Editors

Proceedings of Seventh International Congress on Information and Communication Technology

ICICT 2022, London, Volume 1

Springer

Editors
Xin-She Yang
Middlesex University
London, UK

Simon Sherratt
The University of Reading
Reading, UK

Nilanjan Dey
JIS University
Kolkata, India

Amit Joshi
Global Knowledge Research Foundation
Ahmedabad, India

ISSN 2367-3370 ISSN 2367-3389 (electronic)
Lecture Notes in Networks and Systems
ISBN 978-981-19-1606-9 ISBN 978-981-19-1607-6 (eBook)
https://doi.org/10.1007/978-981-19-1607-6

This Springer imprint is published by the registered company Springer Nature Singapore Pte Ltd.
The registered company address is: 152 Beach Road, #21-01/04 Gateway East, Singapore 189721, Singapore

Preface

The Seventh International Congress on Information and Communication Technology will be held during 21–24 February 2022 in a hybrid mode and organised by Global Knowledge Research Foundation. The associated partners were Springer and InterYIT IFIP, Activate Learning, City of Oxford College, UK. The conference will provide a useful and wide platform both for display of the latest research and for exchange of research results and thoughts. The participants of the conference will be from almost every part of the world, with backgrounds of either academia or industry, allowing a real multinational multicultural exchange of experiences and ideas.

A great pool of more than 1100 papers were received for this conference from across 95 countries among which around 300 papers were accepted and will be presented through digital platforms during the two days. Due to the overwhelming response, we had to drop many papers in the hierarchy of the quality. Total 42 technical sessions will be organised in parallel in 4 days along with a few keynotes and panel discussions in hybrid mode. The conference will be involved in deep discussion and issues which will be intended to solve at global levels. New technologies will be proposed, experiences will be shared, and future solutions for design infrastructure for ICT will also be discussed. The final papers will be published in four volumes of proceedings by Springer LNNS Series.

Over the years. this congress has been organised and conceptualised with collective efforts of a large number of individuals. I would like to thank each of the committee members and the reviewers for their excellent work in reviewing the papers. Grateful acknowledgements are extended to the team of Global Knowledge Research Foundation for their valuable efforts and support.

I look forward to welcoming you to the 7th Edition of this ICICT Congress 2022.

Amit Joshi, Ph.D.
Organising Secretary, ICICT 2022
Director—Global Knowledge Research Foundation
Ahmedabad, India

Contents

Smart Wearable Shoes Using Multimodal Data for Visually Impaired 1
Ann Nosseir

Conceptual Framework of Database Development on Bidong Island: The Case Vietnamese Boat People (VBP) Campsite Facilities for Historical Tourism 19
Dazmin Daud, Nursyamilah Annuar, and Antashah Mohd Nor

An Evaluation of Techniques for Classification of Conditional Sentences and Their Structural Components 29
G. B. Sanjana, Sundar Guntnur, and Shivali Agarwal

Design of an Assistive Low-Cost 6 d.o.f. Robotic Arm with Gripper 39
Vasile Denis Manolescu and Emanuele Lindo Secco

An Improved Method for the Sizing of a Stand-Alone Photovoltaic System: Application at Ngoundiane Village in Senegal 57
Pape Made Diouf, Amy Sadio, Papa Lat Tabara Sow, Ibrahima Fall, and Senghane Mbodji

Blockchain Based Software Engineering Requirements Analysis and Management 75
Bandar Ali Alrami AL Ghadmi, Omar Ahmed Abdulkader, and Ahmad Abdulaziz Alwarhi

Computer Simulation of the Response of a Semiconductor Wafer with a Self-Affine Pattern in the Form of a System of Coupled Ring Grooves to Electromagnetic Radiation 85
Gennadi Lukyanov, Alexander Kopyltsov, and Igor Serov

The Computer Engineering in ECG Analysis Based on Scatter Mapping 93
Svetlana Aleksandrova, Irina Kurnikova, Marina Aleksandrova, Nikolay Kislyy, Tatiana Kochemasova, and Maria Zavalina

Traffic Disturbance Mining and Feedforward Neural Network to Enhance the Immune Network Control Performance 99
Ali Louati, Fatma Masmoudi, and Rahma Lahyani

The Use of ICT in Personalizing Self-learning in Time of Crisis: A Human Computer Interaction Perspective in a Developing Country 107
Ghada Refaat El Said

Advanced Processing and Classification of Plant Disease 127
Sufola Das Chagas Silva E Araujo, V. S. Malemath, and K. Meenakshi Sundaram

Design of an Interactive BB8-Like Robot 137
Mia Innes and Emanuele Lindo Secco

Transfer Learning in Deep Reinforcement Learning 145
Tariqul Islam, Dm. Mehedi Hasan Abid, Tanvir Rahman, Zahura Zaman, Kausar Mia, and Ramim Hossain

A Novel Current Control Scheme for Grid-Connected Single-Phase PWM Bridgeless Power Converters 155
Khalid Javed, Lieven Vandevelde, and Frederik De Belie

Disk Space Management Automation with CSI and Kubernetes 171
Anastasia Shemyakinskaya and Igor Nikiforov

Accuracy of Potentiometric Methods for Measuring Ion Activity in Solutions 181
O. M. Vasilevskyi, V. M. Sevastianov, K. V. Ovchynnykov, V. M. Didych, and S. A. Burlaka

Evaluating Effect of Microsoft HoloLens on Extraneous Cognitive Load During Simulated Cervical Lateral Mass Screw Placement 191
Dmitriy Babichenko, Edward G. Andrews, Stephen P. Canton, Eliza Beth Littleton, Ravi Patel, Dukens Labaze, and Andrew Mills

Network Modeling—A Convenient Way to Study IP Networks 203
Ivan Nedyalkov and Georgi Georgiev

A New Innovation Concept on End-user's Contextual and Behavioural Perspectives 213
Reem Aman, Shah J. Miah, and Janet Dzator

Computational Modelling of the Role of Leadership Style for Its Context-Sensitive Control Over Multilevel Organisational Learning 223
Gülay Canbaloğlu, Jan Treur, and Anna Wiewiora

Enumeration of LCD and Self-dual Double Circulant Codes Over $\mathbb{F}_q[v]/ <v^2 - 1>$ 241
Shikha Yadav and Om Prakash

Autonomous Dysfunction and the Phenomenon of Early Aging of Regulatory Systems 251
Irina Kurnikova, Shirin Gulova, Ramchandra Sargar, and Nikolay Kisliy

Principles for Assurance on Corporate Governance of ICT 257
Petrus M. J. Delport and Rossouw Von Solms

The Role of Telecommunication Technology During COVID-19 Pandemic in Indonesia 275
Vina Fujiyanti, Syifaul Fuada, and Nadia Tiara Antik Sari

Online Purchase Over Pandemic Covid-19: Its Growth and Future in Malaysia 285
Tang Mui Joo and Chan Eang Teng

MAGNeto: An Efficient Deep Learning Method for the Extractive Tags Summarization Problem 297
Hieu Trong Phung, Anh Tuan Vu, Tung Dinh Nguyen, Lam Thanh Do, Giang Nam Ngo, Trung Thanh Tran, and Ngoc C. Lê

Design and Development of a Mobile Outdoor AR Application for On-Site Visualization of Wind Turbines 311
Simon Burkard and Frank Fuchs-Kittowski

An Incorporated Solution to Support Elder People in Staying in Their Familiar Surroundings 323
Dominic Mircea Kristaly and Sorin-Aurel Moraru

Analysis of Indoor Localization Using Beacons for the Visually Impaired: A Systematic Literature Review 339
Juan Surco-Anacleto and Michael Cabanillas-Carbonell

Information and Communication Technologies for Employability in Times of COVID-19, a Systematic Literature Review 349
Jesus Palacios-Loayza, Carlos Ayala-Inca, and Michael Cabanillas-Carbonell

The Acceptance and Challenges of Online Learning over Covid-19 Pandemic 359
Eang Teng Chan and Mui Joo Tang

Learn to Ask What You Don't Know ... 369
Binay Dahal, Sing Choi, and Kazem Taghva

Location-Based Service Discovery for Mobile-Edge Computing Using DNS ... 379
Kurt Horvath, Helmut Wöllik, Uran Christoph, and Valentin Egger

Promoting Viable Supply Chain Management (SCM) in the Nigeria Agro-Allied Industry Using Internet of Things ... 389
Makinde Oluwafemi Ajayi and Opeyeolu Timothy Laseinde

Recovery System of Work Performance by Using Indoor Environmental Changes Based on EEG-Movement Feature Space ... 401
Hinata Serizawa and Yoshihisa Fukuhara

Moroccan Sign Language Video Recognition with Deep Learning ... 415
Abdelbasset Boukdir, Mohamed Benaddy, Othmane El Meslouhi, Mustapha Kardouchi, and Moulay Akhloufi

Home Automation System and Quality of Life in Low-Income Households: A Systematic Review of the Literature from 2010 to 2021 ... 423
Jenifer Diana Bustamante-Gonzales, Hugo Eladio Chumpitaz-Caycho, and Franklin Cordova-Buiza

Detecting Termites in Wood Structure Using Internet of Things Approach ... 431
Nur Zaimah Ahmad, Lutfil Hadi Zaifri, Bazilah A. Talip, and Aznida Abu Bakar Sajak

Performance Evaluation of Boosted 2-Stream TCRNet ... 443
Shah Hassan, Md Jibanul Haque Jiban, and Abhijit Mahalanobis

Design of a Cascaded Single-Phase Multilevel Inverter for Photovoltaic Applications ... 451
Darío Fernando Yépez Ponce, Héctor Mauricio Yépez Ponce, and William Manuel Montalvo López

An IoT Architecture to Enhance Monitoring and Predictive Maintenance for Cultural Heritage Buildings ... 461
Mario Casillo, Massimo De Santo, Marco Lombardi, Rosalba Mosca, Domenico Santaniello, and Carmine Valentino

A BIM-Based Approach for Decision Support System in Smart Buildings ... 471
Francesco Colace, Caterina Gabriella Guida, Brij Gupta, Angelo Lorusso, Francesco Marongiu, and Domenico Santaniello

Deficiencies of Computational Image Recognition in Comparison to Human Counterpart 483
Vladimir Vinnikov and Ekaterina Pshehotskaya

Electronic Health Record's Security and Access Control Using Blockchain and IPFS 493
Md. Yeasin Ali, Suhaib Ahmed, Muhammad Iqbal Hossain, A. B. M. Alim Al Islam, and Jannatun Noor

Implementing Butterfly Key Expansion Using Post-Quantum Algorithms 507
Ahmad Salman and Zachary Blankinship

Perceived Readiness of Information and Communication Technology Policy in Supporting Mobile Learning in Times of COVID-19 at South African Schools 517
Baldreck Chipangura

Application of Random Forest Model in the Prediction of River Water Quality 525
Turuganti Venkateswarlu and Jagadeesh Anmala

Supervised Learning-Based PV Output Current Modeling: A South Africa Case Study 537
Ely Ondo Ekogha and Pius A. Owolawi

Use of Social Networks by Russian Politicians 547
Olga Gris and Anna Sosnovskaya

Design and Implementation of Verifiable Blockchain-Based e-voting System 557
Seiwoong Choi, HeeSeok Choi, and Kwang Sik Chung

DevOps Best Practices in Highly Regulated Industry 567
Ruth G. Lennon

Citizens' Use of Social Media: A Thematic Analysis on Digital Co-Production in Disaster Management 587
Vicente A. Pitogo and Jesterlyn Q. Timosan

Robot Welding Path Planning and Application Based on Graphical Computing 597
Jingjing Lou, Xujiang Yu, Yongfei Chen, Zhubing Sun, and Pengfei Zheng

The Interpolation-Vandermonde Method for Numerical Solutions of Weakly Singular Volterra Integral Equations of the Second Kind 607
E. S. Shoukralla, B. M. Ahmed, Ahmed Saeed, and M. Sayed

Adoption of Cloud-Based Communicable Disease Surveillance in Taiwan: Chief Information Officers' Perspectives of Hospitals 615
Pi-Jung Hsieh and Hui-Min Lai

Design and Evaluation of a Novel and Modular Educational Robot Platform Based on Technology Acceptance Model 633
Avraam Chatzopoulos, Konstantinos Kalovrektis, Apostolis Xenakis, Elefterios Chondrogiannis, Michail Papoutsidakis, Michail Kalogiannakis, and Sarantos Psycharis

Assessing the Effects of Landmarks and Routes on Neuro-Cognitive Load Using Virtual Environment 645
Usman Alhaji Abdurrahman, Lirong Zheng, and Usman Haruna

On the Transposition of Translator Functions in a Digital Communicative Environment 657
Lyudmila Kushnina, Elena Alikina, Irina Perlova, Kristina Permiakova, and Marina Khudaiberdina

Factors Affecting Intelligent Enterprise Resource Planning System Migrations: The South African Customer's Perspective 665
Precious Mushayi and Thembekile Mayayise

Analysis of an Efficient ZnO/GeTe Solar Cell Using SCAPS-1D 677
Mostafa M. Salah, A. Zekry, Mohamed Abouelatta, Ahmed Shaker, Mohamed Mousa, and Ahmed Saeed

Looping Through Color Space: A Simple Augmentation Method to Improve Biased Object Detection 687
Pascal Lampert, Janis Jung, Andreas Hubert, and Konrad Doll

Detection of Retinopathy of Prematurity Stages Utilizing Deep Neural Networks 699
Nazar Salih, Mohamed Ksantini, Nebras Hussein, Donia Ben Halima, Ali Abdul Razzaq, and Sohaib A. Mahmood

Iterative Approach for Reduction of Index-2 Periodic Models Using Generalized Inverse Procedure 707
Atia Afroz, Mohammad-Sahadet Hossain, Musannan Hossain, and Mashrur Wasek

Smart Village Crop Planning: Enhancing Farmer's Decision-Making Culture with Data-Driven Predictive Model 717
Ariza Nordin and Faizah Ahmad Faizar

Closed-Domain Multiple-Choice Question Answering System for Science Questions 729
Kedar P. Vaidya, Sanya A. Chetwani, and Mansi A. Radke

Quality Management Within and Visiting e-cultural Tourist Destinations: Case Study Rural Parish of San Miguelito ... 743
Alicia Porras-Angulo, Alba Hernández-Freire, Johana Porras-Quispe, and Adriana Cuesta-Chiriboga

Use of GeoGebra in Learning to Solve the Problem of Calculating the Root of a Nonlinear Equation ... 753
Judith Keren Jiménez-Vilcherrez, Felicita Marcela Velásquez-Fernández, Araceli Margarita Acevedo-Ruiz, Ricardo Velezmoro-León, and Robert Ipanaqué-Chero

Usability Evaluation Using Unmoderated Remote Usability Testing on Angkasa LMS Website Case Study ... 761
Veronikha Effendy, Dana Sulistiyo Kusumo, Nungki Selviandro, and Kusuma Ayu Laksitowening

Distributed Deep Reinforcement Learning for Resource Allocation in Digital Twin Networks ... 771
Jie Luo, Jie Zeng, Ying Han, and Xin Su

The COVID-Enforced Adoption of Technology for Reluctant Entrepreneurial Businesses: A Systematic Literature Review ... 783
Gareth Mclean and Adriana A. Steyn

Integrated Remote Primary Care Infrastructure: A Framework for Adoption and Scaling of Remote Patient Management Tools and Systems ... 797
Barimwotubiri Ruyobeza, Sara S. Grobbelaar, and Adele Botha

A Comprehensive Virtual Classroom Dashboard ... 819
Amber Kimberling and Sampson Akwafuo

Freddy Render: A Horizontally Scaled Blender-Based Solution for 3D Graphics Rendering ... 829
Mike Peralta and Sampson Akwafuo

Medical X-Ray Image Classification Employing DCGAN and CNN Transfer Learning Techniques ... 839
Md. Asif Talukdar, Ayesha Siddika, Ahasanul Haque Abir, Mohammed Ziad Hassan, and Muhammad Iqbal Hossain

A Survey on Counterfeits in the Information and Communications Technology (ICT) Supply Chain ... 849
Samar Saleh, Rong Lei, Weihong Guo, and Elsayed A. Elsayed

Banking Credit Risk Analysis using Artificial Neural Network ... 871
Charles Maruma, Chunling Tu, and Claude Nawej

Implementation, Analysis, and Emulation of Electric Vehicle Powertrain System with Sensorless Field Controlled PMSM Drive 879
Monika Verma, Mini Sreejeth, and Madhusudan Singh

Verification of the Effectiveness of Learning Materials that Support Self-regulation for Learning Considering Differences in Career Resilience: Acquiring Knowledge of Level 3 Automated Driving Vehicles 889
Maki Arame, Junko Handa, Yoshiko Goda, Masashi Toda, Ryuichi Matsuba, Huiping Zhou, Makoto Itoh, and Satoshi Kitazaki

ICT-Enabled Vehicle Theft Detection and Recovery System 901
Kamlesh Kumawat and Vijay Singh Rathore

Determination of Antibiotic Resistance Level in *Klebsiella* using Machine Learning Models 913
Snehal Gupta, Sreemoyee Chatterjee, Amita Sharma, Marina Popolizio, Vincenzo Di Lecce, Mariantonietta Succi, Patrizio Tremonte, Rita Dario, and Vijay Singh Rathore

Author Index 925

Editors and Contributors

About the Editors

Xin-She Yang obtained his D.Phil. in Applied Mathematics from the University of Oxford, and subsequently worked at the Cambridge University and the National Physical Laboratory (UK) as Senior Research Scientist. He is currently Reader in Modeling and Optimization at Middlesex University London and Adjunct Professor at Reykjavik University (Iceland). He is also elected Bye-Fellow at the Cambridge University and IEEE CIS Chair for the Task Force on Business Intelligence and Knowledge Management. He was included in the "2016 Thomson Reuters Highly Cited Researchers" list.

Simon Sherratt was born near Liverpool, England, in 1969. He is currently Professor of Biosensors in the Department of Biomedical Engineering, University of Reading, UK. His main research area is signal processing and personal communications in consumer devices, focusing on wearable devices and health care. He received the 1st place IEEE Chester Sall Memorial Award in 2006, the 2nd place in 2016 and the 3rd place in 2017.

Nilanjan Dey is Assistant Professor in the Department of Information Technology, Techno India College of Technology, India. He has authored/edited more than 75 books with Springer, Elsevier, Wiley, CRC Press and published more than 300 peer-reviewed research papers. He is Editor-in-Chief of the International Journal of Ambient Computing and Intelligence; Series Co-editor of Springer Tracts in Nature-Inspired Computing (STNIC); and Series Co-editor of Advances in Ubiquitous Sensing Applications for Healthcare, Elsevier.

Amit Joshi is currently Director of Global Knowledge Research Foundation and also Entrepreneur and Researcher who has completed his masters and research in the areas of cloud computing and cryptography in medical imaging. He has an experience of around 10 years in academic and industry in prestigious organizations. He is an active member of ACM, IEEE, CSI, AMIE, IACSIT-Singapore, IDES, ACEEE, NPA and many other professional societies. Currently, he is International Chair of InterYIT at International Federation of Information Processing (IFIP, Austria), He has presented and published more than 50 papers in national and international journals/conferences of IEEE and ACM. He has also edited more than 40 books which are published by Springer, ACM and other reputed publishers. He has also organized more than 50 national and international conferences and programs in association with ACM, Springer, IEEE to name a few across different countries including India, UK, Europe, USA, Canada, Thailand, Egypt and many more.

Contributors

Omar Ahmed Abdulkader Faculty of Computer Studies, Arab Open University, Riyadh, Kingdom of Saudi Arabia

Usman Alhaji Abdurrahman School of Information Science and Technology, Fudan University, Shanghai, China;
Department of Computer Science, Yusuf Maitama Sule University, Kano, Nigeria

Dm. Mehedi Hasan Abid Daffodil International University, Dhaka, Bangladesh

Ahasanul Haque Abir BRAC University, Dhaka, Bangladesh

Mohamed Abouelatta Faculty of Engineering, Ain Shams University, Cairo, Egypt

Aznida Abu Bakar Sajak Computer Engineering Technology Section, Malaysian Institute of Information Technology, Universiti Kuala Lumpur, Kuala Lumpur, Malaysia

Araceli Margarita Acevedo-Ruiz Universidad Tecnológica del Perú, Piura, Peru;
Universidad César Vallejo, Piura, Peru;
Universidad Nacional de Piura, Urb. Miraflores s/n Castilla, Piura, Peru

Atia Afroz Department of Mathematics and Physics, North South University, Dhaka, Bangladesh

Shivali Agarwal IBM Research, Bengaluru, Karnataka, India

Nur Zaimah Ahmad Computer Engineering Technology Section, Malaysian Institute of Information Technology, Universiti Kuala Lumpur, Kuala Lumpur, Malaysia

B. M. Ahmed Faculty of Engineering and Technology, Future University in Egypt, Cairo, Egypt

Suhaib Ahmed Department of Computer Science and Engineering, BRAC University, Dhaka, Bangladesh

Makinde Oluwafemi Ajayi Faculty of Engineering and the Built Environment, University of Johannesburg, Johannesburg, South Africa

Moulay Akhloufi Département d'Informatique, Université de Moncton, Moncton, Canada

Sampson Akwafuo California State University, Fullerton, CA, USA

Marina Aleksandrova Department of Therapy and Endocrinology, RUDN University, Moscow, Russia

Svetlana Aleksandrova Department of Therapy and Endocrinology, RUDN University, Moscow, Russia

Bandar Ali Alrami AL Ghadmi Faculty of Computer Studies, Arab Open University, Riyadh, Kingdom of Saudi Arabia

Md. Yeasin Ali Department of Computer Science and Engineering, BRAC University, Dhaka, Bangladesh

Elena Alikina Perm National Research Polytechnic University, Perm, Russia

A. B. M. Alim Al Islam Department of Computer Science and Engineering, Bangladesh University of Engineering and Technology, Dhaka, Bangladesh

Ahmad Abdulaziz Alwarhi Faculty of Computer Studies, Arab Open University, Riyadh, Kingdom of Saudi Arabia

Reem Aman Ministry of Education, Riyadh, Kingdom of Saudi Arabia, Newcastle Business School, University of Newcastle, Newcastle, NSW, Australia

Edward G. Andrews University of Pittsburgh Medical Center (UPMC), Pittsburgh, PA, USA

Jagadeesh Anmala Birla Institute of Technology and Science, Pilani, Hyderabad, Telangana, India

Nursyamilah Annuar Faculty of Business and Management, Universiti Teknologi MARA, Cawangan Perlis, Kampus Arau, Arau Perlis, Malaysia

Maki Arame Polytechnic University of JAPAN, Tokyo, Japan;
Kumamoto University, Kumamoto-city, Japan

Carlos Ayala-Inca Universidad Autónoma del Perú, Lima, Perú

Dmitriy Babichenko University of Pittsburgh, Pittsburgh, PA, USA

Mohamed Benaddy Laboratory of Engineering Sciences and Energies, FPO, Ibn Zohr University, Ouarzazate, Morocco

Zachary Blankinship James Madison University, Harrisonburg, VA, USA

Adele Botha Council for Scientific and Industrial Research, Pretoria, South Africa

Abdelbasset Boukdir Laboratory of Engineering Sciences and Energies, FPO, Ibn Zohr University, Ouarzazate, Morocco

Simon Burkard Hochschule für Technik und Wirtschaft (HTW) Berlin, Berlin, Germany

S. A. Burlaka Vinnytsia National Agrarian University, Vinnytsia, Ukraine

Jenifer Diana Bustamante-Gonzales Engineering Faculty, Universidad Privada del Norte, Lima, Peru

Michael Cabanillas-Carbonell Universidad Privada del Norte, Lima, Peru

Gülay Canbaloğlu Department of Computer Engineering, Koç University, Istanbul, Turkey;
Center for Safety in Healthcare, Delft University of Technology, Delft, The Netherlands

Stephen P. Canton University of Pittsburgh, Pittsburgh, PA, USA

Mario Casillo DIIn, University of Salerno, Fisciano, SA, Italy

Eang Teng Chan Tunku Abdul Rahman University College, Kuala Lumpur, Malaysia

Sreemoyee Chatterjee IIS (deemed to be University), Jaipur, India

Avraam Chatzopoulos University of West Attica, Aigaleo Attiki, Greece

Yongfei Chen Zhejiang Linix Motor Co., Ltd, Dongyang, China

Sanya A. Chetwani Visvesvaraya National Institute of Technology, Nagpur, India

Baldreck Chipangura University of South Africa, Florida Campus, Johannesburg, South Africa

HeeSeok Choi ATGLab R&D Center, Seoul, Korea

Seiwoong Choi Graduate School, Department of Information Science, Korea National Open University, Seoul, Korea

Sing Choi University of Nevada, Las Vegas, USA

Elefterios Chondrogiannis Agricultural University of Athens, Athens Attiki, Greece

Uran Christoph Carinthia University of Applied Sciences, Carinthia, Austria

Hugo Eladio Chumpitaz-Caycho Engineering Faculty, Universidad Privada del Norte, Lima, Peru

Kwang Sik Chung Department of Computer Science, Korea National Open University, Seoul, Korea

Francesco Colace DIIn, University of Salerno, Fisciano, Italy

Franklin Cordova-Buiza Research and Innovation Department, Universidad Privada del Norte, Lima, Peru

Adriana Cuesta-Chiriboga Facultad de Ciencias Humanas Y de La Educación, Universidad Técnica de Ambato, Ambato, Ecuador

Binay Dahal University of Nevada, Las Vegas, USA

Rita Dario Medical Management, Hospital Medical Management, Polytechnic of Bari, Bari, Italy

Sufola Das Chagas Silva E Araujo Department of Computer Science Engineering, KLE Dr. M.S.S Sheshgiri College of Engineering and Technology, VTU, Belagavi, India

Dazmin Daud Faculty of Business and Management, UCSI University, Cheras, Kuala Lumpur, Malaysia

Frederik De Belie Department of Electromechanical, Systems and Metal Engineering, EEDT Decision and Control, Flanders Make, Ghent University, Ghent, Belgium

Massimo De Santo DIIn, University of Salerno, Fisciano, SA, Italy

Petrus M. J. Delport Nelson Mandela University, Port Elizabeth, South Africa; Noroff University College, Kristiansand, Norway

Vincenzo Di Lecce Department of Electrical and Information Engineering, Polytechnic of Bari, Bari, Italy

V. M. Didych National Pirogov Memorial Medical University, Vinnytsia, Ukraine

Pape Made Diouf Research Team in Renewable Energies, Materials and Laser of Department of Physics, Alioune Diop University, Bambey, Senegal

Lam Thanh Do PIXTA Vietnam, Hanoi, Vietnam;
Hanoi University of Science and Technology, Hanoi, Vietnam

Konrad Doll University of Applied Sciences Aschaffenburg, Aschaffenburg, Germany

Janet Dzator Newcastle Business School, University of Newcastle, Newcastle, NSW, Australia

Veronikha Effendy Telkom University, Bandung, Indonesia

Valentin Egger Carinthia University of Applied Sciences, Carinthia, Austria

Ely Ondo Ekogha Tshwane University of Technology, Pretoria, South Africa

Elsayed A. Elsayed Rutgers University-New Brunswick, Piscataway, NJ, USA

Faizah Ahmad Faizar Faculty of Computer and Mathematical Sciences, Universiti Teknologi MARA, Shah Alam, Malaysia

Ibrahima Fall Research Team in Renewable Energies, Materials and Laser of Department of Physics, Alioune Diop University, Bambey, Senegal

Syifaul Fuada Universitas Pendidikan Indonesia, Bandung, Indonesia

Frank Fuchs-Kittowski Hochschule für Technik und Wirtschaft (HTW) Berlin, Berlin, Germany

Vina Fujiyanti Universitas Pendidikan Indonesia, Bandung, Indonesia

Yoshihisa Fukuhara Faculty of Data Science, Musashino University, Tokyo, Japan;
Asia AI Institute, Musashino University, Tokyo, Japan

Georgi Georgiev South–West University "Neofit Rilski", Blagoevgrad, Bulgaria

Yoshiko Goda Kumamoto University, Kumamoto-city, Japan

Olga Gris RANEPA, Moscow, Russia

Sara S. Grobbelaar Stellenbosch University, Stellenbosch, South Africa

Caterina Gabriella Guida DICiv, University of Salerno, Fisciano, Italy

Shirin Gulova Department of Therapy and Endocrinology, RUDN University, Moscow, Russia

Sundar Guntnur RVCE, Bengaluru, India

Weihong Guo Rutgers University-New Brunswick, Piscataway, NJ, USA

Brij Gupta Department of Computer Science and Information Engineering, Asia University, Taichung, Taiwan;
Research and Innovation Department, Skyline University College, Sharjah, United Arab Emirates;
Staffordshire University, Stoke-on-Trent, UK

Snehal Gupta IIS (deemed to be University), Jaipur, India

Donia Ben Halima CEMLab, ENIS, University of Sfax, Sfax, Tunisia

Ying Han Chongqing University of Posts and Telecommunications, Chongqing, China

Junko Handa Polytechnic University of JAPAN, Tokyo, Japan;
Kumamoto University, Kumamoto-city, Japan

Usman Haruna Department of Computer Science, University of Terengganu, Terengganu, Malaysia;
Department of Computer Science, Yusuf Maitama Sule University, Kano, Nigeria

Mohammed Ziad Hassan BRAC University, Dhaka, Bangladesh

Shah Hassan University of Central Florida, Orlando FL, USA

Alba Hernández-Freire Facultad de Ciencias Humanas Y de La Educación, Universidad Técnica de Ambato, Ambato, Ecuador

Kurt Horvath Carinthia University of Applied Sciences, Carinthia, Austria

Mohammad-Sahadet Hossain Department of Mathematics and Physics, North South University, Dhaka, Bangladesh

Muhammad Iqbal Hossain Department of Computer Science and Engineering, BRAC University, Dhaka, Bangladesh

Musannan Hossain Department of Electrical and Computer Engineering, North South University, Dhaka, Bangladesh

Ramim Hossain Daffodil International University, Dhaka, Bangladesh

Pi-Jung Hsieh Department of Hospital and Health Care Administration, Chia Nan University of Pharmacy and Science, Tainan, Taiwan, ROC

Andreas Hubert University of Applied Sciences Aschaffenburg, Aschaffenburg, Germany

Nebras Hussein Biomedical Engineering Department, Al-Khwarizmi College of Engineering, Baghdad University, Baghdad, Iraq

Mia Innes Robotics Laboratory, School of Mathematics, Computer Science and Engineering, Liverpool Hope University, Liverpool, UK

Robert Ipanaqué-Chero Universidad Nacional de Piura, Urb. Miraflores s/n Castilla, Piura, Peru

Tariqul Islam Daffodil International University, Dhaka, Bangladesh

Makoto Itoh University of Tsukuba, Tsukuba-city, Japan

Khalid Javed Department of Electromechanical, Systems and Metal Engineering, EEDT Decision and Control, Flanders Make, Ghent University, Ghent, Belgium

Md Jibanul Haque Jiban University of Central Florida, Orlando FL, USA

Judith Keren Jiménez-Vilcherrez Universidad Tecnológica del Perú, Piura, Peru

Tang Mui Joo Tunku Abdul Rahman University College, Kuala Lumpur, Malaysia

Janis Jung University of Applied Sciences Aschaffenburg, Aschaffenburg, Germany

Michail Kalogiannakis University of Crete, Rethymnon Crete, Greece

Konstantinos Kalovrektis University of Thessaly, Galaneika, Lamia, Greece

Mustapha Kardouchi Départemement d'Informatique, Université de Moncton, Moncton, Canada

Marina Khudaiberdina Perm National Research Polytechnic University, Perm, Russia

Amber Kimberling California State University, Fullerton, CA, USA

Nikolay Kisliy Department of Therapy and Endocrinology, RUDN University, Moscow, Russia

Nikolay Kislyy Department of Therapy and Endocrinology, RUDN University, Moscow, Russia

Satoshi Kitazaki Automotive Human Factors Research Center, Tsukuba-city, Japan

Tatiana Kochemasova Department of Therapy and Endocrinology, RUDN University, Moscow, Russia

Alexander Kopyltsov Saint Petersburg State University of Aerospace Instrumentation, St. Petersburg, Russia

Dominic Mircea Kristaly Transilvania University of Brasov, Brasov, Romania

Mohamed Ksantini CEMLab, ENIS, University of Sfax, Sfax, Tunisia

Kamlesh Kumawat Department of CS & IT, IIS (Deemed to be University), Jaipur, India

Irina Kurnikova Department of Therapy and Endocrinology, RUDN University, Moscow, Russia;
Department of Aviation and Space Medicine, Federal State Budgetary Educational Institution of Further Professional Education, Russian Medical Academy of Continuous Professional Education, Moscow, Russia

Lyudmila Kushnina Perm National Research Polytechnic University, Perm, Russia

Dana Sulistiyo Kusumo Telkom University, Bandung, Indonesia

Dukens Labaze University of Pittsburgh Medical Center (UPMC), Pittsburgh, PA, USA

Rahma Lahyani Operations and Project Management Department, College of Business, Alfaisal University, Riyadh, Saudi Arabia

Hui-Min Lai Department of Business Administration, National Taichung University of Science and Technology, Taichung, Taiwan, ROC

Kusuma Ayu Laksitowening Telkom University, Bandung, Indonesia

Pascal Lampert University of Applied Sciences Aschaffenburg, Aschaffenburg, Germany

Opeyeolu Timothy Laseinde Mechanical and Industrial Engineering Tech Department, University of Johannesburg, Johannesburg, South Africa

Rong Lei Rutgers University-New Brunswick, Piscataway, NJ, USA

Ruth G. Lennon Atlantic Technological University, Letterkenny, Ireland

Eliza Beth Littleton University of Pittsburgh, Pittsburgh, PA, USA

Marco Lombardi DIIn, University of Salerno, Fisciano, SA, Italy

Angelo Lorusso DIIn, University of Salerno, Fisciano, Italy

Jingjing Lou Yiwu Industrial and Commercial College, Yiwu, China

Ali Louati Department of Information Systems, College of Computer Engineering and Sciences, Prince Sattam bin Abdulaziz University, Al-Kharj, Saudi Arabia; SMART Lab, ISG, University of Tunis, Tunis, Tunisia

Gennadi Lukyanov ITMO University, St. Petersburg, Russia

Jie Luo China Academy of Telecommunications Technology, Beijing, China

Ngoc C. Lê PIXTA Vietnam, Hanoi, Vietnam;
Hanoi University of Science and Technology, Hanoi, Vietnam

William Manuel Montalvo López Instituto Superior Tecnológico, Luis Tello, Ecuador

Abhijit Mahalanobis University of Central Florida, Orlando FL, USA

Sohaib A. Mahmood Ibn AL Haitham Teaching Eye Hospital, Baghdad, Iraq

V. S. Malemath Department of Computer Science Engineering, KLE Dr. M.S.S Sheshgiri College of Engineering and Technology, VTU, Belagavi, India

Vasile Denis Manolescu Robotics Laboratory, School of Mathematics, Computer Science and Engineering, Liverpool Hope University, Liverpool, UK

Francesco Marongiu DIIn, University of Salerno, Fisciano, Italy

Charles Maruma Tshwane University of Technology, Pretoria, South Africa

Fatma Masmoudi Department of Information Systems, College of Computer Engineering and Sciences, Prince Sattam bin Abdulaziz University, Al-Kharj, Saudi Arabia

Ryuichi Matsuba Kumamoto University, Kumamoto-city, Japan

Thembekile Mayayise University of the Witwatersrand, Johannesburg, South Africa

Senghane Mbodji Research Team in Renewable Energies, Materials and Laser of Department of Physics, Alioune Diop University, Bambey, Senegal

Gareth Mclean University of Pretoria, Pretoria, South Africa

K. Meenakshi Sundaram Department of Engineering and Applied Sciences, Botho University, Gaborone, Botswana

Othmane El Meslouhi ENSA–Safi, Cadi Ayyad University, Marrakech, Morocco

Kausar Mia Daffodil International University, Dhaka, Bangladesh

Shah J. Miah Newcastle Business School, University of Newcastle, Newcastle, NSW, Australia

Andrew Mills University of Pittsburgh, Pittsburgh, PA, USA

Antashah Mohd Nor Faculty of Communication and Media Studies, Universiti Teknologi MARA, Shah Alam, Selangor, Malaysia

Sorin-Aurel Moraru Transilvania University of Brasov, Brasov, Romania

Rosalba Mosca DIIn, University of Salerno, Fisciano, SA, Italy

Mohamed Mousa Electrical Engineering Department, Future University in Egypt, Cairo, Egypt

Precious Mushayi University of the Witwatersrand, Johannesburg, South Africa

Claude Nawej Tshwane University of Technology, Pretoria, South Africa

Ivan Nedyalkov South–West University "Neofit Rilski", Blagoevgrad, Bulgaria

Giang Nam Ngo PIXTA Vietnam, Hanoi, Vietnam

Tung Dinh Nguyen PIXTA Vietnam, Hanoi, Vietnam

Igor Nikiforov Peter the Great St. Petersburg Polytechnic University, St. Peterburg, Russia

Jannatun Noor Department of Computer Science and Engineering, BRAC University, Dhaka, Bangladesh

Ariza Nordin Faculty of Computer and Mathematical Sciences, Universiti Teknologi MARA, Shah Alam, Malaysia

Ann Nosseir British University in Egypt, Cairo, Egypt;
Institute of National Planning, El Shourk, Egypt

K. V. Ovchynnykov Vinnytsya National Technical University, Vinnitsya, Ukraine

Pius A. Owolawi Tshwane University of Technology, Pretoria, South Africa

Jesus Palacios-Loayza Universidad Autónoma del Perú, Lima, Perú

Michail Papoutsidakis University of West Attica, Aigaleo Attiki, Greece

Ravi Patel University of Pittsburgh, Pittsburgh, PA, USA

Mike Peralta California State University, Fullerton, CA, USA

Irina Perlova Perm National Research Polytechnic University, Perm, Russia

Kristina Permiakova Perm National Research Polytechnic University, Perm, Russia

Hieu Trong Phung PIXTA Vietnam, Hanoi, Vietnam;
Hanoi University of Science and Technology, Hanoi, Vietnam

Vicente A. Pitogo College of Computing and Information Sciences, Caraga State University, Butuan City, Philippines

Darío Fernando Yépez Ponce Universidad Politécnica Salesiana, Quito, Ecuador; Instituto Superior Tecnológico, Luis Tello, Ecuador

Héctor Mauricio Yépez Ponce Instituto Superior Tecnológico, Luis Tello, Ecuador

Marina Popolizio Department of Electrical and Information Engineering, Polytechnic of Bari, Bari, Italy

Alicia Porras-Angulo Facultad de Ciencias Humanas Y de La Educación, Universidad Técnica de Ambato, Ambato, Ecuador

Johana Porras-Quispe Universidad Técnica Particular de Loja, Loja, Ecuador

Om Prakash Department of Mathematics, Indian Institute of Technology Patna, Bihta, Patna, India

Ekaterina Pshehotskaya Department of Information Security, Moscow Polytechnic University, Moscow, Russian Federation

Sarantos Psycharis ASPETE, Athens Attiki, Greece

Mansi A. Radke Visvesvaraya National Institute of Technology, Nagpur, India

Tanvir Rahman Daffodil International University, Dhaka, Bangladesh

Vijay Singh Rathore Department of CS & IT, IIS (Deemed to be University), Jaipur, India

Ali Abdul Razzaq Ibn AL Haitham Teaching Eye Hospital, Baghdad, Iraq

Barimwotubiri Ruyobeza Stellenbosch University, Stellenbosch, South Africa

Amy Sadio Research Team in Renewable Energies, Materials and Laser of Department of Physics, Alioune Diop University, Bambey, Senegal

Ahmed Saeed Faculty of Engineering and Technology, Future University in Egypt, Cairo, Egypt;
Electrical Engineering Department, Future University in Egypt, Cairo, Egypt

Ghada Refaat El Said Future University in Egypt (FUE), New Cairo, Egypt

Mostafa M. Salah Electrical Engineering Department, Future University in Egypt, Cairo, Egypt;
Faculty of Engineering, Ain Shams University, Cairo, Egypt

Samar Saleh Rutgers University-New Brunswick, Piscataway, NJ, USA

Nazar Salih CEMLab, ENIS, University of Sfax, Sfax, Tunisia

Ahmad Salman James Madison University, Harrisonburg, VA, USA

G. B. Sanjana RVCE, Bengaluru, India

Domenico Santaniello DIIn, University of Salerno, Fisciano, SA, Italy

Ramchandra Sargar Department of Therapy and Endocrinology, RUDN University, Moscow, Russia

Nadia Tiara Antik Sari Universitas Pendidikan Indonesia, Bandung, Indonesia

M. Sayed Faculty of Electronic Engineering, Menoufia University, Shibin Al Kawm, Egypt

Emanuele Lindo Secco Robotics Laboratory, School of Mathematics, Computer Science and Engineering, Liverpool Hope University, Liverpool, UK

Nungki Selviandro Telkom University, Bandung, Indonesia

Hinata Serizawa Faculty of Data Science, Musashino University, Tokyo, Japan

Igor Serov Human Genome Research Foundation, St. Petersburg, Russia

V. M. Sevastianov Vinnytsya National Technical University, Vinnitsya, Ukraine

Ahmed Shaker Faculty of Engineering, Ain Shams University, Cairo, Egypt

Amita Sharma IIS (deemed to be University), Jaipur, India

Anastasia Shemyakinskaya Peter the Great St. Petersburg Polytechnic University, St. Peterburg, Russia

E. S. Shoukralla Faculty of Electronic Engineering, Menoufia University, Shibin Al Kawm, Egypt

Ayesha Siddika BRAC University, Dhaka, Bangladesh

Madhusudan Singh Delhi Technological University, New Delhi, India

Anna Sosnovskaya RANEPA, Moscow, Russia

Papa Lat Tabara Sow Research Team in Renewable Energies, Materials and Laser of Department of Physics, Alioune Diop University, Bambey, Senegal

Mini Sreejeth Delhi Technological University, New Delhi, India

Adriana A. Steyn University of Pretoria, Pretoria, South Africa

Xin Su Tsinghua University, Beijing, China

Mariantonietta Succi Department of Food Microbiology, University of Molise, Campobasso, Italy

Zhubing Sun Zhejiang Linix Motor Co., Ltd, Dongyang, China

Juan Surco-Anacleto Universidad Autónoma del Perú, Lima, Peru

Kazem Taghva University of Nevada, Las Vegas, USA

Bazilah A. Talip Informatics and Analytics Section, Malaysian Institute of Information Technology, Universiti Kuala Lumpur, Kuala Lumpur, Malaysia

Md. Asif Talukdar BRAC University, Dhaka, Bangladesh

Mui Joo Tang Tunku Abdul Rahman University College, Kuala Lumpur, Malaysia

Chan Eang Teng Tunku Abdul Rahman University College, Kuala Lumpur, Malaysia

Jesterlyn Q. Timosan College of Computing and Information Sciences, Caraga State University, Butuan City, Philippines

Masashi Toda Kumamoto University, Kumamoto-city, Japan

Trung Thanh Tran PIXTA Vietnam, Hanoi, Vietnam

Patrizio Tremonte Department of Agricultural, Environmental and Food Science, University of Molise, Campobasso, Italy

Jan Treur Center for Safety in Healthcare, Delft University of Technology, Delft, The Netherlands;
Social AI Group, Department of Computer Science, Vrije Universiteit Amsterdam, Amsterdam, The Netherlands

Chunling Tu Tshwane University of Technology, Pretoria, South Africa

Kedar P. Vaidya Visvesvaraya National Institute of Technology, Nagpur, India

Carmine Valentino DIIn, University of Salerno, Fisciano, SA, Italy

Lieven Vandevelde Department of Electromechanical, Systems and Metal Engineering, EEDT Decision and Control, Flanders Make, Ghent University, Ghent, Belgium

O. M. Vasilevskyi Vinnytsya National Technical University, Vinnitsya, Ukraine

Ricardo Velezmoro-León Universidad Tecnológica del Perú, Piura, Peru;
Universidad César Vallejo, Piura, Peru;
Universidad Nacional de Piura, Urb. Miraflores s/n Castilla, Piura, Peru

Felicita Marcela Velásquez-Fernández Universidad César Vallejo, Piura, Peru

Turuganti Venkateswarlu Birla Institute of Technology and Science, Pilani, Hyderabad, Telangana, India

Monika Verma Delhi Technological University, New Delhi, India

Vladimir Vinnikov Department of Computer Sciences, Higher School of Economics, Moscow, Russian Federation

Rossouw Von Solms Nelson Mandela University, Port Elizabeth, South Africa

Anh Tuan Vu PIXTA Vietnam, Hanoi, Vietnam

Mashrur Wasek Department of Electrical and Computer Engineering, North South University, Dhaka, Bangladesh

Anna Wiewiora QUT Business School, Queensland University of Technology, Brisbane, Australia

Helmut Wöllik Carinthia University of Applied Sciences, Carinthia, Austria

Apostolis Xenakis University of Thessaly, Galaneika, Lamia, Greece

Shikha Yadav Department of Mathematics, Indian Institute of Technology Patna, Bihta, Patna, India

Xujiang Yu Yiwu Industrial and Commercial College, Yiwu, China

Lutfil Hadi Zaifri Computer Engineering Technology Section, Malaysian Institute of Information Technology, Universiti Kuala Lumpur, Kuala Lumpur, Malaysia

Zahura Zaman Daffodil International University, Dhaka, Bangladesh

Maria Zavalina Department of Therapy, Izhevsk State Medical Academy, Izhevsk, Russia

A. Zekry Faculty of Engineering, Ain Shams University, Cairo, Egypt

Jie Zeng Tsinghua University, Beijing, China

Lirong Zheng School of Information Science and Technology, Fudan University, Shanghai, China

Pengfei Zheng Yiwu Industrial and Commercial College, Yiwu, China

Huiping Zhou University of Tsukuba, Tsukuba-city, Japan

Smart Wearable Shoes Using Multimodal Data for Visually Impaired

Ann Nosseir

Abstract The visually impaired people's ultimate goal is to walk freely and comfortability indoors and outdoors. They fear to hit into steps, stones, or uneven floor. Wearable technologies whether image-based or sensors-based provide a solution. However, image-based technologies have issues of detecting an obstacle accurately with no delay. The sensors-based technologies have limitations of the data quality. Therefore, the sensors need to be fitted closer to the obstacles to capture the data, and they require filter to remove the noise data. This work presents a novel wearable, simple, low-cost, user-friendly device. It supports visually impaired to walk in different areas. The system provides accurate data to support visually impaired detecting the obstacles surround them, i.e., front, left, right, and back. It works in multiple environments. The shoes will help them to walk indoors and avoid obstacles on the floor. In outdoors, like pedestrian road, parks, or forests, it will detect pit holes and pumps. The proposed system consists of three parts. The first part, which is a low-cost Internet of things (IoT) system, attaches sensors to shoes to collect data about the context. The second part works like a filter to remove the noise data. Four anomaly machine learning algorithms are applied to choose the most accurate—K-NN, SVM, decision tree, and random forest. The third part is a risk level assessment using fuzzy rules. The results of comparing the anomaly algorithms accuracy show that the random forest is 0.99 with a std. dev $\pm$ 0.01. The fuzzy rules defined the three different ranges for the levels of risk.

Keywords Fuzzy systems · Health and safety · Supervised learning · Internet of things · Wearable computers

A. Nosseir (✉)
British University in Egypt, Cairo, Egypt
e-mail: nosseir12@yahoo.co.uk

Institute of National Planning, El Shourk, Egypt

X.-S. Yang et al. (eds.), *Proceedings of Seventh International Congress on Information and Communication Technology*, Lecture Notes in Networks and Systems 447,
https://doi.org/10.1007/978-981-19-1607-6_1

1 Introduction

Statistics in 2020 [1] by the World Health Organization (WHO) show that there is one billion visually impaired persons globally. They cannot see clearly in nearby distance. This is caused by different factors such as uncorrected refractive errors, cataract, or age-related macular degeneration. The percentage of the visually impaired is four times in the low- and mid-income regions than in the high-income regions. This is due to the lack of appropriate eye healthcare services. Additionally, the percentage of impaired people becomes higher with elderly. There is as well a number of visually impaired in young age people [2]. They face a number of difficulties in indoors and mostly outdoor environments. Riazi et al. [3] research shows that the main difficulties the visually impaired face in the outdoor environment are unsafe sidewalks, existence of obstacles on sidewalks, walking into glass doors, and others. Additionally, visually impaired fear the most tumbling down from stairs, falling into holes or pits, and bumping into objects in footpaths. To avoid these difficulties, they use the conventional navigation aids such as cane stick or dogs.

Recently with the development of technology, a number of assistive devices developed to support impaired vision people. The WHO classifies these assistive devices into three categories: electronic travel aids (ETAs), electronic orientation aids (EOAs), and position locator devices (PLDs). The ETAs devices give information about environment conveyed by the sensory modality. The EOAs devices inform impaired vision with the direction. The PLDs enable sending GPS position for tracking or in case of an emergency [4].

These devices provide vital information for visually impaired. Considering that the visually impaired people first priority is to walk freely in the street; EOAs and PLDs are supportive. In unfamiliar environments, ETAs devices give impaired people mobility and inform them with risks of different outdoors problems like hitting obstacles and others. ETAs devices provide more information than the conventional navigation tools because these devices have sensing input units that receive inputs from the environment. ETAs devices and especially wearable technologies allow hands-free and portability. Blind and impaired people receive information from these technologies about uneven floor, stones, or holes to prevent them from crippling. Having a smart wearable shoes enables recognizing these obstacles immediately as the sensors are located closer to these obstacles.

This work presents a novel smart wearable shoe. It is not only alerting visually impaired with obstacles in the streets, but also it gives details about the location of the of obstacles whether it is in front, back, right, or left of the visually impaired and fires. In other words, it works in multiple environments. It as well uses machine learning techniques to filter noise, i.e., anomaly data from the sensors. It gives also information of the obstacles risk level.

The paper starts with discussing the related work of detecting obstacles for visually impaired. This is followed by the description of the system's framework. In the third section, the details of sensors attached to the shoes are presented. In the fourth section, four anomaly machine learning algorithms, namely K-NN, SVM, decision tree, and random forest, are applied on the sensors data collected by our shoes prototype to filter noise data and distinguish the anomaly data. The accuracies of these algorithms are compared. In the fifth section, the fuzzy algorithm is described. Finally, this work is discussed and the paper ends up with conclusions and future work.

2 Related Work

Because the cane stick is simple and informs visually impaired with static obstacles on the ground, uneven surfaces, or holes, technological advancements such as image processing and sensor technology [5, 6] have been added to this conventional cane stick to provide more details about the obstacles. Research of [3] shows that visually impaired sometimes feel uncomfortable with the canes because a few of these sticks are fragile and break or are heavy to carry for a long time. In some cases, they crash or get stuck in small holes.

Dakopoulos and Bourbakis [7] work concludes that wearability and hands-free tools offer crucial features for the ETAs devices because they allow flexibility for the users, permit them to pay attention and get engaged with the surrounding environment. Gandy et al. [8] emphasized the importance of the physical placement of the devices and their ergonomics of wearable devices. These devices are worn on the body. Examples are head-mounted devices, wristbands, vests, belts, or shoes. Wearable devices are either image-based or ultrasonic-based.

2.1 Image-Based

These image-based are usually located in the head and the upper body such glasses, binocular, gloves jacket. Bai et al. [9] developed glasses that have a camera to give the visually impaired people visual hints. They use a fisheye camera to improve the accuracy of locating important features of the environment. Jiang et al. [10] capture images in a fixed frequency using a binocular vision sensor. The imaged captured is sent to the cloud for additional calculations. Zientara et al. [11] embed camera in a smart glass to allow independency of impaired vision in daily life activities like shopping. The camera is connected to a local wireless routing and uses mainly surf algorithm to identify the product on the shelf in the supermarket. Herghelegiu et al. [12] proposed an algorithm to identify potential risk situations such as negative obstacles, i.e., holes. The recognition is built on an estimation of the ground surface in the stereo images.

The challenges of the images-based techniques are mainly the latency and accuracy. It takes time to run a model to identify an object in an image. Usually, the image-based system is performing the calculations on the cloud that adds more to time. Additionally, the walking pace and the position of the camera affect the detection of the image. Furthermore, the accuracy relies on the algorithm, the images used in the training phase, the condition of the capture image.

2.2 Ultrasonic-Based

The sensors are commonly situated in the lower part of the body, for example, in a jacket, gloves, a leg shin, or shoes. Siddhartha et al. [13] designed a jacket with embedded senor that enables visually impaired to detect obstacles. It has four ultrasonic sensors fixed as follows: two in front and two in the back of the jacket. It buzzes the impaired vison when the sensors detect an obstacle within 200 cm from any of the four sensors. Alvarado et al. [14] installed ultrasonic sensors in a glove. It supports visually impaired when moving and sensing obstacles to support them grabbing an object. It alerts them when the sensors encounter with an obstacle. The sensors are fixed in the fingers, and the system gives guidance for two directions based on the position of the hand. Rahman et al. [15] added ultrasonic sensor on the leg shin aligned in an angle to aim toward the ground to detect humps in the street. When the hump is detected, the visually impaired is notified. There are other systems by Vignesh et al. [16], Xu et al. [17], and Wu et al. [18] that embed sensors are in shoes. Gokalgandhi et al. [19] studied the different sensors embedded in the smart shoes. They found out that ultrasonic sensors are attached to the shoes to give information for blind people about obstacles such as pit holes or uneven floor within a distance 5–200 cm.

2.3 Machine Learning and IoT Devices

Ultrasonic sensors are simple and cheap and allow developing user-friendly devices to improve visually impaired mobility. However, sensors can report faulty data either missing values or outliers [20]. Gaddam et al. [21] point out the importance of detecting anomaly outlier and sensor fault data, i.e., failures using anomaly detection techniques such as SVM, K-NN, and others.

On the other hand, sensors produce enormous amount of data. Developing models that learn the patterns of these data supports predicting risks. Tayyaba et al. [22] and Karakaya and Ocak [23] used fuzzy-based algorithm to provide easy navigation and allow visually impaired to avoid obstacles. Their systems receive data from the sensors and give information about the walking area to make decisions about walking speed such as 'stop,' 'very slow,' 'slow,' 'medium,' and 'fast.'

This paper chooses from the different wearables technologies the shoe to support vision impaired because they provide a compact, lightweight, portable, and hands-free product. Additionally, shoes are conserved by the industry. Nike develops a shoe product to sense the vitals for the athletics (see Fig. 1).

Fig. 1 Nike shoe [24]

3 Methodology

The system attaches different sensors in different parts of the shoes to detect pits holes, pumps, fires, and pavements. Then, the data collected by the sensors is classified to assess the context. Whether it is risky or not. In that respect, this work has experimented four classifiers—K-NN, SVM, decision tree, and random forest—to choose the one with best accuracy. These classifiers will be discussed in the following section. The system as well gives an extra information about the level of risk using fuzzy rules. Figure 2 shows these steps.

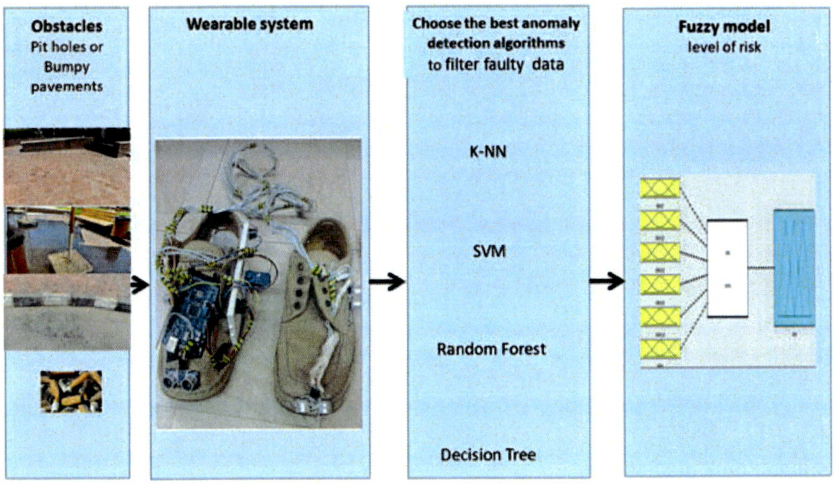

Fig. 2 System steps

4 Implemented System

Based on the objective of the system, the type and the allocation of the sensor are decided. The main sensor attached to the shoes is the ultrasonic sensor to detect obstacles [17, 18].

4.1 Wearable IoT Device

This system uses two types of sensors, namely ultrasonic sensor and flame sensor, to detect obstacle like pumps, uneven floor, pit hole, and fires. The main components of the system are 'At Mega 2560' microcontroller board to connect these sensors, a battery, and a SD memory. The following section has more details of the system, and how the above components are attached (see Figs. 3 and 4).

1. System Component

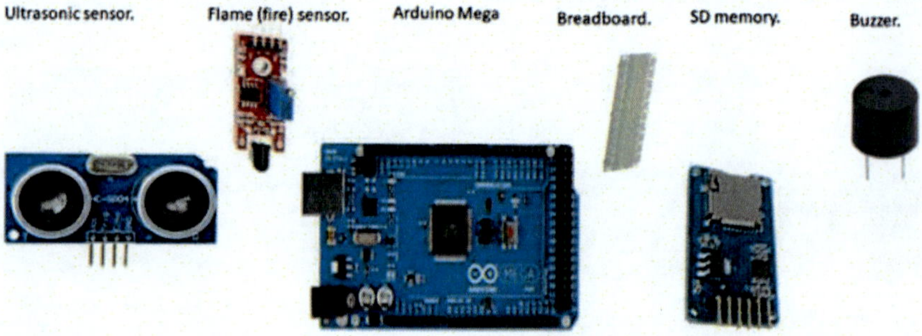

Fig. 3 System components

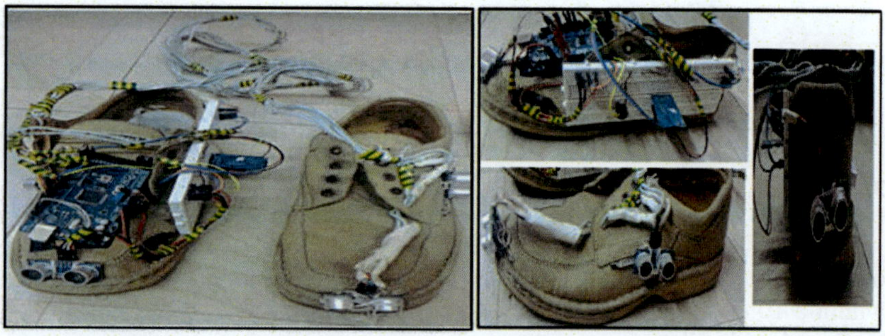

Fig. 4 Sensors attached to the shoes

At Mega 2560 is a microcontroller board-based that is designed for projects that require more I/O lines, more sketch memory, and more RAM. It contains 54 digital input and output pins, 16 analog inputs 4UARTs (hardware serial ports), a16MHz crystal oscillator, a USB connection, a power jack, an ICSP header, and a reset button (see Fig. 3). Ultrasonic sensor measures the distance of a target object by emitting ultrasonic sound waves and converts the reflected sound into an electrical signal. The ultrasonic module sends waves in the shape of pulses, and the angle of waves spread is less than 15°. With a 5 V, the average distance detected is from 1 to 5 m. Flame sensor detects flame or a light of the wave length of light between 760 and 1100 nm with an angle of 60°. It has a comparator chip LM393 that makes module readings stable. It works with 3.3–5 V voltage. The buzzer is a tiny speaker that can be connected directly to an Arduino board. It works when electricity is applied because electricity effects crystals and changes its shape. The buzzer operates around the audible 2 kHz range. It is activated with 12 V. The sound output at 10 cm is 90 dB. Micro-SD card modules allow communication with the memory card and write or read the information on them. It transfers data to and from a standard SD card to add mass storage and data logging to a project. The module interfaces in the SPI protocol. It stores up to 16 GB. A battery is used to support operating the buzzer, memory, and the six sensors (left ultrasonic sensor, right ultrasonic sensor, front ultrasonic sensor, rear ultrasonic sensor, hole ultrasonic sensor, and fire ultrasonic sensor). It is A 9 V battery and made up of small individual 1.5–V cells (see Fig. 3).

2. The Sensors Placement

   The sensors are fixed on both sides of the shoes. On the right shoe, three ultrasonic sensors are fixed: the 'right' sensor is fixed on the right side of the right shoe, the 'front' sensor is fixed on the front of the right shoe, and the 'rear' sensor is fixed on the back of the right shoe. All sensors are facing forward to detect walls or any obstacle.

   The Arduino board is fixed on top of the right shoe, and the flame sensor is fixed on the Arduino board. On the left side of the right shoe, the breadboard is fixed to attach the memory and the wires. On the left shoe, the 'left' ultrasonic sensor is fixed on the left of the left shoe. The 'hole' ultrasonic sensor is fixed on the front of the left shoe, and it is pointing downwards to detect holes (see Fig. 4).

3. Rules of How the Sensors Are Working

First, the left, right, front, and rear ultrasonic sensors are facing forward to detect obstacles and any objects. Then, the buzzer is activated when an obstacle is detected within a 10 cm away from the sensors. Second, the hole ultrasonic activates the buzzer when the distance between the sensor and an obstacle is equal or more than 120 cm. Third, the flame sensor triggers the buzzer when it detects a fire. Fourth, every second, the SD memory saves the reading of the six sensors (left, right, front, rear, and hole ultrasonic sensor and fire sensor). Finally, the buzzer tone varies based on the risk level indicated from smart agent implemented.

4.2 *Developing a Smart Agent*

The sensors data has a marginal error, i.e., a percentage of faults and noise [20] and [24]. It can send inaccurate information to the visually impaired person. These records are annotated to be noise records, and four supervised anomaly algorithms, namely K-nearest neighbors (K-NN), support vector machine (SVM), decision tree, and random forest tree, are applied to select the best and use as a filter to separate the noise from the sensors data. To apply these algorithms and compare among them, the wok followed this methodology.

1. Data Collection

The data collected at the food court of the university which is an area of 500 m' square. It has different obstacles and stairs, umbrella, ramp, and few light cigarettes were added (see Fig. 2). 11,710 records were saved in the memory. Each record contains the readings of the six sensors. The data collected at the food court of the university which is an area of 500 m' square. It has different obstacles and stairs, umbrella, ramp, and few light cigarettes were added (see Fig. 2). 11,710 records were saved in the memory. Each record contains the readings of the six sensors.

2. Data preprocessing

Before applying any of the algorithms, the preprocess has been done in two steps. First, the missing data or noise data where labeled into noise or not noise was encoded into 1 and 0. Second, it is a requirement to normalize the dataset to apply machine learning. The data is scaled to lie between a given minimum and maximum value of zero and one. Minmax normalization equation is below (1).

$$X \text{ normal} = \frac{x - \min(x)}{\max(x) - \min(x)} \tag{1}$$

3. Cross-validation

The cross-validation is applied to enable generalizing the model created from a limited sample data. It presents how a classifier algorithm may perform once the distribution of training data gets changed in each iteration. The data is randomly partitioned into equal size subsamples. The process of the cross-validation is repeated, and the results can be the averaged for a single estimation. The data is split into five subsets or folds.

4. Results and Analysis

To compare the four machine learning algorithms, the ROC curves and AUC values are applied. The receiver operating characteristic (ROC) curve is created to compare the performance of the machine learning-based models. ROC curve shows the performances of the algorithm in each fold. The curve plots the true positive rate (TPR) and false positive rate (FPR). Equations 2 and 3 calculate the TPR and the FPR. A steep slope at the beginning of the curve shows a higher true positive (correct) classification of the algorithm, whereas increasing the FP rate causes the curve to flatten. Area under the ROC curve. (AUC) is the entire two-dimensional area underneath the entire ROC curve. It is from (0,0) to (1,1).

$$TPR = TP/(TP + FN) \quad (2)$$

$$FPR = FP/(FP + TN) \quad (3)$$

Machine learning algorithms were employed to compare the performance. The analysis of the binary classification of noise data results shows that, among the machine learning-based algorithms, random forest attains the highest accuracy 99% with a Std.Dev of ± 0.05 followed by decision tree 93% with a with a Std.Dev of ± 0.03 and SVM and K-NN 89% with a Std.Dev of ±0.05 and ± 0.06, respectively. Moreover, ROC was applied in order to show machine learning algorithms results. Figures 5, 6, 7 and 8 show the ROC curves and AUC values for all four machine learning models in each fold. The AUC accuracy results indicate the robustness and promising performances. The following step is the assessment of the level of risk using fuzzy logic.

Fig. 5 Random forest ROC curve

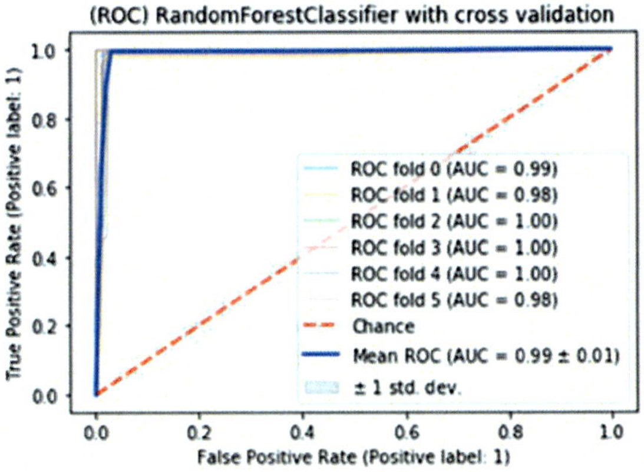

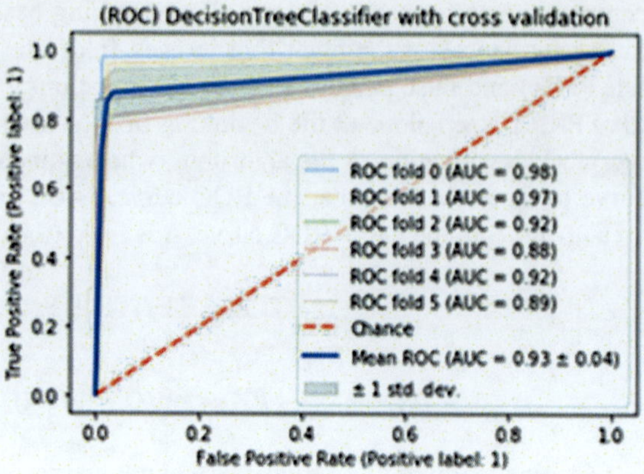

Fig. 6 Decision tree ROC curve

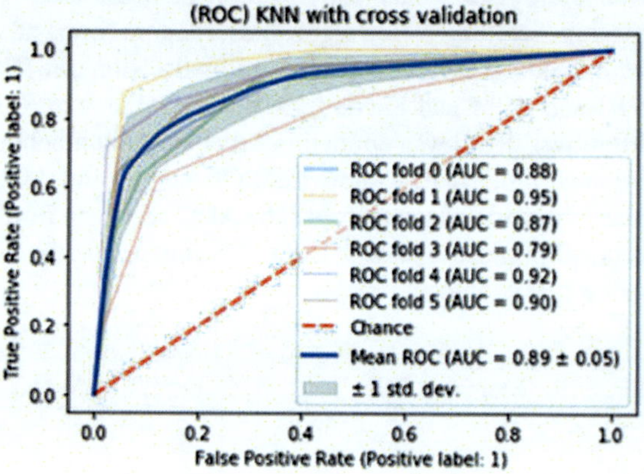

Fig. 7 K-NN ROC curve

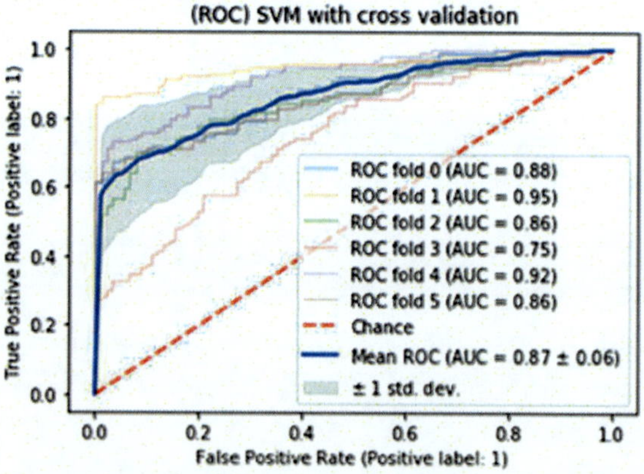

Fig. 8 SVM ROC curve

4.3 Fuzzy Logic

Fuzzy logic (FL) is similar to the human reasoning. It produces an acceptable and definite output. It works on multivariable (0, 0.5, 0 0.6,1, etc.) that involves all intermediate possibilities between digital values YES and NO, i.e., 0 and 1. Fuzzy models deal with problems relating to ambiguous, subjective and imprecise judgments. The fuzzy models recognize and represent the vagueness in the data.

1. Input data

The model has six input. They are the data the six sensors, namely left ultrasonic sensor, right ultrasonic sensor, front ultrasonic sensor, rear ultrasonic sensor, hole ultrasonic sensor, and flame (fire) sensor.

2. Membership function

It is mapping a set to have degrees of membership between the intervals of 0 to 1. In that respect, it gives a grade for each object in the range of 0 to 1. Triangular fuzzy function, which is the shape of membership function, is simplest and most appropriate. It has three points minimum, maximum, and modal. The ranges of the six sensors are as follows: low is between (0, 0.3), medium is between (0.31, 0.7), and the high is between (0.71, 1). Figures 9, 10, 11, 12, 13 and 14 represent the membership function of each sensor input.

3. Fuzzy Rule-Based Model (FRBM)

FRBM or a fuzzy inference system has a number of rules to produce the output. This system supports impaired vision, and identifying the different levels of risk whether low, medium, or high is of a value for these people. The fuzzy inference has 30 rules. There are three scenarios for the risks. The first scenario, in Fig. 15,

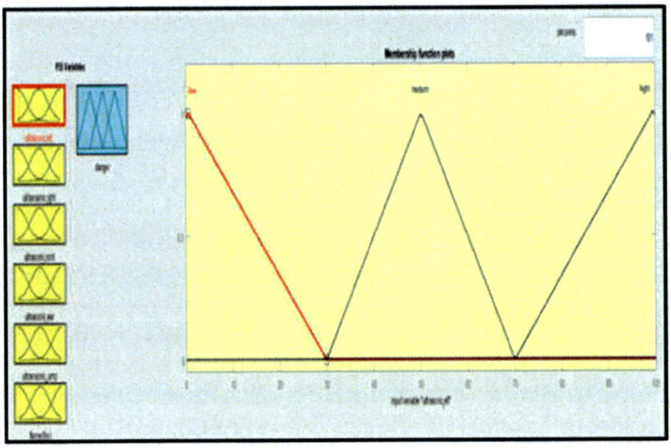

Fig. 9 Left ultrasonic sensor membership function

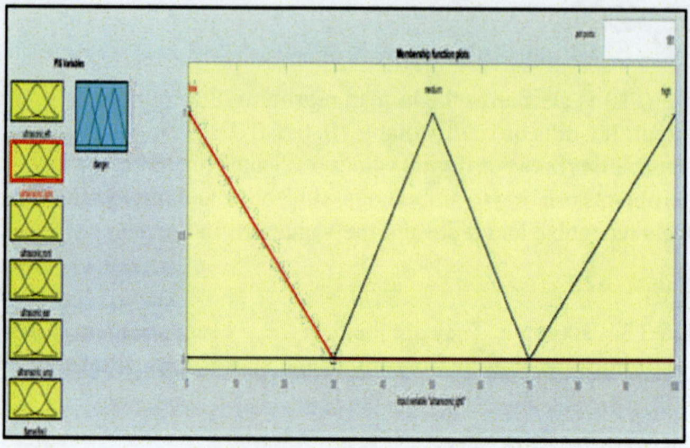

Fig. 10 Right ultrasonic sensor membership

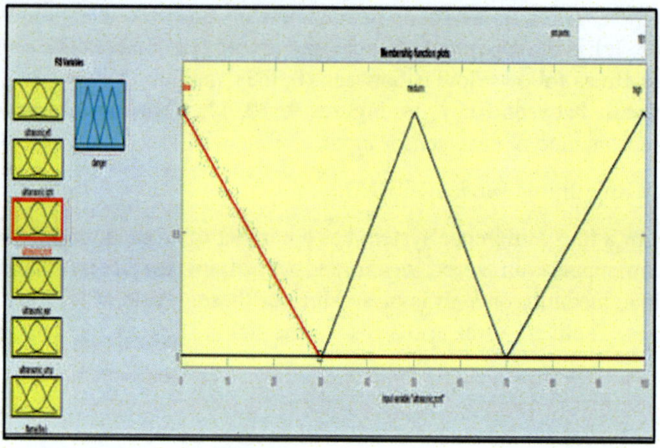

Fig. 11 Front ultrasonic sensor membership

it shows the rules and the combinations of sensors value and it is a low risk. These are the sensors values: left (low) = 22, right (low) = 1.5, front(low) = 26.4, rear (low) = 19.6, hole (low) = 18.8 flame (fire) (low) = 21.4, and the rate of total risk is 13.8 which means the rate of totally risk is (low) because the low risk the rate is between (0, 30). The second scenario where the risk is medium, in Fig. 16, it is the combinations of the sensors values. These are left (medium) = 50, right (medium) = 50, front(medium) = 50, rear (medium) = 50, hole (medium) = 50, flame (fire) (medium) = 50, and the rate of total risk is 50 which means the rate of totally risk is medium because the medium risk the rate is between (31, 70).

The third scenario, in Fig. 17, the values of sensors in the high risk are left (high) = 60.2, right (high) = 90.2, front(high) = 87.1, rear (high) = 90.7, hole (high) =

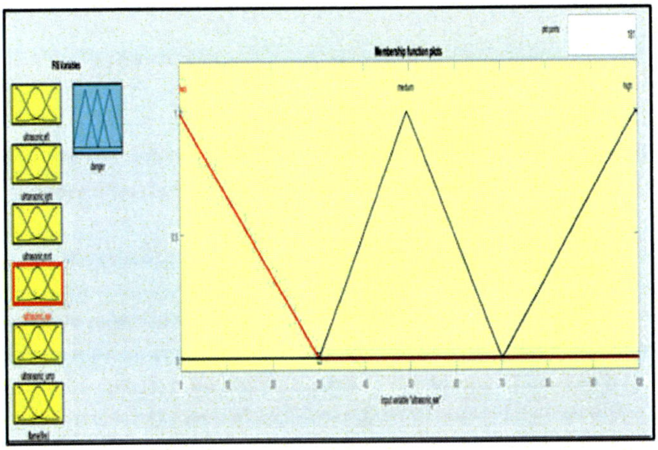

Fig. 12 Rear ultrasonic sensor membership function

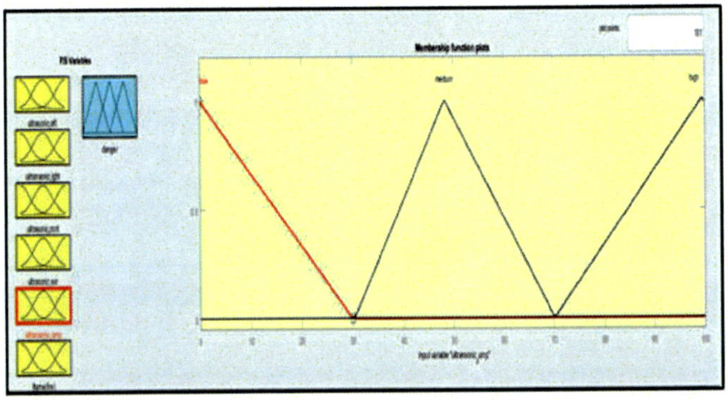

Fig. 13 Hole ultrasonic sensor membership function

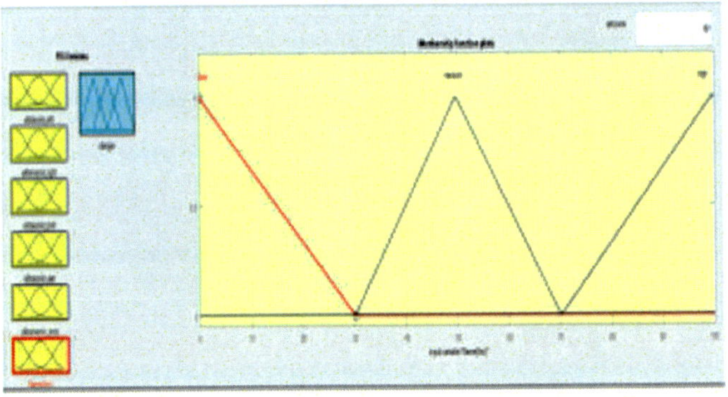

Fig. 14 Flame sensor membership function

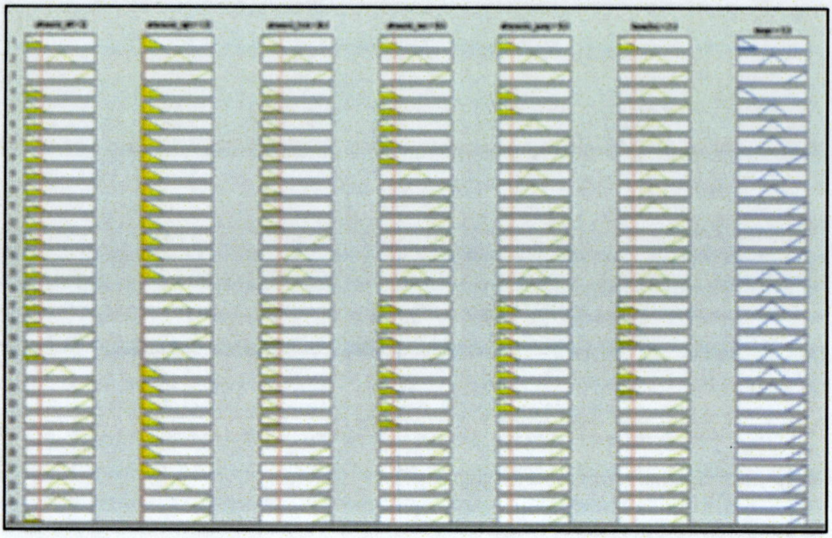

Fig. 15 Rate of risk is low

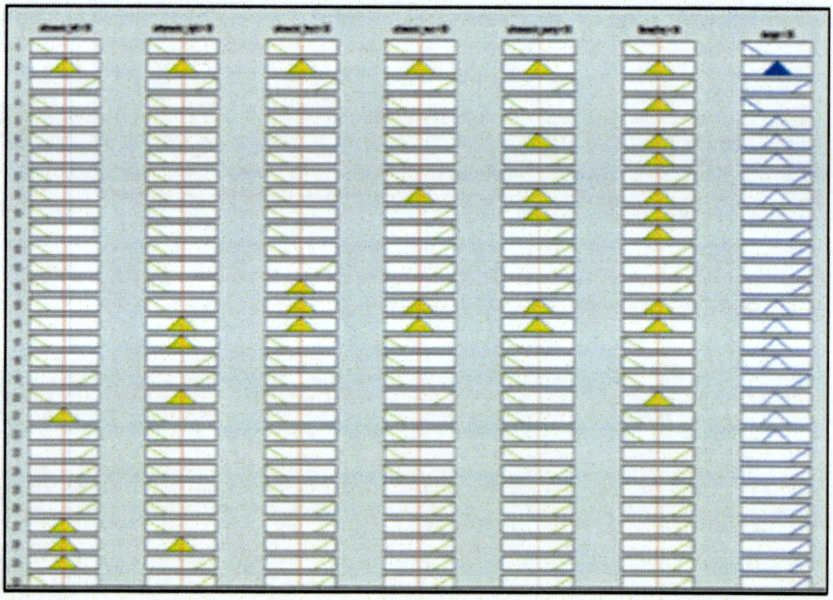

Fig. 16 Rate of risk is medium

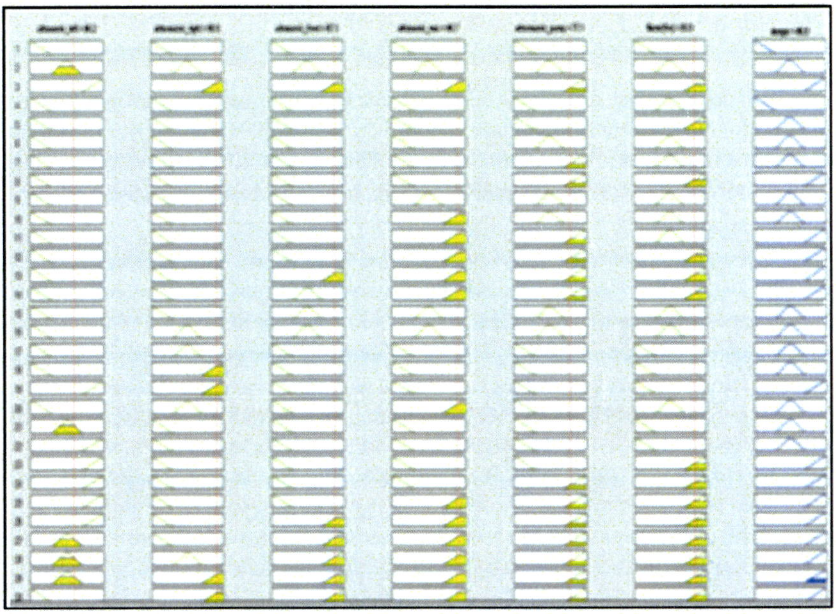

Fig. 17 Rate of risk is high

77.1, flame (fire) (high)-83.9, and the rate of total risk is 86.9 which means the rate of totally risk is (high) because the high risk the rate is between (71, 100).

5 Discussions

This work has developed shoes that detect obstacles and warn impaired vision people. One of the contribution is it supports impaired vision walking in multiple environments. The shoes will help them to walk indoors and avoid obstacles on the floor. In outdoors like pedestrian road, it will detect pit holes and pumps. In case, they walk in parks or forests, and it can notify them against pushes of fires. Elmannai and Elleithy [2] devised a list of the guidelines for designing satisfactory performance of assistive devices [2].

In light of these guidelines, the following session discusses the work contributions. The guidelines are simple, economically accessible, wearable, performance, and reliability.

5.1 Simple and Economically Accessible

This work developed an affordable and simple design. It embeds sensors in the shoes. The design does not confuse the user, and it is understandable because it tells them the location of the obstacle comparatively to where they stand, i.e., in front, rear, left, right of the visually impaired person. The attached sensors are considerably cheap sensors and attainable.

5.2 Performance and Reliability

Work of [17, 18] embed sensors to the shoes. The motivation of their work is to identify obstacles on the floor. They have used either both shoes or only one (see Fig. 18). The general steps of the work of these devices are: getting the data from the sensors [16–18], may process the data then sending an alert to the user.

Wu et al. [18] have attached sensors to shoes and applied a case-based algorithm to give information about any obstacle to the user through vibration and send an emergency call. Vignesh et al. [16] work attached sensors, and users get information on the mobile. Xu et al. [17] applied the SVM to classify the different body posture. In this work to ensure the reliability and the performance of the detecting the surrounding obstacles, it attached the sensors in both shoes (see Fig. 4). Additionally, their systems did not encounter the noise and error [24, 25] of the sensor data. This work considered applying anomaly machine learning algorithm to improve the performance and reliability of the system. The algorithm filters the data and the system sends only information about the obstacles whether it is in a front, rear, left, right of the person or it is a fire. The accuracy result of the random forest is the best. 99%. The integration of fuzzy logic and IoT provides extra information for the visually imparted about the risk level.

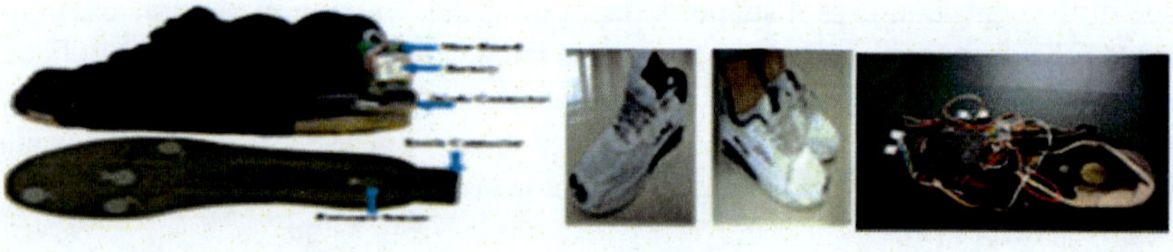

Fig. 18 Sensors attached to different wearable shoes [16–18]

5.3 *Wearable*

Wearable technology allows impaired vision mobility and walking freehand. Lee et al. [25] discuss the challenges and the factors to maintain sustainability of wearable devices. These factors are cost-effective, have a long battery life, durable, i.e., waterproof or shockproof, and scalable, i.e., easily produced/developed. This work's system attached sensors to wearable shoes. It is cost-effective, and there are examples in the market where the sensors have been embedding to the shoes [24]. The scalability and durability of this system are attainable in case it is taken to the industrial level.

6 Limitations

To gain confident in the reliability and usability of our system, different empirical experiments are required to test the following factors. For example, the perception of the users, i.e., young or old. Elderly acceptance and use of technology are different from the younger generation. The second factor is the environment to test detecting the obstacles indoors and outdoors. The third factor is the buzz feedback and the different users' impressions, for instance, the response time to the feedback or the volume of the sound.

7 Conclusions and Future Work

This work presents novel, affordable, cheap, wearable shoes to support visually impaired. It allows them to walk freehand in different environments indoors, outdoors, a park and avoid obstacles. The shoes have sensors embedded around the shoes to detect obstacles in the front, rear, right, and left. The system improves the performances of the sensors by using anomaly algorithm and compares among the accuracy of four algorithms, namely K-NN, SVM, decision tree, random forest. Random forest has the best accuracy. The fuzzy rules inform the visually impaired about the seriousness of the risk and give three levels of high moderate or low risk to help them to take the appropriate action. In the future, more empirical experiments with different users in different environments, and feedbacks are required to gain confident in the system. Adding another feature such as maps to guide for paths and roads enables having an integrated system.

References

1. WHO "No Title." [Online]. Available https://www.who.int/news-room/fact-sheets/detail/blindness-and-visual-impairment
2. Elmannai W, Elleithy K (2017) Sensor-based assistive devices for visually-impaired [1.0] people: current status, challenges, and future directions. Sensors (Basel) 17(3):565–582
3. Riazi A, Riazi F, Yoosfi R, Bahmeei F (2016) ScienceDirect outdoor difficulties experienced by a group of visually impaired Iranian people. J Curr Ophthalmol 28(2):85–90
4. Lin B, Lee C, Chiang P (2017) Visually impaired people. Sensors 17:1–22
5. Nowshin N, Shadman S, Joy S, Aninda S (2017) An intelligent walking stick for the visually-impaired people. pp 94–101
6. Helmy M et al. (2014) Smart cane: assistive cane for visually-impaired people smart cane: assistive cane for visually-impaired people. Int J Comput Sci
7. Dakopoulos D, Bourbakis NG (2010) For blind: a survey. 40(1):25–35
8. Brashear H, Maribeth Gandy TS, Westeyn T (2008) Wearable systems designissues for agingor disabled users. In: Helal BA, Abdelsalam (Sumi), Mokhtari M (ed) The engineering handbook of smart technology for aging, disability, and independence. Wiley, Inc., pp 317–338
9. Bai J, Lian S, Liu Z, Wang K, Liu D (2019) Virtual-blind-road following based wearable navigation device for blind people. IEEE Trans Consumer Electron
10. Jiang B, Yang J, Lv Z (2019) Wearable vision assistance system based on binocular sensors for visually impaired users. IEEE Internet of Things J 6(2):1375–1383
11. Zientara PA, Lee S, Smith GH, Irick KM (2017) Third eye : a shopping assistant for the visually impaired. Computer 50:16–24
12. Herghelegiu P, Burlacu A, Caraiman S (2017) Negative obstacle detection for wearable assistive devices for visually impaired. In: 21st international conference on system theory, control and computing (ICSTCC)
13. Siddhartha B, Chavan AP, Uma BV (2018) ScienceDirect an electronic smart jacket for the navigation of visually impaired society. Mater Today Proc 5(4):10665–10669
14. Alvarado JA, Chavez MDP, Rodriguez RA, Embs I, Universidad P (2016) Design of a wearable technology for the visually impaired people. In: 2016 IEEE ANDESCON, 19–21, Peru, pp 1–4
15. Marufur M, Milon R, Shishir I, Saeed A, Khan A (2020) Obstacle and fall detection to guide the visually impaired people with real time monitoring. SN Comput Sci
16. Vignesh N, Meghachandra Srinivas Reddy P, Nirmal Raja G, Elamaram E, Sudhakar B (2018) Smart shoe for visually impaired person. Int J Eng Technol 7:116–119
17. Xu Q, Gan T, Chia SC, Li L (2016) Design and evaluation of vibrating footwear for navigation assistance to visually impaired people. In: IEEE international conference on internet of things (iThings) Things (iThings) and IEEE green computing and communications (GreenCom) and IEEE cyber, physical and social computing (CPSCom) and IEEE smart data, December, pp 305–310
18. Wu W, Ning L, Junquan T (2017) Smart shoes for obstacle detection. In: The 10th international conference on computer engineering and networks, pp 1317–1326
19. Gokalgandhi D, Kamdar L (2020) A review of smart technologies embedded in shoes
20. The HY, Liehr AWK, Wang KIK (2020) Sensor data quality: a systematic review. J Big Data 1–50
21. Gaddam A, Wilkin T, Angelova M (2020) Detecting sensor faults , anomalies and outliers in the internet of things: a survey on the challenges and solutions. Electron Article 1–15
22. Tayyaba S, Ashraf MW, Alquthami T, Ahmad Z (2020) Fuzzy-based approach using IoT devices for smart home to assist blind people for navigation. Sensors 1–13
23. Karakaya S, Faculty E, Ocak H, Faculty E (2019) Fuzzy logic-based moving obstacle avoidance method. Comput Sci Theory Res 9(1):1–9
24. Nike, "No Title." [Online]. Available https://www.pinterest.it/pin/442478732109652967/
25. Lee J, Kim D, Ryoo H, Shin B (2016) Sustainable wearables: wearable technology for enhancing the quality of human life. Sustainability 1–16

Conceptual Framework of Database Development on Bidong Island: The Case Vietnamese Boat People (VBP) Campsite Facilities for Historical Tourism

Dazmin Daud, Nursyamilah Annuar, and Antashah Mohd Nor

Abstract The lack of maximizing the technology in today's historical tourism is the main concern for the success in this area. Previous studies have proven the success rate of using the current innovation and trends in general tourism sector in which they help to increase the visibility of the tourist attractions and number of visitors in one tourist spot. Terengganu, one of the states in Malaysia, is one of the best tourist attractions in the country. It is well-known for its popular islands, namely Redang Island, Lang Tengah Island, Kapas Island and Tenggol Island. Yet, many local and international tourists are not aware of the high potential historical island called Bidong Island. Not many tourists go there. Many come due to the informal communication such as from mouth to mouth with little formal promotions and marketing activities. Bidong Island was used as Vietnamese Boat People (VBP) refugee camp facilities from 1978 to 1991. Currently, this lonely island has started to show her visibility. This is a good sign of the state's domestic tourism activity that will contribute towards the Gross Domestic Profits in tourism sector. Therefore, the aim of the present study is to provide an overview of database development on the Bidong Island for the Vietnamese Boat People former campsite facilities. A field observation study is proposed for the actual study.

Keywords Database development · Bidong Island · Vietnamese boat people · Historical tourism

D. Daud
Faculty of Business and Management, UCSI University, 56000 Cheras, Kuala Lumpur, Malaysia
e-mail: dazmindaud@ucsiuniversity.edu.my

N. Annuar (✉)
Faculty of Business and Management, Universiti Teknologi MARA, Cawangan Perlis, Kampus Arau, 02600 Arau Perlis, Malaysia
e-mail: nsyamilah@uitm.edu.my

A. Mohd Nor
Faculty of Communication and Media Studies, Universiti Teknologi MARA, 40000 Shah Alam, Selangor, Malaysia
e-mail: antashah@uitm.edu.my

X.-S. Yang et al. (eds.), *Proceedings of Seventh International Congress on Information and Communication Technology*, Lecture Notes in Networks and Systems 447,
https://doi.org/10.1007/978-981-19-1607-6_2

1 Introduction

From 1978 to 1991, the presence of Vietnamese Boat People (VBP) in Bidong Island was increasing. By the time, the Island was closed as a refugee camp in 1991, about 250,000 Vietnamese had passed through the camp according to the United Nation High Commissioner for Refugees website, ohchr.org. The VBP called this island as "Little Saigon", remembering their former capital city of Saigon, Vietnam when they left. In 1979, Malaysia has started to receive thousands of VBP and the numbers were increasing. This alarming number triggered Malaysia to cooperate with the United Nations High Commissioner for Refugees (UNCHR) to set up shelter facilities at Bidong Island. The status of the shelter facilities there remained as a temporary shelter before the VBP was accepted by the third countries.

Since 1979, this island has been used as a safe place for VBP to stay. The Government of Malaysia together with the International Red Cross Society had planned and organized for basic facilities in the island. This included main administrative building, hospital, schools, vocational training centre, church, quarters and even a place for a graveyard. The island also was equipped with a jetty and basic logistics facilities such as freshwater tank and dry food storage. The closure of the island in 1991 which has placed up to 250,000 of VBP was later than accepted in third countries such as America, Canada, Australia and France. Some of them were sent back to their motherland, Vietnam.

The closure of the VBP facilities in 1991 has led to the abandonment of the facilities. In November 2020, our team visited and explored the abandoned VBP facilities on this Island. Findings from this visit show that the facilities are in the status of environmental damage. Despite the continuous effort done by the State Museum of Terengganu as a caretaker for the island, it faces huge challenges in terms of manpower and finance. The facilities in this Island are totally ruined as the main materials to build them are from woods. Only the concrete pillars and bricks remain intact. The effort from the State Museum is tagging these remaining pillars and bricks. From time to time, it needs to be maintained and administrated. The tagging is done in a simple way—printed simple photograph and name/label with laminated plastic. Then it is placed with an iron stick as a signage. Figure 1 shows the signage. When we visited these signages, they were in bad condition due to environmental damage.

There are two issues that trigger a research paper from the above background of the study. First, developing a database for the former VBP facilities at Bidong Island can help the State Museum to update and to complete information that is currently not in a completion mode. Once the database is completed, the Museum can provide the information using a virtual platform for visitors to view virtually the former facilities on the Island. The argument here is that not all public are able to visit the island regularly. The current limited basic facilities and out of sea-route destination make the visit to the island not a loop for tourists. With the virtual information from the Museum, the public can save travelling cost to the island. At the same time, it preserves the island from potential pollution incurred by human being.

Fig. 1 Signage that shows former visual and audio school for the VBP. Behind the signage, there are two concrete pillars of the school. (*Source* Researchers personal photo taken in Bidong Island 2020)

Second, the collaboration between the researchers and the State Museum regarding database development on VBP former facilities in Bidong Island provides a competitive advantage for both parties. The researchers have the equipment and expertise in developing the database. This will assist the lack of manpower and expertise from the State Museum side. By having updated database, it is much easier for the Museum to plan the historical tourism model with other State Government agencies such as State Tourism. Furthermore, the Museum can focus on other historical areas which need further attention.

The above two issues provide these specific research objectives for this paper:

1. To provide an overview of how the database development on former VBP facilities in Bidong Island is done; and
2. To assist State Museum of Terengganu in promoting the former VBP facilities in Bidong Island as historical tourism from the outcome of the database development.

The following research question is addressed. Does database development for the former VBP facilities in Bidong Island provide a solution for the Terengganu State Museum to make the former facilities as historical tourism?

2 Literature Review

2.1 VBP at Bidong Island

Bidong Island is a tiny island off the coast of Terengganu. In 1978, the Malaysian Government designated the island as the authorized refugee camp until it was officially closed in 1991. During this period, the Malaysian government had instructed any arriving boatloads of refugees to land on the island. The island at that time served as a temporary home and transit point for thousands of refugees. At the island, they lived in unsanitary living conditions in basic shelters such as boat timbers, plastic sheeting, flattened tin cans and palm fronds. The last group of VBP left the island in 1991. Since then, it was returned to the State Government of Terengganu. The facilities there remain unoccupied and gradually ruined due to weather factors.

Previous studies have shown various findings about VBP and Bidong Island. However, the previous studies were lacking in the findings related to the VBP and Bidong Island from the perspective of database development. Previous studies have focused on green conservation for flora and fauna [12, 23], general tourism [33], weather impact [28] and sustainability in pure science [32].

Issue about the VBP in Bidong Island has always been the scope of previous studies. Majority of the previous studies focused on issue that relate to international policy on refugees, politics and crisis. Baharuddin and Enh [5] provided insights into the issues related to VBP from the perspective of diplomatic relationship between Malaysia and Vietnam. Through the methodology of content analysis, their findings revealed that the presence of the refugees in Malaysia sovereignty, caused a serious national security issue. From the other side, the VBP issue is able to make Malaysia gain experience in handling refugees. Malaysia is able to assist the UNHCR in handling refugee crisis from Myanmar and Bangladesh by following strictly the Refugee Convention of 1951 [4].

Meanwhile, [7] focused on the output from their qualitative study about the VBP. Their findings were based on the personal experience of former VBP who revisited Bidong Island to remember those who have perished there. The experience of returning was an emotional and uncanny for the former refugees. Both participants and observers reported that the trip was a catalyst for the telling of refugee stories and their memories. A similar personal experience study was done by Quynh Giao [25]. This is a personal case study about female VBP on the Island. The content was about a modest living condition in the facilities before she was transferred to the third country.

2.2 Database Development

Database development simply means plan, construct, build and maintain a system so that an organization can stay organized and use their data effectively. It helps organizations to organize and track information. A literature about database development can be traced back as early as 1990 pertaining its roles and importance [29]. For the past 30 years, studies that involved database development have clearly indicated the importance of database development. Recently, database development is applied in pure science, engineering, operations management, technology, business and social sciences [10, 18, 20, 24, 27, 31].

In a study of an island, database development has been used widely in previous studies. Majority of the previous studies use database development for monitoring volcanic activities on islands and for studying social sciences [6, 19]. From the perspective of tourism, [16] have successfully outlined the geographic information systems (GIS) database into visitor satisfaction in Boston Harbor Islands National Recreation Area (BHINRA). Visitors can easily access the database for information that is needed. From the point of view of the BHINRA operator, the database provides controlling mechanism for visitors coming in and out. Database from the GIS has also been used in selecting site on an island for aquaculture tourism. In addition, a study from [22] also shows the importance of database in sustaining tourism industry in Canary Islands.

Literature about the application of database development for the Malaysia islands has demonstrated the important findings. GIS was used as a base tourism decision by the public for vacation in Langkawi Island [17]. In another study, the geographic (spatial) information technology was used for computing, storing and manipulating spatial information for tourism information by the State Government of Kedah, Malaysia, for tourism in Langkawi Island [3]. The extension from the previous study has allowed for public participation as an information supplier. This has led to collaborative tourism planning for Islands in Malaysia [17].

To execute database development, it needs a tool. To ensure database development is reliable, the 360° media is used. In photography, an omnidirectional camera (from "omni", meaning all), also known as 360° camera, is a camera having a field of view that covers approximately the entire sphere or at least a full circle in the horizontal plane. From the perspective of tourism, embedding 360° videos into a travel application will provide viewers a tour as if they plan an actual vacation. It helps the prospective tourists on a virtual walkthrough of their upcoming vacation. Hence, the use of this application will be beneficial for tourism marketing [1].

Previous studies have investigated the 360° application in tourism sector both locally and internationally. However, studies conducted mainly focused on developed countries. In addition, the studies show how the application is assisting tourists to visualize before their actual trip to the choice of tourist attraction [8, 14, 21].

2.3 *Historical Tourism*

The uniqueness of Bidong Island is that it has a VBP legacy, being a former refugee centre. When it is compared with other Islands in Malaysia, the tourism activity at this island is more to the historical tourism concept. According to [30], historical tourism tends to have a different kind of visitors. The visitors are interested in visiting historical sites. They visit and view places that are related to historical events. Tourists take an interest in historical information which is linked to the respective points of interest. In the case of Bidong Island, it is about the VBP former facilities. Here, it can be linked with a study from [13] where a database was developed for cemeteries in Europe for recording cemeteries and cultural preserves. In Europe, for example in Spain, the concern about the importance of historical tourism has gained much attention from the public [9]. The importance of and the need for historical tourism,

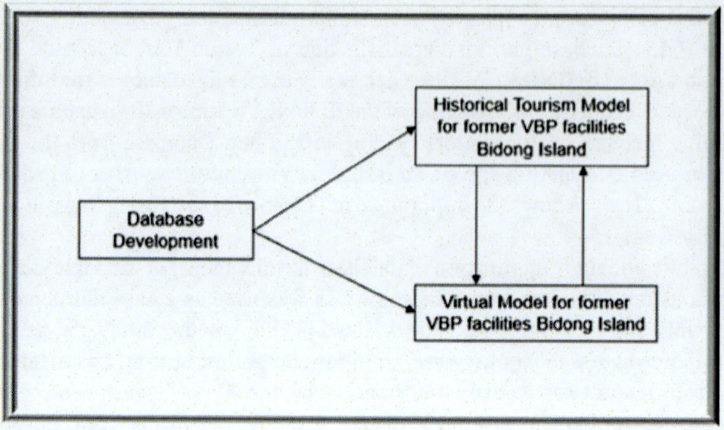

Fig. 2 Conceptual framework for former VBP facilities in Bidong Island, database development and historical tourism

therefore, cannot be denied. With the aid of database development, historical tourism can provide ways of activating sustainable tourism strategies [11].

From the review of the literature, a conceptual framework for showing an association between former VBP facilities at Bidong Island, database development and historical tourism is developed. Figure 2 shows the conceptual framework.

From Fig. 2, database development supports the historical tourism model for the former VBP facilities in Bidong Island. At the same, the State Museum through the database development can develop the virtual former VBP facility for those who are not able to visit the island physically. There is a reciprocal coordination between historical tourism model and virtual model. Three propositions are developed from this conceptual framework:

1. There is an association between database development and historical tourism model for former VBP facility in Bidong Island;
2. There is an association between database development and virtual model for former VBP facility in Bidong Island; and
3. There is an association between historical tourism model and virtual model that is related to the former VBP facility in Bidong Island.

3 Proposed Research Design

A research design that relates to qualitative research is proposed for this study. This study employs qualitative research to understand all the three associations that derive from the conceptual framework. For this purpose, the researchers are observing the status of the former VBP facilities in Bidong Island through in-depth interviews, site visit, application of 360° video–audio image capture and data recording [2,

15]. In addition, secondary data collection method is also being employed to gather documented information.

To reduce the area of possible interest regarding the mapping of the former VBP facility in Bidong Island, a set of in-depth interviews is conducted. This procedure is done because the present geographical area carries very little resemblance of a facility. It is therefore giving a clearer understanding of the causal mechanisms that lead to breakdown of abstract points. The respondents are the Head of Terengganu State Museum and operational supervisor from the Museum.

The aim for the site visit is to gather primary information about the former VBP facilities in Bidong Island. It is to prepare testimony addressing the purpose of the site visit. The site visit enables researchers to understand more clearly the micro-location of the former facility site: the visibility and prominence of the site, the current terrain, the current land use and the access to the site. For a bird's-eye view on the site, a drone (model DJI Phantom 3 Advance) and a high-resolution camera (model Canon 6D) are media tools that will be used. Drone with high-resolution cameras has the advantage of capturing the terrain from a so-called bird's-eye view with high longitudinal and lateral accuracy.

The employment of 360° media on this study will help the Terengganu State Museum enhance traditional applications with immersive content. This will later be used by the tourism agency in promoting the island as one of the historical tourism products in Terengganu. Embedding 360° videos in a travel application will provide viewers a tour as though they plan an actual vacation. It takes the prospective tourists on a virtual walkthrough of their upcoming vacation. Hence, the use of this application will be beneficial for tourism marketing [1]. The 360° media through a stereoscopic virtual reality (VR) can be displayed via desktop/laptop browsers and from mobile applications. Thus, it is much accessible to museum visitors or public even without additional VR hardware.

A dual lens of 360 model camera will be used. The dual lens is the 360 one resolution or R. The lens assists the camera to provide basic functions and features using 360 technology. The outcome is traditional flat video, high dynamic range (HDR) video, time-lapse mode and bullet time video. The camera allows to record up to 5.7 K resolution and is compatible with smartphone to control the camera with the remote-controlled application provided by camera manufacturer.

Database about the former VBP facility at Bidong Island then is developed from the data that derive from in-depth interviews, site visit, 360° media capture and secondary data. Once all the data are documented, the findings will be presented to the related stakeholder such as the Terengganu State Museum.

4 Discussion and Conclusion

This study aims to assist the State Government of Terengganu to promote Bidong Island as a historical tourism. At the same time, the study can provide a database from the 360° technique pertaining to the virtual VBP facilities to the State Museum of Terengganu. It should be noted that not all public visitors can have the opportunity to visit the island physically. Some of them are just visiting the State Museum for seeking knowledge and information. With the aid of the 360° technology, visitors at the museum can virtually "visit" the island and gain knowledge about the historical complex of the former VBP facilities.

The expected outcome from database development on the former VBP facility will further assist as tools of marketing communications. The use of the application gives the actual visual image on the island as a whole. Another expected outcome of the application is that it will be used by the tourism agencies such governmental agencies as well as private agencies. The usage of new media trends as marketing tools will help to reach the potential number of tourists visiting the island. A study conducted by Rahimizhian et al. [26] shows the acceptance of 360° application as a tool of promoting tourism destination which indirectly promotes electronic Word of Mouth (eWOM).

Database development in the tourism sector is not a new technology. The use of Geographical Information Systems (GIS), for instance, has been able to provide maps for users as a source of information or reference. From the scope of historical tourism, a complete mapping and information resources are very important to help domestic and foreign tourists reach their destinations accurately and safely. Regardless of whether the tourists are in Bidong Island or at the State Museum, database development helps to record essential information to the users. Continuous innovation will further be improved with the proper planning of the facility and for recording matters.

This product is a uniqueness to the State Museum to attract more visitors. The 360° can encourage the public to learn about the history of the island in relations to VBP. In addition, the State Government of Terengganu may use the findings from this study to disseminate more information about the VBP complex to local tourist agencies. Thus, it can promote the island as part of the tourism activity in the state.

For knowledge extension, the development of the database for mapping the Bidong Island using 360° technique can provide new area of study in mass communication. It can develop an association between mass communication and tourism for integrating knowledge. The public will perceive a new dimension of knowledge that have elements of mass communications and historical tourism. It can promote the actual tourist fieldwork to attract those who want to learn more about Malaysian history via mass communication. In future, it may create an opportunity for higher learning institutions to have a good collaboration with the museum State Government for other historical projects of mapping. A field observation study is proposed for the future data collection.

In conclusion, this paper presents a conceptual framework for the needs of database development of the former VBP facilities in Bidong Island. The proposed study endeavours to make both practical contribution and knowledge extension in the field of mass communication.

References

1. Adachi R, Cramer EM, Song H (2020) Using virtual reality for tourism marketing: a mediating role of self-presence. Soc Sci J 1–14. https://doi.org/10.1080/03623319.2020.1727245
2. Adnan WH, Daud D, Jamaluddin MF (2021) The forgotten World War II airfield: the case of Morib airfield. J Soc Sci 7(1):10–18. https://doi.org/10.37134/ejoss.vol7.1.2.2021
3. Ahmad A, Masron T, Osman MA, Mohammed B, Marzuki A (2011) Initial studies on web-based tourism decision support system (WBTDSS) case study: Langkawi Island, Kedah. In: Proceedings of 2nd reginal conference on tourism research, sustainable tourism research cluster, University Sains Malaysia, Penang, Malaysia
4. Ahmad AA, Abdul Rahim Z, Mohamed AMH (2016) The refugee crisis in Southeast Asia: the Malaysian experience. Int J Novel Res Humanity Soc Sci 3(6):80–90
5. Baharuddin SA, Enh AM (2018) Vietnamese refugees: a global issue in the history of Malaysia foreign affairs. J Nusantara Stud 3(1):1–18
6. Blahut J, Klimeš J, Rowberry M, Kusàk M (2018) Database of giant landslides on volcanic islands—first results from the Atlantic Ocean. Landslides 15:823–827. https://doi.org/10.1007/s10346-018-0967-3
7. Carruthers A, Huynh-Beattie B (2011) Dark tourism, diasporic memory and disappeared history: the contested meaning of the former indochinese refugee camp at Pulau Galang. The Chinese/Vietnamese Diaspora: Revisiting the Boatpeople. Ed. Yuk Wah Chan. New York, Routledge, https://doi.org/10.4324/9780203813102
8. Corbillon X, De Simone F, Simon G (2017) 360-degree video head movement dataset. In: Proceedings of the 8th ACM on multimedia systems conference. Association for Computing Machinery, New York, Unites States, pp 199–204
9. De Freitas IV, Sousa C, Ramazanova M (2021) Historical landscape monitoring through residents' perceptions for tourism: the World Heritage Porto City. Tourism Plann Develop 18(3):293–313. https://doi.org/10.1080/21568316.2020.1769717
10. Demirguç-Kunt A, Klapper L, Singer D, Ansar S, Hess J (2020) The global findex database 2017: measuring financial inclusion and opportunities to expand access to and use of financial services. The World Bank Econ Rev 34(1):2–8
11. Goussous J, Al-Jaafreh O (2020) Sustainable tourism development in historical cities case study: Karak, Jordan. In: Al-Masri A, Al-Assaf Y (eds) Sustainable development and social responsibility, vol 2. Advances in science, technology and innovation (IEREK Interdisciplinary Series for Sustainable Development). Springer, Cham. https://doi.org/10.1007/978-3-030-32902-0_34
12. Hamza A, David G, Mcafee A, Abdullah MT (2018) Annotated checklist of avifauna in Pulau Bidong, Malaysia. J Sustain Sci Managem 13(1):105–118
13. Johannsson H, Felicori M, Borgatti C, Caraceni S, Garutti L, Vysniauskiene A, Baliulyte I, Zabiela S, Sarris A, Peraki E (2007) An interactive graveyard information management tool and virtual memoriam database. In: Figueiredo A, Velho G (eds) Proceedings of the XXXIII computer applications and quantitative methods in archaeology conference (March 2005, Tomar, Portugal): The World in Your Eyes, pp 145–150
14. Kelling C, Vaataja H, Kauhanen O (2017) Impact of device, context of use, and content on viewing experience of 360-degree tourism video. In: Proceedings of the 16th international conference on mobile and ubiquitous multimedia. Association for Computing Machinery, New York, United States, pp 211–222. https://doi.org/10.1145/3152832.3152872
15. Lawrenz F, Keiser N, Lavoie B (2003) Evaluate site visits: a methodological review. Am J Eval 24(3):341–352. https://doi.org/10.1177/109821400302400304
16. Leung Y, Shaw N, Johnson K, Duhaime R (2002) More than a database: integrating GIS data with the Boston Harbor Islands visitor carrying capacity study. The George Wright Forum 19(1):69–78. https://www.jstor.org/stable/43597789
17. Masron T, Marzuki A, Mohamed B, Ayob NM (2014) Conceptualise tourism support system through web-based GIS for collaborative tourism planning. J Malaysian Instit Planners 13:59–90

18. Motohashi K (2020) Development of patent database in Thailand for assessing local firms' technological capabilities. World Patent Inform 63. https://doi.org/10.1016/j.wpi.2020.101998
19. Naidu S (2016) Does human development influence women's labour force participation rate? evidences from the Fiji Islands. Soc Indic Res 124:1067–1084. https://doi.org/10.1007/s11205-015-1000-z
20. Omidvarkarjan D, Cipriano D, Rosenbauer R, Biedermann M, Meboldt M (2020) Implementation of a design support tool for additive manufacturing using a feature database: an industrial case study. Progress in Additive Manuf 5:67–73. https://doi.org/10.1007/s40964-020-00119-5
21. Pasanen K, Pesonen J, Murphy J, Heinonen J, Mikkonen J (2019) Comparing tablet and virtual reality glasses for watching nature tourism videos. In: Information and communication technologies in tourism 2019, Springer, Cham, pp 120–131. https://doi.org/10.1007/978-3-030-05940-8_10
22. Pérez OM, Telfer TC, Ross LG (2003) Use of GIS-based models for integrating and developing marine fish cages within the tourism industry in tenerife (Canary Islands). Coast Manag 31(4):355–366. https://doi.org/10.1080/08920750390232992
23. Pesiu E, Abdullah MT, Salim J, Salam MR (2016) Tree species composition in Pulau Bidong and Pulau Redang. J Sustain Sci Managem 1:48–60
24. Pham-Duc B, Tran T, Trinh T, Nguyen T, Nguyen N, Le H (2020) A spike in the scientific output on social sciences in Vietnam for recent three years: evidence from bibliometric analysis in Scopus database (2000–2019) (2020). https://doi.org/10.1177/0165551520977447
25. Quynh-Giao NV (2007) Journey of the abandoned: Endless refugee camp and incurable traumas. Sings 32(3):580–584. https://doi.org/10.1086/510157
26. Rahimizhian S, Ozturen A, Ilkan M (2020) Emerging realm of 360-degree technology to promote tourism destination. Technol Soc 63:101411. https://doi.org/10.1016/j.techsoc.2020.101411
27. Reiller PE, Descosters M (2020) Development and application of the thermodynamic database PRODATA dedicated to the monitoring of mining activities from exploration to remediation. Chemosphere 251. https://doi.org/10.1016/j.chemosphere.2020.126301
28. Safuan CDM, Roseli NH, Bachok Z, Xia C, Qiao F (2020) First record of tropical storm (Pabuk–January 2019) damage on shallow water reef in Pulau Bidong, South of South China Sea. Reginal Stud in Marine Sci 35. https://doi.org/10.1016/j.rsma.2020.101216
29. Taddei A, Biagini A, Distante G (1990) The European ST-T database: development distribution and use. Computers in Cardiology 1990, CA, Los Alamitos, IEEE Computer Society Press, pp 177–180. https://doi.org/10.1109/CIC.1990.144191
30. Varfolomeyev A, Korzun D, Ivanovs A, Soms H, Petrina O (2015) Smart space-based recommendation service for historical tourism. Proc Comput Sci 77:85–91. https://doi.org/10.1016/j.procs.2015.12.363
31. Wang Y, Zhang S, Li F, Zhou Y, Zhang Y, Wang Z, Zhang R, Zhu J, Ren Y, Tan Y, Qin C, Li Y, Li X, Chen Y, Zhu F (2020) Therapeutic target database 2020: enriched resource for facilitating research and early development of targeted therapeutics. Nucleic Acids Res 48(1):1031–1041. https://doi.org/10.1093/nar/gkz981
32. Xin TJ, Shaari H, Ghazali A, Ibrahi NB (2020) Monthly physicochemical variation of tropica island groundwater of Pulau Bidong, South China Sea. Groundwater for Sustain Developm 10. https://doi.org/10.1016/j.gsd.2020.100358
33. Zakaria AA, Rahim NAA, Abdullah MT (2018) Reptile diversity as an ecotourism potential in Malaysia. Ecotourism Potentials in Malaysia 42–47

An Evaluation of Techniques for Classification of Conditional Sentences and Their Structural Components

G. B. Sanjana, Sundar Guntnur, and Shivali Agarwal

Abstract Conditional statements are an important part of procedural knowledge as they determine the decision points in the control flow. In order to bootstrap conversation bots and automation tools automatically from natural language procedure documents, it is important to be able to classify the conditional statements accurately and separate the condition and effect correctly. This paper aims at exploring three different techniques to classify and analyze conditional statements and discusses the advantages and drawbacks of each of them. This paper also aims at understanding the drawbacks of the three techniques and overcoming them by building models with better performance.

Keywords NLP · Support vector machine (SVM) · Rhetorical structure theory (RST)

1 Introduction

Many AI applications such as question answering, guided troubleshooting, and conversational agents depend on documents like guides and manuals to bootstrap automatically. A lot of such documents contain procedures and processes written in natural language. While most of the steps are instructional or action-oriented sentences, there are enough instances of conditional sentences which act as decision points. It is important to identify such conditionals and model the flow as per the condition and effect identified from the sentence.

G. B. Sanjana · S. Guntnur (✉)
RVCE, Bengaluru, India
e-mail: sundargs2000@gmail.com

G. B. Sanjana
e-mail: sanjana051198@gmail.com

S. Agarwal
IBM Research, Bengaluru, Karnataka, India
e-mail: shivaaga@in.ibm.com

X.-S. Yang et al. (eds.), *Proceedings of Seventh International Congress on Information and Communication Technology*, Lecture Notes in Networks and Systems 447,
https://doi.org/10.1007/978-981-19-1607-6_3

The paper aims at evaluating state-of-the-art techniques for identifying conditional statements and recognizing the condition, as well as the action or effect part of the statement. The three techniques explored are Rule-based approach, Support Vector Machine (supervised learning), and Rhetorical Structure Theory (discourse-based learning). The research questions that the paper sets out to address are as follows: (RQ1) What kind of conditional sentences can be handled well by each of these techniques? (RQ2) Can a model trained on a single conditional sentence be applied to blocks of conditional sentences? (RQ3) Can the structural components of a conditional sentence, namely condition and consequent (effect), be identified by these techniques?

The rest of the paper discusses how to address these questions and the insights obtained. Section 2 describes the tasks through which the three techniques are evaluated. Section 3 provides dataset overview and Section 4 gives the results of each of the techniques on each of the tasks. In Sect. 5, a detailed analysis of the strengths and weaknesses of each approach is carried out and concluded in Sect. 6. Section 5 also discusses methods adopted in our evaluation to reduce the weaknesses in three techniques based on observations.

1.1 Related Work

The Stanford dependency parser [1] can be used to detect the conditional statements using the SBAR node. The experiment results have an F1 score of 76.4. The main drawback of this model was that it does not provide in detail analysis of the conditional statement. Our model aims to solve this problem by exploring three models to identify the conditional statements, classify them, and extract the condition and effect/action part of the statement.

There is another method proposed in Automatic Generation of Conditional Diagnostic Guidelines [2] based on trigger words. LIBSVM was applied to make a binary decision for each trigger pair as to whether a phrase headed by trigger word is condition. The main drawback of this model was that the condition and the effect parts were not recognized. This is overcome in our models by building a dataset for an SVM model to classify a given part of the sentence as condition or effect.

Identifying condition action sentences using a heuristic-based information extraction [3] also describes a heuristics-based approach that creates a set of rules. Literature in [4, 5] suggested similar approaches that are using patterns based on a set of English connectives. These methods are not effective, because creating efficient patterns is difficult. The paper aims at creating rules that effectively recognizes the conditional statements, classifies them, and also extracts the condition and effect part of the statement. The rule-based model built by us has better performance in all of the above tasks.

Another work [6] deals with using two recurrent neural networks, namely encoder and decoder to mine conditions. A drawback of the above model is that it only identifies the conditions but fails to recognize the actions. The model proposed in

Table 1 Expected output of each task

Example	Task 1	Task 2	Task 3
If prompted, agree to app requests	Conditional	Simple	Condition—If prompted Effect—agree to app requests
If there are no gears and the torque is low, it probably would not allow the robot	Conditional	Complex	Condition 1—No gears Condition 2—torque is low Effect—it probably would not allow the robot to move forward

our paper performs the task of recognizing both condition and the action. The model in the above paper [6] uses a dataset in [7], the observed results have low precision, recall, F1 score, whereas the model proposed by our paper has better results.

2 Problem to Solve

The paper defines three tasks (refer to Table 1 for illustration of the tasks) as follows to evaluate each of the three techniques. The tasks have been defined based on experience in building conversation agents.

Task 1. Classifying the conditional statement: Given a sentence, classifying whether it is a conditional sentence or not.

Task 2. Classifying the type of conditional statement: Given a conditional sentence, this task predicts whether it is a simple or a complex conditional sentence. A conditional sentence which has multiple predicates in its condition is defined as a complex conditional.

Task 3. Identifying the effect and condition part of the sentence: This is used to recognize the action, and the condition governing the action in a sentence.

3 Dataset

The data used in evaluation of techniques was built from scratch by processing WikiHow [1] pages that had instructions to perform technical tasks. The WikiHow procedures were scraped using HTML2text [8] and then were cleaned to get procedures. The data was annotated to obtain the conditional statements and also the condition and the effect part of it. WebAnno [9] was used to annotate conditional sentences as well as the condition and effect part of it. The data has been divided into two parts: single sentence conditional statements and block conditional statements.

The final dataset consisted of single sentences and nonconditional sentences as the features, and they were annotated into three classes as simple, complex, and

Table 2 Dataset for the tasks

Type	Tag	Example sentence syntax	No. of sentences
Single sentence	Simple Complex	If (condition) (action) If (condition1 and/or condition2) (action)	400 20
Block sentence	If-else-if	If (condition1) (action1) Else if (condition n) (action n)	10
Nonconditional	Complex	The next step is as follows	400

nonconditional. The main aim of this paper is to evaluate only single conditional statements. Block sentences are a part of future work. The distribution of the data with respect to different types of conditional sentences is shown in the Table 2.

Apart from WikiHow, a small set of Lenovo troubleshooting documents were also used for model validation.

A dataset of conditions from English sentences with conditions [7] was used as test data for SVM model task 1 and task 2.

4 Approaches for Conditional Sentence Analysis

In this section, an overview of three techniques that were evaluated by the three tasks defined in Sect. 2 are provided.

4.1 Rule Based

In the following, rule-based technique built for the tasks is described. The ideas in literature [2, 3] are used to come up with the new set of rules that could fit the dataset. Corresponding rules are given in Tables 3 and 4.

Task 1. Conditional statements have the following trigger words—if, when, and unless. Words 'to', 'once', and 'on' were also considered as trigger words as these sentences could be converted to conditional statements. It is also observed that the conditional statements also have a verb close to the trigger words.

Table 3 Rules to perform Task 1 and Task 2

Type	Tag	Example sentence syntax
Simple	Left Right	trigger (*) (Verb) (*), (*) (Verb) (*) (*) (Verb) (*) trigger (*) (Verb) (*)
Complex	Left	trigger (*) (Verb) (and/or) (*) (Verb), (*) (Verb) (*)

Table 4 Rules to perform Task 3

Part	Condition location	Rule
Condition	Condition to left Condition to right	trigger (*) (Verb) (*), trigger (*) (Verb) (*)
Effect	Condition to left Condition to right	, (*) (Verb) (*) (*) (Verb) (*) trigger

Task 2. Complex conditional statements usually have and/or in the conditional part. A simple conditional statement has a trigger word and a single condition. Examples can be seen in Table 1.

Task 3. The condition can appear to the right or left of the sentence. In sentences where conditions appear to the left end, the condition and effect are separated by a comma. In sentences where condition appears to the right end, the entire sentence till the trigger word is the effect, e.g., 'Press and hold 1 if you are on AT&T'.

4.2 Support Vector Machine

The second technique evaluated was supervised learning-based approach. SVM was used to train the models. Two SVM models were built, one for classification of conditional statements and other for recognizing part of the sentence as condition or effect.

Task 1 and Task 2. SVM with a linear kernel was built for multiclass classification. It had three labels: 'Simple Conditional', 'Complex Conditional', and 'Nonconditional'. The preprocessing techniques include tokenization, stemming, and removal of stop words. The features had tf-idf vectorization of sentences. The bias in the dataset was overcome by oversampling.

Task 3. SVM with linear kernel was built to train to predict whether a given part of a sentence was condition or effect. It had two labels, 'condition' and 'effect'. The features had tf-idf vectorization of PoS tagging of the part of the sentence.

4.3 Rhetorical Structure Theory

An existing implementation of rhetorical structure theory was used [10] as the third technique. The three tasks were carried out using RST parser as follows:

Task 1. The sentences were passed through Discourse Parser. The parser identified conditional statements as conditional (Fig. 1).

Task 2. From the number of nucleus, whether a conditional statement is simple or complex can be told. From the observations of the output, it is seen that, complex

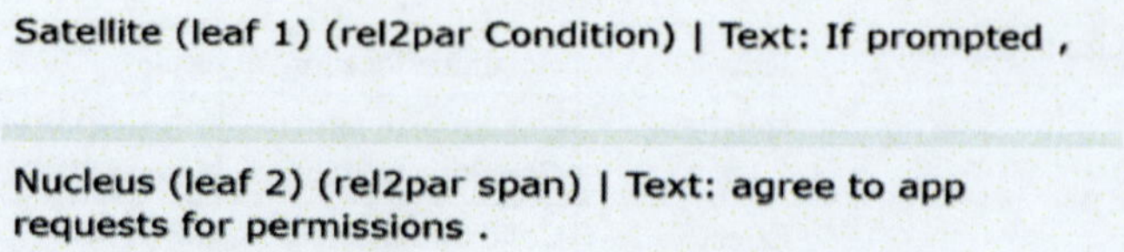

Fig. 1 RST output for simple sentences

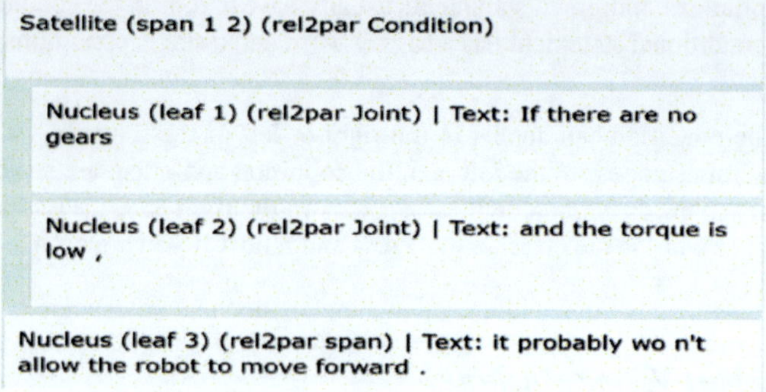

Fig. 2 RST output for a complex sentence

conditional statements have two nucleus and each predicate of the condition is distributed to each leaf and are annotated as rel2par joint as seen in Fig. 2.

Task 3. The discourse parser was successfully able to recognize the condition and the effect part of a conditional statement.

5 Results

5.1 Model Wise Results

Tables 5 and 6 show the results on tasks for Rules-based, SVM, and RST-based parser, respectively. Block sentences have not been considered in building the rule-based and SVM models.

Observations and analysis for Rule-based model:

1. Better understanding of the dataset and the thoroughgoing of the existing research papers helped in coming up with a new set of rules as mentioned in Task 1, Task 2, and Task 3 of Sect. 4.1
2. The model built in this paper outperformed the model in literature [3] which had a baseline efficiency of only 0.50. This can be compared with results achieved by our model as noted in Table 5.

Table 5 Results for rule-based methods

Task	Precision	Recall	Class	Precision	Recall
Task 1	0.86	0.92	Condition	1	0.85
			Nonconditional	0.72	1
Task 2	0.76	0.77	Simple	0.98	0.84
			Complex	0.54	0.70
			Nonconditional	0.72	1
Task 3	0.58	0.54	Condition	0.66	0.61
			Effect	0.51	0.46

Table 6 Results for SVM model

Task	Precision	Recall	Class	Precision	Recall
Task 1	0.96	0.96	Condition	0.97	0.98
			Nonconditional	0.96	0.94
Task 2 (without stop word)	0.88	0.86	Simple	0.84	0.75
			Complex	1.00	1.00
			Nonconditional	0.79	0.83
Task 2 (with stop word)	0.97	0.93	Simple	0.92	0.83
			Complex	1.00	1.00
			Nonconditional	0.94	0.91
Task 3 (without POS)	0.81	0.73	Condition	0.84	0.61
			Effect	0.79	0.85
Task 3 (with POS)	0.85	0.85	Condition	0.82	0.84
			Effect	0.88	0.86

Observations and Analysis for SVM:

1. The results show a dataset without removal of stop words, for classifying the types of statements using SVM provided better results. On further analysis of tf-idf scores, it was observed that some stop words like 'if' (tf-idf 0.096) and 'to' (tf-idf 0.095) has low tf-idf scores and hence were important for recognition of conditional statements.
2. Task 3 results for SVM without PoS tagging were not good as compared to with PoS tagging. The reason is that in condition the word 'to' is followed by a 'verb' whereas in effect it is not followed by verb, example: PoS tagging of effect part of the sentence: ['point_NN', 'to_TO', 'the_DT', and 'corner_NN',']. PoS tagging of condition part of the sentence: ['to_TO', 'install_VB', and 'it_PRP'].
3. The model built by us without stop word removal and with POS tagging gave an accuracy of 84% on the test data from [7]. Whereas the model used in paper [6] has a low precision of 0.676, recall of 0.6086, and F1 score of 0.6405 for the above dataset (Table 7).

Table 7 Results of RST model

Task	Precision	Recall	Class	Precision	Recall
Task 1	0.93	0.94	Condition	1.00	0.88
			Nonconditional	0.87	1.00
Task 2	1.00	0.88	Simple	1.00	0.88
			Complex	1.00	0.77
			Nonconditional	0.87	1.00
Task 3	1.00	1.00	Condition	1.00	1.00
			Effect	1.00	1.00

Observations and Analysis for RST:

1. Complex conditional statements with trigger word 'if' and multiple conditions are detected properly (both conditions to left and right).
2. Complex conditional statements with trigger word 'if' and multiple conditions separated by ',' are detected only as conditional statements and do not detect the different conditions.
3. Some complex conditional statements, despite having multiple conditions separated by words 'and', 'or' are not detected because their multiple conditions can be represented as relational operations (Example: If you are using Windows 8 or later, the Blue Screen is slightly different.
4. RST works good for nonconditional statements as it does not tag any nonconditional statement as conditional, but low precision is due conditional statements with trigger words other than 'if'.

5.2 *Detailed Analysis*

This section digs deeper to answer RQ1, RQ2, and RQ3.

RQ1. It can be seen that SVM works best for Task 1 which is a simple binary classification task. Quantitatively, RST is marginally better than SVM for simple conditionals in Task 2. On doing qualitative analysis as shown in Table 10, it is found that SVM works best for identifying simple conditionals, as it can recognize sentences with all types of trigger words. RST is not effective for triggering words other than 'if'. For complex conditionals, it is seen that RST works best for complex sentences. SVM is not suitable for complex conditionals due to very few data samples. Rule-based approach is not a generalizable approach for complex conditionals.

RQ2. In order to answer RQ2, the models trained on single sentences we run on block conditional sentences. The rule-based and SVM methods did not perform well in Task 1 and could not perform Task 2 and Task 3. RST worked well in recognizing block sentences and connecting them with nucleus as elaboration.

RQ3. RST is a natural fit for Task 3 as it can identify condition and effect components from a sentence. Rule-based technique requires complex rules for working effectively. SVM can identify and classify the pre-split components of the sentence (condition, effect) but cannot separate the components given a sentence, so is partially suitable for Task 3 (Tables 8 and 9).

Our findings can be summarized as:

1. Simple machine learning techniques like SVM can perform well in classification tasks for simple conditionals and even outperform discourse-based models in some cases.

Table 8 Analysis of models for simple conditions

Type	Example	RST result	SVM result	Rule-based
Simple condition with trigger word 'if', 300	1. If prompted, agree to app requests for permissions 2. This will be helpful if you need to restore contacts	Detects as conditional and also detects parts of condition	Detects as simple condition and given parts of it can identify as condition, effect	Detects as conditional and also detects parts of condition
Simple condition with trigger word 'Once' 22 'When' 16 'To' 14	1. Once you have done so, select 'End Task' 2. When they begin acting up, pay attention to the issue 3. To restore your contacts from this backup in the future, n select the.VCF file	Does not detect properly	Detects as simple condition and given parts of it can identify as condition, effect	Detects as conditional and also detects parts of condition

Table 9 Analysis of models for complex conditionals

Sentence	Property	RST result	SVM result	Rule-based
Press and hold 1 if you are on AT&T, T-Mobile, Sprint, Cellular One, or Metro PCS	Condition to the right with multiple conditions separated by comma	Detects as conditional but cannot detect parts of condition	Detects as complex conditional but cannot detect parts of condition	Detects complex conditionals but cannot detect parts of condition
If you're using Windows 7 or down, skip this step	Sentence with condition that can be represented as relational operator	Detects as conditional but cannot detect parts of condition	Detects as complex conditional but cannot detect. parts of condition	Detects both complex conditional and parts of condition

2. SVM can potentially work for complex conditionals with good training; however, the dataset did not have enough samples to prove this hypothesis.
3. SVM works best for classifying nonconditionals. Discourse-based methods are better suited for structural component analysis of conditional sentences.

Based on these findings, it can be seen that if a dataset contains simple conditional sentences and there is no requirement for Task 3, then SVM should be the technique of choice. If the dataset has a higher proportion of block sentences, then RST-based techniques will work better. If the requirement is to extract the condition and effect from a conditional, then RST-based techniques should be used.

6 Conclusion

From the above observation, it can be concluded that each model has its advantages and drawbacks and can be applied depending on the nature of the dataset and task requirements. The above observations have also helped in understanding where the techniques fail, this in turn helped in coming up with solutions that led to creating models that have better performance than existing models.

References

1. Park H, Motahari Nezhad HR (2018) Learning procedures from text: codifying how-to procedures in deep neural networks. In: Published in proceedings of companion of the web conference 2018. https://doi.org/10.1145/3184558.3186347
2. Baldwin T, Guo Y, Syeda-Mahmood T (2017) Automatic generation of conditional diagnostic guidelines. In: Published in AMIA annual symposium proceedings 2016. Published online 2017 Feb 10, pp 295–304
3. Wenzina R, Kaiser K (2013) Identifying condition action sentences using a heuristic-based information extraction method. In: Chapter from book process support and knowledge representation in health care: AIME 2013 joint workshop, KR4HC 2013/ProHealth 2013, Murcia, Spain, June 1, 2013, Revised Selected Papers. https://doi.org/10.1007/978-3-319-03916-9_3
4. Chikersal P, Poria S, Cambria E, Gelbukh AF, Siong CE (2015) Modelling public sentiment in Twitter. In: CICLing. vol 2. pp 49–65
5. Mausam, Schmitz M, Soderland S, Bart R, Etzioni O (2012) Open language learning for information extraction. In: EMNLP-CoNLL. pp 523–534
6. Gallego FO, Corchuelo R (2019) On mining conditions using encoder-decoder networks. In: Proceedings of the 11th international conference on agents and artificial intelligence (ICAART 2019). pp 624–630
7. FernanOrtega. https://www.kaggle.com/fogallego/reviews-with-conditions?select=sentences.csv
8. aaronsw. https://github.com/aaronsw/html2text
9. Muhie Yimam S, Gurevych I, Eckart de Castilho R , Biemann C (2013) WebAnno: a flexible, web-based and visually supported system for distributed annotations
10. Joty S, Carenini G, Ng R, Mehdad Y (2013) Combining intra-and multi-sentential rhetorical parsing for document- level discourse analysis. In: Proceedings of the 51st annual meeting of the association for computational linguistics (ACL 2013), Sofia, Bulgaria

Design of an Assistive Low-Cost 6 d.o.f. Robotic Arm with Gripper

Vasile Denis Manolescu and Emanuele Lindo Secco

Abstract The robotics industry is rapidly evolving driven by better and cheaper computer chips and affordable 3D printing technologies. All these aspects are a catalyst that helps in building new concepts and prototypes at a lower cost, and easier and faster than ever before. This paper presents the entire process of building an articulated robotic arm with 6 degrees of freedom (DOF) and a gripper, all controlled from a designed Arduino command center. The project will go through the 3D designing process and the selection of different actuators. Then, an optimization of the hardware options for controlling the motors and the software to operate the robotic arm is presented. Finally, advantages and drawbacks of the proposed architecture are discussed.

Keywords Robotic arm · Assistive robot · Low-cost robotics

1 Introduction

The meaning of what is known to be a robotic arm has evolved significantly in the past 20 years. The concept was mostly referring to a pre-programmed mechanical arm found in the assembling industries, capable of doing a repetitive task fast and without interruptions for long periods. Mainly, those tasks were actions like welding, painting, palletizing, screwing, picking, and placing or other tasks that involved not very complicated movements [1]. Today, the meaning of a robotic arm is becoming gradually more structured, descriptive, and specialized, while the machine capabilities became a lot more complex with extreme speeds and accuracy and a high level of efficiency. As for the future, relatively new concepts, like artificial intelligence, machine learning, and deep learning, present a promising opportunity for the

V. D. Manolescu (✉) · E. L. Secco
Robotics Laboratory, School of Mathematics, Computer Science and Engineering, Liverpool Hope University, L16 9JD Liverpool, UK
e-mail: 20203547@hope.ac.uk

E. L. Secco
e-mail: seccoe@hope.ac.uk

X.-S. Yang et al. (eds.), *Proceedings of Seventh International Congress on Information and Communication Technology*, Lecture Notes in Networks and Systems 447,
https://doi.org/10.1007/978-981-19-1607-6_4

robotics industry [2–4]. All these data-driven software-side techniques are capable of enhancing the capabilities of the robots more efficiently and offer them full autonomy around humans. When starting to design a robotic arm, the number of joints represents one of the key points. Terminology wise, the "axes movement" or the "degrees of freedom (DOF)" are to robots as the joints are to the human body. The end effector represents the working or the operation tools attached to the upper end of the robot, that effectively performs the task, for example, a griper. To be able to move the end effectors in a 3-dimensional space, at least 3 joints are necessary, while to change the angle of the tools, a minimum of 6 joints are required [5].

The subject of this paper is the design and build of an articulated robotic arm. Articulated robots offer high flexible movements, and it is the most common type of industrial robot. They present at least 3-serial joint linkages and performs only rotary motions. The paper is organized as it follows: a *Design Setup* section will present the main characteristic of the robot and the details of the designing process. The *Hardware* section will present a depth view of all the components integrated into the project and will systematically analyze the joint structure to understand the requirements and the functionality of each connection. Then, a *Software* section will go through the software used to make the 3D design of the robot, the Arduino IDE environment, and the code developed to give control to the robotic arm. *Results and Discussion* will point out the major problems encountered and offer an insight into how they have been managed. All the future improvements that can be applied to the robotic arm will be covered by the *Future Work* section. The final section, *Conclusion,* will create an objective overview of the whole project and reflect on the end result.

2 Design Setup

The overall design inspiration for the robotic arm was taken from Gluon, developed by Innfos Drive Technologies Co (Fig. 1). One of the primary design objectives was to make the robot to be modular and to be able to easily combine the 3D parts in various ways allowing future changes in the joint structure. This feature also offers a great advantage for future upgrades. The picture in Fig. 1 shows the similarities between those two designs. The final version is 75 cm tall with 55 cm working envelop.

All types of actuators have been tested in the pre-building process of the robot. The final version works with 2 bipolar stepper motors, one rotating the base and the other one drives the second joint, while the rest of the joints are each driven by 5 servomotors, the last one being the gripper. The gripper design was not a priority, and therefore, the 3D model was downloaded from the open sources grabcad.com library, re-scaled and 3D printed. All credit goes to its creator [6]. Additionally, a pressure sensor has been integrated into the gripper fingers. The arm is controlled by an Arduino Mega which offers the advantage of a larger number of connections: 54

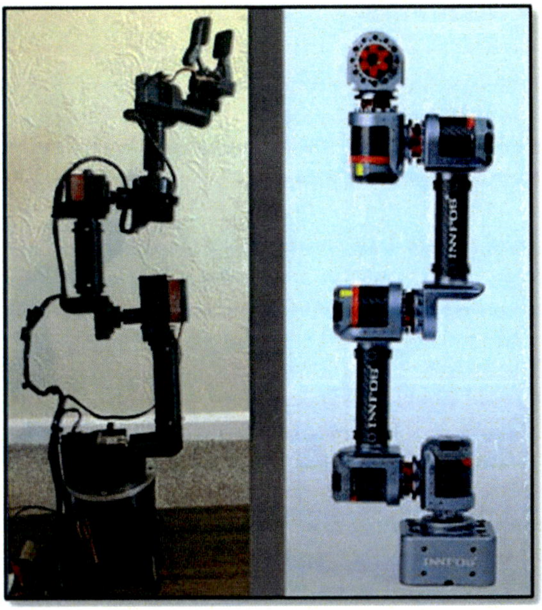

Fig. 1 The proposed design versus Gluon design (left and right panels, respectively)

digital input/output pins and 16 Analog inputs [7]. While in operation, the Arduino board is connected to a ~12 V power supply and is linked to the following:

(1) one CNC shield expansion board holding two DRV8825 stepper motor drivers. This shield is connected to a separate 12 V power supply (Fig. 2).
(2) a total of 6 dual-axis joystick modules which control the motors.
(3) one push-button switch programmed to drive the motor in a pre-set position, defined in the code as home.
(4) one PCA9685 servo driver module that is separately connected to 6 V generated by a DC–DC Voltage Regulator from the 12 V power supply.

More details about each of these components are presented in the next section of the paper.

Fig. 2 The Arduino board with the CNC shield and DVR8825 stepper drivers on the top

3 Hardware

In this section, we will look at the components used in building the robotics arm. The main goal is to analyze their purpose in the project and to justify their use. As a general reference, Fig. 3 illustrates the final degree of movement for each joint of the robotic arm, which are analyzed in depth in the following paragraphs.

3.1 Bipolar Stepper Motors

This section will analyze the stepper motors used to build the first 2 joints and will also talk about the 3D design around them.

Joint 1—The Base. The base of the robotic arm is built using a *NEMA 17*, model *42HD4027*—which is a medium-size bipolar stepper motor. To operate, this actuator needs 3.3 V and a current rate of 1.5 A/phase. This makes it capable of producing 0.4 N/cm of torque. It also rotates with 1.8° per step, generating 200 steps per single revolution (360°). The 360° rotation of the base is a horizontal movement on the x–z-axis. Counting the rest of the other joints and structure going up from the base, which weighs around 1.3 kg, the *42HD4027 stepper motor* generates enough torque to perform a smooth and stable movement.

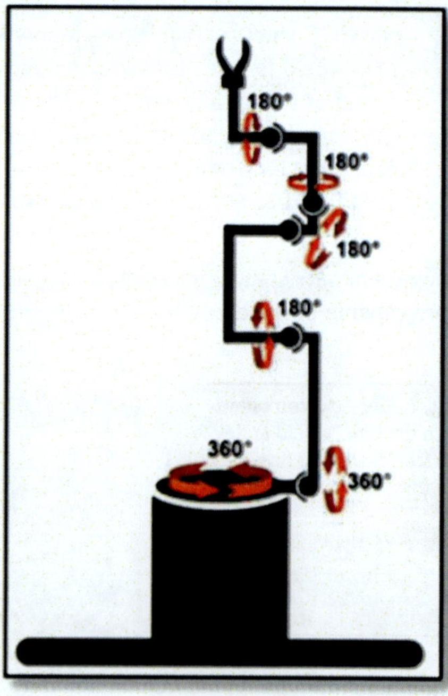

Fig. 3 Overview of the DOF angles and robot main movements

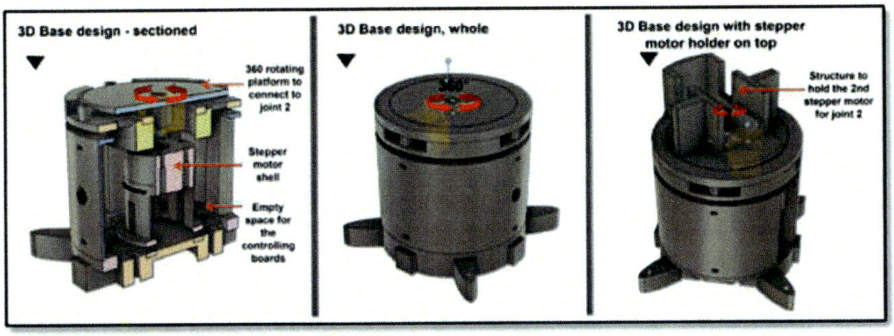

Fig. 4 Design of the robotic arm base (Fusion 360 ®)

The 3D model of the base is built in Autodesk® Fusion 360 and is designed to entirely encapsulate the stepper motor (Fig. 4—left panel).

Inside the base structure, the *42HD4027 stepper motor* holds itself in place by its own shell. This has been designed using one *ABEC-1* deep groove ball bearing, one *AXK* needle roller thrust bearing with vibration pads, one 3D-printed shaft extension, and one 3D-printed outer housing. The bearings are installed on the motor shaft extension, while the housing closes down using the screws of the motor (Fig. 5—left panel).

The new shaft extension goes inside the base, as shown in Fig. 4—left panel, and connects to the rotating platform of the base (Fig. 4—right panel). The rotating platform is built using a heavy-duty turntable bearing of 14 cm in diameter, which is attached to the holding structure of the second stepper motor of the second joint.

Joint 2. The second joint has a 360° rotation on the x–y-axis, but when the arm is installed on a flat surface, the movement needs to be limited to 300° to avoid hitting the ground. Considering the weight of the arm structure and the up and down movement, the arm has a stress point in the stepper motor shaft of around 20 N/cm. Along the building process, the joint has been previously tested with a few other types of motors: high torque brushless geared DC motor, high torque servomotor, and other stepper motors with less torque (see also Problems and Solution paragraph), but none were capable of handling the weight of the arm.

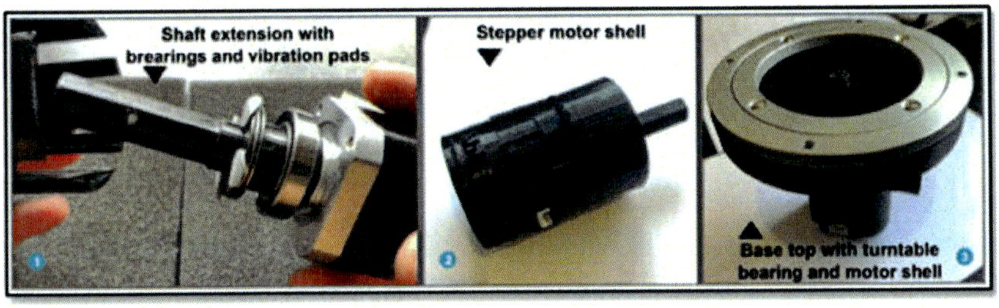

Fig. 5 Shell structure of the stepper motor number 1

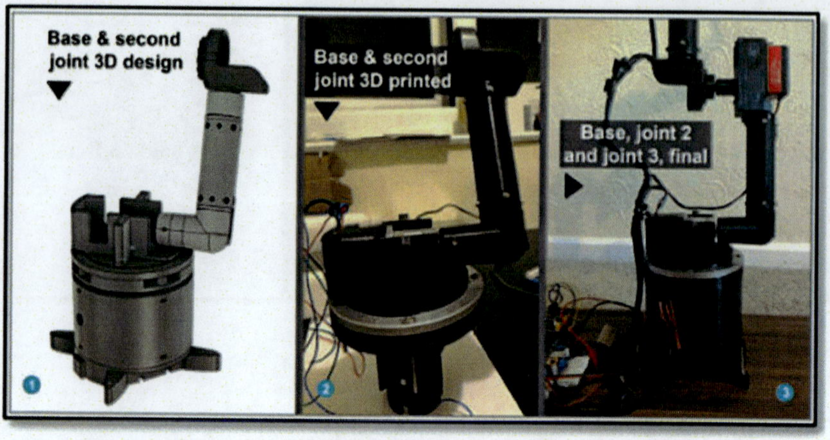

Fig. 6 3D design versus the final 3D-printed parts of the base, of joint 2 and of joint 3 (left, central and right panels, respectively)

The final version of the second joint uses a high torque *NEMA 17* bipolar stepper motor—model *17HS19-1684S-PG51*, having a planetary gearbox attached to it with a ratio of 51:1 (Fig. 6—central panel). The *17HS19 stepper motor* operates at a current rate of 1.68A per phase with a rotation of 0.035° per step, generating approximately 10,285 steps per revolution. This actuator is capable of producing around 400 N/cm of torque. Comparing the requirements of the project with the capacity of the stepper motor, there is an unnecessary big difference in torque, but because of accessibility and reasonable price, the 17HS19 seemed to be the most viable solution available. The second joint connects with the third joint through a 3D-printed 90° tube, which is directly installed on the joint-2 stepper motor shaft (Fig. 6). In the upper end of the tube, there is the joint-3 servomotor holder which will be discussed in the Servo section.

3.2 *Stepper Motors Controller*

This section will continue the talk about the parts that ensure the functionality and control of the 2 bipolar stepper motors presented above. These parts are the CNC shield and the DRV8825 stepper drivers.

CNC Shield V3. The CNC shields are specialized boards created around the Arduino microcontroller that offer integral solutions to build CNC machines. They offer the necessary power to drive multiple stepper motors and include functions like speed/direction control, stop, hold, micro-stepping, I2C, and other more personalized functions [8]. Compared to other more dedicated options to control stepper motors, the CNC shield presents a few other advantages, like robust shape compatible with

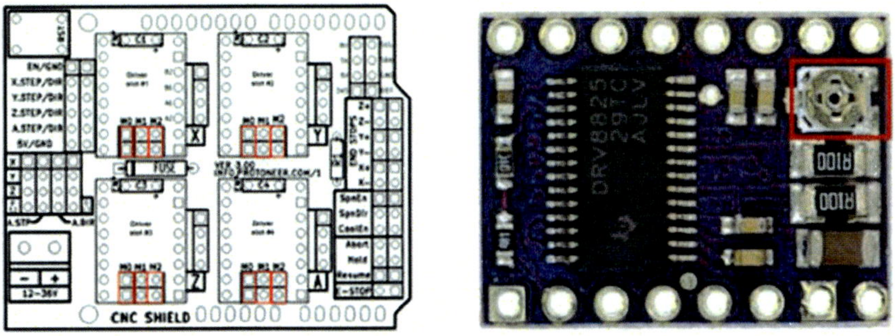

Fig. 7. The CNC shield V3 layout functions [9] and the current regulation potentiometer (red square) on the left and right panels, respectively

Arduino pinout, interface easy to work with, individual customizable controls, open-source design, and reasonable price. According to the datasheet, the shield can operate with an input voltage of 12–36 V; in this project, it is connected to a 12 V DC power supply. On top, the shield can hold up to 4 stepper drivers like A4988, DRV8825, and TCM 2100, etc. In this case, having to control just 2 stepper motors, only the X-slot and the Y-slot will each carry a DRV8825 driver. The *M0*, *M1,* and *M2* pins are directly controlling the micro-stepping indexer of the driver attached to it (Fig. 7, left panel). More about that in the DRV8825 driver section below.

DRV8825 Stepper Motor Driver. The *DRV8825* is an integrated motor driver that uses *N-channel MOSFETs* configured as 2 full *H-bridge* circuits. These are used to switch the flow of the current through the motor windings to control the speed and direction. A single driver can operate one stepper motor or two DC motors [10]. According to the datasheet, the driver operates at 8.2–45 V with a maximum current of 2.5A per phase [11]. In the project, the driver will get its voltage directly from the *CNC shield* which is connected to a 12 V DC power supply. Compared to the *A4988* stepper driver, *DRV8825* supports micro-stepping down to 1/32 which makes it fully compatible with the CNC shield capabilities—that is the main reason why it was the final choice in this project.

After individual testing, both drivers end up being set to 4 micro-steps per step with *M0* and *M2* pins set as low and *M1* set as high. This configuration seems to move the stepper motors more smooth and natural than the other micro-stepping modes.

Setting up the stepper driver. Another very important feature in the driver configuration is the current regulation which needs to be set according to each stepper motor specifications. This is done from a small potentiometer located on top of the module (Fig. 7, right panel). The potentiometer needs to be manually set according to the manufacturer formula:

$$I_{\mathrm{FS}}(A) = V_{\mathrm{REF}}(V)/(A_V \times R_{\mathrm{SENSE}}(\Omega)) \tag{1}$$

where I_{FS} is the stepper motor maximum current rate per winding; I_{max} is the stepper motor total rated current; V_{REF} is the maximum voltage reference corresponding to the maximum current allowed to flow into the stepper motor; A_V is the current sense amplifier gain (according to the datasheet this parameter is a factor of × 5); R_{SENSE} is the sense resistor value of the driver module, which in this case it is 0.2 Ω. After replacing the variables with the known values, given that $I_{max} = 2 \times I_{FS}$, it holds

$$I_{FS}(A) = \mathbf{V_{REF}}(V)/(\mathbf{5} \times \mathbf{0.2}(\Omega)) \tag{2}$$

It is important to notice that:

- V_{REF} is essentially determining the current limit per coil. If it is set too high, it can overheat or burn both the motor and the driver. If it is set too low, the motor loses steps or even stop moving.
- It is a good safety measure to adjust V_{REF} 10% lower.
- The R_{SENSE} variable can have different value, strictly depending on the manufacturer of the stepper driver.

Once the final formula is set, the next step is to determine the $\mathbf{V_{REF}}$ for each stepper motor and adjust the corresponding driver, namely:

$$\text{Joint 1} - \text{Stepper motor 42HD4027}: V_{REF} = (1.5/2)^*90\% = \mathbf{0.675\ V}$$
$$\text{Joint 2} - \text{Stepper motor 17HS19}: V_{REF} = (1.68/2)^*90\% = \mathbf{0.765\ V}$$

The final step is to physically adjust the current limit on the DRV8825 driver according to the calculus results.

3.3 Servomotors

This section aims at discussing the selection of the 5 servomotors which were used for prototyping the joints 3 to 6 and the gripper.

Joint 3. The third joint is a relatively high-tension point in the robotic arm assembly. The minimum torque to move the upper structure is around 15 N/cm. The motion task is even harder when considering that there is an x–y-axis movement, plus counting for a possible load carried by the gripper. The joint is capable of rotating 180° without restrictions. To be able to properly operate this joint, there was a need for a powerful servomotor. A great advantage for choosing a servomotor was the balance between a lightweight actuator and a relatively high torque performance. The most significant disadvantages were the limited motion that can be achieved and the high acquisition cost.

In the end, the third joint was built with the *DS5260ssg servomotor*, which comes with a gearbox ratio of 279:1. According to the datasheet, it operates with a voltage of 6–8.4 V and it is capable of 60 kg/cm of torque. The design for this joint is made

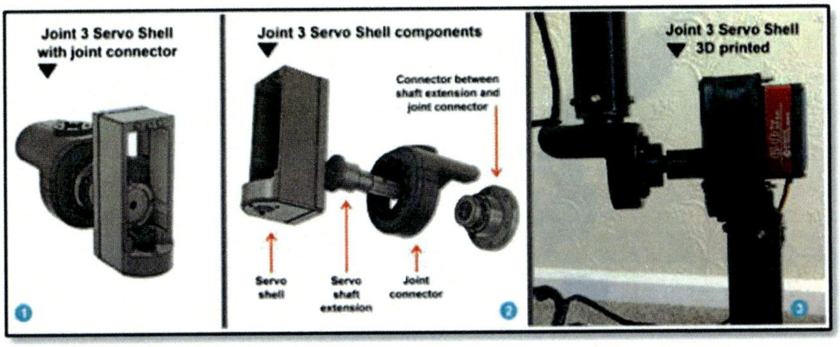

Fig. 8 Design of the joint 3 (left and central panel) and final assembly of the parts (right panel)

to attach the servomotor to a 3D-printed shell while extending the motor shaft from inside the shell to the next joint connector. This is done by using another 3D-printed part—a pawn-shaped shaft extension (Fig. 8). At the same time, between the servo shell and the shaft extension that comes out of it, there is an *ABEC-1* deep groove ball bearing that is used to minimize the bending and the pressure created by the upper joints upon the servo shaft (Fig. 8—left panel).

Joints 4, 5, and 6. Joint 4 is having a structure that weighs approximately 600 g that needs to lift off. Taking into consideration a possible gripper payload and the x–y-axis movement, the necessary torque to ensure a proper motion is approximately 10 N/cm. The joint can rotate 180° with no restrictions. The actuator used is a servomotor *DS3218MG* designed by DSServo with an operating voltage of 4.8–6.8 V and a gearbox ratio of 275:1.

Joint 5 and 6 are identically built, and they use the same type of servomotor, the *Diymore MG996R*. These servo actuators operate with a voltage of 4.8–7.2 V and are capable of generating up to 15 kg/cm of torque.

Similar to joint 3, these three servomotors were chosen because they pack enough torque to operate the robotic arm and they are very light—weighting 55–60 g.

For all the three joints, the 3D design is similar and it is created to integrate the servos into the joint connectors as shown in Fig. 9. The motor connects to a 2-part frame while the shaft is linked to a hexagonal pawn-shaped shaft extension. In the same way as in the third joint, between the shaft extension and the joint connector, there is a ball bearing to avoid bending and decrease the pressure (Fig. 9—left panel).

The end effector or gripper. Because the construction of the griper was not a priority for the robotic arm, its design comes from a free online source [6]. The final 3D-printed version is driven by a *Diymore MG996R* servomotor, the same actuator used in joints 5 and 6. The rotation of the motor is software limited to 120°, which is slightly less than the motion allowed by the mechanical parts.

Both gripper fingers are having anti-slipping pads, and one of them is equipped with the *SF15-600 pressure sensor film* [12]. The sensor can sense and record the

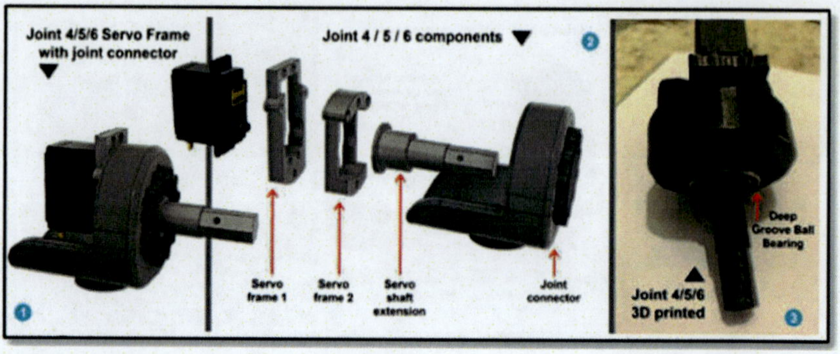

Fig. 9 Top 3 joints—Inside and outside structure

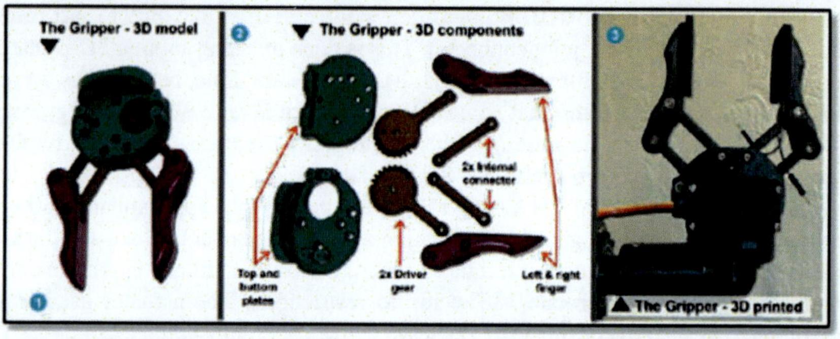

Fig. 10 The gripper design, from the 3D assembly (left and central panels) to the 3D printing manufacturing (right panel)

squeezing force applied to any grabbed object, and it can be used to create gripping profiles for a different type of targets (Fig. 10).

3.4 *Servomotor Controller*

This section of the paper goes through the components used to control the servomotors, which are the *PCA9685 servo driver* and the *DC–DC voltage regulator*.

PCA9685 Servomotor Driver with I2C. The PCA9685 used in this project is a 16-channel I2C-bus protocol controller with a capacity of 12-bit output per channel and a fixed frequency (24–1526 Hz). Because of its features, the board can be used as *Pulse Width Modulation* (PWM) controller or to adjust the duty cycle, to individually drive up to 16 servomotors per chip [13].

Arduino communicates with the PCA board using I2C protocol through *Serial Data Line* (SDA) and *Serial Clock Line* (SCL) pin connection. This is how the direction and speed of the 5 servos are individually controlled (see also the Software section below).

To operate, the *PCA9685* gets 6 V from the 12 V DC power supply using a DC–DC voltage regulator.

DC–DC Voltage Regulator. In order to simplify the development of the device, a single DC power supply was utilized, which is by default set to 12 V to be used by the CNC shield to control the stepper motors.

The *PCA9685 servo driver* is set to use the same power supply, but because it only needs 6 V, a *Buck DC–DC converter* is used to step down the voltage. The step-down switching regulator board is built based on the *XLSEMI XL4015 IC chip*, with 96% efficiency and capable of converting 8–36 V into 1.25–32 V [16]. The board can regulate the voltage using 2 push buttons and a 3-digits LED or from the *W503 trimmer potentiometer*.

3.5 Intuitive Control Panel (i.e., Dual-Axes Joystick Interface)

The centralized control of all the actuators is performed using 6 joysticks and 1 push-down button (Fig. 11). The 4-way directional 2-axis joystick is used as Analog user input and controller, and it is connected to the Arduino. The module consists of two perpendicular 10 kΩ potentiometers controlling the x- and y-axis with the joystick movement. It includes springs to auto-return to the center position. It also has an integrated joystick *push-down button*—which is used in this project as a home button with a pre-defined repositioning of each actuator.

Fig. 11 The 2-axis, 4-direction joystick interface with the push buttons (left panel) and the main control panel details (central and right panels)

4 Software

All the 3D designing and 3D printing have been possible using Autodesk Fusion 360. The programming of the Arduino Mega 2560 board has been developed by using the Arduino IDE interface, which is a powerful, open-source C/C++ Programming Language environment developed for the Arduino products, but not limited to it.

4.1 The Arduino Code

The servomotors controlling is developed using an external library specially developed for the PCA9685 driver board. The library cannot be downloaded and installed using the conventional way from Arduino IDE, the Library Manager [14]. It needs to be manually added to the default library folder of the IDE. Once installed, the library gets called. In practice, the most important aspect done by this library is to partition the internal registers on the PCA9685 chip to allow I2C communication for commands.

Global Variables and Pinouts. The global variables section in the code is split between those related to stepper motor control and those related to servomotor control. The first declared are the Arduino pinouts used to control the direction and the speed for both stepper motors.

The variables *x_dirPin/x_stepPin* are commanding the base direction and speed, while y*_dirPin* and *y_stepPin* are the same commands for the next joint. These first 2 actuators will be controlled by the same joystick, the base is related to the x-axis with up-down joystick movement, while joint-1 is related to the y-axis with left–right joystick movement. The joystick data given by the potentiometer movement is read by the Arduino Analog pins *A0* and *A1*, and it is stored by the *vrx_data* and *vry_data* variables. The variables *x_steps_per_rev* and *y_steps_per_rev* represent the number of steps per a 360° rotation for each stepper motor. These are used together with the joystick data to synchronize in mapping the motion.

Concerning the servomotor global variable declaration, the first step is to define the *Arduino I2C master* address that will be used to communicate with the PCA board, as it follows: the code defines the Arduino Mega pinouts *A8* to *A11* for each joint, which are used to read the positioning data from each joystick. The data from those *Analog pins* are paired with the *Servo_Position* variables to map and send motion commands to the servomotors via the PCA board.

Every joystick press-down button defined by the *homeButton_* variable gets detected by the Arduino pins 22, 24, 26, 28, and 30. This command will automatically drive the related servomotor to its pre-defined initial position. The same thing happens with the *allHome* variable connected to pin 40, which is related to a universal push-down button that drives all servomotors, at the same time, to their initial position.

Setup function. The first four lines of code in the setup function is setting the *_stepPin* and *_dirPin* pins as output (sending data). All home button pins are set to read the data and detect when the buttons are pressed.

Code line 58 is initializing the *PCA9685* by turning the Boolean variable *SERVO_MODE* to true. This variable is located in the library. The next line of code is activating all the library functions and features by turning the sleep mode off (false).

Loop function. One of the primary commands of the main loop is to run another function called *Joystick*. This dictates the two stepper motors rotation based on the joystick motion. The joystick potentiometer registers its movement with values between 0 and 1023, both on the x- and y-axis. When it is not used, it auto-sets itself into a default middle position corresponding to a value of around 512 (Fig. 12). At the same time, using a for loop, it keeps updating the joystick position and checks if it gets released to the default state, which would trigger the command of stopping the motor. Finally, if nothing is detected, the actuator will keep rotating at the same speed (Fig. 12).

The same logic is used to correlate the joystick down-positioning with the movement of the actuator to the left. The joystick function finishes by adapting the same algorithm design to the joystick left–right movement and the rotation of the second stepper motor.

Servomotor controls. For the servomotor control, the main loop checks first if any of the *homeButton* or *allHome* buttons are being pressed. The check is done by 5 if statements having the negation argument *!digitalRead()*, which detects any change of state. If a press of a button gets detected, it will execute one of the *homeJoint*

Fig. 12 The *Joystick* function

```
244 // stepper joystick function
245 void joystick() {
246   vrx_data = analogRead(vrx);
247   vry_data = analogRead(vry);
248
249    ////////// Joint 1 - stepper \\\\\\\\\\
250    // stop when joystick is into middle position
251    if ( (vrx_data > 490)  &&   (vrx_data < 540)   ){ //"Nothing"; }
252    // when joystick is UP move clockwise
253    if ( vrx_data > 540  ){
254     digitalWrite(x_dirPin,HIGH);
255     for (int x = 0; x < x_steps_per_rev; x++) {
256      vrx_data = analogRead(vrx);
257      if ((vrx_data > 490) && (vrx_data < 540)) { break; }
258      digitalWrite(x_stepPin,HIGH);
259      delayMicroseconds(2000);
260      digitalWrite(x_stepPin,LOW);
261      delayMicroseconds(2000);
262     }
263    }
264    // when joystick is down position move counter-clockwise
265    if ( vrx_data < 490) {
266     digitalWrite(x_dirPin,LOW);
267     for(int x = 0; x < x_steps_per_rev; x++) {
268      vrx_data = analogRead(vrx);
269      if ( (vrx_data > 490)  &&   (vrx_data < 540)   ){ break; }
270      digitalWrite(x_stepPin,HIGH);
271      delayMicroseconds(2000);
272      digitalWrite(x_stepPin,LOW);
273      delayMicroseconds(2000);
```

```
// homeJoint_3 method - servo
void homeJoint_3() {
    for (int i = 0; i<600; i++) {
      if (Servo0Position > 10) {
        // brings servo to Home position
        HCPCA9685.Servo(0, Servo0Position--);
        // in library : void Servo(byte Chan, unsigned int Pos
        delay(20);
      }
      if (Servo0Position < 10) {
       // brings servo to Home position
        HCPCA9685.Servo(0, Servo0Position++);
        delay(20);
      }
     // int lastPosition_joint3 = Servo0Position; // TEST: memorize last position
    }
```

Fig. 13 The *Return to Home* function

functions. This function is adapted for each joint, but it generally works the same way: a for loop is used with a given maximum value greater than the servomotor steps per a full revolution—this is determined by testing each servo. No matter in what position the actuator is, it will always loop back to the initial position of the shaft (Fig. 13).

Linking Servomotor with Joystick. The last part of the main loop is a series of *if statements* grouped to define the movement of each servomotor. The logic of the algorithm works the same for each joint. The joystick potentiometer value is read, and the following 2 events may occur:

1. The value leans toward the upper direction (i.e., greater than 600) and the servomotor position is less than its maximum possible value (420), then the servo position is increased by 1 step to the right.
2. The value leans toward the down direction (i.e., less than 400) and the servo position is greater than its minimum (10), then the servomotor position is decreased by 1 step to the left.

In both cases, the loop is delayed by 10 ms and then it sends the new position value to the PCA board in order to be effectively executed by the actuator.

5 Results, Discussion, and Challenges

A set of preliminary trials was performed in order to check the reliability and robustness of the robot. During these preliminary tests, two main problems were noticed:

The first problem regards the fact that, initially, the robotic arm was supposed to be built using a small size stepper motor for all joints, but after testing the actuator on the 3D-printed structure, the results proved that the motor was not capable of handling such a weight. The next option to solve this issue was to try using a 12 V high torque, reversible DC motor, which on paper looked like a viable solution. But

Fig. 14 Selection of the actuators (see details in the Results section below)

again, after testing on the structure of the arm, it also failed to lift the weight of the upper joints. Finally, the solution was to use the *17HS19 bipolar stepper motor,* which proved to be more than capable to handle the joint requirements. Additionally, to use this 600 g actuator, the robotic arm base had to be redesigned (Fig. 14).

A second problem that was noticed during the tests was about the 3rd and 4th joint actuators: finding the proper motor for the next two joints proved to be challenging. After furtherly testing the 12 V DC motor and a couple of other servomotors, none of them were capable of lifting or moving the arm properly because of the weight. Then, a solution was to try the highest torque servo available and gradually go for a lower torque servo for each upper joint.

6 Future Work

One of the near-future plans is to use the pressure sensor that is already been included in the gripper structure and to develop an algorithm capable of using that data to better control the end effectors. Moreover, the PCA9685 cannot properly be used when dealing with various servomotors of different voltages. A new option for a new servo driver needs to be explored as well the development of a Human and Graphical User Interface [15, 17].

The modular design of the robotic arm makes it easy for other future upgrades. The major challenge that will help the project reach its true potential is to develop an encapsulated type of joint around a high torque actuator and incorporate into the same space features like integrated encoder, temperature sensors, high gearbox ratio for high precision motion, feedback with hall sensors, backlash correction, and vibration dissipation housing. These are all planned to be tested and implemented in the near future.

Another important update could be to integrate cameras and proximity sensors into the robotic structure and feed the data into an AI type of algorithm that can help the robot become socially interactable, aware of obstacles, and being able to receive and reproduce commands.

7 Conclusion

This project started with the goal of exploring a variety of actuators and other controlling hardware by building a 6 DOF robotic arm. 3D designing proved to be a fairly creative and satisfying phase, while 3D printing has been more demanding and the most time-consuming task. The electronic side of the project ended being the one that required a lot of research and documentation, done in parallel with intensive testing and problem-solving situations, but one of the most engaging and complex. Meanwhile, the software development has been a great task of logical thinking, testing, debugging, and documenting.

Acknowledgements This work was presented in coursework form in fulfilment of the requirements for the BEng in Robotics for the student Vasile Denis Manolescu from the Robotics Lab, School of Mathematics, Computer Science and Engineering, Liverpool Hope University.

Appendix

A video demonstration of the working robotic arm is available at the following link. https://www.youtube.com/watch?v=NlBRkktJQAw

References

1. Builtin (2021) Available at: https://builtin.com/roboticsgrabcad. Last Accessed 02 Sep 2021
2. Magenes G, Ramat S, Secco EL (2002) Life-like sensorimotor control: from biological systems to artifacts. Curr Psychol Cogn 21(4–5):565–596
3. Magenes G, Secco EL (2003) Teaching a robot with human natural movements. In: XI Winter Universiads, conference on biomechanics and sports. Italy, pp 135–144
4. Secco EL, Visioli A, Magenes G (2004) Minimum jerk motion planning for a prosthetic finger. J Robot Syst 21(7):361–368
5. Kawasaki Robotics (2021) Available at: https://robotics.kawasaki.com/ja1/xyz/en/1803-01/. Last Accessed 01 Sep 2021
6. Gripper design 3D model, Open source use. Available at: https://grabcad.com/library/dripper-4. Last Accessed 02 Sep 2021
7. Arduino, Arduino Mega. Available at: https://www.arduino.cc/en/Main/arduinoBoardMega/. Last Accessed 02 Sep 2021
8. All3dp, Arduino CNC shield guide. Available at: https://all3dp.com/2/arduino-cnc-shield-buyers-guide/. Last Accessed 01 Sep 2021
9. Osoyoo, CNC shield V3 guide. Available at: https://osoyoo.com/2017/04/07/arduino-uno-cnc-shield-v3-0-a4988/. Last Accessed 01 Sep 2021

10. Northwestern_University, Northwestern Robotics and Biosystems (2009) Available at: http://hades.mech.northwestern.edu/index.php/Driving_a_high_current_DC_Motor_using_an_Hbridge. Last Accessed 01 Sep 2021
11. Texas_Instruments (2021) Texas Instruments—DRV8825 driver. Available at: https://www.ti.com/product/DRV8825. Last Accessed 02 Sep 2021
12. Secco EL, Moutschen C (2018) A soft anthropomorphic and tactile fingertip for low-cost prosthetic and robotic applications. EAI Trans Pervasive Health Technol 4:14
13. nxp.com—Datasheet—PCA9685 Servo Driver Available at: https://www.nxp.com/docs/en/data-sheet/PCA9685.pdf. Last Accessed 02 Sep 2021
14. hobby.component.com, HCPCA9685 library. Available at: https://forum.hobbycomponents.com/viewtopic.php?f=58&t=2034, last accessed 2021/09/02
15. Chu TS, Chua AY, Secco EL(2021) A wearable MYO gesture armband controlling sphero BB-8 robot. HighTech Innov J 1(4):179–186. https://doi.org/10.28991/HIJ-2020-01-04-05. Last Accessed 02 Sep 2021
16. LCSC. XL4015 Datasheet (2014) Available at https://datasheet.lcsc.com/szlcsc/1811081616_XLSEMI-XL4015E1_C51661.pdf. Last Accessed 20 Aug 2021
17. Secco EL, Scilio J (2020) Development of a symbiotic GUI for robotic and prosthetic hand. In: Intelligent systems conference (IntelliSys), Amsterdam, The Netherlands

An Improved Method for the Sizing of a Stand-Alone Photovoltaic System: Application at Ngoundiane Village in Senegal

Pape Made Diouf, Amy Sadio, Papa Lat Tabara Sow, Ibrahima Fall, and Senghane Mbodji

Abstract An improved method for the sizing of a stand-alone photovoltaic system in Ngoundiane site is investigated in this paper. From the numerical model, proposed in Sadio et al. (2018), the optimal combination between PV and battery capacities is found out, by taking into account the annual monthly irradiation variation and the simultaneity coefficient. So, all average monthly values of solar irradiation are considered instead of one average value and the maximal value of load demand is used and determined from energy balance and consumers behavior. After generating all PV/batteries combinations for each value of solar irradiation, the best configuration is chosen by using the TLCC. It is shown that the PV and battery capacities decrease by 25.13 and 88.39%, respectively, when compared to the intuitive method. However, the obtained LPSP, assessed to 0.3, is considered as high and is due to the lack of complete data.

Keywords Numerical improved method · Stand-alone photovoltaic system · Optimal combination · Annual monthly irradiation variation · Simultaneity coefficient

1 Introduction

The rapid global depletion of fossil fuel sources has prompted an urgent search for alternative sources to meet current energy demand [1]. The global climate change due to increased greenhouse gases, caused by use of these resources, will have a negative consequence for human health, environment, and ecosystem [2]. In this context, the integration of renewable energies in energy production becomes essential. Thanks to its many advantages, photovoltaic solar energy is one of the most promising solutions in addition to being well shared around the world, harmless to the environment, and inexhaustible on the human scale. It is now widely exploited for the electrical energy production. However, the major problems associated with the

P. M. Diouf · A. Sadio · P. L. T. Sow · I. Fall · S. Mbodji (✉)
Research Team in Renewable Energies, Materials and Laser of Department of Physics, Alioune Diop University, Bambey, Senegal
e-mail: senghane.mobdji@uadb.edu.sn

X.-S. Yang et al. (eds.), *Proceedings of Seventh International Congress on Information and Communication Technology*, Lecture Notes in Networks and Systems 447,
https://doi.org/10.1007/978-981-19-1607-6_5

use of solar energy for power generation are their discontinuous and unpredictable nature and the high cost of some components of photovoltaic systems. Many research works have been conducted to optimize PV systems by using various methods such as classical (intuitive, numerical, and analytical) and artificial intelligence (Genetic Algorithm, Neural Network, etc.) methods. In Khatib et al. [3], different sizing methods have been discussed and it has been showed that numerical methods are more accurate when compared to others methods and require more computing time. A new sizing method of stand-alone PV system, using an improved algorithm is investigated in Sarhan et al. [4]. It takes into account the fluctuation effect of solar irradiation during cloudy days, increasing the precision of the system conception. Eduardo et al. [5] designed a sizing methodology to obtain the optimal combination between the available energy sources and the load demand. The Loss of Power Supply Probability (LPSP) during a consecutive time period and minimal cost are used as technical and economic criteria, respectively. It has been concluded that the increase of the LPSP and consecutive time period lead to the one of energy shortage. A safe technico-economic sizing methodology is studied by Bouabdallah et al. [6]. The optimal reliability of their sizing procedure is based on Clearness Index, which is evaluated by using a great numbers of possible scenarios. This Clearness index is randomly generated from transition Matrix of Markov. It has been showed that LPSP is sensitive to the diversity of the possible configurations of the sunshine. A new reliability indicator, using a probabilistic approach, has been developed by Padilla et al. [7]. The optimization of a hybrid PV/Wind system based on this indicator has been applied by using Barranquilla's meteorological data. Results showed that this approach can generate a system with a high reliability. A new approach has been proposed by Rachchh et al. [8] in order to maximize the total solar capacity and the corresponding energy efficiency. The main optimization parameters considered in this paper are the tilt angle, altitude, profit factor, solar height, and shading. It has been found that the capacity and generated energy can be improved over 25% for a given area. Qayoom et al. [9] presented a new analytical model to optimize the choice of different components of a PV system for the predetermined charge satisfaction. This model has been compared with other numerical and analytical ones. It is proved that this model is more convincing, because it integrates several useful variables such as suitable Loss of Load Probability (LLP), latitude, Clearness Index, absorbed solar radiation, efficiency of PV array, load demand, and unit cost of PV generator and battery. Ibrahim et al. [1] elaborated an improved numerical method for the optimization of stand-alone photovoltaic systems, in remote areas. The authors developed a new technique to model PV output current by using Random Forest Technique and a dynamic model, which reflect the dynamic behavior of the battery. Results showed that the numerical proposed method is optimal for a LLP of 0.01 with a minimum cost.

Literature shows that Artificial Intelligence Techniques are more relevant when some data are missing and lead to more effective results for complex sizing problems handling many parameters ([3, 10, 11]). An optimization method based on Genetic Algorithm (GA) has been expressed to minimize the voltage fluctuations caused by the intermittence of solar PV systems in distribution systems by Babacan et al. [12]. Sensitivity studies showed that results qualitatively follow the conception objectives of the «Objective» function. A novel method for sizing of stand-alone photovoltaic system using multi-objective differential evolution algorithm and hybrid multi-criteria decision making methods is elaborated by Halboot et al. [13]. The first method has been used to optimize a set of system configurations, while the second one was able to select the best configuration. This sizing approach is 27 times faster than numerical method. Mellit et al. [14] reviewed different sizing techniques. They concluded that conventional sizing methods, such as empiric, numerical, analytical, and hybrid, present a good solution, when all required data are available.

The output of PV systems strongly depends on the availability of solar radiation and ambient temperature [1], and is unpredictable and non-linear. Therefore, energy stores, such as batteries, are used to meet the demand, when there is a shortage of available energy to overcome the demand for charging [15]. So most sizing works have been achieved based on the looking for a better configuration between PV and battery capacities. Meanwhile, to our knowledge, we note a low contribution of Senegal in these works. Indeed, studies have been focused in the fundamental. Recently, Sadio et al. [16, 17, 19] have addressed three papers, treating PV system sizing optimization. The first paper proposed a new numerical sizing approach of a stand-alone photovoltaic power at Ngoundiane, Senegal. A new numerical sizing has been developed, based on determinist approach, adapted to the nature of meteorological data and load demand in the Ngoundiane site. First, an intuitive method has been applied, and thereafter, the numerical sizing method is implemented. In the second paper, a comparative study based on Genetic Algorithm is investigated for the optimal sizing of the stand-alone photovoltaic system in the Ngoundiane site. Finally, in the third paper, the sizing of a stand-alone PV water pumping system has been investigated. Results obtained from these papers have been satisfactory. However, all of them have not considered the random nature of solar irradiation and ambient temperature.

Following the same procedure, we propose in this paper an optimal sizing to find out the best configuration between PV and battery capacities. The nominal capacity of different components is computed by using a simple intuitive method. Afterward, a numerical approach which takes into account monthly variation of solar radiation while maintaining constant the ambient temperature is laid. Total life cycle cost is used as economic criterion to make an optimal choice. This sizing method is run from MATLAB software by using meteorological data and load profile of Ngoundiane site and technical and economic parameters of components.

2 Materials and Method

A pattern of PV system studied in this paper is shown in Fig. 1, and it consists of the following:

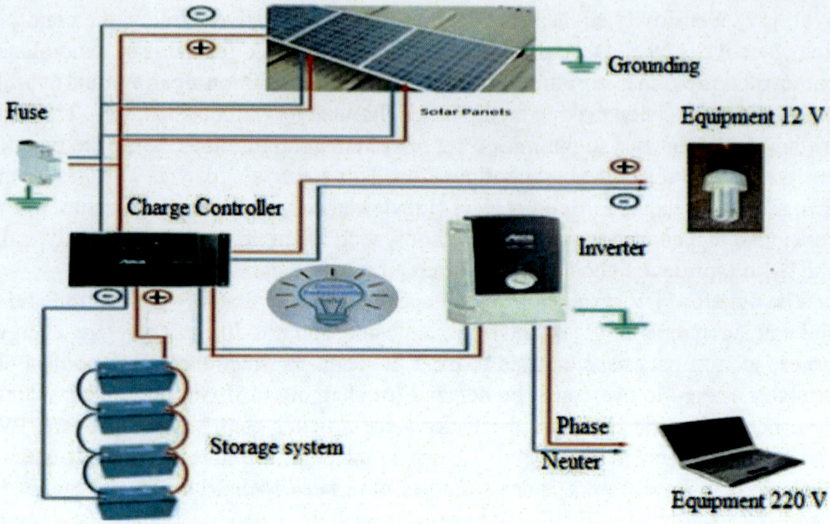

Fig. 1 General configuration of a PV system

- PV panels: it is the main production source of energy in the system; it converts solar energy into direct electric current;
- battery storage: it allows to store the generated energy from PV modules for being used, during night and cloudy days;
- inverter: it converts the direct electric energy from PV modules and/or battery into alternating electric current;
- charge controller: it protects batteries against deep discharge and overload to avoid damage and drastic reduction of their lifetime.

2.1 Solar Radiation and Load Profile in Ngoundiane Site

Ngoundiane site is an annexe of Alioune Diop University, situated in Ngoundiane, Thiès. Its geographical coordinates are 14.723864° north latitude and −16.734604° west longitude. Ngoundiane site consists of social campuses, and pedagogic and administrative buildings.

Solar potential in Ngoundiane. Ngoundiane has an important solar potential. In Sadio et al. [19], the annual profile of monthly solar irradiation average in Ngoundiane is presented as showed in Fig. 2. The highest value of solar irradiation is estimated to 7.05 kWh/m^2/day and is recorded in the month of April, while the lower value of irradiation is observed in the month of December with a value of 4.88 kWh/m^2/day. The average value of irradiation is 5.87 kWh/m^2/day.

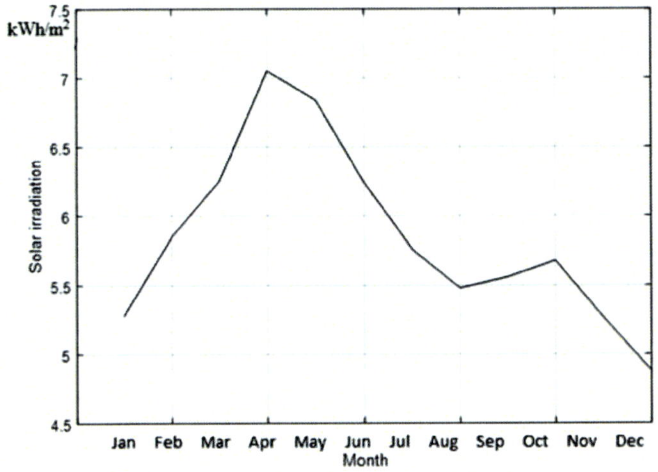

Fig. 2 Annual variation of monthly solar irradiation average

Load profile. The type load profile of Ngoundiane site is established from the energy balance and labor fees of users. Some assumptions have been expressed:

- one-hour time interval is considered;
- some of the electrical equipment are not taken into account;
- three types of load profile corresponding to three periods in the year are defined according to the weather conditions and the needs of the users during these periods. We assume that the three periods correspond to the months of November up to February, March to July, and August to October, respectively.

The three-load profiles resulting from this study are shown in Fig. 3. It shows the power of apparatus which simultaneously work for each one-hour interval.

The maximal power is recorded in the second period, precisely during class hours. This is due to the hot weather that stimulates the use of strong power equipment. The lower load demand is observed in break hours. The maximal power is used in the sizing process to maximize the PV energy production. It is equal to 81869 W, and the maximal energy quantity is 327476 Wh. During the night, between 7 PM and 3 AM, energy consumption drops to 112245 Wh and becomes very weak between 3 and 7 AM, around 13576 Wh.

2.2 *Simple Empiric Model*

A simple empiric model (intuitive method) is used to find the nominal approximate size of different components of the PV system. Equations 1–10 give the nominal capacities of different components.

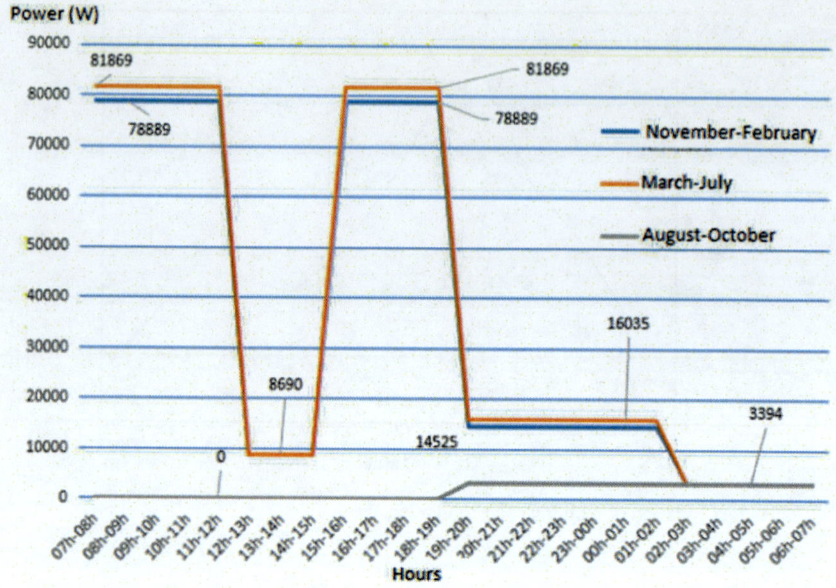

Fig. 3 Load profile for the periods

$$C_{\text{PV}} = \frac{1000 \cdot E_L}{\eta \cdot E_I} \tag{1}$$

$$C_{\text{Wh}} = \text{DOD} \cdot U_{\text{bat}} \cdot C_{\text{bat}} \tag{2}$$

C_{PV} is the total PV capacity, 1000 in W/m^2 represents the reference irradiance in the optimal conditions of sunniness and temperature, η is the global efficiency of PV system, estimated to 0.6, E_I corresponds to the average energy received by modules, it is equal to 5.85 kWh/m^2/day in Ngoundiane's zone, C_{Wh} in Wh is the total capacity of storage system, **DOD** corresponds to the deep of discharge, U_{bat} is the voltage of storage system, and C_{bat} is the total capacity of storage system in Ah, and it is expressed by Eq. 3:

$$C_{\text{bat}} = \frac{E_L \cdot N}{\text{DOD} \cdot U_S \cdot \eta_{\text{bat}}} \tag{3}$$

N is the number of autonomy days, U_S in V is the voltage of the system, and η_{bat} represents the battery efficiency.

Input and output currents of charge controller are computed from following equations, respectively:

$$I_I = \frac{P_{\text{CC}} \cdot N_{\text{CC}}}{U_S} \tag{4}$$

$$I_O = I \cdot N_{CC} \quad (5)$$

I in A is the operating current of the charge controller, P_{CC} in W and N_{CC} are the total capacity and the number of charge controller, respectively, and are given by Eqs. 6 and 7, respectively:

$$P_{CC} = U_{CC} \cdot I \quad (6)$$

$$N_{CC} = \frac{P_{CC}}{C_{PV}} \quad (7)$$

U_{CC} in V is the voltage of a charge controller. I_I and I_O in A must be greater than the currents produced by PV array and consumed by charge, respectively. Both the currents are expressed from Eqs. 8 and 9, respectively:

$$I_{PV} = \frac{C_{PV}}{U_S} \quad (8)$$

$$I_L = \frac{P_L}{U_L} \quad (9)$$

P_L in W and U_L in V are the power and voltage of alternating charges. The nominal capacity of inverter is calculated from Eq. 10:

$$P_{In} = 1.2 \cdot P_L \quad (10)$$

2.3 *Sizing Approach Based on Monthly Solar Radiation Average*

This approach is focused on the operating optimization of PV array and batteries. Indeed, the PV array generates the energy to make loads work and to charge batteries. This energy stored in batteries is used later, when PV energy is insufficient to meet load demand. The generation of energy from PV array strongly depends on the availability of many meteorological parameters such as solar irradiation and temperature. These last have an unforeseeable nature, which make them difficult to check. So, it is important to find the best configuration between PV array and storage system in relation to meteorological data for an optimal PV production which satisfies the load demand and minimize the use of batteries and the energy deficiency. In this study, the configurations between PV and battery capacities are looked for, according to monthly solar radiation average and the optimal one is selected based on TLCC. The different stages of the sizing approach are detailed in this section.

Energy models for PV and batteries. The model used to simulate the PV output is given by Eq. 11:

$$E_{PV} = A_{PV} \cdot E_I \cdot \eta_r \cdot \eta_{inv} \cdot \eta_w \cdot \left[1 - \beta\left(T_{amb} + \frac{G}{800}(T_{NOC} - 20)\right) - T_{ref}\right] \quad (11)$$

A_{Pv} is the total area of PV array, η_r, η_{inv}, and η_w are the reference, inverter, and wire efficiencies, respectively, T_{amb}, T_{NOC}, and T_{ref} are ambient, optimal operating, and reference temperatures, respectively.

The PV energy depends on solar irradiation and ambient temperature, which change according to the time, but in this study, only the monthly variation of solar irradiation is considered. The temperature has a constant value, and it corresponds to the highest monthly value to maximize the PV energy.

If the energy generated by PV array is greater than load demand and the energy stored in the battery storage is lower than the maximal capacity, then batteries are been charging in accordance with Eq. 12:

$$E_{bat} = E_{batmin} + \left(E_{PV} - \frac{E_I}{\eta_{inv}}\right) \cdot \eta_{bat} \quad (12)$$

E_{batmin} is the minimal energy allowed in the battery.

If the PV energy is lower than load demand and the maximal capacity of batteries is greater than E_{batmin}, then the energy stored is used to compensate the missing energy from Eq. 13:

$$E_{bat} = E_{batmin} + \left(\frac{E_I}{\eta_{inv}} - E_{PV}\right) \quad (13)$$

In order to protect batteries against damages and drastic reduction of their lifetime, they are subject to the following restriction:

$$E_{batmin} \leq E_{bat} \leq E_{batmax} \quad (14)$$

E_{batmax} is the maximal energy allowed in the batteries.

Excess and deficit of energy. If the PV energy is higher than load demand and the capacity of batteries is equal to its maximal value, there will be an excess of energy. If the PV energy is weaker than load demand and the capacity of batteries is insufficient, an energy shortage occurs. In this study, only the energy shortage is considered to evaluate the technical relevance of the proposed sizing. It is quantified by the Loss of Power Supply Probability (LPSP), which is defined by the ratio between the energy shortage and the load demand. It is expressed by Eq. 15:

$$\text{LPSP} = \frac{\text{LPS}}{E_L} \quad (15)$$

LPS corresponds to the shortage energy, and it is computed by using Eq. 16:

$$\mathrm{LPS} = E_L - (E_{\mathrm{PV}} + E_{\mathrm{bat}} + E_{\mathrm{batmin}}) \cdot \eta_{\mathrm{bat}} \quad (16)$$

The LPSP varies between 0 and 1. A LPSP close to 0 means a good satisfaction of load demand and the one close to 1 is equivalent to a bad satisfaction of load requirements.

Description of the different stages of sizing algorithm. The different stages of sizing algorithm are described below.

Step 1: Define input parameters:

- the load demand is equal to the maximal consumption estimated from load profile;
- all monthly values of solar irradiation are used in the sizing process;
- the temperature is constant and corresponds to the highest monthly value;
- the technical characteristics of chosen component models are set.

Step 2: The time, PV, and battery capacity ranges are set. Then, input parameters are initialized and the energy generated by PV array is calculated by using the first value of solar irradiation. From this calculation, two situations can appear:

- the situation corresponding to the battery charge in accordance with Eq. 12:
- the situation corresponding to battery discharge expressed by Eq. 13:
- if neither of these two situations occurs, then there is an excess production or an energy shortage defined by Eq. 16.

Step 3: Increment the technical parameters:

- PV capacity is less than maximal capacity; in this case, it is incremented and the calculation of PV energy is repeated with the same value of irradiation;
- PV capacity is equal to the maximum.

Step 4: In this last case, the time is compared to the maximal time. There are two possibilities:

- the time is less than maximal time, and in this case, it is incremented and **step 2** is repeated;
- the time is equal to maximal, and then the different configurations of PV and battery capacities are generated for all values of monthly irradiations. Figure 4 shows all step of sizing algorithm.

3 Analysis of Results and Discussion

3.1 Intuitive and Numerical Results

Table 1 summarizes results given by intuitive method. The total PV capacity required by system is 93298. W_P corresponds to 374 PV modules of 250 W_P, while the total

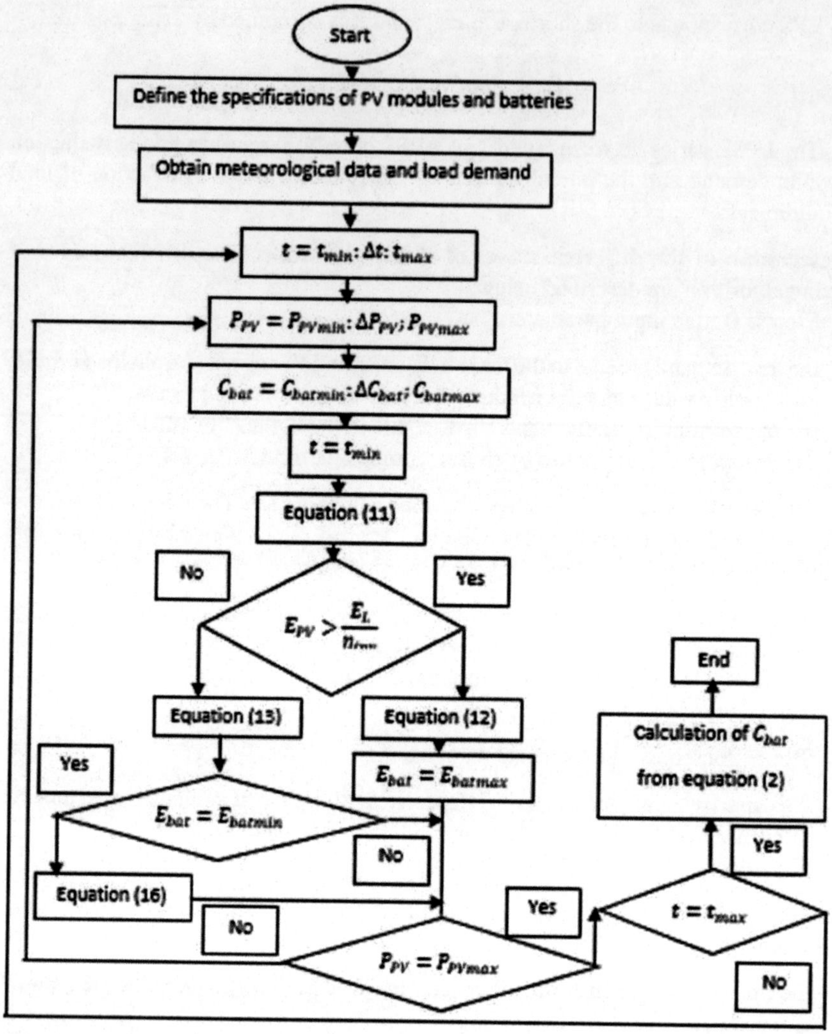

Fig. 4 Flowchart of proposed sizing algorithm

Table 1 Results provided with intuitive method

System components	Results of intuitive method
Total PV capacity	93500 W_P
Total number of PV modules	374
Total capacity of storage system	25840 Ah
Total number of batteries	408
Total number of charge controller	3
Total number of inverter	11

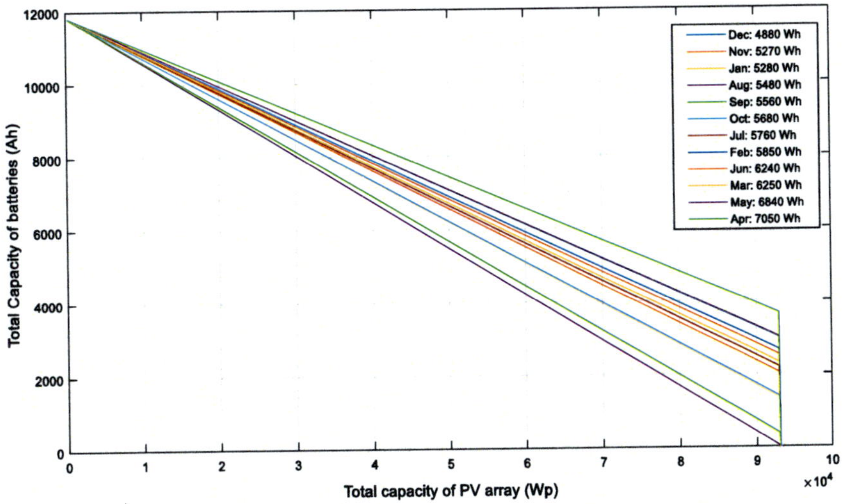

Fig. 5 PV/battery combinations for different monthly values of irradiation

battery capacity is 25840 Ah. A number of 240 batteries are appropriate to yield this energy quantity. In regard to charge controller, only three of which characteristic are 250 V/100 A with an input voltage of 48 V being sufficient. The total number of inverter is 11, with a power of 10000 VA and input and output voltages of 48 and 225 V, respectively.

The different combinations of PV and battery capacities for all values of solar irradiations provided with numerical method are shown in Fig. 5:

From Fig. 5, we notice that a lower irradiation value requires a higher capacity to install. While a higher irradiation value gives smaller capacities. Indeed, the energy quantity produced is more important when the irradiation is high, and in this case, PV and battery capacities are restricted. On the contrary, a small solar irradiation leads to decreased output energy, so capacity of components increase. Figures 6 and 7 illustrate this phenomenon. They give different combinations between PV and battery capacities provided with higher and smaller values of irradiation, respectively. They show when a low value of irradiation is used for sizing, a great value of component capacities is necessary to meet load demand. So, the optimal configuration of system will be chosen from the worst month (Fig. 6). Since all configurations from this figure have the same reliability, the best one is selected based on an economic criterion.

3.2 *Economic Analysis*

Different concepts and approaches have been established to assess the economic cost of PV systems. In this study, the Total Life Cycle Cost (TLCC) is used to find the

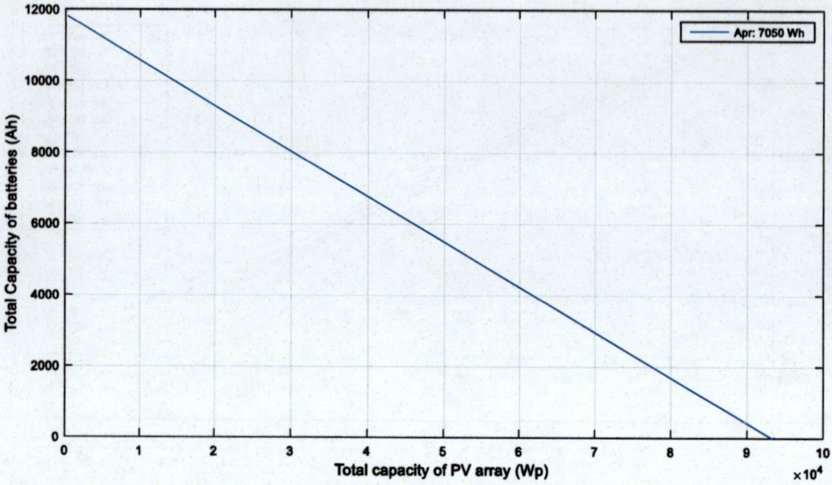

Fig. 6 PV/battery combinations for greatest value of irradiation

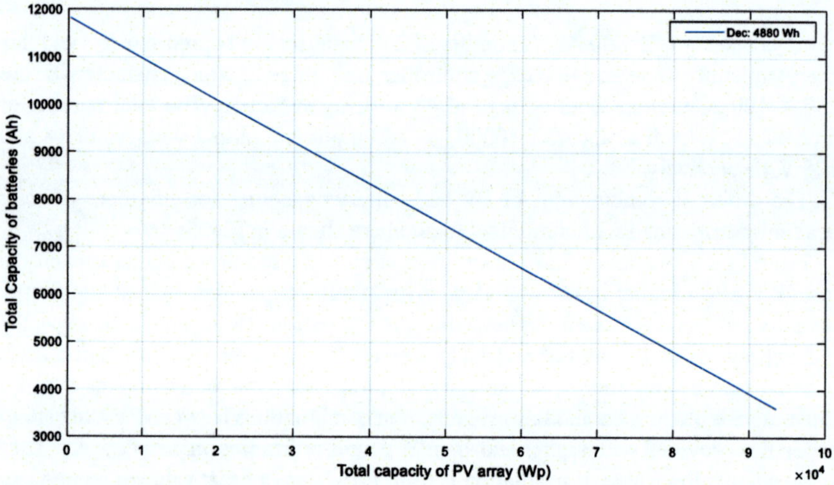

Fig. 7 PV/battery combinations for smallest value of irradiation

optimal PV/battery configuration. It is mathematically expressed by Eq. 17:

$$\text{TLCC} = K_{IC,S} + K_{RE,S} + K_{M,S} \tag{17}$$

$K_{IC,S}$ is the initial capital cost of the system in FCFA, it consists of component prices, labor, design, and fitting costs, and $K_{RE,S}$ and $K_{M,S}$ are the net present value of replacement and maintenance costs, respectively. Initial capital cost is calculated

Table 2 Cost per units of components

Components	Cost per unit
PV module	553 FCFA/W_P
Battery	62.2 FCFA/Ah
Charge controller	276.6 FCFA/W
Inverter	276.5 FCFA/W

from the following equation:

$$K_{IC,S} = K_{PV}.P_{PV} + K_{bat}.C_{bat} + K_{CC}.P_{CC} + K_{inv}.P_{inv} + K_{ins} \tag{18}$$

K_{PV}, K_{bat}, K_{CC}, and K_{inv} are the costs per unit in FCFA, given in Table 2, of PV, battery, charge controller, and inverter components, respectively, K_{ins} is the cost of installation, and it represents 10% of the initial capital cost.

We assume that the PV module lifetime is the same as the system and need no replacement but battery, charge controller, and inverter require to be changed. The replacement cost is calculated from Eq. 19:

$$K_{RE,S} = K_{RE,bat} + K_{RE,CC} + K_{RE,inv} \tag{19}$$

$K_{RE,S}$, $K_{RE,CC}$, and $K_{RE,inv}$ are the replacement costs of battery, charge controller, and inverter, respectively, and are computed from Eqs. 20 to 22, respectively.

$$K_{RE,bat} = K_{bat,T}.\left(\frac{1}{1+i}\right)^N \tag{20}$$

$$K_{RE,CC} = K_{CC,T}.\left(\frac{1}{1+i}\right)^N \tag{21}$$

$$K_{RE,inv} = K_{inv,T}.\left(\frac{1}{1+i}\right)^N \tag{22}$$

N is the lifetime of components, $K_{bat,T}$, $K_{CC,T}$, and $K_{inv,T}$ are the total expended costs for acquisition of battery, charge controller, and inverter, respectively, i represents the market interest rate, and it is calculated from Eq. 23:

$$i = i' + \overline{f} + i' + i'\overline{f} \tag{23}$$

$\overline{f}$ means the inflation rate, i' corresponds to the reel interest rate, and it is determined by the local bank and expressed from Eq. 24:

$$i' = \text{BRL} - 2\% \tag{24}$$

Table 3 Values of economic parameters

Designation	Values (%)
$\overline{f}$	0.5
BRL	3.5

Table 4 Values of TLCC

PV/batteries combinations	TLCC
10000W$_P$/ 10700 Ah	124012660 FCFA
20000 W$_P$/9500. Ah	130969349.1 FCFA
30000 W$_P$/8000 Ah	137829197.4 FCFA
40000 W$_P$/7000 Ah	144850447 FCFA
50000 W$_P$/5500 Ah	151710295.5 FCFA
60000 W$_P$/4500 Ah	158731545.1 FCFA
70000 W$_P$/3000 Ah	165591393.5 FCFA
80000 W$_P$/1700 Ah	172515802.3 FCFA
90000 W$_P$/500 Ah	179472491.5 FCFA

BRL is the Base Lending Rates.

The operating and maintenance cost for 20 years are calculated from the following Eq. 25:

$$K_{M,S} = K_M\left[\frac{(1+i)^N - 1}{i(1+i)^N}\right] \tag{25}$$

K_M is the maintenance cost for each year (1% of the initial total cost). The values of economic parameters are given in Table 3.

The calculation of the TLCC corresponding to each combination in Fig. 6 led to results, given in Table 4:

From Table 4, we remark that the combination corresponding to the greatest and smallest PV and battery capacities, respectively, gives the highest TLCC, while the combination which gives the weakest and most large PV and battery capacities, respectively, achieves the smallest TLCC. In the first case, the PV capacity would produce a great quantity of energy but the storage system is very limited to store the requisite energy. While in the second one, the PV capacity is very restricted and would be not able to generate an important energy quantity. As all these combinations are the same reliability, the best one will be selected, based on the smallest storage capacity, which is able to satisfy the energy consumption during night, assessed to 112245 Wh. According to Table 5, this combination corresponds to 70000 W$_P$/3000 Ah. Indeed, with this combination, the PV capacity would sufficiently produce energy to meet load requirements and load the storage system. Regarding this last, they could store the necessary energy to supply electricity for night and cloudy days. The LPSP for this combination is assessed to 0.3.

Table 5 Energy stored in batteries corresponding to each combination

PV/batteries combinations	Energy stored in batteries
10000 W_P/ 10700 Ah	410880 Wh
20000 W_P/9500 Ah	218880 Wh
30000 W_P/8000 Ah	307200 Wh
40000 W_P/7000 Ah	268800 Wh
50000 W_P/5500 Ah	211200 Wh
60000 W_P/4500 Ah	172800 Wh
70000 W_P/3000 Ah	115200 Wh
80000 W_P/1700 Ah	65280 Wh
90000 W_P/500 Ah	19200 Wh

This numerical method generates a PV capacity, less than the one provided with intuitive method. On the contrary, the storage capacity has drastically dropped, compared to the empirical method. In terms of percentage, the PV and battery capacities are decreased by 25.13 and 88.39%, respectively.

In [16], the proposed numerical method reduced the capacity of the storage system by 75% and the TLCC by 65% compared to the intuitive method, with only 0.01 of ALPSP, while the PV capacities are the same for intuitive and numerical method. The same variation is observed for battery capacity. Concerning the PV capacity, it was kept constant in [16] and has decreased in the model proposed in this paper. Differences with the results in this paper is due to the fact that all average monthly values of solar irradiation are considered and the load profile has been defined from energy balance of the site and the consumer's behavior. However, the average LPSP found in this paper is considered as relatively high but must be further optimized. So, it is very important to use very accurate input data which reflects their real variations in order to optimize the final system operating by minimizing the deficit and excess energy.

4 Conclusion

In this paper, an improvement of the numerical method, proposed in Sadio et al. [16], is made. This improvement is based on two criteria. The first one concerns the solar irradiation for which all average monthly values are considered instead of only one value. The second is about the load demand for which the maximum value is used instead of the total value and is determined from the energy balance and the behavior of consumers. After generating all PV/batteries configurations for each value of solar irradiation, the optimal combination is selected by using the TLCC. It is shown that the PV and battery capacities are decreased by 25.13% and 88.39%, respectively, when compared to the intuitive method. These results show, in this improved method, PV and battery capacities dropped much, compared to intuitive method, like [16], where the battery capacity was strongly restricted by the model, while the PV capacity was kept constant in [16] and has decreased in the model proposed in this paper. Because of the unavailability of data and the non-integration of some parameters, the accuracy of results can be affected; this explains the high value of LPSP. So more effective improvements to generate more high-performance PV systems must be made. It could be interesting to develop new techniques and methodologies to predict with precision the solar energy availability and to design good sizing and optimization models leading to better results.

References

1. Ibrahim IA, Khatib T, Azah M (2017) Optimal sizing of a standalone photovoltaic system for remote housing electrification using numerical algorithm and improved system models. Energy 126:392–403
2. El Mghouchi Y, Chhamc E, Zemmouria EM, El Bouardib A (2019) Assessment of different combinations of meteorological parameters for predicting daily global solar radiation using artificial neural networks. Build Environ 149:607–622
3. Khatib T, Ibrahim IA, Mohamed A (2016) A review on sizing methodologies of photovoltaic array and storage battery in a standalone photovoltaic system. Energy Convers Managem 120:330–348
4. Sarhan A, Hizam H, Mariun N, Ya ME (2018) An improved numerical optimization algorithm for sizing and configuration of standalone photovoltaic system components in Yemen. Renew Energy
5. Nogueira CEC, Vidotto ML, Niedzialkoski RK (2014) Sizing and simulation of a photovoltaic/wind energy system using batteries, applied for a small rural property located in the South of Brazil. Renew Sustain Energy Rev 29:151–157
6. Bouabdallah A (2015) Safe sizing methodology applied to a standalone photovoltaic system. Renew Energy 80:266–274
7. Padilla R, Mercado A (2018) Measuring reliability of hybrid photovoltaic-wind energy systems.: a new indicator. Renew Energy 106:68–77
8. Rachchh R, Kumar M, Tripathi B (2016) Solar photovoltaic system design optimization by shading analysis to maximize energy generation from limited urban area. Energy Convers Managem 146:244–252
9. Qayoom A, Othman A, Ragai A, Rigit H (2012) A novel analytical model for optimal sizing of standalone photovoltaic systems. Energy 46:675–682
10. Mellit A (2010) ANN based GA for generating the sizing curve of standalone photovoltaic system. Adv Eng Softw 41:687–693
11. Prakash P, Khatod D (2016) Optimal sizing and siting techniques for distributed generation in distribution systems.: a review. Renew Sustain Energy Rev 57:111–130
12. Babacan O, Torre W, Kleissl J (2017) Siting and sizing of distributed energy storage to mitigate voltage impact by solar PV in distribution systems. Solar Energy 156:199–208
13. Halboot D, Nabil M, Khatib T (2019) A novel method for sizing of standalone photovoltaic system using multi-objective differential evolution algorithm and hybrid multi-criteria decision-making methods. 174:1158–1175
14. Melit A (2007) Sizing of photovoltaic systems: a review. Revue des Energies Renouvelables 10:463–472
15. Tan CW, Ayop R, Isa NM (2018) Components sizing of standalone photovoltaic system based on loss of power supply probability. Renew Sustain Energy Rev
16. Sadio A, Fall I, Mbodji S (2017) New numerical sizing approach of a standalone photovoltaic power at Ngoundiane; Senegal. Endorsed Trans Energy Web Info Technol 5:122017–012018

17. Sadio A, Fall I, Mbodji S, Sow PLT (2018) A comparative study based on the Genetic Algorithm (GA) method for the optimal sizing of the standalone photovoltaic system in Ngoundiane site. Endorsed Trans Energy Web Info Technol 5:7–18
18. Sadio A, Mbodji S, Dieng B, Ndiaye A, Fall I, Sow PLT Sizing of standalone photovoltaic water pumping system of a well in Ngoundiane site. Springer. (Publication in progress)
19. Sadio A, Fall I, Mbodji S, Sissoko G (2017) Analysis of meteorological data for photovoltaic applications in Ngoundiane site. Endorsed Trans Energy Web and Inform Technol 3:2312–8623

Blockchain Based Software Engineering Requirements Analysis and Management

Bandar Ali Alrami AL Ghadmi, Omar Ahmed Abdulkader, and Ahmad Abdulaziz Alwarhi

Abstract Requirements analysis and engineering is a vital phase in any software project's lifecycle, and the success or failure during this phase mainly determines the entire project's outcome. Recently the challenges incredibly increased in the software industry either in the technology, project management, or requirements engineering, drive-by leveraging the diversity of the tremendous tools and techniques that aim to avoid failure in requirements engineering and analysis. However, the tools and techniques will not always tackle the most significant challenge: validate, align, and confirm the needs and outcomes accordingly to both parties, the customer and the vendor, ensuring the confirmation and validation process is trustworthy. A unique process that ensures the requirements reliability is by introducing a framework that will authenticate the requirements analytics and engineering by Ethereum blockchain technology smart contract for the customers and vendor to guarantee substantial agreement on all requirements aspects through the project's lifecycle.

Keywords Smart contract · Ethereum · Blockchain · Decentralized application

1 Introduction

Requirements analysis is a statement that all stakeholders agreed about how the service, component, or part of the software will become [1]. This statement would emphasize the indispensable of this phase of any software life cycle, and any failure at this phase will affect the whole project to unknown consequences [2]. Researches reveal that the software project fails mainly due to the poorly requirements management and unachievable project expectations. Brenda Whittaker state [3, 4] that 45% of failed projects did not succeed to deliver particular advantages. In [5], Kamata and Tamai report that "Clear statement of requirements" is represented as a third rank

B. A. A. AL Ghadmi · O. A. Abdulkader (✉) · A. A. Alwarhi
Faculty of Computer Studies, Arab Open University, Riyadh, Kingdom of Saudi Arabia
e-mail: o.abdulkader@arabou.edu.sa

B. A. A. AL Ghadmi
e-mail: b.alrami@arabou.edu.sa

X.-S. Yang et al. (eds.), *Proceedings of Seventh International Congress on Information and Communication Technology*, Lecture Notes in Networks and Systems 447,
https://doi.org/10.1007/978-981-19-1607-6_6

with 13.0% of project success factors responded and "incomplete requirements" is represented as a first rank with 13.1% of project impaired factors responded according to the Standish Group during 1994 [6]. This leads to the establishment that customer expectation is a bundle of views on many aspects of the final product, which many projects fail to capture in an early stage of the project life cycle. That's encouraging growing several requirements analysis and project management approach today [7], trying to fill the gap by employ tools and techniques to ensure covering all customer's requirements. Regardless of those approaches are assisting in capturing the customer's requirements or not. Many projects are still failing to deliver the customer expectations in the final products, which draws attention to that customer's awareness and organization awareness are evolved during the project's life cycle [8]. However, relatively the customer's requirements could be developed and change during the project's stages. And it is challenging to address the customer's needs. Even if the customer's requirements are documented in an early stage, the gap will persist, which will be hard to discover until the project progress evolves and reaches the point which will be hard to accept change request this could be carried from one side. On the other side, even if the requirements are managed well, this does not mean the project expectations are met because it has been observed from a different perspective, a technical mindset against a business mindset. That is why requirements management needs to securely address both views during the project's lifecycle to obtain the project's goal and meet customer expectations. Gotel and Finkelstein [9] define the requirements traceability, which aims to explain and comply with the life of a requirement in backward and forward way. Blaauboer et al. investigate five different factors, which comprise: process flow, customer awareness, development organization awareness, return on investment and stakeholder preferences. Most of the leaders in the software development project who are interviewed are unfamiliar to the traceability aspect. Due to the advantages of the RT smart contractual enforcement of traceable trustworthiness and objectives through various tamper-proof transactions [10], this project aims to align outcomes between vendors and customers in every aspect during the project's life cycle. Therefore, building a trustworthy requirements management and engineering platform is an essential step. This project is based on blockchain technology that traces and validates the requirements and the process between the parties without the need to reply to a reliable third party to organize given transactions. The following sections deep dives into details to show how the approach will fill the requirements management and engineering based on blockchain technology fill the gap [11].

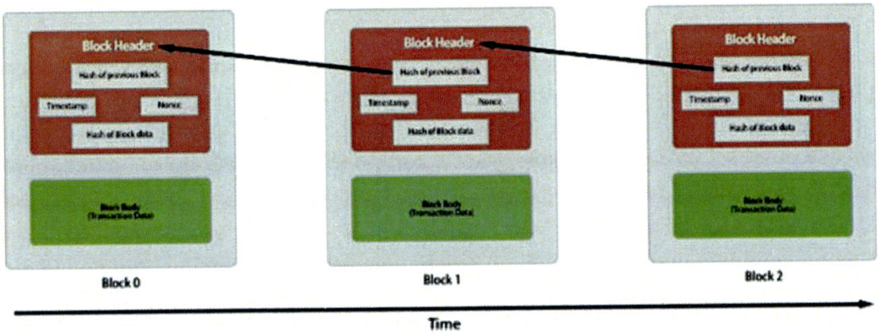

Fig. 1 The structure of a blockchain

2 Related Research

2.1 *Blockchain*

Nowadays, cryptocurrency is probably the favorite buzzwords in media, industry, and academia. Currently, the market cap of Bitcoin is $935.44B [12, 13]. That paid more attention even for people who never decided to cryptocurrency or understand how it works are talking about Bitcoin, Ethereum, and other digital currencies and blockchain is the underlying technology, which was initially proposed in 2008 [14] and implemented in 2009. Blockchain is defined as a disseminated immutable public ledger, which is able to record various transactions between two given parties in an efficient, permanent and identified manner within blocks' sequence. In particular, each block possesses a block header where the block body consists of a validated list and authentic transactions, which are applied to the network that is related to a blockchain [15]. A block header includes a block body of hash representation where the former hash value pertaining to the block header is also included (see Fig. 1).

The pre-configure block represents the primary block, namely, the genesis block, which includes a 0 value that forms a former address. Each blockchain network consists of a published genesis block. Additionally, each block should be attached along to the closest blockchain according to the agreed-upon consensus approach. Nonetheless, every block should be valid, and hence, should be validated in an independent manner through every blockchain network user.

2.2 *Ethereum*

In [16] Ethereum highlights many drawbacks of the Bitcoin such as the sizes of the blockchain, bandwidth and throughput [17]. Further, Ethereum develops the scripting aspects, including the on-chain meta-protocols, which permits the developers to generate different features as interoperability, standardized, feature-completeness, applications scalable and ease of development [18]. The Ethereum's architecture aims at complying with the following ideologies: **Simplicity, Universality, Modularity, Agility, and Non-discrimination**, and **non-censorship**. In fact, these ideologies represent a philosophy that can provide a robust base [19]. The Ethereum account consists of four different domains, which comprises: the nonce to validate where every transaction is managed one time only, the contract code, the storage of an account that is by default empty and the current ether balance. In Ethereum, the code of the pattern execution contract indicates to the "Ethereum virtual machine code", which can also indicate to the "**EVM** code". As declared by Grishchenko, "*This is quasi Turing complete as the otherwise Turing complete execution is restricted by the upfront defined resource gas that effectively limits the number of execution steps*" [20].

2.3 Smart Contract

In 1994, Szabo produced smart contracts [21], which represent a simple bench of codes written by computer programs. Smart contracts run independently without intervention from a third party, which and will be executed automatically once the contract's predefined conditions are met [22]. Deploying Smart Contracts on blockchain eliminates the role of a trusted third party which is required in traditional contracts. Figure 2 shows a transaction example for a smart contract delivery service executed on a blockchain. Service Requester will append delivery requests through the blockchain network. Smart Contract Payment trigger produces for the requester,

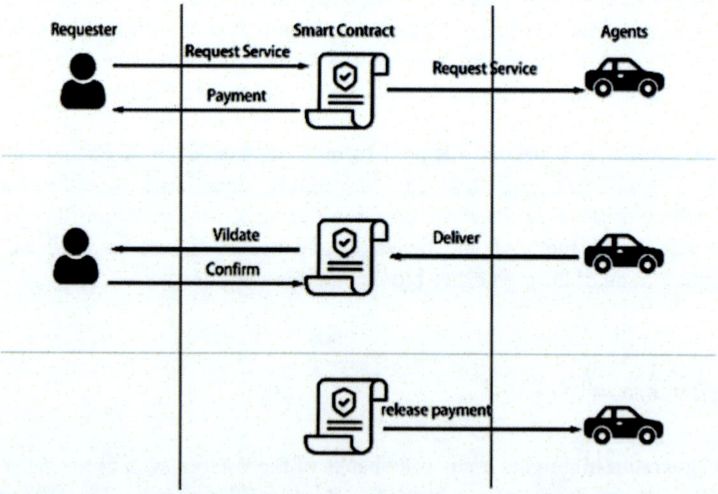

Fig. 2 An example of smart contract delivery service

and then the agent can claim verified delivery request. When the agent confirms the delivery request, the service requester has to confirm it, and Smart Contract will trigger to release the payment to the agent.

The Smart Contract process will be executed as mentioned without a third party intervention like a central bank or one point only that is related to verification transactions, which is represented into a peer-to-peer form. In fact, smart contracts are documented in high-level languages. Solidity is defined as a high-level language and an object-oriented that can be used to implement smart contracts. Solidity represents an important part of the Ethereum platform that operated on the EVM, which is a curly-bracket language that is affected by the C++ programming language, JavaScript and Python [23].

2.4 Decentralized Applications

Decentralized applications or dApps are a set of wares that communicates with the blockchain, which manages the state of all network actors. dApps is front-end use blockchain for data storage and smart contracts for their app logic. Once the decentralized application is deployed on the blockchain network, it will remain immutable and cannot change. It uses JavaScript, HTML, CSS, where the business logic is lying on a database or back-end server in central traditional front-end apps. The same scenario applies to decentralized applications, but the business logic will be on the decentralized blockchain network [24]. This empowers dApps which avoid the single point of failure.

3 Model Implementation

3.1 Generic Context Structure

To satisfy the intent is to build a framework that can validate the requirements along with expected outcomes. The ultimate architecture is to fulfill the need is to design the framework to validate the requirements management process itself and make it simple, accessible, reliable, and scalable. This solution could be achieved by design a smart contract to be deployed to validate every requirement deliver from vendor to customer through the Ethereum blockchain network. Few constraints need to get attention shown below:

- Privacy: Requirements Information details that need to remain private which will be saved off-chain and will be saved on central server back-end
- Abstraction layer Ethereum service retrieval: To guarantee to retrieve the transactions from the Ethereum network.

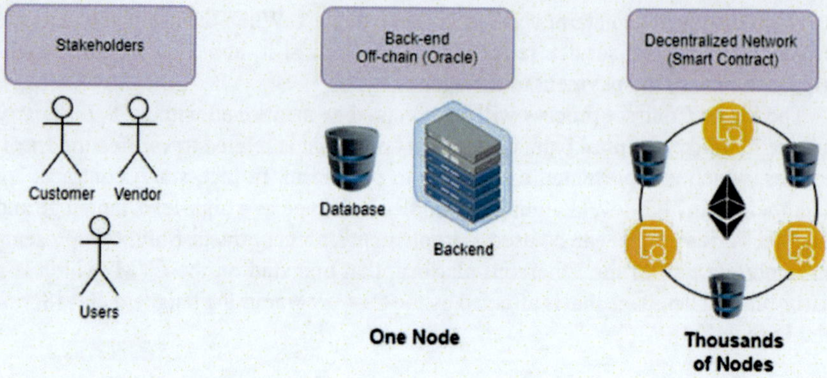

Fig. 3 Framework architecture

- Store requirements assets: dApp and Ethereum are not designed to store data which will be costly because 1 GB will cost around $76,000.

The framework will be semi-centralized applications to achieve the needs mentioned above.

3.2 *Framework Architecture*

Figure 3 shows the proposed framework architecture, whereas mentioned in the section above fulfill the need to sign every requirement proposed to sign through the smart contract by Ethereum blockchain and save extra data off-chain. This will empower the framework to do necessary calculations on the private back-end and save and manipulating the data that does not need to be relayed on the Blockchain network. Each verified transaction by Ethereum will be saved on the Ethereum alongside a reference indicating the transaction itself on the back-end to lock it and checking the piece of the requirement and outcome between the stakeholders and refer it is approved.

3.3 *Use Case*

Table 1 shows the functional requirements of our framework. The framework's process requires both Actors to do validating and confirm transactions on any listed requirements, ensuring both parties are aligned and agreed.

Table 1 Use case of the framework

Requirement ID	Framework use case	
	Requirement definition	Actor
R1.0	Create new project	Vendor
R1.1	Add new requirement and expected outcome	Vendor
R1.1	Add new requirement and expected outcome	Vendor
R1.1.1	Update requirement	Vendor–customer
R1.1.2	Review and comment On requirement	Vendor–customer
R.1.2	Confirm project	Vendor–customer

Table 2 Framework functional requirements

Actor	Actors
	Actor definition
Vendor	Business analyst, developer, project manager, product manager
Customer	Project owner, end user

3.4 Actors

The actors that will interact with the framework and their primary function are shown below in Table 2.

3.5 Workflow Process

Figure 4 shows the workflow that explains how the framework will process every single requirement request verification through the layers and stakeholders.

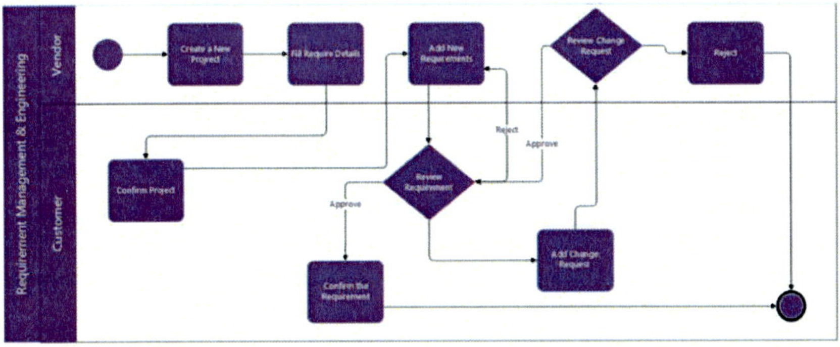

Fig. 4 Verification workflow process

4 Experimental Analysis and Results

The proposed framework is stressful to design the best approach solution for ensuring the process of requirements management and engineering, which is obtained efficiently. Accordingly, the following findings are comprised of:

- The ability to deal with enormous papers and researches, which create the awareness of the needs related to the business experience to be understood and be aware of the latest research within the field of expertise.
- Blockchain and smart contracts can be tricky in the development stage. The development of the Truffle framework is applied to easily simulate the blockchain and smart contract within a development mode without the need for deploying the produced code in each time.
- The Truffle Ganache Ethereum framework is extremely advantageous as it assists in generating a personal Ethereum blockchain that is applied for running different involved tests, executing different provided commands, and checking different involved states when handling the way, a chain is being processed.

5 Conclusion and Future Research

To identify the problem, it is essential to aware that blockchain and cryptocurrency have become a significant research problem. This was a huge favor for the project to rely on the correct concrete, which what is draw the attention of the smart contracts was started in the 1990s as paper by Nick Szabo, and after ten years later blockchain is shine. This paper address some aspects and challenges that are facing today's software industry in requirements management and requirements engineering. The experimental results indicate that the proposed model could significantly impact the project's outcomes for the stakeholders. The most significant result is to verify the outcomes from both parties with a confidant. In the future, the framework could be an API call that can be embedded and called from another off-chain party to help the industry to build such function internally with little effort.

Acknowledgements The authors would like to thank the Arab Open University for supporting this research paper.

References

1. Azham Hussain EOOCM (2016) Requirements: towards an understanding on why software projects fail. 06010 Sintok, Malaysia
2. Dhirendra Pandey USAKR (2010) An effective requirement engineering process model for software development and requirements management. In: International conference on advances in recent technologies in communication and computing

3. Whittaker B (1999) What went wrong? Unsuccessful information technology projects. In: Information management and computer security
4. Diana White JF (2002) Current practice in project management—an empirical study. Int J Project Manag
5. Mayumi Itakura Kamata TT (2007) How does requirements quality relate to project success or failure? 15th IEEE—international requirements engineering conference
6. International SG (1994) [Online]. Available: http://www.standishgroup.com
7. Demi S (2020) Blockchain-oriented requirements engineering: a framework. In: IEEE 28th international requirements engineering conference (RE)
8. Floris Blaauboer KSAMNA (2007) Deciding to adopt requirements traceability in practice. Springer-Verlag Berlin Heidelberg
9. OC, Finkelstein ACW (1994) An Analysis of the requirements traceability problem. IEEE
10. EA, SG, Wenli Yang DHLDABK (2019) A survey on blockchain-based internet service architecture: requirements, challenges, trends, and future. In: IEEE access
11. SXHDXCAHW Zibin Zheng (2017) An overview of blockchain technology: architecture, consensus, and future trends. In: IEEE 6th international congress on big data
12. Coin Market Cap (2021) [Online]. Available: https://coinmarketcap.com/
13. Coindesk (2021) [Online]. Available: https://www.coindesk.com/coindesk20
14. Nakamoto S (2008) Bitcoin: a peer-to-peer electronic cash system
15. Dylan Yaga PMNRKS (2018) Blockchain technology overview. National Institute of Standards and Technology
16. Buterin V, Ethereum White Paper. [Online]. Available: https://blockchainlab.com/pdf/Ethereum_white_paper-a_next_generation_smart_contract_and_decentralized_application_platform-vitalik-buterin.pdf
17. Jesse Yli-Huumo DKSCSPKS (2016) where is current research on blockchain technology?—A systematic review. PLOS ONE
18. Dejan Vujičić DJSR (2018) Blockchain technology, Bitcoin, and Ethereum: a brief overview. In: 17th international symposium INFOTEH-JAHORINA
19. ethereum.org. "Ethereum Whitepaper", [Online]. Available: https://ethereum.org/en/whitepaper/
20. MMaCS Ilya Grishchenko (2018) Foundations and tools for the static analysis of ethereum smart contracts. In: Chockler H, Weissenbacher G (eds)
21. Szabo N (1996) Smart contracts: building blocks for digital markets. [Online]. Available: https://www.fon.hum.uva.nl/rob/Courses/InformationInSpeech/CDROM/Literature/LOTwinterschool2006/szabo.best.vwh.net/smart_contracts_2.html
22. Zibin Zheng SXH-NDWCXCJWMI (2019) An overview on smart contracts: challenges, advances and platforms. In: Future generation of computer system
23. Weili Chen ZZJCENPZYZ (2018) Detecting ponzi schemes on ethereum: towards healthier blockchain technology. In: Web economics, monetisation, and online markets, France
24. Ziechmann K (2021) Introduction to dapps. [Online]. Available: https://ethereum.org/en/developers/docs/dapps/

Computer Simulation of the Response of a Semiconductor Wafer with a Self-Affine Pattern in the Form of a System of Coupled Ring Grooves to Electromagnetic Radiation

Gennadi Lukyanov, Alexander Kopyltsov, and Igor Serov

Abstract We simulated the response of a surface with a circular relief on a semiconductor wafer constructed using self-affine transformations to the effect of an incident electromagnetic wave. The study assumed that the main mechanism leading to the reaction of the plate to incident radiation is electric polarization, which, for example, is the basis for the functioning of a number of electronic components, such as a MOS FET or a CCD. Since silicon belongs to polarizable materials, a spatial separation of charges occurs in a changing electric field in the volume of a silicon crystal in accordance with the law of change of field. If a silicon wafer is used, on one of the surfaces of which a certain relief is created, then the distribution of charges under the relief will be uneven in space in accordance with the pattern of this relief.

Keywords Regular self-affine microrelief · Ring-shaped grooves · Electric field · Computer simulation · Wave structure

1 Introduction

Regular microrelief surfaces have been used for a long time. These include, for example, diffraction gratings. Widely used are devices on surfactants, the basis of which is also a regular surface relief. Now there is a study of new areas of application of devices with a regular microrelief on the surface. For example, there are studies [1] where the effect of thermal radiation from such heated surfaces is investigated. It was shown [1] that in the near field this radiation has spatial coherence.

G. Lukyanov (✉)
ITMO University, Kronverksky Pr. 49, 197101 St. Petersburg, Russia
e-mail: gen-lukjanow@yandex.ru; gn_lukyanov@itmo.ru

A. Kopyltsov
Saint Petersburg State University of Aerospace Instrumentation, Bolshaya Morskaya 67, 190000 St. Petersburg, Russia

I. Serov
Human Genome Research Foundation, Bolsheokhtinsky prospect, 16, bldg. 1, lit. A, 195027 St. Petersburg, Russia
e-mail: director@aires.fund

X.-S. Yang et al. (eds.), *Proceedings of Seventh International Congress on Information and Communication Technology*, Lecture Notes in Networks and Systems 447,
https://doi.org/10.1007/978-981-19-1607-6_7

In our study, we present an object that also has the properties of self-similarity and scale invariance, but so far it is not widely represented in the studies. It is obtained in the process of scaling and rotation of the circle taken as a basis and further transformations of the aggregates thus obtained [2].

An affine transformation of a vector whose origin coincides with the origin and the end has coordinates (x_1, y_1), into a vector whose origin is at a point with coordinates (b_1, b_2), and the end at a point with coordinates (x_2, y_2) has the form [3]:

$$\begin{cases} x_2 = a_{11}x_1 + a_{12}y_1 + b_1 \\ y_2 = a_{21}x_1 + a_{22}y_1 + b_2 \end{cases} \tag{1}$$

\System (1) can be represented in the form of the matrix:

$$\begin{bmatrix} x_2 \\ y_2 \end{bmatrix} = \begin{bmatrix} a_{11} & a_{12} & b_1 \\ a_{21} & a_{22} & b_2 \end{bmatrix} \begin{bmatrix} x_1 \\ y_1 \\ 1 \end{bmatrix} \tag{2}$$

and to illustrate by Fig. 1.

Also, using affine transformations, you can assign the operation of rotation through the angle α.

$$T_1 = \begin{bmatrix} \cos\alpha & -\sin\alpha & 0 \\ \sin\alpha & \cos\alpha & 0 \end{bmatrix} \tag{3}$$

and scaling.

$$T_2 = \begin{bmatrix} m & 0 & 0 \\ 0 & m & 0 \end{bmatrix} \tag{4}$$

For $m > 1$, the distance from the origin occurs; for $m < 1$, it approaches the origin.

With an increase or decrease in the scale of the figure by a factor of m, an increase or decrease in its size by a factor of m occurs.

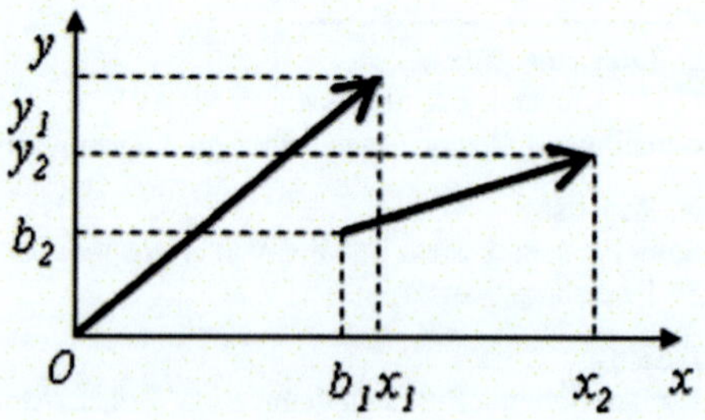

Fig. 1 Affine transformations of the vector

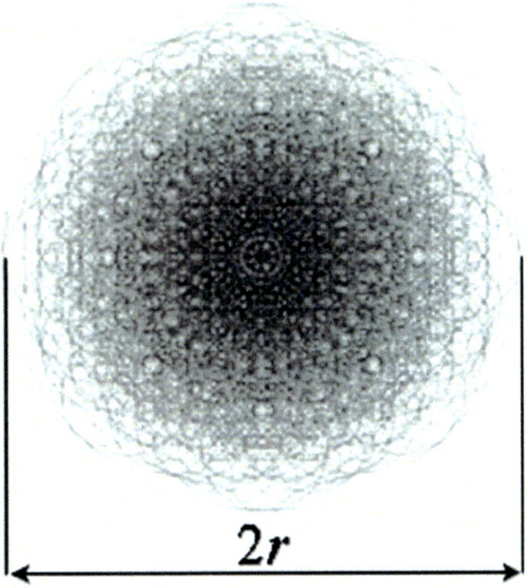

Fig. 2 The LIFETUNE resonator

2 Object of Study

The behavior of a silicon wafer was studied, on the surface of which a pattern of a large number of ring-shaped grooves was etched by plasma-chemical etching (see Fig. 2). The studied object, the LIFETUNE resonator, is a silicon wafer on the surface of which there are annular grooves 0.2 μm wide and 0.8 μm deep, the pattern of which obeys the laws of self-similarity and scale invariance, and is based on affine transformations that is, this surface is self-affine by construction [2].

This figure was obtained as a result of the implementation of affine transformations, the initial stages of which are illustrated in Fig. 3.

3 Experiment

We considered the interaction of an electromagnetic wave with a plate surface for a non-stationary case, for a two-dimensional model. A change in the distribution of tension with time over the surface of the resonator was simulated for various boundary conditions.

An electric field interacting with a semiconductor causes a charge displacement phenomenon and, due to the fact that the plate has a smaller thickness in the "groove" region, the concentration of charge carriers in the groove region will be higher than in neighboring regions. Then most of the charge carriers are concentrated in the regions under the grooves (see Fig. 4).

Let the charge density of two adjacent grooves be q_1 and q_2, respectively, and the potentials φ_1 and φ_2 (see Fig. 5).

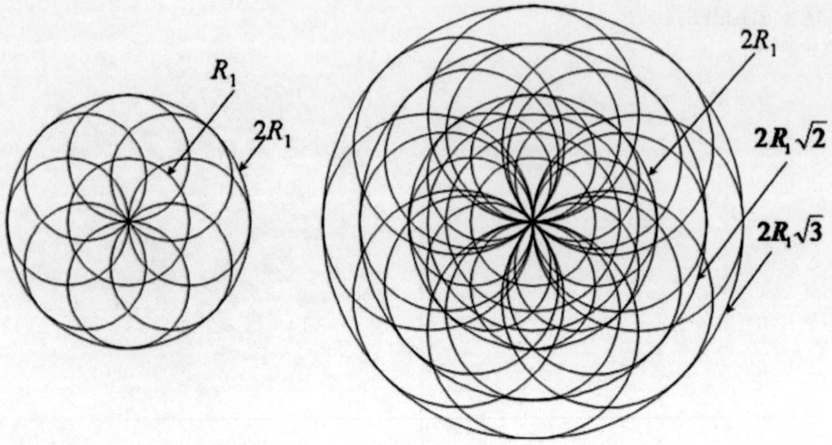

Fig. 3 Create a self-affine relief

Fig. 4 The charges

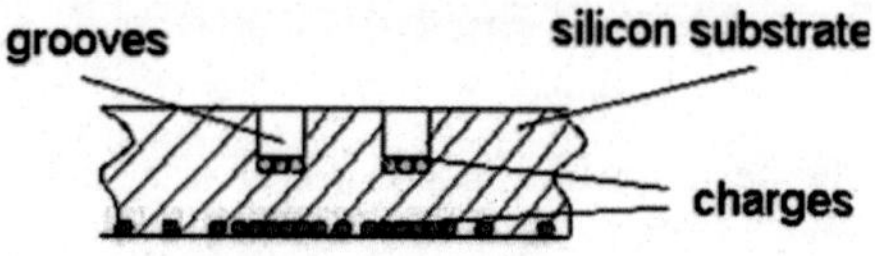

Fig. 5 The potentials φ_1 and φ_2

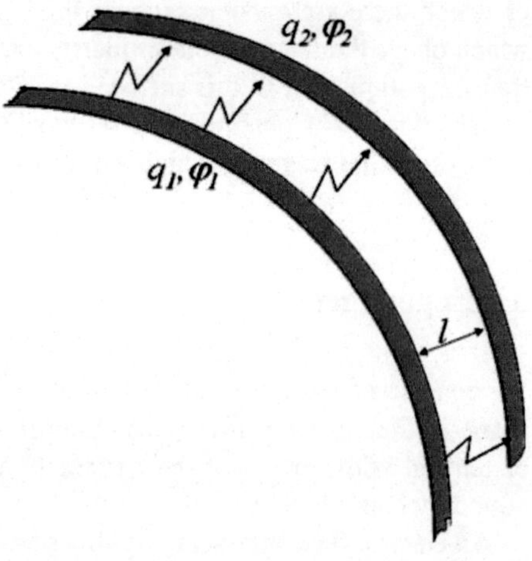

When the potential reaches some critical value φ_c, a current arises along the shortest distance between the grooves. The induced electric field strength E_{ind} then has the form $E_{ind} = (\varphi_1 - \varphi_2)/l$.

The mathematical model for this case has the form:

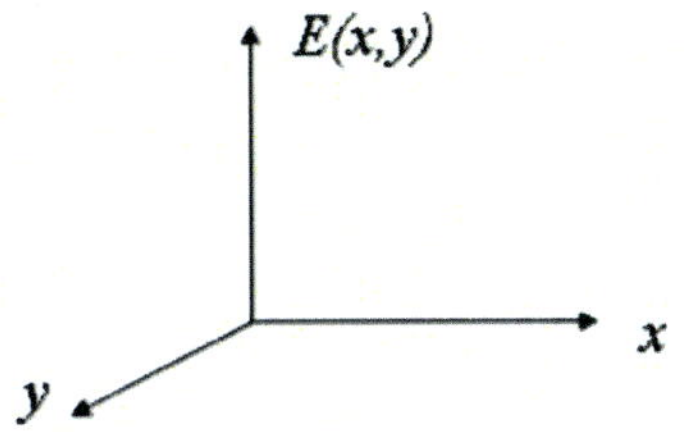

Fig. 6 The distribution of $E(x, y)$ over resonator in plane (x, y)

$$\frac{\partial E}{\partial t} = \alpha_1 \left(\frac{\partial^2 E}{\partial x^2} + \frac{\partial^2 E}{\partial y^2} \right) - \frac{E}{\alpha_2}. \tag{5}$$

where: E—electric field strength, t—time; α_1, α_2 are the coefficients, x, y are the coordinates. In the simulation, it was assumed that the law of current change when the potential reaches the value $i = 1 - (e^{-\beta t}\cos(\omega_0 t))$.

The condition at the cavity boundary: $E = 0$ for $r > \sqrt{x^2 + y^2}$. In the last expression, r is the radius of the resonator (Fig. 2). In addition, the results were compared with the sink in the center of the cavity ($E = 0$) and without it, when the value of E in the center is obtained as a result of calculation by model (5).

4 Results and Discussion

The simulation results in the form of the distribution of the value of E (x, y) (see Fig. 6) are shown in Fig. 7. Since model (5) is dynamic, the figures show different stages of the process at different times, in which waves of different lengths and orientations are visible.

The sizes of the plate along the x and y axes are 20 × 20 mm. Waves with different lengths and orientations arise due to the complex structure of the resonator surface, which creates an "orchestra" of interconnected wave processes.

5 Conclusions

Regardless of the conditions at the surface boundary, after some time t_s, a stable multi-frequency distribution of the electric field strength over the resonator surface is established.

The surface under consideration acts as a transducer of the radiation incident on it and gives a response in the form of a set of waves. When the period of incident electromagnetic radiation changes, the distribution of the electric field on the surface retains its character.

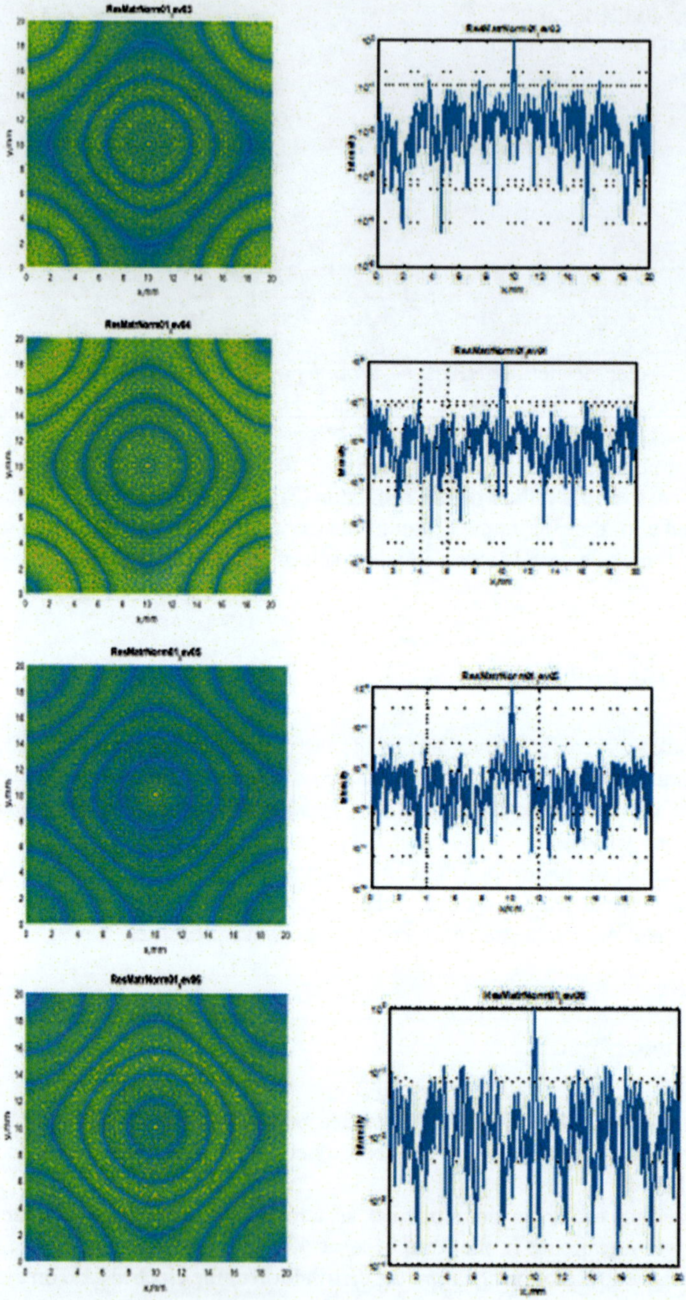

Fig. 7 $E\,(x, y)$. Change over time

References

1. Greffet JJ, Carminati R, Joulain K, Mulet JP, Mainguy S, Chen Y (2002) Coherent emission of light by thermal sources. Nature 416:61–64. www.nature.com
2. Kopyltsov A, Lukyanov G, Serov I (2007) Coherent emission of electromagnetic radiation from the surface of semiconductor plate with the self-affine relief. In: The 3rd international IEEE scientific conference on physics and control (PhysCon 2007). Potsdam, Germany, pp 63–67
3. Peitgen HO, Jurgens H, Saupe D (2004) Chaos and fractals. In: New Frontiers of Science, 2nd edn. Springer-Verlag

The Computer Engineering in ECG Analysis Based on Scatter Mapping

Svetlana Aleksandrova, Irina Kurnikova, Marina Aleksandrova, Nikolay Kislyy, Tatiana Kochemasova, and Maria Zavalina

Abstract The article presents data on the effectiveness of computer analysis of cardiac activity using the method scatter mapping of ECG. Based on the assessment of the integral index of the myocardium and other parameters in different age groups, it is proved that the method allows early detection of signs of cardiovascular diseases.

Keywords Cardioversion device · Scatter mapping of ECG · Cardiovascular diseases · Myocardial index · Heart rate variability

1 Introduction

Cardiovascular diseases are an urgent health problem in most countries of the world, and experts from the World Health Organization (WHO) predict a further increase in cardiovascular morbidity and mortality.

Cardiovascular disease ranks first among diseases and causes of death in modern society.

S. Aleksandrova (✉) · I. Kurnikova · M. Aleksandrova · N. Kislyy · T. Kochemasova
Department of Therapy and Endocrinology, RUDN University, Miklukho-Maklayast. 6, 117198 Moscow, Russia
e-mail: information@rudn.ru

I. Kurnikova
e-mail: information@rudn.ru

M. Aleksandrova
e-mail: information@rudn.ru

N. Kislyy
e-mail: information@rudn.ru

T. Kochemasova
e-mail: information@rudn.ru

M. Zavalina
Department of Therapy, Izhevsk State Medical Academy, Kommunarov Str, 281426034 Izhevsk, Russia
e-mail: asmi2@yandex.ru

X.-S. Yang et al. (eds.), *Proceedings of Seventh International Congress on Information and Communication Technology*, Lecture Notes in Networks and Systems 447,
https://doi.org/10.1007/978-981-19-1607-6_8

[WHO, 2018 https://www.who.int/cardiovascular_diseases/ru/].

The problem of high mortality is directly related to late diagnosis; therefore, special attention in modern scientific research is paid to identifying predictors of cardiovascular diseases and improving methods of early (screening) diagnosis. For a long time, electrocardiography has been the main method used for screening cardiovascular diseases [1]. However, the sensitivity of this method does not allow detecting violations at the preclinical stage.

The development of computer technologies, modern methods of digital data processing have led to the emergence of new diagnostic computer electrocardiographic (ECG) systems. And to identify predictors of cardiovascular diseases (CVD), methods for monitoring indicators of functional and metabolic adaptation (dispersion mapping, heart rate variability, photoplethysmography, multifrequency segmental bioimpedansometry) have appeared. In our study, we used the method of scatter (dispersion) mapping (SM) of the ECG (SM ECG) using the "Cardioversion-06s" device based on a new computer technology for analyzing the ECG signal [2–4].

2 Purpose

To identify early predictors of high cardiovascular risk in older and elderly people based on a computer assessment of the integral myocardial index and the state of the body's functional reserves.

3 Materials and Methods

The study was conducted on the basis of Clinical Hospital No. 7 (Moscow) and approved by the Ethics Committee of RUDN University (No. 27 dated December 18, 2018). 2019 to 2020 145 patients were examined: 30 patients with CVD—age up to 65 years (group I), 34 patients with CVD—age over 65 years(group II), 29 patients—without diagnosed CVD under the age of 65 years (group III), 23 patients without CVD over 65 years old (group IV) and control group—29 healthy individuals under the age of 45 (group V). SM ECG was performed on days 1–3 after hospitalization. The follow-up period after discharge from the hospital was 6–8 months.

The examination of each patient included: anamnesis, physical and laboratory instrumental examination, electrocardiography (ECG) at rest using the Alton-106-S apparatus (Russia), monitoring of functional metabolic adaptation using the computer analysis of the Cardioversion device data and analysis of indicators heart rate variability (HRV). To objectify the state of the heart, the patients underwent echocardiography (EchoCG) on the Aplio 400 apparatus. Statistical processing of the study data was performed using the statistical packages STATISTICA 6.0, SPSS 17.0 and Microsoft Excel 2013 software.

4 Method of Registration of SM CG

The device "Cardioversion-06 s" was used. According to the test results, based on the DB-PTB ECG data of the German Institute of Metrology, when dividing the groups "norm"—"pathology", the sensitivity of the indicator "Myocardium" was 84% and the specificity—73% [6], when tested on a prospective sample, the sensitivity and specificity was 75.6 and 81.4%, respectively [4].

Within 30–60 s, the ECG signal is recorded. For each part of the myocardium, an automatic analysis of low-amplitude oscillations of the ECG signal in successive cardiac contractions (analysis of micro-alternations) is carried out, which is fundamentally different from the standard contour ECG analysis. Micro-alternations are sensitive indicators of the total influences of the physiological systems of the body involved in the mechanisms of regulation of the heart. The cardioversion device reacts to changes in the ion balance in myocytes, shifts in sympathoadrenal activity and other metabolic changes that are not manifested by an ECG or ultrasound of the heart. Similarly, the cardioversion device reacts to the latent dynamics of the compensatory reaction of the left ventricle, which makes it possible to timely diagnose the state of cardiac overload (see Fig. 1) and evaluate two indicators: the integral index "Myocardium" (IMI), which characterizes the severity of electrophysiological

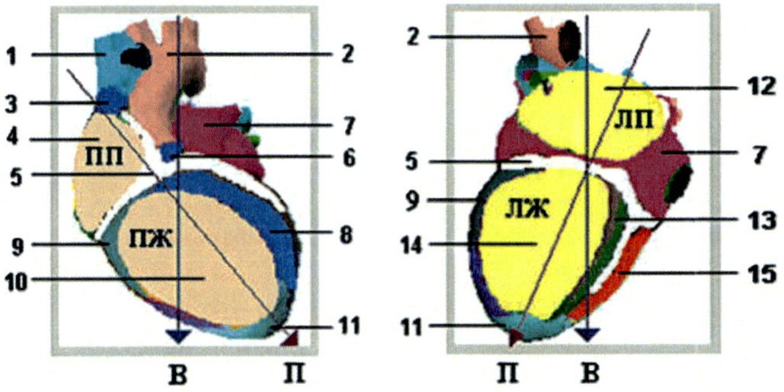

Fig. 1 The results of the analysis are presented in the form of a three-dimensional picture (scattering map or "portrait of the heart" ("Cardioversion-06 s"). Visual structure of the right and left parts of the portrait of the heart. ПП—right atrium (RA), ПЖ—right ventricle (RV), ЛП—left atrium (LA), ЛЖ—left ventricle (LV), B—vertical axis of the heart, П—longitudinal axis of the heart, 1—vena cava superior, 2—the aorta, 3—integral rhythm indicator, 4—right atrial myocardial status indicator (depolarization variance), 5—,indicator of anomalies of the interval P-Q, 6—indicator of the stability of the AV-conducting, 7—integral indicator of the state of two atria (common properties due to a common source of excitation), 8—indicator of the final phase of depolarization of the right ventricle (projection in the area of the interventricular septum), 9—indicator of the duration of ventricular repolarization (correlates with Q-T), 10—indicator of the status of the right ventricular myocardium (scattering of repolarization), 11—ventricular repolarization duration indicator (QRS duration), 12—left atrial myocardial status indicator (depolarization scattering), 13—indicator of the final phase of depolarization of the left ventricle, 14—indicator of the state of the left ventricular myocardium (scattering of repolarization), 15—indicator of the final phase of depolarization of the right ventricle (projection on the posterior wall)

abnormalities; indicator "Rhythm"—an integral index of deviations from the norm of indicators of rhythm variability. Both indicators change in the range from 0 to 100% (the larger the indicator value, the greater the deviation from the norm).

"Heart portrait" presents a schematic image of the heart in two projections, divided into zones (Fig. 1), the intensity of staining in color from green to red, depending on the severity of pathological changes in the corresponding zone.

For the general characteristic of the activity of regulatory systems, an indicator was used in the form of a sum of estimates (modulo) of individual states and characteristics of the heart rate regulation system (criteria)—the indicator of the activity of regulatory systems (IARS). IARS characterizes the activity of regulatory systems in general, which depends on the overall response of the body to environmental factors. The IARS value is determined in conditional points (in the range from 0 to 10).

5 Results and Discussion

Angina pectoris is more commonly diagnosed in CVD patients with less than 65 years of age. In patients with CVD over 65 years of age, myocardial infarction (MI) was more often diagnosed, which was complicated by heart failure I, II FC (88.5; 11.5%). (Table1).

According to the results of the indicators of the monitor of functional-metabolic adaptation at the time of examination, in patients with CVD over 65 years of age, the maximum changes were observed in terms of IMI, RHYTHM (Table 2).

The Rhythm rate was significantly higher in patients with CVD in both age groups and significantly differed from the same indicator in the comparison and control groups. In the patients of the comparison group, the mean values of the

Table 1 Comparative characteristics of patients with CVD ($n = 64$)

Indicator	<65 years ($n = 30$)	>65 years ($n = 34$)
Sex (m\f), *n* (%)	13 (43,3)/17 (56,7)	9 (26,5)/25 (73,5)
ASCVD *n* (%)	19(67,9)	26(76,5)
MI *n* (%)	11(36,7)	16(47,1)
CAD *n* (%) I, II, III *n* (%)	15(48,3) 1 (6,7)/10(66,7)/4 (26,7)	22(64,7) 0 (0)/8(38,1)/13(61,9)
PVD 1,2 *n* (%)	19(86,4)/3(13,6)	24 (96)/1(4)
CHF I, II *n* (%)	12(54,5)/10(45,5)	23(88,5)/3(11,5)
HTN I, II, III *n* (%)	2(13,3)/4(26,7)/9 (60,0)	0 (0)/6(30,0)/14(70,0)
Smokers/non-smokers *n* (%)	7(23,3)/23(76,7)	7(20,6)/27(79,4)

Note MI—myocardial infarction; ASCVD—atherosclerotic cardiovascular disease; CAD—coronary artery disease; PVD—peripheral arterial disease; CHF—chronic heart failure; HTN—Hypertension

Table 2 Comparative characteristics of patients with CVD, without CVD and the control group in terms of IMI, IARS, and Rhythm

Groups	Indicators		
	IMP (%)	IARS	RHYTHM
CVD < 65 years (n = 30)	18 (17;19)	7 (6;7)	32 (21;60)
CVD > 65 years (n = 34)	24 (17;32)	6,5(6;7)	48 (42;63)
p 1 (between I and II groups)	0,020	0,938	0,015
p 2 (between I and III groups)	0,000	0,169	0,002
Without CVD < 65 years (n = 29)	16 (15;17)	6 (6;7)	14 (14;23)
Without CVD > 65 years (n = 23)	17 (16;17)	6 (6;7)	16 (14;23)
p 3 (between III and IV groups)	0,319	0,722	0,287
p 4 (between II and IV groups)	0,000	0,297	0,000
Group of healthy people up to 45 years old	14 (14;15)	5 (4; 6)	14 (13;14)
p 5 (between V and I groups)	0,000	0,002	0,000
p 6 (between V and II groups)	0,000	0,11	0,01
p 7 (between V and III groups)	0,000	0,02	0,00
p 8 (between V and IV groups)	0,000	0,003	0,000

Note p 1—reliability of differences between indicators within the observation group (I and II); p 2—the reliability of differences between the indicators of the observation group CVD < 65 years and the comparison group without CVD < 65 years; p 3—reliability of differences between indicators within the group of comparisons; p 4—the significance of differences between the indicators within the CVD observation group 65 years and the comparison group without CVD > 65 years; p 5—reliability of differences between the parameters of the control group and the observation group CVD < 65 years; p 6—reliability of differences between the indicators of the control group and the observation group CVD > 65 years; p 7—reliability of differences between the indicators of the control group and the comparison group without CVD < 65 years; p 8—reliability of differences between the indicators of the control group and the comparison group without CVD > 65 years

analyzed parameters fully corresponded to those obtained in healthy individuals, which allowed us to consider that the IMI and RHYTHM indices are CVD markers.

The IMI index was 40% higher in the patients of the observation group over 65 years old than in the comparison group and in healthy people. The RHYTHM index in patients of the observation group up to 65 years old was 2 times higher, in patients over 65 years old it was 3 times higher than in the comparison groups. The IMI index was significantly higher in the patients of the observation group over 65 years old compared not only with the control group, but also with the comparison group.

6 Conclusion

The main conceptual idea of the study was to diagnose changes in the cardiovascular system as early as possible on the basis of computer analysis of integral indicators, to trace the condition of patients at the stages of disorders in the cardiovascular system, starting with a group of patients who did not have cardiovascular diseases in as the root cause of the disease, as well as at the stage of clinical disorders in patients with cardiovascular pathology. The data obtained indicate that using the SM ECG method, it is possible to identify groups of people with ischemic changes, left ventricular hypertrophy and cardiac arrhythmias at the screening stage (physical examination), which will allow timely in-depth examination and prescribe or adjust treatment.

Control values for assessing the risk of developing CVD should be considered indicators of IMI and RHYTHM. With an increase in the IMI indicator by more than 30%, and the rhythm indicator is more than 2 times, the patient should be included in the risk group and sent for examination to a cardiologist. The parameters of dispersion mapping of the ECG, characteristic of significant changes in the heart, were determined: left ventricular hypertrophy, myocardial ischemia, and cardiac arrhythmias.

References

1. Gracheva SV, Ivanova GG, Syrkina AL (eds) (2007) New methods of electrocardiography. Technosphere pp 552
2. Ivanov GG, Sulla AS (2008) Method of dispersion (scaterring) mapping of ECG in clinical practice. Moscow
3. Software for screening studiesheart Cardioversion-06s: user manual, medical computer systems. (26.10.2003, edition of 25.06.2004). 51
4. Materials of the site "Medical computer systems". URL: http://www.mks.ru/dev/KardioVisor-6C/?picture=3. Ddate accessed 17 Aug 2012

Traffic Disturbance Mining and Feedforward Neural Network to Enhance the Immune Network Control Performance

Ali Louati, Fatma Masmoudi, and Rahma Lahyani

Abstract Traffic disturbance in urban cities challenges the most advanced traffic signal control systems (TSCS). The challenge is mainly related to the capability of TSCS to ensure a quick detection and to suggest suitable decisions. Neural network has shown great potential in predicting traffic disturbance. In addition, smart clustering could be beneficial to ensure fast disturbance reaction while TSCS are providing control decisions. Moreover, the immune network approach has succeeded in controlling interrupted intersections. Motivated by these assumptions, we propose in this paper a disturbance mining approach based on the occurrence of traffic disturbances to ensure optimal signals control that minimizes traffic delay. Initially, the queue delay is calculated based on mutual information of different traffic scenarios. At that point, within the maximum traffic delay constraint, the feedforward neural network is considered to find the optimal traffic delay and maximize traffic fluidity. As a result, disturbances and related control decisions are clustered based on the calculated traffic delay. Our approach helped the immune network control system (INCS) by prompting it with faster reaction and lower traffic delay compared to its classical version.

Keywords Feedforward neural network · Clustering · K-means · Artificial immune network · Traffic signal control systems · Disturbance

---

A. Louati (✉) · F. Masmoudi
Department of Information Systems, College of Computer Engineering and Sciences, Prince Sattam bin Abdulaziz University, Al-Kharj 11942, Saudi Arabia
e-mail: a.louati@psau.edu.sa

F. Masmoudi
e-mail: f.masmoudi@psau.edu.sa

A. Louati
SMART Lab, ISG, University of Tunis, Tunis 2000, Tunisia

R. Lahyani
Operations and Project Management Department, College of Business, Alfaisal University, Riyadh 11533, Saudi Arabia
e-mail: rlahyani@alfaisal.edu

X.-S. Yang et al. (eds.), *Proceedings of Seventh International Congress on Information and Communication Technology*, Lecture Notes in Networks and Systems 447,
https://doi.org/10.1007/978-981-19-1607-6_9

1 Introduction

Traffic control for interrupted intersections remains a tough problem for both engineers and researchers. This toughness is mainly due to the large number of disturbance types. In addition, while developing TSCS, two major difficulties could be raised, one is related to the system modeling [1] while the second one is to the optimization task [2–4]. As pointed out in [5–7], several recent TSCS are developed to predict future states of traffic states and provide in advance an appropriate control decision. To deal with the pointed out issues, two principle kinds of approaches could be found [8]. The first one is called the flow model-based approaches while the second one is known as the simulation-based approaches. The flow model-based approaches aim to formulate analytical models capable to describe the macroscopic traffic flow dynamics measured at multiple locations [9]. Authors in [9] claimed that due to the complexity of traffic scenarios, the modeling errors and costs must be carefully considered. The simulation-based approaches estimate and predict future traffic flow states based either on simulations or artificial intelligence learning [10]. Artificial intelligence has shown its capability to perform several traffic control-related tasks including modeling, learning, and reproducing macroscopic traffic flow dynamics by using measurements of recorded traffic flow [11]. In contrast, simulations reproduce and describe the actions taken by individual microscopic traffic participators.

When an intersection is disturbed, the parameters of traffic situation change from their normal levels, which launch the disturbance detection function inside the TSCS. The disturbance symptoms generated via this function were stored inside a case base [6, 12]. Over time, an extensive number of disturbance symptoms are progressively accumulated inside the TSCS. As we progress toward a more automated "smart" system, the identification of the disturbance events can lead to what we call a "situational awareness." This awareness would be helpful particularly whenever the system is stressed, or even while facing the initial phases of a potential blackout.

To test TSCS, indicators including traffic delay, volume, density, and speed are crucial. These parameters are estimated periodically via an evaluation based on real-time and historical data [13]. The prediction of such indicators would aid TSCS to act fast during the management of the traffic network [5, 14]. On that note, the neural networks show recently the ability to model traffic's nonlinear behavior [7]. Moreover, they reflect on spatiotemporal property and easily could be integrated into different data sources [5].

Among fundamental modes of learning and understanding, we can highlight the approach of organizing data into sensible groupings. Analysis-based clustering is considered as the formal study of objects grouping-based algorithms and performed according to a measured proximity. K-means is a popular clustering algorithm and is still widely used based on several ways including dealing with heuristics. It has two well-known variants in the literature of pattern recognition, which are ISODATA and FORGY [15]. The K-means assigns each data point to one single cluster. Authors in [16] reduce data through replacing groups with their related centroids before performing clustering. This approach has made the K-means and fuzzy C-means faster.

Fig. 1 Overview of the suggested system

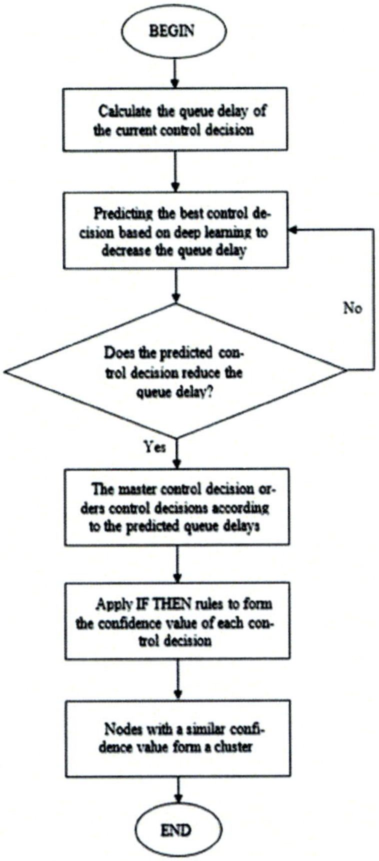

Moreover, K-means has been considered in text mining. For example, authors in [17] analyzed chapter keywords for the anaerobic digestion filed. Prevailing directions of e-government research by analyzing keywords included in e-government publications have been considered [18].

In this study, we propose a novel disturbance mining approach (DMA) based on the occurrence of disturbances to ensure an optimal signals control that minimizes traffic delay in intersections, through minimizing the computational time necessary for finding suitable control decisions. To achieve this objective, we focus on analyzing types, causes, and possible decisions associated with disturbances. To do so, the feedforward neural network predicts disturbances, while the DMA utilizes K-means clustering algorithm to classify these disturbances.

2 Proposed Model

The deep learning technique is implemented to decrease vehicles queue length. The resulting vehicles delay is compared with a predefined threshold. If the resulting delay is below this threshold, the control decision is assigned a maximum confidence. Hence, the master control decision starts forming the pair control decision based on the order of queue delay values. Therefore, the predicted queue delays are utilized to form the cluster. The confidence is calculated for each control decision based on rewards and penalties. In this work, we adopted the rewards and penalties process developed in [15, 16]. The deep feedforward neural network (DFNN) is found as a suitable method to solve the disturbance classification problem by approximating any measurable parameter or function [19]. The DFNN has three layers, which are the input, hidden, and output layers. Each layer has a set of neurons and a nonlinear function of activity of the previous layer [20, 21]. The Sigmoid is considered as an activation function. The number of hidden layers was set to 3. In this work, the ultimate aim of the learning process is to train the neurons weights to output queue delay as minimum as possible compared to the training set extracted from [5–7]. To measure the learning performance, we used the mean square error [22]. The smaller the value provided by the latter, the more accuracy is provided to the experimental data and the prediction model. Figure 1 illustrates the proposed model.

3 Disturbance Prediction-Based Clustering

In this section, we represent a descriptive analysis and the suggested disturbance mining approach for clustering.

3.1 Descriptive Analysis

To conduct the cluster and frequency analysis, we have appended the predicted disturbance data into a text file. We have estimated the distribution of disturbances based on the period of day (rush hour or normal hour). We noticed that disturbances during rush hours boomed compared to those during normal hours, which seems natural based on a traffic expert's point of view. In addition, it is also reasonable to state that a few disturbances were founded between 11 p.m. and 5 a.m.

3.2 Disturbance Mining Approach

This work proposes novel mining terms called disturbance mining approach (DMA). DMA is applied as a document clustering approach inspired from a text mining model. In particular, the k-means clustering algorithm is adopted, through which the predicted disturbances are classified into subgroups to form what is called clusters. As the latter is considered as an unsupervised learning approach, there are no predefined classes or labels. Therefore, clusters are built based on disturbances similarity. DMA provides a summary of traffic situation represented as vectors, which are used to predict control decision. We have adopted a diversity of disturbances and related control decisions considered in [5–7]. The K-means has the following steps:

Step 1: Define the "K" number of clusters.
Step 2: The initial cluster centers were established by generating k random points.
Step 3: K-means assigns each point to the closest cluster center.
Step 4: The new cluster centers are recomputed.
Step 5: If no more changes occur (a convergence criterion is met), the algorithm stops and clusters are built.

4 Methodology and Experimentation

In this work, DMA is implemented using Python and its related libraries including Pandas, Numpy, and Scikit-learn. DMA classifies the disturbances into clusters by merging all queue delays considered in [5–7] into one single data file, and traits each queue delay as one single disturbance. As a result, a file with 424 disturbances is generated and appended into one dataset. This dataset is divided into two separated corpora based on their correlated period of day. The rush hours data is associated with the first corpus, while the normal hours data is associated with the second corpora. Then, the k-means clustering algorithm is considered. We adopted the trial-and-error approach [23] to define the number of clusters (k). This technique is capable to specify the best number of clusters. For different values of k, trial-and-error approach compares the obtained clustering results and decides which one represents the appropriate value of k. In our case, the number of clusters provided by the trial-and-error approach is 5. Next, each cluster is named based on its assigned disturbances and unique association. This operation allows the distinguishing between clusters while ensuring meaningful description of disturbances. The obtained results represent a clue to the concepts' focus for each cluster based on the period of the day. This will help to assess future disturbances, which could be appended, to avoid redundancy, and solve more complex traffic problems. In addition, this could be beneficial for researchers of the field to raise gaps and non-covered types of disturbances. Finally, the results of the clustering algorithm revealed a few continuous types of disturbances, where similar value of queue length and delay appears in the two periods of day. The disturbances for rush and normal hour's periods are as follows (Table 1):

- **Congestion,**
- **Low traffic volume,**
- **High traffic volume,**
- **Special events,**

Table 1 Sample of disturbances distribution during 2 periods of the day

Disturbance	Rush hour	Normal hour
Congestion	55	15
High traffic volume	45	12
Low traffic volume	25	35
Special events	10	5
Accidents	29	12
Emergency vehicles	30	20
Balanced traffic distribution	45	28
Unbalanced traffic distribution	26	32
Total	265	159

Fig. 2 System performance based on traffic delay

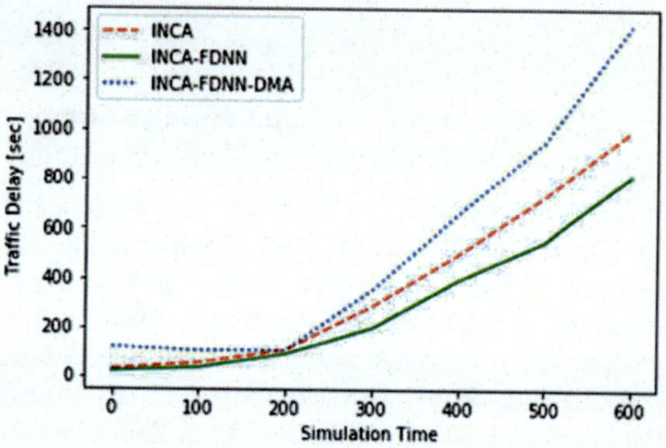

- **Accidents,**
- **Emergency vehicles,**
- **Balanced traffic distribution,**
- **Unbalanced traffic distribution.**

Figure 2 illustrates the traffic delay benchmarking results. One single intersection is considered in the experiments with four approaches. Each approach has 400 m of length and has two lanes. More details regarding the simulated intersection could be found in [24, 25]. SUMO has been adopted as traffic simulator. During the simulation, a set of scenarios are assigned one by one. Each scenario has one hour duration. The vehicles injected into the intersection vary from 500 to 1800 vehicles. More information regarding the scenarios could be found in [26]. The Immune Network Controller Algorithm (INCA) [24] is considered to assess our suggested system. In Fig. 2, the blue-dashed line illustrates the traffic delay provided by INCA, the red-dashed line represents the traffic delay provided by INCA coupled with FDNN called INCA-FDNN, while the green line represents the performance of INCA coupled with FDNN and DMA, called INCA-FDNN-DMA. Despite INCA-FDNN shows better performance compared to the classic INCA, the INCA-FDNN-DMA outperforms both controllers and improves the traffic delay as the simulation time progresses. Since disturbances are associated with predefined clusters and each cluster represents a specific type of disturbance, all of these have helped INCA-FDNN-DMA to reduce traffic delay by improving the computational time necessary for fetching the appropriate control decision. We conclude that we have enhanced the performance of the classical INCA by reducing the computational time to access the relevant control decision and thus is due to DMA approach.

5 Conclusion

This work combines different artificial intelligence techniques to minimize traffic delay at interrupted traffic flow. The feedforward neural network approach has been considered to enhance the traffic delay using multiple traffic scenarios. These scenarios were varied based on real-life parameters. The unsupervised learning approach is used to find optimal control decisions. Disturbances and control decisions are both clustered according to the traffic delay. The experimentation shows that the suggested approach has enhanced the performance of the immune network control algorithm by reducing the traffic delay, and minimizing the computational time necessary to fetch for the appropriate control decision. This work is limited to one single intersection. Therefore, in the future, it is recommended to test the proposed system on multiple intersections network.

Acknowledgments This project was supported by the Deanship of Scientific Research at Prince Sattam bin Abdulaziz University through the research project No. 2021/01/18209. This research was partially supported by grant IRG 19102 from Alfaisal University. This support is gratefully acknowledged.

References

1. Mirchandani P, Head L (2001) A real-time traffic signal control system: architecture, algorithms, and analysis. Transp Res Part C Emerg Technol 9(6):415–432
2. Hammami M, Bechikh S, Louati A, Makhlouf M, Ben Said L, Feature construction as a bi-level optimization problem. Neural Comput Appl
3. Said R, Bechikh S, Louati A, Aldaej A, Ben Said L (2020) Solving combinatorial multi-objective bi-level optimization problems using multiple populations and migration schemes. IEEE Access 8:141674–141695
4. Louati A, Lahyani R, Aldaej A, Mellouli R, Nusir M (2021) Mixed integer linear programming models to solve a real-life vehicle routing problem with pickup and delivery. Appl Sci 1
5. Louati A (2020) A hybridization of deep learning techniques to predict and control traffic disturbances. Artif Intell Rev 53:5675–5704. https://doi.org/10.1007/s10462-020-09831-8
6. Louati A, Louati H, Li Z (2020) Deep learning and case-based reasoning for predictive and adaptive traffic emergency management. J Supercomput 77(5):4389–4418. https://doi.org/10.1007/s11227-020-03435-3

7. Louati A, Louati H, Nusir M, Hardjono B (2020) Multi-agent deep neural networks coupled with LQF-MWM algorithm for traffic control and emergency vehicles guidance. J Ambient Intell Hum Comput 11(11):5611–5627. https://doi.org/10.1007/s12652-020-01921-3
8. Li L, Wen D, Yao D (2014) A survey of traffic control with vehicular communications. IEEE Trans Intell Transp Syst 15(1):425–432
9. Timotheou S, Panayiotou CG, Polycarpou MM (2015) Distributed traffic signal control using the cell transmission model via the alternating direction method of multipliers. IEEE Trans Intell Transp Syst 16(2):919–933
10. Houli D et al (2020) Multiobjective reinforcement learning for traffic signal control using vehicular ad hoc network. EURASIP J Adv Signal Process 2010(1):724035
11. Hajidavalloo MR, Li Z, Chen D, Louati A, Feng S, Qin WB (2020) A mechanical system inspired microscopic traffic model: modeling, analysis, and validation. arXiv preprint arXiv:2012.02948
12. Louati A, Elkosantini S, Darmoul S, Ben Said L (2016) A case-based reasoning system to control traffic at signalized intersections. IFAC-PapersOnLine 49:149–154
13. Nagy AM, Simon V (2018) Survey on traffic prediction in smart cities. Pervasive Mob Comput 50:148–163
14. de Gier J, Garoni TM, Rojas O (2010) Traffic flow on realistic road networks with adaptive traffic lights
15. Jain AK (2010) Data clustering: 50 years beyond K-means. Pattern Recogn Lett
16. Eschrich S, Ke J, Hall LO, Goldgof DB (2003) Fast accurate fuzzy clustering through data reduction. IEEE Trans Fuzzy Syst 11(2):262–270
17. Wang L, Wu Y, Chen T, Wei C (2013) The interactions of phenanthroline compounds with DNAs: preferential binding to telomeric quadruplex over duplex. Int J Biol Macromol 52(1):1–8
18. Abu-Shanab E, Harb Y (2019) E-government research insights: text mining analysis. Electron Commer Res Appl 38:100892
19. Louati H, Bechikh S, Louati A, Aldaej A, Ben Said L (2021) Joint design and compression of convolutional neural networks as a bi level optimization problem. Neural Comput Appl
20. Louati H, Bechikh S, Louati A, Hung CC, Ben Said L (2021) Deep convolutional neural network architecture design as a bi-level optimization problem. Neurocomputing 439:44–62. https://doi.org/10.1016/j.neucom.2021.01.094
21. Louati H, Bechikh S, Louati A, Aldaej A, Said LB (2021) Evolutionary optimization of convolutional neural network architecture design for thoracic X-ray image classification. In: Fujita H, Selamat A, Lin JCW, Ali M (eds) Advances and trends in artificial intelligence. Artificial intelligence practices. IEA/AIE 2021. Lecture notes in computer science, vol 12798. Springer, Cham. https://doi.org/10.1007/978-3-030-79457-6_11
22. Louati A, Lahyani R, Aldaej A, Aldumaykhi A, Otai S (2022) Price forecasting for real estate using machine learning: a case study on Riyadh city. Concurrency Comput Pract Experience 34(6):6748
23. Pham DT, Dimov SS, Nguyen CD (2005) Selection of K in K-means clustering. Proc Inst Mech Eng Part C J Mech Eng Sci 219(1):103–119
24. Louati A, Darmoul S, Elkosantini S, Ben Said L (2018) An artificial immune network to control interrupted flow at a signalized intersection. Inf Sci (NY) 433–434:70–95
25. Louati A, Elkosantini S, Darmoul S, Ben Said L (2019) An immune memory inspired case-based reasoning system to control interrupted flow at a signalized intersection. Artif Intell Rev 52(3):2099–2129
26. Louati A, Elkosantini S, Darmoul S, Louati H (2018) Multi-agent preemptive longest queue first system to manage the crossing of emergency vehicles at interrupted intersections. Eur Transp Res Rev 10(2):52

The Use of ICT in Personalizing Self-learning in Time of Crisis: A Human Computer Interaction Perspective in a Developing Country

Ghada Refaat El Said

Abstract As schools and universities being shut across the world during the COVID-19 pandemic, a timely call for research on effective self-online learning approaches is emerged. To identify gaps in the field, this paper analyzes sample of existing personalized self-learning platform in terms of features and limitations. The paper highlights key review studies, underlining the research trends, potentials, and challenges of the use of advanced ICT techniques, in personalized education over a decade. Aiming to identify challenges, opportunities, and new trends in this research area, focus group sessions were conducted with educational technology experts from Egypt, to discuss opportunities and challenges of ICT-based personalized self-learning tools in the Egyptian education context in the time of pandemic. The thematic analysis of the collected qualitative data suggested some generic aspects for future trends, such as: Intelligent Tutoring, Content Generation, User Control, Career Path Advising, and Integration with web2.0. However, results of this research uncover the importance of additional aspects, which seems specific for developing countries context, such as: ICT Literacy and Digital Equity. Other Cultural Consideration were reveals, such as Language and Collectivism considerations. The results of this study revealed a collection of best practices, potential challenges, current limitations, and future scope for the use of ICT in personalized self-learning platforms in a developing country context. These results might be useful for education platform developers, and education policy makers, for an effective self-learning experience in the time of crisis. These results would provide a baseline for future research in the domain.

Keywords Personalized self-learning platforms · Learning analytics · E-learning cultural considerations

G. R. E. Said (✉)
Future University in Egypt (FUE), 90th Street, Fifth Settlement, New Cairo, Egypt
e-mail: ghada.refaat@fue.edu.eg

X.-S. Yang et al. (eds.), *Proceedings of Seventh International Congress on Information and Communication Technology*, Lecture Notes in Networks and Systems 447,
https://doi.org/10.1007/978-981-19-1607-6_10

1 Introduction

In previous decades, there were high hopes for education technology to evolve and revolutionize learning systems, but the results were mostly disappointing [12]. An OECD report in 2020 found no link between what countries spend on ICT in schools and student's performance [13]. In the past few years, advanced ICT tools, such as artificial intelligence (AI) has come to public attention through successful applications in speech and facial recognition and search engines, and there is little doubt that AI would have a huge impact in solving complex challenges in education [1]. The term AI is used broadly in this research as an umbrella term that subsumes methods, algorithms, and systems that learn from data. In general, AI is the use of data acquired from users to build models, understand its complexity, and using these models to solve relevant problems. Artificial intelligence can transform education by putting learners in charge of their own education, by shaping learning experience to varying needs and through compensating the traditional roles of schools and universities, in situation such as the current pandemic.

On the other hand, the benefits of personalized education have been research-tested and proven long ago [4]. Individually tutored students would achieve better outcomes than 98% of their peers, namely, in math and reading [14]. Personalized learning ensures every student fulfills their potential, regardless of their initial proficiency [20]. However, delivering that education has been difficult due to time and resources limitation inside classrooms. Most of developing countries are adopting a cost-effective traditional mass education system that has a one system fits for all, hindering effective learning chances and opportunities for economic growth [8]. While the COVID-19 pandemic offers an opportunity to revisit the essential conditions for adapting personalized online learning; even technologically advanced countries are facing challenges concerning the effectiveness of online self-learning [10, 22]. Challenges are augmented in developing countries, with un-even distribution of IT infrastructure, lack of financial resources, and modest IT skills for educators and learners [17].

This paper aims to investigate gaps in the tools and platforms developed in the market, and aims to reveal challenges, opportunities, and future trends of the use of ICT in supporting personalized self-learning in a developing country. The paper started, in Sect. 2, by reviewing sample of AI-based personalized self-learning tools. The section also highlights key review studies, underlining the research trends, potentials, and challenges of the use of AI in education over a decade. The main contributions of the study is the overview of current limitations, future scope, and market opportunities within the context of a developing country, based on experts feedback collected in focus group online session. Section 3 reports the administration of the focus group sessions, with comprehensive analysis of the collected data. Section 4 concludes the research and provides basic insights on potential directions of future research and open issues in the domain in developing countries.

2 Literature Review

2.1 Personalized Learning

Learner have different needs in the way they acquire knowledge, hence, contemporary learning methods converge to learner-focused, tailored models that serve each group of learners with better engagement and personalized interaction [2]. Recent studies identified several learning advantages to the personalized learning approach, including better engagement, improved dropout rates, and higher academic performance [16, 18]. Technology plays vital roles in the design, implementation, and operations of tailored learning experiences, including analytics and recommender systems. By analyzing learners' behaviors, preferences, and learning style, recommender systems provides remedial actions including additional exercises and practice examples. Recommendations are also given on the administrative level, such as changes in prerequisite, identifying students at risk, and actions to improve retention, and dropout prediction [2]. The adaptability of the learning tools to cater for multiple learning styles would not have been feasible just a few years ago before the wide accessibility of ICT in the area of education.

2.2 ICT-Based Personalized Learning Platform

Currently, AI is being expensively applied to personalize education in various ways. In most cases, personalization of content is based on individual's learning characteristics such as: learning style, study hours, knowledge level, and behavior. Different techniques are used to assess for example, Mindspark intelligent tutoring platform draws on a set of 45,000 questions per day to identify patterns in incorrect answers, based on which specific remedial exercises are recommended. Another intelligent tutoring tool (iTalk2Learn), remedial tasks are recommended based on previous performance and behavior patterns detected via student's speech in tutoring sessions. In foreign language training platforms, such as Dueolingo, learners are provided with an AI-powered virtual assistant capable speech recognition and synthesis, leading to personalize content based on skills, level, test performance, and available study hours. Other tools, such as Dream box and Quizalize, gamify the learning experience through a points-based reward system to keep learners engaged and progressing toward proficiency. Other tools for personal teaching coach and tutor for corporate training (e.g.: Fulcrum laboratory, Elevate laboratories) tailor real-world skills content based on previous experience and learning curve with assess skills and predict performance, with a customize daily training focus. In most of K-12 personalized self-learning platforms (e.g.: Dream box, Scoop-pad, SmartEd, and Socrative) assessments are conducted in and between lessons to provide the right next lesson at the right time, while strategically increases the learning velocity, while also adapting the plans to match the specific learning style of each student. Table 1 lists sample

Table 1 Sample of AI-based personalized self-learning tools/platform

AI-based self-learning tools/platform	Key features	Personalized learning aspects								
		Learning style	Study hours	Test performance	Topic weighting	Skills assessment	Collective intellect	Gamification	Voice recognition	Behavior
Surgent CPA review: (surgentcpareview.com) CPA exam preparation	Automated study plans that direct student to the specific topic where they have weakness, while also adapting the plans to match the specific learning style of each student	✓	✓	✓	✓	✓				
Alta Knewton (knewton.com) Higher education courses	Instructional content (including text, video, examples, and assessments) selected based on learning style and skills	✓		✓		✓				

(continued)

Table 1 (continued)

AI-based self-learning tools/platform	Key features	Personalized learning aspects								
		Learning style	Study hours	Test performance	Topic weighting	Skills assessment	Collective intellect	Gamification	Voice recognition	Behavior
Fulcrum laboratory (fulcrumlabs.ai) Personal teaching coach and tutor for corporate training	Performance-driven content tailored to improve real-world skills based on previous experience and learning curve with assess skills and predict performance	✓	✓	✓	✓	✓	✓			
Scoop-pad (scootpad.com) K-8 grader remedial lesson	Deliver personalized enrichment, remediation and mastery based on automatically detect knowledge gaps in real-time and scaffold learning to prerequisite skills	✓		✓		✓	✓	✓		

(continued)

Table 1 (continued)

AI-based self-learning tools/platform	Key features	Personalized learning aspects								
		Learning style	Study hours	Test performance	Topic weighting	Skills assessment	Collective intellect	Gamification	Voice recognition	Behavior
Dream box (dreambox.com) K-8 Education	Assessment in and between lessons, to provide the right next lesson at the right time, while strategically increases the learning velocity	√		√		√	√	√		
Quizalize (quizalize.com) K-12 Classroom test with 100,000 quizzes	Provide instant data on student mastery and automatically assign differentiated follow-up activities with game-changing results	√		√		√	√	√	√	√

(continued)

Table 1 (continued)

AI-based self-learning tools/platform	Key features	Personalized learning aspects								
		Learning style	Study hours	Test performance	Topic weighting	Skills assessment	Collective intellect	Gamification	Voice recognition	Behavior
Socrative (socrative.com) K-12 class room assessment	On the fly assessment in classes to evaluate learning and provide individual's remedy	√		√	√	√	√	√		√
Elevate Labs (elevateapp.com) Soft skills training provider	Customize daily training focus and choose between 35 + games with collective assessment	√	√	√		√	√	√		
SmartEd (smartlearning.ca) Smart learning resources for elementary students	Customizes learning materials to suit the needs and styles of students with gamification tool to enhance real-time students engaging	√		√		√		√	√	√
Matific (matific.com) K-6 grader math learning site	Personalized math learning based on local curriculum	√		√	√	√	√	√	√	

(continued)

Table 1 (continued)

AI-based self-learning tools/platform	Key features	Personalized learning aspects								
		Learning style	Study hours	Test performance	Topic weighting	Skills assessment	Collective intellect	Gamification	Voice recognition	Behavior
Duolingo (duolingo.com) Language-learning	Auto generation of daily goal and content based on skills, level, test performance, and available study hours		✓	✓		✓		✓		
Brainly (brainly.com) Assignment help tool	Help learners to explore questions and develop concept-based answers through collective knowledge						✓			
iTalk2Learn (italk2learn.com) Tutoring platform for Math	Lesson recommender based on performance and behavior patterns detected via student's speech in tutoring sessions			✓					✓	✓

(continued)

Table 1 (continued)

AI-based self-learning tools/platform	Key features	Personalized learning aspects								
		Learning style	Study hours	Test performance	Topic weighting	Skills assessment	Collective intellect	Gamification	Voice recognition	Behavior
Thinskter Math (thinkster.com) Math tutoring program	Tracking how students arrive at answers for different questions and suggest question-targeted feedback for a better approach			✓						

of AI-based self-learning platform offering personalized learning with key features and learning aspects, based on which personalization is done.

2.3 *Review Studies*

Recent review studies targeting specific application areas of AI in education have been proposed. Many of these reviews conducted bibliometric analysis for AI in education for the last decade. For example the work of Baker and Yacef [3] Review the history and current trends in the field of educational data mining, categorizing the diversity of research, and listing the most cited papers in between 1995 and 2005. Most recently, the work of Zhai et al. [21] conducted a content analysis of 100 papers from 2010–2020 and identified potential research trends and challenges in adoption of AI in education. The work of Ahmad et al. [1] is particularly comprehensive, providing a detailed bibliometric analysis over six years in various related sub-domains such as learning analytics, educational data mining, and big data in education. Table 2 lists sample of key review studies on AI-based personalized learning with scope an contribution of each review study.

3 Research Method

Focus group online sessions were conducted to collect qualitative data, around the ICT-based personalized learning platforms in the Egyptian education context. In the past years, the use of focus group for collecting qualitative data, has emerged in social science research [19]. Focus groups are an economical, fast, and efficient method for obtaining data from multiple participants simultaneously [11]. The sense of belonging to a group can increase the participants' sense of cohesiveness, and help them to feel safe to share information [11]. The interactions that occur among the participants can lead to more spontaneous responses, with opportunities to discuss perceptions, ideas, opinions, and solutions [6]. The focus group in this study consisted of seven educational technology Egyptian experts. The group met twice in July 2021, via online Zoom sessions, moderated by the researcher of this paper. Sessions were one week apart and each session lasted for two hours and this was when data saturation was reached. Data saturation is reached when information occurs so repeatedly that the researcher can anticipate it and whereby the collection of more data appears to have no additional interpretive worth [6]. The moderator started each session with the following question: *"What are the ICT opportunities, challenges, and future directions, in the Egyptian context, for self-online learning in the time of crisis"*. As sessions were recorded for the purpose of data collection, the moderator mainly facilitated the discussion, encouraged all the members to participate, and take notes. Experts were selected based on convenience sampling, and their participation was on volunteer bases, Table 3 lists the profile of participants.

Table 2 Sample of key review studies on AI-based personalized self-learning

Review study	Scope	Contributions
Zhai et al. [21]	Content analysis of 100 papers including 63 empirical papers and 37 analytic papers focusing on the use of artificial intelligence in education, from 2010–2020	Identifying potential research trends of AI in education (Internet of things, swarm intelligence, deep learning, and neuroscience). Highlighting challenges (techniques, changing roles of teachers and students, and ethical issues)
Ahmad et al. [1]	Review of various applications of AI in education including student grading and evaluations, students' retention and drop out prediction, sentiment analysis, intelligent tutoring, and recommendation systems	Highlighting the existing tools deployed in several applications of AI in education. A detailed bibliometric analysis over six years (2013–2019) in various related sub-domains such as learning analytics, educational data mining, and big data in education
Baker and Yacef [3]	Review the history and current trends in the field of Educational Data Mining, considering methodological shifts, with emphasis on prediction	Categorizing the diversity of research in educational data mining research. Listing the most cited papers from 1995–2008
Romero and Ventura [15]	Survey of the most relevant studies carried out in this field of Educational Data Mining from 2005–2010	List the most typical/common tasks in the educational environment that have been resolved through data mining techniques
Fischera et al. [7]	Data mining operationalization, and behavioral processes to personalize and enhance instruction and learning. Institutional data analysis to improve decision making through early warning systems	Challenges for AI use in education were identified, as following: balancing data privacy and protection with data sharing and research, training researchers in educational data science methodologies

3.1 Thematic Analysis

The Zooms sessions were transcribed and coded through thematic analysis. Thematic analysis exceeds counting words and rather identify patterns and themes, describing the problem under investigation [9]. Relevant words/sentences were labeled, then coded, then combined later into overarching themes. Themes were then grouped or split based on focus group participants' citations and the researcher interpretations. Thematic analysis, through the process of six phases, creates meaningful patterns to identify unfolded themes emergent from the data [5]. In phase1, the researcher became familiar with the data by reading and re-reading the data set to look for

Table 3 Focus group participants profile

Demographic variable		Count (percentage) $N = 7$
Gender:	Male	4
	Female	3
Age:	Below 40 years old	2
	Between 41–50	3
	Above 50	2
Education technology working experience:	5–10 years	2
	More than 10 years	5
Education technology specialization:	HCI free lancer	1
	Usability engineer	2
	MOOCs designer	1
	Content developer	1
	ICT educator	1
	Post graduate learner	1

meanings, patterns, and repeating issues [5]. In phase2, initial codes were done as a means of data simplification, by labeling relevant words, phrases, or sentences. Following the guidelines of Braun and Clarke [5], data were found relevant to code if it is repeated in several places, it surprises the researcher, and/or focus group member (s) explicitly states that it is important. In phase3, relevant codes were combined overarching themes, which are reviewed in phase4, as some themes were grouped together, others were split into subthemes. This process is repeated until full satisfaction with the thematic map. In phase5, themes are defined and given names, going beyond surface meanings of the data. Finally, in phase6, results are reported and quotes from focus groups were contained. At the end of the thematic analysis procedures, themes illustrated in the next section were identified.

4 Results

Thematic analysis of the focus group sessions resulted in four main themes and nine subthemes, identified as ICT opportunities, challenges, and future directions, in the Egyptian context, for personalizing self-online learning, in the time of crisis, Table 4 summarizes themes, subthemes, and themes definitions.

Table 4 ICT opportunities, challenges, and future directions in self-online learning

main theme	Subtheme	Definition
Personalized learning	Intelligent tutoring/mentoring Content generation User control	The extent to which tool analyses learners' learning styles and weaknesses to provide customized content and remedial feedback
Long term learning management	Career path advising Integration with Web2.0	The supporting tools, learners can get, to design their own learning tracks based on long term career objectives
ICT support	ICT literacy Digital equity	The extent to which ICT infrastructure, accessibility, technical support, and IT skills, are available
Cultural consideration	Collectivism Translation tool	The extent to which learning tool can accommodate learning interactivity cultural and social aspects, and language

4.1 *Theme 1: Personalized Learning*

Consensus was reached within the focus group participants that personalization is one of the main contributions of ICT in education. It was agreed that by analyzing learners existing knowledge, customized learning materials with proper content and context, targeting areas that require focus and skipping already familiar materials, and by providing control to user to select tools and sequence, all lead to increased productivity. Within the broad theme of personalized learning, more specific contributing categories emerged relating to: intelligent tutoring/mentoring, content generation, and user control.

4.1.1 Intelligent Tutoring/Mentoring

Participants praised the ability to track learners' work, adjust feedback and provide hints along the way, leading to a more fruitful learning experience and enhance learners' engagement.

> With intelligent tutoring, …. answers could be graded and feedback could be given directly after the end of the test….. Even in quick quizzes, errors could be corrected instantly…keeping learners more engaged. (Participant: P#1, male: M, HCI Free Lancer: FL).
>
> Learner's individual weaknesses can be highlighted and additional guidance could be offered, leading to a more productive learning experience. (Participant: P#4, male: M, MOOC Designer: DS).

Intelligent tutoring systems recognize individuals' patterns in incorrect answers, based on which errors are corrected instantly, and tailored remedial exercises are provided, without criticizing learners or judging them.

> By correcting error instantly, and proposing ways to avoid them in the future, students are encouraged to learn without fear or shame of making mistake". (Participant: P#7, female: F, Post Graduate Learner: LR).
>
> As an educator, I know that in class when students make mistakes or fail to answer questions, they feel ashamed… self-online learning does not criticize learners in front of the whole class, these smart tools evaluate learners without judging them. (Participant: P#6, female: F, Educator: ED).

Participants suggested e-mentoring system that can simulate the communication with a real mentor. Customized answers in response to learner messages can be generate, giving tips on what learners need to improve.

> Intelligent and personalized mentoring system can continuously assess learner performance, determining knowledge and skill levels, and create a portfolio that follows the learner along their lifelong learning journey (Participant: P#5, female: F, Content Developer: CD).

4.1.2 Content Generation

As explained by an educator in the focus group, one of the main problems with the conventional classroom is the issue of sequence. Instructors, must teach the same topics to all students in the same order, regardless of their previous knowledge and academic level. Adapting content and sequences to individual needs, focusing on learners' weak point in each topic, would be a major benefit. In addition, the tool can use the collected feedback on learner's progress to create next tasks, tailored project work, or individual instruction that target individual's demonstrated areas of need. While content generation can be done for a particular learner, and recommendations for institutions regarding syllabus modifications, can be given.

Consensus was reached that learner's interests can be boosted by tailoring content based on identified learning style for individual learners, content relevance to their previous knowledge, learning curve, and available time for learning.

> With real-time assessments of learners' needs, this technology could track performance and continuously adjust content accordingly. (Participant: P#2, male: M, Usability Engineer: UE).
>
> Performance assessment could be done between small chunks of content to suggest the right next content while increasing the learning curve toward the targeted level of knowledge. (Participant: P#5, female: F, Content Developer: CD).

4.1.3 User Control

A main advantage of online self-learning is that each individual can study at his/her own pace. User can skip sessions, revise content, replay videos, and revise materials. Participants suggested more learners control in terms of selection of learning tool, learning sequence, and learning times, schedule, and slots.

> Such tools can provide individual learner control to set amount of time to learn per week, to select learning times and slots, and select preferred instruction language. (Participant: P#6, female: F, Educator: ED).
>
> Individual should be able to select preferred learning tools such as games, quizzes, and other learning and exploratory activities that combine programs of study with learners' interests. (Participant: P#7, female: F, Post Graduate Learner: LR).

4.2 Theme 2: Long Term Learning Management

Participants of the focus group raised the issue to support learners to design their own learning path, based on their professional learning development objectives. With the help of a career path e-advising, learners can track their progress along their career learning track. It can also guide learners in any area of weakness by providing helpful resources available online and web2.0. Ideas for allowing learners to follow a tailored learning track, toward a lifelong learning portfolio or a targeted career path, were discussed. Within the broad theme of long term learning management, more specific contributing categories emerged relating to: career path advising, and integration with web2.0.

4.2.1 Career Path Advising

The idea of career path prediction was raised by the designer in the group and was praised by other participants. Learners need to be able to design their own aimed learning tracks, based on their learning professional development career path. By selecting between the content modules, an assessment after each milestone can allow learners to skip modules and redesign the learning track.

> ICT in education should aim to personalize lifelong learning by helping individual learners to set their own goals, follow a customized learning path, and create a tailored learning sequence toward targeted skills. (Participant: P#4, male: M, MOOC Designer: DS).
>
> Learners should be given the possibility to design their own learning track based on desired careen, including selection of specific topics, with specific sequence, with a quick assessment after each topic to review the learning track. (Participant: P#1, male: M, HCI Free Lancer: FL).
>
> Learners can share their professional development profile with recruiters and potential employers. (Participant: P#3, male: M, Usability Engineer: UE).

4.2.2 Integration with Web2.0

Participants suggested that an online course should not be limited to a single tool, the use of integrative feature of the web, guiding learners to other helpful resources available online, such as web2.0 resources, can better support learners to progress in their own learning path.

Institutions/Employers can set professional development tracks on for their current or potential Students/Employees. (Participant: P#2, male: M, Usability Engineer: UE).

By integrating other resources, like the learner's wiki, Google drive and drop-box, they can share their educational and professional development profile with potential employers. (Participant: P#4, male: M, MOOC Designer: DS).

Based on assessment for learners, tools can help predicting the best career paths and specialization areas for individual learners, providing a clear focus of what skills they need to enhance, and incorporating resources from various web2.0 resources and storage repositories, in an integrative online learning environment. (Participant: P#1, male: M, HCI Free Lancer: FL).

4.3 *Theme 3: ICT Support*

ICT support is especially crucial for online learning in developing countries, with diverse IT infrastructure capabilities from one region to another within the same country. Lack of financial resources and modest IT skills for some learners are main challenges for e-learning in general in developing communities. Within the broad theme of ICT support, more specific contributing categories emerged relating to: ICT literacy, and digital equity.

4.3.1 ICT Literacy

All participants praised the idea suggested by the freelancer in the group that online-self tools should provide instructional videos on trouble shooting simple problems, and to educate new online learners on network setup tips as well as ergonomic tips.

Self-online learning tools should include instructional videos to support new learners, or learners with weak technological skills, solving Wi-Fi home network issues, ... telling them how to configure voice and video, and helping them troubleshoot simple problems. (Participant: P#1, male: M, HCI Free Lancer: FL).

Synchronous and asynchronous technical support should be given to learners… forms for such support could be: instantaneous respond to queries, replies to reported problems, instructional videos, tips and tutorials. (Participant: P#2, male: M, Usability Engineer: UE).

Tools should include tips on best practice for setting up a workstation, ensuring a safe work environment, and ergonomic tips for protecting physical health. (Participant: P#4, male: M, MOOC Designer: DS).

4.3.2 Digital Equity

In developing countries, such as Egypt, ICT infrastructure, in terms of Internet reliability and computer devices affordability vary from one place to another. Due to this technological limitations and divide, online learning materials should be accessed in multiple ways where captions and a transcript of the video could be a solution for poor Internet connectivity and limited computer capabilities.

> Not all Egyptian learners have the appropriate hardware and the stable Internet connectivity for online learning….. during the lock down, students experienced connection problems with e-learning platforms in many areas of Egypt. (Participant: P#4, male: M, MOOC Designer: DS).
>
> Policy makers need to ensure all learners access to Supportive Technologies, such as: high bandwidth connectivity, seamless multimedia services, reliable cloud services. (Participant: P#2, male: M, Usability Engineer: UE).

4.4 *Theme 4: Cultural Considerations*

Group discussed that self-online learning tools made a fundamental transformation in replacing in class collaborative learning, where learners interact, to a one-way learning, where participants independently learn materials with minimal interaction with peers. Such transformation might not be convenient with some learners' culture such as collectivist culture, where interaction with instructors and peers is in the center of the learning process. Within the broad theme of cultural consideration, more specific contributing categories emerged relating to: Collectivism, and translation tool.

4.4.1 Collectivism

The focus group participants discussed the collectivist culture of Egyptian learners, where relationships and the inter-connectedness between learners play a central role in the learning process. Participants emphasizes the important for online self-learning tool to facilitate regular remote communication with peers, live group discussion, and virtual social gathering, in order to reduce the sense of social isolation, and match with Egyptian learners expectations.

> Egyptian education system tend to teach for the whole group and allow students to learn one from the other, whereas in other societies, the individualistic societies, tend to teach by focusing on the individual, emphasizing personal responsibility for learning. (Participant: P#6, female: F, Educator: ED).

Focus group participants agreed that self-learning in general, is lacking Collaborative Learning. The poor opportunities for peer communication, the boring nature of videos, and the lack of the on-the spot comments provided by learners, are all representing advantages of face-to-face learning over the self-online learning and may cause a feel of isolation for learners, especially in the time of lock down.

> Lacking interaction with peers and instructors is a main challenge in online learning in general…with the lockdown conditions, this challenge is amplified. (Participant: P#1, female: F, content developer: CD).
>
> It feels lonely… it is hard to make friends and engage with peers….. furthermore, it is hard to consult with instructors and receive immediate feedback. (Participant: P#2, male: M, postgraduate IT learner: PG).

Innovations in online learning should emphasis on peers interaction and collaboration to co-construct knowledge and learning experience in this virtual space. (Participant: P#4, male: M, MOOC Designer: DS).

4.4.2 Translation Tool

Participants discussed that the level of English language used in some self-online learning tools may be sophisticated for Egyptian and other non-English native language learners. To accommodate various cultures, there is a need for simple terminologies, short sentences, and simple sentence structures, and computer-aided translation tool for video subtitles and text.

5 Conclusion

This paper studied the role of ICT in supporting education at the time of pandemic, through online self-learning tools. Online focus group sessions, with seven Egyptian experts in the field were conducted, and feedback were analyzed around suggestions for future directions of the use of ICT in education to support self-learning in the times of pandemic in Egypt. Intelligent tutoring systems were suggested to be able to track learners' work, adapt content and sequences to individual needs, correct errors instantly, and tailored remedial exercises, without criticizing learners, leading to a more fruitful learning experience and enhancing learners' engagement. Recommendations for more learners' control in terms of selection of learning tool, learning sequence, language, and learning times, schedule, and slots. Control is suggested to be given to learners to design their own learning path, based on their professional learning development objectives. With the help of a career path e-advising, learners can track their progress along their career learning track. Career path advising can guide learners in any area of weakness by providing helpful resources available online and web2.0. Learners can share their professional development profile with recruiters and potential employers; Institutions/Employers can set professional development tracks on for their current or potential Students/Employees. Self-learning tool should provide instructional trouble shooting videos to educate new users on network setup tips as well as ergonomic tips. Online self-learning tool need to promote collaborative learning and facilitate regular remote communication with peers, live group discussion, and virtual social gathering, in order to reduce the sense of social isolation, and match with Egyptian learners expectations.

Based on this study results, the employment of ICT in education needs to be complemented by a wider digital transformation, especially in developing communities with current limited ICT infrastructure that would lead to major changes in education. Policy makers in developing countries need to give priority to strengthen components of this ecosystem, such as: seamless broadband delivery, high performance computing affordability, cloud computing utility, and technology literacy for learners and educators. It is only with this wider digital ecosystem that ICT can lead to fundamental shift in education systems.

6 Limitations and Future Research

The results of this study are based on the opinions of seven education technology experts, educator, and learners, within an online focus group discussion during the COVID-19 pandemic. As with any research there are some limitations to the current study. In this study, the sample is one of these limitations. Focus group participations were selected based on convenient sampling. Participants were confined to one geographical and cultural context. Furthermore, the focus group method does present limitations in terms of relatively small sample size and possible bias of moderator. On the other hand, while completion rates is an important issue in online self-learning courses, however, completion rate is out of the scope of the current study. Findings of this qualitative exploratory study can usefully form the basis of mature hypotheses which can be further tested quantitatively. It would be interesting to explore whether some of the newly identified themes, arising from this research, will be validated by future quantitative examination, and on a larger sample.

References

1. Ahmad K, Qadir J, Al-Fuqaha A, Iqbal W, Elhassan A, Benhaddou B, Ayyash M (2021) Artificial intelligence in education: a panoramic review. https://www.researchgate.net/publication/342323576_Artificial_Intelligence_in_Education_A_Panoramic_Review. Last accessed 21 July 2021
2. Al-Samarraie H, Shamsuddin A, Alzahrani A (2019) A flipped classroom model in higher education: a review of evidence across disciplines. Educ Technol Res Develop 68(6). https://doi.org/10.1007/s11423-019-09718-8
3. Baker RS, Yacef K (2009) The state of educational data mining in 2009: a review and future visions. J Educ Data Min 1(1):3–17. https://doi.org/10.5281/zenodo.3554657
4. Bloom BS (1984) The 2 sigma problem: the search for methods for group instruction as effective as one-to-one tutoring. Educ Res 13(6):4–16
5. Braun V, Clarke V (2006) Using thematic analysis in psychology. Qual Res Psychol 3(2):77–83. https://doi.org/10.1191/1478088706qp063oa
6. Duggleby W (2005) What about focus group interaction data? Qual Health Res 15:832–840
7. Fischera C, Pardosb A, Bakerc R, Williamsd J, Smythe P, Yue R, Slaterc S, Bakere R, Warschauere M (2020) Mining big data in education: affordances and challenges. Rev Res Educ 44(1):130–160
8. Gómez-Jordana MR (2021) The current COVID-19 pandemic in Africa: economic effects of the pandemic on the continent- actions to be taken. https://atalayar.com/en/content/current-covid-19-pandemic-africa. Last accessed 22 July 2021
9. Guest G, MacQueen K, Namey E (2012) Applied thematic analysis, sage publications. ISBN: 9781483384436, https://doi.org/10.4135/9781483384436.n1

10. Havergal C (2021) Refund students if online teaching has fallen short, say MPs, Times Higher Education. https://www.timeshighereducation.com/news/refund-students-if-online-teachinghas-fallen-short-say-mps. Last accessed 23 Jun 2021
11. Krueger RA (2000) Focus groups: a practical guide for applied research, 3rd edn. Sage, Thousand Oaks, CA
12. Moorhouse BL (2020) Adaptations to a face-to-face initial teacher education course 'forced' online due to the COVID-19 pandemic. J Educ Teach 46(4):609–611. https://doi.org/10.1080/02607476.2020.1755205
13. OECD (2020) Organization for economic co-operation and development: education at a glance, ISSN: 19991487, https://doi.org/10.1787/19991487
14. Pane JF, Steiner ED, Baird MD, Hamilton LS (2015) Continued progress: promising evidence on personalized learning. rand corporation
15. Romero C, Ventura S (2010) Educational data mining: a review of the state of the art. IEEE Trans Syst Man Cybern Part C Appl Rev 40(6):601–618
16. Sein-Echaluce M, Fidalgo-Blanco A, García-Peñalvo F (2019) Preface of the book innovative trends in flipped teaching and adaptive learning. In: Sein-Echaluce M, Fidalgo-Blanco A, García-Peñalvo F (eds) Advances in educational technologies and instructional design, AETID. IGI Global, pp xiii–xxiv
17. UNESCO (2021) United nations educational, scientific and cultural organization: education: from disruption to recovery: COVID-19 impact on education. https://en.unesco.org/COVID19/educationresponse. Last accessed 12 Jun 2021
18. Uzir N, Gašević D, Matcha W, Jovanović J, Pardo A (2020) Analytics of time management strategies in a flipped classroom. J Comput Assist Learn 36(1). https://doi.org/10.1111/jcal.12392
19. Wilkinson S (2004) Focus group research. In: Silverman D (ed) Qualitative research: theory, method, and practice, Thousand Oaks, CA: Sage, pp 177–199
20. Wolf R, Armstrong C, Ross SM (2018) Study of Knewton online courses for undergraduate students: examining the relationships among usage. In: Assignment completion, and course success. Johns Hopkins University Center for Research and Reform in Education
21. Zhai X, Chu X, Chai C, Jong M, Istenic A, Spector M, Liu J, Yuan J, Li Y (2021) A review of artificial intelligence (AI) in education from 2010 to 2020. Complexity 8812542
22. Zhong R (2021) The coronavirus exposes education's digital divide. https://www.nytimes.com. Last accessed 10 May 2021

Advanced Processing and Classification of Plant Disease

Sufola Das Chagas Silva E Araujo, V. S. Malemath, and K. Meenakshi Sundaram

Abstract Weather, pests and various other factors cause a lot of crop yield to decrease. Crop losses are more in countries which are tropical and, knowledge and investments in crop health management is very less (Sufola Das Chagas Silva Araujo, Meenakshi Sundaram Karuppaswamy. Comparative Analysis of K-Means, K-Nearest Neighbor Segmentation Techniques, IEEE (2016) [15]). Manual detection are taxing as our eyes have to perceive the indications of the disease based on shape and color. A model of Guntur-4 variety of chili plants was developed that can classify particular diseases. There has been made use of multiple models to train and detect such diseases to figure out which model is more accurate. Each model uses object detection techniques to recognize certain features on leaves and categorize them into different diseases. Different infected leaf images of whitefly, Yellowing, Curled, and Healthy were collected and tested on different models built to try and find which model is best suited for this particular data set. Time complexity, accuracy, and resource usage were computed to build the best automatic leaf image disease detection model.

Keywords Infection · Classifier · G-4 chili leaf

1 Introduction

The models built were used to identify the diseases infecting crops, which lessen produce. I.C.A.R Goa, observed that the production of Guntur-4 variation of chilies in the state of Goa had lessened drastically due to infection and virus attacks. This research is aimed at improving food security by overall reducing crop losses caused due to plant diseases, especially for small-income farmers [1, 2].

S. Das Chagas Silva E Araujo (✉) · V. S. Malemath
Department of Computer Science Engineering, KLE Dr. M.S.S Sheshgiri College of Engineering and Technology, VTU, Belagavi, India
e-mail: sufolachagas100@rediffmail.com

K. Meenakshi Sundaram
Department of Engineering and Applied Sciences, Botho University, Gaborone, Botswana

X.-S. Yang et al. (eds.), *Proceedings of Seventh International Congress on Information and Communication Technology*, Lecture Notes in Networks and Systems 447,
https://doi.org/10.1007/978-981-19-1607-6_11

2 Disease Detection Networks

Leaf analyzer using neural networks to detect plant disease were built and tested. It was needed to find the best neural network technique by performing operations on a few good neural network techniques.

2.1 *Faster R-CNN*

Faster R-CNN was implemented for disease detection by (a) producing boxes bounding positions of disease in the image; (b) extracting features using CNN; (c) predicting the different classes by using classifying layer and (d) building regression layer to bring more accuracy in disease identification [3].

2.2 *Single Shot MultiBox Detector (SSD)*

Single Shot MultiBox Detector was used for recognition and classification of objects in real-time. SSD is precise with regards to accuracy and gives assistances to scale up the accuracy for low resolution images, which increases the speed [4]. SSD was implemented by creating feature maps and then convolution filters were used to identify the disease class. Localization loss was computed amidst the ground truth box and predicted boundary box. For negative predictions three losses were corrected depending on the confidence score of the predicted classes [5]. The final loss function was computed as [5]:

$$L(x, c, l, g) = \frac{1}{N}(L_{\text{conf}}(x, c) + \alpha L_{\text{loc}}(x, l, g) \tag{1}$$

where N is count of positive matches and weight α for localization loss [5].

2.3 *YOLOv3*

YOLO was trained by learning features with the neural network to classify the disease [6]. YOLO used a deep neural network. The objectness score is 1, when the bounding box greatly overlapped a ground truth object as compared to other boxes. YOLOv3 targeted and used a threshold of 0.5 [7].

The YOLO network was trained using following error functions,

$$\text{loss} = \lambda_{\text{coord}} \times \text{loss}_{\text{regression}} + \text{loss}_{\text{classification}} \tag{2}$$

$$\text{Loss}_{\text{regression}} = \sum_{i=0}^{s^2}\sum_{j=0}^{B} 1_{ij}^{\text{obj}}\left[(x_i - x_i^*)^2 + (y_i - y_i^*)^2\right] + \sum_{i=0}^{s2}\sum_{j=0}^{B} 1_{ij}^{\text{obj}}\left[\left(\sqrt{w_i} - \sqrt{w_i^*}\right)^2 + \left(\sqrt{h_i} - \sqrt{h_i^*}\right)^2\right] \tag{3}$$

$$\text{loss}_{\text{classification}} = \sum_{i=0}^{s^2}\sum_{j=0}^{B} 1_{ij}^{\text{obj}}\left[c_i - c_i^*\right]^2 + \lambda_{\text{noobj}}\sum_{i=0}^{s^2}\sum_{j=0}^{B} 1_{ij}^{\text{noobj}}\left[c_i - c_i^*\right]^2 + \sum_{i=0}^{S^2} 1_{i}^{\text{obj}} \sum_{c\in\text{classes}} \left(p_i(c) - p_i^*(c)\right)^2 \tag{4}$$

The first among the last three terms disciplines the objectness and assumes that the bounding boxes will predict the correct disease [8]. The second term for bounding boxes tells about not having the object, and the last term disciplines the class for the bounding box which predicts the objects [8]. Logistic regression is used to calculate the confidence score and class predictions in YOLO v3.

3 Building and Implementing Disease Detection Models

Training and testing was performed by building various models like Faster R-CNN, SSD and YOLOv3 to detect their efficiency [9]. Dataset collected belonged to four different classes: Whitefly, Yellowing, Curled and Healthy and were labeled using labelImg and stored pascal_voc xml format.

3.1 Splitting and Preparing Binary Data for Training and Testing

7464 images were split as dataset for training, testing for the class: Healthy, Yellow, White fly and Curled [10]. Feature maps extraction and disease detection model is as shown (see Fig. 1).

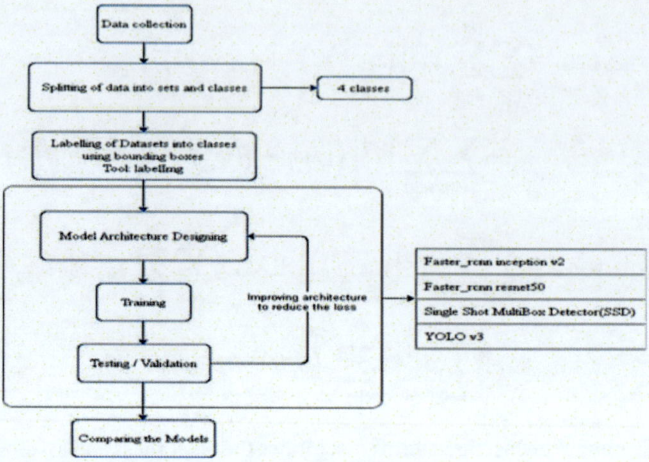

Fig. 1 Feature maps extraction and disease detection model

3.2 Data Acquisition and Labeling

Datasets of chili leaves containing various attributes were collected from the ICAR—Central Coastal Agricultural Research Institute, Goa. Images collected belonged to four different classes: Whitefly, Yellowing, Curled and Healthy. Data for training and testing was segregated from the whole dataset [11–13]. 7464 images were split as TrainData (6000, 13,656), TestData (1464, 3192) [11, 14] amounting to 18,848 boxes.

3.3 Model Architecture Design

Faster R-CNN, SSD and YOLOv3 were built and configured to different settings for G-4 chili leaf disease detection. Training was performed on the various models. Testing was incorporated to find a suitable model for disease detection [9].

Faster R-CNN inception model: Configuration during training involved setting the config files as follows.

• num_classes = 4 • grid_anchor_generator – scales: [0.25, 0.5, 1.0, 2.0] – aspect_ratios: [0.5, 1.0, 2.0] • training_steps: 1000 • initial learning rate: 0.0002	• scheduled_learning_rate: 0.00002 • momentum_optimizer_value: 0.9 • Score_converter: SOFTMAX • iou_threshold: 0.6 • Epochs: 1000 • Loss; 0.2–0.3

Training was monitored using the tensorboard tool for loss and accuracy monitoring (see Fig. 2a). Shows confusion matrix obtained by testing the model and shows (see Fig. 2b) Faster R-CNN inception model tested image.

Observation done on testing the image (see Fig. 2b) showed that the model took 15–20 s to predict with a prediction confidence accuracy of 90–92%. Faster R-CNN inception model detected white fly and yellow for leaves in the image with an accuracy of 94 and 81%, respectively.

Faster R-CNN resnet50 model: Training process config files setting were as below. Training was monitored using the tensorboard tool for loss and accuracy. The number of iterations were less and the performance was better than the previous mode. Increasing the training iteration more than 910 would lead to over-fitting.

• num_classes = 4 • grid_anchor_generator • scales: [0.25, 0.5, 1.0, 2.0] • aspect_ratios: [0.5, 1.0, 2.0] • training_steps: 910 • initial learning rate [15]: 0.0003	• scheduled_learning_rate [15]: 0.00003 • momentum_optimizer_value [15]: 0.9 • Score_converter: SOFTMAX • iou_threshold: 0.6 • Epochs: 910 • Loss: 0.09–0.1 (better than other models)

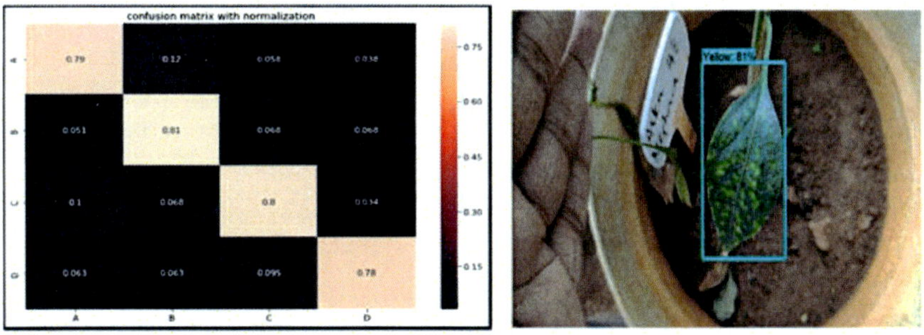

(a) (b)

Fig. 2 Faster R-CNN inception performance **a** Confusion matrix using for A-Healthy, B-Curl, C-Yellow, D-Whitefly. **b** Faster R-CNN inception model tested image

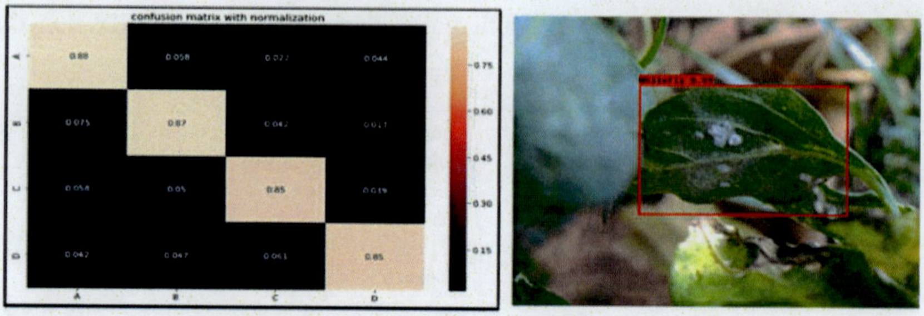

(a) (b)

Fig. 3 Faster R-CNN resnet50 performance. **a** Confusion matrix for A-Healthy, B-Curl, C-Yellow, D-Whitefly. **b** Faster R-CNN resnet50 model tested image

The confusion matrix obtained by testing model is shown (see Fig. 3a) below. On testing the model, the following observations were made on the image shown (see Fig. 3b). The model took 30–40 s for prediction. Prediction confidence accuracy obtained was 96%. The model was slightly computer intensive, but its CPU usage was similar to the previous model.

The above image (see Fig. 3b) was tested on Faster R-CNN resnet50 model, and it detected a class White fly of the given chili leaf with an accuracy score of 94%. Resnet50 model had a higher accuracy, compared to inception_v2 model. Even the loss obtained on training was very less in the Resnet50 model compared to the other. But it was observed that both models were computer intensive and this created a problem of running the model on the handheld devices. Therefore, to avoid this issue other models were trained and created using SSD and YOLOv3 techniques.

SSD (Single Shot MultiBox Detector): The pre-trained model used for this process is "ssd_mobilenet_v2_quantized_coco", which has a speed of 29 ms, COCOmAP [^1]: 22 outputs Boxes. In SSD, train.record and test.record were used as feed for training and testing. Training process was configured by setting config files as follows:

• num_classes: 4	• kernel_size: 3
• activation: RELU_6	• dropout_keep_probability: 0.8
• iou_threshold[16]: 0.6	• iou_threshold: 0.6 (for prediction)
• loss_type: CLASSIFICATION[16]	• score_converter: SIGMOID
• initial_learning_rate: 0.004	• Epochs: 1000
• decay_factor: 0.97	• Loss: 0.9–1.0

(a) (b)

Fig. 4 SSD performance. **a** Confusion matrix for A-Healthy, B-Curl, C-Yellow, D-Whitefly. **b** Single Shot MultiBox detector model tested image

Training was monitored and the model was implemented for 1000 iterations to avoid over-fitting. Average loss obtained on training the model was about 0.9–1.0 which was not satisfactory. The confusion matrix computed on testing is shown (see Fig. 4a). Prediction confidence accuracy obtained was not satisfactory when the image was tested (see Fig. 4b). The model CPU usage was considerably less compared to faster R-CNN. The image tested detected class yellow but the accuracy obtained was very less, i.e., 66%. SSD was not suitable for disease detection, as the model did not perform well.

YOLOv3: For YOLOv3, darknet53 was used and the entire model was built using pyTorch. Default yolov3 weights file were used as the initial weight input to the model. To improve the training process, the following configurations setting was done:

• decay = 0.0005	• Classes = 4
• width and height = 416	• num = 9
• learning rate = 0.001	• jitter = 0.3
• momentum = 0.9	• ignore_threshold = 0.5
• Hue = 0.1, Saturation = 1.5	• truth threshold = 1

The leaky activation function was used with the convolutional layers, with varying filters, setting the stride, size and pad. Three yolo layers, were configured by setting, anchors as 10,13, 16,30, 33,23, 30,61, 62,45, 59,119, 116,90, 156,198, 373,326. K-means was used for anchor box prediction and x, y values were used for configuration of anchor boxes [17]. For K-means clustering, IOU metric is used [17]. Initially, K value was chosen randomly which got optimized on training [17] 0.102 epochs trained the model in 7 h, which was implemented on google colab. The model's final learning rate was set to 9.99e-11.

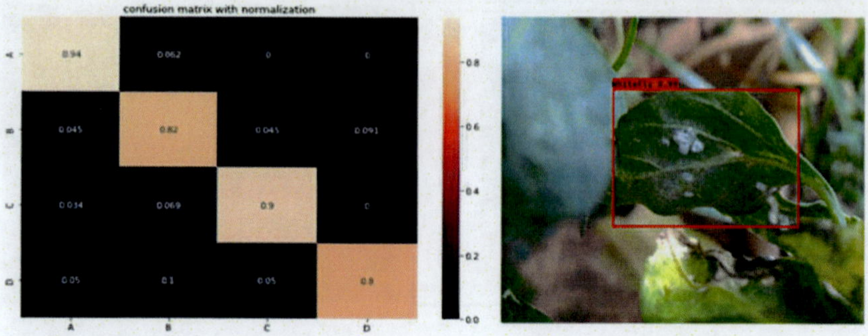

Fig. 5 YOLOv3 performance. **a** Confusion matrix for A-Healthy, B-Curl, C-Yellow, D-Whitefly. **b** YOLOv3 model tested image

The model was trained and tested on the image shown (see Fig. 5b). It was observed that the model detected the three classes- healthy, yellow, white flies and curl- with a confidence of 97, 98, 99 and 98%. The model took the least amount of time of 1–2 s to predict. The result was concluded using confusion matrix as shown (see Fig. 5a) below.

After testing YOLOv3 we observed that the model's prediction accuracy was about 98–99% and it took least amount of time to predict, i.e., 1–2 s. It used less of the CPU. YOLOv3 also made it possible to predict multiple classes for a particular predicted anchor box, since it made use of the SIGMOID function. YOLOv3 took more time for training compared to other models, but prediction time and CPU usage was the least.

4 Result Analysis

The built models enable users to perform leaf analysis, retrieve past analysis data and add extra data to the analytical information. Using object detection techniques, four models were built, trained and tested to detect disease in chili crops. The leaves were of four classes, i.e., healthy, yellow, white fly and curl. Table 1 shows the precision, sensitivity, specificity and accuracy computed by testing each model on the collected data set.

After performing various machine learning disease detection techniques, the best disease detection technique was YOLOv3 which had Precision of 99.27%, Sensitivity of 99.32%, Specificity of 83.59% and Accuracy of 98.661%. This model is the most ideal compared to others used for chili leaf disease detection. Thus, it was concluded that YOLOv3 is the best model for disease detection.

Model conversion to.tflite: The trained and tested models were run on handheld device such as android or iOS phones. This was performed by converting the models to.tflite format. Tensorflow Lite interprets the model in Flatbuffer file format (.tflite)

Table 1 Performance analysis of different models

Model Name	TP	FN	FP	TN	Precision (%)	Sensitivity (%)	Specificity (%)	Accuracy (%)
Faster R-CNN Inception_V2-4	5736	696	816	216	87.55	89.18	20.93	79.74
Faster R-CNN Resnet50-4	6480	312	456	216	93.42	95.40	32.14	89.71
SDD	5928	816	1080	240	83.14	86.72	18.18	74.60
YOLOv3	7099	48	52	265	92.27	99.32	83.59	98.66

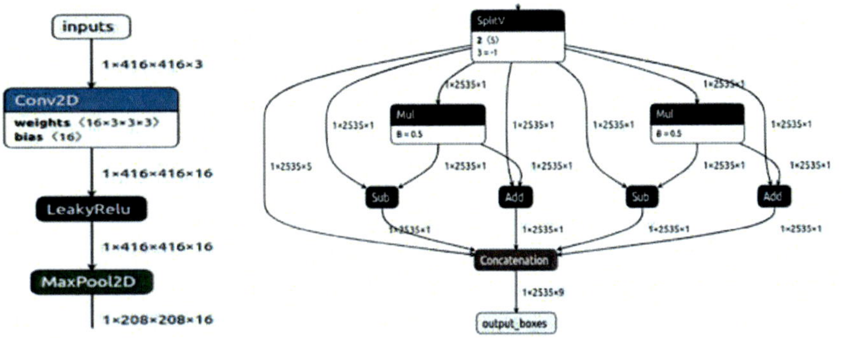

(a) (b)

Fig. 6 Tflite converted model. **a** Input section. **b** Output section

which is generated by the Tensorflow Converter. The model was converted to.pb format before converting to.tflite format.

(see Fig. 6a) shows the input section and (see Fig. 6b) shows the output section of the.tflite converted model. These can be used on mobile device to predict and identify diseases.

5 Conclusion

The model enabled users to perform leaf analysis, retrieve past analysis data and add extra data to the analytic information on their mobile devices. Four models were built, trained and tested to detect disease in chili crop using this disease detection technique. YOLOv3 was tested to be the best disease detection technique as it gave most accurate results. Model conversion and integration with mobile devices using Tensorflow Lite was accomplished to bring added compatibility.

Acknowledgements We would like to thank ICAR-Central Coastal Agricultural Research Institute for their support in allowing us to use their facility to acquire data. We would also like to thank Dr. R. Ramesh, Principal Scientist (plant pathology) and section in-charge (Crop Science) for helping us collect the dataset and use their labs for the research.

References

1. Araujo SDCS, Karuppaswamy MS (2016) Vegetable-fruit identification based on intensity and texture segmentation. Int J Control Theor Appl [15]
2. Araujo SDCS, Karuppaswamy MS (2018) Recognition and detection of object using graph-cut segmentation. In: 7th world conference on applied sciences, engineering & management. international journal of engineering &technology [15]
3. Tanzi L, Piazzolla P, Enrico V (2020) Surgery room management: a deep learning perspective. Int J Med Robot Comput Surg
4. Alipio MI, Peñalosa KT, Unida JC (2020) In-store customer traffic and path monitoring in small-scale supermarket using UWB-based localization and SSD-based detection. Journal of Ambient Intelligence and Humanized Computing. https://doi.org/10.1007/s12652-020-02236-z
5. Liu X, Du J, Yang J, Xiong P, Liu J, Lin F (2020) Coronary artery fibrous plaque detection based on multi-scale CNN. International Journal of Environmental Research and Public Health Vols. 2 to 19
6. Araujo SDCS, Karuppaswamy MS, Malemath VS (2020) Disease identification in chilli leaves using machine learning techniques. Int J Eng Adv Technol (IJEAT) ISSN: 2249–8958
7. Djebbar K, Mimi M, Berradja K, Taleb-Ahmed A (2019) Deep CNN for detection and classification of tumours in mammograms. In: 6th International conference on image and signal processing and their applications
8. Zhong M, Meng F (2019) A YOLOv3-based non-helmet-use detection for seafarer safety aboard merchant ships. Journal of Physics Conference Series 1325(1):012096 Follow journal https://doi.org/10.1088/1742-6596/1325/1/012096
9. Abdullah A, Oothariasamy J (2020) Vehicle counting using deep learning models: comparative study. International Journal of Advanced Computer Science and Applications, https://doi.org/10.14569/IJACSA.2020.0110784
10. ICT Innovations 2019. Big Data Processing and Mining, Springer (2019)
11. Computational vision and bio-inspired computing, Springer (2020)
12. Artificial intelligence and evolutionary computations in engineering. Springer (2020)
13. Advances in Computational Intelligence. Springer, (2019)
14. Advanced computing technologies and applications, Springer (2020)
15. Castillo A, Tabik S, Pérez F, Olmos R, Herrera F (2018) Brightness guided pre-processing for automatic cold steel weapon detection in surveillance videos with deep learning. Neurocomputing
16. Ansari S (2020) Building computer vision applications using ANN. Springer Professional "Wirtschaft+Technik", Springer Professional "Technik", Springer Professional "Wirtschaft"
17. Classification techniques for plant disease detection. Int J Recent Technol Eng (2020)
18. de Luna RG, Dadios EP, Bandala AA (2018) Automated image capturing system for deep learning-based tomato plant leaf disease detection and recognition. https://doi.org/10.1109/TENCON.2018.8650088 Conference: TENCON 2018 - 2018 IEEE
19. Araujo SDCS, Karuppaswamy MS (2016) Comparative Analysis of KMeans, KNearest neighbour segmentation techniques. IEEE [15]

Design of an Interactive BB8-Like Robot

Mia Innes and Emanuele Lindo Secco

Abstract Inspired by the famous Star Wars movie, we decide a moving and interactive robot which is similar to the BB8 character of the sequel. The proposed system is based on a low-cost set of components allowing to control the device wirelessly by means of a mobile app. The robot incorporates an mp2 module and a visual interactive system, and it could be used for human–robot interactive applications.

Keywords Low-cost robotics · Interactive robotics · Human–robot interaction · Hamster drive mechanism

1 Introduction

BB8 is a very famous robotic device from the movie sequel "Star Wars" [1]. This robot is made up of a rolling ball-shaped trunk combined with a spherical head on its top.

In 2015, a miniature toy version of the character came out that replicated the characters movements and sounds [2]. Such a toy has an interesting design since its main spherical body or trunk is equipped with a gyroscope, to detect the direction and movement of the robot trunk, and two wheels that are controlled by motors to cause the body of the robot to move from the inside [2, 3]. There is also a base plate inside that is used as a counterweight to keep the wheels at the bottom of the sphere, and there is a vertical bearing which keeps the wheels in contact with the floor. Finally, the head is connected on to the rolling body through magnets that are placed in the body, and at the bottom of the head, the magnets keep a connection between the head and body so that the head stays on top, while the body roles [2].

M. Innes (✉) · E. L. Secco
Robotics Laboratory, School of Mathematics, Computer Science and Engineering, Liverpool Hope University, Liverpool L16 9JD, UK
e-mail: 19003867@hope.ac.uk

E. L. Secco
e-mail: seccoe@hope.ac.uk

X.-S. Yang et al. (eds.), *Proceedings of Seventh International Congress on Information and Communication Technology*, Lecture Notes in Networks and Systems 447,
https://doi.org/10.1007/978-981-19-1607-6_12

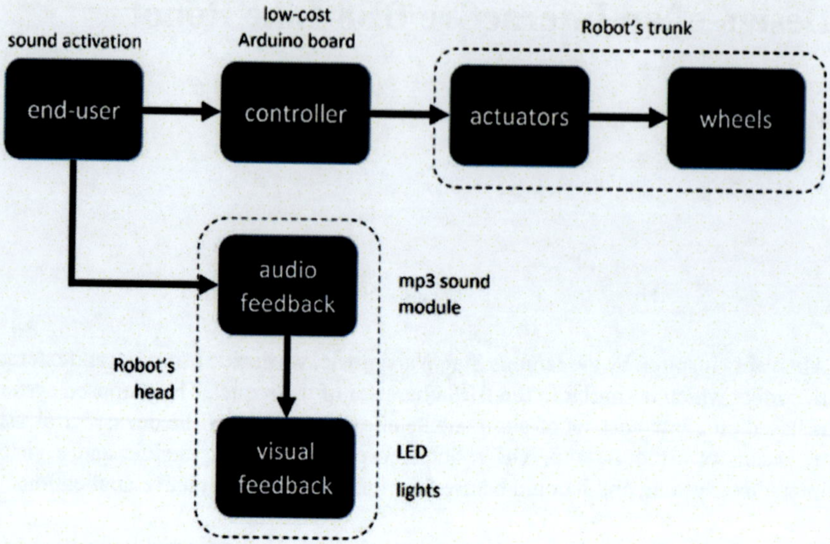

Fig. 1 Functional diagram of the proposed robotics' design

This robot represents an interesting opportunity to enhance the interaction of toys with the end user of the toys, giving it nice look and appealing [4]. However, such a design needs to integrate further components which can enhance the interaction capability of the device, such as:

(i) wireless communication and control system
(ii) sensors integration for proper human–robot interaction
(iii) use of low-cost components and open-source software to decrement cost and increment use

Here, we propose to customize the BB8 robot with a novel design which allows the wireless control of the robot and the integration of low-cost components as well as low-cost sensors and devices for a better human–robot interaction. The robot is made of six main functional elements which are shown in Fig. 1.

2 Materials and Methods

2.1 Interaction Points

The project has two interaction points. The first point is the control of the movement. The project is controlled by an app called *Elegoo BLE tool*; the app connects to the Bluetooth module attached to the electronics inside the body of the project [5]. Once connected, the app allows the user to control the movements of the project

Fig. 2 Overview of the inner side of the robot trunk

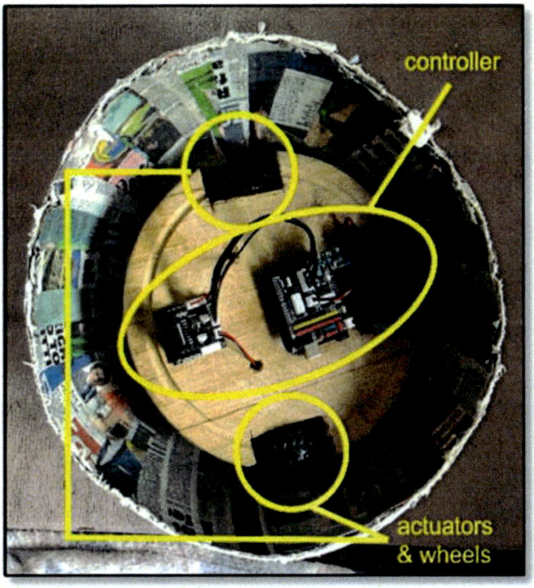

by commanding it to move either forward, backward, or turns. The wireless control allows the device to move freely of up to a 4 m distance from the app, and the user can also determine how fast they want the project to move. The second interaction point is through the mp3 sound module. The module outputs sound through a speaker once commanded. There is a button connected to the top of the head of the project that can be pressed for the sound to play. This also switches the interactive lights on through the mic module that is connected to the end of the LED lights. The mic detects sounds nearby and outputs the lights to flash in time with the sound (Fig. 2).

2.2 Design, Assembly, and Integration

The main part of the robot is made of three layers of paper *mache* newspaper. This one is placed on a 51-cm beach ball. A further layer of cotton canvas and all-purpose filler is positioned on the mache and sanded; then, a further layer of varnish is deposited on top. Once the varnish had dried, the body was then coated in black spray paint, and silver and gold spray paint were then added for detail. The head consists of a 20-cm polystyrene semi-sphere that was cut down to 17 cm and the covered in all-purpose filler, sanded, and coated in layers of varnish. Once the varnish had dried, the head was then coated in spray paint, and the details and designs were then added. At the bottom of the head, there are three makeshift rollers that are used bearings for the head to allow the head to roll on the body. Three neodymium magnets were attached onto the structure inside the body, and another three was attached onto to the makeshift rollers at the bottom of the head.

2.3 Electronics

The main electronics in the robot's body combines a robotics car kit [6] which works with the mobile app. Wireless communication is provided by means of a Bluetooth module and an Arduino Uno board [7]. This board manages the signals to a bridge driver connected to the Arduino, and the bridge driver will then command the motors which will move the wheels in the direction instructed. Arduino Uno-embedded system was adopted because this board can be easily connected to another computer system via a USB port, which allows quick prototyping and the code to be constantly modified and updated accordingly [2]. However, it is also a way for power to be supplied to the board (Fig. 3).

In order to actuate the device, *two DC motors* and a *H-bridge driver* were attached onto a wooden circular board using screws. The wires from each of the DC motors were then connected to their corresponding spaces on the H-bridge driver. The Arduino Uno was connected onto the board along with the shield. Finally, the battery was then attached to the board and connected to the Arduino along with the H-bridge drive.

Another electronic device that was implemented for this project was an *MP3 sound module*. This worked by pressing a button that was attached to the sound module containing a micro-SD card to play a different sound out of the speaker each time the button was pressed. The device was made by attaching a mini speaker with

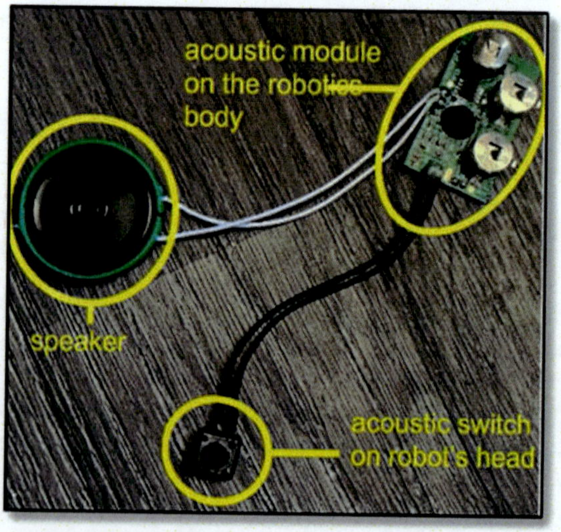

Fig. 3 Acoustic module with the switch, the speaker, and the mp3 processing unit [8]

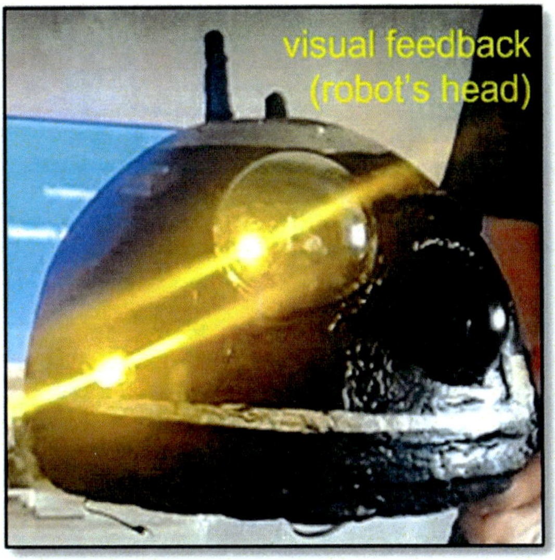

Fig. 4 LED lights inside the project

two small jumper cables and a button switch cable wire to the mp3 sound module [8]. A micro-SD card was then used to import six different sounds onto the mp3 module. To test, the button was pressed 12 different times to ensure each sound worked twice.

The final electronic device implemented in the project was sound-reactive LED lights. These lights worked by using a mic module as a sound sensor to detect any audio nearby, and once the audio has been detected, the sensor then transmits that audio to the lights. These were made using a BC547 transistor. The C end of the transistor was attached to the BRG section on the yellow LED light strip.

A 68 kΩ was then attached to the B end of the transistor, and the other end of the ohm was attached to the positive section of the LED lights. The positive end of the mic module was then attached to the E end of the transistor, and then, the negative end was connected to a jumper cable that was attached to the negative end of a power supply. Another jumper cable was used to attach the positive end of the power supply to the transistor. The device was tested by placing a phone next to the mic module and playing sound at three different levels to see if the mic module was able to sense all the different sound levels (Fig. 4).

2.4 *Software*

The programming that has been develop in this project has been written in Arduino using an Arduino Uno. The code contains seven different sections. The first section of the code defines the pins for each of the sections, and the second section of the code is for the forward command that will move the card forward. The third section

Fig. 5 Overview of the main function of the Arduino IDE program

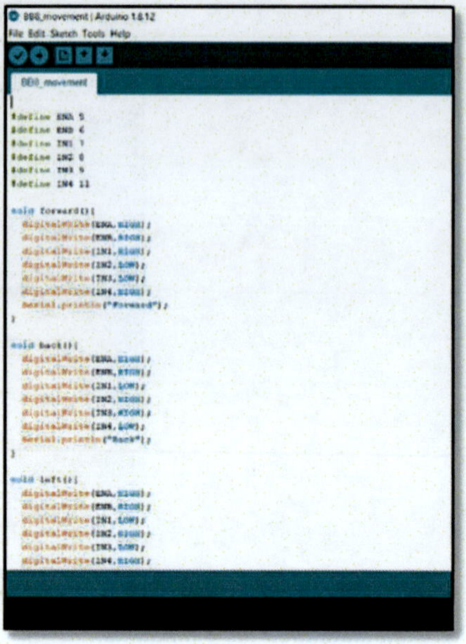

of the code is for the backward command that will move the car backward, and there are two sections for left and right commands as well. The final section is for the loop function that allows the command and movements to be repeated until the device has reached its required destination.

2.5 *Wireless Communication*

The project works by an app wirelessly sending commands to an Arduino via IEEE 802.15.1 protocol, namely using a HC-08 Bluetooth 4.0 module. The HC-08 module uses the BLE protocol as it is suitable for transferring small amounts of data between nearby devices and allows the device to operate for a longer period (Fig. 5).

3 Results and Discussion

When given a command, the project was capable of moving accordingly. The motors inside the body of the project will cause the body to roll across the floor. The Bluetooth module keeps connected to the device and app for up to 4 m which gives the project a lot of distance to move. The magnets placed inside both the head and body cause a pull that forces the head to remain on top of the body as the project moves. Table 1 shows the light display for the mic sensor based on the sound level of the audio; as

Table 1 Preliminary testing of the LED light sensor

	Preliminary tests setup	
	Sound level	*LED light display outcome subhead*
Light sensor test	High	Strong light display
	Medium	Strong light display
	Low	Weak light display

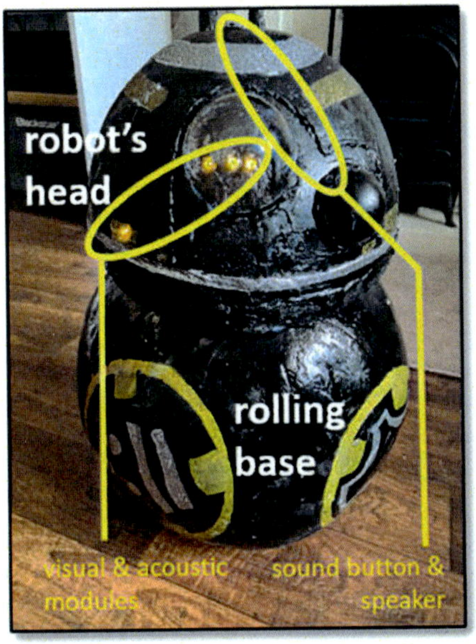

Fig. 6 Whole robot design with the low-cost controller, the wireless communication module, and the visual and acoustic systems

seen, the mic sensor has the strongest output on a high or medium level. To ensure the strongest light display, the mic sensor has been placed next to the speaker from the sound module in the project (Fig. 6).

4 Conclusion

While the project has been built successfully and met all its requirements, there are many ways in which it could be developed and improved in the future. A future development for the project could be to add a sensor that would detect sound and automatically output sound as a response to create a human–robot interaction [9–12]. This could then be used as a companion for the elderly or lonely. A motion sensor could also be attached to the project to avoid the robot from crashing into any obstacles; this could be added with a self-drive alteration, in which a user could enter a specific location for the project to get to, and the project will take itself to that location, much like the delivery robots that are used by some shops in the UK to deliver items to those who are unable to leave their house [12]. Advantages for these future developments include accessibility for those unable to leave their houses; it would provide comfort for those who live by themselves, and due to the low cost of the project, it could be accessible for a lot of people to buy. However, a disadvantage of these future developments could be that as the material is cheap and homemade, it might not last long, so a different lightweight material might need to be used which might change the cost of the project [10].

Acknowledgements This work was presented in coursework form in fulfillment of the requirements for the BEng in Robotics of the student Mia Innes from the Robotics Laboratory, School of Mathematics, Computer Science, and Engineering, Liverpool Hope University.

References

1. Sánchez C (2021) How does BB-8 work?. [online] How does BB-8 work?
2. Sphero (2021) Sphero BB-8 Robot and R2-D2 Robot | Star Wars Robot Toys | Sphero. [online]
3. Chu TS, Chua AY, Secco EL, A wearable MYO gesture armband controlling sphero BB-8 robot. HighTech Innov J 1(4):179–186. http://dx.doi.org/https://doi.org/10.28991/HIJ-2020-01-04-05
4. Chu TS, Chua AY, Secco EL (2021) Performance analysis of a neuro fuzzy algorithm in human centered and non-invasive BCI. In: Lecture notes in networks and systems. Springer
5. Github APK (2021). https://github.com/keopx/Arduino/blob/master/ElegooTool.apk. [online]
6. Elegoo (2021) https://www.elegoo.com/products/elegoo-smart-robot-car-kit-v-4-0 [online]
7. Unknown W (2021) What is Arduino UNO? | Amazing 11 Features of Arduino UNO. [online] EDUCBA
8. HALJIA (2021) www.HALJIA ISD1820 Audio Sound Voice Recording Playback: Amazon.co.uk: Electronics [online]
9. Maereg AT, Lou Y, Secco EL, King R (2020) Hand gesture recognition based on near-infrared sensing wristband. In: Proceedings of the 15th international joint conference on computer vision, imaging and computer graphics theory and applications (VISIGRAPP 2020), pp 110–117. ISBN: 978-989-758-402-2, https://doi.org/10.5220/0008909401100117
10. McHugh D, Buckley N, Secco EL (2020) A low-cost visual sensor for gesture recognition via AI CNNS. In: Intelligent systems conference (IntelliSys) 2020. Amsterdam, The Netherlands
11. Chu TS, Chua AY, Secco EL (2021) Performance analysis of a neuro fuzzy algorithm in human centered and non-invasive BCI. In: Sixth international congress on information and communication technology (ICICT), 2021—lecture notes in networks and systems. Springer
12. Howard AM, Secco EL (2021) A low-cost human-robot interface for the motion planning of robotic hands. In: Intelligent systems conference (IntelliSys) 2021, advances in intelligent systems and computing, lecture notes in networks and systems, vol 3, no 30. Springer, pp 296. 978-3-030-82198-2

Transfer Learning in Deep Reinforcement Learning

Tariqul Islam, Dm. Mehedi Hasan Abid, Tanvir Rahman, Zahura Zaman, Kausar Mia, and Ramim Hossain

Abstract Reinforcement learning has quickly risen in popularity because of its simple, intuitive nature, and its powerful results. In this paper, we study a number of reinforcement learning algorithms, ranging from asynchronous q-learning to deep reinforcement learning. We focus on the improvements they provide over standard reinforcement learning algorithms, as well as the impact of initial starting conditions on the performance of a reinforcement learning agent.

Keywords Deep learning · Transfer learning · Reinforcement learning · Convolutional neural networks · Q-networks

1 Introduction

Reinforcement learning is a class of machine learning algorithms that are designed to allow agents provided with only the knowledge of the states it visits and the actions available to the agent to learn how to maximize its reward function, quite similar to the trial-and-error approach. There are different techniques used for reinforcement learning, one of the most popular ones being Q-learning where an agent develops a policy that chooses the action that is estimated to lead to the greatest total

T. Islam (✉) · Dm. M. H. Abid · T. Rahman · Z. Zaman · K. Mia · R. Hossain
Daffodil International University, Dhaka, Bangladesh
e-mail: tariqul15-2250@diu.edu.bd

Dm. M. H. Abid
e-mail: mehedi15-226@diu.edu.bd

T. Rahman
e-mail: tanvir15-2245@diu.edu.bd

Z. Zaman
e-mail: zahura15-1381@diu.edu.bd

K. Mia
e-mail: kausar15-2248@diu.edu.bd

R. Hossain
e-mail: ramim15-2246@diu.edu.bd

X.-S. Yang et al. (eds.), *Proceedings of Seventh International Congress on Information and Communication Technology*, Lecture Notes in Networks and Systems 447,
https://doi.org/10.1007/978-981-19-1607-6_13

future rewards. Reinforcement learning has seen great recent success, particularly in Playing Atari with Deep Reinforcement Learning [1] and Mastering Chess and Shogi by Self-Play with a General Reinforcement Learning Algorithm [2] as it is a relatively simple yet extremely powerful algorithm, making it an interesting class of learning algorithms to study. Furthermore, the training of reinforcement learning agents is extremely slow, since the information it is provided is minimal, which means that there is a lot of room for improvement with reinforcement learning algorithms. Transfer learning on the other hand is a class of machine learning algorithms that seek to transfer knowledge gained from solving one problem and applying it to another problem, so transfer learning can solve the problem of speed for reinforcement learning agents. In this paper, we discuss the impact of initial conditions with transfer learning on the convergence of reinforcement learning agents. In real life, we know that initial starting conditions matter. Consider a person who chooses to learn a sport: the athletic ability, age, equipment, training, and instructor will all influence the time it takes for the person's skill to peak. If it were at all possible, we would want to transfer the traits of high-performing athletes to the beginner to provide better chances at performing well. Based on this intuition, we want to experiment with transferring the models of trained reinforcement learning agents as the initial starting conditions of reinforcement learning algorithms and confirm that this hypothesis does indeed apply here too. That is, we want to show that given better initial conditions, an agent will likely achieve high performance faster than an agent with worse initial conditions. This is reasonable, and we can easily produce simple examples that illustrate the point. Consider an extreme example where an agent uses a neural network to model its policy, and all the weights in the network are initialized to zero. Then all the weights follow the same gradients, and the policy will likely perform poorly. Conversely, an agent with a policy model that has been trained for extremely long periods of time will likely be much closer to optimality: Hence, it will likely take much less time to converge. Intuitively, it makes sense that better initial conditions lead to optimal performance faster, and we wish to establish this for reinforcement learning, by means of a simple form of transfer learning.

2 Related Work

An interesting improvement to Q-learning is asynchronous Q-learning (AQL). This technique involves one central, shared neural network. Then each asynchronous agent copies the shared network as its own individual network, learns on its own, and periodically shares its accumulated updates with the shared neural network. Furthermore, each agent will periodically copy the shared neural network as its own individual neural network, making use of the learning that other agents have done. In effect, an AQL agent searches across multiple locations in the state space while sharing information with other agents, speeding up its learning process. Wang et al. [3] represent the ride dispatching problem and suggest suitable solutions which are based on deep Q-networks. Nowadays, the GPS authorization applications are getting more popular and are also used in ride-sharing. To get the result, they use a window of 100 circumstances for counting the reward curve and the total number of training duration is 40,000 circumstances. Our work has suggested a procedure which has based on DQN for this dispatching platform. They are successful to show that CFPT is most successful and better than other methods. Victoria et al. [4] aim to establish a method for deep reinforcement learning that will refine the effectiveness and capacity of this advisable method by structural perceptivity and relational argument. They advisable relational model has gained favorable performance and solved more than 98% of levels. Lample et al. [5] focus on representing a structure to face 3D infrastructure in FPS games. In recent times, deep reinforcement education has shown much success to achieve human-level control. In this paper, they describe a procedure to increase the efficiency of the model to utilize the information of game features. They apply the DQRN model because of its good performance accuracy. This model is instructed and used to shorten Q-learning. Our advisable structure is trained to permit various models in various phases of FPS games. This paper [6] aims to establish an efficient model that will repetitively store the results of a chemical reaction and select new exploratory conditions to upgrade feedback outcomes. Here they take random 5000 functions, and the DRO takes 32 steps to arrive at the standard of regret, where some other algorithms such as CMA-ES takes 111 steps, SNOBFIT takes 187 steps, and Nelder–Mead fails. Our established DRO model has shown its remarkable performance to optimize chemical reactions. This model has already shown its ability to optimize and also increase the speed of reaction. Baldazo et al. [7] aim to suggest a new model for the mean embedding of distribution which is based on DRL. Nowadays, DRL has widely used for solving various multi-agent collaboration problems. In our advisable model, they use the agents as a sample and as input use mean embedding. Besides they describe various features of the mean embedding by using radial basis functions and training neural networks. The paper [8] aims to establish some effective methods to upgrade exploration conjunctional optimization based on DRL. In recent times, DRL has successfully shown an excellent improvement to solve different kinds of control problems. The paper [9] aims to explore mobile edge computing for smart (IoT) based on deep reinforcement learning. In incent times, there has been tremendous advancement in developing IoT. Basically, Kiran et al. [10] aim to show a classification of automated driving activity where can apply DRL methods. With the advancement of the DRL network, the autonomous driving system has gained high fidelity.

3 Background

3.1 Reinforcement Learning

The reinforcement learning task is often formulated as a Markov decision process, a modeling framework useful for partially random, partially controlled environments, which is certainly the case in reinforcement learning where the environment may behave randomly, but the agent has control over its own actions. In the reinforcement learning task, a Markov decision process consists of the following elements:

1. S_E: The set of states that the environment (with the agent in it) E can be in.
2. A_E: The set of possible actions that the agent can take in the environment.
3. W_E: $S_E \times A_E \rightarrow S_E$: The function that determines the resulting state given a starting state and an action.
4. R_E: $S_E \times S_E \rightarrow R$: The function that gives the immediate reward for a state transition.

The agent constructs a policy π_E: $S_E \rightarrow A_E$ that maps a state in the state space to an available action that leads to the highest total immediate and future rewards. So we formulate a utility function $U_{\pi E}$: $S_E \rightarrow R$ that determines all rewards received by following the policy given a starting point s_0:

$$U\pi_E(s_0) = \sum_{t=0}^{\infty} \gamma^t R_E(s_t, W_E(s_t, \pi_E(s_t))) \tag{1}$$

Then the policy for our reinforcement learning agent can be defined as follows:

$$\pi_E(s_t) = \text{argmax}_{a\epsilon AE} U_{\pi E}(W_E(st, a)) \tag{2}$$

3.2 Q-learning

Often times, an agent does not have access to $W_{E,}$ and in such cases, the agent's policy is said to be model-free. The agent must, then, estimate the utility function by its internal Q-value function $Q_{\pi E}$: S_E ! R. A simple way to represent $Q_{\pi E}$ is to use a table in which each possible state and action pair is listed, and the estimated cumulative reward is the entry. To learn the optimal Q-value function which we denote by Q_E, we use the Q-learning algorithm on our Q-value function $Q_{\pi E}$. In one-step Q-learning, the algorithm takes one step at every training iteration t from state s_t, observes the reward received r_t and the new state $s_t + 1$, and updates the policy as follows:

$$Q_{\pi E}(s_t) = Q_{\pi E}(s_t) + \alpha(r + \gamma Q_{\pi E}(s_t + 1) - Q_{\pi E}(s_t)) \tag{3}$$

with α as the learning rate, typically a real number between 0 and 1. This algorithm sets the target value to be the discounted sum of all the future rewards estimated by $\gamma Q_{\pi E}(s_t + 1)$ added to the observed immediate reward r. The difference between the target value and output value is then a weighted by the learning rate and used to update the Q-function. To avoid settling for a non-optimal policy (premature convergence of policy), an exploration factor is introduced: Is the probability that the agent will ignore its policy and execute a random action, to diversify its experiences and to avoid local minima in its policy. As time progresses, the exploration rate is decayed, so that the agent relies more (but not completely) on its policy. However, it is often the case that the exploration rate is not allowed to decay to 0 and is instead held at some fixed minimum exploration rate, to discourage the policy from sinking into a local minimum.

3.3 *Q-networks*

It becomes hard to maintain such a Q-table when the size of the state space increases: for example, consider an agent playing a video game, using the screen's pixel values as its state space. If the state is a $84 \times 84 \times 3$ array of 8 bit pixels, and there are four actions available, the q-value table will hold $284 * 84 * 3 + 2 \approx 106{,}351$ entries! A popular solution to the problem of poorly scaling tables is the use of artificial neural networks in Q-learning termed Q-networks [11, 12]. Q-networks map states in the state space, represented by frames from the game, to q-values for each possible action. Q-networks learn to approximate Q in a way intuitively similar to the update formula for the Q-table, by computing gradients for the network based on the output of the network (determined without knowing the next states) and target Q-values (determined using the next states) for the network [1, 13]. Q-networks are far more powerful than Q-tables because they can also approximate Q-values for states it has not yet seen and scales much better in terms of size. However, they come with the downside of being harder and slower to train.

4 Approach and Experiments

We first describe the infrastructure available to us for our experiments. For high numbers of independent experiments, we use a distributed high-throughput computing resource through the Center for High Throughput Computing (CHTC) available at UW-Madison. For our guaranteed convergence experiment, we tested using a custom maze environment with a state size of 4 and an action space size of 4. For our initial conditions experiment, we describe how we done operational convergence criteria for our problem setting [14, 15]. We say that the agent's learning has stopped if the winning rate over the last 100 evaluations averages to a value greater than 78, the same stopping criteria for the environment FrozenLake.

4.1 Guaranteed Convergence Given Infinite Time

We know from Even-Dar et al. [16] that using the action elimination algorithm, our reinforcement learning agent will converge given infinite time. Our hypothesis was that the algorithm would work for a reinforcement learning agent in an environment with an extremely simple problem with an extremely small state space. We expected to see the algorithm converge given a couple month's time. Unfortunately, the algorithm's progress exponentially diminished, and we never saw the convergence (or anything even close) after 2 months of running the algorithm on a high-throughput computing cluster. As such, we affirm that "Infinite time" really does mean some enormous time quantity that is infeasible. We ran this experiment on a Google Cloud Compute Engine instance with 8 cores and 32 GB RAM.

4.2 Impact of Initial Conditions on Convergence

We hope to find that given better initial conditions, our DQN agent will converge faster. We provided these initial conditions as trained DQN models, saved after various periods of pre-training. We hypothesize that models that have had more pre-training will require less time to converge, while models that have had little pre-training. We first show the baseline performances of each initial condition in Fig. 3. Then we show training times until convergence starting from each initial condition in Fig. 4. Our hypothesis is affirmed through this experiment as we can see that indeed agents with more pre-training had faster times to convergence. The pre-training was done on a system with an Intel i7-7700 k overclocked to 4.9 GHz with 32 GB DDR4 3200 MHz SDRAM on a Samsung 960 EVO M.2 drive. When testing each initial condition, we used CHTC. Each job was run on a system with 8 CPUs and 10 GB memory (Figs. 1 and 2).

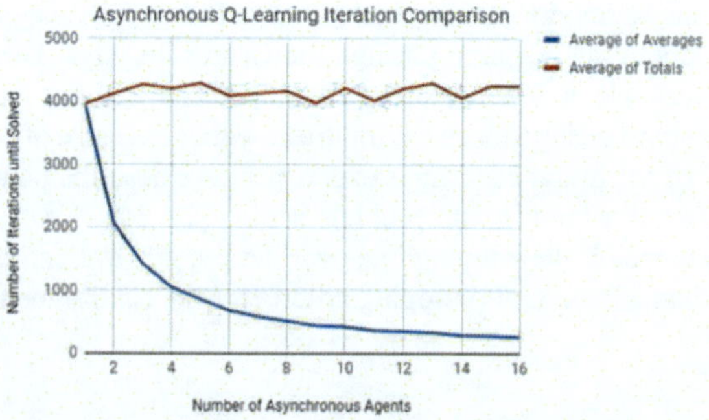

Fig. 1 Comparison of average total number of iterations over all agents until the task was solved and the average of the number of iterations of each agent

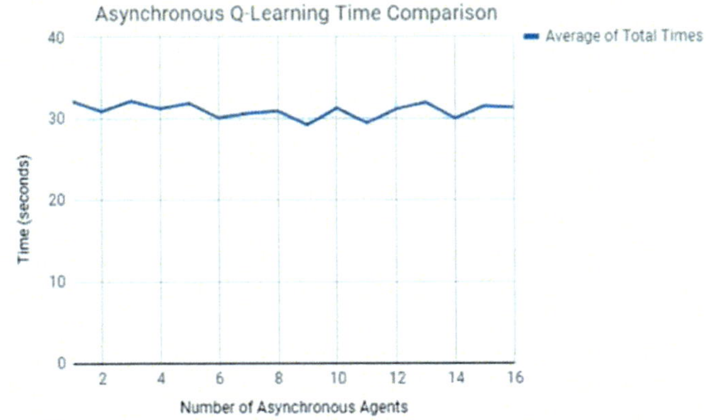

Fig. 2 Comparison of average total time over all agents until the task was solved

Fig. 3 Baseline performances of the DQN agent. The *x*-axis shows the number of iterations that had passed when the agent was saved while the *y*-axis shows the winning percentage of the agent over 1000 games

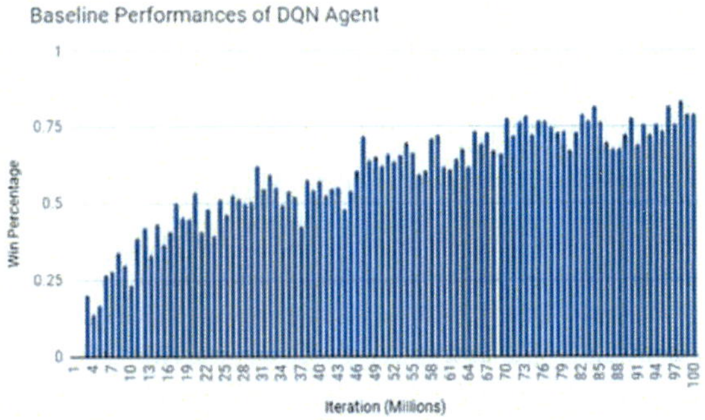

Fig. 4 Comparison of time to convergence for different amounts of pre-training for DQN agents

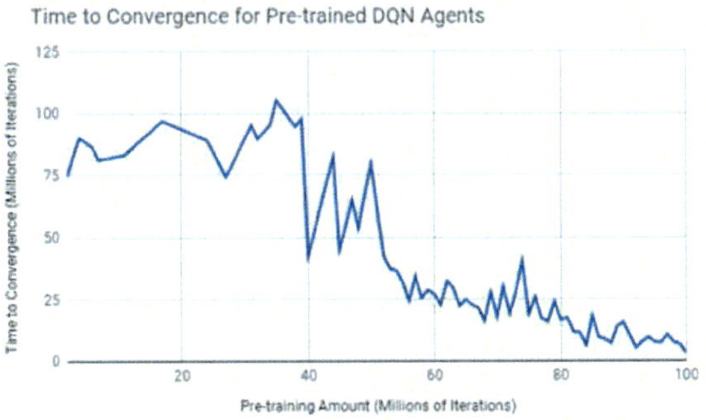

5 Conclusion and Future Work

We have shown that initial conditions greatly impact the rate of convergence for reinforcement learning. As a result, transfer learning shows great potential for accelerating the convergence rate of reinforcement learning agents. Transfer learning has already seen great success in deep reinforcement learning, and we hope that this research is now further motivated. In the future, we would want to study adversarial learning in the reinforcement learning setting [17]. Intuitively, presenting challenges allows humans to learn better, and we believe that this translates to reinforcement learning agents as well. In fact, it has already been shown that this adds robustness [18]. Furthermore, adversarial learning is perfectly suited for two-player games like many of the Atari games. Hence, our future work should include studies in adversarial learning in the reinforcement learning setting. We would also like to study learning models for multiple games and use transfer learning to apply these models to different reinforcement learning tasks.

References

1. Mnih V, Kavukcuoglu K, Silver D, Graves A, Antonoglou I, Wierstra D, Riedmiller M (2013) Playing atari with deep reinforcement learning
2. Silver D, Hubert T, Schrittwieser J, Antonoglou I, Lai M, Guez A, Lan-tot M, Sifre L, Kumaran D, Graepel T, Lillicrap T, Simonyan K, Hassabis D (2017) Mastering chess and shogi by self-play with a general reinforcement learning algorithm
3. Wang Z et al (2018) Deep reinforcement learning with knowledge transfer for online rides order dispatching. In: 2018 IEEE International Conference on Data Mining (ICDM). IEEE
4. Manfredi V et al (2021) Relational deep reinforcement learning for routing in wireless networks. In: 2021 IEEE 22nd international symposium on a World of Wireless, Mobile and Multimedia Networks (WoWMoM). IEEE
5. Lample G, Chaplot DS (2021) Playing FPS games with deep rein-forcement learning. In: Thirty-first AAAI conference on artificial intelligence
6. Zhou Z, Li X, Zare RN (2017) Optimizing chemical reactions with deep reinforcement learning. ACS Cent Sci 3(12):1337–1344
7. Baldazo D, Parras J, Zazo S (2019) Decentralized multi-agent deep reinforce-ment learning in swarms of drones for flood monitoring. In: 2019 27th European Signal Processing Conference (EUSIPCO). IEEE
8. Landajuela M et al (2021) Discovering symbolic policies with deep reinforcement learning. In: International conference on machine learning, PMLR
9. Zhao R et al (2020) Deep reinforcement learning based mobile edge computing for intelligent Internet of Things. Phys Commun 43:101184
10. Kiran BR et al (2021) Deep reinforcement learning for autonomous driving: a survey. IEEE Trans Intelligent Transp Syst
11. Simonyan K, Zisserman A (2014) Very deep convolutional networks for large-scale image recognition
12. Szegedy C, Liu W, Jia Y, Sermanet P, Reed S, Angue-lov D, Erhan D, Van-houcke V, Rabinovich A (2015) Going deeper with convolutions
13. Mnih V, Kavukcuoglu K, Silver D, Rusu AA, Veness J, Bellemare MG, Graves A, Riedmiller M, Fidjeland AK, Ostrovski G, Petersen S, Beattie C, Sadik A, Antonoglou I, King H, Kumaran

D, Wierstra D, Legg S, Hassabis D (2014) Human-level control through deep reinforcement learning
14. Yin H, Pan SJ (2017) Knowledge transfer for deep reinforcement learning with hierarchical experience replay
15. Mnih V, Badia AP, Mirza M, Graves A, Harley T, Lillicrap TP, Silver D, Kavukcuoglu K (2016) Asynchronous methods for reinforcement learning
16. Even-Dar E, Mannor S, Mansour Y (2006) Action elimination and stopping conditions for reinforcement learning
17. Lin Y-L, Hong Z-W, Liao Y-H, Shih M, Liu M, Sun M (2017) Tactics of adversarial attack on deep reinforcement learning agents
18. Pinto L, Davidson J, Sukthankar R, Gupta A (2017) Robust adversarial reinforcement learning

A Novel Current Control Scheme for Grid-Connected Single-Phase PWM Bridgeless Power Converters

Khalid Javed, Lieven Vandevelde, and Frederik De Belie

Abstract In this paper, a new bridgeless single-phase AC–DC power factor corrector is presented. The proposed scheme is based on the interleaved topology of buck–boost converter without a bridge rectifier at the input stage for AC–DC conversion. The required AC–DC conversion is done with the help of the interleaved topology of the buck–boost converter. The absence of rectifier bridge at the input side of the circuit means the reduced numbers of the diodes in the circuit which results in less conduction and power losses. The advantages of such topology includes less power losses in the circuit, low voltage stress on the switch, and improved efficiency of the overall system. Two circuits are used in this paper which are connected in parallel to each other, each circuit comprising the two interleaved boost converters for AC–DC conversion to reduce the power losses and to provide an alternative approach for efficiency improvement. This Power Factor Correction PFC converter will control the output voltage to provide a regulated DC voltage at the user end, and at the same time, it will draw a sinusoidal input current from the power supply source to maintain the power factor of the system. MATLAB is used as a software for results, and simulation results are performed to present the feasibility of the proposed technique.

Keywords Bridgeless interleaved connected converters · Power factor correction in parallel connected modules · Efficiency improvement

K. Javed (✉) · L. Vandevelde · F. De Belie
Department of Electromechanical, Systems and Metal Engineering, EEDT Decision and Control, Flanders Make, Ghent University, Ghent, Belgium
e-mail: Khalid.Javed@UGent.be

L. Vandevelde
e-mail: Lieven.Vandevelde@UGent.be

F. De Belie
e-mail: Frederik.DeBelie@UGent.be

X.-S. Yang et al. (eds.), *Proceedings of Seventh International Congress on Information and Communication Technology*, Lecture Notes in Networks and Systems 447,
https://doi.org/10.1007/978-981-19-1607-6_14

1 Introduction

With the development of new electronic equipment, the demand for power quality improvement has been also increased in recent years. Power quality improvement is all about reducing the harmonic factor of the power lines and efficiency improvement of the power system. So for this purpose, Power Factor Correction PFC is an active topic nowadays in power electronics with some great development already achieved in this regards [1–4]. As some of the more strict regulations are implemented in [5] on all types of the power systems regarding the Total Harmonic Reduction THD of the input current and power quality regulation, for this purpose power factor correction in single-phase power supply sources are mandatory. There are different types of PFC but among them active PFC is preferably used.

In active PFC, the load behaves like a resistor due to which irregular currents are not drawn from the power supply which results in a high power factor for the system with negligible harmonics in the line current [6]. In a conventional PFC converter setup, there is a rectifier stage at the input end which is then followed by a DC–DC converter, i.e., Buck, Boost, Buck–Boost, Cuk, and a Flyback converter, etc. These types of setups are suitable for a low power range or somehow for medium power ranges also but when there is increase in the power levels then these types of setups are not much useful as it has high conduction losses at the diodes of bridge rectifier which alternatively effects the overall efficiency of the system. Sometimes, the diodes of the bridge rectifier can also be destroyed because of the heat generated due to the power losses in bridge rectifier. To prevent the breakdown of the diodes at the bridge rectifier, some high-efficient heat dissipating and current handling techniques are introduced into the system which results in the increase of size along with the price of the power supply.

To deal with such types of the problems, researchers developed different types of the bridgeless PFC topologies presented in [7–13]. All these techniques are working on improving the efficiency of the system. Most of the techniques are working with boost converter topology because of its characteristics, i.e., low cost, easy to control, and ease of implementation. At the same time, it also has some implementation draw backs as for boost converter the output voltage is always higher than the peak value of input supply voltage due to that the isolation between the input and the output is not easily implemented. The other drawback includes the current limitation problems during the overload conditions and the high start-up inrush current.

The concept of interleaving topology is quite old. Reference [14] implements the use of interleaved topology in 1972. The number of stages used in that article was eight. The advantages of this interleaved topology includes the high switching frequency with low switching losses and high power density with high efficiency of power conversion. The drawback of this [14] was an increase in the size and complexity of the system which leads to the reduced reliability of the system. Some other interleaved techniques of converter is presented in [15, 16], in which the size of the input EMI filter is reduced with increase in switching frequency because of this interleaving technique.

The control technique presented in this paper consists of both bridgeless and interleaved topologies. The PWM converter of the proposed topology operates in voltage follower method in which the current is forced to follow the sinusoidal envelop. This type of configurations has low harmonic order which results in a significant THD. Variable duty cycle is implemented in this control technique to overwhelm the lower order harmonic distortion [17]. Buck–boost topology is used as a DC–DC power converter as it is useful for both purposes, i.e., either bucking the voltage or for boosting it. The rest of the paper is organized as follow: i.e., in Sect. 2, a bridgeless interleaved topology of two boost converter is presented along with their operation. Section 3 is about the detail control technique of this interleaved structure. In Sect. 4, two identical structures of interleaved topology are connected in parallel and their operation along with control is discussed. The improvement in the efficiency by using parallel connected devices is presented in Sect. 5 of this paper. Simulation results of MATLAB are presented in Sect. 6 of this paper while at the end the conclusion is made in Sect. 7.

2 Bridgeless Interleaved Topology of Single-Circuit Buck–Boost Converter

Figure 1 illustrates a block diagram of the bridgeless PFC with interleaved topology of buck–boost converter. An input EMI filter along with an inrush control relay is placed in between the input supply source and the PFC converter while a capacitor filter and load is at the outer end of the circuit. The control of such type of converter works on the correction of the input supply current waveform by controlling the inductor current to follow the sinusoidal voltage with the help of high frequency.

Two buck–boost converters are connected in interleaved scheme for each half line cycle. The input voltage of this circuit is symmetrical in two half line cycles while the line current of the circuit flows only through the two diodes of the circuits which results in the low conduction losses. The space utilization and the thermal

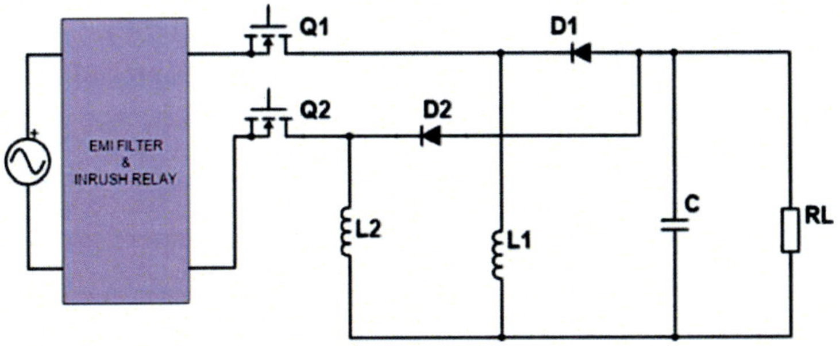

Fig. 1 Interleaved connected scheme of buck–boost converter topology

performance of the circuit also get improved because of using two inductors as compared to using a single conventional inductor. So it can be helpful in improving the efficiency of the system even for high power applications. The sum of both inductor currents, i.e., I_{L1} and I_{L2} represents the total current from the input supply source. The ripple currents will be out of phase in these inductors due to which they will cancel the effect of each other. This will also lead to the reduction of EMI filter size by mitigating the high frequency ripple current which is generated because of the high speed switching of the buck–boost converter. Output capacitor high frequency ripple is also reduced in this interleaving technology.

The detailed operation of bridgeless interleaved buck–boost converter is performed in two different cycles of the input voltage, i.e., positive and negative half cycle as drawn by red line in Fig. 2. During the positive half cycle of the input supply source, the power is transmitted from input to the output with the help of L_1. For the negative half cycle, the power is supplied with the help of L_2. For the first positive cycle when the switch Q_1 is ON, then path is provided for the current between the load and the input supply source with the help of inductor L_1. The circuit Q_1, L_1, and D_1 acts as a single buck–boost converter switching at high frequency for the first positive cycle as drawn by a red line in Fig. 2a while the same is the case for negative half cycle in which a link is provided between the input supply

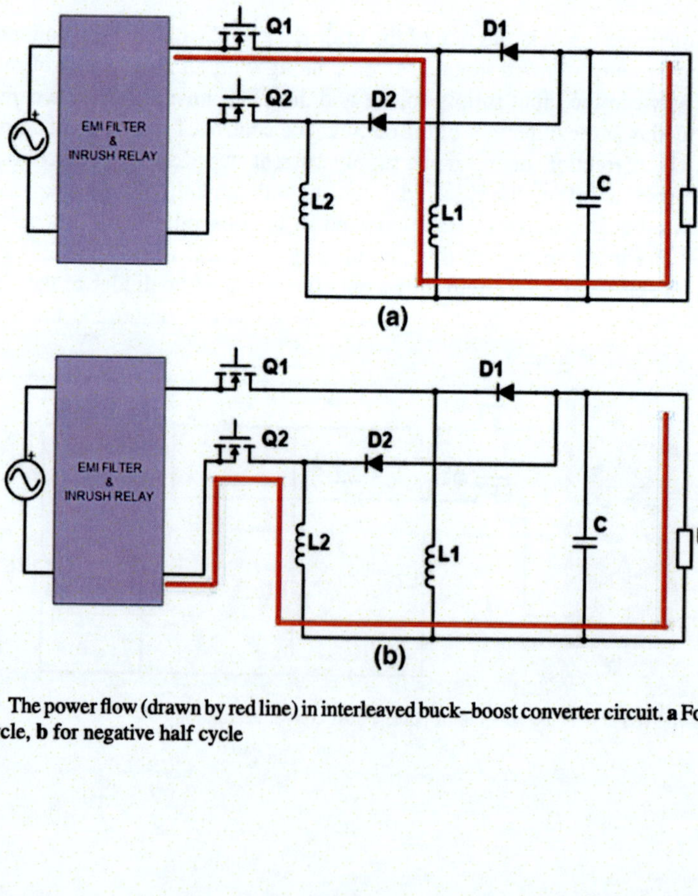

Fig. 2 The power flow (drawn by red line) in interleaved buck–boost converter circuit. **a** For positive half cycle, **b** for negative half cycle

source and the output load with the help of inductor L_2 as drawn again by a red line in Fig. 2b. Here, Q_2, L_2, and D_2 reacts as a single buck–boost converter switching at high frequency for the other cycle.

3 Control Technique for a Single-Circuit Interleaved Topology

Power Factor Correction is one of the most integral part of the circuits with the implementation of IEC 620000-3-2 standards for input current harmonic content, which defines the limitation on any types of the electronic loads for injecting harmonic components into the supply line. So for any AC–DC-rectified system, the PFC is a front-end stage as shown in Fig. 3. The performance factor of this PFC stage is to extract a pure sinusoidal form of input current from the supply source and also regulates the DC voltage at the output end. The stage after PFC in Fig. 3 is Phase-Shifted Full Bridge PSFB converter. The voltage which is coming from the PFC stage is a high DC voltage, so this PSFB converter will convert it into a low voltage of about 12 or 48 V. This stage will also provide high frequency isolation and voltage translation to this rectifier system. In this paper, we are focusing on the implementation of PFC stage with a bridgeless interleaved buck–boost converter along with their control scheme. Figure 4 shows the control scheme for the bridgeless interleaved topology of two buck–boost converters.

The above setup is designed specifically to improve the efficiency of the system by reducing the power losses at the bridge rectifier diodes. This will also be helpful in the power factor correction of the power system. As discussed earlier, buck–boost converter is used in interleaved structure as an AC–DC conversion stage as no bridge rectifier is used for AC–DC conversion. So it means that both voltage rectifying and voltage regulations is performed by the interleaved buck–boost converters. The control for this setup is provided digitally which is shown in Fig. 4. The control structure consists of two main loops, i.e., inner current control loop and outer voltage control loop. The outer loop is responsible for voltage control and regulation while the inner loop is current control loop. The voltage control loop is taken commonly for both converters while current control loop contains two sub-loops, i.e., the current control loop for converters 1 and 2.

The input supply voltage is applied to the converter circuit with the help of an EMI filter as shown in Fig. 4. Two buck–boost converters are connected in interleaved

Fig. 3 Block diagram of bridgeless power factor corrector converter

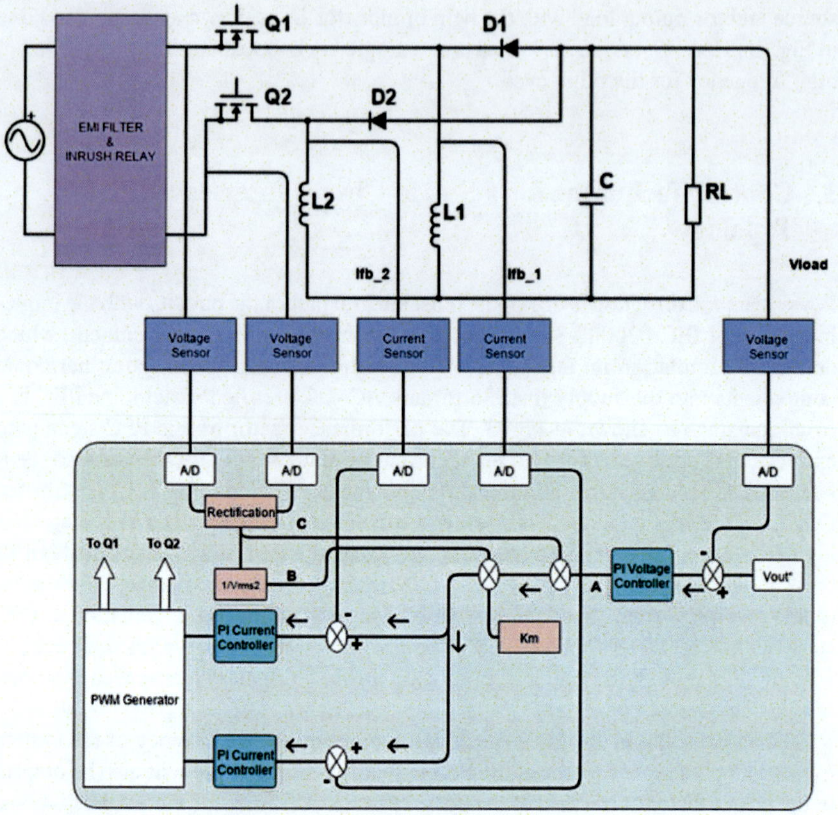

Fig. 4 Control scheme of single circuit of two interleaved PFC converters

topology which acts as a PFC stage, each one operating in buck–boost mode alternately during each half of input AC line cycle. The interleaved buck–boost converters will convert the AC line voltage into DC line voltage during each half cycle, i.e., during the first half cycle first converter will provide the DC voltage to the load while during the second half cycle, the second converter will be responsible for voltage supply to the load. So the first buck–boost converter is formed by the switch Q_1, inductor L_1, and reverse diode D_1, while similarly the switch Q_2, inductor L_2, and reverse diode D_2 combines together to form the second buck–boost converter. At the output end, a capacitor C_{out} and a load R_L is connected across these converters. This capacitor is responsible for the voltage storage and then supplying a regulated voltage to the load. The control scheme is carried in a digital environment and all the signals which are needed from the circuit are sensed and can be seen from the Fig. 4. In total, 5 different signals are sensed from the circuit which are input voltage line V_{in_L} signal and input voltage neutral V_{in_N} signal, two current sensors each one at inductor of each buck–boost represented by I_{fb_1} and I_{fb_2} and at the output end the load voltage V_{load}. All these signals are used inside the digital control schemes for voltage and current control loops for each buck–boost converter.

The bus voltage, which is sensed at the output load, is compared with a certain reference value for output voltage V_{ref} and then their error is fed into PI voltage regulator. This voltage controller is responsible for the voltage regulation according to the reference voltage. The output from this voltage controller represents the amount of power transferred to the load from the input supply with the help of this converter. It is denoted by letter A. Furthermore, this output from voltage controller A is multiplied by three further signals, i.e., B, C, and K_m which are done to provide the reference current for the inner current control loop.

Signal B represents the inverse of square of root means square RMS value of the input voltage while the signal C actually represents the rectified input voltage. Signal B is responsible for fast feed-forward control of the PFC system while the signal C is responsible for shaping of the line/input current waveform according to the input supply voltage waveform. The range of the reference current I_{ref} is adjusted with the help of gain factor K_m. So the output after the multiplication of these three factors, i.e., B, C, and K_m, is now producing a reference value for the average of inductor current, i.e., I_{L1} and I_{L2}.

From here, we have two current control loops for controlling the current of each converter correspondingly while the reference current remains the same for both. In each current control loop, this reference current I_{ref} is compared with the sensed I_{fb_1} and I_{fb_2} in both loops, respectively. The output from this comparison is an error signal which is then fed into the PI controller at current control loop. In this current control loops, both the currents of inductors, i.e., L_1 and L_2, are forced to follow the waveform of the reference current which is purely sinusoidal in accordance with input supply voltage. For providing the PWM signals to both the switches of interleaved converters, the output from current PI controllers is compared with a certain reference signal of certain frequency.

4 Connecting Two Parallel Identical Structures of Interleaved Topology

Connecting the switching mode converters in parallel is a well-known technique nowadays for delivering stable and reliable power to the user end. Both the circuits of parallel connected converters are operating with the same switching frequency in order to minimize the conduction losses, current ripple, and switching frequency of each phase. Two circuits are connected in parallel, each one comprising two buck–boost converters which are connected in interleaved topology as shown in Fig. 5. Both of the circuits are feeding a common output load at the user end.

Two circuits are connected in parallel for the purpose of providing an uninterrupted energy supply to the load. Each circuit contains buck–boost converter topology as a PFC stage. Both the circuits are supplied by the same input supply source at the input.

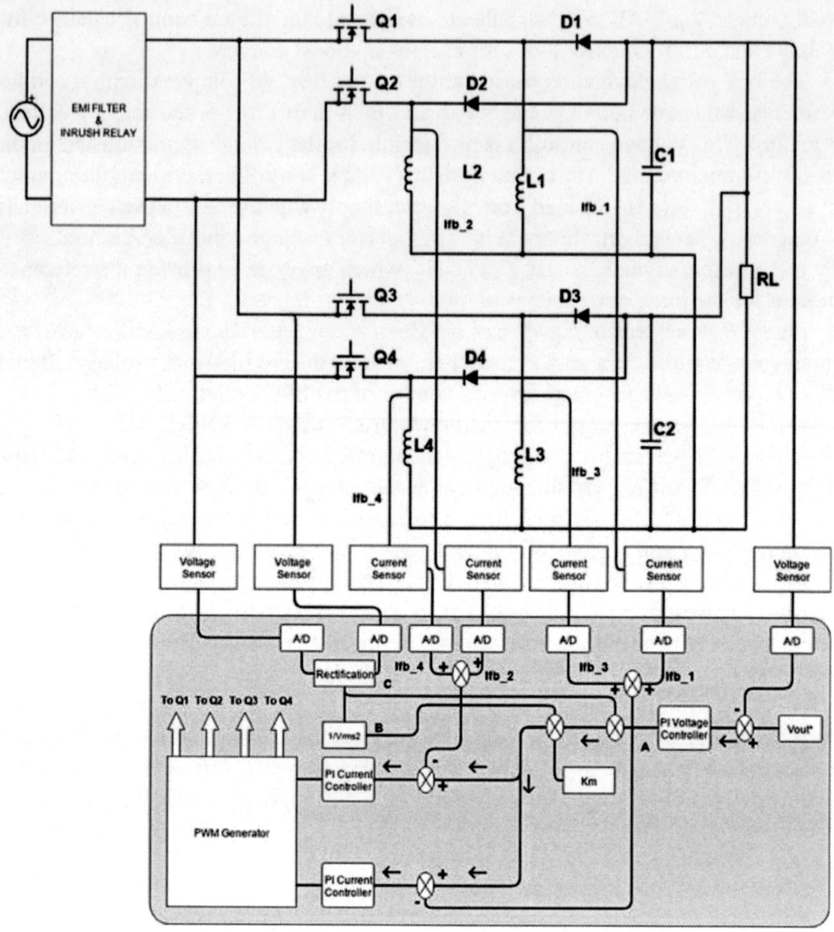

Fig. 5 Control scheme of two parallel circuit of interleaved PFC converters

Rectification is performed with the help of interleaved technique as already discussed in the previous section. A common output from both of the circuit is provided to a load at the user end. The connection need of parallel technique is to provide a regular power supply to the load, i.e., if any of the circuit has some faults in energy supply, then the load energy will be shifted to the load by the second circuit.

The control structure for two converters connected in parallel is almost the same as that for the single converter with interleaved topology. The only difference is that the number of current sensors is increased for the second circuit. Two extra current sensors are placed at the interleaved inductors of second circuit converter, which means that I_{fb_3} is for first inductor of second converter, i.e., L_3 while I_{fb_4} is current sensed from second inductor of second converter, i.e., L_4. The rest of the scheme remains the same for the rest of the sensors. These two sensors will be added into the digital control scheme environment for current and voltage regulations of both of the circuits.

From Fig. 5, it is obvious that the control scheme comprises two main loops, i.e., outer voltage loop which is also known as a slower loop and inner current control loop also referred as faster loop. Outer voltage loop will be operated at a very low frequency as compared to the current control loop in order to give enough time for the inductor currents to track the input voltage waveform. The current sensed I_{fb_3} and I_{fb_4} are provided into the current control loops for controlling the line currents of the second circuit in a manner that I_{fb_3} is provided into first current control loop while I_{fb_4} is provided into the second control loop. The first current control loop has the signal from I_{fb_1} and I_{fb_3} while the second loop has the signals from I_{fb_2} and I_{fb_4}.

The main difference between the single circuit and the parallel circuit is mainly at current control loops. The reference current I_{ref} coming from the voltage control loop is further supplied into the two current control loop where it is compared with the summation of inductor currents present in that loop, i.e., in first current control loop, we have I_{fb_1} and I_{fb_3} while the second control loop has I_{fb_2} and I_{fb_4}. So the summation of I_{fb_1}, I_{fb_3} and I_{fb_2}, I_{fb_4} is compared with the reference current to generate the error which is tackled then by their respective current PI controller. The rest of the process will remain the same. The output PWM signals will be the output from this digital control scheme which will be provided to the respective switches of the converters with the help of gate signals.

5 Efficiency Improvement in Parallel Connected Circuits

The above-discussed scheme is used mostly in the chargers of electric vehicles. Parallel connected devices are always helpful in improving the efficiency of the high power battery chargers. Such type of chargers are widely used in the industries for obtaining high efficiency over the load range. The advantages of parallel connected AC–DC converter includes obtaining peak efficiency, and smaller and cheaper power components for most of the battery chargers. Parallel connected converter scheme has low current ripples which in turns increases the efficiency of the system and also reduces the conducted EMI noise. These current ripples are the main sources of voltage distortion, heating, and noise which cause a gradual decrease in the efficiency of the system.

Parallel connection of the switches or diodes has positive impact on the efficiency improvement of the circuit. The efficiency of the two diodes connected in parallel will be greater than the single diode and IGBT switch connection. The same is the case in two IGBTs connected in parallel will provide more efficiency to the system. In this paper, four IGBT switches are connected in parallel in the set of two in each circuit so it has a positive impact on the efficiency of the overall system.

The efficiency improvement of the buck–boost converter has also some bright effects on the output voltage of the system. As efficiency and voltage have direct proportionality, increasing the efficiency means a gradual increase in the output voltage of the system. The switches connected in parallel have more positive impact on the output voltage of the circuit which shows that the parallel connection of the switching converters also have impact on decreasing the ON resistance of the switches plus improving the efficiency of the system [18]. It also improves the efficiency of the front-end PFC stage by reducing the forward voltage drop in line current path.

Both of the circuits used in the scheme are bridgeless interleaved converters with no bridge rectifier at the input stages of both circuits so it means that there are no voltage drops or power losses at the input stage. Both the circuits are using their interleaved topologies of converters for AC–DC conversion so which means that less power losses in the circuit as compared to those with bridge rectifiers at the input stage. The low power losses means that more energy is transferred to the load which leads to the higher efficiency of the system.

6 Simulations and Results

Two circuits are connected in parallel to provide a most cost-effective design. Each circuit contains two buck–boost converters which are connected in interleaved form to provide a most efficient and reliable system. Each circuit contains bridgeless power converters to improve the power losses in the system. The AC–DC conversion is performed with the help of this interleaved scheme of the converters as discussed earlier. The load power is shared equally between the converters to reduce the stress on a single circuit. With the absence of bridge diodes, the power dissipation is reduced to supply more efficient power to the load. Such type of parallel schemes are also efficient in improving the energy storage of the circuit which is also helpful in the output power estimation capabilities. The proposed scheme is verified with the help of MATLAB/Simulink to improve the efficiency and power factor of the system, the output voltage, and line current control. All these results are obtained from Simulink to provide the effectiveness of proposed scheme. The output reference voltage is set to be 400 V which can be seen in Fig. 6 along with the output current of the circuit.

Figure 6 shows clearly that the reference voltage which is specified in the voltage control loop of the circuit is achieved successfully. So the output voltage regulator is working finely, and a pure DC output voltage is available at the output end for the load. After voltage control loop, as discussed that output from voltage loop is multiplied by the rectified supply voltage and then divided by the inverse of square RMS to provide a reference for both of the inner current control loops. So the reference signal for both current control loops are shown in Fig. 7. This signal is purely sinusoidal according to the rectified inner supply voltage which will be latterly followed by the line currents of each converter to remain sinusoidal and to remove the harmonics from the line current.

The line currents for first circuit, i.e., for converters 1 and 2 are shown in Fig. 8 while the line currents for second circuit, i.e., converter 3 and 4 are shown in Fig. 9. Before giving these currents to the control loops, they are added to each other as

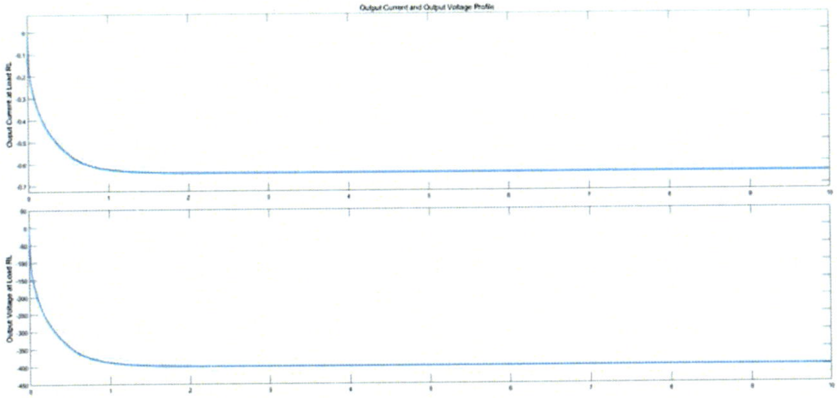

Fig. 6 Output current and voltage profile of bridgeless parallel connected interleaved converters

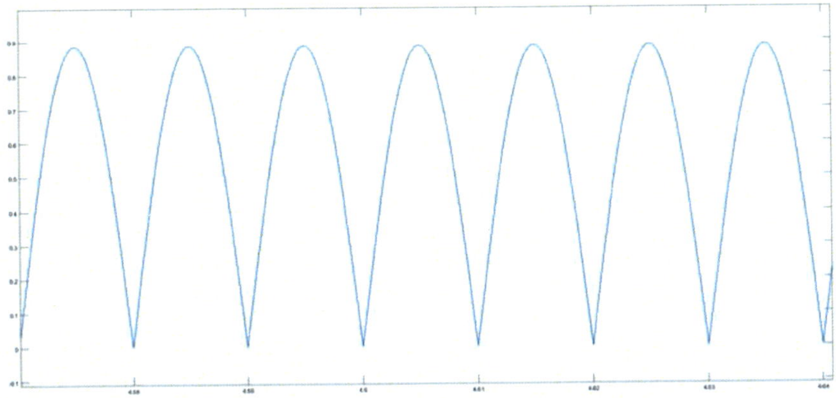

Fig. 7 Reference current waveform for current control loops after voltage control loop

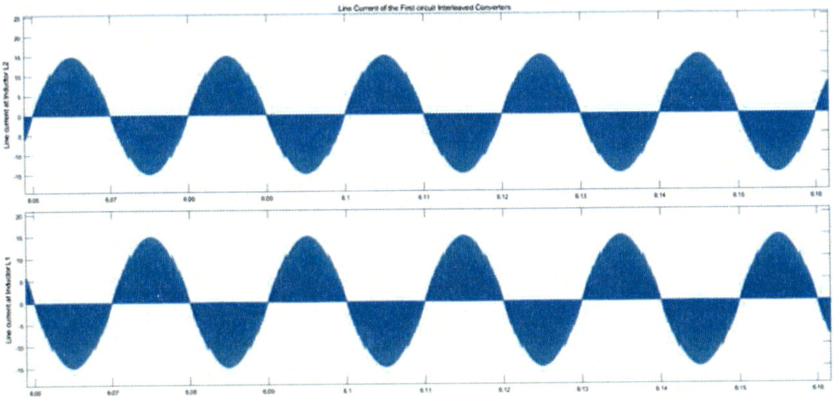

Fig. 8 Line current profile for first and second converter

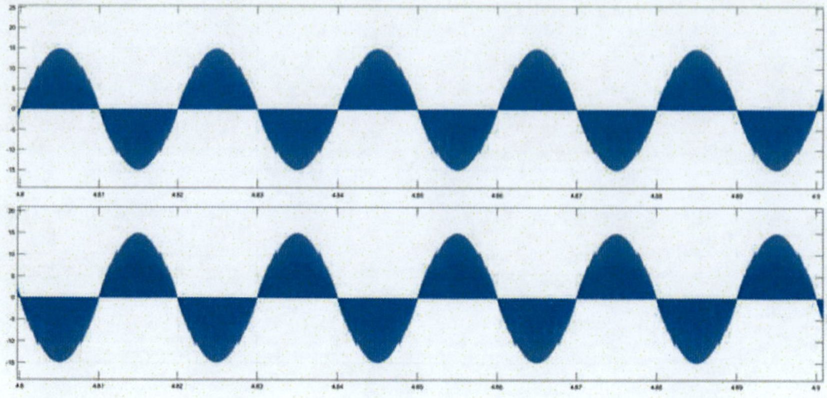

Fig. 9 Line current profile for third and fourth converter

shown in Fig. 5. These converter currents are then forced to follow the reference current in a pre-defined manner, as shown in Fig. 10, in each control loop to have a pure sinusoidal line current which will provide the sinusoidal inner current for the system which will be in phase with the input supply voltage. The current voltage PI regulator is responsible for the error omission between the reference and the line current. The current waveform after current control loop along with the reference current is compared in Fig. 11, and it is visible that current control loop is focusing completely on the current waveforms of the inductors to follow the reference current waveform. After current control loops, the DC PWM generator is used to generate the gate signals for the switches of each converters. The gate signals for the switches are shown in Fig. 12. The supply current and voltage are purely sinusoidal and they are

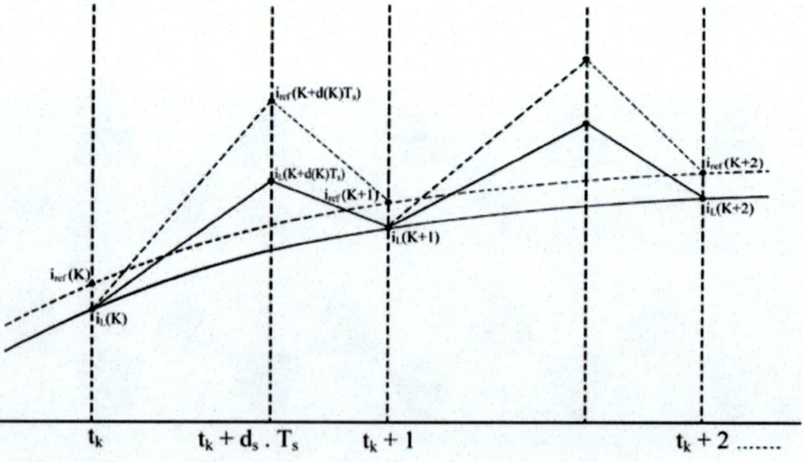

Fig. 10 Pre-defined scheme for inductor current following the reference current

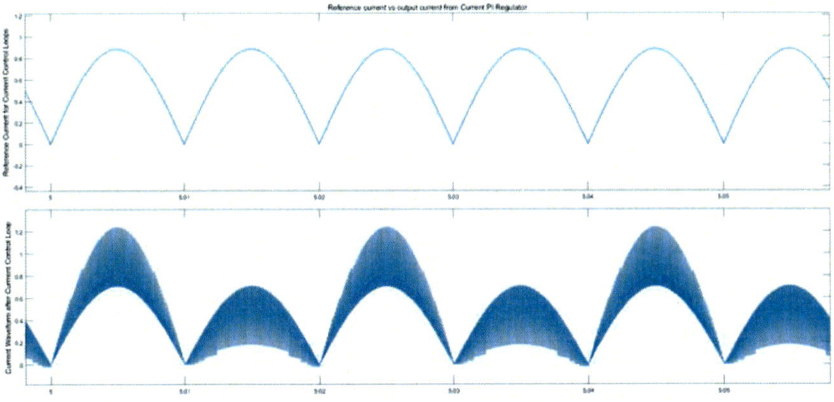

Fig. 11 Reference current along with the current waveform after current PI regulator

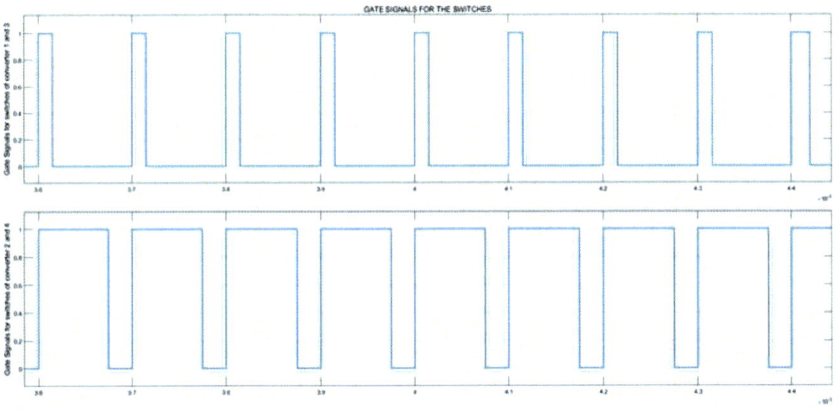

Fig. 12 Gate signals for the switches

also in phase, as shown in Fig. 13. This will results in a low value of Total Harmonic Distortion THD of the supply/line current which is shown in Fig. 14. The efficiency of the whole system with respect to the input supply voltage is provided in Fig. 15.

7 Conclusion

Interleaved topology of the converter is introduced in this research work in order to provide a system which can perform AC–DC conversion with less power dissipation inside the system. Two identical circuits of the interleaved converters are connected for the purpose of having less stress over the single converter components and to provide a continuous power supply to the system. Control scheme is performed

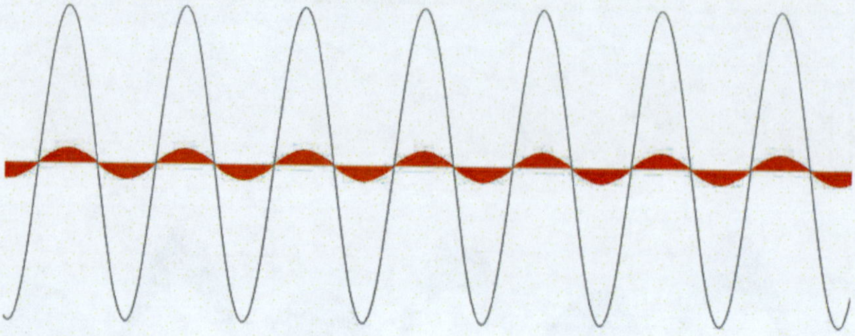

Fig. 13 Input voltage and current profile

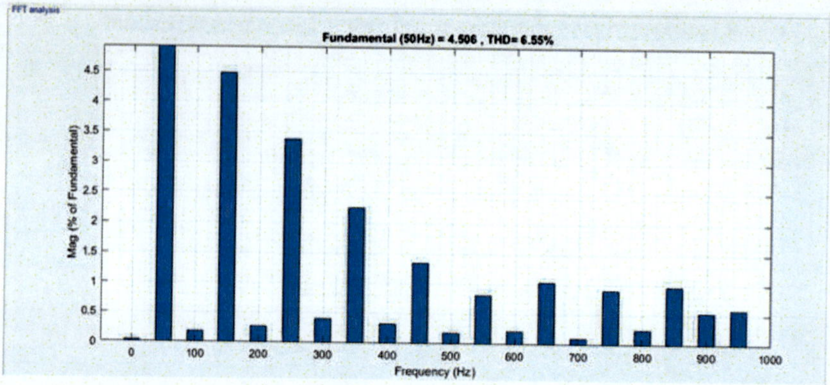

Fig. 14 THD value of the whole system

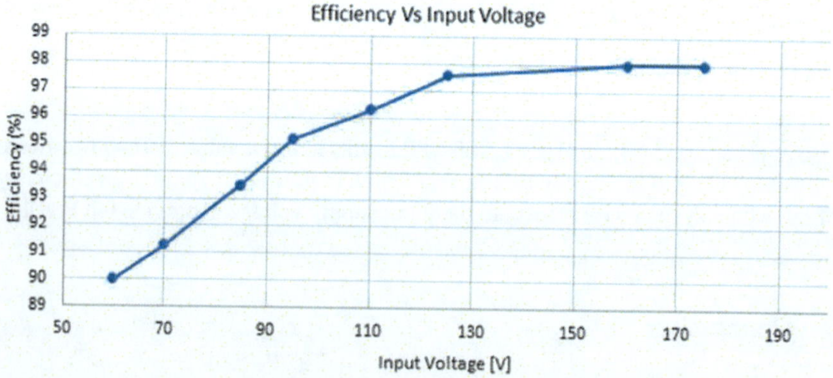

Fig. 15 Efficiency of overall system with respect to the input supply voltage

successfully for single circuit with two interleaved buck–boost converters and then the same control scheme is extended for two identical circuits which are connected in parallel. So in total, we have 4 converters and with the help of the control scheme the gate signals are provided to the gate terminal of each switch in order to perform the output voltage and line current control. MATLAB/Simulink results are provided in the end to verify the effectiveness of the scheme in detail.

References

1. Qiao C, Smedley KM (2001) A topology survey of single-stage power factor corrector with a boost type input-current-shaper. IEEE Trans Power Electron 16(3):360–368
2. Gracia O, Cobos JA, Prieto R, Uceda J (2003) Single phase power factor correction: a survey. IEEE Trans Power Electron 18(3):749–755
3. Jovanovic MM, Jang Y (2005) State-of-the-art, single-phase, active power-factor-correction techniques for high-power applications—an overview. IEEE Trans Ind Electron 52(3):701–708
4. Villarejo JA, Sebastian J, Soto F, de Jódar E (2007) Optimizing the design of single-stage power-factor correctors. IEEE Trans Ind Electron 54(3):1472–1482
5. Limits for Harmonic Current Emissions (Equipment Input Current ≤ 16 A per Phase), EMC Part 3-2, 3rd edn. IEC 61000-3-2, Nov 2005
6. Erickson RW, Maksimovic D (2001) Fundamentals of power electronics, 2nd edn. Kluwer, Norwell, MA
7. Mitchell DM (1983) AC–DC converter having an improved power factor. U.S. Patent 4 412 277, 25 Oct 1983
8. Salmon JC (1992) Circuit topologies for single-phase voltage-doubler boost rectifiers. In: Proceedings of IEEE applied power electronics conference, pp 549–556
9. Tollik D, Pietkiewicz A (1992) Comparative analysis of 1-phase active power factor correction topologies. In: Proceedings of INTELEC, pp 517–523
10. Enjeti PN, Martinez R (1993) A high performance single phase AC to DC rectifier with input power factor correction. In: Proceedings of IEEE applied power electronics conference, pp 190–195
11. Souza AF, Barbi I (1995) A new ZVS-PWM unity power factor rectifier with reduced conduction losses. IEEE Trans Power Electron 10(6):746–752
12. Salmon JC (1995) Circuit topologies for PWM boost rectifiers operated from 1-phase and 3-phase ac supplies and using either single or split dc rail voltage outputs. In: Proceedings of IEEE applied power electronics, pp 473–479
13. Souza AF, Barbi I (1999) A new ZVS semiresonant high power factor rectifier with reduced conduction losses. IEEE Trans Ind Electron 46(1):82–90
14. Miwa BA, Otten DM, Schlecht MF (1992) High efficiency power factor correction using interleaving techniques. In: Proceedings of APEC '92 (IEEE Catalog no. 92CH3089-0), pp 557–568
15. Lee P-W, Lee Y-S, Cheng DKW, Liu XC (2000) Steady-state analysis of an interleaved boost converter with coupled inductors. IEEE Trans Ind Electron 47(4):787–795
16. Canesin CA, Goncalves FAS, Melo GA, de Freitas LCG (2009) DCM boost interleaved converter for operation in AC and DC to trolleybus application. In: EPE '09, 13th European conference on power electronics and applications, Barcelona, pp 1–10
17. Athab HS (2008) A duty cycle control technique for elimination of line current harmonics in single-stage DCM boost PFC circuit. In: TENCON 2008—2008 IEEE Region 10 conference, Hyderabad, pp 1–6
18. Mohammed MF, Ahmad AH, Humod AT (2018) Efficiency Improvement of dc/dc Boost converter by parallel switches connection. Int J Appl Eng Res 13(9):7033–7036

Disk Space Management Automation with CSI and Kubernetes

Anastasia Shemyakinskaya and Igor Nikiforov

Abstract Kubernetes is a container orchestration system, which is used in production-ready platform as a service such as OpenShift. To manage and provision storage for application, the container storage interface (CSI) exists in container orchestration systems. The article contains comparison analyze of bare-metal CSI implementations that shows advantages and disadvantages of existing CSI implementations, based on which bare-metal implementation is thought to be a most powerful one, but not free from a gap for improvement. So, there is a need to extend CSI for bare-metal storage provisioning to avoid virtualization and cloud overhead and minimize manual storage management operations. The paper considers bare-metal CSI extension for automation of local disk management as well including ephemeral volume support.

Keywords Container storage interface · Kubernetes · Ephemeral volumes

1 Introduction

Container orchestration (CO) systems gain their popularity for simplifying the process of application deployment on clusters [1]. CO systems make application that run on cluster more independent from each other and safe. The following main CO system exists: Kubernetes [2], Mesos [3], Docker Swarm [4]. Kubernetes is one of the most popular from them and de facto standard of orchestration systems [5]. It is an open-source software that is used for managing and deploying a Docker container cluster.

Every application, that is, runs on cluster, requires to store data like records, databases, log files, big data [6] and internet of things devices [7], structured and unstructured content on storages and disks.

A. Shemyakinskaya (✉) · I. Nikiforov
Peter the Great St. Petersburg Polytechnic University, Polytechnicheskaya 29, 195251 St. Peterburg, Russia
e-mail: a.shemyakinskaya@edu.spbstu.ru

X.-S. Yang et al. (eds.), *Proceedings of Seventh International Congress on Information and Communication Technology*, Lecture Notes in Networks and Systems 447,
https://doi.org/10.1007/978-981-19-1607-6_15

By default, Kubernetes supports integration with a limited number of storage systems [8] that leads to an integration difficulty during adding support of new storage systems into Kubernetes and that difficulties are described in detail in Sect. 2.

To simplify integration of different storages to the Kubernetes infrastructure, generalized interface called container storage interface (CSI) exists. Storage vendors can implement CSI on their side, so that orchestration systems can use new storage systems.

There are many implementations of CSI for local disks provisioning, which are considered and compared in Sect. 2, and based on the comparison, bare-metal implementation is thought to be a most powerful one, but not free from a gap for improvement:

- Lack of ephemeral volumes support in CSI, what leads to manual disk space removal after the pod deletion.
- Manual volume expansion in Kubernetes, what leads to spending more time on routine operation.
- End-to-end test automation absence.

Section 3 describes theoretical basis about CSI ephemeral volumes and principles of their realization.

Section 4 provides detailed realization of ephemeral volumes and results achieved by their introduction in CSI.

Volume expansion and end-to-end test automation will be covered in the future.

2 Container Storage Interface

CSI is designed as a standard to provide integration of arbitrary storage systems, like Ceph, Portworx, NetApp, and other with container orchestration (CO) systems such as Kubernetes [9].

Adding new storage systems (called "volume plugins") to a Kubernetes system is a tedious task, because the storage implementation functionality is the main part of the Kubernetes code, and vendors who would like to add support for their storage to Kubernetes (or even fix a bug in an existing storage plug in) are forced to:

- Make changes to the Kubernetes source code, which requires them to have additional knowledge about the internal structure of the Kubernetes system source code.
- Follow all policies for the development, testing, and release of new versions of Kubernetes, which introduces additional dependencies and time delays in the storage development process on vendor's side.

The presence of storage support code on the Kubernetes side causes additional difficulties not only for storage vendor's development, but also for the Kubernetes developers themselves:

- Violation of the reliability and integrity of the system due to external dependencies on the features of the storage system implementation.
- Laboriousness of testing, support, and approval of new versions.

Introduction of CSI interface simplifies the development process for third-party storage support in Kubernetes for both storage vendors and Kubernetes developers, because there is no necessity to change Kubernetes code, storage vendors do not need to understand Kubernetes release lifecycle and it is more option to customize storage. CSI driver is deployed as a separate application, consisting of CSI identity, controller, and node. Identity is responsible for identification of driver and provides information about it. Controller handles request for creation, deletion, and extension of volume. Node usually mounts created volumes on node [10]. All components run with sidecar containers in their pods in Kubernetes cluster. Sidecar containers watch Kubernetes components: volumes, claims, etc. CSI is handling request from these sidecars.

The paper considers following main CSI realization:

- Bare-metal CSI [11] is a CSI implementation that allows container orchestration systems (such as Kubernetes) to use storage for applications that run on that system. Bare-metal CSI manages locally attached drives on a cluster.
- CSI driver-LVM [12] uses local Kubernetes node storage to provide persistent storage for pods, creates an LVM logical volume on local disks.
- Minio Direct CSI [13] uses local Kubernetes node storage directly.
- TopoLVM [14] uses LVM for persistent local storage.

To compare CSI implementations, following criteria are introduced:

- Open-sourced criteria, which show if the code of the tool is hosted on public servers.
- Ephemeral volume support shows if the CSI implementation supports work with ephemeral volumes (more details can be found in Sect. 3).
- Automatic volume expansion shows if the CSI implementation supports automatic volume expansion with no or little manual work.
- Test automation shows if CSI tests can be run against implementation in automatic way.
- Support of raw-block volumes [15]. A raw-block volume is a volume that appears as a block device inside the container.
- Disk replacement procedures. That criterion shows if there is a capability of CSI to monitor disk status and health and indicates if disk is needed to be replaced. In addition, that criteria also show if there is program support for adding or removing disks.
- Scheduler extender support for Kubernetes. The scheduler distributes pods to nodes. The default scheduler can be replaced entirely, or multiple schedulers can run at the same time [16]. That criterion shows if CSI can use custom scheduler algorithm based on capacity on Kubernetes nodes.
- Support for drive types such as HDD, SSD, LVM, NVMe.

Based on the comparative analysis, presented in Table 1, it is possible to conclude that bare-metal CSI project is the most powerful CSI implementation. But, it is not free from downsides, so it is necessary to eliminate them.

Bare-metal CSI needs to have such features as:

- Support for working with ephemeral volumes in Kubernetes, so that there is no more need for manual disk space removal after the pod deletion.
- Automatic volume expansion, what eliminates tedious manual work and reduces time on Kubernetes cluster management.
- Support for automation of integration testing, so that all new features are tested against the existing requirements to avoid regression in CSI implementation.

Ephemeral volumes are covered in the paper in below sections. Other topics will be covered in the future paper.

3 Ephemeral Volumes Support

3.1 Persistent Volumes Versus Ephemeral Volumes

Typically, the volumes provided by the external storage driver in Kubernetes are persistent, and their lifecycle is completely independent. Persistent volume is independent of pod and can exist without application using it. So, if pod is deleted, the according persistent volume remains in the system with kept data.

The mechanism for requesting and defining such volumes in Kubernetes is persistent volume claim (PVC) [17] and persistent volume (PV) objects. Initially, volumes supported by the container storage interface (CSI) driver could only be used through this PVC/PV mechanism.

But, there are also use cases for data volumes, whose content and lifecycle are tied to a pod (a group of containers that run as a single unit), for example, caches. That data volumes are called ephemeral, and they should be deleted within the removal of the pod.

There is no information in the CSI specification if the volume should be removed on physical level or not, CSI specification only talks about that relation should be broken. That leads to the situation, where the developers of the storage should take care by themselves removal of the ephemeral storage after pod removal.

Table 1 CSI implementation comparison table

Tool name	Open-sourced	Ephemeral volume	Volume expansion	Test automation	Raw-block volumes	LVM support	Disk and node replacement procedures	Scheduler extender	Drive type support
Bare-metal CSI	+	–	–	–	+	+	+	+	+
Topolvm	+	+	+	+	+	+	–	–	–
Csi driver-LVM	+	+	+	+	+	+	–	–	–
Minio	+	–	–	–	–	–	–	–	–

("+" denotes CSI satisfies criteria, "–" denotes criteria does not apply for CSI)

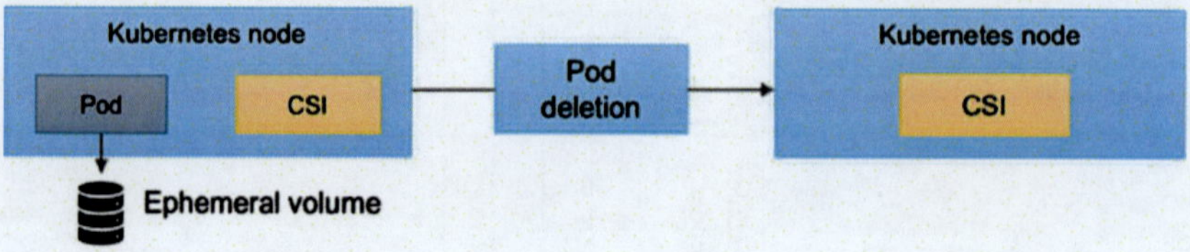

Fig. 1 Ephemeral volumes conceptual scheme

4 Ephemeral Volumes Realization

4.1 Conceptual Scheme

Figure 1 illustrates the conceptual scheme of ephemeral volumes. Firstly, Kubernetes has one running application in pod, which uses ephemeral volume to store data and CSI. After pod deletion, CSI also deletes volume. So, after pod has been removed, CSI remains in the system, but ephemeral volume has been deleted.

4.2 Implementation

Bare-metal CSI treats persistent and ephemeral volumes the same way. When a request is received, bare-metal CSI creates a partition and file system on the disk and mounts it in a container. Therefore, to implement ephemeral volumes.

There is a necessity to organize code so that it can be reused for different types of volumes. For this purpose, new common interface "volume operations" was introduced. Its implementation is used for both ephemeral and persistent volumes to avoid duplicated code.

RPC requests for NodePublishVolume and NodeUnpublishVolume have been implemented to handle ephemeral volumes. NodePublishVolume and NodeUnpublishVolume are functions used to mount or unmount volume in container. For ephemeral volume, it was expanded. Request contains indicator if volume is ephemeral. If this parameter is provided, functions also execute volume creation and deletion in operating system before mounting it in container.

Figure 2 shows the flow of ephemeral volumes work. Persistent volume is created and deleted in special requests "CreateVolume" and "DeleteVolume." These requests are called by provisioner. But, created volumes are mounted and unmounted in another requests "NodePublishVolume" and "NodeUnpublishVolume." Both these requests are called by kubelet. Kubelet is a special daemon running on each node and managing containers in pods [18]. Ephemeral volumes are created, deleted, mounted, and unmounted in the same requests "NodePublishVolume" and "NodeUnpublishVolume." So, it is managed by kubelet daemon.

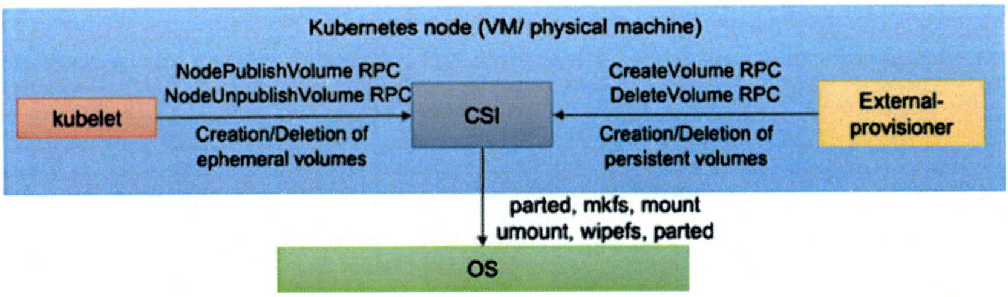

Fig. 2 Implementation of ephemeral volumes in bare-metal CSI

4.3 Results

Prior adding ephemeral volumes support to bare-metal CSI, it was necessary to also delete the volume after removing the pod, which required the use of additional Kubernetes commands:

```
kubectl delete pod <pod_name>
kubectl delete pvc <pod_pvc_name>
```

After adding support of ephemeral volumes, only one Kubernetes command is remaing. It removes both the Pod and its volume:

```
kubectl delete pod <pod_name>
```

There are two metrics can be measured:

Automatic execution time of command sequence: two independent commands and single commands. The below Fig. 3 illustrates that.

We can see that when we execute single command, we have benefit in 5% in time in comparison to two commands. The time difference is not so huge when working in cluster with single node, but if to consider cluster of 1000 nodes, then those seconds become hours in benefit.

Fig. 3 Automatic execution of commands

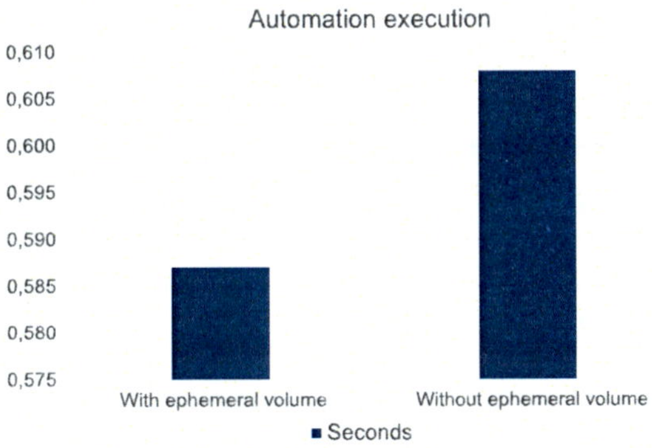

Fig. 4 Manual execution of commands

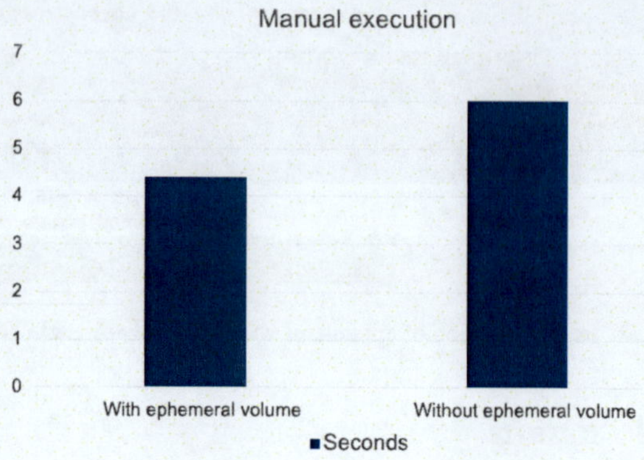

Manual (not scripted) command execution time. In this case, cluster administrator needs to: manually identify pod_name and pod_pvc_name and then execute two commands. With the support of ephemeral volumes, cluster administrator needs to only find a pod_name and execute single command. Time difference and benefices here are about 40% (Fig. 4).

5 Conclusion

A study of existing CSI implementations is performed. Bare-metal CSI is the most powerful implementation for local disk provisioning, and its disadvantages have been eliminated in the work by adding new automation features—added support for working with ephemeral volumes in Kubernetes, which reduced the time and the number of commands used to remove pods.

And as a general result of the work—bare-metal CSI become more automated and reliable. It obtained new functionality; therefore, it can satisfy more requirements from users and attracts new customers. In the future paper, test automation and volume expansion will be covered.

Acknowledgements The research is partially funded by the Ministry of Science and Higher Education of the Russian Federation under the strategic academic leadership program "Priority 2030" (Agreement 075-15-2021-1333 dated 30.09.2021).

References

1. Sithiyopasakul J, Archevapanich T, Purahong B, Sithiyopasakul P, Benjangkaprasert C (2021) Automated resource management system based on kubernetes technology. In: 2021 18th international conference on Electrical Engineering/Electronics, Computer, Telecommunications

and Information Technology (ECTI-CON), pp 1146–1149. https://doi.org/10.1109/ECTI-CON51831.2021.9454911
2. Shemyakinskaya AS (2020) Hard drives monitoring automation approach for Kubernetes container orchestration system. In: Shemyakinskaya AS, Nikiforov IV (eds) Proceedings of the institute for system programming of the RAS, vol 32, no 2, pp 99–106. https://doi.org/10.15514/ISPRAS-2020-32(2)-8
3. Marathe N, Gandhi A, Shah JM (2019) Docker swarm and kubernetes in cloud computing environment. In: 2019 3rd International Conference on Trends in Electronics and Informatics (ICOEI), pp 179–184. https://doi.org/10.1109/ICOEI.2019.8862654
4. Li X, Jiang Y, Ding Y, Wei D, Ma X, Li W (2020) Application research of Docker based on Mesos application container cluster, In: 2020 International Conference on Computer Vision, Image and Deep Learning (CVIDL), pp 476–479. https://doi.org/10.1109/CVIDL5123ßß3.2020.00-47
5. Pereira Ferreira A, Sinnott R (2019) A performance evaluation of containers running on managed Kubernetes services. In: 2019 IEEE international conference on cloud computing technology and science (CloudCom), pp 199–208. https://doi.org/10.1109/CloudCom.2019.00038
6. Voinov N, Rodriguez Garzon K, Nikiforov I, Drobintsev P (2019) Big data processing system for analysis of GitHub events. In: 2019 XXII international conference on Soft Computing and Measurements (SCM)), pp 187–190. https://doi.org/10.1109/SCM.2019.8903782
7. Kochovski P, Sakellariou R, Bajec M, Drobintsev P, Stankovski V (2019) An architecture and stochastic method for database container placement in the edge-fog-cloud continuum. IEEE Int Parallel Distributed Process Symp (IPDPS) 2019:396–405. https://doi.org/10.1109/IPDPS.2019.00050
8. In-tree storages. Available at https://kubernetes.io/docs/concepts/storage/volumes/#volume-types. Accessed 04.07.2021
9. CSI. Available at https://kubernetes-csi.github.io/docs/introduction.html/. Accessed 04.07.2021
10. CSI components. Available at https://kubernetes-csi.github.io/docs/developing.html. Accessed 04.07.2021
11. Bare-metal CSI. Available at https://github.com/dell/csi-baremetal. Accessed 04.07.2021
12. CSI driver lvm. Available at https://github.com/metal-stack/csi-driver-lvm. Accessed 04.07.2021
13. Minio direct CSI. Available at https://github.com/minio/direct-csi. Accessed 04.07.2021
14. CSI specification. Available at https://github.com/container-storage-interface/spec/blob/master/spec.md. Accessed 04.07.2021
15. Topolvm. Available at https://github.com/topolvm/topolvm. Accessed 04.07.2021
16. Raw block volumes. Available at https://kubernetes-csi.github.io/docs/raw-block.html. Accessed 04.07.2021
17. Scheduler extender. Available at https://kubernetes.io/docs/concepts/extend-kubernetes/#scheduler-extensions. Accessed 04.07.2021
18. Goethals T, DeTurck F, Volckaert B. Extending Kubernetes clusters to low-resource edge devices using virtual Kubelets, IEEE Trans Cloud Comput. https://doi.org/10.1109/TCC.2020.3033807

Accuracy of Potentiometric Methods for Measuring Ion Activity in Solutions

O. M. Vasilevskyi, V. M. Sevastianov, K. V. Ovchynnykov, V. M. Didych, and S. A. Burlaka

Abstract Structural realizations of the digital ion selective transducers, constructed on different rutotom principles are proposed: ADC of the successive approximation, time-pulse conversion and voltage to frequency conversion. Corresponding conversion equations are obtained, their static characteristics are constructed, measurements errors, emerging as a result of using one or another construction principles are investigated. As a result of the research, it was found that in order to improve the measurement accuracy, it is advisable to introduce an additional measuring temperature channel. To ensure high accuracy of ion activity measurement in the lower measurement range of 0.3 pX and to take into account the temperature deviation by 1 degree Celsius, it is necessary to construct a temperature measuring channel with a relative error of 0.05%.

Keywords Ion selective electrodes · Static characteristics · Errors · Measuring

1 Principles of the Digital Ion Selective Transducer Realization and Their Mathematical Models

1.1 Ion Selective Transducer Built on the Principle of ADC of Sequential Approximation

Realization of a digital potentiometric ion selective transducer for measuring the activity of substance ions can be performed on the base of ADC of the successive

O. M. Vasilevskyi (✉) · V. M. Sevastianov · K. V. Ovchynnykov
Vinnytsya National Technical University, 95 Khmelnitskoye Shose, Vinnitsya 21021, Ukraine
e-mail: o.vasilevskyi@gmail.com

V. M. Didych
National Pirogov Memorial Medical University, Vinnytsia 21000, Ukraine

S. A. Burlaka
Vinnytsia National Agrarian University, 3, Solnyschna St, Vinnytsia 21008, Ukraine

X.-S. Yang et al. (eds.), *Proceedings of Seventh International Congress on Information and Communication Technology*, Lecture Notes in Networks and Systems 447,
https://doi.org/10.1007/978-981-19-1607-6_16

approximation in its structure. Structural diagram of such transducer is presented in Fig. 1.

As it is seen from Fig. 1 digital ion selective transducer comprises: ion selective electrode pX, reference electrode pX_C, scale converter (SC) that performs function of non-inverted amplification, analog-to-digital converter (ADC), central processing unit (CPU), reprogrammed read only memory (EEPROM), controller for data transfer via the serial port RS232 to the computer (PC) and liquid crystal display (LCD). Signal from the output of ion selective electrode pX is amplified by the scale converter to level of reference value, set by the reference electrode, where ADC functions and further it passes to (HI) input of ADC. ADC transforms the voltage value into binary code.

Model transformation equation of the suggested digital transducer of ions activity, built on the principle of successive approximation, using analog-to-digital converter, has the form:

$$N_{\mathrm{ADC}} = \left(U_0' - \alpha(273.15 + t)n_a^{-1} pX_i\right)k(U_{\mathrm{ref}})^{-1}2^m, \quad (1)$$

where U_{ref}—is the value of the reference voltage of ADC, set by the reference electrode; *m*—is the ADC bit rate; *k*—is amplification factor of the scale converter; α—is temperature coefficient of steepness S that equals $198.4 \times 10^{-3}/°C$; *t*—is the temperature of the environment being analyzed (°*C*); U_0'—graduation voltage, which is determined by the selection of the reference point; pX_i—concentration of ions; n_a—is the charge of ion [1–4].

Characteristic of the conversion equation of the digital ion selective transducer of ions activity, built on the principle of the successive approximation is shown in Fig. 2.

It is seen from the obtained characteristics of the dependence of ADC code change on the activity of ions that the function of the transformation of the digital potentiometric ion selective transducer of ions activity is linear.

Taking into account the conversion function (1) mathematical model of the quantization error of the digital potentiometric transducer of ions activity is described by the expression:

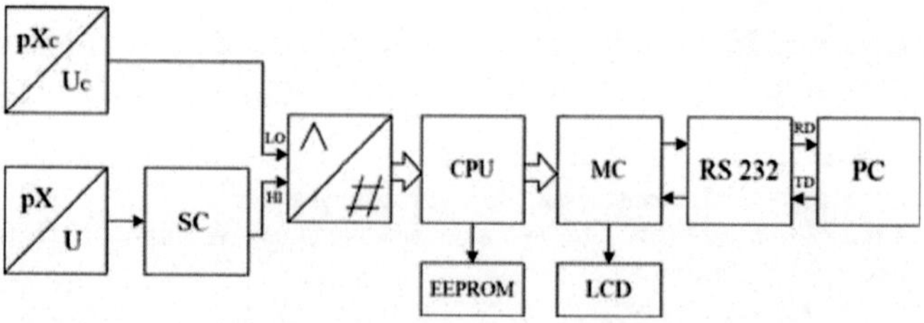

Fig. 1 Structural diagram digital potentiometric transducer of ions activity, built on the principle of ADC of serial approximation

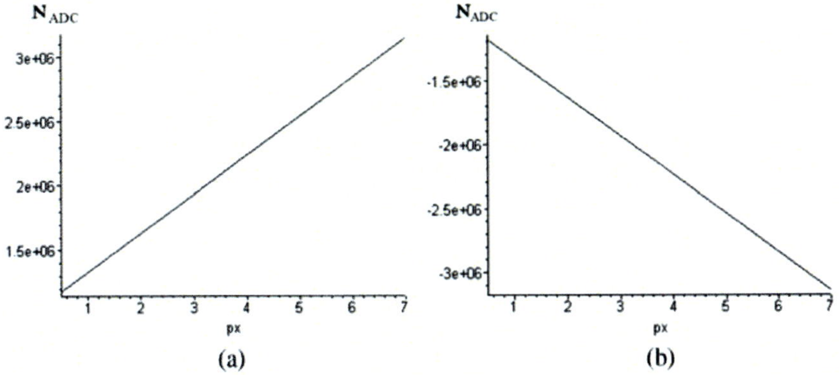

Fig. 2 Characteristics of ions activity change while realization of the digital transducer, built according to the principle of the analog-to-digital conversion of the successive approximation: **a** while measuring of negatively changed ions; **b** while measuring of positively changed ions

$$\delta_{\text{ADC}} = U_{\text{ref}}\left(U_0' - \alpha(273.15 + t)n_a^{-1} pX_i\right)^{-1}\left[k2^m\right]^{-1} 100\%. \quad (2)$$

Analyzing the obtained mathematical model of the error (4) it is seen that it decreases while measuring greater values of pX ions activity and its change characteristic is nonlinear (Fig. 3).

As it is seen from the obtained characteristic of the digital transducer of ions activity error change (Fig. 3) the methodical component of the error does not exceed $8.6 \times 10^{-5}\%$ and is of nonlinear character. Ways of the methodical error decrease is the increase of the number of bits, but this will lead to the increase of digital transducer cost.

Determination of the amplitude changes of the analytical signal of ion selective electrode by means of ADC causes a number of problems, dealing with the necessity of the simultaneous provision of high capacity, fast acting, accuracy and cost. That

Fig. 3 Characteristic of the error change of the ions activity transducer, based on the principle of analog-to-digital conversion of the successive approximation

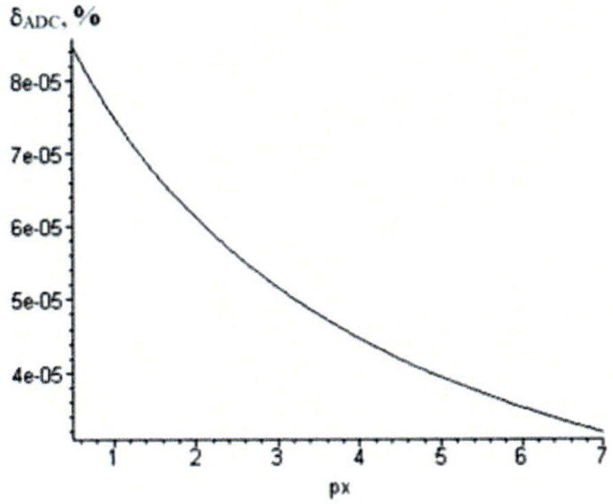

is why, the second variant of the circuit realization of digital ion selective transducer of ions activity, based on the principle of time-pulse conversion is proposed.

1.2 Ion Selective Transducer, Built on the Principle of Time-Pulse Conversion

For measuring the instantaneous voltage values from the outputs of ion selective electrodes instead of analog–digital conversion unit the principle of time-pulse conversion, based on the sawtooth generator and comparators could be used. Also in order to improve the accuracy the additional temperature measuring channel can be added [1–4]. Structural diagram of the digital ion selective transducer of ions activity, built on the principle of time-pulse conversion is shown in Fig. 4. In such transducer of ions activity the measured voltage is converted in time interval T_x with further quantization by the pulses of the reference frequency f_0 of the quartz-crystal resonator of the microcontroller. Temperature measuring channel is realized according the similar principle (time-pulse conversion), as the primary converter thermoresistive converter will be used, and non-inverting operation amplifier will be used as the scale converter.

Basic element of the structural realization of time-pulse conversion method is comparison device (CD), realized on two comparators, sawtooth generator G_U and RS-trigger T. Quantization of the time interval by the pulses of the reference

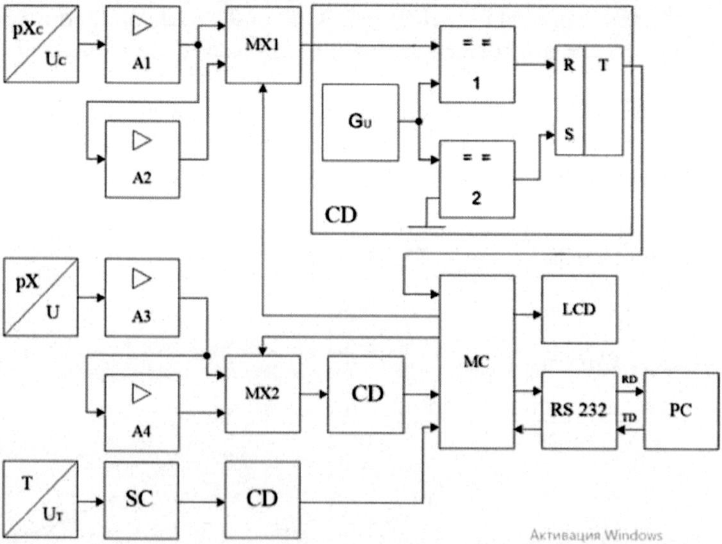

Fig. 4 Structural diagram of digital transducer of ions activity, built on the principle of time-pulse conversion

frequency f_0 is performed in the microcontroller by means of the built-in analog comparator and coincidence circuit.

Digital transducer of ions activity, built on the principle of time-pulse conversion (Fig. 4) comprises: ion selective electrode pX, reference electrode pX_C, operational amplifiers (A1–A4), multiplexers (MX1, MX2), comparison devices (CD), micro-controller (MC) and liquid crystal display (LCD). Comparison devices consist of the sawtooth generator, two comparators and RS-trigger. Additional temperature measuring channel comprises thermoresistive converter, scale converter (SC) and CD.

Signals from the outputs of the reference electrode pX_C, ions selective electrode and thermoresister reach the operational amplifiers (OA), it should be noted that the OA in the measuring channel of ion activity are provided in two variants—for the measurement of both positive and negative voltages from the outputs of ion selective electrodes. This is connected with the fact that in the process of measuring negatively charged ions the voltage at the output of the electrodes will be positive and in the process of measuring positively charged ions the voltage is negative. Process of measuring positively and negatively charged ions is controlled by microcontroller using multiplexors. After the amplification, voltage signals enter the comparison device (CD), where they are converted into the time interval T_x. Further by means of the built-in analog comparators, the allocated time interval is filled with pulses of the reference frequency f_0. Number of pulses, which entered in the allocated by means of the comparison device (CD) time interval is determined by the expression:

$$N_{U/T} = k\left(U_0' - \alpha(273, 15 + t)n_a^{-1}pX_i\right)Kf_0 \tag{3}$$

where k—is the amplification factor of the operational amplifier (OA); K—is proportionality coefficient, which depends on the steepness of linearly varying voltage of the generator G_U; f_0—are pulses of the reference frequency of the quartz-crystal resonator of the microcontroller.

Static characteristics of the digital transducer of ions activity, built on the principle of time-pulse conversion are presented in Fig. 5.

Error of the measuring channel of ions activity, taking into account model conversion Eq. (3) is described by the expression:

$$\delta_{U/T} = \left[k\left(U_0' - \alpha(273.15 + t)n_a^{-1}pX_i\right)Kf_0\right]^{-1}100\%. \tag{4}$$

Analysis of the obtained expression of the error of the measuring channel of ions activity, built on the principle of time-pulse conversation shows that the error decreases with the increase of pX ions activity and its changing characteristic is nonlinear (Fig. 6).

Possible ways of the quantization error decrease (4) is the increase of the reference frequency f_0 value, proportionality coefficient K and amplification factor k.

As it is seen from the obtained characteristic of the error change (Fig. 6) its maximum value does not exceed $2.45 \times 10^{-5}\%$ and is 3.5 times less than the error of

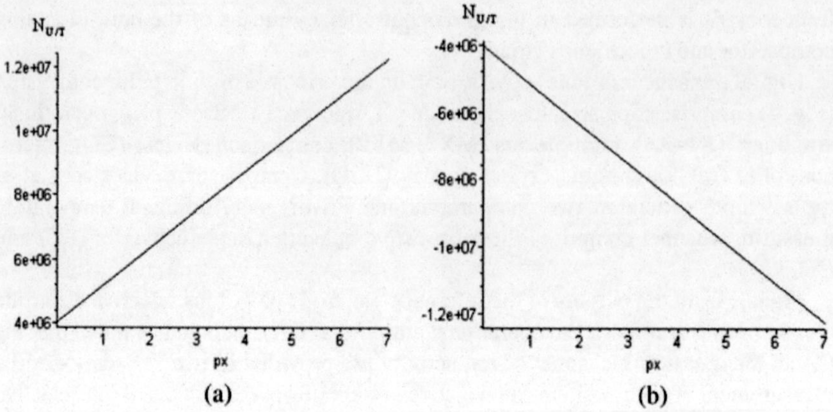

Fig. 5 Static characteristics of digital transducer of ions activity, built on the principle of time-pulse conversion: **a** while measuring negatively charged ions; **b** while measuring positively charged ions

Fig. 6 Characteristic of the error variation of the ions activity measuring channel, built on the principle of time-pulse conversion

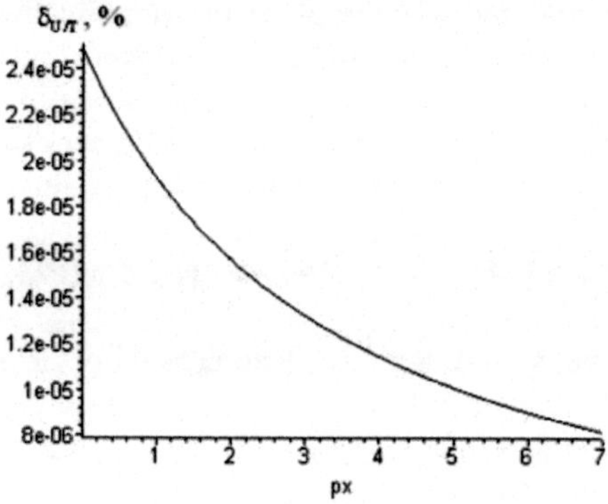

the digital transducer, built on the base of ADC of the successive approximation [1, 2]. The drawbacks of this principle of construction are low noise immunity, caused by the nonlinearity of the varying voltage of the generator G_u and instability of the comparator response level. That is why, we will investigate the third variant of the digital transducer of ions activity, based on the principle of voltage to frequency conversion.

1.3 Ion Selective Transducer, Built on the Principle of Voltage into Frequency Conversion

For the comparison with the previous realizations of the digital transducers of the ions activity the third variant of the digital potentiometric transducer, built on the

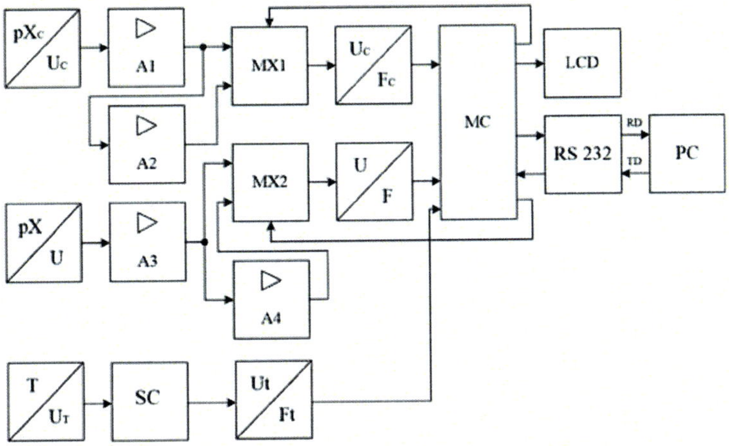

Fig. 7 Structural diagram of the digital transducer of ions activity, built on the base of voltage into frequency convertors

principle of voltage to frequency conversion is suggested. Structural diagram of such digital transducer is shown in Fig. 7.

The device comprises: ion selective converter (pX/U), reference electrode (pX_C/U_C); two operation amplifiers in each of measuring channels (MC) of ions activity (A1–A4) to provide measuring of both positive and negative values of pX; multiplexors (MX1 and MX2); voltage into frequency converters (VFC) (U_C/F_C and U/F) for the conversion the potentials of the reference electrode U_C/F_C and ion selective electrode U/F into frequency; microcontroller (MC); liquid crystal display (LCD); voltage levels converter (RS232) for data transfer to the computer (PC). Also temperature measuring channel (MC), consisting of the thermoresistive converter, scale converter (SC) and voltage into frequency converter (VFC). Conversion equation of the suggested digital transducer of ions activity, built on the principle of U/F conversion has the form:

$$N_{\mathrm{U/F}} = U_{\max} f_0 \tau \left[\left(U_0' - \alpha(273,15 + t) n_a^{-1} pX_i \right) k \right]^{-1} \tag{5}$$

where $U_{\max}$—is the value of the reference voltage of UFC (10 V); $\tau = RS$—constant of UFC time, used for setting the full-scale output frequency of the quartz-crystal resonator of the microcontroller ($R = 1$ kOhm, $C = 47$ mF); f_0—is the frequency of the quartz-crystal resonator of the microcontroller (20 MHz); k—is amplification factor of OA [2].

Representations of the conversion Eq. (5) of the digital transducer of ions activity, built on the principle of voltage into frequency conversion, are shown in Fig. 8.

From the obtained characteristics of the number of pulses change dependence on the ions activity it is seen that the conversion function of the given transducer is

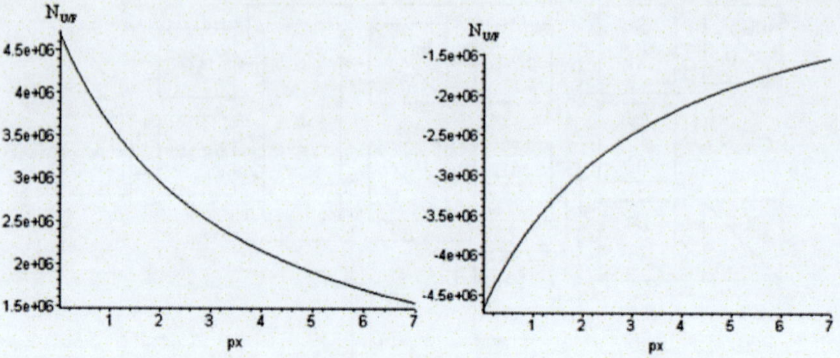

(a) (b)

Fig. 8 Static characteristics of the measuring channel of ions activity, built on the principle of voltage into frequency conversion: **a** while measuring negatively charged ions; **b** while measuring positively charged ions

not linear but nonlinearity of VFC in the wide range of frequencies change does not exceed $2 \times 10^{-3}\%$.

Error of the digital transducer of ions activity, built on the base of voltage into frequency converter with the account of the conversion Eq. (5) is described by the expression:

$$\delta_{\mathrm{U/F}} = k\big(U_0' - \alpha(273.15 + t)n_a^{-1} pX_i\big)[U_{\max}\tau f_0]^{-1}100\%. \quad (6)$$

Characteristic of the transducer error change (6) is shown in Fig. 9.

Analysis of the obtained equation of the relative error of the ions activity transducer, built on the principle of voltage into frequency conversion shows that it increases with the increase of the measuring range of ions activity pX, and its change characteristic is linear (Fig. 9). Linearing of the error characteristic of the digital

Fig. 9 Characteristic of the error variation of ions activity measuring channel, built on the principle of voltage into frequency conversion

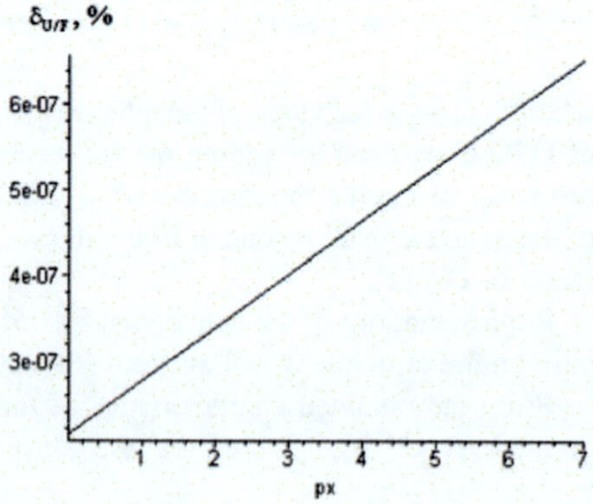

transducer of ions activity enables to introduce easily the corrections if necessary. As it is seen from Fig. 9, maximum error of the digital transducer does not exceed $6.5 \times 10^{-7}\%$ which is 37 times less than the error of the digital transducer, built on the principle of time—pulse conversion and 130 times smaller than the maximal error of the digital transducer, built on the principle of analog-to-digital conversion of the successive approximation. Possible ways of decreasing the relative error of the digital transducer (6) is the increase of the reference frequency value f_0 and time constant τ.

All the characteristics of the errors changes are built in the range of ions activity change from 0.3 to 7 pX at the temperature of 20 °C. As it is seen from the obtained characteristics of the relative errors change (Figs. 3, 6 and 9) the smallest value of the error $7.7 \times 10^{-7}\%$ has the third variant of the realization of the digital ion selective transducer of ions activity, based on the principle of voltage into frequency conversion. The only drawback of such realization is nonlinearity of the static characteristic. However, the range of binary code change is great and this nonlinearly is of minor importance for measuring ions activity.

2 Conclusions

Thus, on the base of the above-mentioned, the conclusion can be made that the best variant of realization of the unified system of the automated control of humus constituents in the soil with increased methodical component of the control reliability will be the system, built on the base of the improved method of the ionometry, using in the structure of measuring channel (MC) method and means of voltage into frequency conversion. Also in the process of construction of the system of the automated control the combination of the method of voltage into frequency conversion with the method of time-pulse conversion is possible, for instance, usage of one of the methods in temperature MC, and other method—in MC of ions activity or vice versa.

References

1. Vasilevskyi O, Didych V et al (2018) Method of evaluating the level of confidence based on metrological risks for determining the coverage factor in the concept of uncertainty. In: Proceedings Volume 10808
2. Vasilevskyi OM, Kulakov PI et al (2017) Vibration diagnostic system for evaluation of state interconnected electrical motors mechanical parameters. In: Proceedings of SPIE 104456C
3. Fuchigami T (eds) (2014) Fundamentals and applications of organic electrochemistry
4. Flow Analysis with Spectrophotometric and Luminometric Detection (2012)

Evaluating Effect of Microsoft HoloLens on Extraneous Cognitive Load During Simulated Cervical Lateral Mass Screw Placement

Dmitriy Babichenko, Edward G. Andrews, Stephen P. Canton, Eliza Beth Littleton, Ravi Patel, Dukens Labaze, and Andrew Mills

Abstract The use of augmented reality (AR) is widely accepted as a feasible training, planning, and prototyping tool. Unlike virtual reality (VR), which implies a complete immersion in a virtual world, AR adds digital elements to a live view by using a headset or camera on a smartphone. The ability to project digital elements into the physical world, combined with the Federal Food and Drug Administration (FDA) approval to use the Microsoft HoloLens in surgical procedures, presents a unique opportunity to explore and develop novel neurosurgical and orthopedic surgery training applications of AR, specifically in spine surgery. The potential of AR in spine surgery training lies in its ability to project CT-generated 3D models of the simulated patient's bony anatomy with overlaid pre-planned screw trajectories, thus allowing learners to practice with real-time guidance. As AR technologies become more mature, numerous research studies have identified AR's potential detriments to learning, including distraction and increased extraneous cognitive load. In this paper, we present our work on evaluating the effect of the presence of a Microsoft HoloLens 1 AR headset on extraneous cognitive load and on task performance during a simulated surgical procedure. A matched crossover trial design was used in which a combined group of 22 neurosurgery and orthopedic surgery residents, ranging in their training from the second postgraduate year (PGY-2) to chief resident (PGY-7 for neurosurgery and PGY-5 for orthopedic surgery, respectively. Participants were asked to place cervical lateral mass screws in a standardized, 3D-printed cervical spine with and without the Microsoft HoloLens 1 headset worn. Lateral mass screws were placed bilaterally at C4 to C6, with six cervical lateral mass screws placed by each participant in each trial, totaling 12 total screws placed. Overall time to drill six pilot holes, time for placement of each individual screw, pilot hole proximity to a predetermined entry point as defined by the Magerl method, and the presence of medial/lateral breaches were assessed and used as surrogate measures of mental

D. Babichenko (✉) · S. P. Canton · E. B. Littleton · R. Patel · A. Mills
University of Pittsburgh, Pittsburgh, PA 15260, USA
e-mail: dmb72@pitt.edu

E. G. Andrews · D. Labaze
University of Pittsburgh Medical Center (UPMC), Pittsburgh, PA 15260, USA

X.-S. Yang et al. (eds.), *Proceedings of Seventh International Congress on Information and Communication Technology*, Lecture Notes in Networks and Systems 447,
https://doi.org/10.1007/978-981-19-1607-6_17

taxation. The SURG-TLX questionnaire, a validated measure of extraneous cognitive load, was also used to compare cognitive strain of the task with and without the HoloLens 1.

Keywords Augmented reality · Mixed reality · Microsoft HoloLens · Surgical training · Intraoperative navigation · Extraneous cognitive load

1 Introduction

Clinical procedure training is a critical component of surgical education. The expertise required to demonstrate reproducible superior skills at specific surgical task comes from performing deliberate practice where supervision, guidance, and feedback from an expert are the necessary components of learning and improvement. Traditionally, surgical resident training has been approached using the apprenticeship model, which remains the gold standard today [1]. In this traditional model, a surgical resident spends five to seven years encountering a wide range of clinical conditions and gradually developing the necessary clinical judgment and operative skill under the supervision of expert surgeons. However, changes in workplace practices and patient privacy standards have led to a decline in training opportunities for novice clinicians [2]. Proliferation of minimally invasive surgical techniques has reduced the number of inpatient encounters [3]. Moreover, the potential risks to surgical patients associated with "learning on the job" present a number of ethical concerns and are becoming increasingly unacceptable [4].

Over the past decade, a large number of task trainers and simulators have been developed to allow surgical trainees to practice specific procedures in safe simulated environments. Recent improvements in immersive media technologies, such as virtual (VR) and augmented reality (AR), have created new opportunities to develop more even more realistic surgical task trainers, ones that are more portable, less expensive, and are capable of providing real-time anatomical navigation guidance and feedback.

Before we continue, it is important to note the key differences between VR and AR. VR headsets completely obscure the user's vision of the physical world while the displays within the headset project a virtual (computer-generated) environment. Most often, users interact with the virtual environment through the use of handheld controllers, or, in several advanced implementations, through the use of hand gestures or facial expressions [5–7]. AR headsets rely on either semitransparent heads-up displays (HUD) or external cameras to allow the users to see the physical world while overlaying and projecting digital objects onto the physical environment.

The current leading candidates for AR systems to be applied in the surgical training include Magic Leap One and Microsoft HoloLens (the Microsoft HoloLens was recently approved by the FDA for intraoperative use [8]). As these technologies are becoming less expensive and more accessible, they are finding more and more applications in surgical training, in preoperative planning, and in specific surgical procedures such as dental implant surgery, pedicle screw placement, and neuroendoscopy [9–11].

As AR technologies become more mature, numerous research studies have identified AR's potential detriments to learning, including distraction and increased extraneous cognitive load [12–15]. Before developing AR-based surgical training systems, it is important to explore the possible shortcomings of these technologies. The mere presence of an AR headset could potentially be considered a hindrance to surgical training that could potentially distract surgical resident from the task at hand due to the limited capacity of working memory and taxation on cognition. In their 1994 and 1998 work, Sweller et al. proposed the cognitive load theory and suggested that there were three different subtypes of cognitive load: intrinsic cognitive load (ICL), germane cognitive load (GCL), and extraneous cognitive load (ECL) [16, 17]. ECL, which is relevant to this study, refers to cognitive activities that are not directly related to the task at hand, but ultimately distract from this task. Even though Sweller and Mayer described the implications of ECL in relation to learning technologies, many have adapted this theory to cognitive analysis of physical task performance, such as surgical procedures [18–21].

The overarching aim of this research is to determine whether the mere presence of a Microsoft HoloLens AR headset would prohibitively affect ECL during a simulated surgical procedure involving cervical lateral mass screw placement as measured by (1) self-reported extraneous cognitive load and (2) surrogate measures of ECL represented by specific objective components of the surgeon's performance.

2 Materials and Methods

2.1 Participants

This study was approved by UPMC institutional internal review board. A total of 22 subjects were recruited for this study. All subjects were neurosurgery or orthopedic surgery residents ranging from postgraduate year two (PGY-2) to chiefs (postgraduate year five (PGY-5) or seven (PGY-7) for orthopedic surgery and neurosurgery, respectively).

2.2 Model Design

Existing studies of AR applications in neurosurgery and orthopedic surgery have been performed either on cadaveric spines or on live patients [10, 22]. Anatomical variances in cadaveric and live patient spines introduce additional variables into the experiments that could potentially affect the validity of research. In 2019, a study by Bohl et al. assessed the use of 3D-printed spines in surgical training and concluded that it is possible to 3D print a spine that would provide the surgical and mechanical feel of the bony structures on pedicle cannulation and pedicle screw placement like that of cadaveric or live patient spines [23]. Thus, this study was conducted with 3D printed spines to ensure that spine segments used by all study subjects were identical.

To create 3D-printed spine segments, a de-identified cervical CT scan was converted to a 3D model using a combination of an open-source Python script and 3DSlicer [24] (Figs. 1 and 2). From this model, 56 identical cervical spine segments spanning C2-T1 were 3D-printed using a MakerGear 2 3D printer with Overture 1.75 mm polylactic acid (PLA) thermoplastic aliphatic polyester filament.

A single trial-and-error assessment was conducted with a PGY-4 neurosurgery resident and a PGY-2 orthopedic surgery resident to determine the amount of infill

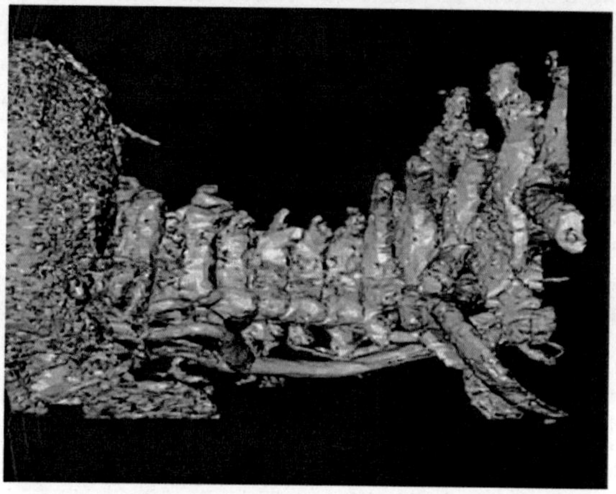

Fig. 1 3D model converted from CT DCOM stacks using an open-source Python script

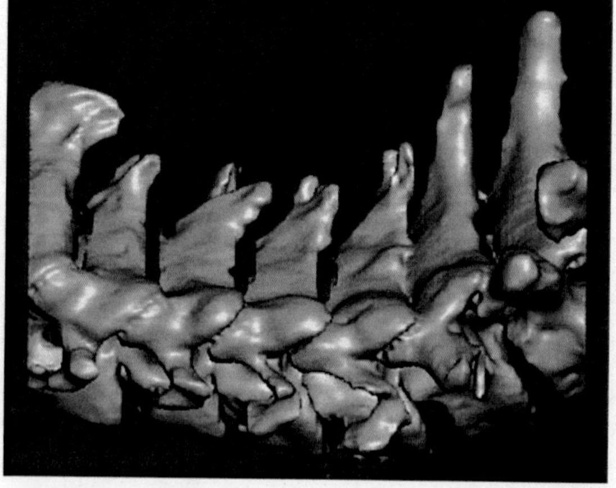

Fig. 2 Cleaned 3D model of a cervical spine section prepared for printing

required for structural integrity of the print and to ensure that the drilling resistance and density of the model closely resemble that of human bone. Infill is a set of repetitive structures automatically created during the 3D printing process to take up space inside an otherwise empty 3D print. Multiple copies using 40, 50, 60, and 70% infill were fabricated, and the residents were asked to drill a single pilot hole and place a single lateral mass screw. The surgeons then reported how well the density and the resistance of each version of the print resembled that of human bone. Both surgeons reported that 70% infill resulted in the closest approximation of human bone density, and thus all spines were printed at this infill percentage.

The 3D-printed cervical spines were then placed in a custom-made mannequin simulator that allowed for easy exchange of the 3D-printed spines in between trials (Figs. 3 and 4). The mannequin was constructed from the upper torso of a CPR task trainer with an attached section of 4-inch diameter PVC pipe fitted with metal brackets to guarantee the same positioning of spine segments during all trials. Artificial skin from the original CPR task trainer was refitted to cover the cervical attachment in such a way that the lateral edges of the 3D-printed spine were partially obscured to best resemble a typical posterior exposure of the subaxial cervical vertebrae from C2 to T1. Finally, surgical drapes were used to cover the mannequin in an effort to reduce distracting visual stimuli by isolating the surgical field.

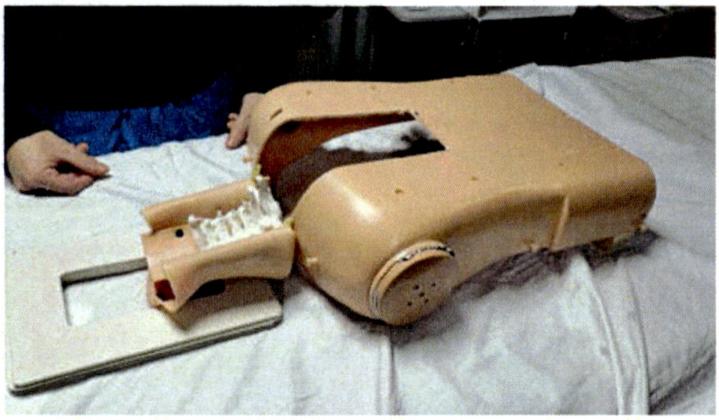

Fig. 3 A mannequin simulator with interchangeable cervical spine segments

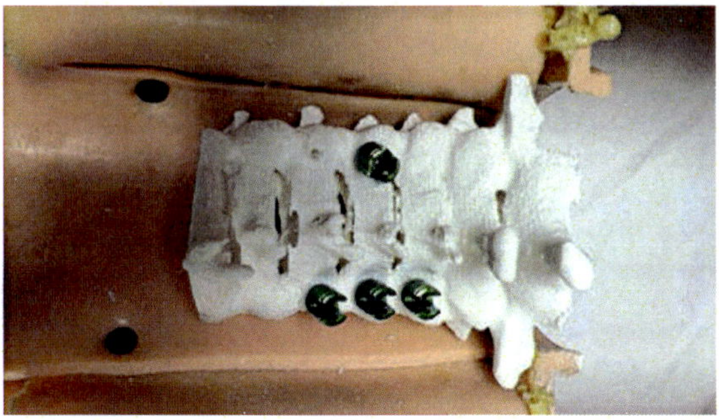

Fig. 4 Interchangeable cervical spine segment with four lateral mass screws already placed

2.3 Study Design

A case-crossover (within subjects) study design was used to evaluate the impact of the Microsoft HoloLens 1 on ECL. To reduce priming bias, subjects were randomly selected into two groups. Subjects from Group A first completed the tasks while wearing the Microsoft HoloLens 1 and then repeated the tasks with the headset off. Subjects from Group B first completed the tasks without wearing the Microsoft HoloLens 1 and then repeated the tasks with the headset on. All subjects completed the first five questions of the SURG-TLX questionnaire twice, once after each trial [21]. Question six was omitted from analysis because several different rooms were used during the study.

For each study session, the simulator was secured in a prone position, affixed to a table using compression straps and draped with surgical drapes, leaving C2-C7 vertebrae readily visible. Researchers reviewed the Magerl technique for placement of cervical lateral mass screws with all participants, regardless of training level [25]. All subjects were then asked to drill practice pilot holes on a spare 3D-printed cervical spine to familiarize themselves with the material's response to the drill bit. They were subsequently asked to place six cervical screws into the lateral masses of C4-C6 vertebra (Table 1). All subjects used the same Stryker drill (model CD4) with a 0.25-inch bit for drilling pilot holes and Depuy-Synthes OCT hextip minipolyaxial screwdriver to place six Depuy-Synthes Mountaineer minipolyaxial cervical lateral mass screws.

The subjects were asked to drill all six pilot holes before placing any screws. Once the pilot holes were drilled, the subjects placed all six cervical screws following the instructions in Table 1. Two researchers used timer applications on smartphones to record the time duration for each subject to drill the pilot holes and then time duration to place each screw, resulting in seven time measurements per subject per trial. The time for screw placement was defined as the time from seating the screw tip in the pilot hole to disengaging the screwdriver. A researcher always reloaded the screwdriver for the subject for efficiency, and this time was not excluded from the measurement.

Table 1 Cervical lateral mass screw placement tasks

Sequence	Task description
1	Identify key bony landmarks on the dorsal surface of the cervical spine to assist with screw trajectory planning
2	Use a drill to create six (6) pilot holes at the desired entry points on the C4, C5, and C6 lateral masses
3	Load the screw onto the screwdriver (pre-loaded)
4	Place the cervical screw
5	Repeat steps 2–4 until all six lateral mass screws are placed
6	Complete SURG-TLX questionnaire

The subjects completed the SURG-TLX questionnaire and repeated the entire procedure again in the crossover arm with or without the Microsoft HoloLens 1 headset, depending into which group they were randomized initially.

Researchers collected each spinal segment after each trial and marked it with the subject's unique identifier and with either an H or NH to signify the Microsoft HoloLens 1 was worn or not worn. Thus, each subject had two spines linked to their performance by the end of the study, one completed with, and one completed without the HoloLens on. Spines were analyzed for accuracy of initial pilot hole placement as well as for lateral mass breaches. To assess pilot hole accuracy, the dorsal surface of the lateral mass was subdivided into four even quadrants. Accuracy was determined by placement of the subject's pilot hole in the correct quadrant along with distance from Magerl's standard entry point 1 mm medial and 1 mm cephalad from the center of the lateral mass. To assess breaches, direct visualization of lateral mass wall violation was identified using a steel probe inserted in the screw tract; only medial/lateral breaches were recorded given technical constraints of the model. As a final measure to assess the effect of the HoloLens on extraneous cognitive load, the Surgical Task Load Index was computed based on participants' answers to the first five SURG-TLX questions on a Likert scale of 1 (low demand) to 20 (high demand).

2.4 Statistical Analysis

Stata/SE 15.1 (StataCorp, College Station, TX) was used for statistical analyses. The Wilcoxon signed rank test was used to compare the effect of condition (presence or absence of the Microsoft HoloLens headset) on (1) median frequency of lateral mass medial/lateral breaches, (2) median times to drill six holes, (3) median time to place individual screws, (4) median distance of trial entry point to Magerl entry point, (5) correct quadrant placement percentage, and (6) SURG-TLX scores.

3 Results

A total of 11 neurosurgery and 11 orthopedic residents participated in the study (Table 2). Overall, no significant difference between the two subject groups in the variables of interest was seen. Only the median drilling times and screw placement times were greater for the NH compared to the H intervention, but this difference was not statistically significant (Table 3). The median values for the SURG-TLX workload questionnaire were identical for all items, except for situational stress and distraction which were slightly greater for H intervention (Table 4), but again this was not statistically significant.

A total of 6 medial breaches (3 H, 3 NH) and 2 lateral breaches (0 H, 2 NH) were observed. Of note, two participants accounted for the 8 total breaches; one participant performed all 6 medial breaches, and another participant performed the 2

Table 2 Subjects' characteristics ($n = 22$)

PGY	% of Cohort	PGY	% of Cohort
PGY-2	9.1	PGY-5	22.7
PGY-3	9.1	PGY-6	9.1
PGY-4	40.9	PGY-7	9.1

Note PGY-6/7 were exclusively neurosurgery residents as a natural byproduct of Orthopedic Surgery/Neurosurgery program structuring

Table 3 Evaluation results of lateral mass screw placement outcome measures

	Intervention	Median	IQR	*p*-value
Drilling duration (s)	Headset	65.6	47.7–100.8	0.86
	No headset	77.7	53.2–105.7	
Screw placement duration (s)	Headset	21.5	15.3–26.4	0.39
	No headset	23.1	17.7–29.3	
Distance from target (mm)	Headset	1.0	0.5–1.6	0.12
	No headset	1.0	1.0–1.6	
Correct quadrant placement (%)	Headset	0.8	0.67–0.8	0.81
	No headset	0.8	0.5–0.8	

Table 4 Results of self-reported SURG-TLX extraneous cognitive load evaluation

	Intervention	Median	IQR	*p*-value
Mental demands	Headset	2	1.0–3.0	0.32
	No headset	2	1.0–3.0	
Physical demands	Headset	2	1.0–3.0	0.48
	No headset	2	1.0–3.0	
Temporal demands	Headset	2	1.0–3.0	0.71
	No headset	2	1.0–3.0	
Task complexity	Headset	2	1.0–3.0	0.71
	No headset	2	1.0–3.0	
Situational stress	Headset	2.5	1.0–3.0	0.37
	No headset	2	1.0–3.0	

lateral breaches. The participants were both PGY-4, with one neurosurgery resident and one orthopedic surgery resident. The data were subsequently stratified into a "junior" or "senior" category, defined as PGY-3 or below and PGY-4 and above for junior and senior, respectively. There was no statistically significant effect observed for any of the outcome measures when the groups were stratified by level-of-training (data available). Finally, all trial 1 s were compared to trial 2 s to see if there was a learning effect. As with the other comparisons, no statistically significant difference was seen.

4 Limitations

In discussing this work, it is important to acknowledge its limitations. We were not able to conduct this study in an actual operating room (OR); the differences in the space used for this study and an OR may have unexpected effects on the validity of this research. Furthermore, during surgical procedures surgeons wear facemasks, visors, lights, and other devices. In our study, the only device that the subjects wore was a Microsoft HoloLens MR headset. The 3D-printed spines used in the study could constitute another potential limitation. While the similarity between the 3D-printed spines and cadaveric spines based on feedback from two surgeons was assessed, evaluation by a larger group of surgeons may have produced different feedback/results regarding the 3D-printed spines' fidelity. This limitation presents further opportunity for research as the tool of 3D-printing offers significant customization and scale in reproducibility. The costs of this reproducibility would need to be addressed by comparing the products of 3D printing and other physical materials used in surgical education. Last, but not least, our study had 22 subjects. While similar studies have been conducted with smaller numbers of participants [10, 22], the lower statistical power of the study with this smaller subject group can falsely produce a lack of statistical significance.

5 Discussion and Future Work

In this paper, we have presented our work on exploring the effect of the mere presence of a Microsoft HoloLens AR headset on (1) self-reported extraneous cognitive load during cervical spine lateral mass screw placement and (2) surrogate measures of surgeons' performance as represented by time to place individual lateral mass screws, total time to drill pilot holes, gross accuracy of pilot hole placement, and absolute number of medial/lateral breaches. Overall, there was no statistically significant difference in outcomes in all measures when the trials with the HoloLens 1 were compared to the trials without the HoloLens 1. There was also no observed difference in results when the groups were further stratified based on level of seniority, which addressed the effect of technical prowess and familiarity on performance, or when trial 1 and trial 2 were compared, which addressed task learning effect. Furthermore, subjects did not report any difference in extraneous cognitive load as determined by the analysis of responses to the SURG-TLX questionnaire. These findings indicate that the presence of a Microsoft HoloLens 1 AR headset does not impact a surgeons' extraneous cognitive load during a standardized and familiar procedure such as cervical lateral mass screw placement. The importance of this

Fig. 5 A screenshot from a Unity3D implementation of a 3D spinal segment overlay with pre-planned screw placement trajectories

finding becomes apparent when looking at the future directions of AR research in surgical training and intraoperative use in spine surgery.

Given these findings, we began work on developing an AR-based surgical navigation tool that utilizes computed tomography (CT)-generated 3D models to project an augmented overlay onto a simulated surgical field, align the 3D model with the simulated patient's physical anatomy, and project pre-planned lateral mass screw trajectories (Fig. 5).

References

1. Castanelli DJ (2009) The rise of simulation in technical skills teaching and the implications for training novices in anaesthesia. Anaesth Intensive Care 37(6):903–910. https://doi.org/10.1177/0310057X0903700605
2. Rodriguez-Paz JM et al (2009) Beyond 'see one, do one, teach one': toward a different training paradigm. BMJ Qual Saf 18(1):63–68. https://doi.org/10.1136/qshc.2007.023903
3. Kneebone R (2003) Simulation in surgical training: educational issues and practical implications. Med Educ 37(3):267–277
4. McGaghie WC, Issenberg SB, Petrusa ER, Scalese RJ (2010) A critical review of simulation-based medical education research. Med Educ 44(1):50–63
5. Lu W, Tong Z, Chu J (2016) Dynamic hand gesture recognition with leap motion controller. IEEE Signal Process Lett 23(9):1188–1192. https://doi.org/10.1109/LSP.2016.2590470
6. Zhang F, Chu S, Pan P, Ji N, Xi L (2017) Double hand-gesture interaction for walk-through in VR environment. In: 2017 IEEE/ACIS 16th International Conference on Computer and Information Science (ICIS), pp 539–544. https://doi.org/10.1109/ICIS.2017.7960051
7. Hickson S, Dufour N, Sud A, Kwatra V, Essa I (2019) Eyemotion: classifying facial expressions in VR using eye-tracking cameras. In: 2019 IEEE Winter Conference on Applications of Computer Vision (WACV), pp 1626–1635. https://doi.org/10.1109/WACV.2019.00178
8. LinkedIn. FDA approves first HoloLens augmented reality system for surgicalse. https://cacm.acm.org/news/232744-fda-approves-first-hololens-augmented-reality-system-for-surgical-se/fulltext. Accessed 06 Mar 2020
9. Katić D et al (2015) A system for context-aware intraoperative augmented reality in dental implant surgery. Int J Comput Assist Radiol Surg 10(1):101–108
10. Elmi-Terander et al (2018) Feasibility and accuracy of thoracolumbar minimally invasive pedicle screw placement with augmented reality navigation technology. Spine 43(14):1018

11. Finger T, Schaumann A, Schulz M, Thomale U-W (2017) Augmented reality in intraventricular neuroendoscopy. Acta Neurochir (Wien) 159(6):1033–1041
12. Chu H-C (2014) Potential negative effects of mobile learning on students' learning achievement and cognitive load—a format assessment perspective. J Educ Technol Soc 17(1):332–344
13. Liu T-C, Lin Y-C, Tsai M-J, Paas F (2012) Split-attention and redundancy effects on mobile learning in physical environments. Comput Educ 58(1):172–180
14. McKnight RR, Pean CA, Buck JS, Hwang JS, Hsu JR, Pierrie SN (2020) Virtual reality and augmented reality—translating surgical training into surgical technique. Curr Rev Musculoskelet Med 1–12
15. Pfandler M, Lazarovici M, Stefan P, Wucherer P, Weigl M (2017) Virtual reality-based simulators for spine surgery: a systematic review. Spine J 17(9):1352–1363
16. Sweller J (2010) Element interactivity and intrinsic, extraneous, and Germane cognitive load. Educ Psychol Rev 22(2):123–138. https://doi.org/10.1007/s10648-010-9128-5
17. Sweller J (1994) Cognitive load theory, learning difficulty, and instructional design. Learn Instr 4(4):295–312. https://doi.org/10.1016/0959-4752(94)90003-5
18. Hoffman R (2005) Protocols for cognitive task analysis, p 109
19. Keller J, Leiden K, Small R, Goodman A, Hooey B (2003) Cognitive task analysis of commercial jet aircraft pilots during instrument approaches for baseline and synthetic vision displays. In: NASA aviation safety program conference on human performance modeling of approach and landing with augmented displays, p 15
20. Hart SG, Staveland LE (1988) Development of NASA-TLX (Task Load Index): results of empirical and theoretical research. In: Advances in psychology, vol 52. Elsevier, pp 139–183
21. Wilson MR, Poolton JM, Malhotra N, Ngo K, Bright E, Masters RSW (2011) Development and validation of a surgical workload measure: the surgery task load index (SURG-TLX). World J Surg 35(9):1961. https://doi.org/10.1007/s00268-011-1141-4
22. Elmi-Terander et al (2019) Pedicle screw placement using augmented reality surgical navigation with intraoperative 3D imaging: a first in-human prospective cohort study. Spine 44(7):517
23. Bohl MA et al (2019) The barrow biomimetic spine: face, content, and construct validity of a 3D-printed spine model for freehand and minimally invasive pedicle screw insertion. Glob Spine J. https://doi.org/10.1177/2192568218824080
24. Pieper S, Halle M, Kikinis R (2004) 3D Slicer. In: 2004 2nd IEEE international symposium on biomedical imaging: nano to macro (IEEE Cat No. 04EX821), vol 1, pp 632–635. https://doi.org/10.1109/ISBI.2004.1398617
25. Lateral mass screw insertion (Magerl technique), site name. https://surgeryreference.aofoundation.org/spine/trauma/occipitocervical/basic-technique/lateral-mass-screw-insertion-magerl-technique. Accessed 02 Nov 2020

Network Modeling—A Convenient Way to Study IP Networks

Ivan Nedyalkov and Georgi Georgiev

Abstract The present paper proposes the use of platforms for modeling of communication networks, in particular modeling of IP networks. The study of IP networks, by using modeled IP networks, has many advantages, which are described in the work. The study of IP networks through modeled ones is shown in practice, through the GNS3 platform. Different capabilities of the proposed platform are presented when using modeled IP networks.

Keywords GNS3 · IP networks · Monitoring tools · Modeled IP networks · Wireshark

1 Introduction

The IP networks are now constantly entering people's daily lives. Almost all household appliances have network (communication) functionality and the ability to connect to the Internet for remote control. Almost any device can now be connected to the Internet. In addition, telecommunications operators are constantly providing new services, higher speeds for Internet access, IP TV with hundreds of programs, the number of high-definition channels (HD channels) is constantly increasing, channels with 4 k resolution are added, and interactive IP television is offered, increasing the number of streaming platforms for series and movies and many, many other services.

In order for these services to be provided to users, IP networks must be "prepared" for them. This is achieved through continuous research in the field of IP networks [1–5]. The research may be related to study of time delays, study of network loading when using different dynamic routing protocols, study of network convergence time, using different dynamic routing protocols, researches related to network monitoring, researches related to the creation of new working algorithms, and other.

I. Nedyalkov (✉) · G. Georgiev
South–West University "Neofit Rilski", 66 Ivan Mihailov str, Blagoevgrad 2700, Bulgaria
e-mail: i.nedqlkov@gmail.com

X.-S. Yang et al. (eds.), *Proceedings of Seventh International Congress on Information and Communication Technology*, Lecture Notes in Networks and Systems 447,
https://doi.org/10.1007/978-981-19-1607-6_18

The aim of the paper is to present the convenience of using platforms for modeling of IP networks in their research. Modeling platforms offer almost unlimited possibilities for modeling of IP networks of any size. The paper presents several different topologies of modeled IP networks and for each a brief analysis of the results will be made.

2 The Need for Modeling of IP Networks

In order to be able to perform the above researches, it is necessary to implement experimental IP networks with the appropriate equipment. The problem comes from the fact that such network equipment is expensive, and not every school or educational institution can afford it. Free access to working network equipment for research purposes is not possible—no one will allow reconfiguration of working equipment for research purposes. Therefore, programs for modeling of IP networks come to the rescue. The advantages of using platforms for creation of modeled IP networks are

- No costs for the purchase of network equipment for the creation of the experimental IP networks;
- Ability to continuously reconfigure the modeled network, which is impossible to happen in a working and operating IP network;
- Ability to design and test the IP network in a secure virtual environment;
- Work with disk images on real network devices. A modeled network built with such images will be almost 100% closer to a real one working with such real network devices;
- Ability to "transfer" an already built and working IP network in a virtual environment for testing purposes, without the need to perform these tests on a real, working network.

Of course, the use of a virtual environment for studying IP networks also has its drawbacks:

- The workstation on which the modeling of IP networks will be implemented must have serious computing capabilities. This also leads to the investment of a lot of financial resources in the purchase of the necessary hardware for the assembly of such a workstation;
- Some IP platforms for modeling are not free, and it is necessary to purchase the appropriate licenses to use certain platform functionality.

3 Results from IP Network Modeling

The proposed platform for modeling of IP networks is GNS3. GNS3 is a free and open platform for modeling of IP networks. GNS3 allows to model a small IP network consisting of only a few devices hosted on a personal computer or to model a large, corporate IP network consisting of multiple devices [6–12]. The platform allows network specialists to virtualize real hardware devices. Virtualization is accomplished through the use of software called Dynamips [13]. GNS3 supports many devices from many network manufacturers and providers.

GNS3 supports simultaneous device emulation and simulation [14]. GNS3 also offers additional features such as collaboration with real communication networks. The modeled network can be easily connected to another real, working IP network or be connected to the Internet. This allows studying the response of the network. In this way, it will be possible to detect in advance possible problems that can be eliminated before the physical implementation of the modeled network and its connection to other IP networks. The platform also allows working with tools for traffic monitoring in the modeled network. This will allow network engineers to monitor what is happening in the modeled network. This feature is very convenient and necessary. It can be used to check if the network is functioning as planned. Figure 1 shows a modeled IP network implemented in GNS3.

The purpose of the created model is to study whether the use of the MPLS technology improves the performance for this specific IP network. The modeled IP network of Fig. 1 consists of five high-end routers (R1–R5). These routers are disk image emulations of real routers operating systems. The modeled IP network of Fig. 1 consists of five high-end routers (R1–R5) and five switches (Switch 1 to Switch 5). The routers are disk image emulations of real routers operating systems and the switches are simple simulation models of switches created by the GNS3 developers. Four virtual machines used for hosts in the modeled network (VM–VM3). IP telephone exchange—asterisk and the module Router_Firewall, which is used to connect to real IP networks or the Internet. The studies were performed first without activated and then with activated MPLS. The dynamic routing protocol is OSPF. If this network is to be physically implemented, it will be necessary to invest large sums of

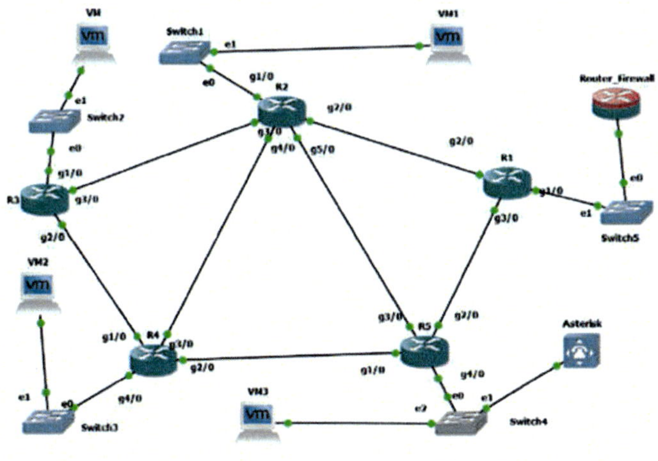

Fig. 1 Topology of the modeled IP network

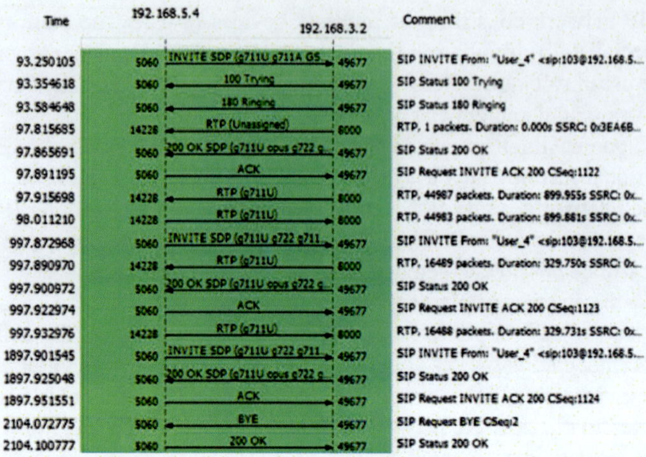

Fig. 2 Successful SIP handshake between asterisk and VM

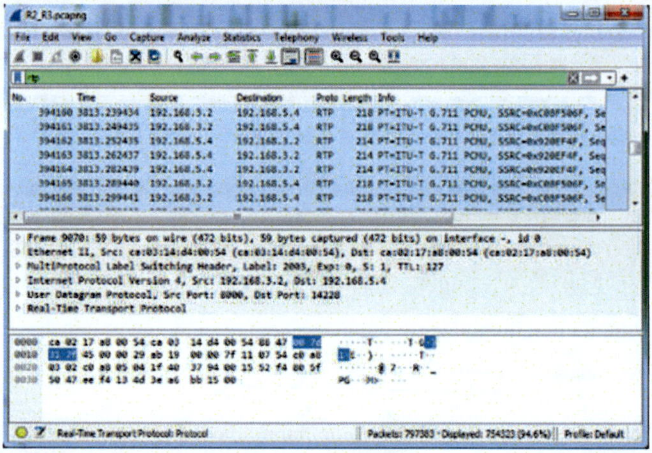

Fig. 3 RTP packets exchange between VM and asterisk

money only in the purchase of routers and switches. Another convenience is that the entire network, together with the virtual machines, is "located" in the platform, in the workstation where the network is modeled. An unlimited number of users (virtual machines) can be created. In the modeled network from Fig. 1, mainly VoIP traffic is exchanging, as each virtual machine is a subscriber to the asterisk. During the study, virtual machines build telephone conversations with each other. Wireshark's built-in IP telephony research functionality is used for voice flow parameter analysis.

Fig. 4 Summarized results for the R2–R3 link without MPLS

Forward		Reverse	
192.168.3.2:8000 → 192.168.5.4:15432		192.168.5.4:15432 → 192.168.3.2:8000	
SSRC	0xdb2c560e	**SSRC**	0x3098e788
Max Delta	220.03 ms @ 316813	**Max Delta**	100.01 ms @ 316808
Max Jitter	18.43 ms	**Max Jitter**	16.30 ms
Mean Jitter	8.04 ms	**Mean Jitter**	4.05 ms
Max Skew	-449.14 ms	**Max Skew**	-146.62 ms
RTP Packets	75599	**RTP Packets**	75619
Expected	75599	**Expected**	75619
Lost	0 (0.00 %)	**Lost**	0 (0.00 %)
Seq Errs	0	**Seq Errs**	0
Start at	3133.028178 s @ 316796	**Start at**	3133.121189 s @ 316800
Duration	1512.31 s	**Duration**	1512.45 s
Clock Drift	41 ms	**Clock Drift**	14 ms
Freq Drift	8000 Hz (0.00 %)	**Freq Drift**	8000 Hz (0.00 %)

Figure 2 shows the successful SIP handshake for the telephone connection between asterisk (192.168.5.4) and VM (192.168.3.2). Figure 3 shows the exchange of RTP packets between VM and asterisk.

Figure 4 presents the summarized results for the voice stream that passes through VM, R3, R2, R5, asterisk, and vice versa, without the use of the MPLS technology. Figure 5 shows the same values, but when using the MPLS technology. As can be seen from the values of the jitter parameter for both directions, its average value is well below the maximum allowable value of 30 ms [15, 16]. As it can be seen from the obtained results for such a small network and when exchanging only voice traffic, the use of the MPLS technology does not lead to any significant improvements in the voice flow parameters.

Figures 6 and 7 present the results for the instantaneous values of the jitter in both directions for each moment of the call duration. As can be seen from the obtained results, the use of the MPLS leads to a slight deterioration of the instantaneous values of the jitter in the reverse direction.

The following can be briefly summarized for the presented modeled network: The modeled network from Fig. 1 is working, and voice streams are successfully and smoothly exchanged in it. The use of the MPLS technology together with OSPF

Fig. 5 Summarized results for the R2–R3 link with MPLS

Forward		Reverse	
192.168.3.2:8000 → 192.168.5.4:18078		192.168.5.4:18078 → 192.168.3.2:8000	
SSRC	0xc0bf506f	**SSRC**	0x0920ef4f
Max Delta	100.01 ms @ 217810	**Max Delta**	220.03 ms @ 217801
Max Jitter	16.21 ms	**Max Jitter**	18.60 ms
Mean Jitter	8.08 ms	**Mean Jitter**	8.08 ms
Max Skew	-288.54 ms	**Max Skew**	-450.05 ms
RTP Packets	114302	**RTP Packets**	114291
Expected	114302	**Expected**	114291
Lost	0 (0.00 %)	**Lost**	0 (0.00 %)
Seq Errs	0	**Seq Errs**	0
Start at	2127.455914 s @ 217773	**Start at**	2127.676942 s @ 217787
Duration	2286.22 s	**Duration**	2286.06 s
Clock Drift	34 ms	**Clock Drift**	30 ms
Freq Drift	8000 Hz (0.00 %)	**Freq Drift**	8000 Hz (0.00 %)

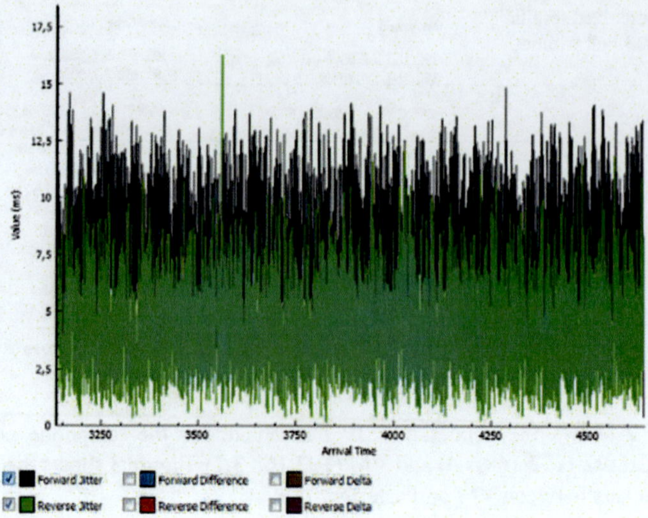

Fig. 6 Instantaneous values for the jitter in both directions without MPLS

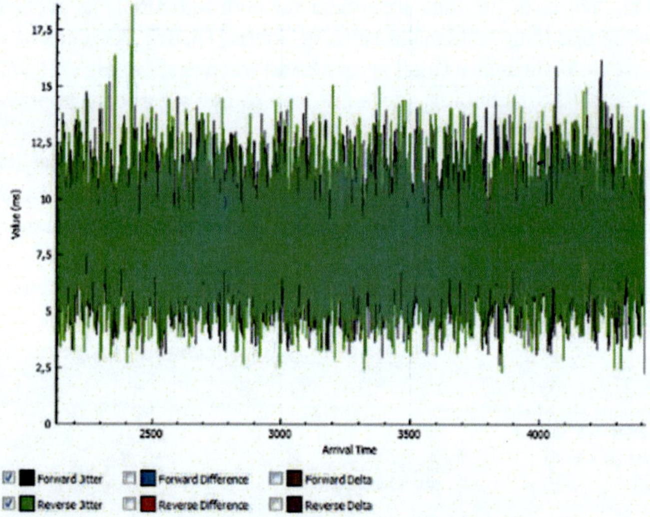

Fig. 7 Instantaneous values for the jitter in both directions with MPLS

does not lead to any significant improvements in the modeled network, made of only several routers. GNS3 offers the ability to integrate with various monitoring tools for IP networks, which allows great convenience in the study of modeled IP networks.

Figure 8 presents a topology of another modeled network with two users—VM1 and VM2, four routers (R1–R4), the same as those used in the network of Fig. 1,

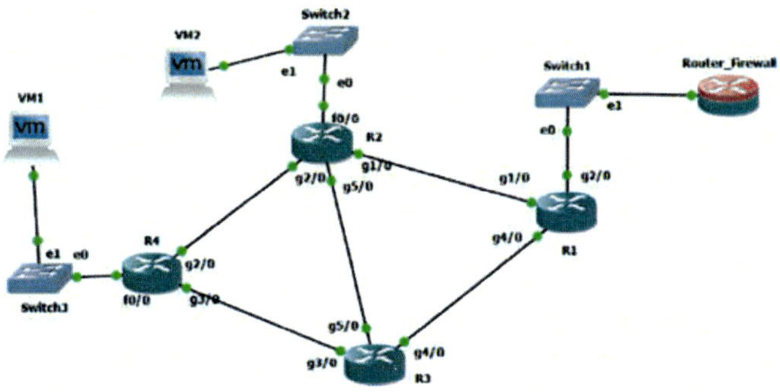

Fig. 8 Topology of the modeled network

three switches and again the module for Internet access or connection to other real networks. OSPF is used again. The new in this modeled network is the connection to the Internet. As such, another functionality and advantage are shown when using platforms for modeling of IP networks—connecting and working with real IP networks.

Figure 9 presents a Wireshark result that shows stream of HTTP packets to VM1. This stream is a download of a very large file—a few gigabytes. Because the downloaded file is larger than 1400 bytes, a huge number of packets will need to be sent from the server to the client to produce the file downloaded from the server.

Figure 10 shows the generated traffic that is monitored at the Router_Firewall interface. The result is obtained by using the Colasoft Capsa Free Network Analyzer. For greater accuracy, the sample interval is set to 1 s. As it can be seen, the average value is about 40 Kbytes/s. This is only HTTP traffic, and it is heterogeneous.

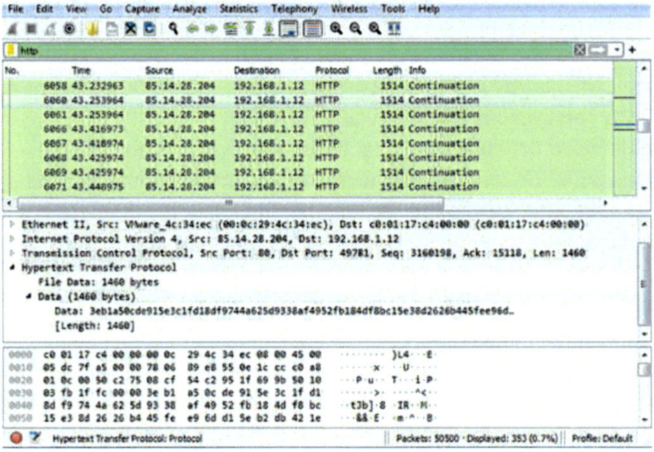

Fig. 9 HTTP continuation file transfer

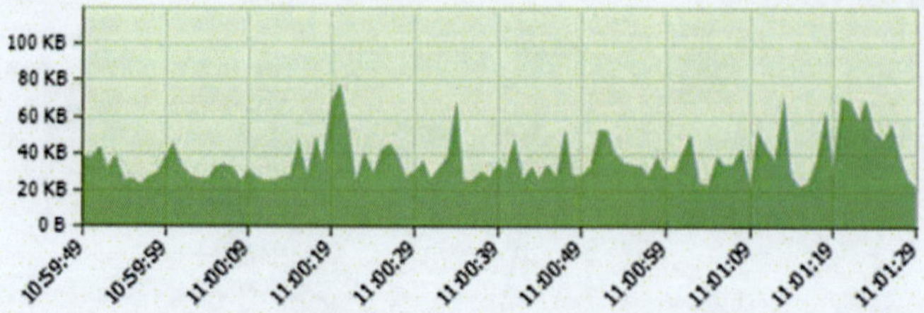

Fig. 10 Total generated traffic by bytes

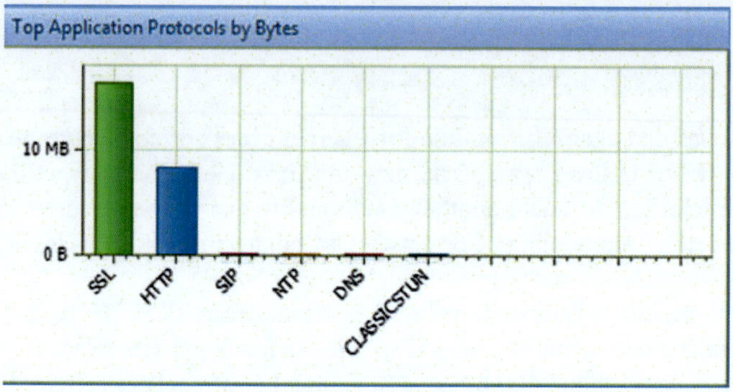

Fig. 11 Most used protocols and the traffic they generate

Figure 11 shows the most used network protocols, and what traffic was generated from them. The results were obtained again from the monitoring of the Router_Firewall interface. As it can be seen from the results, the secure sockets layer (SSL) protocol is the most used. This indicates that the Internet server from which the file is downloaded uses an old cryptographic protocol for the client–server connection instead of transport layer security (TLS). The presence of session initiation protocol (SIP) traffic is due to testing a creation of a VoIP telephone connection with an IP telephone exchange in a real network.

For the modeled network from Fig. 8, the following can be briefly summarized: a working modeled IP network has been created in which data are exchanged. The modeled network is connected to the Internet, and as a result, there is a successful download of data from a remote server, which is proofed by the results of Wireshark. The ability to connect modeled networks to other real IP networks is one of the great advantages of using IP platforms for modeling in the study of IP networks.

4 Conclusions

The use of platform for modeling of IP networks is proposed. Using modeled networks instead of building real test ones has many advantages, such as less financial costs; working with disk images of real operating systems of real network devices; convenience; possibility for continuous reconfiguration and testing of the modeled network, which is impossible for a working real IP network; ability to modeling an entire working real IP network for testing and research. The drawbacks are very high-computing capabilities of the workstation that will be used for the modeling of the IP networks; some platforms for modeling and the additional tools for them are not free; achieving a working IP network model takes a long time, and this time depends on the complexity of the network.

Practical examples show some of the opportunities offered by the platforms for modeling of communication networks. The GNS3 platform is presented. The proposed platform offers the possibility of integration with many different tools for monitoring IP networks.

The ability of IP network platforms for modeling to connect the modeled network with real IP networks and in particular with the Internet, allows to expand the research on IP networks. This allows checking how the modeled IP network reacts to the Internet connection.

Platforms for modeling are very useful for professionals who design IP networks. As such, they will be able to test their network before it is built and put into operation.

The main challenges that can be faced while pursuing this kind of research can be additional system settings of the platform in order to achieve successful connection to real networks/Internet and sometimes the need of using additional tools and modules.

Last but not least, these platforms are very suitable and convenient for use in training, because not every educational organization can afford to purchase network equipment.

References

1. Mirtchev ST (2019) Study of preemptive priority single-server queue with peaked arrival flow. In: 2019 X national conference with international participation (ELECTRONICA), pp 1–4
2. Mirtchev ST (2020) Investigation of queueing systems with a polya arrival process. In: 2020 28th national conference with international participation (TELECOM), pp 29–32
3. Sapundzhi F, Popstoilov M (2019) C # implementation of the maximum flow problem. In: 2019 27th national conference with international participation (TELECOM), pp 62–65
4. Ganev B, Marinov MB, Nikolov D, Ivanov A (2021) High-resolution particulate matter monitoring and mapping in urban environments. In: 2021 12th national conference with international participation (ELECTRONICA), pp 1–4. https://doi.org/10.1109/ELECTRONICA52725.2021.9513728
5. Cherneva GP, Hristova VI (2020) Evaluation of FHSSS stability against intentional disturbances. In: 2020 28th national conference with international participation (TELECOM), pp 14–16

6. Imran M, Khan MA, Abdul Qadeer M (2018) Design and simulation of traffic engineering using MPLS in GNS3 environment. In: 2018 second international conference on computing methodologies and communication (ICCMC), pp 1026–1030
7. Biradar AG (2020) A comparative study on routing protocols: RIP, OSPF and EIGRP and their analysis using GNS-3. In: 2020 5th IEEE international conference on recent advances and innovations in engineering (ICRAIE), pp 1–5
8. Korniyenko B, Galata L, Ladieva L (2019) Research of information protection system of corporate network based on GNS3. In: 2019 IEEE international conference on advanced trends in information theory (ATIT), pp 244–248
9. Castillo-Velazquez J, Ramirez-Diaz E, Niño WRM (2019) Use of GNS3 cloud environment for network management emulation when comparing SNMP vs syslog applied over an advanced network. In: 2019 IEEE 39th Central America and Panama convention (CONCAPAN XXXIX) pp 1–6
10. Uramová J, Segeč P, Papán J, Brídová I (2020) Management of cybersecurity incidents in virtual lab. In: 2020 18th international conference on emerging elearning technologies and applications (ICETA), pp 724–729
11. Liu S, Wang H, Liu J, Xian M (2019) Feasibility analysis of network security teaching platform based on KVM and GNS3. In: 2019 international conference on information technology and computer application (ITCA), pp 310–313
12. Angelescu N, Puchianu DC, Predusca G, Circiumarescu LD, Movila G (2017) DMVPN simulation in GNS3 network simulation software. 2017 9th international conference on electronics, computers and artificial intelligence (ECAI), pp 1–4
13. https://www.iteasypass.com/Dynamips.htm
14. Getting Started with GNS3: https://docs.gns3.com/docs/
15. Szigeti T, Hattingh C (2004) End-to-End QoS network design: quality of service in LANs, WANs, and VPNs. In: Cisco Press. part of the networking technology series, ISBN-10: 1-58705-176-1
16. Cisco—understanding delay in packet voice networks, white paper. https://www.cisco.com/c/en/us/support/docs/voice/voice-quality/5125-delay-details.html

A New Innovation Concept on End-user's Contextual and Behavioural Perspectives

Reem Aman, Shah J. Miah, and Janet Dzator

Abstract This research study explores this original phenomenon for proposing a new concept that will act as an overarching descriptor of innovation types: idea, object and behaviour. This proposed concept, relating to intangible innovation, will explain the sequence within one or many connected intangible activities that provide novelty to its end user relative to previous activities and practices. Using a design science research approach, the study comprises two goals: (a) identifying opportunities and issues to measure intangible inputs to the innovation and (b) proposing a framework for extending the existing innovation theories that to better capture intangible end-user innovation and its diffusion insights in their online environment across nations.

Keywords Diffusion of innovation theory · User innovation · Visual analytics · Big data · Design science research

1 Introduction

This study looks at an alternative paradigm of ICT application development for end user in their own context of innovation. We look at the purpose of technology to generate knowledge from information relevant to innovation support. This innovation focuses on informed actions by concentrating on method and task [1]. The entire design and inductive tasks would be considering as an exploratory and end-user-oriented design science research that highlight end-user's particular contextual situation approach for designing conceptual solution framework [2–4].

R. Aman (✉)
Ministry of Education, Riyadh, Kingdom of Saudi Arabia, Newcastle Business School, University of Newcastle, Newcastle, NSW, Australia
e-mail: reem.aman@uon.edu.au

S. J. Miah · J. Dzator
Newcastle Business School, University of Newcastle, Newcastle, NSW, Australia
e-mail: shah.miah@newcastle.edu.au

J. Dzator
e-mail: janet.dzator@newcastle.edu.au

X.-S. Yang et al. (eds.), *Proceedings of Seventh International Congress on Information and Communication Technology*, Lecture Notes in Networks and Systems 447,
https://doi.org/10.1007/978-981-19-1607-6_19

Innovation by individual end users in the household sector (HHS) is a widespread phenomenon on a national scale. Individual end users in the household sector modify and develop tangible and intangible innovations in household projects context. There are multiple conceptualizations for intangible innovations, which indicate the importance of conceptual clarity to distinguish between tangible and intangible innovations. It is important for research to develop conceptual clarity that could extend academic knowledge on understanding the phenomena, the nature of it as an individual project instead of idea, object or practice, its diffusion and the impact of it on a national scale. Therefore, the exploration of studying this phenomenon will have theoretical and empirical implication as academic research.

Utilizing the end-user-oriented design understanding, this study proposes an overarching conceptualization of innovation types by end users in the household sector, to illustrate and communicate the complexity of intangible innovation. It will explore end-user innovation scope [5–7], through visual analytics of big data to discuss a new type of intangible innovation, scope innovation within a household project context. This conceptual framework will give a design basis of a fully functional end-user innovation support solution that provides decision support in relation to different intangible innovations in household sector.

2 Research Background

2.1 *Intangible Innovation*

Intangible innovations can be difficult to distinguish, especially when it comes to its diffusion in end-users' context. Therefore, it was suggested that operating innovation at a process level will assist its identification and capturing its diffusion [8]. This can be challenging in a survey study, or impossible. Although, it is important to understand intangible innovation by end users in the household sector, It was argued that traditional methods might not be reliable [9]. For example, the lack of understanding by the individual of the nature of innovation, especially in a household sector setting, because they might associate it with a daily practice [9].

2.2 *Intangible Innovation by HHS*

There are multiple conceptualizations for intangible innovations, which indicate the importance of conceptual clarity to distinguish between tangible and intangible innovations. This would be of paramount task to assist the classification and the measurement of individual end-users' innovation impact in the household sector. Innovation by end users in the household sector received wide attention in the literature, since it was first captured in a national survey, which proposed it scale and scope [10].

End-user innovation in the household sector is the modification or the development of innovation by individuals who are not commercial groups [10].

Multiple national survey studies illustrated the phenomena in several developed and developing countries [10–18]. Furthermore, it was argued that there are value propositions of tangible and intangible types of innovation developed and modified by individual end users.

2.2.1 Technique and Service Innovation

Technique innovation by end users was identified in an adventure sports context. It included novel movements in the water, physical movements and environmental movements [19]. Technique innovation by individual end users was defined as a "new way of doing", done by an individual, as a skilful and planned activity, either involves a physical object, or operates in an environment with physical objects [20]. In the banking sector, it was found that innovation in service was first initiated by end users from the household sectors, in developed [21] and developing countries [22]. Service innovation by end users is to meet self needs [21] and not for third party like in commercial settings [23].

2.2.2 Behavioural Innovation

Another important dimension of end-user innovation is the behavioural innovation which can be seen as a combined form of technique and service innovation. Behavioural innovation was defined as "one or a connected sequence of intangible problem-solving activities that provide a functionally novel benefit to its user developer relative to previous practice" [24]. However, a particular focus of this innovation is limited to develop systematic procedure for further guide and assistance.

This study will propose an overarching conceptualization of behavioural innovation, to illustrate and communicate the complexity of intangible innovation, when discussed in future research. The proposed conceptualization in this study will discuss a new type of intangible behavioural innovation, scope innovation, which include innovation types: idea, object and practice. Also, it will act in contrast with tangible innovation. This contribution to the literature will need the development of a new design science model, to facilitate the study in capturing the scope and significant of intangible innovation by individuals in the household sector on a national scale, in detail.

2.2.3 HHS Innovation Scope

The exploration of scope innovation will add to our knowledge about end-user innovation and extend the adopters categories in the diffusion of innovation theory [24]. For example, the topics that were discussed in relation to individual innovators characteristics include gender [25], personality [26] and capabilities [12]. The pilot study will enable the inclusion of additional variables that will be more relevant to characterized innovators based on extend of novelty. This will create a new set of innovators/adopter's categories that will extend the diffusion of innovation theory. Therefore, it is proposed in this study that the identification of tangible and intangible innovation diffusion will aid the exploration of innovators novelty based on the scope of their process innovation in a project context. Therefore, we develop the research question as: what attributes of the individual innovation are vital for facilitating diffusion and adoption of technological artefact?

3 Research Methodology

Improvement of the artefact design knowledge is an essential component of design science research [27]. Design research "...seeks to create innovations that define the ideas, practices, technical capabilities and products through which the analysis, design, implementation, management and use of information systems can be effectively and efficiently accomplished" [27, p. 76]. Therefore, this paradigm, going beyond the traditional qualitative and quantitative methods, provides supportive knowledge and guide for the artefact design (in this study, the artefact is the conceptual framework for scope innovation).

Design science also goes beyond the traditional sequential phase or step-oriented design methods ensuring that the problem-solving method is to be designed is fully problem informed and enables creation of new generalizable problem-solving knowledge (e.g. prescriptive solution (design) knowledge) for similar issues [27–29]. Design studies have introduced various approaches such as action design research—ADR [30]; design science research methodology [31] and participatory action design research [32], although there are some limitations on each of them. Analysing different approaches, we found that [31] six activity-oriented approaches have been somehow driven by the seven design guidelines of [27] together provide us a suitable synergy for supporting the objectives of our study. We therefore rebuild a suitable design approach that meets our goals both in articulating end-user innovation realities and designing a problem-solving method that may integrate computational techniques. A conceptual framework is defined as a design research artefact that provides specifications of ways of performing tasks as well as for illustrating relationship between components (attributes, causes or factors that creates effect) within the particular problem context.

Different variables and the assumed relationships between those variables are included in the framework model and reflect the innovation expectation's goal directed activities in order to solve or address particular problem called wicked problems (e.g. comprises technical, human and organizational elements in the problems [27]). In our study, our approach is based on three phases (illustrated in Table 1):

Table 1 Three design phases for conducting the study

Phases	Tasks in our study
Design phase 1: Problem realization and artefact types	Literature review; data analysis; gap analysis; end-users' provisions; context analysis; problem articulation; artefact type selection
Design phase 2: Artefact creation and evaluation, and	Identifying components of framework; framework design; framework validation; framework justification and reformation parameter identification
Design phase 3: Research contributions of the artefact and communication of results	Dissemination of design ideas; design artefact and identification of generalizable parameter of the proposed framework; applicability analysis

The three design phases for conducting the study are informed through [27] and articulated on [31] six activities. Hevner et al. [27] suggested that design science research must talk about the creation of an innovation and purposeful development that may capture the problem situation, reality, and the key demands of the purpose in a specific problem domain. This implies that a collection of innovative conceptual artefact that can reinforce quality by creating effective design to meet the needs of the end users as well as being able to fulfil the process, users' and situational requirements within a problem space (e.g. household project contexts). The definitions of [27] establish two useful views that can help define a purposeful artefact design and its properties.

4 Theoretical Framework

In the diffusion of innovation theory (DOI), diffusion is the spreading of properties by the penetration of invention through a process of imitation, and the properties resulting from invention, transmit from an individual to another individual [33]. Therefore, diffusion is a phenomenon operating on a micro-level. The essence of imitation between individuals in the diffusion process is a repetition that is conditioned by universal laws [33]. There are multiple models of innovation development [34]. Some of innovation development models that include invention and diffusion were based on the diffusion of innovation theory [35]. These models were used by economics and managers [36]. However, these innovations are bounded by market failure [14].

Innovations by end users diffuse in two forms: consumer to businesses or consumer to consumer (C2B or C2C) [37]. In the first path, the innovation is diffused to the market through firm's adoption. In the second path, the innovation is either diffused through a community and into the market or be adopted by peers. In this path, the innovation ends up creating a new market [38]. In individual's innovation projects, end users develop innovation in a collaborative mode and in an individual mode [9]. In a collaborative mode, the innovation project has a distributed nature [39]. The end user re-innovates and diffuse the innovation within a community, where another member may require supportive information to adopt or rectify the innovation, which will then diffuse to another member for modification in their own context. This complex phenomenon is represented in open-source software development [39], but appropriate support framework are yet to be developed for different stakeholders in this sector.

5 Proposed Conceptualization

Our proposed conceptual framework embraces components of intangible innovation beyond technique, service or behaviour as a scope innovation. These can connect parts of most theoretical and empirical studies in this problem domain on an individual level, ambiguously. End-users' scope is one of the dimensions of process innovation diffusion and how it would be autonomous and systemically nurtured for end user enhancement. The proposed framework could provide supportive environment for the nature of innovation outcome. Figure 1 illustrates some initial components of the framework that will be a primary basis of our design study.

In the proposed framework, we revised the scope was defined as "the extent of the area or subject matter that something deals with or to which it is relevant" that can be viewed as an opportunity. Furthermore, a scope can take the form of an investigation or evaluation [40] and defined in organizational development as the number of units

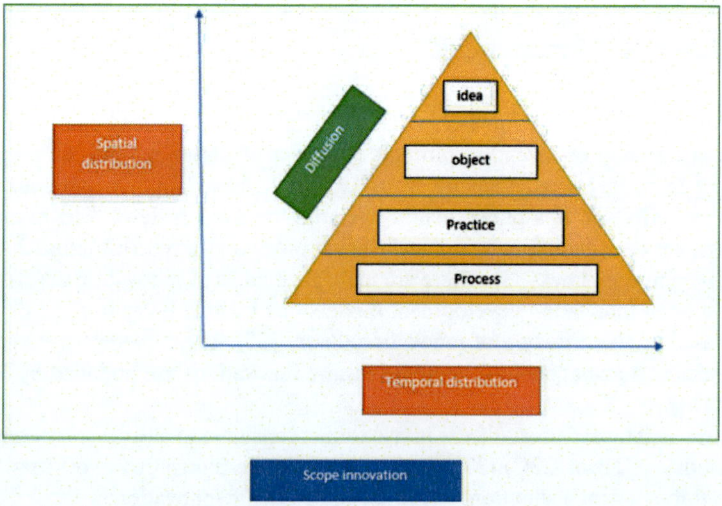

Fig. 1 Key components of the proposed end-user innovation framework

associated with the innovation through diffusion and change [41]. For example, the scope of user innovation was discussed in a process innovation context, and its magnitude was empirical demonstrated nationally [5] In the household sector, the view of individual end-user innovation phenomena [42] shows promises that was firstly identified in modifications done by individuals. Also, it was argued that most individual end users modify or develop process equipment and software to meet their needs [43], and the two properties of scope: autonomous and dichotomous [41], which are used as a fundamental benchmarking of the entire framework design.

6 Discussion and Conclusion

The study was to describe a new methodological foundation for designing an end-user innovation design framework in the household sector. Design science research becomes the central element that would provide systematic guide and assistance for the entire design study in capturing end-user innovation contextual details and converting them into a purposeful artefact (e.g. regarding intangible innovation). The three phases in the design methodology will enable us in designing relatively new concept for promoting intangible components of innovation that offers both further reinvention and invention support to end users. We deployed an openly online environment within peer-to-peer communication setting, to capture intangible and tangible innovation diffusion. This method will not replace traditional methods such as surveys, it will act as a complement to it. The filtering process of individual scope innovation will be captured to develop meta data of innovation types, innovation needs, nature of novelty, diffusion activities and innovators characteristics, which will be initially based on secondary data. Our extended understanding would contribute to rectify the current form of DOI theory, including new components in relation to individual innovation needs in their own context.

This study recreated an alternative paradigm of ICT application development for end user in their own context of innovation. We redefined the context as "end user own context of design innovation" in this paper for the first time in the literature. We believe that the purpose of technology is to assist end users for intervention support and produce new knowledge for innovation support. It will focus on informed actions by focusing on method and task [1]. The entire design and inductive tasks would be considering as an exploratory and end-user-oriented design science research that highlight end-user's particular contextual situation approach for designing conceptual solution framework [2, 3].

The literature of individual innovation in the household sector is at its emergent stage at a national scale. New understanding and knowledge should be reproduced in this sub-domain for new researchers and industry practitioners for more provisions of modifying and developing tangible and intangible innovations, such as in household projects context. There are multiple conceptualizations for intangible innovations, which indicate the importance of conceptual clarity to distinguish between tangible and intangible innovations. We attempt to develop required conceptual clarity that will extend academic knowledge on understanding the phenomena, the nature of it as an individual project instead of idea, object or practice, its diffusion and the impact of it on a national scale. Therefore, the exploration of studying this phenomenon will have theoretical and empirical implication as academic research.

The application of big data analytics approach can be utilized as they are growing in other sectors (e.g. in higher education [44]). The findings of this study in future would be used to develop innovative ICT application design for end users following design guidelines that are established in other associative areas of functionalities for empowering end users [2, 45, 46] and highlighting features of building end-user specific service-based systems [2, 47, 48].

References

1. Kawalek JJP (1998) User innovation in information systems practice, Doctoral dissertation, Sheffield Hallam University, United Kingdom
2. Miah SJ (2008) An ontology based design environment for rural business decision support, Doctoral dissertation, Griffith University, Australia
3. Miah S, Kerr D, von Hellens L (2014) A collective artefact design of decision support systems: design science research perspective. Inf Technol People 27(3):259–279
4. Miah SJ, Gammack JG, McKay J (2019) A metadesign theory for tailorable decision support. J Assoc Inf Syst 20(5):570–603
5. de Jong J (2016) The empirical scope of user innovation. In: Revolutionizing innovation. MIT Press, Cambridge, MA
6. de Jong J, Ben-Menahem SM, Franke N, Füller J, von Krogh G (2021) Treading new ground in household sector innovation research: Scope, emergence, business implications, and diffusion. Res Policy 50(8)
7. von Hippel E (201) The broad scope of free innovation. In: Free innovation. 1st edn. MIT Press, Cambridge, MA
8. Bogers M, West J (2012) Managing distributed innovation: strategic utilization of open and user innovation. Creat Innovat Manage 21(1):61–75
9. von Hippel E (2016) Free innovation, 1st edn. The MIT Press, Cambridge, MA
10. von Hippel E, de Jong JPJ, Flowers S (2012) Comparing business and household sector innovation in consumer products: Findings from a representative study in the United Kingdom. Manage Sci 58(9):1669–1681
11. Bengtsson L (2016) How big and important is consumer innovation in Sweden?—A comparison with five other countries. In: the 2016 open and user innovation conference. Harvard Business School Press, Boston, MA
12. Chen J, Su YS, de Jong J, von Hippel E (2020) Household sector innovation in China: impacts of income and motivation. Res Pol 49(4)
13. de Jong J (2013) User innovation by Canadian consumers, analysis of a sample of 2021 respondents
14. de Jong J, von Hippel E, Gault F, Kuusisto J, Raasch C (2015) Market failure in the diffusion of consumer-developed innovations: patterns in Finland. Res Pol 44(10):1856–1865
15. Fursov K, Thurner T, Nefedova A (2017) What user-innovators do that others don't: a study of daily practices. Technol Forecast Soc Chang 118:153–160
16. von Hippel E, DeMonaco H, de Jong J (2017) Market failure in the diffusion of clinician-developed innovations: The case of off-label drug discoveries. Sci Publ Pol 44(1):121–131
17. Kim Y (2015) Consumer user innovation in Korea: an international comparison and policy implications. Asian J Technol Innov 23(1):69–86

18. Ogawa S, Pongtanalert K (2011) Visualizing invisible innovation continent: evidence from global consumer innovation surveys
19. Hienerth C, Von Hippel E, Jensen M, Berg Jensen M (2014) User community vs. producer innovation development efficiency: a first empirical study. Res Pol 43(1):190–201
20. Hienerth C (2016) Technique innovation. In: Revolutionizing innovation, 1st ed. MIT Press, Cambridge, MA, pp 331–352
21. Oliveira P, von Hippel E (2011) Users as service innovators: the case of banking services. Res Policy 40(6):806–818
22. van der Boor P, Oliveira P, Veloso F (2014) Users as innovators in developing countries: the global sources of innovation and diffusion in mobile banking services. Res Policy 43(9):1594–1607
23. Lusch R, Nambisan S (2015) Service innovation: a service-dominant logic perspective. MIS Q 39:155–176
24. von Hippel CD, Cann AB (2021) Behavioral innovation: Pilot study and new big data analysis approach in household sector user innovation. Res Pol 50(8)
25. Mendonça J, Reis A (2020) Exploring the mechanisms of gender effects in user innovation. Technol Forecast Soc Change 155
26. Stock RM, von Hippel E, Gillert NL (2016) Impacts of personality traits on consumer innovation success. Res Pol 45(4):757–769
27. Hevner AR, March ST, Park J, Ram S (2004) Design science in information systems research. MIS Q 28(1):75–105
28. Gregor S, Hevner AR (2013) Positioning and presenting design science research for maximum impact. MIS Q 37(2):337–355
29. Persson JG (2016) Current trends in product development. In: The 26th CIRP design conference, Procedia CIRP, vol 50, pp 378–383 (2016)
30. Sein MK, Henfridsson O, Purao S, Rossi M, Lindgren R (2011) Action design research. MIS Q 35(1):37–56
31. Peffers K, Tuunanen T, Rothenberger MA, Chatterjee S (2007) A design science research methodology for information systems research. J Manag Inf Syst 24(3):45–77
32. Bilandzic M, Venable J (2011) Towards participatory action design research: adapting action research and design science research methods for Urban informatics. J Commun Informat 7(3)
33. Kinnunen J (1996) Gabriel tarde as a founding father of innovation diffusion research. Acta Sociolog 39(4):431–442
34. Rothwell R (1994) Towards the fifth-generation innovation process. Int Mark Rev 11(1):7–31
35. Rice RE, Rogers EM (1980) Reinvention in the innovation process. Knowl Creat , Diffus Utilizat 1(4):499–514
36. von Hippel E (2017) Free innovation by consumers—how producers can benefit. Res Technol Manage 60(1):39–42
37. Gambardella A, Raasch C, von Hippel E (2017) The user innovation paradigm: impacts on markets and welfare. Manage Sci 63(5):1450–1468
38. Hyysalo S (2009) User innovation and everyday practices: micro-innovation in sports industry development. R and D Manage 39(3):247–258
39. von Hippel E (2002) Open source software projects as user innovation networks
40. Oxford Dictionary. https://www.lexico.com/definition/scope?locale=en. Last accessed 2021/10/01
41. Sisaye S, Birnberg J (2010) Extent and scope of diffusion and adoption of process innovations in management accounting systems. Int J Account Inf Manag 18(2):118–139
42. von Hippel E (1977) Dominant role of the user in semiconductor and electronic subassembly process innovation. IEEE Trans Eng Manage EM-24(2):60–71
43. Gault F, von Hippel E (2011) The prevalence of user innovation and free innovation transfers: implications for statistical indicators and innovation policy. Working paper, Alfred P. Sloan School of Management, MIT, Cambridge, MA
44. Miah SJ, Miah M, Shen J (2020) Editorial note: Learning management systems and big data technologies for higher education. Educ Inf Technol 25:725–730

45. Miah SJ (2009) End user as application developer for decision support. In: The 5th American conference on information system. San Francisco, California, pp 1088–1095
46. Miah SJ, Debuse J, Kerr D, Debuse V (2010) A practitioner-oriented decision support process for forestry pest management. In: The 21st Australasian conference on information systems. Brisbane, Australia
47. Miah SJ, Gammack J, Kerr D (2007) Ontology development for context-sensitive decision support. In: The 3rd international conference on semantics, knowledge, and grid (SKG 2007), pp 475–478. IEEE, Xi'an, China
48. Miah SJ, Ahamed R (2011) A cloud-based DSS model for driver safety and monitoring on Australian roads. Int J Emerg Sci 1(4):634–648

Computational Modelling of the Role of Leadership Style for Its Context-Sensitive Control Over Multilevel Organisational Learning

Gülay Canbaloğlu, Jan Treur, and Anna Wiewiora

Abstract This paper addresses formalisation and computational modelling of context-sensitive control over multilevel organisational learning and in particular the role of the leadership style in influencing feed forward learning flows. It addresses a realistic case study with focus on the role of managers for control of multilevel organisational learning. To this end a second-order adaptive self-modelling network model is introduced and an example simulation for the case study is discussed.

Keywords Organisational learning · Leadership style · Context-sensitive control · Computational modelling · Self-modelling networks

1 Introduction

Organisational learning is a shared knowledge development process involving individuals, groups and the organisation. Organisational learning occurs through formation of shared mental models and common believes developed by organisational members and institutionalised for future use. Intermediary agents such as projects or teams are also involved in the process of learning [7, 9, 23, 22]. The team level occurs through discussion and developing of shared understanding at the team level,

G. Canbaloğlu (✉)
Department of Computer Engineering, Koç University, Istanbul, Turkey
e-mail: gcanbaloglu17@ku.edu.tr

G. Canbaloğlu · J. Treur
Center for Safety in Healthcare, Delft University of Technology, Delft, The Netherlands
e-mail: j.treur@vu.nl

J. Treur
Social AI Group, Department of Computer Science, Vrije Universiteit Amsterdam, Amsterdam, The Netherlands

A. Wiewiora
QUT Business School, Queensland University of Technology, Brisbane, Australia
e-mail: a.wiewiora@qut.edu.au

X.-S. Yang et al. (eds.), *Proceedings of Seventh International Congress on Information and Communication Technology*, Lecture Notes in Networks and Systems 447,
https://doi.org/10.1007/978-981-19-1607-6_20

achieved through collective actions, dialogue, shared practices and mutual adjustment. Although organisational members are involved in the process of organisational learning, organisational level learning can be people-independent and captured in routines and practices, even if the organisation loses some of its members. The process of organisational learning is non-linear, dynamic and context specific. It can be influenced by contextual factors such as leadership style, organisational culture or structure [23].

The diversity of the involved individuals and contextual factors brings an abundance of possible learning scenarios. Even in a project run by a single team, there may be multiple learning scenarios and contextual factors affecting decisions related to the organisational learning process. The multilevel and context-dependent characteristic of organisational learning makes it hard to observe and analyse.

Computational modelling and in particular the self-modelling network modelling approach introduced in [19] and explained in Sect. 3 in this paper, offers a useful tool to comprehend and represent the complex process of organisational learning; e.g. [4–4]. A detailed real-world learning scenario, explained in the in Sect. 2, is used to observe and analyse the process of organisational learning, with a focus on a context specific control of a leadership style. Using self-modelling networks with different context factors allows to incorporate a variety of management contexts, which enriches the possible learning scenarios, and provide better understanding of the effects of these contexts on the learning outcomes. The designed computational model is described in more detail in Sect. 4. Simulation results of the model follow in Sect. 5 with added images for a simulation scenario and a discussion part is included in Sect. 6.

2 Multilevel Organisational Learning

In this section, it is briefly discussed how multilevel organisation learning works and by an example scenario it is illustrated how leadership style can play an important role in it.

2.1 Multilevel Organisational Learning

Organisations operate as a system or organism of interconnected parts. Similarly, organisational learning is considered a multilevel phenomenon involving dynamic connections between individuals, teams and organisation [7, 9]. Due to the complex and changing environment within which organisations operate, the learning constantly evolves and some learning may become obsolete. Organisational learning is a vital means of achieving strategic renewal and continuous improvement, as it allows an organisation to explore new possibilities as well as exploit what they have already learned [15]. Organisational learning is a dynamic process that occurs in

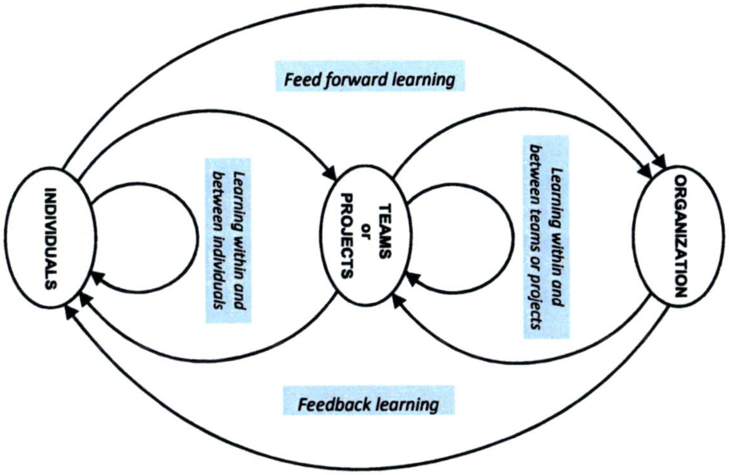

Fig. 1 Multilevel organisational learning: multiple levels and nested cycles (with depth 3)

feedback and feed forward directions. Feedback learning helps in exploiting existing and institutionalised knowledge, making it available for teams and individuals to utilise. Feed forward learning assists in exploring new knowledge by individuals and teams and institutionalising this knowledge at the organisational level [7]. As such, organisations may learn from individuals and teams via feed forward learning. Institutionalised on the organisational level can subsequently be accessed and used by the teams and individuals via feedback learning. This dynamic and adaptive process is depicted in Fig. 1. There are number of ways by which individuals, teams and organisations learn. For example, individuals can learn by reflecting on their past experiences and observing others. Teams can learn via joint problem solving or sharing their mental models. Organisations can learn from individuals and teams by capturing learning and practices into organisational manuals, policies or templates, which are then made available for teams and individuals to utilise. Recent research has pointed to the role of leaders in influencing learning flows between individuals, teams and organisations, which is discussed in the following section.

2.2 *The Influential Role of Leaders in Facilitating Multilevel Learning*

Management research found that leaders influence organisational learning [8, 11]. Leaders have been described as social architects of organisational learning [6, 11] who can either inhibit or facilitate learning flows [22]. For example, findings from [16] suggest that leaders facilitate feed forward learning by creating an environment for open and transparent communication. Edmondson [8] demonstrated that those leaders who purposefully obliterate power differences and encourage input and debate promote an environment conducive to learning, whereas leaders who choose to retain their status and power tend to tighten control at the expense of learning. Such leaders provide environment in which individuals are discouraged to share ideas or be open to others.

More recently, research identified the role of leaders in facilitating learning linkages between individuals, teams and organisation. A case study on multilevel learning in the context of a global project-based organisation revealed that senior leaders facilitate individual to individual and team to team (same level), as well as individual to team and individual to organisation (feed forward) learning flows [22]. This is because senior leaders have access to different parts of the organisation, well developed networks and a position of power to influence the transfer of learning between the levels. As such, senior leaders can facilitate an environment in which individuals can exchange knowledge, bounce ideas off each other, discuss ideas and engage in joint problem solving. Furthermore, [22] research demonstrated that by using a position of influence, leaders can either restrict or promote individual ideas for organisational improvement, hence affect institutionalisation of learning. Overall, existing research found evidence that leaders and their leadership styles impact the flow of learning within organisations. Systematic literature review of mechanisms facilitating multilevel organisational learning revealed that although research begins to identify the role of leaders in facilitating learning flows, more studies are required to better understand this phenomena and uncover the specific contextual factors and connections that leaders can influence to enable multilevel learning flows [23].

2.3 The Example Scenario Used as Illustration

In this section, we illustrate a real learning scenario that occurred in a project-based setting. An experienced project manager—Tom (pseudo name) was recently employed in an established large, and highly hierarchical organisation—Alpha (pseudo name). For the scenario description, see the left hand column of Table 1. Variables identified in this example are: learning from past experiences, role of leaders to effectively transfer learning and institutionalise learning, and a resistance to learn from a novice. This example also demonstrates that leaders have a powerful role in the organisation to promote ideas and institutionalise them. In Table 1 further analysis of the scenario can be found in terms of the conceptual mechanisms involved and how they can be related to computational mechanisms.

3 The Self-modelling Network Modelling Approach Used

In this section, the network-oriented modelling approach used is briefly introduced. A temporal-causal network model is characterised by; here X and Y denote nodes of

Table 1 Further analysis of the example scenario

Scenario	Conceptual mechanisms	Computational mechanisms
(1) Tom brought with him to the organisation learnings and insights that he acquired in his previous roles as a project manager. One of the insight he acquired was about the use of tollgates in projects	**Individual learning** His individual mental models based on individual learning from past experiences	**Learning by observation within project teams** World states with mirroring links to mental states and hebbian learning for the mental model • Team 1 without tollgates → weaker world states • Team 2 with tollgates → stronger world states • Tom decides to learn the mental model of Team 2
(2) During a team meeting, he shared one of his past learnings with other project managers. It was an idea to implement tollgates (approval points before proceeding to the next stage of the project that allows to review reediness and progress of the project). The organisation did not use tollgates or other processes to monitor progress of the project	**Feed forward learning from individual Tom to Team 3** Individual to team learning / sharing mental models based on past experiences	**Controlled individual and feed forward learning from individual mental model of Tom to mental models of the other individuals in the team and to team mental model formation** • Controlled decision of Tom to communicate his individual mental model to the other individuals in Team 3 • Shared mental model formation within Team 3
(3) The idea was well received by the team, but no call for action to implement the idea was requested by Tom's immediate boss	**No feed forward learning from Team 3 to the organisation** Lack of approval for a novice. Organisational learning stops, due to lack of interest from the immediate leader/or reluctance to take on board insights suggested by a novice. Illustrating the influential role of leaders	**Controlled feed forward learning from team to organisation** • Controlled by a control state for the immediate manager's approval • Due to the lack of this approval no feed forward learning to the organisation level takes place • This depends on the context factor that Tom is a novice in the new organisation • This also depends on the leadership style of Tom's immediate boss. That leader choose to retain their power to tighten control in expense of learning

(continued)

Table 1 (continued)

Scenario	Conceptual mechanisms	Computational mechanisms
(4) After several months, Tom raised the idea again with a higher level manager, who liked the idea and discussed it with others in the organisation who supported the idea and requested that tollgates are implemented as a new process to manage projects	**Feed forward learning from Team 3 to the organisation** By approval from higher level manager: institutionalisation of learning takes place illustrating the influential role of leaders	**Controlled feed forward learning from team to organisation based on communication with others** Controlled by a control state for the higher maneger's approval • This control state does not depend on being a novice • Instead it depends on feedback of some others in the organisation • This is obtained by communication channels back and forth to them • This depends on the leadership style of the higher level manager who displayed openness and welcomed the new idea from the employee

the network that have activation levels that can change over time, also called states [19]:

- *Connectivity characteristics*: Connections from a state X to a state Y and their weights $\boldsymbol{\omega}_{X,Y}$
- *Aggregation characteristics*: For any state Y, some combination function $\mathbf{c}_Y(..)$ defines the aggregation that is applied to the single causal impacts $\boldsymbol{\omega}_{X,Y}X(t)$ on Y from its incoming connections from states X
- *Timing characteristics*: Each state Y has a speed factor $\boldsymbol{\eta}_Y$ defining how fast it changes for given causal impact.

The following canonical difference (or related differential) equations are used for simulation purposes; they incorporate these network characteristics $\boldsymbol{\omega}_{X,Y}$, $\mathbf{c}_Y(..)$, $\boldsymbol{\eta}_Y$ in a standard numerical format:

$$Y(t + \Delta t) = Y(t) + \boldsymbol{\eta}_Y\left[\mathbf{c}_Y\left(\boldsymbol{\omega}_{X_1,Y}X_1(t), \ldots, \boldsymbol{\omega}_{X_k,Y}X_k(t)\right) - Y(t)\right]\Delta t \quad (1)$$

for any state Y and where X_1 to X_k are the states from which Y gets its incoming connections. The above concepts enable to design network models and their dynamics in a declarative manner, based on mathematically defined functions and relations. The available dedicated software environment described in [19, Chap.9], includes a combination function library with currently around 50 useful basic combination functions. Some examples of combination functions that are applied here can be found in Table 2.

Table 2 Examples of combination functions for aggregation available in the library

	Notation	Formula	Parameters
Advanced logistic sum	$\mathbf{alogistic}_{\sigma,\tau}(V_1, \ldots, V_k)$	$\left[\frac{1}{1+e^{-\sigma(V_1+\cdots+V_k-\tau)}} - \frac{1}{(1+e^{\sigma\tau})}\right](1+e^{-\sigma\tau})$	Steepness $\sigma > 0$ Excitability threshold τ
Steponce	$\mathbf{steponce}_{\alpha,\beta}(..)$	1 if time t is between α and β, else 0	Start time α; end time β
Complement identity	$\mathbf{comp\text{-}id}(V_1, \ldots, V_k)$	$1 - V_1$	
Hebbian learning	$\mathbf{hebb}_{\mu}(V_1, V_2, V_3)$	$V_1 * V_2(1 - V_3) + \mu V_3$	V_1,V_2 activation levels of the connected states; V_3 activation level of the self-model state for the connection weight; persistence factor μ

Combination functions as shown in Table 2 are called *basic combination functions*. For any network model some number m of them can be selected; they are represented in a standard format as $\mathrm{bcf}_1(..)$, $\mathrm{bcf}_2(..)$, …, $\mathrm{bcf}_m(..)$. In principle, they use parameters $\pi_{1,i,Y}$, $\pi_{2,i,Y}$ such as the λ, σ and τ in Table 2. Including these parameters, the standard format used for basic combination functions is (with V_1, …, V_k the single causal impacts): $\mathrm{bcf}_i\,(\pi_{1,i,Y}, \pi_{2,i,Y}, V_1 \cdots, V_k$. For each state Y just one basic combination function can be selected, but also a number of them can be selected; this will be interpreted as a weighted average of them with *combination function weights* $\gamma_{i,Y}$ as follows:

$$\mathbf{c}_Y(\pi_{1,1,Y}, \pi_{2,1,Y}, \ldots, \pi_{1,m,Y}, \pi_{2,m,Y}, \ldots, V_1, \ldots, V_k) = \left(\frac{\gamma_{1,Y}\,\mathrm{bcf}_1(\pi_{1,1,Y}, \pi_{2,1,Y}, V_1, \ldots, V_k) + \cdots + \gamma_{m,Y}\,\mathrm{bcf}_m(\pi_{1,m,Y}, \pi_{2,m,Y}, V_1, \ldots, V_k)}{\gamma_{1,Y} + \cdots + \gamma_{m,Y}}\right) \quad (2)$$

Selecting only one of them for state Y, for example, $\mathrm{bcf}_i(..)$, is done by putting weight $\gamma_{i,Y} = 1$ and the other weights 0. This is a convenient way to indicate combination functions for a specific network model. The function $\mathbf{c}_Y(..)$ can then just be indicated by the weight factors $\gamma_{i,Y}$ and the parameters $\pi_{i,j,Y}$, according to (2).

Realistic network models are usually adaptive: often not only their states but also some of their network characteristics change over time. By using a *self-modelling network* (also called a *reified* network), a similar network-oriented conceptualization can also be applied to *adaptive* networks to obtain a declarative description using mathematically defined functions and relations for them as well; see [19]. This works through the addition of new states to the network (called *self-model states*) which represent (adaptive) network characteristics. In the graphical 3-D format as shown in Sect. 4, such additional states are depicted at a next level (called *self-model level* or *reification level*), where the original network is at the *base level*. As an example, the weight $\omega_{X,Y}$ of a connection from state X to state Y can be represented (at a next self-model level) by a self-model state named $\mathbf{W}_{X,Y}$. Similarly, all other network characteristics from $\boldsymbol{\omega}_{X,Y}$, $\boldsymbol{\gamma}_{i,Y}$, $\boldsymbol{\pi}_{i,j,Y}$, $\boldsymbol{\eta}_Y$ can be made adaptive by including self-model states for them. For example, an adaptive speed factor $\boldsymbol{\eta}_Y$ can be represented by a self-model state named $\mathbf{H}_Y$, an adaptive combination function weight $\boldsymbol{\gamma}_{i,Y}$ can be represented by a self-model state $\mathbf{C}_{i,Y}$.

As the outcome of such a process of network reification is also a temporal-causal network model itself, as has been shown in ([19], Chap. 10), this self-modelling network construction can easily be applied iteratively to obtain multiple orders of self-models at multiple (first-order, second-order, …) self-model levels. For example, a second-order self-model may include a second-order self-model state $\mathbf{H}_{\mathbf{W}_{X,Y}}$ representing the speed factor $\eta_{\mathbf{W}_{X,Y}}$ for the dynamics of first-order self-model state $\mathbf{W}_{X,Y}$ which in turn represents the adaptation of connection weight $\boldsymbol{\omega}_{X,Y}$. Similarly, the weight $\omega_{Z,\mathbf{W}_{X,Y}}$ of an incoming connection from some state Z to a first-order self-model state $\mathbf{W}_{X,Y}$ can be represented by a second-order self-model state $\mathbf{W}_{Z,\mathbf{W}_{X,Y}}$.

This self-modeling network modeling approach has successfully been used to obtain computational models for dynamics, adaptation and control of mental models (e.g., [20, 17, 21]). The approach to multilevel organisational learning described in the current paper builds further on that previous work.

4 The Adaptive Computational Network Model Designed

In this section the designed computational network model will be explained. Based on the analysis of the learning scenario presented in Sect. 2 and Table 1 in particular the following computational mechanisms were considered for the different phases in the example scenario. Note that teams T1 and T2 are from Tom's previous organisation and team T3 is in his current organisation.

(1) **Team learning by observation for teams T1 and T2 and feedback learning from team T2 to individual A**

Team mental model learning for teams T1 and T2 is assumed to be based on observation of world states and mirroring them in the mental model combined with Hebbian learning. Here team T1 without tollgates shows weaker world states (lower activation levels), whereas team T2 with tollgates shows stronger world states. Based on this difference in success, A (which is Tom) makes a controlled decision to individually learn the shared team mental model from team T2 (feedback learning). The control is based on the good performance of team T2.

(2) **Individual and team learning from the individual mental model of A to mental models of the other individuals B and C in team T3 and to (feed forward) mental model formation by team T3**

Tom makes a controlled decision to communicate his individual mental model (learnt in the past from team T2) to the other individuals B and C in team T3 (individual learning from other individuals); control is based on providing space and time for team T3 to exchange knowledge. Within this team, this is followed by shared mental model formation which makes the mental model of Tom the shared team mental model of team T3 (feed forward learning).

(3) **Feed forward learning from team T3 to organisation controlled by A's immediate manager D**

Control is modelled by a control state related to the immediate manager D's approval. Due to the lack of this approval no feed forward learning to the organisation level takes place. D's non-approval depends on the context factor that Tom is a novice in the new organisation, and D's leadership style which is based on retaining power at the expense of learning.

(4) **Feed forward learning from team T3 to organisation controlled by a higher manager E**

Control is modelled by a control state related to the higher manager E's approval. This control state does not depend on being a novice; instead it depends on feedback of some other individuals F and G in the organisation. This feedback is obtained by communication from E (after having received the proposal) to F and G and from F and G back to E's state for approval. The leadership style of E also played a role. E displayed openness to ideas from others, hence he was more receptive welcomed the tollgate idea as a way to improve organisational processes.

In previous work [3, 4], a number of the involved computational mechanisms already have been described and used. In other work [5] some other mechanisms have been pointed out but not used yet. For example, learning by observation is only briefly described for individuals in [4], but in the current paper it is actually applied in (1) at the level of teams T1 and T2. Furthermore, the mechanisms for the control decisions for Tom (a) to decide to adopt the mental model of team T2 in (1), and (b) to decide to share this mental model with the members of team T3 in (2), are new too. Moreover, an important focus of the current paper is the control of the feed forward learning by managers in (3) and (4) above; this also is new here as that was not addressed in previous work such as Canbaloğlu et al. [3, 4].

4.1 Team Learning by Observation for Teams T1 and T2 and Feedback Learning from T2 to Individual A

In Fig. 2a adopted from Canbaloğlu et al. [5] it is shown how internal simulation of a mental model by any individual B (triggered by context state con_1) activates subsequently the mental model states a_B to d_B of B and these activations in turn activate Hebbian learning of their mutual connection weights. Here for the Hebbian learning [12], the self-model state $\mathbf{W}_{X,Y}$ for the weight of the connection from X to Y, uses the combination function $\mathbf{hebb}_{\mu}(V_1, V_2, W)$ shown in Table 2, last row. More specifically, this function $\mathbf{hebb}_{\mu}(V_1, V_2, W)$ is applied to the activation values V_1, V_2 of X and Y and the current value W of $\mathbf{W}_{X,Y}$. To this end upward (blue)

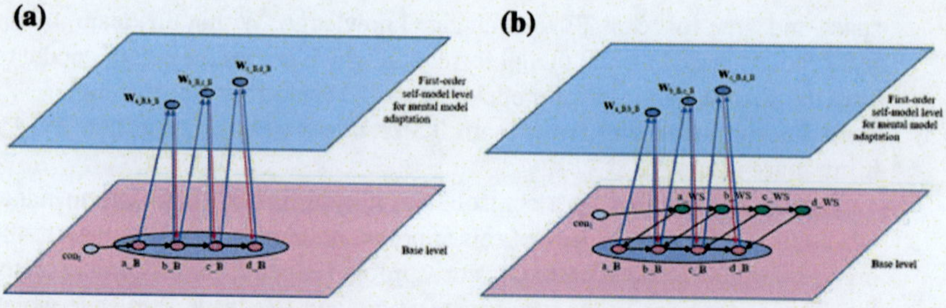

Fig. 2 **a** Left. Learning by internal simulation: Hebbian learning during internal simulation **b** Right. Learning by observation: Hebbian learning after mirroring of the world states

connections are included in Fig. 2a (also a connection to $\mathbf{W}_{X,Y}$ itself is assumed but usually such connections are not depicted). The (pink) downward arrow from $\mathbf{W}_{X,Y}$ to Y depicts how the obtained value of $\mathbf{W}_{X,Y}$ is actually used in activation of Y. Thus, the mental model is learnt by individual B. If the persistence parameter μ is 1, the learning result persists forever; if $\mu < 1$, then forgetting takes place. For example, when $\mu = 0.9$, per time unit 10% of the learnt result is lost.

For learning by observation as used as an individual learning mechanism in the current paper, the above is applied in conjunction with mirroring links (e.g., [13, 14, 18, 10]) to model the observation; see Fig. 2b. Here mirror links are the (black) links from World States a_WS to d_WS to the corresponding mental model states a_B to d_B. When the world states are activated because things are happening in the world, through these mirror links they in turn activate B's related mental model states which in their turn activate Hebbian learning like in the case of pure internal simulation pointed out above. In the network model introduced in the current paper, the mechanism of learning by observation as described is applied at the team level to teams T1 and T2.

4.2 Abstracted Overall View on the Process

After the teams T1 and T2 have learned their (shared) mental models, individual A (Tom) decides to adopt the team mental model of team T2 as his own individual mental model: feedback learning. This is shown in Fig. 3 which gives an abstracted view of the overall process. This time, the ovals stand for groups of states: in the base plane groups of base states for one mental model and at the first-order self-model level (blue middle plane) for groups of self-model **W**-states for one mental model.

The arrow from the first-order self-model for the shared mental model for team T2 indicated by SMM-W-T2 to the first-order self-model for the individual A's mental model indicated by IMM-W-A-T2 depicts the adoption by A of the team mental model of team T2 as individual mental model (feedback learning).

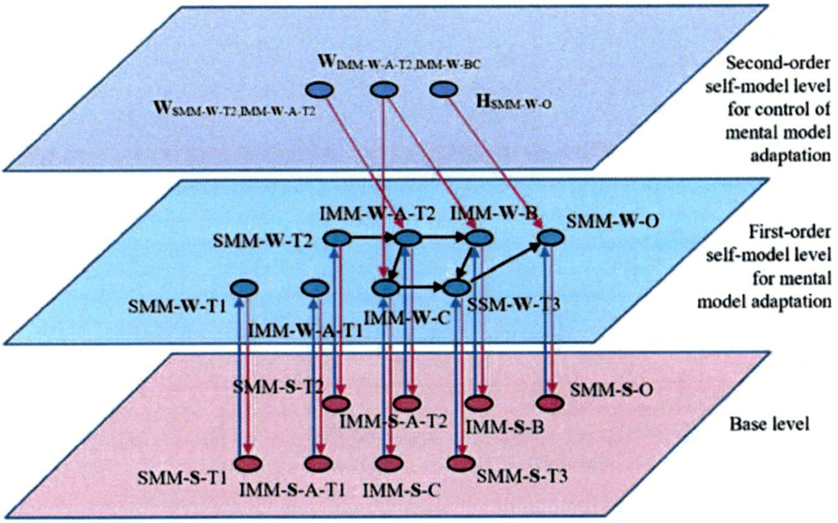

Fig. 3 Abstract view on the connectivity of the first- and second-order self-model of the mental models: SMM = Shared Mental Model, IMM = Individual Mental Model, **S** = States of mental model, **W** = Connection Weights of mental model, T1, T2 = teams from previous organisation, T3 = team in current organisation, A = Tom, B and C = team members in T3, O = organisation

This happens (via the pink downward link) under control of second-order self-model state $\mathbf{W}_{\text{SMM-W-T2,IMM-W-A-T2}}$ that gives the weight of this connection from SMM-**W**-T2 to IMM-**W**-A-T2 values 0 or 1. This second-order self-model state $\mathbf{W}_{\text{SMM-W-T2,IMM-W-A-T2}}$ depends on a sufficiently high activation value (>0.6) of team T2's world state d_WS_T2, which makes the control of the adoption decision for feedback learning context-sensitive.

In the new organisation, Tom decides to communicate this individual mental model to the team members (B and C) in team T3. This is depicted in Fig. 3 by the arrows from IMM-**W**-A-T2 to IMM-**W**-B and IMM-**W**–C. This is learning for one individual from another individual. Note that also here control is used for the decision to actually do this: the two pink downward arrows from the control state $\mathbf{W}_{\text{IMM-W-A-T2,IMM-W-BC}}$ at the second-order self-model level to IMM-**W**-B and IMM-**W**-C. This control state $\mathbf{W}_{\text{IMM-W-A-T2,IMM-W-BC}}$ is context-sensitive as it depends on a suitable time within team T3's meetings. After that, within team T3 this model is chosen as shared mental model, depicted in Fig. 3 by the arrows from IMM-**W**-B and IMM-**W**-C to SMM-**W**-T3 (feed forward learning).

Note that in the model, for the sake of simplicity no explicit control over this shared mental model formation for T3 was included. As a next step it is shown in Fig. 3 how this shared mental model of T3 is institutionalised and becomes shared mental model for the organisation O. However, for this step control is needed by an authorised manager D or E (the pink downward connection from state $\mathbf{H}_{\text{SMM-W–O}}$,

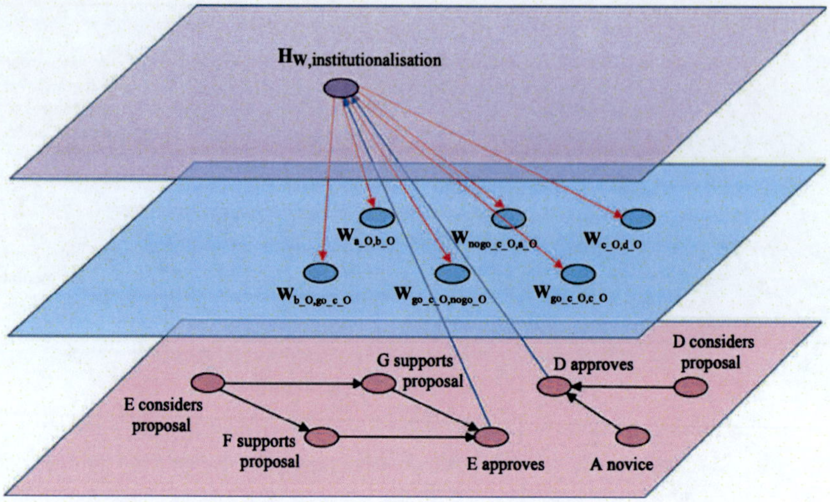

Fig. 4 Network for approval of institutionalisation and the context-sensitive control over it

which is also context-sensitive as it depends on approval by D or E. This form of context-sensitive control is described in more detail in Sect. 4.3 and Fig. 4.

4.3 Context-Sensitive Control of Institutionalisation of the Shared Mental Model by Managers D and E

For the control of the process of institutionalisation some context factors are crucial: approval from an authorised manager is required; see Fig. 4. In this scenario two of such managers act: D and E. As can be seen in the base plane, manager D makes approval dependent on the fact that Tom is a novice because of his leadership style focused on retaining control and power.

This makes that D does not approve. However, manager E whose leadership style was based on distributed power and opennes to ideas lets approval depend on whether some knowledgeable persons F and G from the organisation are in support of approval, and then follows their suggestion; see the four (two incoming and two outgoing) arrows in the base plane related to the states called F supports proposal and G supports proposal. The (blue) upward links from base plane to upper plane activate the control for effectuation of the decision for approval, by making the adaptation speed of the related **W**-states in the middle plane nonzero. It is by these mechanisms that the control of the institutionalisation is addressed in a context-sensitive manner. For more details of the model, see the Linked Data at https://www.researchgate.net/publication/355186556.

5 Simulation Results

In this section, the simulation results are discussed for the example scenario described in Sect. 2. In Fig. 5 the world states for team T1 (the team not using tollgates) and for team T2 (the team using tollgates) are shown. The monotonically increasing curves show the world states for team T1; here the value for the last task d stays under 0.4 (the thicker red curve). The curves for the world states for team T2 show increasing pattern with initially a slight fluctuation due to the nogo-feedback loop. Although a bit slower than for team T1, in the end the last task d for T2 reaches a value above 0.6 (the thicker green curve). This value satisfies the criterion of Tom for adopting the team mental model of T2 as his own individual mental model (feedback learning).

Figure 6 shows an overview of the adaptations of the different mental models learnt together with the context-sensitive control over these adaptations. In the different phases the following can be seen:

Time 0–150: Within the previous organisation, teams T1 and T2 learn their (shared) mental models (team learning by observation as discussed in Sect. 4.1).

Time 150–200: Within the previous organisation (because of good results in the world), at this point Tom decides to adopt the shared mental model of team T2 (time 150) and actually adopts it (around time 200). This control decision is based on the good results of the outcomes of team T2 (here state X_{10} is used as an indicator; see the green line in Figs. 5 and 6) together with an appropriate timing for, which is a context factor modelled by state X_{63} (pink curve occurring at time 150); these two together activate second-order self-model state $\mathbf{W}_{\mathbf{W},\mathbf{TomsChoice}}$ for control of the effectuation of this decision (the orange curve starting at time 150).

Time 300–400: In the new organisation, Tom decides to share (time 300) and actually shares (around time 350) his mental model of team T2 with team members B and C in team T3. This control decision is based on another context factor modelled

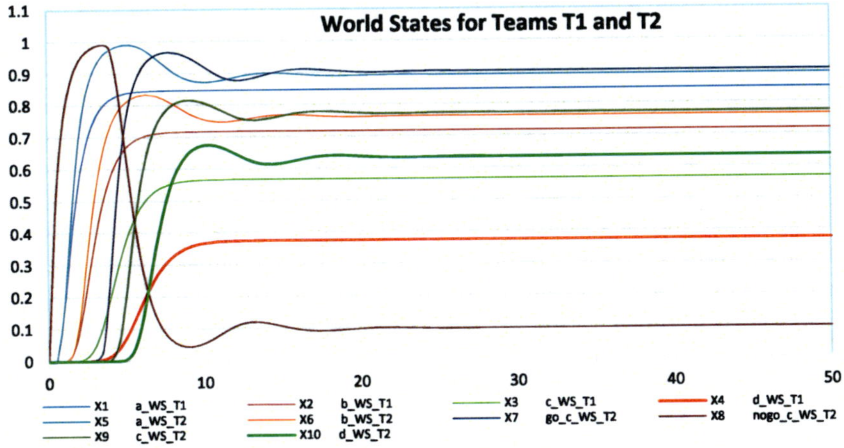

Fig. 5 Simulation outcomes for the world states for teams T1 and T2

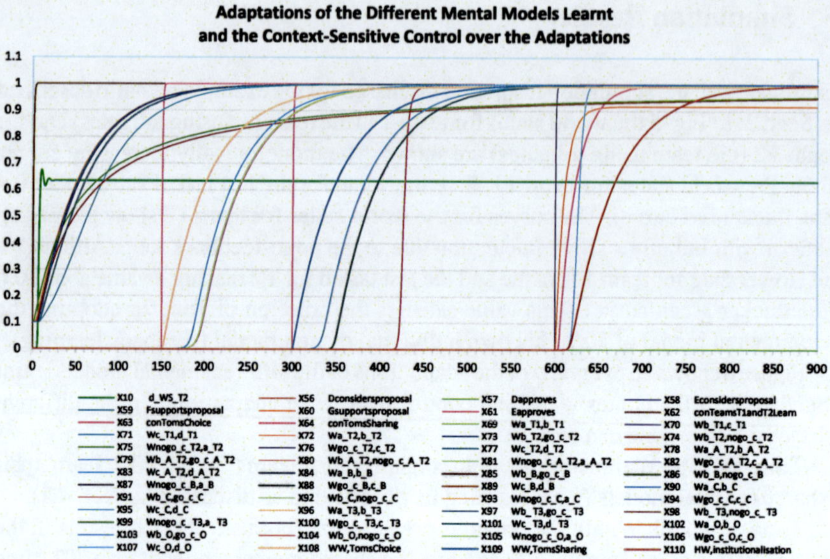

Fig. 6 Simulation outcomes for the first-order self-model states for adaptation of the various mental models learnt, the second-order self-model states for control over the adaptation, and the context factors making this control context-sensitive

by state X_{64} (pink curve occurring at time 300). This context factor activates second-order self-model state $\mathbf{W}_{\mathbf{W},\mathbf{TomsSharing}}$ for control of the effectuation of this sharing decision (the blue curve starting at time 300), so that the communication actually takes place. After having received it, B and C adopt the communicated individual mental model as shared mental model for team T3 (at time 350).

Time 400–450: The just formed shared mental model of team T3 is proposed for institutionalisation to manager D (between time 350 and 400); D does not approve it, due to Tom being a novice and due to D's leadership style of retaining power (yet another context factor, modelled by state X_{55}).

Time 600–650: Tom happens to meet manager E and uses the occasion to propose the institutionalisation (this is a last context factor, modelled by state X_{61}; see the curve occurring at time 600). Manager E communicates this proposal to F and G, and they give supportive feedback. Upon this E approves the proposal (state X_{61}, the pink curve occurring between times 600 and 625). This activates second-order self-model state $\mathbf{H}_{\mathbf{W},\mathbf{institutionalisation}}$ (the brown curve starting around time 615) which controls that the institutionalisation is actually realised (around time 650).

6 Discussion

As holds for many social processes, multilevel organisational learning suffers from the many context factors that often influence these processes and their outcomes in decisive manners. For example, an important decision for the organisation may depend on unplanned and occasionally meeting a higher manager like in the example case study addressed in this paper. Due to such context factors it may seem that sustainable generally valid laws or models will never be found, as any proposal often can easily be falsified by putting forward an appropriate context factor violating it. Given this perspective of highly context-sensitive processes, in particular when addressing a serious challenge like mathematical formalisation and computational modelling of multilevel organisational learning, it makes sense to explicitly address relevant context factors within such formalisations. This already has been done in particular for the specific process of aggregation of mental models in feed forward multilevel organisational learning in [1, 2]. In the current paper, this idea of context-sensitivity also has been addressed for other subprocesses in multilevel organisational learning.

Analysing a realistic case study, context factors have been identified that play a role in a number of steps within the example multilevel organisation learning process covered by the case study. These context factors have decisive effects on different parts of the process, such as adopting mental models for proven good practices and managers with their specific world view that need to give approval to institutionalisation. In the case study, the context factors had a proper setting so that in the end institutionalisation actually took place. However, if only one of the chain of these context factors would have had a different setting, that outcome would not have been achieved. The developed computational model explicitly addresses these decisive context factors and is able to explore for any setting of them what the outcome will be, thus covering both successful and less successful outcomes.

In previous work Canbaloğlu et al. [3, 4], a number of the computational mechanisms involved in the case study addressed here already have been introduced. However, there are a number of new ones too. For example, learning by observation is briefly pointed out for individuals in Canbaloğlu et al. [5] but here it is actually applied in (1) in Table 2 at the level of Team 1 and Team 2 for simulation. Furthermore, the mechanisms for the control decisions for Tom (a) to decide to adopt the mental model of Team 2 in (1) and (b) to decide to share this mental model with the members of Team 3 in (2) in Table 2, are new too. Moreover, an important focus of the current paper is the control of the feed forward learning by managers in (3) and (4) in Table 2; this indeed is new here as that was not addressed in previous work such as Canbaloğlu et al. [3, 4].

The presented findings have important implications for management studies, suggesting that computational modelling is a promising tool to predict changes in learning over time and demonstrate, via modelling, how different leadership styles can facilitate or inhibit organisational learning, hence further expand on findings by Edmundson [8] and [22].

Using computational tools for modelling learning scenarios has many benefits for practice. Modelling learning is a cost-effective decision making tool that helps predict learning outcomes and select best mechanisms for learning without investing time and money on implementing untested solutions. Using computational modelling enables to forecast different scenarios, which then provide basis for more informed decisions about the best possible mechanisms for implementation in the real world. For example, as demonstrated in our study, computational modelling can help organisations make more informed decisions on the most suitable leadership styles that will promote organisational learning. In doing so, organisations can invest in leadership training to achieve greater learning outcomes.

References

1. Canbaloğlu G, Treur J (2021a) Context-sensitive mental model aggregation in a second-order adaptive network model for organisational learning. In: Proceedings of the 10th international conference on complex networks and their applications. studies in computational intelligence, Springer Nature
2. Canbaloğlu G, Treur J (2021b) Using boolean functions of context factors for adaptive mental model aggregation in organisational learning. In: Proceeding of the 12th international conference on Brain-Inspired Cognitive Architectures, BICA'21. Studies in computational intelligence, Springer Nature (2021b)
3. Canbaloğlu G, Treur J, Roelofsma PHMP (2021) Computational modeling of organisational learning by self-modeling networks. Cog Sys Res 73:51–64
4. Canbaloğlu G, Treur J, Roelofsma PHMP (2022) A self-modeling network model for the role of individual and team learning in organisational learning. In: Proceeding of the 7th International Congress on Information and Communication Technology, ICICT'22. Lecture notes in networks and systems, Springer Nature
5. Canbaloğlu G, Treur J, Wiewiora A (2022a) Computational modeling of multilevel organisational learning: from conceptual to computational mechanisms. In: Proceeding of computational intelligence: automate your world. The Second International Conference on Information Technology, InCITe'22. Lecture Notes in Electrical Engineering, Springer Nature
6. Chang A, Wiewiora A, Liu Y (2021) A socio-cognitive approach to leading a learning project team: a proposed model and scale development. Int J Project Manage 39(6):646–657
7. Crossan MM, Lane HW, White RE (1999) An organizational learning framework: from intuition to institution. Acad Manag Rev 24:522–537
8. Edmondson AC (2002) The local and variegated nature of learning in organizations: a group-level perspective. Organ Sci 13:128–146
9. Fiol CM, Lyles MA (1985) Organizational learning. Acad Manag Rev 10:803–813
10. Van Gog T, Paas F, Marcus N, Ayres P, Sweller J (2009) The mirror neuron system and observational learning: Implications for the effectiveness of dynamic visualizations. Educ Psychol Rev 21(1):21–30
11. Hannah ST, Lester PB (2009) A multilevel approach to building and leading learning organizations. Leadersh Q 20(1):34–48
12. Hebb DO (1949) The organization of behavior: a neuropsychological theory. John Wiley and Sons, New York
13. Iacoboni M (2008) Mirroring people: the new science of how we connect with others. Farrar, Straus & Giroux, New York
14. Keysers C, Gazzola V (2014) Hebbian learning and predictive mirror neurons for actions, sensations and emotions. Philos Trans R Soc Lond B Biol Sci 369:20130175

15. March JG (1991) Exploration and exploitation in organizational learning. Organ Sci 2(1):71–87
16. Mazutis D, Slawinski N (2008) Leading organizational learning through authentic dialogue. Manag Learn 39(4):437–456
17. Van Ments L, Treur J (2021) Reflections on dynamics, adaptation and control: a cognitive architecture for mental models. Cogn Syst Res 70:1–9
18. Rizzolatti G, Sinigaglia C (2008) Mirrors in the brain: how our minds share actions and emotions. Oxford University Press
19. Treur J (2020) Network-oriented modeling for adaptive networks: designing higher-order adaptive biological, mental and social network models. Springer Nature, Cham
20. Treur J, Van Ments L (eds) (2022) Mental models and their dynamics, adaptation, and control: a self-modeling network modeling approach. Springer Nature
21. Van Ments L, Treur J, Klein J, Roelofsma PHMP (2021) A second-order adaptive network model for shared mental models in hospital teamwork. In: Nguyen NT et al (eds) Proceeding of the 13th International Conference on Computational Collective Intelligence, ICCCI'21. Lecture Notes in AI, vol 12876. Springer Naturem, pp 126–140
22. Wiewiora A, Chang A, Smidt M (2020) Individual, project and organizational learning flows within a global project-based organization: exploring what, how and who. Int J Project Manage 38:201–214
23. Wiewiora A, Smidt M, Chang A (2019) The 'How' of multilevel learning dynamics: a systematic literature review exploring how mechanisms bridge learning between individuals, teams/projects and the organization. Eur Manag Rev 16:93–115

Enumeration of LCD and Self-dual Double Circulant Codes Over $\mathbb{F}_q[v]/ < v^2 - 1 >$

Shikha Yadav and Om Prakash

Abstract Consider a finite field $\mathbb{F}_q$ having cardinality $q = p^n$, for an odd prime p and a natural number n. In this chapter, we consider $S = \mathbb{F}_q + v\mathbb{F}_q$, where $v^2 = 1$ and obtain conditions for double circulant (DC) codes over S to be LCD and self-dual. Using these conditions, we enumerate LCD and self-dual DC codes over S of length $2m$. Further, we obtain some examples of DC codes by defining a Gray map ψ from the ring S^m to $\mathbb{F}_q^{2m}$. Finally, some bounds are presented on the relative distances of the families of LCD and self-dual DC codes.

Keywords Double circulant code · LCD code · Self-dual code · Gray map

1 Introduction

Double circulant (DC) codes have been an interesting topic for many researchers. These codes belong to a subclass of quasi-cyclic (QC) codes. In 2001, the algebraic structure of QC codes was presented in [6] and later this study was extended over chain rings [7]. Based on these theories, many studies were carried on DC and double negacirculant codes over finite field, various rings such as $\mathbb{Z}_4$, $\mathbb{Z}_{p^2}$ and Galois ring [1, 2, 4, 10–12]. Recently, these codes were studied in terms of generator polynomials over some non-chain rings such as $\mathbb{F}_q + u\mathbb{F}_q$, $\mathbb{F}_q + u\mathbb{F}_q + u^2\mathbb{F}_q$, $\mathbb{F}_q + u\mathbb{F}_q + v\mathbb{F}_q$ in [13–15]. Following these studies, we enumerate LCD and self-dual DC codes over S. We also define a Gray map and study some of its properties. Further, we present distance bounds on the families of Gray images of LCD and self-dual DC codes.

S. Yadav (✉) · O. Prakash
Department of Mathematics, Indian Institute of Technology Patna, Bihta, Patna 801106, India
e-mail: 1821ma10@iitp.ac.in

O. Prakash
e-mail: om@iitp.ac.in

X.-S. Yang et al. (eds.), *Proceedings of Seventh International Congress on Information and Communication Technology*, Lecture Notes in Networks and Systems 447,
https://doi.org/10.1007/978-981-19-1607-6_21

2 Preliminaries

Throughout the chapter, we consider q to be an odd prime power such that $q = p^b$, when $p \equiv 1 \pmod 4$ and $q = p^{2b}$, when $p \equiv 3 \pmod 4$, for some natural number b. It is required for the existence of square root of -1 (see [7]).

For $0 \neq \gamma \in \mathbb{F}_q$ with $2\gamma \equiv 1 \pmod p$, we consider $\epsilon_1 = \gamma(1+v)$ and $\epsilon_2 = \gamma(1-v)$, where $v^2 = 1$. Then, using CRT decomposition, we can write $S \cong \epsilon_1\mathbb{F}_q \oplus \epsilon_2\mathbb{F}_q$ (see [9]). That is, an arbitrary element $r \in S$ is of the form $r = \epsilon_1 a_1 + \epsilon_2 a_2$, where $a_1, a_2 \in \mathbb{F}_q$. Any element $\epsilon_1 a_1 + \epsilon_2 a_2 \in S$ is a unit if and only if a_1, a_2 both are nonzero elements in $\mathbb{F}_q$.

Recall that a linear code $\mathcal{L}$ of length m over S is defined as an S-submodule of the module S^m. The Hamming weight w_H is defined as $w_H(y) = |\{i : y_i \neq 0\}|$, for $y = (y_0, y_1, \ldots, y_{m-1}) \in \mathcal{L}$. The minimum Hamming distance of $\mathcal{L}$ is the minimum of the weights of nonzero codewords. We define the Euclidean inner product of $r = (r_0, r_1, \ldots, r_{m-1}), s = (s_0, s_1, \ldots, s_{m-1}) \in S^m$ as $r \cdot s = \sum_{i=0}^{m-1} r_i s_i$. Then the Euclidean dual $\mathcal{L}^\perp$ of the code $\mathcal{L}$ is

$$\mathcal{L}^\perp = \{r \in S^m : r \cdot s = 0, \forall s \in \mathcal{L}\}.$$

We say that $\mathcal{L}$ is an LCD (resp. self-dual) code if $\mathcal{L} \cap \mathcal{L}^\perp = \{0\}$ (resp. $\mathcal{L} = \mathcal{L}^\perp$). Recall that a double circulant code is a linear code having the generator matrix (I, B), for an identity matrix I and a circulant matrix B of same order.

For a polynomial $q(z) = a_0 + a_1 z + \cdots + a_m z^m$, its reciprocal polynomial is defined by $q^*(z) = z^m q(1/z)$. If $q^*(z) = q(z)$, we say that $q(z)$ is a self-reciprocal polynomial.

3 Enumeration of Double Circulant Codes for Odd Value of *m*

We consider the factorization into irreducible polynomials of $z^m - 1$ over S of the form

$$z^m - 1 = b(z-1)\prod_{i=2}^{s} f_i(z)\prod_{j=1}^{t} p_j(z)p_j^*(z), \tag{1}$$

where $b \in S^*$, i.e., a unit element in S, $f_i(z)$ are irreducible self-reciprocal polynomials of even degree $2e_i$, $p_j^*(z)$ are reciprocal polynomials of $p_j(z)$ with $\deg(p_j(z)) = \deg(p_j^*(z)) = d_j$, for $2 \leq i \leq s$ and $1 \leq j \leq t$. Using CRT, we get

$$\frac{S[z]}{\langle z^m-1\rangle}\cong\frac{S[z]}{\langle z-1\rangle}\oplus_{i=2}^{s}\frac{S[z]}{\langle f_i(z)\rangle}\oplus_{j=1}^{t}\left(\frac{S[z]}{\langle p_j(z)\rangle}\oplus\frac{S[z]}{\langle p_j^*(z)\rangle}\right)$$
$$\cong S\oplus_{i=2}^{s}\left(S_{2e_i}\right)\oplus_{j=1}^{t}\left(S_{d_j}\oplus S_{d_j}\right),$$

where $S_r=\mathbb{F}_{q^r}+v\mathbb{F}_{q^r}$, $v^2=1$ for $r=2e_i$ or d_j. Infact, for any linear code $\mathcal{L}$ over $\frac{S[z]}{\langle z^m-1\rangle}$, we have

$$\mathcal{L}\cong\mathcal{L}_1\oplus_{i=2}^{s}\mathcal{L}_i\oplus_{j=1}^{t}(\mathcal{L}_j^{'}\oplus\mathcal{L}_j^{''}) \tag{2}$$

for some linear codes $\mathcal{L}_1$ over S, $\mathcal{L}_i$ over S_{2e_i}, for $2\le i\le s$ and $\mathcal{L}_j^{'}$, $\mathcal{L}_j^{''}$ over S_{d_j}, $1\le j\le t$. The following lemma can be proved similar to, Lemma 3.1 of [13].

Lemma 21.1 *For a DC code $\mathcal{L}$ over S as given in (1), consider $\alpha_1=(1,c_{e_1})$, $\alpha_i=(1,c_{e_i})$, $\alpha_j^{'}=(1,c_{d_j}^{'})$, $\alpha_j^{''}=(1,c_{d_j}^{''})$ as generators of the codes $\mathcal{L}_1,\mathcal{L}_i,\mathcal{L}_j^{'},\mathcal{L}_j^{''}$ $(2\le i\le s, 1\le j\le t)$. Then the necessary and sufficient conditions for a DC code over S to be self-dual and LCD are as follows:*

1. *For self-dual: $1+c_{e_1}^2=0$, $1+c_{e_i}^{q^{e_i}+1}=0$ and $1+c_{d_j}^{'}c_{d_j}^{''}=0$.*
2. *For LCD: $1+c_{e_1}^2\in S^*$, $1+c_{e_i}^{q^{e_i}+1}\in S_{2e_i}^*$ and $1+c_{d_j}^{'}c_{d_j}^{''}\in S_{d_j}^*$.*

Using this lemma, we now obtain some enumeration results.

Theorem 21.1 *Assume that for an odd integer m, the factorization of z^m-1 over S be as given in (1), where $m=1+\sum_{i=2}^{s}2e_i+2\sum_{j=1}^{t}d_j$. Then, there are*

$$4\prod_{i=2}^{s}(q^{e_i}+1)^2\prod_{j=1}^{t}(q^{d_j}-1)^2$$

self-dual DC codes over S of length $2m$.

Proof The enumeration of self-dual DC codes can be obtained by listing the codes $\mathcal{L}_1,\mathcal{L}_i,\mathcal{L}_j^{'},\mathcal{L}_j^{''}$. We have 4 options for $\mathcal{L}_1$, having the generators $(1,\mu)$, $(1,-\mu)$, $(1,\mu v)$ and $(1,-\mu v)$, where $\mu^2=-1$.

For listing the second constituent codes $\mathcal{L}_i$, we need to find options for $c_{e_i}\in S_{2e_i}$ satisfying $1+c_{e_i}c_{e_i}^{q^{e_i}}=0$. Let $c_{e_i}=x\epsilon_1+y\epsilon_2$, for some $x,y\in\mathbb{F}_{q^{2e_i}}$. Then

$$1+(x\epsilon_1+y\epsilon_2)(x\epsilon_1+y\epsilon_2)^{q^{e_i}}=(1+x^{1+q^{e_i}})\epsilon_1+(1+y^{1+q^{e_i}})\epsilon_2=0$$

if and only if $x^{1+q^{e_i}}=-1$ and $y^{1+q^{e_i}}=-1$. Therefore, the total number of options for c_{e_i} are $(q^{e_i}+1)^2$.

For listing the third constituent codes pairs $\{\mathcal{L}_j^{'},\mathcal{L}_j^{''}\}$, it is sufficient to find options for the pairs $\{c_{d_j}^{'},c_{d_j}^{''}\}$ in S_{d_j} satisfying $1+c_{d_j}^{'}c_{d_j}^{''}=0$. The following cases arise:

- For each $c_{d_j}^{'}\in S_{d_j}^*$, there is only one option for $c_{d_j}^{''}$ and we have $|S_{d_j}^*|=(q^{d_j}-1)^2$ options for the pairs.

- If $c'_{d_j} \in S_{d_j} \setminus S^*_{d_j}$, then $c'_{d_j} = x\epsilon_1 + y\epsilon_2$, for some $x, y \in \mathbb{F}_{q^{d_j}}$ such that either x or y or both are 0. If $c''_{d_j} = \beta_1\epsilon_1 + \beta_2\epsilon_2 \in S_{d_j}$, where $\beta_1, \beta_2 \in \mathbb{F}_{q^{d_j}}$, we have

$$1 + c'_{d_j}c''_{d_j} = (1 + x\beta_1)\epsilon_1 + (1 + y\beta_2)\epsilon_2 = 0,$$

which implies that $x\beta_1 = -1$ and $y\beta_2 = -1$. It leads to a contradiction as either x or y is zero.

From all the above cases, we conclude our result.

Now, we count LCD DC codes of length $2m$ over S.

Theorem 21.2 *If $z^m - 1$ have the factorization as given in (1), then there are*

$$(q-2)^2 \prod_{i=2}^{s} (q^{2e_i} - q^{e_i} - 1)^2 \prod_{j=1}^{t} (q^{4d_j} - 2q^{3d_j} + 3q^{2d_j} - 2q^{d_j} + 1).$$

LCD DC codes of length $2m$ over S.

Proof The enumeration of LCD DC codes can be obtained by listing the codes $\mathcal{L}_1, \mathcal{L}_i, \mathcal{L}'_j, \mathcal{L}''_j$ as done for the self-dual codes. For listing the first constituent code $\mathcal{L}_1$, we need to find options for $c_{e_1} \in S$ such that $1 + c_{e_1}^2 \in S^*$. We have the following possibilities:

- If $c_{e_1} = 0$, then $1 + c_{e_1}^2 = 1 \in S^*$
- If $0 \neq c_{e_1} \in \langle\epsilon_1\rangle$ and $c_{e_1} = x\epsilon_1$, for some $x \in \mathbb{F}_q^*$, then $1 + c_{e_1}^2 = 1 + x^2\epsilon_1 \in S^*$ if and only if $x \neq \pm\mu$, where $\mu^2 = -1$. Therefore, we have $(q-3)$ options. Similarly, when $0 \neq c_{e_1} \in \langle\epsilon_2\rangle$, we have $q-3$ options.
- If $c_{e_1} \in S^*$ then $c_{e_1} = x\epsilon_1 + y\epsilon_2$, where $x, y \in \mathbb{F}_q^*$ and

$$1 + c_{e_1}^2 = (1 + x^2)\epsilon_1 + (1 + y^2)\epsilon_2 \in S^*$$

if and only if $x, y \neq \pm\mu$, where $\mu^2 = -1$. It gives $(q-3)^2$ options for c_{e_1}.

Adding all of these, we have $(q-2)^2$ options for c_{e_1}.

For listing the second constituent codes $\mathcal{L}_i$ $(2 \leq i \leq s)$, we need to find options for $c_{e_i} \in S_{2e_i}$ such that $1 + c_{e_i}^{1+q^{e_i}} \in S^*_{2e_i}$. We have the following possibilities:

- If $c_{e_i} = 0$, then $1 + c_{e_i}^{1+q^{e_i}} = 1 \in S^*_{2e_i}$
- If $0 \neq c_{e_i} \in \langle\epsilon_1\rangle$ and $c_{e_i} = x\epsilon_1$, for some $x \in \mathbb{F}^*_{q^{2e_i}}$, then $1 + c_{e_i}^{1+q^{e_i}} = 1 + x^{1+q^{e_i}}$ $\epsilon_1 \in S^*_{2e_i}$ if and only if $x^{1+q^{e_i}} \neq -1$. Therefore, we have $(q^{2e_i} - q^{e_i} - 2)$ options. Similarly, when $0 \neq c_{e_i} \in \langle\epsilon_2\rangle$, we have $(q^{2e_i} - q^{e_i} - 2)$ options.
- If $c_{e_i} \in S^*$ then $c_{e_i} = x\epsilon_1 + y\epsilon_2$, for some $x, y \in \mathbb{F}^*_{q^{2e_i}}$ and

$$1 + c_{e_i}^{1+q^{e_i}} = (1 + x^{1+q^{e_i}})\epsilon_1 + (1 + y^{1+q^{e_i}})\epsilon_2 \in S^*_{2e_i}$$

if and only if $x^{1+q^{e_i}} \neq -1$ and $y^{1+q^{e_i}} \neq -1$. Therefore, we have $(q^{2e_i} - q^{e_i} - 2)^2$ options.

So, in this case we have $1 + 2(q^{2e_i} - q^{e_i} - 2) + (q^{2e_i} - q^{e_i} - 2)^2 = (q^{2e_i} - q^{e_i} - 1)^2$ options for $\mathcal{L}_i$, where $2 \leq i \leq s$.

Now, for listing the third constituent codes pairs $\{\mathcal{L}_j^{'}, \mathcal{L}_j^{''}\}$, it is sufficient to find options for the pairs $\{c_{d_j}^{'}, c_{d_j}^{''}\}$ in S_{d_j} satisfying $1 + c_{d_j}^{'} c_{d_j}^{''} \in S_{d_j}^*$. The following cases arise:

- If $c_{d_j} = 0$, then $1 + c_{d_j}^{'} c_{d_j}^{''} \in S_{d_j}^*$ for all $c_{d_j}^{''} \in S_{d_j}$. So, we have $|S_{d_j}| = q^{2d_j}$ options for the pairs.
- If $c_{d_j}^{'} \in S_{d_j}^*$ then $c_{d_j}^{''} \in S_{d_j}^* - \frac{1}{c_{d_j}^{'}}$ and $|S_{d_j}^* - \frac{1}{c_{d_j}^{'}}| = |S_{d_j}^*|$. We have $|S_{d_j}^*|^2 = (q^{d_j} - 1)^4$ options for the pairs.
- If $0 \neq c_{d_j}^{'} \in \langle \epsilon_1 \rangle$ then $c_{d_j}^{'} = x\epsilon_1$, for some $x \in \mathbb{F}_{q^{d_j}}^*$. Assume that $c_{d_j}^{''} = \beta_1\epsilon_1 + \beta_2\epsilon_2$, for some $\beta_1, \beta_2 \in \mathbb{F}_{q^{d_j}}$. Then,

$$1 + c_{d_j}^{'} c_{d_j}^{''} = 1 + (x\beta_1)\epsilon_1 = (1 + x\beta_1)\epsilon_1 + \epsilon_2 \in S_{d_j}^*$$

  if and only if $x\beta_1 \neq -1$. Therefore, we have $(q^{d_j} - 1)^2 q^{d_j}$ options. Similarly, when $0 \neq c_{e_i} \in \langle \epsilon_2 \rangle$, we have $(q^{d_j} - 1)^2 q^{d_j}$ options.

In this case, we have $q^{2d_j} + (q^{d_j} - 1)^4 + 2(q^{d_j} - 1)^2 q^{d_j} = (q^{4d_j} - 2q^{3d_j} + 3q^{2d_j} - 2q^{d_j} + 1)$ options for the pairs $\{c_{d_j}^{'}, c_{d_j}^{''}\}$ and hence for $\{\mathcal{L}_j^{'}, \mathcal{L}_j^{''}\}$. Now, adding all of the above options we get our result.

4 Gray Image

In this section, we first define a Gray map from S^m to $\mathbb{F}_q^{2m}$ and study some of its properties. Further, we obtain generator matrix of the Gray image of a double circulant code $\mathcal{L}$ of length $2m$ over S and present several examples of LCD and self-dual DC codes $\mathcal{L}$ over S along with their corresponding Gray images.
We define a map $\psi : S^m \to \mathbb{F}_q^{2m}$ by

$$\begin{aligned}\psi(y_1 + vy_1^{'}, y_2 + vy_2^{'}, \ldots, y_m + vy_m^{'}) \\ = (y_1 - y_1^{'}, y_2 - y_2^{'}, \ldots, y_m - y_m^{'}, y_1 + y_1^{'}, \ldots, y_m + y_m^{'}).\end{aligned}$$

Then ψ is a Gray map (distance preserving isometry) and preserves the duality property (see [9]). This map also preserves self-duality as given below.

Lemma 21.2 *A linear code $\mathcal{L}$ over S of length m is a self-dual code if and only if $\psi(\mathcal{L})$ is a self-dual code over $\mathbb{F}_q$ of length $2m$.*

Table 1 LCD and self-dual codes obtained as Gray images of DC codes over S of length $2m$

m	$a_1(x)$	$a_2(x)$	$\psi(\mathcal{L})$	Remark	Comparison
2	44	411	$[8, 4, 2]_5$	self-dual	
3	221	121	$[12, 6, 3]_5$	LCD	$[12, 6, 2]_5$ [14]
3	041	322	$[12, 6, 4]_5$	self-dual	$[12, 6, 4]_5$ [13]
4	2011	2120	$[16, 8, 4]_5$	LCD	$[16, 8, 4]_5$ [13]
4	0041	1100	$[16, 8, 4]_5$	self-dual	$[16, 8, 4]_5$ [13]
6	123134	201340	$[24, 12, 4]_5$	LCD	
9	121110210	211323120	$[36, 18, 6]_5$	LCD	$[36, 18, 6]_5$ [14]
12	124010321224	111341440102	$[48, 24, 7]_5$	LCD	$[48, 24, 6]_5$ [14]

Proof Let $\mathcal{L}$ be a linear code over S and $\mathcal{L}^\perp$ be its dual code. Then, by Theorem 3.1 of [9], we have $\psi(\mathcal{L})^\perp = \psi(\mathcal{L}^\perp)$. Therefore, $\mathcal{L} = \mathcal{L}^\perp$ if and only if $\psi(\mathcal{L}) = \psi(\mathcal{L})^\perp$ as ψ is one-one. This completes the proof.

The following lemma provides the generator matrix for Gray image of a double circulant code over S.

Lemma 21.3 *Let $\mathcal{L}$ be a DC code with the generator matrix $G = (I, B)$, where $B = B_1 + vB_2$ for some $m \times m$ matrices B_1, B_2 over $\mathbb{F}_q$ and the identity matrix I. Then the generator matrix of $\psi(\mathcal{L})$ is given by*

$$M = \begin{pmatrix} I & I & B_1 - B_2 & B_1 + B_2 \\ -I & I & B_2 - B_1 & B_1 + B_2 \end{pmatrix}_{2m \times 4m}.$$

Proof We can obtain the matrix M by vertically joining the matrices $\psi(G)$ and $\psi(vG)$.

Note that the above lemma also provides a method for obtaining the generator matrix for Gray image of any linear code over S.

In the following table, we obtain some examples of DC codes over S along with their corresponding Gray images. The first column of Table 1 indicates m such that $\mathcal{L}$ is a $[2m, m]$ code over S. Second and third columns represent polynomials $a_1(x), a_2(x)$ such that $(1, a_1(x) + va_2(x))$, is a generator of the code $\mathcal{L}$, where the coefficients of polynomial are written in decreasing order, i.e., 411 stands for $4x^2 + x + 1$. The fourth column represents the Gray image of code $\mathcal{L}$ while fifth column specifies its nature. In the sixth column, we compare our obtained codes from the already available double circulant codes in other papers. The minimum distance of all the codes has been calculated by Magma software [3].

5 Distance Bound

For this section, we consider m to be an odd prime. Let q be a primitive root modulo m and the factorization of $z^m - 1$ into irreducible factors over S be given by

$$z^m - 1 = (z-1)(1+z+\cdots+z^{m-1}) = (z-1)r(z), \tag{3}$$

where $r(z) = 1+z+\cdots+z^{m-1}$. Then, by using the CRT decomposition, we get

$$\frac{S[z]}{\langle z^m - 1\rangle} \cong \frac{S[z]}{\langle z-1\rangle} \oplus \frac{S[z]}{\langle r(z)\rangle} \cong S \oplus S',$$

where $S' = \mathbb{F}_{q^{m-1}} + v\mathbb{F}_{q^{m-1}}$, $v^2 = 1$. We set the ring $\mathcal{S} = \frac{S[z]}{\langle r(z)\rangle}$. A nonzero codeword of a cyclic code which is generated by $r(z)$ is said to be a constant vector. Now, we state two lemmas which provide us the maximum number of DC and self-dual DC codes having a particular kind of element (g, h), where g is not a constant vector. We omit their proofs as these can be proved on the similar lines to, Lemma 4.1 of [12], and Lemma 4.2 of [12], respectively. These two lemmas will be used for establishing our main result (Theorem 21.3).

Lemma 21.4 *Let $0 \neq y = (g, h) \in S^{2m}$, where g is not a constant vector. Then the number of DC codes $\mathcal{L}_a = (1, a)$ containing y, where $a \in \mathcal{S}$ are at most q^{m+1}.*

Lemma 21.5 *Let $0 \neq y = (g, h) \in S^{2m}$, where g is not a constant vector. Then the number of self-dual codes $\mathcal{L}_a = (1, a)$ containing y, where $a \in \mathcal{S}$ are at most* $4(1 + q^{\frac{m-1}{2}})$.

Now, under the assumptions on m and q mentioned in the beginning of this section, we can approximate the number of LCD and self-dual DC codes (using Theorems 21.1 and 21.2) for very large m approaching to infinity, by the following proposition.

Proposition 21.1 *Assume that for an odd prime m, q be an odd prime power which is also a primitive root modulo m and the factorization of $z^m - 1$ is given by (3). Then, for m approaching to infinity, the number of self-dual (resp. LCD) DC codes over S can be approximated to $4q^{m-1}$ (resp. q^{2m}).*

For a family $\mathcal{L}_{<m>}$ of codes of length m over $\mathbb{F}_q$ having parameters $[m, k_m, d_m]$, the *rate* ρ is defined by $\rho = \limsup_{m\to\infty} \frac{k_m}{m}$ and the *relative distance* δ is defined by $\delta = \liminf_{m\to\infty} \frac{d_m}{m}$. Now, there are infinitely many primes m for a fixed non-square q such that q is primitive root mod m, using Artin's conjecture [8]. Therefore, we have an infinite family of DC codes over S associated with the factorization (3). Recall that Hilbert entropy function [5] is defined as $H_q(0) = 0$ and

$$H_q(s) = s\log_q(q-1) - s\log_q(s) - (1-s)\log_q(1-s),$$

if $0 < s \leq 1 - \frac{1}{q}$. Now, we prove our main result for this section.

Theorem 21.3 *Let $\delta > 0$ be given and q be an odd prime power. Then, we have families of self-dual (resp. LCD) DC codes over S having code rate $\frac{1}{2}$ whose Gray images have relative distance δ, whenever $H_q(\delta) < \frac{1}{8}$ (resp. $H_q(\delta) < \frac{1}{4}$).*

Proof Let s_m, l_m denote the sizes of the families of self-dual and LCD DC codes (associated with the factorization (3)) over R, respectively. Then, by Proposition 21.1, we have $s_m \approx 4q^{m-1}$ and $l_m \approx q^{2m}$, for m tending to infinity. The number of elements in S^{2m} whose image under ψ have $< d$ Hamming weight is denoted by $B(d)$. For very large m, assume that we have

$$s_m > \lambda_m B(d_m) \text{ and } l_m > \lambda'_m B(d_m), \tag{4}$$

where $\lambda_m = 4(1 + q^{\frac{m-1}{2}})$ and $\lambda'_m = q^{m+1}$. Then, we have codes of length $2m$ in the family of LCD and self-dual DC codes whose ψ image have minimum Hamming distance $\geq d_m$, by using Lemmas 21.4 and 21.5. We denote the relative distance of this family of codes by δ.

For the truthfulness of the inequality (4), we make some arguments now. Let d_m be the largest positive integer such that $s_m > \lambda_m B(d_m)$, and assume that the growth be of the form $d_m \approx 4\delta m$. Then, using the estimate $B(d_m) \approx q^{4mH_q(\delta)}$ (Ref. [5], Lemma 2.10.3), we have

$$4q^{m-1} > 4(1 + q^{\frac{m-1}{2}})q^{4mH_q(\delta)} \text{ (for self-dual codes).}$$

This holds, for large enough m, if $H_q(\delta) < \frac{1}{8}$ for the family of self-dual DC codes. Similarly, we get that $H_q(\delta) < \frac{1}{4}$ for the family of LCD double circulant codes to satisfy inequality (4).
From the above theorem, it is straight forward that we get families of DC codes with code rate $\frac{1}{2}$ whose Gray images have $\delta < H_q^{-1}(\frac{1}{8})$, for self-dual DC codes and $\delta < H_q^{-1}(\frac{1}{4})$, for LCD DC codes.

6 Conclusion

Here, we presented the enumeration of LCD and self-dual DC codes over S. Moreover, we studied the Gray images of such codes by defining a Gray map from S^m to $\mathbb{F}_q^{2m}$ and obtained few examples of these codes. Later, we presented some bounds on the relative distance with the help of entropy function $H_q(\delta)$. In the future, it would be interesting to investigate double negacirculant codes over S and obtain codes with good parameters.

References

1. Alahmadi A, Güneri C, Ozkaya B, Shoaib H, Solé P (2017) On self-dual double negacirculant codes. Discrete Appl Math 222:205–212
2. Alahmadi A, Özdemir F, Solé P (2018) On self-dual double circulant codes. Des Codes Cryptogr 86(6):1257–1265
3. Bosma W, Cannon J (1995) Handbook of magma functions. University of Sydney
4. Huang D, Shi M, Solé P (2019) Double circulant self-dual and LCD codes over $\mathbb{Z}_{p^2}$. Int J Found Comput Sci 30(3):407–416
5. Huffman WC, Pless V (2003) Fundamentals of error correcting codes. Cambridge University Press
6. Ling S, Solé P (2001) On the algebraic structure of quasi-cyclic codes I: finite fields. IEEE Trans Inf Theory 47(7):2751–2760
7. Ling S, Solé P (2003) On the algebraic structure of quasi-cyclic codes II: chain rings. Des Codes Cryptogr 30(1):113–130
8. Moree P (2012) Artin's primitive root conjecture a survey. Integers 12(6):1305–1416
9. Prakash O, Yadav S, Verma RK (2020) Constacyclic and LCD codes over $\mathbb{F}_q + u\mathbb{F}_q$. Defence Sci J 70(6):626–632
10. Shi M, Huang D, Sok L, Solé P (2019) Double circulant LCD codes over $\mathbb{Z}_4$. Finite Fields Appl 58:133–144
11. Shi M, Qian L, Solé P (2018) On self-dual negacirculant codes of index two and four. Des Codes Cryptogr 86(11):2485–2494
12. Shi M, Huang D, Sok L, Solé P (2019) Double circulant self-dual and LCD codes over Galois ring. Adv Math Commun 13(1):171–183
13. Shi M, Zhu H, Qian L, Sok L, Solé P (2020) On self-dual and LCD double circulant and double negacirculant codes over $\mathbb{F}_q + u\mathbb{F}_q$. Cryptogr Commun 12(1):53–70
14. Yadav S, Islam H, Prakash O, Solé P (2021) Self-dual and LCD double circulant and double negacirculant codes over $\mathbb{F}_q + u\mathbb{F}_q + v\mathbb{F}_q$. J Appl Math Comput 67(1–2):689–705
15. Yao T, Zhu S, Kai X (2019) On self-dual and LCD double circulant codes over a non-chain ring. Chin J Electron 28(5):1018–1024

Autonomous Dysfunction and the Phenomenon of Early Aging of Regulatory Systems

Irina Kurnikova, Shirin Gulova, Ramchandra Sargar, and Nikolay Kisliy

Abstract The article presents the study of the relationship between metabolic and regulatory disorders which is one of the most promising areas of research, especially in patients with systemic disorders such as diabetes mellitus. Modern hardware and computer technologies make it possible, at a fairly subtle level, to assess the effect of infringement of systemic regulatory mechanisms on morphological changes at the level of the vascular bed. We come to the explanation of morphological changes in the body of patients from the organism level of regulation through the methods of mathematical analysis and computer modeling.

Keywords Diabetes mellitus · Autonomic dysfunction · Cardiac autonomic neuropathy · Functional body reserves · Autonomic regulation

1 Introduction

Diabetic autonomic neuropathy (DAN) is quite common [1, 2] and accounts for 3.5–6.0% of patients already at the onset of the disease and 100%in diabetic patients [3]. Cardiac autonomic neuropathy (CAN) is one of the most life-threatening complications in diabetes, as it can lead to reduced heart rate variability (HRV) and sudden death [4, 5]. But, first of all, disturbances in the structure and function of the autonomic nervous system are reflected in the regulatory processes. Heart rate variability impairment is associated with cardiac autonomic neuropathy. A decline in the tone and reactivity of the nervous system worsens the course of the underlying disease and can lead to the progression of vascular complications, but the mechanism of this process - the question is open. It is to this question that we tried to find answers in

I. Kurnikova (✉) · S. Gulova · R. Sargar · N. Kisliy
Department of Therapy and Endocrinology, RUDN University, Miklukho-Maklayast 6, 117198 Moscow, Russia
e-mail: curnikova@yandex.ru

I. Kurnikova
Department of Aviation and Space Medicine, Federal State Budgetary Educational Institution of Further Professional Education, Russian Medical Academy of Continuous Professional Education, Barrikadnaya st., h.2/1, b.1, 125993 Moscow, Russia

X.-S. Yang et al. (eds.), *Proceedings of Seventh International Congress on Information and Communication Technology*, Lecture Notes in Networks and Systems 447,
https://doi.org/10.1007/978-981-19-1607-6_22

our study. Modern methods of mathematical analysis, technologies, and computer assessment programs allow you to do this.

1.1 Purpose

To evaluate the effects of the cardiac form of diabetic autonomic neuropathy on the indicators of autonomic regulation of the cardiovascular system function.

2 Materials and Method

The study was conducted on the basis of the Endocrinology Department of City Clinical Hospital. A.K. Eramishantseva (Moscow). An observational cross-sectional study was conducted. Parameters of heart rate variability were studied in 101 patients with cardiac autonomic neuropathy according to the results of daily monitoring of heart rhythm. The control group consisted of 56 patients without diabetes mellitus.

Instruments and Data Collection Procedure

The parameters of carbohydrate and lipid metabolism of all the patients were studied in dynamics. Heart rate variability was assessed during 24-h ECG monitoring using the "Valenta" MN-08 (Russia) apparatus. Spectral analysis data (frequency characteristics) were evaluated in four groups. Indicators of spectral analysis: GP (total spectrum power), LF, VLF, ULF, HF (mc^2) or HF (%), LF (%), VLF (%), ULF (%). The analysis was carried out with the calculation of the power spectrum of slow oscillations in four frequency ranges. Very low frequencies – VLF (reflects the functional state of the super segmentary structures); low frequencies – LF (the activity of the sympathetic system (increased normal, lower stress, diseases of the cardiovascular system)); high frequencies – HF (the activity of parasympathetic cardio inhibitory center of the medulla oblongata).

Index of vegetative balanced (IVB) = LF/HF – vegetative sympathetic balance relationship is a power ratio of waves of low-frequency (LF) power to high-frequency (HF) waves. Increase-predominance of sympathetic activity, decrease – activated parasympathetic system. The average absolute value in healthy people: *0,7–1,5*. Index of center (IC) = (LF + VLF)/HF. IC (index of centralization) – shows the ratio of the activity of the central contour of regulation to autonomous.

The observation group consisted of 42 patients with type 1 diabetes and 59 patients with type 2 diabetes. The patients who were involved in the age range from 20 to 60 years (20–30 years old - 19 people, 31–40 years old - 24 people, 41–50 years old - 21 people, 51–60 years old - 37 people).

The results were compared with the data which were obtained in 56 healthy individuals, also divided into similar age groups.

For statistical analysis, the program was used STATISTIC 10,0 (Matematica®, Matlab®, HarvardGraphics®) StatSoft).The basic methods of statistical research were linear descriptive statistics (DescriptiveStatistics) with a calculation of the correlation of average standard deviations (corrs / means / SD).

Ethical Consideration

The presented study was carried out in accordance with the scientific direction of the department - "optimization of a systemic approach to the treatment and rehabilitation of patients with diabetes" and was approved in Ethics Committee of the Medical Institute of RUDN University (Protocol №. 8 by February 18, 2016).

3 Results

When comparing the frequency characteristics of heart rate variability between groups of patients with diabetes accompanied by CAN (Table 1) and healthy volunteers (Table 2), an increase in the low-frequency wave spectrum (VLF, ULF) was revealed, which indicates the activation of the central mechanisms of autonomous regulation and the transition from the control level to the control level, which is associated with significantly higher energy costs. Differences in indicators in patients of the compared groups start from the age group of 30 years and then can be traced in all older age groups.

In diabetic patients under the age of 41, an increase in frequency characteristics was observed across the entire spectrum, but hyper sympathetic tone prevailed (the LF/HF coefficient was increased), however, an increase in the power of the spectra characterizing sympathetic and parasympathetic activity makes it possible to understand why the transition to more high-and energy-consuming level of regulation. In the healthy group, the predominance of parasympathetic activity persisted up to 41 years.

In groups 41–50 years old and 51–60 years old, there was a further increase in the activity of central regulatory mechanisms (increase in VLF), and in the group

Table 1 Spectrum of frequency characteristics in age groups of patients with DT1

Parameters	Age 30 yrs (n = 19)	Age 40 yrs (n = 24)	Age 50 yrs (n = 21)	Age 60 yrs (n = 37)
CI	129,2 ± 10,8**	131 ± 7,7*	122,2 ± 4,1	120,4 ± 8,7
HR avg.day	78,2 ± 1, 8*	86,9 ± 6,9*	82,1 ± 6,3*	83,7 ± 11,9*
TP(MS^2)	12,155,6 ± 54,6	3978,3 ± 41,9*	5621,3 ± 67,4*	4437,8 ± 27,4*
ULF (%)	21,2 ± 2,9*	55,9 ± 6,8	48,0 ± 2,9*	40,8 ± 3,8*
VLF (%)	32,3 ± 6,9	27,1 ± 1,9	40,1 ± 8,0	42,7 ± 3,6
LF (%)	27,7 ± 3,5	17,0 ± 0,9**	27,8 ± 3,7*	28,1 ± 6,8*
HF (%)	27,9 ± 2,0	9,2 ± 0, 8*	15,1 ± 1,9*	16,9 ± 4,2*
LF/HF	1,8 ± 0,6	1,5 ± 0,3	2,0 ± 0,1*	1,7 ± 0,1*

Table 2 Spectrum of frequency characteristics in age groups of healthy individuals

Parameters	Age 20–30 yrs (n = 9)	Age 31–40 yrs (n = 14)	Age 41–50 yrs (n = 11)	Age 51–60 yrs (n = 22)
CI	121,5 ± 9,2	125,3 ± 7,5	128,5 ± 7,4	129,1 ± 11,3
HR avg.day	70,1 ± 1, 3	73,8 ± 6,4	76,9 ± 5,8	74,3 ± 9,9
TP(ms^2)	10,211,3 ± 22,6	7734,8 ± 24,5	6524,0 ± 32,3	5418,1 ± 31,3
ULF (%)	16,4 ± 2,1	16,8 ± 4,4	26,3 ± 2,7	25,5 ± 2,6
VLF (%)	20,7 ± 3,2	19,2 ± 2,8	26,6 ± 5,7	32,0 ± 4,1
LF (%)	14,1 ± 2,5	11,5 ± 1,0	26,5 ± 2,9	21,3 ± 3,8
HF (%)	27,6 ± 2,2	25,4 ± 0, 8	33,1 ± 1,7	31,3 ± 2,5
LF/HF	0,7 ± 0,3	0,7 ± 0,2	1,3 ± 0,3	1,6 ± 0,1

*Note*Marked indicators are with deviations from the normal age*
* –p < 0.05; ** –p < 0.01.
CI—circadian index; TP – total spectrum power in the range 0,003–0,40 Hz; ULF—ultra-low-frequency component of the spectrum (waves up to 0,0033 Hz); VLF—very low-frequency component of the spectrum (0,0033–0,04 Hz); LF—low-frequency component of the spectrum (0,04–0,15 Hz); HF—high-frequency component of the spectrum (0,15 to 0,40 Hz); LF/HF—ratio of the low-frequency component of the spectrum to the high-frequency. ms^2*—milisecond; day.avg—the average frequency in a daytime*

of healthy people, similar changes were observed only after 10–15 years. When comparing the spectra of frequency characteristics in the compared groups, it was found that the structure of the frequency spectrum in each age group of patients with diabetes was similar to the frequency spectrum of the healthy group, but in the older age group, that is, the changes in each group of patients with diabetes were comparable to those of the older age group of people without diabetes, which allowed us to put forward the hypothesis about "early aging of regulatory systems" in patients with diabetes mellitus - occurs 10 years earlier than biological age. And this phenomenon makes it possible to explain the early occurrence of atherosclerotic changes in the vessels of diabetic patients (in young patients, similar morphological changes in the vessels are often called macroangiopathy). The formation of cardiac autonomic neuropathy in diabetic patients leads to early aging of regulatory systems, which can be considered as one of the significant mechanisms for the development of macrovascular complications and together with endothelial dysfunction, leads to the development of cardiovascular diseases at a younger age.

We also conducted a comparative analysis of heart rate variability (HRV) indicators in patients with different types of diabetes. In patients with T2 type, autonomic regulation disorders were more pronounced, which was confirmed by a decrease in the circadian index to 1.19 ± 0.08.

Against the background of SD decompensation, the waves of the low-frequency spectrum (LF –25.1 ± 1.4%, VLF -–2.9 ± 3.3%, ULF –29.6 ± 6.9%) prevailed over high-frequency waves (HF-21, 5 ± 1.8%), which should be considered as an unfavorable sign indicating the depletion of the body's functional reserves. In these cases, central mechanisms are connected to the regulation, and the prevalence of the ULF spectrum (more than 50%) is a prognostic sign of the development of an unfavorable cardiovascular event in the coming months. In patients with T1DM, the circadian index did not have such a relationship with the quality of diabetes compensation, but unsatisfactory compensation also led to a significant expansion of the low-frequency spectrum (ULF−35.9 ± 6.8%, VLF−34.0 ± 3.9%, LF−16.9 ± 4.8%) to high frequency (HF−2.0 ± 1.7%). The vegetative balance coefficient in patients with TD1 was higher than the average values −1.58 ± 0.4, which was associated with hypersympathetic activity. A negative balance was also associated in patients with compensation syndrome DT2, but against the background of a highly activated parasympathetic nervous system (LF / HF −0.92 ± 0.1).

The heart rate in the observed patients varied widely and did not depend on the type or compensation of diabetes. The absolute values of the indicators of the intensity of the wave spectrum varied over a wide range and duplicated the relative indicators, but with much less accuracy.

4 Conclusion

As a result of the study, it has been demonstrated that autonomic dysregulation and hyper sympathetic action with centralization of management processes have promoted the manifestation of macrovascular complications of diabetes. Structural changes in the vascular wall in patients with diabetes (macroangiopathy) are similar to atherosclerotic changes, however, occur at younger age. The role of endothelial dysfunction and oxidative stress in the pathogenesis of these disorders is well known. The obtained data of the study show that the "phenomenon of early aging of regulatory systems" is one of the mechanisms of the appearance of macrovascular changes in patients with diabetes mellitus, which explains their early appearance.

References

1. Balcıoglu AS, Muderrisoglu H (2015) Diabetes and cardiac autonomic neuropathy: clinical manifestations, cardiovascular consequences, diagnosis and treatment. World J Diabetes 6(1):80–91
2. Vinik AI, Nevoret ML, Casellini C, Parson H (2013) Diabetic neuropathy. EndocrinolMetabClin North Am 42:747–787. https://doi.org/10.1016/j.ecl.2013.06.00130
3. Dimitropoulos G, Tahrani AA, Stevens MJ (2014) Cardiac autonomic neuropathy in patients with diabetes mellitus. World J Diabetes 5:17–39. https://doi.org/10.4239/wjd.v5.i1.1715
4. Tang ZH, Zeng F, Li Z (2014) Zhou L (2014) Association and predictive value analysis forresting heart rate and diabetes mellitus on cardiovascular autonomic neuropathy in general population. J Diabetes Res 2014:215473. https://doi.org/10.1155/2014/21547317
5. Karayannis G, Giamouzis G, Cokkinos DV, Skoularigis J, Triposkiadis F (2012) Diabetic cardiovascular autonomic neuropathy: clinical implications. Expert RevCardiovascTher 10:747–752
6. Makarov LM (2003) Holter monitoring. 2nd ed. M .: PH, Medpraktika-M, pp 340
7. Rogoza AN, Agaltsov MV, Sergeeva MV (2005) Daily monitoring of blood pressure: options for medical opinions and comments. Nizhny Novgorod, pp 38
8. Ryabykina GV, Sobolev AV (2005) ECG monitoring with rhythm variability analysis. M.: Publishing House Medpraktika-M, pp 224

Principles for Assurance on Corporate Governance of ICT

Petrus M. J. Delport and Rossouw Von Solms

Abstract ICT is critical to the well-being of any modern enterprise and should therefore also be governed and managed appropriately. As such, it is imperative that corporate governance of ICT (CGICT) be implemented and supported by the Board. It is quite clear that the Board remains ultimately accountable for CGICT. Consequently, the Board should have peace of mind regarding its fiduciary duties on CGICT. In other words, the Board must be provided with assurance on the overall efforts of CGICT within the enterprise, in order to provide stakeholder value. Therefore, the focus of this paper is twofold in an attempt to address the relationship between assurance and CGICT. Firstly, this paper will identify important principles and criteria from an assurance point of view, which can ideally be used to assist assurance professionals with understanding the underlying factors of assurance and its relationship with CGICT. Secondly, the principles and criteria identified in this paper can be used to develop a framework for assurance on CGICT, so as to formalise a process for providing the Board with peace of mind on their efforts towards good CGICT.

Keywords Assurance of IT governance · Corporate governance of ICT · Assurance principles · Assurance framework · The board of directors

1 Introduction

It is unclear when the true debate around corporate governance started [1]. However, the idea of the separation of ownership and control, as discussed by Berle and Means [2], is undoubtedly an early contribution to the prominence of this debate.

P. M. J. Delport (✉) · R. Von Solms
Nelson Mandela University, Port Elizabeth 6019, South Africa
e-mail: Pieter.Delport@noroff.no; Petrus.Delport@mandela.ac.za

R. Von Solms
e-mail: Rossouw.Vonsolms@mandela.ac.za

P. M. J. Delport
Noroff University College, 4612 Kristiansand, Norway

X.-S. Yang et al. (eds.), *Proceedings of Seventh International Congress on Information and Communication Technology*, Lecture Notes in Networks and Systems 447,
https://doi.org/10.1007/978-981-19-1607-6_23

Consequently, the concept of corporate governance has been around for quite some time; however, the term "*corporate governance*" was rarely used before the 1980s [3]. Nevertheless, the importance of corporate governance is unquestionable in these modern times as it introduces important concepts that any modern well-run enterprise should aim to follow.

At the forefront of leading, an enterprise is the Board of Directors (hereafter referred to as the Board). For the Board to exercise authority in an enterprise, it would have to direct *what* must be done in the enterprise. Typically, management would then be responsible for implementing these directives and the Board would control or monitor its implementation. Supporting this is the ASX Corporate Governance Principles and Recommendations [4], which can be summarised in that the Board, which fulfils corporate governance, acts on behalf of the shareholders and stakeholders. This is further supported by Justice Owen [5], who states that "*the Board is in control of the assets of an enterprise; however, the Board does not own those assets. They control the assets on behalf of the enterprise and, through the enterprise, others having an interest in the well-being of the entity*". As such, the Board's main responsibility is clear and involves looking after the well-being of the enterprise. For this reason, it is necessary for the Board to consider information and communication technology (ICT) when overseeing the well-being of an enterprise. This is due to ICT having been acknowledged as a core element to the success of any well-run modern enterprise [6]. As a result, ICT has become pervasive in the sense that ICT is now "*built*" into the strategy of most enterprises [7, 8]. This integration results in ICT demanding to be properly governed and managed.

Thus, it is important that the Board's corporate governance mandate extend from a general point of view to include ICT, which is nowadays generally termed the corporate governance of ICT.[1]

1.1 Corporate Governance of ICT

Corporate governance of ICT (CGICT) is defined as "*the system by which the current and future use of I[C]T is directed and controlled*" [10]. Additionally, CGICT involves not only evaluating the needs of ICT but also directing the use of ICT in order to support the enterprise's strategic objectives. After direction has been provided, the implementation and use of ICT should then be monitored, which facilitates the achievement of objectives. It is thus clear from the foregoing that CGICT typically

[1] It should be noted that, in the context of this paper, the terms *corporate governance of ICT* as well as *enterprise governance of IT* represent the same idea. De Haes and Van Grembergen [9] clearly state that "*Enterprise Governance of IT goes beyond the IT-related responsibilities and expands towards (IT-related) business processes*". They continue by stating that the standardisation organisation ISO has also moved in this direction, as represented by the 2008 release of ISO/IEC 38500 [10] as well as the 2016 release [11]. Therefore, in order to align this paper with the ISO/IEC 38500, the term *corporate governance of ICT* will be used as an all-encompassing term including both corporate governance of ICT and enterprise governance of IT.

has three definite tasks which should be addressed [10], which are firstly to evaluate, secondly to direct and lastly to monitor. These three tasks collectively provide the foundation of CGICT. Furthermore, these tasks also form a core part of the fiduciary duties of the Board. Consequently, the Board remains ultimately accountable for CGICT as well [8] and should be seen to exercise these tasks to effectively govern and manage ICT in general. This is essential, as when ICT is governed properly, it adds value to the enterprise. The value that ICT adds to the enterprise allows shareholders or stakeholders to receive maximum benefit. As such, it is important that the functioning of CGICT be measured for effectiveness. As a consequence, this will provide the Board with peace of mind regarding the fulfilment if its fiduciary duties on CGICT. In other words, the Board must be provided with assurance on the overall efforts of CGICT within its enterprise.

Unfortunately, corporate governance and subsequently CGICT are inherently complex and nuanced [12, 13]. As a result, this complexity is extended with the attempts to provide assurance in this regard. For this reason, this paper aims to address this complexity by providing two main perspectives. Firstly, this paper will identify important principles and criteria from an assurance point of view, which can ideally be used to assist assurance professionals with understanding the underlying factors of assurance and its relationship with CGICT. Secondly, the principles and criteria identified in this paper can ideally be used to develop a framework for assurance on CGICT, so as to formalise a process for providing the Board with peace of mind on their efforts towards good CGICT. As such, it is essential to consider the concept of assurance as a starting point.

2 Assurance Defined (What?)

According to the International Auditing and Assurance Standards Board [14], assurance may be defined as "*an engagement in which a practitioner expresses a conclusion designed to enhance the degree of confidence of the intended users, other than the responsible party, about the outcome of the evaluation or measurement of a subject matter against criteria*". In essence, assurance aims to provide "*an assertion about the effectiveness of internal controls*" [14]. COBIT 5 further supports this definition and adds that assurance is pursuant to an accountability relationship between two or more parties [13]. For instance, assurance professionals can be engaged to produce and disseminate a report on the findings or outcome of an assurance engagement to the accountable party [13].

It is clear from the definition that the context of the assurance engagement will determine who the role-players are, as well as what the subject matter is. Hence, one assurance engagement could differ from another assurance engagement. However, it is important to understand the various principles that constitute any typical assurance engagement. Therefore, the five core principles of a typical assurance engagement, as described by the International Auditing and Assurance Standards Board [14] and COBIT 5 For Assurance [13], should be considered.

2.1 Principle 1—Three-Party Relationship

A fundamental prerequisite of providing assurance is the concept of an accountability relationship. In essence, when conducting an assurance engagement an accountability relationship exists [13]. This accountability relationship implies that one party (the auditee) is responsible to another party (the user) for a specific subject matter [13].

Furthermore, the accountability relationship is a core concept of assurance. Although the role-players may differ from engagement to engagement, the accountability relationship typically arises as a result of contractual obligations, or because a user (of the assurance report) is expected to have an interest in how the accountable party (auditee) exercised its responsibilities in the provided delegation or directives [13].

In this instance, the first party is the accountable party. Typically, the accountable party represents the individual, group or entity that is ultimately responsible for governing or managing a specific subject matter, processes or scope [13]. In most assurance engagements, the accountable party is referred to as the *auditee* and typically involves some of the senior or executive management. Furthermore, the auditee typically has to provide some assurance on the fact that it is exercising its responsibility regarding the subject matter. In order to provide any assurance, the second party must be considered.

The second party consists of the assurance professional. Typically, the assurance professional can be referred to as the *practitioner, auditor or assessor* of the specific subject matter. The assurance professional is the person or persons who have overall responsibility for the performance of the assurance engagement [13]. It is important that the assurance professional have adequate experience in conducting a typical assurance engagement, as well as adhere to the principles contained in the Code of Ethics for Professional Accountants [15]. Furthermore, the assurance professional will continually utilise appropriate auditing techniques in order to conduct the assurance engagement. These techniques are standard auditing techniques but, in their entirety, they are sufficiently generic to be applied to any type of assurance engagement [13]. Typically, these appropriate auditing techniques consist of the following summarised tasks, as contained in COBIT 5 for Assurance [13]:

- **Enquire and confirm**—*investigate the context of the assurance engagement, by asking questions and corroborating management statements*
- **Inspect**—*review required elements within the context of the assurance engagement, by searching logs and walk-trough plans, and comparing expected findings*
- **Observe**—*observe all elements relating to the subject matter within the context of the assurance engagement*
- **Re-perform and/or recalculate**—*compare actual elements relating to the subject matter within context of the assurance engagement against expected values, then re-perform and recalculate*
- **Review automated evidence collection**—*collect sample data, analyse the data and extract exceptions.*

Ultimately, these aforementioned appropriate auditing techniques enable the assurance professional to draft and diffuse a report on the subject matter. This report should then be analysed by the third party, the *user* of the assurance report.

The final participant in the three-party relationship is the user of the assurance report. Typically, the user of the assurance report could include a variety of stakeholders, such as shareholders, creditors, customers, the Board, the audit committee, legislators and regulators [13]. In contrast to the first party, who had to provide assurance, the third party requests assurance. This is because the third party has an active interest in monitoring and controlling whether the auditee (first party) exercised its responsibilities or duties appropriately. It should also be noted that for some types of assurance engagement, the auditee (first party) and the user (third party) may be identical. This is very much the case during an internal or self-assessment.

In view of the preceding discussion, it is clear that the three-party relationship is a fundamental principle of assurance, as it clearly indicates the core role-players in a typical assurance engagement. Supporting the three-party relationship is the second principle, which relates to the context of the assurance engagement.

2.2 *Principle 2—Subject Matter*

The subject matter constitutes the specific topic or area over which assurance is to be provided [13]. Typically, the assurance professional will have to evaluate the specific subject matter against agreed criteria or metrics in order to construct an assertion about the effectiveness of the subject matter. Furthermore, it is important to note that the subject matter of an assurance engagement is considered appropriate when it is [14]:

- "*Identifiable, and capable of consistent evaluation or measurement against the identified criteria*".
- "*Such that the information about it can be subjected to procedures for gathering sufficient appropriate evidence to support a reasonable assurance or limited assurance conclusion, as appropriate*".

Once the subject matter has been identified, the assurance professional has to evaluate it. This involves having to identify suitable criteria.

2.3 *Principle 3—Suitable Criteria*

Suitable criteria play an important part in providing consistent grounds for confidence or assurance. In essence, the suitable criteria provide the assurance professional with a means of comparing the existing state of the subject matter against a standard or baseline [14]. It could be said that the assurance professional compares the current state of the subject matter with an envisaged state of the subject matter. Within this context the envisaged state of the subject matter could be based on legislation, baselines, standards and quantitative or qualitative performance metrics [14].

Furthermore, COBIT 5 for Assurance [13] stipulates that the suitable criteria must have the necessary information quality goal attributes. This is supported by the International Auditing and Assurance Standards Board [14]. Accordingly, suitable criteria should typically include the following information quality goal attributes [13]:

- **Objectivity**—*the suitable criteria should not be biased.*
- **Measurability**—*the suitable criteria must allow for reasonably consistent measurement of the subject matter, whether in a quantitative or qualitative manner.*
- **Understandability**—*the suitable criteria should be communicated appropriately and should negate any misinterpretations by the users of the report.*
- **Completeness**—*the suitable criteria should be adequately complete in order to make a conclusion on the subject matter.*
- **Relevance**—*the suitable criteria should be relevant to the identified subject matter.*

These information quality goal attributes of suitable criteria are important for conducting a reasonable assurance evaluation. In line with these, the assurance professional could now execute the assurance evaluation.

2.4 Principle 4—Execute Assurance Evaluation

During the execution of the assurance evaluation, the assurance professional will continually refer to the subject matter and the suitable criteria. In order to conduct the assurance evaluation, COBIT 5 for Assurance [13] highlights that the assurance professional has to follow a structured approach to reach a verdict on the specific subject matter. This structured approach (which will be highlighted in Sect. 4) is critical to ensure reproducibility and the gathering of sufficient appropriate evidence [14].

On completion of the assurance evaluation, it is important that the outcome of the assurance engagement be disseminated to the intended users.

2.5 Principle 5—Reporting and Conclusion on Assurance

With regard to the outcome or the results of the assurance engagement, it is important that the assurance professional communicate the findings and conclusions on the subject matter to both the accountable party (auditee) and the intended users of the assurance report. This is typically done when the assurance professional constructs a report, containing a conclusion, that expresses the assurance obtained about the subject matter and the underlying supporting evidence [14]. In essence, the report aims to either provide or withhold grounds for confidence on the specific subject matter. As such, both the accountable party (auditee) and the intended user of the assurance report could acquire reasonable assurance on the effectiveness of the subject matter.

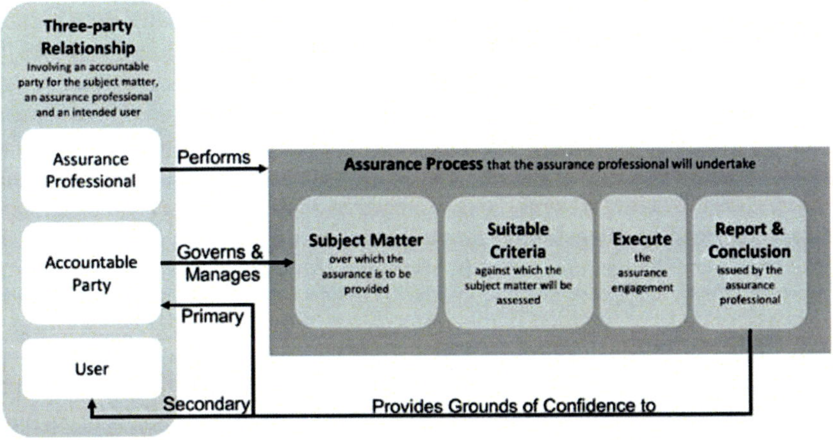

Fig. 1 Critical principles of assurance. Adapted from COBIT 5 for Assurance [13]

Taking the above discussion into consideration, Fig. 1 summarises the relationship among the important principles of a typical assurance engagement.

It is essential that assurance professionals consider these principles as the basis for any assurance engagement. Equally, any attempt at a framework for the assurance of CGICT should be seen to include these five previously mentioned principles of three-party relationship, subject matter, suitable criteria, execution of assurance evaluation and reporting on assurance.

The concept of assurance is critically important for the governance and management efforts in any modern enterprise. In addition, assurance equally applies to the CGICT efforts within the same environment (as highlighted in Sect. 1.1). Therefore, it is important that a clear understanding be gained on the importance of assurance on CGICT.

3 Importance of Assurance on Corporate Governance of ICT (Why?)

As mentioned previously (Sect. 1.1), the Board is ultimately accountable for the well-being of the enterprise. As most business processes in modern enterprises are totally dependent on ICT, this implies that the Board remains ultimately accountable for the overall efforts to implement CGICT in the enterprise [8]. Furthermore, it was also highlighted that the Board typically provides direction and delegates responsibility for implementing CGICT to management. Because the Board remains ultimately accountable, it is essential that the Board monitor the delegation of responsibility, as this is the foundation for providing assurance.

This foundation for providing assurance is evident from the ISO/IEC 38500 Standard [11] which clearly states that the Board "*should monitor, through appropriate measurement systems, the performance of I[C]T. They should reassure themselves that performance is in accordance with strategies, particularly with regard to business objectives*". It is clear that a definite component within the action of monitoring is the *reassurance*. In order for the Board to "reassure" themselves, an accountability relationship exists between the Board and, for instance, executive management (the Chief Executive Officer [CEO] in this case). Consequently, the Board requests grounds for confidence or assurance from executive management on its efforts regarding CGICT. As such, executive management must therefore be able to provide the Board peace of mind or assurance that it is exercising its delegated duties in terms of implementing effective CGICT.

Another instance of accountability in the same environment may be mentioned here. Because the CEO delegated the implementation and execution of CGICT to management (probably the Chief Information Officer [CIO] in this case), management must now provide assurance to the CEO on the delegated directives. As such, the internal auditor (assurance professional) will engage in a typical assurance evaluation, as well as evaluating and comparing the overall efforts to implement CGICT against acceptable criteria. Consequently, the internal auditor will produce an assurance report with the aim of either providing or withholding grounds for confidence (or assurance) on the overall efforts to implement CGICT. Finally, the assurance report, showing whether management has executed the directives, will be handed to the CEO, as the CEO has an active interest in whether management has exercised its responsibilities regarding the implementation of CGICT. This assurance report, in turn, can be issued to the Board as a means of providing assurance. In this context, it is clear that the accountability of the CEO for CGICT is supported by providing assurance on CGICT.

Considering the preceding discussion, it is clear that providing assurance is a means of supporting various role-player's accountability for CGICT. The question at this stage is, how should the accountable party (whether the Board, the CEO or the CIO) provide grounds for confidence or assurance on CGICT?

4 Providing Assurance (How?)

The process of evaluating the results of an assurance engagement on a specific subject matter, and constructing conclusions and recommendations, can be an extensive and quite complex task [13]. The following discussion will focus on identifying two critical frameworks that are key to providing assurance on CGICT. These two frameworks are, firstly, the Common Criteria [16] and, secondly, COBIT 5 for Assurance [13].

4.1 The Common Criteria

The Common Criteria for Information Technology Security Evaluation, typically referred to as the Common Criteria, is a framework contained in the ISO/IEC 15408-1 Standard [16]. The Common Criteria (CC) is a framework for ICT product security certification [16]. Typically, these ICT products are implemented using hardware, firmware and/or software and may include, amongst other things, access control devices and systems, databases, operating systems, key management systems, and network devices [17]. Furthermore, the CC allows for the results of various independent ICT product security evaluations to be compared. This is possible because the CC "*provides a common set of requirements for the security functionality of various I[C]T products and for assurance measures applied to these I[C]T products during a security evaluation.*". In essence, the evaluation of these ICT products provides a level of confidence or assurance for consumers [16]. The assurance will assist consumers to ascertain whether the ICT products fulfil their security needs and uphold their intended purpose [16].

With the foregoing in mind, it is clear that the CC provides a means for consumers (or users) to gain assurance on a specific ICT product (or subject matter). Although the CC is utilised within the domain of assuring ICT products, the ISO/IEC 15408-1 [16] also states that the CC could potentially be extended to other areas of ICT, which include CGICT. As such it is important to identify core criteria from the CC. These core criteria are considered to be the building blocks that can be utilised by an assurance professional to understand CGICT assurance; and more so the building blocks that a typical framework for the assurance of CGICT could incorporate.

Identified Core Criteria: A total of eight core criteria has been identified within the CC. It is important that each of these criteria be discussed in more detail.
Functionality, Correctness and Effectiveness: The CC is based on the construct that any typical ICT asset in an enterprise must be protected. As such, the owners of the enterprise will introduce various countermeasures, typically in the form of controls, to reduce any associated risk [16]. Furthermore, it is important that these countermeasures demonstrate their applicability and justification within the specific risk environment. Consequently, this will allow the owners to gain assurance on the countermeasure. In order to provide assurance on a countermeasure, three critical aspects should be considered. Firstly, the critical aspect of *functionality*, secondly, the critical aspect of *correctness* and, lastly, the critical aspect of *effectiveness* (also referred to as sufficiency). In order to explain functionality, correctness and effectiveness reference is made to both Fig. 2 and the following analogy.

Consider a storage room that contains a single door through which the room is entered. The room contains a valuable asset, such as gold for instance. It is also apparent that the asset (gold) is at risk. Consequently, countermeasures should be put in place to adequately protect the asset. There are various ways in which the asset can be protected; in this analogy, a door lock will be considered as a countermeasure or control.

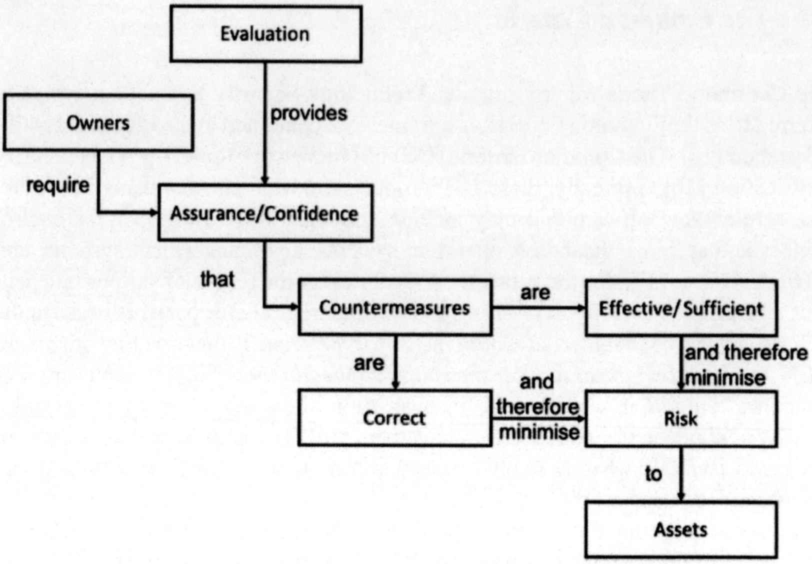

Fig. 2 Common criteria construct. Adapted from the ISO/IEC 15408-1 Standard [16]

It is logical that the lock must be acquired and installed or implemented. If the lock has been acquired but not installed, it can be said that the lock is not functional. In order for the lock to be considered functional, it has to be implemented or installed. This typically relates to the *functionality* of the countermeasure, in that it not only exists but also has been implemented (or installed in this case) within the environment.

While it is important that the lock has functionality, this will be of no use if it is not installed correctly (for instance, installed on the floor instead of the door). It is therefore important that the lock be correctly installed by a certified locksmith, who will install the lock in the correct manner in order to achieve its objective of protecting the asset. This is typically referred to as the *correctness* of the countermeasure or control.

Furthermore, it may be stated that even if the lock is functional and is installed and used in the correct manner, it would not necessarily provide assurance on the effective and sufficient protection of the asset behind the door. As such, the fact that the lock should be an adequately certified lock for protecting the asset is also critical. In other words, the quality or grade of lock should be effective and efficient enough to protect a high value asset. This is typically referred to as the *effectiveness* of the countermeasure or control.

In line with the above discussion, the general concept can be summarised. In order to provide assurance to the owners, the countermeasure must be evaluated. Firstly, it is important that the countermeasure exists and is implemented within the environment. As such, it can be said that the countermeasure is functional. Secondly, it is important to confirm that the countermeasure is implemented correctly, in that the countermeasure is installed by following best practices for instance [16]. It can then be said that the countermeasure is both functional and correct. Lastly, the countermeasure must be both adequate and sufficient to achieve its intended objective of protecting the ICT asset [16]. As such, it can then be stated that the countermeasure is functional, correctly implemented and effective enough to achieve the intended goals. Only then can assurance truly be provided to the owners. The three aspects of functionality, correctness and effectiveness together form an essential core criterion for assurance professionals and that should be included in the development of a framework for the assurance of CGICT.

Target of Evaluation: It has been stated previously (Sect. 4.1) that the CC is a framework for ICT product security certifications. In order to provide certification, the evaluation should target a specific ICT product. As such, the CC refers to the targeted ICT product in the evaluation process, as being the *target of evaluation* (TOE). In essence, the TOE is defined as a set of firmware, software and/or hardware consisting of ICT products or ICT product types [16]. For instance, a TOE could include, amongst other things, a software application, an operating system and a database [16].

Furthermore, the TOE could also be linked to the principle of a subject matter, as discussed in Sect. 2.2. In essence, the TOE becomes the subject matter that assurance is to be provided on.

Three-Party Relationship: Sect. 2.1 discusses the principle of a *three-party relationship*. The CC also highlights the existence of this relationship.

The three-part relationship consists firstly of the party which is referred to as the *consumer*. The CC aims to ensure that the evaluation of the TOE fulfils the needs of consumers [16] and that the consumers can utilise the results or outcomes of the evaluation to assist in making the decision as to whether the TOE fulfils their security needs. In other words, consumers can gain grounds for confidence or assurance on a TOE. It is thus obvious that the consumer may be directly linked to the user of the assurance report, as discussed in Sect. 2.1.

Secondly, the CC refers to a party called the *developer*. The developers are typically responsible for creating or developing a TOE. The CC aims to assist the developers with the development of their TOE, as well as preparing to evaluate their TOE [16]. The developers, in this case, constitute the party that is ultimately accountable, as discussed in Sect. 2.1. In this case, both the developers and accountable party may be referred to as the auditee.

Lastly, the final participant in the three-party relationship is the *evaluator*. The CC provides criteria to be used by evaluators when conducting an evaluation and constructing a conclusion on whether a TOE conforms to specific security requirements [16]. Again, the evaluators can be directly linked to the assurance professional as discussed in Sect. 2.1.

In summary, it is essential that the three-party relationship are understood by assurance professionals and form an integral part of any framework for assurance on CGICT.

Protection Profile: An essential part of the CC is the criterion referred to as a *protection profile* (PP). The PP is an implementation-independent statement of security needs for a TOE type [16]. For instance, the PP provides a generic or general baseline for what a typical network device (TOE type) should look like regarding security.

Furthermore, the PP provides guidance for the three-party relationship. For instance, the consumer can use the PP to understand what a typical TOE type should look like and align it with their needs. The developer, on the other hand, can use the PP to make sure that any new TOE of the same type is built to align with the general baseline, as showcased in the PP. In addition, the evaluator can use the PP as a criterion; for example, the evaluator could compare a new TOE of its corresponding PP and produce reasonable conclusions in the form of assurance.

In essence, the PP can be seen as an overarching general guiding specification for what a typical TOE type should look like.

Security Functional Requirements: Another criterion to consider is the *security functional requirements* (SFR). The CC states that the SFR are "*requirements, stated in a standardised language, which are meant to contribute to achieving the security objectives for a TOE*" [16]. In essence, the SFR is closely linked to the PP. Where the PP provides a generic or general baseline for what a typical TOE should look like, the SFR in contrast considers the environment of the TOE and provides more specific guidance on how a TOE should achieve its security goals in that particular context.

It should be noted that both the PP and the SFR are closely linked and provide input to a typical TOE. Nevertheless, both of which are core and should be considered by assurance professionals and subsequent attempts at developing a framework for assurance on CGICT.

Security Target: Regarding the CC, another core criterion can be identified, which is the *security target* (ST). The CC defines a ST as an "*implementation-dependent statement of security needs for a specific identified TOE*" [16]. From the definition, it is clear that the ST is linked to an instance of a specific TOE. In other words, the ST is implementation-dependent, meaning it is not generic to all TOEs.

In addition, the ST is essential from the perspective of both the developer and evaluator. For instance, the developer will consider the PP and decide to develop a new TOE. The developer will then make various claims on the fact that the new TOE adheres to or aligns with the PP. These claims are then documented or captured in the ST [16]. In contrast, the evaluator will refer to the claims made in the ST when evaluating the TOE.

Accordingly, the ST is an important criterion for providing reasonable assurance on a typical TOE.

Evaluation Assurance Level: In order for an evaluator to provide assurance, the CC refers to an *evaluation assurance level* (EAL). In essence, the EAL is required to provide a degree of confidence on a specific TOE [16]. After an evaluation of a TOE has been conducted, the evaluator can indicate on a predetermined scale the level of confidence regarding the TOE.

Common Criteria Process: In reference to the three-party relationship, it is clear that the CC follows a definite process or methodology which consists of two parts, a preparation and implementation part, as well as an assurance and evaluation part. With this in mind, the first part requires the developers to typically gather information on an intended TOE, by referring to the PP. Furthermore, the developers will make various claims about the TOE, which is documented in the ST, and typically continue to develop or implement the actual TOE. The aforementioned constitutes the *preparation and implementation part*.

Once the preparation or implementation part has been concluded, the TOE can be truly evaluated and assurance provided. At this point, the evaluator can evaluate the TOE by considering both the ST for various claims made and the PP containing a baseline for the TOE. The evaluator conducts the evaluation process and pronounces judgement in the form of a degree of confidence or assurance on the particular TOE. This entire process constitutes the *assurance and evaluation part*. In essence, this part aims to provide grounds for confidence or assurance that a TOE meets the SFR and aligns with the PP [16].

In summary, the aforementioned eight core criteria apparent in the CC are essential for any typical framework for providing assurance. As such, these identified core criteria can be expanded to the realm of CGICT in order to enlighten assurance professionals on CGICT assurance, as well as constructing a framework for the assurance of CGICT. However, it is also important to consider another framework, namely COBIT 5 for Assurance.

4.2 COBIT 5 for Assurance

COBIT 5 and the newly released COBIT 2019 provide a best practice framework for the governance and management of enterprise ICT, which aims to guide role-players on various components for the governance and management of enterprise ICT [18, 19]. In support, COBIT 5 for Assurance builds on the COBIT 5 framework, focusing on assurance [13]. Furthermore, COBIT 5 for Assurance provides detailed guidance for assurance professionals on how COBIT 5 can support different assurance engagements [13].

Therefore, it is essential that COBIT 5 for Assurance (CFA) be considered to further identify possible core criteria in order to understand CGICT assurance and which could be used to construct a framework for the assurance of CGICT.

Identified Core Criteria: Because CFA [13] is closely aligned with the International Framework for Assurance Engagement [14], most of the core criteria have already been discussed in Sect. 2, albeit in the form of principles. In addition, however, two further criteria can be identified. These two further criteria are: firstly, the ability of assessing the enablers/components that support CGICT and secondly, the usage of a defined assurance process. Nonetheless, these additional two criteria should be discussed in detail.

Table 1 Seven enablers of CGICT

Enabler	Description
Processes	Structured set of practices and activities that support the achievement of certain objectives and produce a set of outputs in support of achieving overall IT-related goals. Typically relates to any of the 40 processes (also called objectives) in COBIT 2019
Organisational structures	Typical key structures within an enterprise. Linked to various important roles and responsibilities
Culture, ethics and behaviour	Typically relates to the culture that should be created within an enterprise, as well as the actual behaviour of various role-players
Principles, policies and frameworks	Various documents and frameworks that should be in place in a typical enterprise. Also translates desired behaviour into practical guidance for management
Information flows and items	Various inputs and outputs, also known as work products, that should be leveraged within an enterprise to support communication between processes
Service infrastructure and application	Various systems and applications that should be in place in order to provide the enterprise with ICT processing and services
People, skills and competencies	Typically relates to the existence of key role-players and their underlying skills and competencies required in order to perform their duties effectively

Adapted from COBIT 5 & 2019 [18, 19]

Assessing the Enablers/Components Supporting CGICT: It is clear from previous discussions that CGICT is an essential part of any modern enterprise, and COBIT 5 [19] and COBIT 2019 [18] can play a pivotal role in supporting the efforts of the enterprise. As such, COBIT 5 has introduced the concept of an *enabler model*, which has been further expanded in COBIT 2019, is summarised in Table 1.

The enabler model aims to support the implementation or achievement of an enterprise's efforts to govern and manage CGICT [19]. To support the implementation of CGICT, COBIT 5 has identified seven main categories, referred to as *enablers*, or *components* in COBIT 2019 [18],[2] that support the achievement of effective CGICT. These seven enablers are essential to any attempt at implementing CGICT in an enterprise and, as such, must be evaluated adequately. Therefore, CFA builds on the foundation of these seven enablers and provides a means of assessing or evaluating the seven enablers.

In essence, these seven enablers should be understood by assurance professionals and also be present in a typical framework for assurance on CGICT. Furthermore,

[2] It should be noted that the term *enabler* refers to COBIT 5 terminology. In contrast, COBIT 2019 uses the term *component*. Accordingly, in the context of this paper, the term "*enabler*" will be used to represent both COBIT 5 and COBIT 2019 terminologies.

the core criterion identified is the enablement of assessing or evaluating these seven enablers within a typical CGICT assurance engagement. Without the ability to evaluate these seven enablers, true assurance on CGICT is not quite possible.

COBIT 5 for Assurance Process: It is clear that the CFA allows for the provision of assurance over the aforementioned seven enablers [13]. As such, a final core criterion that can be identified from the CFA is the assurance process that CFA applies to conduct an assurance evaluation of the seven enablers. This particular assurance process consists of three main phases, hereafter referred to as the "*Three-phased approach*". These three main phases include the following:

- **Phase 1**—*Scoping and determining the objective of the assurance engagement on the specific subject matter*
- **Phase 2**—*Understanding the environment and subject matter, identifying and agreeing on criteria, and conducting the assurance evaluation*
- **Phase 3**—*Reporting the outcome or findings of the assurance engagement.*

The three-phased approach is fundamental to providing assurance. As such, Fig. 1 can be seen to include the three-phased approach [13]. Therefore, Fig. 3 elucidates the three-phased approach as linked to the basic principles of assurance.

It is clear from the foregoing discussion that the three-phased approach provides more detail and guidance on the aforementioned CC process, specifically with regard to the assurance and evaluation part (as discussed in Sect. 4.1). Furthermore, the three-phased approach essentially addresses the *how* context of providing assurance. Consequently, the three-phased approach (Fig. 3) should be utilised by assurance professionals in order to provide assurance on CGICT. More so, the three-phased approach (Fig. 3) should be incorporated as a core criterion in a framework for the assurance of CGICT.

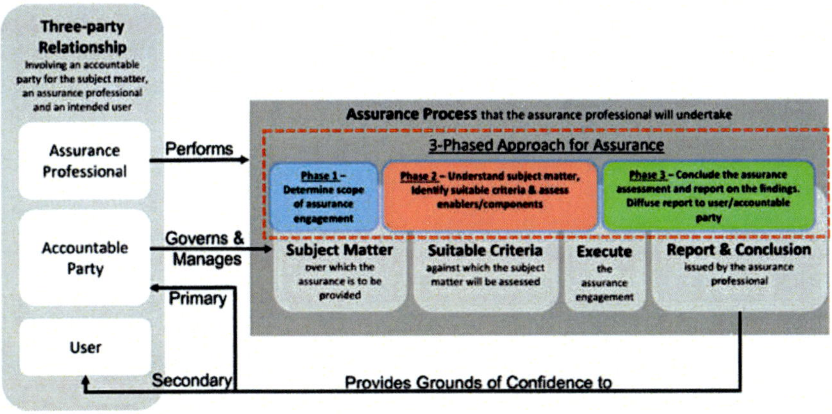

Fig. 3 Three-phased approach to assurance. Adapted from COBIT 5 for Assurance [13]

5 Conclusion

It is clear that corporate governance of ICT is critical to the well-being of any modern enterprise. As such, it is within the fiduciary duties of the Board to assume accountability for CGICT. In order for the Board to exercise its accountability, it must have peace of mind regarding the implementation of CGICT in the enterprise.

Accordingly, after management has followed the directives of the Board and implemented CGICT, it is essential that management provide assurance that it has exercised its responsibilities. In order for management to provide reasonable assurance, various principles of assurance must be considered. These principles form the foundation of any typical assurance engagement.

In this paper, two main frameworks were investigated that could potentially enlighten assurance professionals on CGICT assurance; and more so, to provide insight into the creation of a framework for the assurance of CGICT. Firstly, the Common Criteria framework was critically analysed, and a total of eight core criteria were identified which form the building blocks of an adequate framework for assurance. Secondly, the COBIT 5 for Assurance was also consulted, leading to the identification of two more core criteria. Again, these two criteria should be adopted by assurance professionals and are critical in creating a typical framework for the assurance of CGICT.

In summary, the main construct and principles from the Common Criteria can be combined with the three-phased approach from COBIT 5 for Assurance to summarise the twofold focus of this paper. Firstly, to identify important principles and criteria from an assurance point of view, which can ideally be used to assist assurance professionals with understanding the underlying factors of assurance and its relationship with CGICT. Secondly, identifying principles and criteria that can ideally be used to develop a framework for assurance on CGICT, so as to formalise a process for providing the Board with peace of mind on their efforts towards good CGICT.

In the light of the discussion, it is clear that in order for the Board to perform its fiduciary duties, it must be provided with assurance. The aforementioned principles and criteria can ideally be utilised to assist assurance professionals to provide reasonable assurance on CGICT. More importantly, these principles and criteria can be used to create a framework for assurance on CGICT.

References

1. Du Plessis JJ, Hargovan A, Bagaric M, Harris J (2014) Principles of contemporary corporate governance, 3rd edn. Cambridge University Press
2. Berle AA, Means GC (2017) The modern corporation and private property. Taylor & Francis, New York, USA
3. Tricker B (2015) Corporate governance: principles, policies, and practices, 3rd edn. Oxford University Press, USA
4. ASX Corporate Governance Council (2019) Corporate governance principles and recommendations, 4th edn. Technical report, Australian Securities Exchange

5. Western Australia Supreme Court: The Bell Group Ltd. (in liq) v Westpac Banking Corporation (2008). https://www.allens.com.au/pubs/insol/foinsol14sep09.htm
6. Von Solms S, Von Solms R (2008) Information security governance. Springer, Johannesburg, SA, USA
7. Van Grembergen W, De Haes S (2009) Enterprise governance of information technology: achieving strategic alignment and value. Springer Science & Business Media
8. Institute of Directors South Africa: King Report on Corporate Governance for South Africa (King IV) (2016) Technical report, Institute of Directors South Africa. https://www.iodsa.co.za/page/AboutKingIV. Accessed 2021-10-01
9. De Haes S, Van Grembergen W (2015) Enterprise governance of information technology: achieving alignment and value, featuring COBIT 5, 2nd edn. Springer Publishing Company, incorporated
10. ISO/IEC 38500: International Standard ISO/IEC 38500:2008—Corporate Governance of Information Technology (2008) Technical report, International Organization for Standardization
11. ISO/IEC 38500: International Standard ISO/IEC 38500:2016—Information Technology—Governance of IT (2016) Technical report, International Organization for Standardization
12. Niles S (2021) USA: corporate governance laws and regulations. https://iclg.com/practice-areas/corporate-governance-laws-and-regulations/usa
13. ISACA: COBIT 5 for Assurance (2013) Information systems audit and control association. Rolling Meadows, IL, USA
14. International Auditing and Assurance Standards Board (2008) International framework for assurance engagements. Technical report, International Auditing and Assurance Standards Board
15. International Federation of Accountants (2006) Code of ethics for professional accountants. Technical report, International Federation of Accountants
16. ISO/IEC 15408-1: International Standard ISO/IEC 15408:2009—Information Technology—Security Techniques—Evaluation Criteria for IT Security—Part 1: Introduction and General Model (2009) Technical report, International Organization for Standardization
17. Common Criteria Recognition Arrangement: Common Criteria for Information Technology Security Evaluation (2021). https://www.commoncriteriaportal.org/pps/. Accessed: 2021-10-02
18. ISACA: COBIT 2019 (2018) Framework introduction & methodology. Information Systems Audit and Control Association, Schaumburg, IL, USA
19. ISACA: COBIT 5 (2012) Framework. Information Systems Audit and Control Association, Rolling Meadows, IL, USA

The Role of Telecommunication Technology During COVID-19 Pandemic in Indonesia

Vina Fujiyanti, Syifaul Fuada, and Nadia Tiara Antik Sari

Abstract COVID-19 pandemic gives a very significant impact on many aspects of life in a global scale. In Indonesia, as in many countries in the world, public activities involving many people are restricted and regulated to prevent the virus' broader spread. Technology which still enables activities to be conducted individually and remotely is needed, because during the pandemic, they can only be done very limitedly, if not stopped. Therefore, people are utilizing telecommunications to assist activities during the COVID-19 pandemic. Several innovations involving telecommunication technology are applied, e.g., distance learning (e-learning), online business (e-commerce), and digital health services (e-health). However, there are many challenges in using telecommunication technology during a pandemic in developing countries, especially in Indonesia. This article examines (1) the role of telecommunication technology in education, economy, and health sectors, (2) the challenges, and (3) how telecommunication technology as the solution can be more effective in future. This article provides an outlook and in-depth overview of how the impact of recent technology, especially telecommunication, played a vital role in the three crucial sectors in mid-low countries like Indonesia during the COVID-19 outbreak.

Keywords Telecommunication technology · E-learning · E-commerce · E-health · COVID-19 pandemic · Indonesia

1 Introduction

Currently, the world is facing complications caused by coronavirus disease (COVID-19). Accumulatively globally, an increase in positive cases of COVID-19 occurs every day. In Indonesia, as of April 26, 2021, positive cases of COVID-19 have reached 1.64 million cases (see the real-time updates obtained at Google data: https://www.google.com/search?client=firefox-b-d&q=kasus+covid+di+Indonesia). To prevent this virus comprehensively, World Health Organization (WHO) urges people in the world

V. Fujiyanti · S. Fuada (✉) · N. T. A. Sari
Universitas Pendidikan Indonesia, Bandung, Indonesia
e-mail: syifaulfuada@upi.edu

X.-S. Yang et al. (eds.), *Proceedings of Seventh International Congress on Information and Communication Technology*, Lecture Notes in Networks and Systems 447,
https://doi.org/10.1007/978-981-19-1607-6_24

to stay at home, avoiding places that highly possible to create the masses gather. The Indonesian government implements social distancing and physical distancing policies to prevent more COVID-19 infection. However, the policies taken by the Indonesian government have made a great impact on the use of technology in many areas of life including education, economy, and healthcare sectors. In education sector, this policy makes the learning process in schools (e.g., elementary, secondary, and vocational schools) and universities is run on limited basis. They are fully utilizing technology to implement distance learning. All school and campus officials (faculty members and staffs) do "work from home" (WfH). In the *economy* sector, the WfH has made some Indonesians take the opportunity to utilize the existing technology in buying and selling goods through online platform; e-commerce. E-commerce is an option, and a direct seller (as well as other occupations) could take to improve their income which has got significantly decreased due to the pandemic. While in the healthcare sector, as nowadays, various online applications have been widely employed, e.g., information systems based on doctor services, pharmacy, and so on [1]. Many doctors have chosen to use technology in treating their patients without face-to-face interactions to limit contact between doctor and patient that could possibly lead to COVID-19 infection.

The Indonesian government has collaborated with several major cellular operators in Indonesia, such as Telkomsel, Indosat Ooredoo, Three, XL Axiata, Axis, and Smartfren, in Internet quota assistance policies for distance learning needs [2, 3]. Mobile operators help by providing competitive rates for Internet data quota, meaning that the company continues to run its business to reach customers. In line with that, the conditions between companies, which are competition, technology encouragement, the economic stage that tends to increase, and the COVID-19 pandemic stimulus, have made operators also improve service quality [4, 5]. Society is still required to understand that telecommunication operators in sub-substituting this program are also for social-oriented purposes and the next generation of children. Because after all, the telecommunications industry needs to establish good relationships with stakeholders, including customers, the central government, local governments [6], and all related parties. In this case, the leadership factor of a telecommunications company also determines the stability of its business to survive this pandemic so that telecommunication company employees remain comfortable working to serve loyal customers [7, 8].

In other words, telecommunication technology has become increasingly crucial in helping people's activities during the pandemic. In addition, the world has entered industry 4.0 era, where the situation demands modern society to operate the recent technology fluently in all life activities. However, technology is not enough to bring rapid change; it requires environmental support that can adapt quickly [9]. The COVID-19 pandemic condition causes the telecommunication industry and tele-conferencing system providers grow rapidly. This article will describe the role of telecommunication in the fields of education, economy, and healthcare in more detail. We found that there is no review article discusses the three essential pillars (i.e., education, economy, and healthcare) in Indonesia during the COVID-19 associated with the role of telecommunication industry in one comprehensive study.

2 Methodology

This paper is a survey study in which the literatures are obtained from reliable sources related to the COVID-19 pandemic impact, especially in education, economy, and health sectors in Indonesia, a low-middle country income. We also review the Indonesia's situation during the pandemic which was observed from March 2020 to March 2021. Since the specific case is observed based on the Indonesia territory only so that the references used of this paper as the fundamental framework are almost 80% written in Indonesian, mainly surveyed from domestic literature databases (e.g., National journals and conferences).

3 Results and Analysis

3.1 Education Sector

During this pandemic, telecommunications have an even more crucial role in education sector, where the learning system is transformed into an online learning system. The online learning system is a process of interaction between educators and students using Internet access and network connection [10]. It is one solution to overcome face-to-face learning problems, given the problems of distance, time, location, and costs that have become obstacles during this pandemic [11]. According to reference [12], limiting physical contact between college students to cut the virus transmission/spread is the reason why schools closed. Closing the schools' access and switching the education system from offline to online are *considered* the best solution to reduce the death rate and the spike of COVID-19 positive cases. With this education system transition, the role of technology is becoming even more needed and crucial. Zhang et al. in reference [10] stated that one of the alternatives in modifying the learning in class is Internet access and information technology. According to Wulandari and Almenda in reference [13], technology is very important for teachers in the teaching process. In fact, there has been growing study of technological pedagogical content knowledge (TPACK) since 2006 which shows the need and importance of technology as a helping tool to deliver learning materials [14]. In the pandemic situation, it is pivotal. One of the most important technologies in this time is telecommunication technology.

An example of the uses of telecommunications with Internet technology is in the form of mobile learning (M-learning). It is a collaboration of mobile computing and e-learning used for learning and teaching without being hindered by space and time [15]. M-learning refers to IT tools in the learning process because it can be accessed by electronic gadgets such as laptop, smartphones, and tablets. M-learning is part of e-learning, and they are the primary means in the continuation of distance learning. E-learning can make students can learn independently from various sources [16–18]. In a study of e-learning during COVID-19 pandemic conducted by reference [19], out of

175 respondents, 70.86% of them agreed that e-learning is able to broaden horizons. Furthermore, 82.86% of them admitted that e-learning improves their independence learning abilities.

Online platforms as the technology assistance in online learning are quite a lot, including *WhatsApp group*, *Zoom*, *Google Classroom*, *Google Meet*, *Telegram*, *Line,* etc. Using these learning platforms, teachers and lecturers deliver learning materials through videos, synchronous (real-time) meeting, asynchronous videos, text-based teaching employing PowerPoint media in the *WhatsApp groups,* etc. Technology makes the learning process done without face-to-face interaction since COVID-19 outbreak on March 2020 [20–22]. As an example, in our campus, *Universitas Pendidikan Indonesia* (Indonesia University of Education), in giving quizzes (exams), our lecturers take advantage of several online media (e.g., *Google forms*, *Quizizz*, *Kahoot*, etc.). It proves that the use of technology in learning can create variations that make students do not feel bored during the online learning process. However, online media selection needs to be considered carefully so that students get the impression that the platforms used are proper and effective in supporting their learning activities.

There are many challenges in implementing technology in education. According to reference [23], online education drawbacks include material access, course quality, the relationship between educators and students, and students' honesty. Although online learning is considered the best solution in conducting education during a pandemic, not all institutions can provide online learning process fast. Many institutions are still seeking ways to develop their online learning system so that it runs effectively [23]. One of the challenges in e-learning is educators' and students' skill to operate the existing technology [24]. It is not uncommon for lecturers or teachers and even students to find difficulty using learning support devices because technology continues to develop every time that the devices' operation differs from every version. It can make students less optimal in understanding the material taught by the educator. Reference [25] stated that the lack of students mastering the material made them depressed and anxious in facing exams. Students are also worried about dropping grades and cannot keep up with learning well at the next level. This condition gives anxiety until graduation arrives. Therefore, reference [23] suggested to form study groups that students can support each other so that online learning satisfaction can be fulfilled.

The other challenge is Internet connection. Students are worried that they cannot follow the learning process well because of bad Internet access and the difficulty in getting big Internet quota. Unstable network makes students afraid of experiencing unexpected technical problems during the online learning process. Educators change their teaching style to be effective in providing material online. It is not uncommon for educators not to understand technology. Hence, it is challenging to help them to be able to do the online learning process well, e.g., creating material content, operating meeting applications, etc. During online learning, educators and students have difficulty in communicating two ways.

Often, teachers' explanation could not be received well by the students, and students have relatively more limited time to interact with the teachers. Therefore, distance learning during this pandemic must be continuously developed to be more effective in future.

Big Internet quota is also needed to support e-learning during this pandemic. The contribution of Internet operators in providing free big Internet quota is a great support for education during this situation. Many cellular telephone operators provide educational packages (Internet quotas) to support online learning at affordable prices. In addition, Indonesian Ministry of Education and Culture (*Kemendikbud RI*) also provides free education quota assistance to students and educators every month with different packages in each education level. For example, on December, a college student received an education Internet quota for the next three months (December 2020 to February 2021) only 90 GB.

In summary, telecommunication technology holds an even more crucial role in enabling education in this pandemic situation by e-learning process. However, teachers and students' technology literacy, Internet access, and the price of Internet quota are still becoming challenges. Therefore, e-learning research and development need to be done to investigate strategies to conduct it more effectively. Subsequent section will elaborate similar explanation in the field of economy. It includes the role and challenges of telecommunication in economy sector.

3.2 *Economy Sector*

Telecommunication also has a crucial role in the economy during pandemic. It is the time where more global community uses the Internet in buying and selling activity. The Indonesian Minister of Finance said that when COVID-19 broke out, Indonesian economy declined to minus about 5.3% [26]. This case does not only occur in Indonesia but also in other ASEAN countries such as Malaysia, Thailand, and Philippines. Therefore, the Internet is a vital resource to counter the decreased economy. With the Internet, producers and consumers can communicate without being constrained by distance and time. Many Internet users in the twenty-first century also open up business opportunities with a broader scope. The *We Are Social* launched digital data in 2020. It explains that 2.42 billion of the 4.3 billion people in Asia Pacific have Internet access, and the population that has used social media is around 2.14 billion people. In the 2020 digital data, it is stated that 88% of online transaction players are Indonesian citizens [27]. Industry 4.0 has strong correlation to technology. The role of industry 4.0 in economy sector is as follows: to ensure more businesses survive, to help restoring business to normalcy, and to provide a platform in developing various new medium to long term businesses [9]. It certainly good for future business opportunities through e-commerce utilization.

E-commerce is exchanging products, services, or information via the Internet [28–30]. With an e-commerce system, buyers and vendors not only view and display information about the goods being sold but also negotiate prices, track shipment status, make and receive payments, send and receive orders [31]. According to reference [9], in working with buyers, sellers have to concern about six aspects, i.e., technology, capabilities, firm support, consumption models, flexible growth, and understanding the target market. During this pandemic, various goods are traded according to the sellers' creativity by understanding the trend nowadays. Sellers take advantage of various platforms, such as social media and e-commerce. According to reference [32], social media's role is to connect the interactions of vendors and buyers. Also, social media has the opportunity to find buyers and build the image of a selling brand. Reference [33] stated that the indicators of the survival of an e-commerce company are by viewing the existence of good management team, good network infrastructure, security, and attractive site design. Therefore, e-commerce producers should be reliable and smart in managing their products. Thus, they have quality selling value in consumers' perspectives/needs. Online stores have many advantages for both vendors and buyers. Reference [33] suggests several benefits of e-commerce, for example, online consumers do not have to worry about buying something from a distant store because all of e-commerce will provide the best services to meet a certain standard (quality control, refund guarantee, etc.). In certain events, e-commerce often provides many attractive bonuses such as discounts, cashback, and various vouchers, which are very beneficial for online media users. E-commerce sites in Indonesia such as *Shopee*, *Tokopedia*, *Bukalapak*, and *Akulaku* provide promos and flash sales on several products. The vendors also have many benefits in selling their goods online (unlike offline stores), and the vendors only need relatively low effort in keeping the shop or promoting their products.

With technology, vendors do not need a real shop. Renting cost is not needed. Furthermore, marketing can be done by making digital flyers or pamphlets with persuasive sentences. Online stores also currently make use live streaming feature to conduct promotions. Vendors and buyers can interact directly through the comments feature. The existence of e-commerce as a means of buying and selling can also make customers consider the products to be purchased because several online stores must have displayed much feedback from other customers. It certainly makes it easier for buyers to choose quality products.

However, there is a challenge in using this e-commerce. Personal data are needed to create the e-commerce accounts to be able to do online transactions. It consists data leak risk. Privacy violation may happen. Some people can use our personal data to make a fraud which is criminal and can make our name bad.

In summary, telecommunication brings many benefits in rolling the wheel of economy in this pandemic. The existence of e-commerce offers many advantages both for the vendors and buyers. The challenge of data leak risk is still there but technology is increasingly more sophisticated, and cyber policymakers can guarantee higher security. Subsequent section explains the role and challenge of telecommunication in health sector.

3.3 Health Sector

Not only affect education and economy sectors, COVID-19 pandemic also certainly affects health sector. Health services should be conducted even more carefully by considering a very strict health protocol to avoid infection. By using technology, medical personnel can examine patients remotely. A well-known innovation to the public is an online health consultation system through smartphone application. This innovation is undoubtedly beneficial during this pandemic, considering Indonesia policy that requires people to make distance and to avoid mass gatherings.

Besides, technology can help all people to share information about the COVID-19 virus. However, this certainly raises many challenges where it is not uncommon for the spread of hoaxes (invalid news). To avoid this, one of the major platforms, *Facebook*, announced that they are working with third parties to verify facts to avoid spreading invalid news on social media [34], as Indonesia has numerous of telemedicine platforms that affiliated with the Ministry of Health: Halodoc, YesDoc, Alodokter, KlikDokter, SehatQ, Good Doctor, Klinikgo, Link Sehat, Milvik, Prosehat, and Getwell.

Industrial technology 4.0 applied to digital health services is the beginning of current health industry development. Another innovation in the health industry is artificial intelligence (AI) combined with point of care (POC) diagnostics to perform independent testing after someone exposed to COVID-19 [35]. In addition, there are AI-based cameras developed in China that can identify places and people infected with the virus. It shows that industry 5.0 starts to develop and is promising enough to be applied in health. With this AI technology, doctors may not need to perform surgery directly. It is conducted by specially designed robots. The challenge in e-health is in the degree of accuracy both in online health consultation and robotic health treatment. Some patients might still feel the need of human-to-human interaction to care for their body. Further, research on robot assisted treatment/surgery is also needed to convince more people to be willing to do it.

4 Conclusion

Telecommunication technology is pivotal to many activities during this COVID-19 pandemic, especially to education, economy, and healthcare. It allows education to be carried out remotely and individually through e-learning which is important to realize the social distancing needed in this situation. Likewise, online business/e-commerce serves save and easy interaction between vendors and buyers because all activities are done through technology-based specific platforms. In health sector, many innovations are created to address health problems during this pandemic. One of the examples is online consultation. Quick adaptation is needed in applying the technology, especially in Indonesia. However, challenges still exist from the use of technology in the aforementioned three sectors. The extent of digital literacy to operate the technology, the risk of personal data leak, and the accuracy of diagnosis through online consultation is some of them. Therefore, the utilization of telecommunication technology still needs to be socialized and developed. Thus, its application will be more effective because it still enables and even improves the education, economy, and healthcare services after COVID-19 pandemic. Finally, it is safe to say that "*where there is human activity during a pandemic that requires Internet connectivity that is where telecommunications come to play.*" This paper only explains the common things. Telecommunication technology has many application areas such as wireless, wired, networking, cellular, mobile communication, IoT, and WSN. It should be compared by each specified application. For future work, we will describe the existing technology specifically for each sector. Then, add some several comparisons of the advantage and disadvantage.

References

1. Rohmah MK, Wahyuni KI, Ambari Y (2021) Edukasi dan Pendampingan Dalam Pencegahan COVID-19 Memulai Aplikasi E-Health Pada Mahasiswa Stikes Rumah Sakit Anwar Medika dan Keluarga. J Abdikarya J Karya Pengabdi Dosen Dan Mhs 4:12–18
2. Admin (2021) Siaran Pers : Operator Seluler Berkomitmen Sukseskan Bantuan Kuota Data Internet 2021, https://pusdatin.kemdikbud.go.id/siaran-pers-operator-seluler-berkomitmen-sukseskan-bantuan-kuota-data-internet-2021/. Last accessed 16 Aug 2021
3. Mursid F, Yolandha F (2021) Ini Operator Seluler yang Ikut Program Bantuan Internet, https://www.republika.co.id/berita/qh7gb2370/ini-operator-seluler-yang-ikut-program-bantuan-internet. Last accessed 17 Aug 2021
4. Cristie (2013) Pengaruh Kualitas Pelayanan Terhadap Kepuasan Nasabah Pada BPR Syariah Rinjani Batu. J Ilm Mhs Fak Ekon Dan Bisnis 1:1–10
5. Cristie PA (2013b) Pengaruh Kualitas Pelayanan Terhadap Kepuasan Nasabah Pada Bpr Syariah Rinjani Batu, http://repository.ub.ac.id/106706/s
6. Yuliandani T (2017) Integrasi Kepentingan Stakeholder Dalam Pembangunan Berkelanjutan di Kabupaten Bojonegoro. http://repository.unair.ac.id/67979/
7. Fonda B (2015a) Pengaruh Gaya Kepemimpinan Situasional Terhadap Budaya Organisasi Dan Kepuasaan Kerja Karyawan (Studi Pada Karyawan Pt. Telekomunikasi Indonesia Tbk. Wilayah Malang). http://repository.ub.ac.id/117467/
8. Fonda B, Utami HN, Ruhana I (2015b) Pengaruh Gaya Kepemimpinan Situasional Terhadap Budaya Organisasi dan Kepuasan Kerja Karyawan (Studi pada Karyawan PT. Telekomunikasi Indonesia Tbk. Wilayah Malang). J Adm Bisnis 25:1–8
9. Czifra G, Molnár Z (2020) Covid-19 and industry 4.0. Res Pap Fac Mater Sci Technol Slovak Univ Technol 28:36–45. https://doi.org/10.2478/rput-2020-0005
10. Sadikin A, Hamidah A (2020) Pembelajaran Daring di Tengah Wabah Covid-19. BIODIK 6:109–119. https://doi.org/10.22437/bio.v6i2.9759
11. Kusuma JW, Hamidah (2020) Perbandingan Hasil Belajar Matematika Dengan Penggunaan Platform Whatsapp Group Dan Webinar Zoom Dalam Pembelajaran Jarak Jauh Pada Masa Pandemik Covid 19. J Ilm Pendidik Mat 5:97–106. https://doi.org/10.26877/jipmat.v5i1.5942
12. Viner RM, Russell SJ, Croker H, Packer J, Ward J, Stansfield C, Mytton O, Bonell C, Booy R (2020) School closure and management practices during coronavirus outbreaks including COVID-19: a rapid systematic review. Lancet Child Adolesc Health 4:397–404. https://doi.org/10.1016/S2352-4642(20)30095-X

13. Pujiasih E (2020) Membangun Generasi Emas Dengan Variasi Pembelajaran Online Di Masa Pandemi Covid-19. Ideguru J Karya Ilm Guru 5:42–48. https://doi.org/10.51169/ideguru.v5i1.136
14. Koehler MJ, Mishra P, Cain W (2013) What is technological pedagogical content knowledge (TPACK)? J Educ 193:13–19
15. Husna R (2020) Efektivitas Pembelajaran Turunan Pada Masa Pandemi Covid-19 Melalui Media Mobile Learning Ditinjau Dari Hasil Belajar Mahasiswa. 7:324–333. https://doi.org/10.46244/numeracy.v7i2.1187
16. Abdul Majid NW, Fuada S (2020) E-learning for society: a great potential to implement education for all (EFA) movement in Indonesia. Int J Interact Mob Technol IJIM 14:250–258. https://doi.org/10.3991/ijim.v14i02.11363
17. Majid NWA, Fuada S, Fajri MK, Nurtanto M, Akbar R (2020) Progress report of cyber society v1.0 development as a learning media for Indonesian society to support EFA. Int J Eng Pedagogy 10:133–145. https://doi.org/10.3991/ijep.v10i4.13085
18. Pramudita R, Fuada S, Majid NWA (2020) Studi Pustaka Tentang Kerentanan Keamanan E-Learning dan Penanganannya. J Media Inform Budidarma 4:309–317
19. Radha R, Mahalakshmi K, Kumar DVS, Saravanakumar DA (2020) E-learning during lockdown of Covid-19 pandemic: a global perspective. Int J Control Autom 13:1088–1099
20. Handarini OI (2020) Pembelajaran daring sebagai upaya study from home (SFH) selama pandemi Covid 19. J Pendidik Adm Perkantoran JPAP 8:496–503
21. Astini NKS (2020) Tantangan Dan Peluang Pemanfaatan Teknologi Informasi Dalam Pembelajaran Online Masa Covid-19. Cetta J Ilmu Pendidik 3:241–255. https://doi.org/10.37329/cetta.v3i2.452
22. Purwanto A, Pramono R, Asbari M, Santoso PB, Mayesti L, Hyun CC, Putri RS (2020) Studi Eksploratif Dampak Pandemi COVID-19 Terhadap Proses Pembelajaran Online di Sekolah Dasar. 2:1–12
23. Hassan M (2021) Online teaching challenges during COVID-19 pandemic. Int J Inf Educ Technol 11:41–46. https://doi.org/10.18178/ijiet.2021.11.1.1487
24. Dabbagh N (2007) The online learner: characteristics and pedagogical implications 7:217–226
25. Oktawirawan DH (2020) Faktor Pemicu Kecemasan Siswa dalam Melakukan Pembelajaran Daring di Masa Pandemi Covid-19. J Ilm Univ Batanghari Jambi 20:541–544. https://doi.org/10.33087/jiubj.v20i2.932
26. Putra DA (2021) Sri Mulyani Gambarkan Brutalnya Dampak Covid-19 ke Ekonomi Dunia. https://www.liputan6.com/bisnis/read/4386109/sri-mulyani-gambarkan-brutalnya-dampak-covid-19-ke-ekonomi-dunia. Last accessed 13 Feb 2021
27. Ramadhan B (2020) Data Internet di Indonesia dan Perilakunya Tahun 2020. https://teknoia.com/data-internet-di-indonesia-dan-perilakunya-880c7bc7cd19. Last accessed 25 Dec 2020
28. Irmawati D (2011) Pemanfaatan e-commerce dalam dunia bisnis. J Ilm Orasi Bisnis 6:95–112
29. Khasanah FN, Rofiah S, Setiyadi D (2019) Metode User Centered Design Dalam Merancang Tampilan Antarmuka Ecommerce Penjualan Pupuk Berbasis Website Menggunakan Aplikasi Balsamiq Mockups. JAST J Apl Sains Dan Teknol 3:14–23. https://doi.org/10.33366/jast.v3i2.1443
30. Samsiana S, Herlawati, Nidaul Khasanah F, Trias Handayanto R, Setyowati Srie Gunarti A, Raharja I, Maimunah B (2020) Pemanfaatan Media Sosial dan Ecommerce Sebagai Media Pemasaran Dalam Mendukung Peluang Usaha Mandiri Pada Masa Pandemi Covid 19. J Sains Teknol Dalam Pemberdaya Masy 1:51–62. https://doi.org/10.31599/jstpm.v1i1.255
31. Molla A, Licker PS (2001) E-commerce systems success: an attempt to extend and respecify the delone and maclean model of is success 2:131–141
32. Vernia DM (2017) Optimalisasi Media Sosial Sebagai Sarana Promosi Bisnis Online Bagi Ibu Rumah Tangga Untuk Meningkatkan Perekonomian Keluarga. Util J Ilm Pendidik Dan Ekon 1:105–118
33. Wibowo EA (2014) Pemanfaatan Teknologi E-Commerce Dalam Proses Bisnis. 1:95–108. http://dx.doi.org/https://doi.org/10.33373/jeq.v1i1.222

34. Kapoor A, Guha S, Kanti Das M, Goswami KC, Yadav R (2020) Digital healthcare: the only solution for better healthcare during COVID-19 pandemic? Indian Heart J 72:61–64. https://doi.org/10.1016/j.ihj.2020.04.001
35. Sarfraz Z, Sarfraz A, Iftikar HM, Akhund R (2021) Is COVID-19 pushing us to the Fifth Industrial Revolution (Society 5.0)? Pak J Med Sci 37:591–594. https://doi.org/10.12669/pjms.37.2.3387

Online Purchase Over Pandemic Covid-19: Its Growth and Future in Malaysia

Tang Mui Joo and Chan Eang Teng

Abstract Ever since Malaysian Government imposed movement control order (MCO) due to covid-19, there is a surge in online purchase. Covid-19 pandemic has affected shopping and purchasing behavior of consumers. It is in the view that shifting consumer habits are changing Malaysia's future for retail. Such scenario brought forth by covid-19 pandemic to online purchase, its growth and future. Consumers' buying behavior has been the interest in this research. This research intends to determine the factors of online purchase during the pandemic. It is to identify the common problems encountered by online consumers. It also intends to find out whether it is a temporary behavioral change of the consumer because of the benefits of online shopping during this covid-19 pandemic. Technology acceptance model and the theory of planned behavior used are to study the factors of online purchase and consumers' behavior while adopting online purchase during covid-19 pandemic period and in the future. The subjects for this research are 150 volunteering adults, aged between 20 and 30 years old with at least one-time experience of online purchase during covid-19 pandemic. Google Form has been used, and snowballing is used to reach the subjects. This research has concluded that online purchase is resulting in a positive way. Online sales and promotion are the main reason for the surge of online purchase. Besides, consumers also want to avoid crowds during covid-19. More importantly is that most respondents will continue to purchase online even after the pandemic despite problems encountered in the process.

Keywords Online purchase · Online shopping · Online shopping behavior · Consumers' buying behavior · Covid-19 pandemic · Technology acceptance · Planned behavior

T. M. Joo (✉) · C. E. Teng
Tunku Abdul Rahman University College, 53300 Kuala Lumpur, Malaysia
e-mail: muijoo@hotmail.com

X.-S. Yang et al. (eds.), *Proceedings of Seventh International Congress on Information and Communication Technology*, Lecture Notes in Networks and Systems 447,
https://doi.org/10.1007/978-981-19-1607-6_25

1 Introduction

Covid-19 has changed online shopping forever. The pandemic has accelerated the shift toward a more digital world and triggered changes in online shopping behaviors that are likely to have lasting effects [1]. The scenario happens the same in United States that though its economy continues to reopen, the consumers have no intention of reverting back to their pre-pandemic shopping habits. Instead, consumers continue to shop from the safety of their home through online [2].

In Malaysia, there are over two-thirds of Malaysians now more comfortable shopping online after covid-19 compared to pre-pandemic. There were only 30% of Malaysians preferred to shop online [3]. It is also in the view that shifting consumer habits is changing Malaysia's future for retail. Malaysian consumers have changed their purchasing behaviors for greater convenience and value [4]. Covid-19 pandemic affects not only general behavior but also shopping and purchasing behavior of consumers. The scenario brought forth by covid-19 pandemic to online purchase, its growth and future and consumers' buying behavior, has been the interest in this research.

The purposes of this research have been as below:

(a) To determine the factors of online purchase
(b) To identify the common problems encountered by online consumers
(c) To find out whether it is a temporary behavioral change of the consumer because of the benefits of online shopping during this covid-19 pandemic.

To achieve the purposes of this research, this paper discusses the factors that lead to online purchase and the common problems encountered by online consumers. It is mainly to find out whether online buying behavior is a temporary change due to covid-19 pandemic or a permanent change in the behavior of consumers. This is also the motivation that drives the researchers into studying the trend of public in adopting technology to predict the growth and future of online purchase in Malaysia. Technology acceptance model and the theory of planned behavior are used to support this research. For data collection, voluntary online survey and snowballing have been used upon the samples of this research who are adults aged between 20 and 30, with at least one-time experience of online purchase during covid-19 pandemic. Conclusion is then drawn from the data collected. The challenge of this research is to perceive covid-19 pandemic as an enhancer to technology acceptance. It is assumed that with the much exposure to the use of technology for online purchase during lockdown, work from home and stay at home, people have been molded into the permanent behavior of online purchase.

2 Literature Review

2.1 Factors to Online Purchase During Covid-19

When online purchase is studied, pre-purchase actions are to look at to understand how the process is completed. Therefore, terms and actions like consumers' buying behavior and online shopping will be inter-discussed. There are many factors that can contribute to a consumer's buying behavior. Bashir [5] has identified price, time saving, and convenience as important factors which lead to certain buying behavior in online shopping and online purchase. Online consumers are always seeking trendy products, new attractiveness, and new products that are trending at the period of time and most importantly products with prices that are compatible with their budget. Online consumers do not have limits for online shopping, and the best way to save time and money is through the Internet that allows you to purchase anything online wherever you are. Other key factors that may affect changes in a consumer's online shopping and purchasing behavior is commonly on the trust of the Web site. Trust can be in the aspects of the Web site itself, the counterparts of the transactions and the trust to make online purchase [6].

Online purchase has been popular and successful for many reasons even before MCO. First, online purchase is more convenient than traditional shopping because people can buy anything online anytime due to the operation of an online shop for 24 h [7]. Besides that through online shopping, people can easily compare the products from its details to prices and online stores. All it takes is to search the name of the specific product then can see all the prices from different stores than to drive to many shops or malls and find parking at many places to compare those prices [8]. In addition, online purchases can provide us with more discounts and cashbacks on different holidays [9]. When the shoppers reach the purchase price target which is set by the online seller, they will get the extra vouchers or cash backs, and these vouchers will be given out anytime due to some holidays. This will help us to save more money to get the same quality of the products.

Malaysia has 16.53 million online shoppers which are the 50% of the population, and 62% of users uses their mobile devices for online shopping [10]. Due to the covid-19 pandemic, the online purchase rates in Malaysia have increased as well. More than two-thirds or 73% of Malaysians said that they are more positive about online purchase due to the covid-19 pandemic [3]. Prior to the pandemic, only, 30% of Malaysians said they preferred shopping online compared with 70% who preferred shopping in person. Online shopping and online purchase can reduce the number of trips to reduce the chance of getting the virus, as compared to traditional shopping, during the movement control order (MCO) period. According to a survey conducted by Vodus Insights [11] based on a sample of 15,000 Malaysian adults, MCO has compelled a significant number of non-shoppers to start shopping for groceries, meals, and other non-food items online. This surge in online shopping is observed to be driven by necessity rather than convenience or discounts. 25–40% of online shoppers has only started to shop online after MCO has started to avoid crowded places, and many of these new online shoppers are from the senior age groups (45 years old and above). Safety concern is the primary reason for ordering food and grocery online, while convenience is the primary reason for purchasing non-food items online.

This research is to look at the factors of online shopping and online purchase during pandemic in order to study also the possibility of future trend.

2.2 *Problems Faced by Consumers in Online Purchase*

Though online purchase has its advantages and strengths, there are concerns and disadvantages to be considered. Delivery fee is one of the concerns in online purchase. If the items purchased are large and heavy, more fees occurred for its delivery. Same as distance, it may increase the fee of delivery. On top of that delay in delivery is an issue. It may be due to weather conditions, lost consignments, or other logistic issues. This aspect has become a problem of delivery risk. Online shoppers often worry that they won't receive the product after they buy it. The loss or damage of the goods is associated with the potential loss of delivery. In this, there are also other concerns in delivery risks. For instance, whether customers receive goods and improper handling of goods in the delivery process. Somehow by providing correct shipping status updates, consumers should expect the arrival schedule of the product, allowing customers to reduce their thoughts about shipping and undeliverable [12].

Product risk is defined as the probability that a product fails to meet the initially anticipated performance requirements. It is the most common reason why people do not shop online. Product risk was found to have a significant impact on the frequency of online purchases. Online shoppers are unable to check and test product online; relatively, high levels of product risk are expected when buying online, especially for certain product categories, suggesting that risks associated with product uncertainty may negatively impact online purchase intentions [13].

Online shoppers cannot have a clear idea about the shape and quality of the products and also cannot try the products before buying. Therefore, they need to bear the risk that maybe the items they receive are different from their expectations [14].

Another problem faced by online shoppers is fraud and security issues. As consumers have no chance to inspect goods before purchasing, they are at higher risks of fraud on the part of goods than in physical stores. Online purchase also can be made using stolen credit. Besides that security and privacy are another issue faced. For instance, spams, scams, and telemarketing that come with providing contact information to online merchants [15].

Although there are risks and disadvantages of online purchase, the increase in online purchase has shown that consumers are still willing to overcome the risks looking at the advantages gained. This research is not only studying the benefits and risks of online purchase; it is also to study the trend change in consumer purchases from offline to online.

2.3 *The Changes of Consumers' Buying Behavior in Online Purchase During Covid-19*

When Malaysian government imposed lockdowns and movement control orders (MCO) during covid-19, it has directly affected online businesses positively. Non-shoppers start to shop for groceries, meals, and other non-food items online. This surge in online shopping is observed to be driven by necessity rather than convenience or discounts. 25–40% of online shoppers has only started to shop online after MCO has started. It is to avoid crowded places, and many of these new online shoppers are from the senior age groups (45 years old and above). There are few reasons that lead to the surge of online purchase. Among the reasons are to avoid crowded places, for its conveniences, for its cheaper prices, for its rewards offered, for its ability to pay with e-wallet, for its wider selection of products, and others [11].

Furthermore, during the covid-19 pandemic, the public health sector issued the fear appeals to the public. When consumers linked to social interpretations of perceived fear and risks, impulsive buying will generate. Stay at home, wash your hands to save your loved ones are involved in social interpretations. These fear appeals are able to develop consumers' impulse buying and cause consumers' buying behavior [16]. When people feel panic gradually, consumers cannot keep making decisions rationally. This has led consumers to buy extra things when they need to stay at home during covid-19 pandemic [17].

Furthermore, the motives of online purchase have changed during covid-19 pandemic. It has been found that there is a positive relationship between consumers' external subjective norms and their behavioral intention. This is because consumers are impacted by their own emotional instead of rational motives. Besides that the intention of consumers to purchase clothes online is not only influenced by the media reports from media contents but also the expert opinions. Therefore, the consumers may change their opinions because of the recommendations from people or information around them. Moreover, because of the pandemic, the consumers have less chances to engage in some leisure activities during the lockdown period. This causes them to engage online shopping for the purpose of enjoyment and distraction. The conveniences of online shopping such as no limited time for online shopping, reduce the risk of getting infected by coronavirus also cause the consumers to purchase online [18]. Consumers' buying behavior and interest are changing dramatically due to the limited movement during the lockdown because of covid-19. Consumers are leaning more toward the online shopping methods via few online platforms as well as Web sites and paying via online. In other words, online shopping is important for consumers not only in Malaysia but worldwide [19].

Looking at the trend of online purchase during covid-19 pandemic, it is also a concern in this research whether online purchase will last long, or it is merely a trend to overcome the period.

2.4 Technology Acceptance Model (TAM) and Theory of Planned Behavior (TPB)

This research is applying technology acceptance model (TAM) and the theory of planned behavior (TPB) to study the factors of online purchase and consumers' behavior in adopting online purchase during covid-19 pandemic period and in the future. The function of the model is used in this research to determine the acceptability of the technology among Malaysians to make transformation from offline purchase to online purchase. The acceptability of the users is based on two key factors which are perceived as easy to use and perceived usefulness [20]. Perceived as easy to use refers to people believing that they are able to use the technology easily without using any effort, whereas perceived usefulness indicates people believe that the usage of the technology will be increased. The two key factors in the context of online purchase describe the convenience and ease of how online purchase works. Both factors will influence the behavior of users by having the intention and positive attitude toward online purchase and end up buying the product or service [21].

Theory of planned behavior is built out of some constructs. First is the individual's attitude toward the behavior which is determined by their beliefs and personal evaluation on how a certain behavior is able to make either positive or negative outcome to their life [22]. Secondly, subjective norms focus on everything around the individual such as his or her cultural background, social network, and group belief. The opinions from the individual's reference groups like family members and peers are also able to influence him or her to either engage in the behavior or not. In addition, the motivation from the reference group to meet their expectation can be a positive subjective norm. Third is about perceived behavioral control in TPB. This construct refers to the beliefs of an individual's ability to perform certain behavior and also the ease and difficulty of one performing the behavior [23]. With this theory, we are able to determine whether a person has a positive or negative attitude toward online purchase. Further to this, it will reflect the future trend and possible use of online purchase in daily lives.

3 Methods

The subjects are 150 adults, aged within 20–30 years old based on the rational that Malaysian adults aged between 20 and 30 years old have become the age group with the largest increase for non-food items online shopping with an increased by 83% during the MCO period [24]. The subjects must have at least one-time experience of online purchase during covid-19 pandemic. This is a voluntary-based online survey using Google Form and snowballing from the volunteers.

The researchers use quantitative design, online survey to collect data. Online survey has been chosen as it allows the researchers to collect data from a big number of respondents. Online survey is at a zero cost. Especially over pandemic period, there is no face-to-face contact needed. The replies from respondents can be much more instant. The questionnaire in the format of Google Form will be distributed through emails and social media platforms to targeted respondents though snowballing among volunteered friends and families. The questionnaire is divided into three sections. Section one is on demography; section two is on factors of online purchase during covid-19; section three is on consumers' buying behavior during covid-19. All the questions are close ended and partly Likert. The timeframe of the survey is a month, starting from January 15, 2021 to February 14, 2021.

4 Results

4.1 Demographics

This section comprises demographic profiles of the 150 respondents. The demographic profiles include gender, age group, occupation status, and the income level of the respondents.

Table 1 shows the number and percentage of respondents based on gender, age, occupation status, and income level. The data show that there are 70% female respondents over 30% male respondents out of 150 of them. Among them, 88.7% of the respondents is from the age group of 20–24 years old. This age group has become one of the age groups with the largest increase for non-food items online shopping

Table 1 Demographic profile. *Source* Online survey conducted from January 15, 2021 to February 14, 2021

Demographic profile	Number	Percentage (%)
Gender		
Male	45	30
Female	105	70
Age		
20–24	133	88.7
25–30	17	11.3
Occupation Status		
Students	125	83.3
Workers	24	16
Unemployed	1	0.7
Income level		
Below RM 1000	122	81.3
RM 1000–RM 2000	10	6.7
RM 2001–RM 3000	10	6.7
RM 3001–RM 4000	5	3.3
Above RM 4000	3	2

during the MCO period [24]. Most of the respondents are students with the respond rates of 83.3%. Among the respondents, there are 122 respondents from the income level which is below RM 1000. There are only 2% of the respondents with income above RM 4000.

4.2 *Factors of Online Purchase During Covid-19*

In this section, the factors that lead to online purchase during covid-19 period will be presented and analyzed. Details of the responses will be shown in Table 2.

Table 2 shows the reasons why respondents prefer to have online purchases during covid-19 period. Among 150 respondents, 30.7% of the respondents performs online purchases mainly for online sales and promotions. Other than that online purchases have been performed as it is able to avoid crowds during pandemic at the same time they are able to get products not sold at their places. Comparing prices though ranked the least, the number is not far below. It shows that the reasons to purchase online are rather close to each other.

Other factors that may lure Malaysians to purchase online would be due to the factors summarized in Table 3.

Table 3 shows the online purchase performed by the respondents attracted to marketing strategies of social media, online live video streaming, and online advertisements. All three marketing strategies have reflected their success in attracting people to purchase online. All three have shown high number of strongly agree with agree from 132, 137 to 134.

There are other personal factors of online purchase being studied in this research. The data have shown that 72 of respondents purchase online as they want to try new things. There are 89 of them feel that their lifestyles have been improved by purchasing new stuff online.

Overall result for this part shows that the respondents prefer online purchase during covid-19 period because of the online sales and promotions. Additionally, the online marketing strategies have also contributed to the factors of online purchase during covid-19 period. Social media, online live video streaming, and online advertisements

Table 2 Factors of online purchase during covid-19. *Source* Online survey conducted from January 15, 2021 to February 14, 2021

Factors of online purchase during covid-19	Number	Percentage (%)
Able to buy products not sold at my place	37	24.7
Able to compare prices of the products	30	20
Able to avoid crowds	37	24.7
Able to get online sales and promotion	46	30.7

Table 3 Other factors to online purchase during covid-19. *Source* Online survey conducted from January 15, 2021 to February 14, 2021

Questions	Strongly agree	Agree	Neutral	Disagree	Strongly disagree
Attracted to online selling through social media	90	42	15	3	0
Attracted to product selling through online live video streaming	85	42	16	6	1
Attracted to online advertisements	80	54	13	3	0
I will try new things when purchasing online	20	52	61	15	2
I have improved my lifestyle by purchasing new stuff	28	61	49	10	2

have become the trendy platform for the sellers to sell their products. It is especially effective as there are more people required to stay at home and work from home during pandemic period. There are respondents willing to try new things through online purchase. During pandemic period, online shopping and online purchase are ways to improve their lifestyle.

4.3 *Consumers' Buying Behavior During Covid-19 Pandemic*

The data have been collected from 150 respondents where 89 of them are long-term online purchasers and 61 of them have started online purchase. The data of buying behavior during covid-19 pandemic have been summarized into a few tables to be discussed as below (Table 4).

The table reflects that covid-19 pandemic has affected consumers' buying behavior. The influence of buying behavior in the categories of product has been summarized into a comparison in Table 5.

Based on the response, it is clear that the number of online purchase for grocery has increased tremendously as compared to other categories of purchase. It is also

Table 4 Influence of covid-19 to consumers' buying behavior. *Source* Online survey conducted from January 15, 2021 to February 14, 2021

Questions	Yes	No	Maybe
Is online shopping easy to find what you are looking for during the pandemic?	140	9	1
Is covid-19 pandemic affecting your buying behavior?	105	38	7

Table 5 Products purchased online before and during the movement control order. *Source* Online survey conducted from January 15, 2021 to February 14, 2021

Products	Before movement control order	During movement control order
Grocery shopping	23	54
Clothes	43	17
Skin care	8	16
Health supplies	6	7
Accessories	70	56

from the response that 99% of them feel that online shopping is useful during the pandemic period.

When it is asking about how often the respondents shop and purchase online, there are 12.4% of them purchased very often; 36.2% of them are often, and 35.2% of them are rather often. Looking at such high percentage of them purchase online frequently, it is making sense where 60% of them has spent more money online during covid-19. On the other hand, the high frequency of online purchase reflects that they are at ease to adopt the technology to perform purchases.

The last question in this section is whether the respondents will continue to purchase online when the pandemic is over. The responses are very positive. There are 73.3% will continue to purchase online as they find it convenient. 21% of them will continue to purchase online because they always can get what they want online. There are 2.9% of them will not purchase online as they have no trust in online shops. Another 2.9% of them will not continue to purchase online as they find the quality of the product is bad.

5 Conclusion

In conclusion, the role of online purchase is resulting in a positive way during covid-19. Online sales and promotion are the main reason for the surge of online purchase. Besides, consumers also want to avoid crowds during covid-19; it is to reduce the possibility of exposure to covid-19. Other than that online purchase may improve their lifestyle with the new stuff purchased. Though it is not an overall impact to online purchase, there is a significant surge in it. More importantly is that most respondents will continue to purchase online even after the pandemic despite problems encountered in the process.

Referring to technology acceptance model and the theory of planned behavior, this research has reflected that respondents are at the point of utilizing and adopting online purchase at ease during covid-19 pandemic period. Even though some of them may not adopt online purchase, most of them are into it due to its convenience and ability to provide what they are looking for online. Many of the respondents' feedback show that they will continue to purchase online. In long term, online business is the key concern. At the same time, offline business shall need extra care for its performance.

The authors acknowledged the raw materials from Cheng Huey Shin, Kick Chia Miam, Yeo Pek Hwa, Lee Pei Err, Mak Yi Lin, Lai Wai Kuan, Dominique Chong Cheng Yi, Ng Ling Huei, and Bryen Aden Aeria.

References

1. United Nations Conference on Trade and Development. https://unctad.org/news/covid-19-has-changed-online-shopping-forever-survey-shows
2. PYMNTS. https://www.pymnts.com/digital-payments/2020/why-financial-institutions-must-build-digital-future-avoid-losing-customers/
3. The Edge Markets. https://www.theedgemarkets.com/article/over-twothirds-malaysians-now-more-comfortable-shopping-online-after-covid19-%E2%80%94-stanchart
4. Microsoft Malaysia News Center. https://news.microsoft.com/en-my/2021/07/14/shifting-consumer-habits/
5. Bashir A (2013) Consumer behavior towards online shopping of electronics in Pakistan. Thesis, Seinäjoki University of Applied sciences
6. Karayanni D (2003) Web-shoppers and non-shoppers: compatibility, relative advantage and demographics. Eur Bus Rev. https://doi.org/10.1108/09555340310474640
7. Heather Abrafi Agyapong. https://core.ac.uk/download/pdf/161426026.pdf
8. Cheema U, Rizwan M, Jalal R, Durrani F, Sohail N. http://www.aessweb.com/pdf-files/131-141.pdf
9. Vijay TS, Prashar S, Sahay V. https://scielo.conicyt.cl/pdf/jtaer/v14n1/0718-1876-jtaer-14-01-00102.pdf
10. International Trade Administration. https://www.trade.gov/country-commercial-guides/malaysia-ecommerce
11. Vodus Insights. https://vodus.com/article/impact-on-e-commerce-since-covid-19-began
12. Tham KW, Dastane O, Johari Z, Ismail NB. https://www.koreascience.or.kr/article/JAKO201915658234374.page
13. Bo D, Sandra F, Kwon WS. https://web.csulb.edu/journals/jecr/issues/20141/Paper2.pdf
14. Woolf A (2014) Let's think about the Internet and social media. Raintree, United Kingdom
15. Sunitha CK, Edwin Gnanadhas M (2018) Problems towards online shopping. Int J Emerg Technol Eng Res 6(1):14–17
16. Naeem M (2021) Understanding the customer psychology of impulse buying during COVID-19 pandemic: implications for retailers. Int J Retail Distrib Manage 49(3):377–393. https://doi.org/10.1108/IJRDM-08-2020-0317
17. Mehta S, Saxena T, Purohit N (2020) The new consumer behaviour paradigm amid COVID-19: permanent or transient? J Health Manag 22(2):291–301
18. Koch J, Frommeyer B, Schewe G (2020) Online shopping motives during the Covid-19 pandemic—lessons from the crisis. Center for Management of Muenster University, p 10247
19. Hashem TN (2020) Examining the Influence of Covid 19 Pandemic in changing customers' orientation towards e-shopping. Mod Appl Sci 14(8):59–76
20. Bhattacharjee J, Chetty P. https://www.projectguru.in/online-consumer-behaviour-theory-model/
21. Zeba F, Ganguli S (2016) Word-of-mouth, trust, and perceived risk in online shopping. Int J Inform Syst Serv Sector 8(4):17–32
22. Glanz K, Timer B, Viswanath K (2015) Health behaviour: theory, research and practice, 5th edn. Jossey-Bass, San Francisco
23. Kautonen T, van Gelderen M, Fink M (2015) Robustness of the theory of planned behavior in predicting entrepreneurial intentions and actions. Entrepreneurship Theor Pract 39(3):655–674
24. Vodus Insights. https://insights.vodus.com/article/covid-19-mco-impact-on-malaysia-e-commerc

MAGNeto: An Efficient Deep Learning Method for the Extractive Tags Summarization Problem

Hieu Trong Phung, Anh Tuan Vu, Tung Dinh Nguyen, Lam Thanh Do, Giang Nam Ngo, Trung Thanh Tran, and Ngoc C. Lê

Abstract In this work, we study a new image annotation task named Extractive Tags Summarization (ETS). The goal is to extract important tags from the context lying in an image and its corresponding tags. We adjust some state-of-the-art deep learning models to utilize both visual and textual information. Our proposed solution consists of different widely used blocks like convolutional and self-attention layers, together with a novel idea of combining auxiliary loss functions and the gating mechanism to glue and elevate these fundamental components and form a unified architecture. Besides, we introduce a simple but effective data augmentation technique dedicated to alleviate the effect of outliers on the final results. Last but not least, we explore a self-supervised pre-training strategy to further boost the performance of the model by making use of the abundant amount of available unlabeled data. Our model shows the good results as 90% F_1 score on the public NUS-WIDE benchmark, and 50% F_1 score on a noisy large-scale real-world private dataset. Source code for reproducing the experiments is publicly available at: https://github.com/pixta-dev/labteam.

Keywords Deep learning · Language and vision · Self-supervision

H. T. Phung (✉) · A. T. Vu · T. D. Nguyen · L. T. Do · G. N. Ngo · T. T. Tran · N. C. Lê
PIXTA Vietnam, 8th Floor, Truong Thinh Building, Phung Chi Kien, Cau Giay District, Hanoi, Vietnam
e-mail: hieu.phungtrong@pixta.co.jp

A. T. Vu
e-mail: tuananh.vu@pixta.co.jp

T. D. Nguyen
e-mail: tung.nguyendinh@pixta.co.jp

L. T. Do
e-mail: lam.dothanh@pixta.co.jp

G. N. Ngo
e-mail: giang.ngonam@pixta.co.jp

T. T. Tran
e-mail: trung.tranthanh@pixta.co.jp

H. T. Phung · L. T. Do · N. C. Lê
Hanoi University of Science and Technology, 1 Dai Co Viet Road, Hanoi, Vietnam

X.-S. Yang et al. (eds.), *Proceedings of Seventh International Congress on Information and Communication Technology*, Lecture Notes in Networks and Systems 447,
https://doi.org/10.1007/978-981-19-1607-6_26

1 Introduction

The goal of Extractive Tags Summarization (ETS) task is to shorten the list of tags corresponding to a digital image while keeping the representativity. The process provides not only an effective way to quickly understand the image's content but also possibly serves other downstream tasks such as image retrieval [25], caption generation [26], object detection [7], etc. A practical application of ETS is supporting tag-based image search engines, which use tags information of images and users' queries to activate the search logic. These systems could return better search results if know the relevance of each image to its corresponding tags. Thankfully, this information can be easily and automatically extracted by using an ETS system.

Undoubtedly, deep learning is a powerful tool for numerous problems these days with the backing of abundant data as well as the advances in hardware technology. In this research, we compose some deep learning models to combine computer vision (CV) and natural language processing (NLP) tasks to employ both visual and textual information. Convolutional neural network (CNN) [15] has shown to be the best tool for extracting visual features from an image. In fact, it can be applied to most CV tasks as long as having enough training data and computational power. To the NLP side, in [24], the authors introduced the Transformer architecture, an attention-based model, which has beaten recurrent models like [3, 9] to become a very effective solution for NLP tasks.

In this work, we proposed a unified architecture that employs both convolutional layers and the Transformer encoder with the help of auxiliary loss function [30] and a gating mechanism to solve the ETS task in an efficient manner to employ visual features from the image as well as textual features from tags. The following is the description of the ETS framework.

Assuming that each image has already been assigned all the possible tags to describe the content of this image, our job is to "summarize" the original set of tags to a more compact form which still being able to help understand the main point presented; i.e., only important tags kept. Along with that, we propose an effective deep learning method to deal with ETS task that can extensively exploit the information in both image and tag.

In Sect. 3, we describe the model architecture as well as the auxiliary loss function. Section 4 is devoted to the data and conducted experiments. Some details about the setting of experiments are left in Sect. 5. In Sect. 6, we discuss the results as well as some potential considerations in the future. But first, in Sect. 2, we describe some researches related to the issue.

2 Related Work

Deep learning in CV. Since the publication of LeNet-5 [16] in the 1990s, CNNs have gone through a long journey to become an effective, or rather the best, general architecture for automatically extracting useful features from images. Due to the excessive computational cost and the disinterests of researchers in this kind of architecture, there was not any significant research on CNN in many years after LeNet-5. It is not until AlexNet [13] that popularized convolutional networks in CV field by beating the second runner-up of the ImageNet ILSVRC 2012 challenge [20] by a huge margin. Year after year, new model architectures [8, 28] have been proposed to surpass the existed solutions.

Deep learning in NLP. Coming to the NLP field, deep learning also has a great influence on its development. It has been a long time since recurrent-based models, such as long short-term memory (LSTM) [9] and gated recurrent unit (GRU) [3], dominated existed solutions in various NLP tasks like language modeling and machine translation [1, 23]. With the newborn of attention-based models like Transformer [24], the boundary of deep learning in NLP field has been pushed further.

Self-supervised learning. These days, self-supervised learning has been widely used to support different tasks [12, 14] because of the abundance of unlabeled data when compared with the nicely labeled one. Self-supervised learning aims at finding a good representation of data with a pretext, or auxiliary, task. The pretext tasks to learn the visual representation of the image data can be grayscale photograph colorization [29], solving jigsaw puzzles [19], predicting image rotations [6], etc. When the target is learning representation from a text corpus, the pretext task can be formed as masked language modeling [5, 14, 18], next sentence prediction [5], sentence order prediction [14], etc. The learned representation can then be used to solve the downstream tasks, e.g., object detection, language translation, and even ETS task.

Mixture of experts (MoE). Since the first introduction in 1991 by Jacob et al. [11], MoE has gone through a long journey [27] to become one of the most flexible method to combine a wide range of different models to achieve what called ensemble learning. Besides the studies in statistical properties, researchers also care about the applicability of MoE in real-world problems. The method has gone to extremes with the help of sparsely-gated MoE layer [21] that can achieve greater than 1000x improvements in model capacity while keeping the computational performance in control. In this work, we practice the MoE idea with a simple gating mechanism to combine the outputs of the tag-stream and the image-stream to form the final results; the details will be discussed in Sect. 3.

3 Methods

3.1 Model Architecture

Our proposed solution, named Multi-Auxiliary with Gating Network (MAGNeto), is composed of two streams: (1) tag-stream which only uses tag's information to solve ETS task and (2) image-stream which requires both image's and tag's information. Above all is a gating mechanism taking the role of fusing the outputs of two streams (Fig. 1).

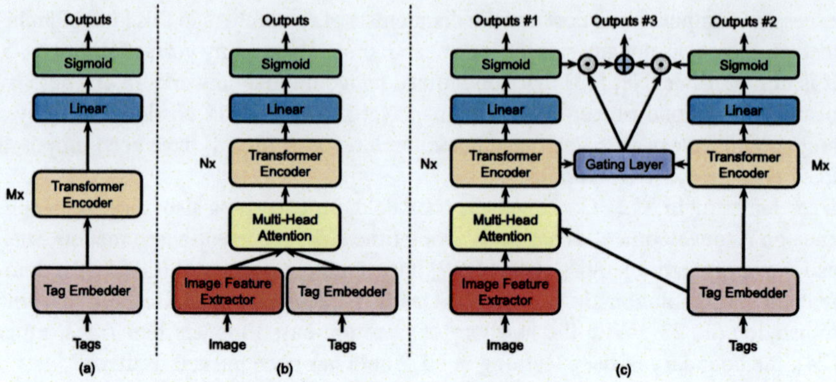

Fig. 1 From the left to the right: **a** the baseline model that utilizes the Transformer encoder to handle tags' features, **b** the baseline model that utilizes the Transformer encoder to handle tags' and image's features, **c** MAGNeto architecture

Tag embedder. The tags are vectorized here, and it could be a lookup table.
Image feature extractor. Any CNN architecture can be used to extract the input image's features. In this research, we adopt ResNet [8] and modify it as the following. First, we drop the top global average pooling [17] and fully connected layers. Second, a 1×1 convolutional layer is plugged on top of the network to get the desired shape of the output feature maps, followed by a batch-norm [10] layer. Finally, the (N, N, d_{model}) output tensor corresponding to $N \times N$ grid regions with d_{model} features is flattened across the first two dimensions.
Multi-head attention layer. In this subsection, we describe the method combining visual features from the image and textual features from tags. Since the target of the ETS task is to determine which tags should be kept, the tags' feature vectors should be projected on the image feature vector space ($\mathbb{R}^{d_{\text{model}}} \rightarrow \mathbb{R}^{d_{\text{model}}}$) by utilizing the attention (scaled dot-product attention [24]) mechanism. The tags' feature vectors are *queries* while image regions' vectors play as *keys* and *values*; i.e., each tag vector "looks" at the whole grid regions of the image feature maps and the duty of the attention mechanism is to guide the tag to which regions should be attended. The desired outputs would be the fusion of the image's and tags' information, which then fed into the Transformer encoder to extract more abstract features.
Transformer encoder. This is one of the most important parts of the whole architecture, composed of identical layers built with multi-head self-attention mechanism and position-wise fully connected feed-forward network. This component figures out the relations among feature vectors, e.g., the vectors that corresponding to tags or grid regions in the image's feature maps. As the MAGNeto architecture described in Fig. 1, there are two blocks of the Transformer encoder: (1) stacked of M identical layers for the tag-stream in the right and (2) stacked of N identical layers for the image-stream in the left.

Table 1 Gating layer's units (the flow of the data is from top to the bottom)

Layer	In shape	Out shape
Dropout	$(bs, l, 2 \times d_{\text{model}})$	$(bs, l, 2 \times d_{\text{model}})$
FC	$(bs, l, 2 \times d_{\text{model}})$	(bs, l, d_{ff})
ReLU	(bs, l, d_{ff})	(bs, l, d_{ff})
Dropout	(bs, l, d_{ff})	(bs, l, d_{ff})
FC	(bs, l, d_{ff})	$(bs, l, 1)$
Squeeze	$(bs, l, 1)$	(bs, l)
Sigmoid	(bs, l)	(bs, l)

bs Batch-size
l The maximum number of tags per item. To serve the batching purpose, all the items must have a fix number of tags; here is l tags. Empty slots are filled up with zero values; i.e., we use padding for items that do not have long enough sets of tags
d_{model} The number of expected features in the encoder inputs
d_{ff} The dimension of the feed-forward network

Gating mechanism. The gating layer is a special part of our model merging the outputs of the tag-stream and the image-stream to form the final result. The sub-layers of this special block are listed in Table 1.

Let $\mathcal{O}_{it} = [o_{it}^0, o_{it}^1, \ldots, o_{it}^{l-1}]^{\mathrm{T}}, o_{it}^i \in [0, 1]$ be the output of the image-stream and $\mathcal{O}_t = [o_t^0, o_t^1, \ldots, o_t^{l-1}]^{\mathrm{T}}, o_t^i \in [0, 1]$ be of the tag-stream. The output of each stream can play standalone as an answer to the problem. However, fusing these outputs to get the ultimate decision usually results in a better outcome. The gating layer uses the intermediate feature vectors, the outputs of two Transformer encoders concatenated along the last axis, which results in $(2 \times d_{\text{model}})$-dimensional feature vectors, in both streams of the network to return a vector $\mathcal{A} = [\alpha_0, \alpha_1, \ldots, \alpha_{l-1}]^{\mathrm{T}}, \alpha_i \in [0, 1]$ for each item. The final output of the model is calculated as follows.

$$\mathcal{O}_{\text{Final}} = \mathcal{A}^{\mathrm{T}} \odot \mathcal{O}_{it} + (\mathbb{1} - \mathcal{A}^{\mathrm{T}}) \odot \mathcal{O}_t, \tag{1}$$

where $\odot$ is the element-wise (or point-wise) multiplication.

3.2 *Auxiliary Loss Function*

In general, since the tags often contain less information than the image, the tag-stream tends to converge much faster than the image stream and it would be difficult for the model to converge. Hence, besides the main loss function guides the training process of the whole model, two auxiliary loss functions for the image-stream and tag-stream have been added to the model. When one stream learns faster than the other, the α values tend to be skewed. In some extreme cases, the slower-converging stream can be ignored altogether. Figure 2 depicts the learning process of the two versions, with and without auxiliary loss functions.

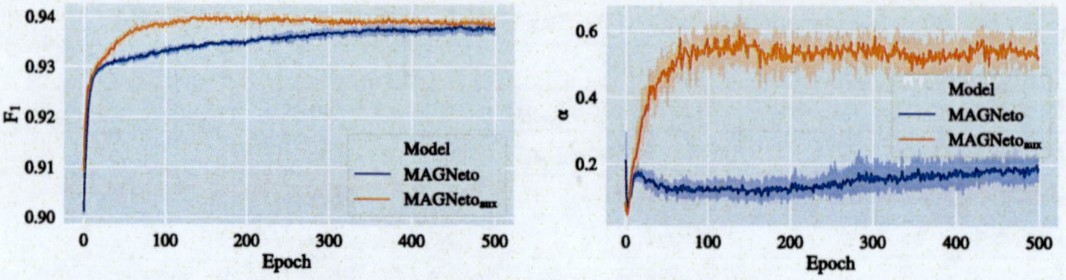

Fig. 2 Without and with auxiliary loss functions

For simplification, the final objective of MAGNeto is computed as:

$$\begin{aligned}\mathcal{L}_{\text{Total}} &= \mathcal{L}_g + \mathcal{L}^1_{\text{aux}} + \mathcal{L}^2_{\text{aux}} \\ &= \mathcal{L}_g + \mathcal{L}_{it} + \mathcal{L}_t,\end{aligned} \tag{2}$$

where $\mathcal{L}_g$ is the loss function for the final output after going through the gating mechanism, two auxiliary loss functions $\mathcal{L}^1_{\text{aux}}$ ($\mathcal{L}_{it}$) and $\mathcal{L}^2_{\text{aux}}$ ($\mathcal{L}_t$) are employed to guide the training processes of the image- and tag-stream, correspondingly.

4 Data and Experiment Results

4.1 Training Data

The NUS-WIDE dataset [2] is chosen for our experiments. To serve the ETS task, we perform some pre-processing steps to the public dataset as follows.

1. For each item (image), only keeping the tags included in the 81 concepts that can be used for evaluation provided by the National University of Singapore.
2. Filtering out all the items that cannot satisfy the condition: having the set of tags that cannot cover all the annotated concepts in the ground-truth.
3. Filtering out all the items that do not have any tag left after applying the two above pre-processing steps.

From 269,648 images, after going through all the pre-processing steps, we have obtained 26,559 satisfied items for the ETS task.

4.2 Baseline Models

Before coming up with the multi-auxiliary with gating idea, we have implemented several architectures from the simplest that only uses textual information as input to more complex ones fed with visual information (Fig. 1).

The Transformer encoder for tag features. This is the most straightforward architecture using only tag information to determine the output. We name this architecture TF-t for later references. Examining this architecture shows us that the relations among tags are extremely significant to find the important concepts for each image. **The Transformer encoder for image with tag features.** Until now, we only use the tag information to do ETS task, but how about the image, the good source of information to decide which tags should be kept? To utilize the image information, we combine it with tag features by projecting the tags' feature vectors to the image feature vector space with the attention mechanism. Then, we use the Transformer encoder to learn the relations among these intermediate features to get more meaningful ones before classifying the input tags. We name this architecture TF-it for later references.

4.3 Proposed Solution

Our proposed architecture has two streams and is utilized with the gating mechanism. Despite the importance of image information, using tag information alone can boost the performance of the model significantly. Obviously, having two experts in hands seem much better; one entirely relies on tag features, and one makes use of all available resources. After having had the outputs of both streams, the gating layer comes into play, fusing the outputs of these two experts, image-stream and tag-stream, to form the final decision on which tags should be marked as important. Based on the F_1 score, MAGNeto beats all the baseline models. The results are described in Fig. 3 and Table 2.

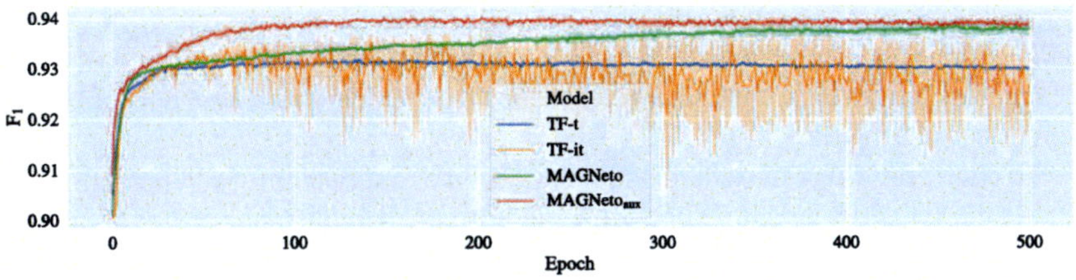

Fig. 3 F_1 comparison among models

Table 2 Evaluation metrics of different architectures

Model	Prec (%)	Recall (%)	F_1 (%)
TF-t	93.4	98.1	93.2
TF-it	94.4	**100.0**	93.8
MAGNeto	93.9	97.6	93.9
MAGNeto$_{aux}$	**94.5**	96.7	**94.1**

5 Experiment Settings

5.1 Outliers and Tag Data Augmentation

Although the powerful self-attention mechanism in the Transformer encoder tremendously helps the MAGNeto architecture to extract useful features from the textual information provided by the input tags, it cannot defend the model from the outliers, i.e., inappropriate tags. In some extreme cases, these unrelated, to the image content, tags could be classified as the important ones. This effect might be the result of the differences between outliers' vectors and the rest. To address this issue without performing any pre-filtering methods to the raw input tags, we propose an augmentation technique for the textual content which can be efficiently performed during the training phase. We name it "tag adding and dropping," or TAD for short, augmentation. Not only does this technique help overcome the outlier problem but also improves the generalization of the model by reducing overfitting. (As the name suggests, TAD augmentation includes two subprocedures: Tag-adding and tag-dropping.)

Tag-adding. This is the procedure of randomly picking unique tags from the vocabulary and employing these to extend the original list of tags corresponding to each item. We have the β coefficient responsible for the maximum ratio between the number of new adding tags and of the original tags that labeled as not important. For example, with $\beta = 0.3$, if an item has a set of 50 tags in which five of them are marked as important tags and the rest, 45 tags, are unimportant, the maximum number of tags we can add up is $A = \lfloor 0.3 \times 45 \rfloor = 13$. After having the β upper bound, we uniformly sample an integer between $[0, A]$ and use it as the true number of tags should be randomly[1] selected from the vocabulary.[2]

Tag-dropping. In contrast to the tag-adding procedure, tag-dropping randomly picking unimportant tags from the original set to remove them from the list. We also have the $\hat{\beta}$ coefficient to limit the number of tags being dropped. In practice, we tend to use $\hat{\beta}$ equal to β; however, these are two independent hyperparameters; thus, different values could be used.

To check the effect of outliers on the model, we first randomly add inappropriate tags to the validation set and evaluate the model; then, we count the number of outliers picked as important tags in the final outputs. The results are described in Table 3.

Please be aware that when working with NUS-WIDE dataset, the results may not accurately reflect the true performance of the model in reality because of (1) the small number of concepts (only 81 concepts), (2) the small number of tags per item after using the pre-processing and filtering strategies (mainly one tag and 13 tags at most), (3) and the small number of images (26,559 items). The experimental results of the model when working with a large-scale real-world dataset can be found in Appendix.

---

[1] Uniformly sampling.

[2] All original tags must be excluded from the vocabulary prior to executing the tag sampling process.

5.2 *Self-supervised Pre-training Strategy*

These days, self-supervised learning has been widely used to support different tasks [12, 14] because of the abundance of unlabeled data when compared with the nicely labeled one. Self-supervised learning aims at finding a good representation of data with a pretext, or auxiliary, task. The learned representation can then be used to solve the downstream tasks, e.g., object detection, language translation, and even ETS task.

To further improve the performance of the MAGNeto model and make use of unlabeled data available, we propose a two-stage training strategy (Fig. 4) which is a combination of self-supervised pre-training and supervised fine-tuning.

Self-supervised pre-training. The model architecture used in this first stage is slightly modified to better fit the purpose of finding a good initialization point for the target model in the supervised fine-tuning stage. In detail, we remove the gating mechanism in the original MAGNeto architecture entirely and the model becomes a two-input two-output architecture. (It means the model still has visual and textual inputs and both outputs are on target to attain the same self-supervised learning objective with the guidance of the loss function $\mathcal{L}_u = \mathcal{L}_{it} + \mathcal{L}_t$.) The objective must be accomplished is detecting outliers, i.e., solving the outlier detection problem; here, outliers referred to the inappropriate tags. It is simply achieved by randomly adding irrelevant tags to the unlabeled training data, and the model's job is to classify relevant and irrelevant tags for each item in the dataset. After being fitted with the pretext task, we are ready for the second stage, fine-tuning the model for the downstream task—ETS.

Supervised fine-tuning. At this stage, we come back to our original MAGNeto architecture which is now initialized with the parameters of the self-supervised pre-trained model in the first stage. Nevertheless, only some specific building blocks are copied to the target model, which are image feature extractor, tag embedder, multi-head attention layer, and two Transformer encoders, while the rest are randomly initial-

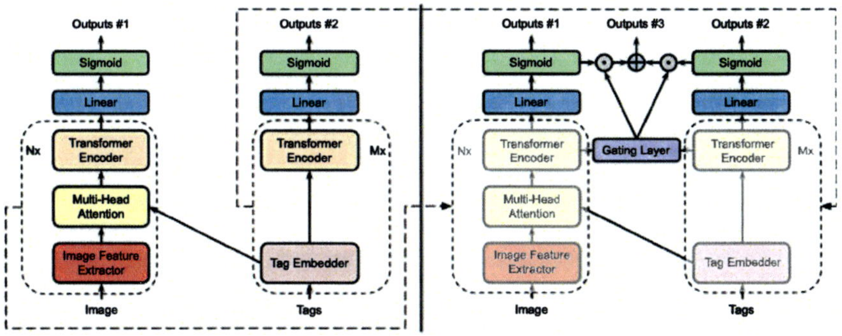

Fig. 4 Training pipeline composed of self-supervised pre-training (left) and supervised fine-tuning (right)

Table 3 Performance comparison among configurations

Pre-trained[a]	β	$\hat{\beta}$	Outlier[b] (%)	F_1 (%)
No	0.0	0.0	39.34	93.24
No	0.5	0.5	28.60	93.25
Yes	0.0	0.0	24.98	94.27
Yes	0.5	0.5	**20.53**	**94.30**

[a]Whether or not using self-supervised pre-training weights
[b]The percentage of items affected by outliers

ized and trained from scratch. Note that we do not freeze the pre-trained parameters yet use labeled data for ETS task to fine-tune them together with the whole model.

The self-supervised pre-training strategy alone alleviates the effect of outliers on the target model (Table 3). When being coupled with TAD augmentation, the model can be even more robust in ignoring inappropriate tags.

6 Conclusion

In this work, a new image annotation task called Extractive Tags Summarization (ETS) is studied. Besides taking advantage of advancements in the deep learning field, we specifically devote our effort to combine these fundamental components together and form a complete model dedicated to solving ETS task. After having ablation studies with various baseline models, the final proposed solution named MAGNeto has shown the effectiveness of the gating mechanism when being cooperated with auxiliary loss functions. Additionally, data augmentation techniques and self-supervised pre-training strategies are also taken care of to further boost the performance of the model.

The gating mechanism presented in this work is just our first attempt to fuse the outputs of the two streams. We have not performed any hyperparameter tuning for this specific layer and even have not decided which are the best input features for it. Therefore, MAGNeto architecture definitely could be further improved just by leveraging a better gating mechanism. Another aspect that can improve the model performance is by preventing the imbalance among the concepts' frequencies. The last thing that should be paid attention to in future works is the mapping mechanism from the tag feature vector space to the image one, e.g., using a multi-head attention layer as proposed in this work.

Acknowledgements This project is sponsored by PIXTA Inc. We are grateful to Minh Huu Nguyen, Duong Thuy Do, and Huyen Thi Le for their fruitful comments, corrections, and inspiration. The last author also received the support from Vietnam Institute for Advanced Study in Mathematics in Year 2020.

Appendix

Testing Model on a Large-Scale Real-World Dataset

Large-Scale Dataset. To test the capacity of the proposed solution, we have generated a large-scale real-world dataset, abbreviate as LS for convenience, which composed of about 700,000 fully annotated images with 24,056 concepts.

Experiments. When working with a large-scale real-world dataset, the result is always imperfect when compared with the experiments conducted in the laboratory. Additionally, the gaps among experiments are usually much more significant, a few percent of F_1 score, when compared to the results while working with the NUS-WIDE dataset. However, despite the horrible noise presented in the dataset, the model is still capable of learning the hidden pattern from the data. Again, the F_1 scores in Fig. 5 show the effectiveness of the gating mechanism.

Outlier Problem. When working with NUS-WIDE dataset, the results of using TAD augmentation to prevent outlier problem may not accurately reflect the true performance of the model and effectiveness of the proposed augmentation method. However, in LS dataset, when we have (1) thousands of concepts (24,056 concepts), (2) each item have about 43 tags on average, (3) and having much more labeled training data (about 700,000 fully annotated images), the effect of outliers (about 8–10% of validation items affected by the presentation of the outliers and 0.5–0.8% after using TAD augmentation) is not severe as in NUS-WIDE dataset.

Training Configurations

Training Data and Batching. We performed various experiments with our models on 2 datasets: the public NUS-WIDE benchmark and LS—a large-scale real-world private dataset. For the demonstrating purpose, we configured *batch-size* = 32 for NUS-WIDE. Yet, when dealing with LS, we used *batch-size* = 256 to make the most of an RTX-2080ti GPU. Moreover, we use $l = 16$ for training with NUS-WIDE dataset, and $l = 64$ while working with LS.

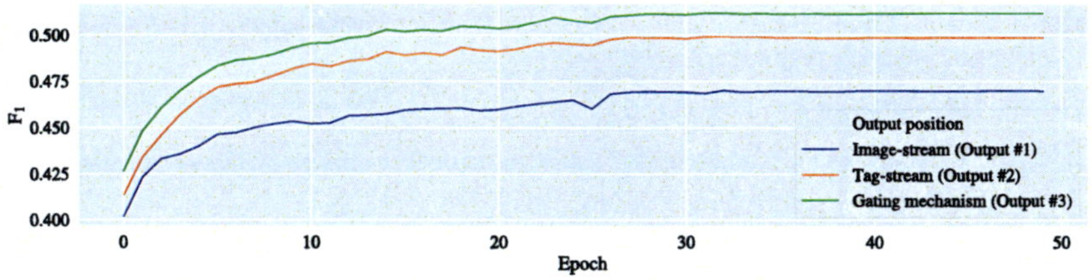

Fig. 5 F_1 comparison of different output positions

Image Feature Extractor. Since the small size of NUS-WIDE dataset after being applied to some pre-processing and filtering steps, ResNet18 was used as the backbone for the image feature extractor. Additionally, all parameters of the backbone were entirely frozen during the training phase in all experiments. With a big dataset like LS, we adopted ResNet50 and froze the first 3 residual blocks of the backbone. **Transformer Encoder**. Through different empirical studies, it is shown that the Transformer encoder is very easy to fit the data in ETS task; thus, depending on the size of the dataset, the hyper-parameter configurations would be varied.

- NUS-WIDE: We used one block for the encoder of the image-stream ($N = 1$) and two for the tag-stream ($M = 2$). All others hyper-parameters are the same for both streams: *d-model* = 128, *heads* = 4, *dim-feedforward* = 512, *dropout* [22] = 0.3, where *d-model* is the number of expected features in the input, *heads* is the number of heads in the multi-head attention model, *dim-feedforward* is the dimension of the feed-forward networks, and *dropout* is simply the dropout value used for the whole architecture.
- LS: We set $N = 2$, $M = 6$, *d-model* = 512, *heads* = 8, *dim-feedforward* = 2048, *dropout* = 0.1.

Optimizer. We used the stochastic gradient descent (SGD) optimizer to train MAG-Neto model with $momentum = 0.9$, *learning-rate* = 10^{-2}; when dealing with LS, *learning-rate* = 3×10^{-2} can be used for faster convergence rate. Furthermore, we also used reduce-lr-on-plateau scheduler with $factor = 0.1$ and $patience = 5$ for the training process with the large-scale dataset.
Image Data Augmentation. Besides frequently used augmentation techniques for images such as flipping and cropping, we also adopted the ImageNet augmentation policies learned by AutoAugment [4]. For image size, we used *input-size* = 112 for NUS-WIDE, and *input-size* = 224 for LS.
Tag Data Augmentation. For TAD augmentation, we used $\beta = 0.5$ and $\hat{\beta} = 0.5$ for NUS-WIDE since the small scale of the dataset. When training with LS, the two coefficients are smaller with $\beta = 0.3$ and $\hat{\beta} = 0.3$.
Hardware. We trained our model on one machine with a single RTX-2080ti GPU.

References

1. Bahdanau D, Cho K, Bengio Y (2014) Neural machine translation by jointly learning to align and translate. arXiv preprint arXiv:1409.0473
2. Chua TS, Tang J, Hong R, Li H, Luo Z, Zheng YT (2009) NUS-wide: a real-world web image database from national university of Singapore. CIVR. Santorini, Greece
3. Chung J, Gulcehre C, Cho K, Bengio Y (2014) Empirical evaluation of gated recurrent neural networks on sequence modeling. arXiv preprint arXiv:1412.3555
4. Cubuk ED, Zoph B, Mane D, Vasudevan V, Le QV (2019) Autoaugment: learning augmentation strategies from data. In: CVPR
5. Devlin J, Chang MW, Lee K, Toutanova K (2018) BERT: pre-training of deep bidirectional transformers for language understanding. arXiv preprint arXiv:1810.04805

6. Gidaris S, Singh P, Komodakis N (2018) Unsupervised representation learning by predicting image rotations. arXiv preprint arXiv:1803.07728
7. Hariharan B, Arbeláez P, Girshick R, Malik J (2014) Simultaneous detection and segmentation. In: ECCV. Springer, Berlin, pp 297–312
8. He K, Zhang X, Ren S, Sun J (2016) Deep residual learning for image recognition. In: CVPR, pp 770–778
9. Hochreiter S, Schmidhuber J (1997) Long short-term memory. Neural Comput 9(8):1735–1780
10. Ioffe S, Szegedy C (2015) Batch normalization: accelerating deep network training by reducing internal covariate shift. arXiv preprint arXiv:1502.03167
11. Jacobs RA, Jordan MI, Nowlan SJ, Hinton GE (1991) Adaptive mixtures of local experts. Neural Comput 3(1):79–87
12. Kolesnikov A, Zhai X, Beyer L (2019) Revisiting self-supervised visual representation learning. In: CVPR, pp 1920–1929
13. Krizhevsky A, Sutskever I, Hinton GE (2012) ImageNet classification with deep convolutional neural networks. In: Advances in neural information processing systems, pp 1097–1105
14. Lan Z, Chen M, Goodman S, Gimpel K, Sharma P, Soricut R (2019) ALBERT: a lite BERT for self-supervised learning of language representations. arXiv preprint arXiv:1909.11942
15. LeCun Y, Bengio Y et al (1995) Convolutional networks for images, speech, and time series. In: The handbook of brain theory and neural networks 3361(10)
16. LeCun Y, Bottou L, Bengio Y, Haffner P (1998) Gradient-based learning applied to document recognition. Proc IEEE 86(11):2278–2324
17. Lin M, Chen Q, Yan S (2013) Network in network. arXiv preprint arXiv:1312.4400
18. Liu Y, Ott M, Goyal N, Du J, Joshi M, Chen D, Levy O, Lewis M, Zettlemoyer L, Stoyanov V (2019) RoBERTa: a robustly optimized BERT pretraining approach. arXiv preprint arXiv:1907.11692
19. Noroozi M, Favaro P (2016) Unsupervised learning of visual representations by solving jigsaw puzzles. In: ECCV. Springer, Berlin, pp 69–84
20. Russakovsky O, Deng J, Su H, Krause J, Satheesh S, Ma S, Huang Z, Karpathy A, Khosla A, Bernstein M, Berg AC, Fei-Fei L (2015) ImageNet large scale visual recognition challenge. IJCV 115(3):211–252. https://doi.org/10.1007/s11263-015-0816-y
21. Shazeer N, Mirhoseini A, Maziarz K, Davis A, Le Q, Hinton G, Dean J (2017) Outrageously large neural networks: the sparsely-gated mixture-of-experts layer. arXiv preprint arXiv:1701.06538
22. Srivastava N, Hinton G, Krizhevsky A, Sutskever I, Salakhutdinov R (2014) Dropout: a simple way to prevent neural networks from overfitting. J Mach Learn Res 15(1):1929–1958
23. Sutskever I, Vinyals O, Le QV (2014) Sequence to sequence learning with neural networks. In: NIPS, pp 3104–3112
24. Vaswani A, Shazeer N, Parmar N, Uszkoreit J, Jones L, Gomez AN, Kaiser Ł, Polosukhin I (2017) Attention is all you need. In: NIPS, pp 5998–6008
25. Wu L, Jin R, Jain AK (2013) Tag completion for image retrieval. TPAMI 35(3):716–727
26. You Q, Jin H, Wang Z, Fang C, Luo J (2016) Image captioning with semantic attention. In: CVPR
27. Yuksel SE, Wilson JN, Gader PD (2012) Twenty years of mixture of experts. IEEE Trans Neural Networks Learn Syst 23(8):1177–1193
28. Zhang H, Wu C, Zhang Z, Zhu Y, Zhang Z, Lin H, Sun Y, He T, Mueller J, Manmatha R et al (2020) ResNeSt: aplit-attention networks. arXiv preprint arXiv:2004.08955
29. Zhang R, Isola P, Efros AA (2016) Colorful image colorization. In: ECCV. Springer, Berlin, pp 649–666
30. Zhao H, Shi J, Qi X, Wang X, Jia J (2017) Pyramid scene parsing network. In: CVPR, pp 2881–2890

Design and Development of a Mobile Outdoor AR Application for On-Site Visualization of Wind Turbines

Simon Burkard and Frank Fuchs-Kittowski

Abstract This paper presents a mobile outdoor augmented reality application that enables a realistic three-dimensional on-site visualization of planned wind turbines in their surroundings. For this purpose, the requirements for such an application are presented, which were obtained from the discussion with representatives of the relevant target groups. In addition, the essential aspects of the graphical user interface and the technical concept (architecture) as well as the implementation and evaluation of the application are presented.

Keywords Mobile outdoor augmented reality · On-site visualization · Wind turbine

1 Introduction

In order to achieve climate protection goals, a steady expansion of renewable energy plants (RE plants) and thus also the planning and construction of new wind turbines (WT) is necessary. However, parts of the (local) population are often rather opposed to concrete plans for new wind power plants, especially due to a feared change of the landscape as well as potential impairments by acoustic (noise) and visual (lighting, shadow cast) emissions [1]. However, since the actual effects of new wind turbines (e.g., on the landscape) often remain unclear for many non-experts, a realistic visualization of the effects of planned wind turbines is necessary in order to achieve a higher acceptance of planned wind turbines among the population. In addition to traditional visualization methods (static photomontage/simulation/sketch), the technology of mobile augmented reality (mAR) offers an innovative visualization method to make planned projects tangible on site in the real landscape. In a mobile AR application,

S. Burkard (✉) · F. Fuchs-Kittowski
Hochschule für Technik und Wirtschaft (HTW) Berlin, Wilhelminenhofstr. 75a, 12459 Berlin, Germany
e-mail: s.burkard@htw-berlin.de

F. Fuchs-Kittowski
e-mail: frank.fuchs-kittowski@htw-berlin.de

X.-S. Yang et al. (eds.), *Proceedings of Seventh International Congress on Information and Communication Technology*, Lecture Notes in Networks and Systems 447,
https://doi.org/10.1007/978-981-19-1607-6_27

3D models of planned wind turbines could be superimposed on the camera image of a mobile device in an accurate and realistic manner. In this way, a realistic picture of the effects of new wind turbines on the landscape could be conveyed from any position.

This paper presents the development of a mobile augmented reality application that enables the visualization of planned wind turbines in real time on-site in the camera image of the mobile device in the immediate vicinity of the user. The paper is structured as follows: In the following Sect. 2, the state of the art in research and technology for the visualization of wind turbines as well as for the realistic representation of virtual objects in outdoor environments using mobile augmented reality is presented. Then, in Sect. 3, the requirements for a mAR application for 3D visualization of wind turbines in the landscape are presented, which were obtained in workshops with potential end-users. Section 4 presents the user interface concept for the mAR application, and Sect. 5 presents the technical concept (architecture) of the mAR application system. Then, finally, Sect. 6 presents key aspects for the implementation and Sect. 7 for the evaluation of the application. The paper ends with a summary and an outlook in Sect. 8.

2 State of Research and Related Work

2.1 Visualization of Planned Wind Turbines

Photorealistic visualizations of such construction projects are usually created in advance with the help of "special software" on the PC, for example by photomontages, e.g., with photoshop (a virtual model is retouched into a real landscape photograph), or by 3D simulations in virtual landscape models, either by visualization tools within established planning software (e.g., 3D animator of the planning software "WindPro") or by specially developed special software solutions (e.g., 3D analysis [2]). Such visualizations allow a relatively realistic view of planned construction projects, but the creation of these graphics usually has to be done by "experts" in advance. Furthermore, the visualization is limited to certain previously defined viewpoints (e.g., in the case of photomontage) or purely virtual environments (e.g., in the EnergyAtlas Bavaria [2]).

2.2 Mobile Augmented Reality for Outdoor Applications

Mobile AR applications and SDKs for the realistic AR representation of virtual objects or information in the close-up range ("indoor") are established and work quite reliably and robustly, for example for the representation of virtual pieces of furniture in one's own home (e.g., the application IKEA Place). The idea of mAR-based representation of virtual objects outdoors (mobile outdoor augmented reality) has also been analyzed and implemented in several other projects (see, e.g., [3, 4]). Many of these applications visualize virtual content outdoors but within a known small-scale environment, e.g., for AR-based representation of flood hazards [5] or for 3D visualization of historical [6] or planned buildings [7]. For precise registration of the mobile device (e.g., smartphone) within the known local environment as a prerequisite for correct placement of virtual AR content in the camera image, this can be done by relying on artificial markers or natural reference images (e.g., house facades) known in advance [5, 6] or 3D models (3D point clouds) of the environment created in advance [7].

In contrast, the realistic representation of information or objects at a specific geographic position within a large-scale environment unknown in advance (e.g., planned wind turbines in the landscape) is particularly challenging: This requires not only local tracking of the mobile camera for a stable AR representation, but also precise localization with respect to a global geo-coordinate system (geo-localization; global registration).

2.3 *Systems for AR Visualization of Wind Turbines*

For mAR-based visualization of planned wind turbines, no systems are available on the market so far. Only two prototype implementations from research projects are known: One is an implementation by LandPlan OS GmbH (research project "MoDal-MR," https://www.landplanos.de/forschung.html), in which the global device registration is performed by manual calibration on the basis of point-shaped reference markers (e.g., church towers) in the landscape, which are defined manually. On the other hand, a prototypical visualization tool exists from the project "Linthwind" of Echtzeit GmbH and ZHAW Switzerland (https://echtzeit.swiss/index.html#projects_AR). In this case, a manual global device registration is carried out on the basis of previously defined mountain peaks of the environment. Both systems are prototype applications with basic functionalities whose requirements and concepts have not yet been documented or examined in more detail.

3 Requirements

In the following section, the requirements for the functionalities and the design of a mAR application for 3D visualization of wind turbines in the landscape will be described. These requirements were identified during potential analysis workshops (see [8]) with potential end-users in cooperation with the Bavarian State Office for the Environment as well as the Bavarian State Ministry for Economic Affairs, Regional Development and Energy.

- **R1 Placement of planned wind turbines on a map and in the camera image**: It should be possible for the user to add or place a virtual WT in his environment in the application. On the one hand, the placement of a WT should be done separately via a map-based display (classic 2D map overlay). On the other hand, the placement of a WT model should also be possible in the camera image, if necessary.
- **R2 mAR visualization of planned wind turbines as a 3D model**: The placed virtual wind turbine should be able to be viewed in the landscape as a mAR visualization from any position. Ideally, the WT should be visualized as a textured 3D model in a representation as realistic as possible.
- **R3 Precise and correct placement of AR content in the camera image**: The representation of mAR content should be as precise as possible; i.e., AR content should be placed correctly in the camera image. This requires precise determination of user location and viewing direction despite imprecise localization sensors in mobile devices. However, the use of external, very precise localization sensors (GPS receiver with external antenna and correction signals) is not desired. For a realistic representation of the virtual 3D models in the environment, occlusions by terrain, vegetation or buildings should also be taken into account.
- **R4 mAR visualization of the wind turbine rotor movement in the 3D model (animation)**: The 3D model can either be static or animated with a fixed rotor speed (can be switched on/off). The implementation of further visualization functions (shadow casting simulation, night marking, etc.) can be omitted.
- **R5 Simultaneous mAR visualization of several planned WTs**: Not only individual WTs, but also several planned WTs (wind park) should be virtually representable. When a WT has been placed, the user shall have the possibility to add one (or more) further WTs to the camera image in the same way.
- **R6 mAR display of meta-data of a planned WT**: In addition to the display of the WT, it should be possible to display corresponding meta-data of a WT, the scope of which should be adapted to the respective users (no excessive demands). For all users, only important parameters (height, rotor diameter, position) should be displayed. For special user groups (e.g., planners), additional information could be provided (power, yield, etc.).
- **R7 Modification of the appearance of the planned WTs (model type and height)**: It should be possible to modify the appearance of the WTs. For this purpose, some standard models of different manufacturers shall be available for selection in order to be able to change the WTG type at runtime. These models can then be scaled arbitrarily to simulate any hub height (minor "distortions" due to proportional scaling would be acceptable).
- **R8 Modification of the orientation and location of the planned WT**: The location and orientation of the virtual WT should also be modifiable. As in the case of placement, moving the WTs should be possible separately via a map-based display as well as in the camera image. The orientation (nacelle orientation) of the virtual WT should also be variable; i.e., the WT should be able to be rotated in any cardinal direction.

- **R9 Consideration of non-permitted areas when placing WTs**: When placing or moving WTs, permitted and non-permitted areas shall be considered and displayed. In this way, placement in "not possible" places shall be prevented (protected areas, minimum distances to buildings and other wind turbines).
- **R10 mAR visualization of planned WTs as POI markers**: An abstract visualization of planned WTs as point-of-interest markers (POI markers) shall also be possible if the WT is located at a greater distance from the viewer.
- **R11 mAR visualization of already existing WTs as POI markers**: The mAR application shall not only offer the possibility to virtually represent newly planned WTs, but also to show already existing WTs as well as other existing WTs via AR representation as POI visualization.[1] This representation can also be bundled for an entire location ("This location X supplies itself with Y percent renewable energy, thereof Z percent wind energy.").
- **R12 Local import/export of the current mAR planning configuration**: The current planning configuration (location and appearance of the WTs placed in the camera image) is to be saved as a project file and can be reloaded at another location. A specific configuration can thus be viewed virtually from different locations. The saved file can also be shared with other users and loaded in another application instance.
- **R13 Video recording of the current AR view**: It should also be possible to make a short video recording of the current AR view so that the planned configuration can be shared with other people as a video.
- **R14 Mobile AR hardware (commercially available smartphones/tablets)**: The mAR application should be developed for use on modern, commercially available smartphones or tablets without special sensor technology.
- **R15 Server-based data provision and online capability**: The mAR application must access various data sources (e.g., WT models, WT meta-data, meta-data of other RE plants, terrain models, property areas, etc.). The provision of these data should be server-based, e.g., via a CMS or a geoserver/GIS. Large amounts of data (e.g., surface models) should be provided in an optimized way to keep the transmitted data volume as low as possible.

4 Design of the graphical user interface

This section describes the designed graphical user interface (GUI) of the mAR application for visualization of planned WTs, which is used to realize all essential functional requirements. The designed screen designs as well as their interrelationships are shown in Fig. 1.

Accurate determination of global camera position and orientation (global camera registration) is not only a requirement for this mAR application (see R3), but a key requirement for accurate AR visualizations in outdoor mobile AR applications

[1] Users can get an impression of where renewable energy systems are already installed in the immediate vicinity and how far the expansion of renewable energies has already progressed.

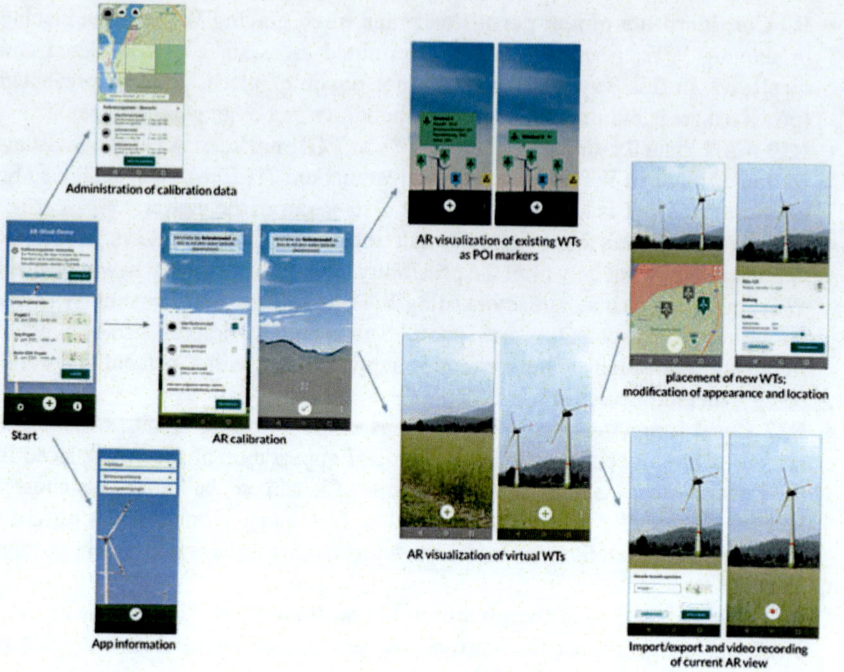

Fig. 1 GUI designs for a mobile application for mAR visualization of wind turbines

in general. In the system presented here, a novel user-assisted registration method (developed by the authors of this paper [9]) was used, which uses georeferenced data to accurately register mobile devices with respect to a global geo-reference system (geo-coordinate system). In this method, the calibration of the device pose is based on visible objects, e.g., on the terrain. Digital 3D terrain, building, and surface models are integrated and used for this purpose. The user manually moves—via two common mobile touch gestures (drag-touch and pinch-zoom gesture)—the projected model of the environment (e.g., terrain model) on the screen so that it matches the actual real view of the world in the live camera video. This user-controlled shift of the virtual environment thus leads to a correction of the global device position and orientation.

The following screen designs have been created:

Start screen: This is the first screen after launching the application. It provides a quick selection of the last saved AR representations (AR projects), the option to open the calibration data management and view application information, as well as the option to start the main AR functions (representing virtual WTs as well as AR information about existing RE plants).

Calibration data management: Before AR rendering of virtual content is possible, AR calibration data of the environment (e.g., 3D geospatial models) must be loaded.

A separate screen provides expert options for custom display, loading, and activation of individual calibration data.

AR calibration: Precise AR display requires manual calibration for precise global localization of the mobile AR application. During calibration, the AR view is to be aligned via user interaction so that the virtual view and the real view match. This is done by moving the virtual calibration objects (3D geomodels). The user can also open a settings menu in the AR calibration view to determine the type of calibration data to be displayed. After manual alignment is complete, the user confirms the calibration, which launches the main AR view.

View virtual WT models in the landscape (AR main view): On this main screen, one or more WTs are displayed using AR visualization at specified geo-positions in the camera view. Depending on the distance, the display is either a 3D model or a POI. From this main view, further functionalities related to AR visualization of WTs in the landscape can be started via button.

Adding a new virtual WT to the AR view: In the AR main view, a new WT model can be added in the user's field of view via button. Clicking the "New WT" button will automatically add a standard WT model in the user's field of view to the AR view.

Changing the location (position) of the WT model: Via a map view, the position of the placed WTs can be adjusted manually. Unsuitable areas are marked accordingly on the map via map overlay. By moving a WEA icon in the map view, the global position of the virtual model changes. This is immediately reflected in the AR view.

Changing the appearance of the WT model: The user has the possibility to change the size (height), the orientation (rotation), and the model of the virtually placed WTs in order to view or compare the effects of WTs with different appearance or height. For this purpose, the user selects a placed WTG model in the AR view or the map view, so that a WTG settings menu appears. There, a 3D WT model can be selected from a list of predefined 3D models and the size (hub height) and orientation can be specified.

Loading and saving AR projects: The current AR configuration, i.e., the WT models (location and appearance) currently placed in the AR view, can be saved as a project file in order to restore and view this AR configuration at another time and/or location.

View existing RE plants as AR-POI representation: Via a button in the start screen, the user has the possibility to view meta-information about relevant renewable energy plants in the surrounding area as a POI-AR representation to get an impression of the availability and type of renewable energy plants in the surrounding area. Clicking on an AR-POI marker expands the marker view with additional meta-information about the selected plant. Clicking on a "list button" opens a (non-AR-based) list view of the RE plants in the vicinity.

Additional information on the mobile application: This screen provides a display of relevant additional (technical and legal) information about the application, e.g., terms of use, background information, etc.

5 Architecture of the mAR System

This section describes the (technical) architecture of the mAR system for visualizing planned WTs and important design decisions are explained. The (technical) architecture of the designed mAR system is roughly sketched in Fig. 2.

The architecture consists of the following main components:

- **mAR client**: The mobile AR client represents all components of the mobile AR application on the mobile device, these include:
  - **mAR application**: This application-specific component implements all the UI functionalities needed to operate and control the mobile mAR application (see GUI design). It also controls the server communication for retrieving and administering the required geospatial data. To realize these functions, the application component uses several application-independent components (local storage, GeoCMS as well as GeoAR Library).

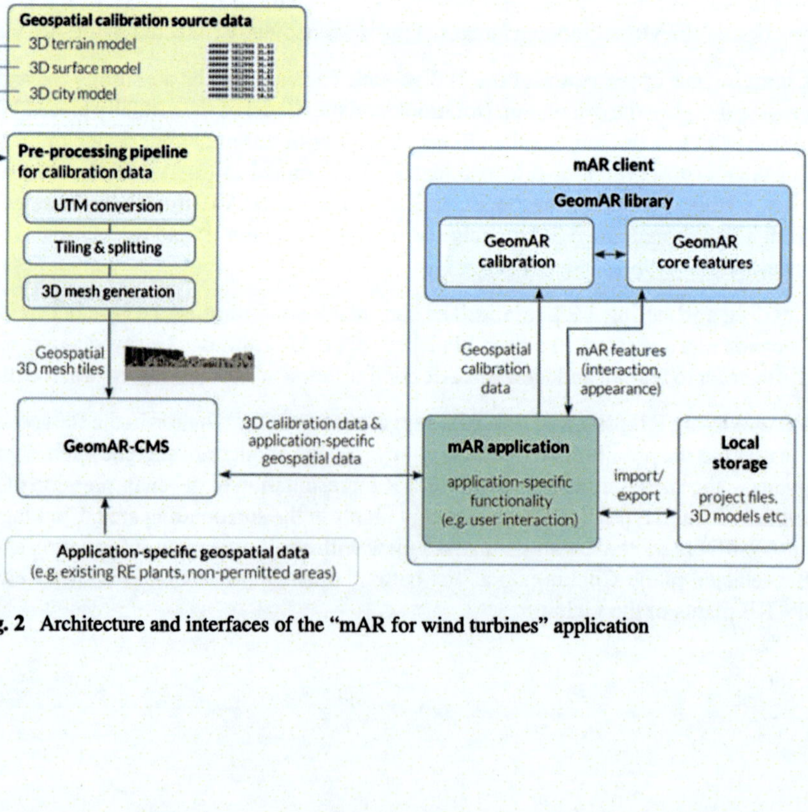

Fig. 2 Architecture and interfaces of the "mAR for wind turbines" application

  - **Local storage**: Stored AR views, 3D models of the wind turbines and temporarily stored calibration data are stored locally within the AR client and can be loaded from there into the mAR application.
  - **GeomAR library**: This client-side library provides necessary geospatial mAR (GeomAR) core functionalities (placement and manipulation of wind turbine models) as well as functions for device calibration (geo-localization). The used technology for device calibration was developed by the authors of this paper [9].

- **GeomAR-CMS**: The main task of this server-based component is the storage and administration of geospatial data needed to run the mAR application. For this purpose, this server component provides:

  - On the one hand, **interfaces** to add, modify, and retrieve application-specific geodata (e.g., WT data) as well as geodata needed for device calibration (e.g., 3D terrain models) (REST API and GUI frontend for administration), and
  - On the other hand, a **database** for persistent storage of these geodata.

- **Geospatial data pre-processing pipeline**: Since 3D geospatial data for device calibration is often not initially available in formats suitable for immediate server-based storage and delivery, the overall system also includes an offline component to convert the 3D geospatial data into smaller-scale 3D tiles that can be efficiently displayed as a 3D model (3D mesh) within the mobile AR client. To convert to 3D models suitable for AR display within the mobile AR client, several conversion steps are required:

  - **UTM conversion**: First, the source data is transformed into a Universal Transverse Mercator (UTM) coordinate system that uses a metric grid (meter).
  - **Tiling and splitting**: For more efficient data handling, the source data is then split into smaller parts with defined square dimensions. This way, the mobile client only needs to load and process smaller tiles with smaller file sizes when rendering the virtual user environment.
  - **3D mesh generation**: Using incremental Delaunay triangulation [10], the generated 3D tiles are finally transformed into optimized triangulated irregular network (TIN) surface meshes with different levels of detail. Wavefront OBJ is used as the target file format for the textureless 3D meshes.
  - **Storage in GeomAR-CMS**: These obj files are finally stored together with the associated meta-data (position, size, and type of 3D tile) in the server-based management system (GeomAR-CMS). From there, they can be provided to the mobile AR client on demand.

Fig. 3 Screenshots from the mAR application—select calibration data, move 3D geomodels, display WT as AR visualization, place WT, modify WT

6 Implementation

The presented components for the realization of the mobile application "mAR for wind turbines" were implemented for practical use on commercially available smartphones and tablets within predefined test areas (Berlin and Augsburg region). Users of the application thus receive an impression of the potential impact of newly planned wind turbines on the landscape within these areas of use.

The **application** was implemented as a native Android mobile application that can be run on standard mobile devices with built-in IMU and GNSS sensors and using Google ARCore SDK as a local VIO tracking system. The AR content in the camera image is rendered using the OpenGL-based Sceneform SDK. The raw data of the 3D geospatial models for AR calibration were provided by the Bavarian Surveying Administration (LDBV) for testing purposes within the scope of the research work. Alternatively, the—often freely available—raw data of other state surveying offices can be integrated in the same way.

The **preprocessing pipeline** was realized based on open-source tools for geospatial data processing, in particular based on the GDAL library. These tools were encapsulated in Docker containers and automatized using Python scripts. The GeomAR-CMS for managing, storing, and providing geospatial data was implemented based on the GeoServer software and using a Ruby-on-Rails web framework. Screenshots from the implemented mobile application can be seen in Fig. 3.

7 Evaluation

The functionality of the mAR application as well as the UI were developed from the beginning in cooperation with representatives of the user group. In initial tests with potential end-users, the application was evaluated positively. Experimental results have shown that the user-driven calibration approach—combined with a robust local tracking system—enables efficient and accurate global registration of mobile devices in various outdoor environments and can determine device orientation with less than one degree deviation. Moreover, due to the positionally accurate integration of 3D environment models thus achieved, this approach is suitable for correctly handling occlusions of virtual AR content caused by vegetation, terrain or buildings. As a result, the realistic visualization through the possible consideration of obscurations of the turbine as well as the correct, robust positioning of the wind turbines was emphasized during the experimental tests achieving a high degree of realism.

The additional manual effort for the user required for calibration, i.e., correcting the position and orientation of the wind turbines, was rated as acceptable. This manual effort does not make the application as easy to use as users would like, but due to the inaccuracies of the sensors on commercially available mobile devices, such additional calibration effort is mandatory to ensure correct visualization. While this provides a source of error due to incorrect use, similar user errors can occur with other visualization methods (e.g., photomontages). Therefore, in next steps, a more intensive user evaluation is planned to identify and implement optimization approaches for better usability—especially with respect to the geospatial data-based calibration methodology.

8 Summary and Outlook

The aim of this paper was the development of a mobile mAR application for the representation of planned wind energy turbines. For this purpose, the requirements for such an application were determined, which were obtained from discussions with representatives of the relevant target groups. Based on this, the user interface concept and the technical concept (architecture) of the application were developed and the mAR application was implemented and evaluated.

The great advantage of an mAR application—especially in comparison to classic visualization processes—is that visualization is easier and faster as well as possible directly on site. This more flexible visualization method can already be clearly demonstrated and experienced by the developed prototype application using the example of the visualization of planned wind energy plants. This advantage could also be used for further mAR applications in the renewable energy sector, e.g., for the mAR visualization of planned power lines.

Limitations of the proposed solution can be seen on the one hand in the fact that complex geodata must be available as well as an infrastructure to manage and provide this data. In addition, the application has only been tested in a small test area. A test in a larger area would place higher demands on the management of even larger amounts of data as well as the infrastructure for data provision. In this context, the focus was on the technical feasibility and the implementation of the functional requirements. In order to achieve greater practicality and reach broader user groups, a stronger focus on usability—especially for non-experts—would have to be placed in the next step.

Acknowledgements The authors gratefully acknowledge financial support from the German Federal Ministry of Education and Research (BMBF) within the research project "mARGo" (grant number 01IS17090B) and the German Federal Ministry for Economic Affairs and Energy (BMWi) within the research project "AR4WIND" (grant number 03EE3046B).

References

1. Hübner G, Pohl J, Warode J, Gotchev B, Nanz P, Ohlhorst D, Krug M, Salecki S, Peters W (2019) Naturverträgliche Energiewende. Akzeptanz und Erfahrungen vor Ort. Bundesamt für Naturschutz (BfN), Bonn
2. Nefzger A (2018) 3D-Visualisierung von Windenergieanlagen in der Landschaft—Webanwendung "3D-Analyse". In: Freitag U, Fuchs-Kittowski F, Hosenfeld F, Abecker A, Reineke A (eds) Umweltinformationssysteme 2018—Umweltbeobachtung: Nah und Fern., CEUR-WS.org, vol 2197, pp 159–177. http://ceur-ws.org/Vol-2197/paper12.pdf. Last accessed 2021/11/05
3. Burkard S, Fuchs-Kittowski F, Abecker A, Heise F, Miller R, Runte K, Hosenfeld F (2021) Grundbegriffe, Anwendungsbeispiele und Nutzungspotenziale von geodatenbasierter mobiler Augmented Reality. In: Freitag U, Fuchs-Kittowski F, Abecker A, Hosenfeld F (eds) Umweltinformationssysteme - Wie verändert die Digitalisierung unsere Gesellschaft?. Springer Vieweg, Wiesbaden, pp 243–260. https://doi.org/10.1007/978-3-658-30889-6_15
4. Rambach J, Lilligreen G, Schäfer A, Bankanal R, Wiebel A, Stricker D (2021) A survey on applications of augmented, mixed and virtual reality for nature and environment. In: Chen JYC, Fragomeni G (eds) Virtual, augmented and mixed reality. HCII 2021. Lecture notes in computer science, vol 12770. Springer, Cham. https://doi.org/10.1007/978-3-030-77599-5_45
5. Haynes P, Hehl-Lange S, Lange E (2018) Mobile augmented reality for flood visualisation. Environ Model Softw 109:380–389.https://doi.org/10.1016/j.envsoft.2018.05.012
6. Panou C, Ragia L, Dimelli D, Mania K (2018) Outdoors mobile augmented reality application visualizing 3D reconstructed historical monuments. In: 4th international conference on geographical information systems theory, applications and management (GISTAM). Scitepress, pp 59–67. https://doi.org/10.5220/0006701800590067
7. Zollmann S, Hoppe C, Kluckner S, Poglitsch C, Bischof H, Reitmayr G (2014) Augmented reality for construction site monitoring and documentation. Proc IEEE 102(2):137–154. https://doi.org/10.1109/JPROC.2013.2294314
8. Fuchs-Kittowski F, Burkard S (2019) Potential analysis for the identification of application scenarios for mobile augmented reality technologies—with an example from water management. In: Schaldach R, Simon KH, Weismüller J, Wohlgemuth V (eds) Environmental informatics—computational sustainability: ICT methods to achieve the UN sustainable development goals. Shaker, Aachen, pp 372–380
9. Burkard S, Fuchs-Kittowski F (2020) User-aided global registration method using geospatial 3D data for large-scale mobile outdoor augmented reality. In: 2020 IEEE international symposium on mixed and augmented reality adjunct (ISMAR2020), pp 104–109. https://doi.org/10.1109/ISMAR-Adjunct51615.2020.00041
10. Heckbert PS, Garland M (1997) Survey of polygonal surface simplification algorithms. Technical report, Carnegie-Mellon University, Pittsburgh

An Incorporated Solution to Support Elder People in Staying in Their Familiar Surroundings

Dominic Mircea Kristaly and Sorin-Aurel Moraru

Abstract The SAVE system is an incorporated solution whose main purpose is to support elder people in staying in their familiar surroundings for as long as possible, whilst still be safe and optimally cared for. From an architectural point of view, the SAVE system is designed on the model of microservices-based architecture, which exposes a *Representational State Transfer* (REST) communication interface, making use of the HTTP/S protocol. The user interfaces (the frontend) of the web applications are developed using the Angular framework and the data is retrieved using the REST APIs of the backend (built on the Spring Boot framework). The messages they exchange with each other are formalized using the *JavaScript Object Notation* (JSON) format. To ensure data persistence, a relational database management system (Oracle's MySQL) was employed, providing a more efficient solution for organizing information in the system. We created a web application—*SAVE Admin Centre*—to manage data efficiently and securely within the SAVE system, which consists of several modules: Dashboard, RO Data, Data Kit, Kits, Devices, Device Types and Users. The data collection sub-system was design to be able to connect to different data sources through data adapters. The security of the SAVE system comes from the use of *public key infrastructure* (PKI), through HTTPS, whilst using standard protocols to ensure maximum interoperability.

Keywords Microservices · Admin Centre · Data collection system

1 SAVE Solution Overview

The SAVE system is an incorporated solution whose main purpose is to support elder people in staying in their familiar surroundings for as long as possible, whilst still be safe and optimally cared for. Secondarily, SAVE supports informal caregivers,

D. M. Kristaly (✉) · S.-A. Moraru
Transilvania University of Brasov, B-dul Eroilor 29, 500036 Brasov, Romania
e-mail: dominic.kristaly@unitbv.ro

S.-A. Moraru
e-mail: smoraru@unitbv.ro

X.-S. Yang et al. (eds.), *Proceedings of Seventh International Congress on Information and Communication Technology*, Lecture Notes in Networks and Systems 447,
https://doi.org/10.1007/978-981-19-1607-6_28

like relatives, in providing optimal care for their wards, whilst maintaining their professional and private life. Additionally, SAVE enables professional caregivers in the development of an optimal support planning and achievement, involving also volunteering associations.

The SAVE solution is built using the latest technologies for applications designed to run in cloud environments (in an IAAS context) and taking advantage of what containerization has to offer (less overhead, increased portability, more consistent operation, greater efficiency in terms of scalability and velocity, better application development [1]).

An overview of the SAVE solution is depicted synthetically in Fig. 1.

The sensors included in the SAVE kit (wearable, ambient and biological) provide raw data about the elderly (end-user) (indicators of his/her well-being, activity, environment, location) by connecting through Internet to dedicated services of the SAVE cloud application [2]. The users of the SAVE solution have dedicated user interfaces, in form of responsive web applications and mobile applications, for accessing its features:

- For the end-users: the SAVE smartwatch face and application and the *SAVE web application*;
- For the caregivers: the *SAVE web application*;
- For the SAVE solution maintenance staff and for the SAVE researchers: the *SAVE Admin Centre* web application.

All web applications are responsive, so they can be viewed from a vast range of devices (desktops, laptops, tablets, smartphones).

The SAVE cloud application is split into several functional units (microservices), that deal with particular aspects of the business logic:

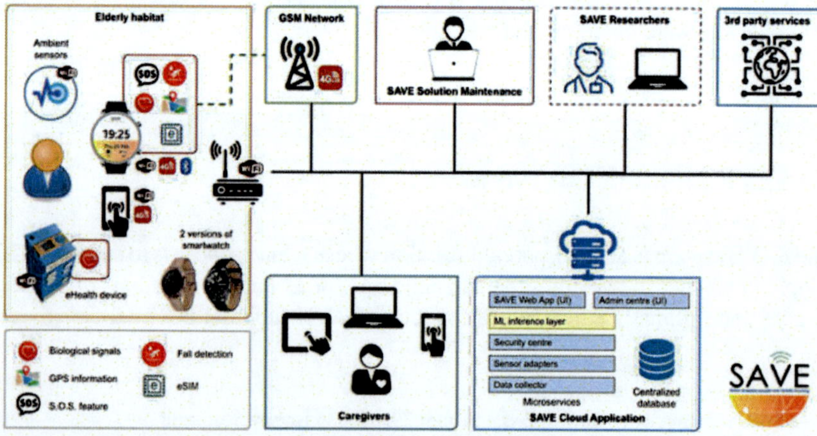

Fig. 1 SAVE solution overview

- *Data collector*: collects data coming from sensors and stores them inside the database. It also offers a communication interface (REST API) for retrieving this data.
- *Sensor adapters*: array of microservices that connect directly to the sensing devices and relay their data in a standardized format to the *Data collector* (using REST calls).
- *Security centre*: offers REST APIs for authentication and authorization operations.
- *SAVE Web App*: is the web-based user interface for end-users and caregivers.
- *SAVE Admin Centre*: is the web-based user interface for maintenance staff and researchers.
- ML inference layer: is a machine learning powered service that analyses the data collected by the sensors and assesses the well-being of the end-users (by conformity with the current profile of the end-user). The development of this layer it is forecasted for the immediate period.

A relational database management system (Oracle's *MySQL*) was used to persist both the configuration and data of the SAVE software solution. To assure high speeds of access to the data, a partitioning algorithm was implemented at cloud application level (based on the kit identifier).

The SAVE solution was designed as a maximum inclusive system, so it can incorporate other sensor kits, developed by 3rd parties, with their own services, as long as they do not require permanent maintenance and configuration. Tests were made with the *Xiaomi Smart Home Sensor Set* in this context, with good results. Also, the *Data Collector* service offers an inclusive REST API for connecting with ease any other sensor systems to the SAVE solution. Work is in progress to include other sensors in the SAVE kit that will make use of this API.

2 SAVE System Architecture

From an architectural point of view, the SAVE system is designed on a service-based architecture. This involves splitting the source code for services or features into independent, smaller software components that need to perform very specific tasks. These components—microservices—are independent components, but they communicate with each other through a common, message-based language using a standardized interface [3]. Usually, the communication protocol used by microservices is the *Hypertext Transfer Protocol* (HTTP) or *Hypertext Transfer Protocol Secure* (HTTPS)—www protocol, and the messages exchanged between services are independent of the technology used, the *JavaScript Object Notation* (JSON) format being usually employed.

Microservices are an alternative to the monolithic application development system. The latter implies developing an application as a single unit. In the case of a monolithic application, all the components are strongly interconnected and thus the application becomes difficult to modify and maintain, but also to move into a cloud environment. Any changes made to the application will cause a process of creating another version of the system (see Fig. 2) [4].

The use of a microservice-based architecture has both technical and implementation logic advantages, which make it easier and more efficient to develop applications in a cloud context, considering the implementation, availability, efficient management, possibility of modification and adaptation, reliability and scalability [5].

The general architecture of SAVE system (see Fig. 3) is composed of several independent components that communicate with each other to serve certain functionalities.

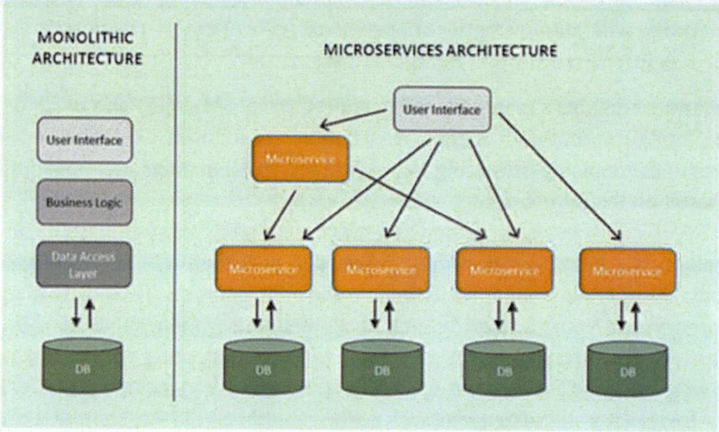

Fig. 2 Monolithic architectures—microservices [8]

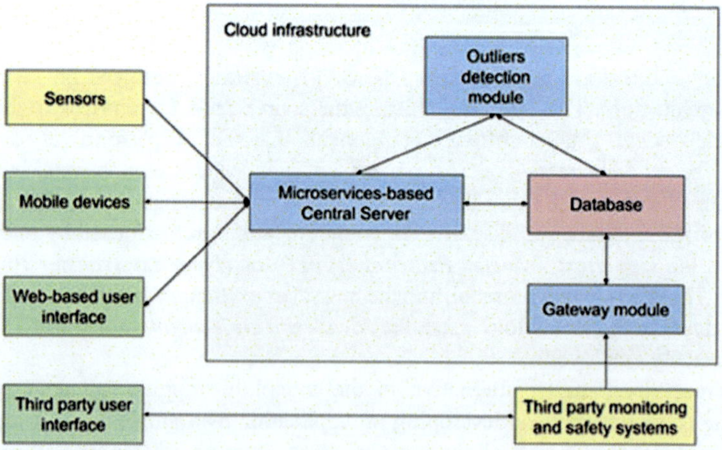

Fig. 3 General architecture of the SAVE system

In this context, the microservices-based architecture of the SAVE system allows running on a cloud infrastructure, scalability, and speed in developing new features [6].

The system has several main components that are hosted into the cloud and perform specific tasks; it also allows interaction with external systems, provided by third parties, such as sensors, mobile devices or web applications that provide data and other monitoring applications that can process data retrieved from the SAVE system.

Looking in detail, the main components of the system are represented by:

- The main microservices, which constitute a virtual central server;
- Microservices for data processing (a machine learning algorithm finds deviations from a pattern, anomalies, trends, etc.);
- Microservices for user interaction;
- Microservices for interaction with external systems;
- User interfaces;
- Database management system.

In terms of implementation, the essential components are developed in the Java programming language using the Spring Boot framework. This solution offers advantages for the development of microservices applications, their running and maintenance over time. Also, updating them is more efficient; it can be done quickly and independently, without affecting in any way other components.

All these microservices expose REST communication interfaces for interconnection, so they use the HTTP or HTTPS protocol. The messages they exchange with each other are standardized using the JSON format.

Thus, communication between internal components and communication with external systems is easy to ensure.

3 The SAVE Database

To ensure data persistence, a relational database management system (Oracle's *MySQL*) is used, providing a more efficient solution for organizing information in the system. Figure 4 presents the organization of the collected data database tables.

The Kits are registered into the *kits* table, and they receive a user-friendly name (field *kt_name*).

The recognized device types are stored in the *device_types* table; they receive an acronym and a default description. The acronym can be used by the user interfaces to identify the component that must be used to handle the data coming from a particular type of device. Also, this acronym can be used for localizing the user interfaces; if the language dictionary does not contain the acronym, the default description will be used.

The SAVE devices are registered into the *devices* table (see Fig. 4). They can be included in one kit and are linked to a device type. The short and long descriptions

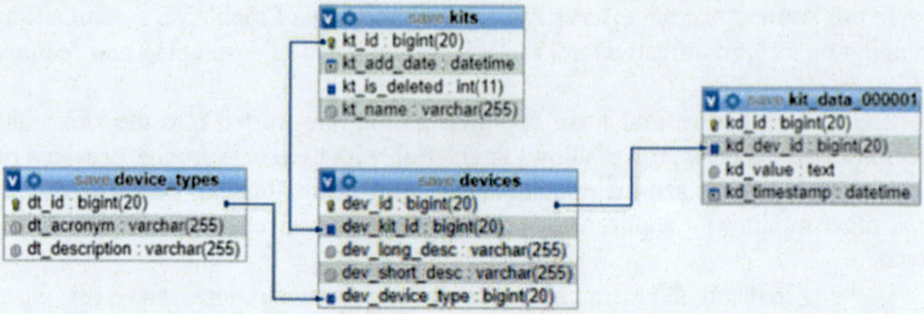

Fig. 4 Tables for the data collecting system

are used on the user interface and are provided by the end-users, to make easier the identification and to discriminate between multiple sensors of the same type.

The values read from the sensors (the "data") are persisted, partitioned at kit level, in the "*kit_data_**" tables; there is one table for each kit. The *kd_value* field contains the sensor data, as transmitted by the sensors itself or a sensor adapter.

Figure 5 presents the structure of users table and several other tables used by the microservices that implement the user interfaces and the link to the smartwatch applications. For the passwords only the SHA-1 hash values are stored.

The *not_data* table stores the scheduled notifications of the end-users. These are transferred periodically (30 s to 1 min or soon as the smartwatch connects to the SAVE cloud application) to the smartwatch. The SAVE web application user interface manages these records.

It must be highlighted that the data coming from the sensors are not linked directly to the end-users, but to the kits. Accessing only this data does not reveal anything about the users' identities.

The *config_packages* table persists the settings of the smartwatch application. The records are automatically generated when an end-user is created in the *SAVE Admin Centre* web application and contains the information needed to "pair" the smartwatch with the right user. When first installed, the SAVE smartwatch application asks for a pin, and based on this it loads the corresponding configuration package from this table.

The temporary GPS data is collected in the *gps_data* table. The maximum duration for which the records are kept is 10 days. Work is in progress to reduce the number of the records, by eliminating consecutive records that have the coordinates closer than 2 m.

4 The Technologies

Figure 6 depicts the simplified technology stack for the microservices included in the SAVE cloud application.

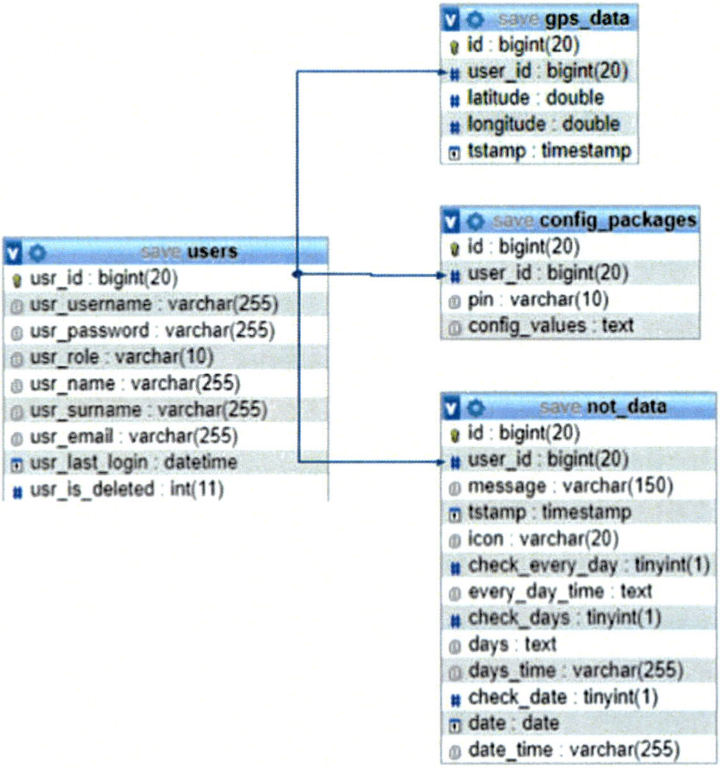

Fig. 5 Tables for the users, notifications, smartwatch app configuration, GPS data

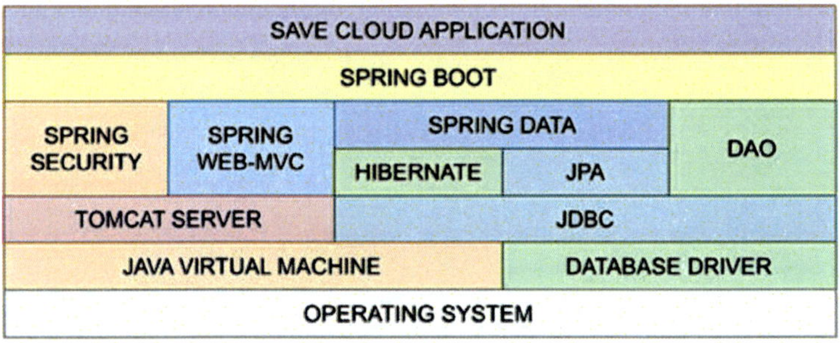

Fig. 6 Technology stack for the SAVE cloud application

The user interfaces (the frontend) of the web applications [7] are developed using the Angular framework and the data is retrieved using the REST APIs of the backend (developed using the Spring Boot framework).

The smartwatch face application and the smartwatch application are Tizen web application, developed in HTL, CSS and JavaScript.

5 The Data Collecting System—Data Collector

One of the most important components of the SAVE system is the data collection service. This service is implemented as a microservice, to be easily scalable with the increase of number of sensors. The communication interface is a REST API (over HTTPS) that accepts JSON formatted data coming from devices that are registered inside SAVE's database.

The minimal structure of the JSON message accepted by the data collecting API is:

```
{
  "kdDevId": <integer_value>,
  "kdValue": <string_value>
}
```

When the sensors cannot provide the data in the accepted format, a data adapter must be implemented, so it will wrap the sensor's native format (binary, text, JSON, XML, etc.) into the accepted structure.

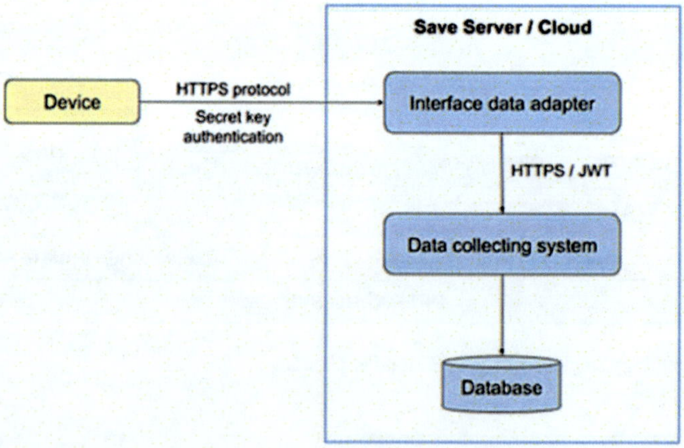

Fig. 7 System communication model

As illustrated in Fig. 7, the devices connect via the HTTPS protocol to a device type-specific adapter interface that transmits the data to the collection system, which will persist it in the database.

The data adapter interface is a software module that is developed specifically for each type of device recognized by the system. Its role is to expose a communication interface with external devices and to convert the format of the received data into a standard format recognized by the data collection component and to send it to the latter.

The adapter streamlines the communication between devices and system, abstracting the data format and providing it to be persisted. In this sense, the adapter is specific to the type of device with which it interacts and can be developed using any technology; it must, however, maintain the standard communication with the central system.

The communication between the adapter and the *Data Collector* is done via the HTTPS protocol, doubled by an authentication based on a secret key (*API Key*).

The adapters are implemented as microservices, using the Java language and the Spring Boot framework.

Once in the system and transposed into a standard format, data from the devices are organized (see Fig. 8) and stored.

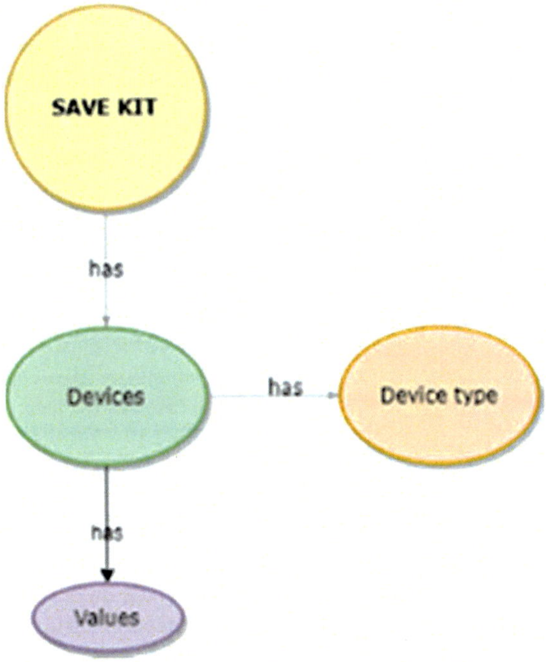

Fig. 8 Data organization in the SAVE system

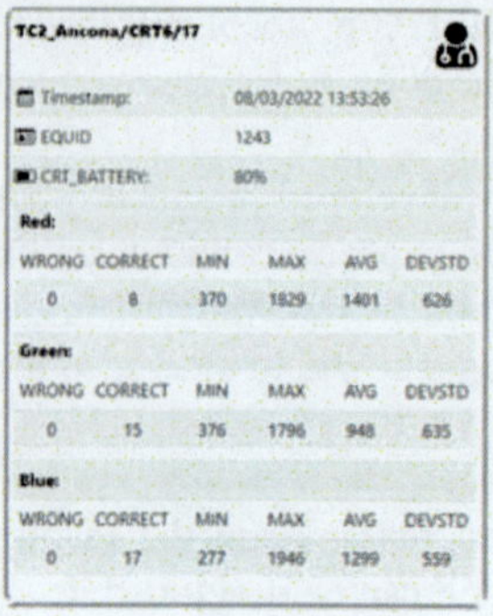

a. Card – version 2

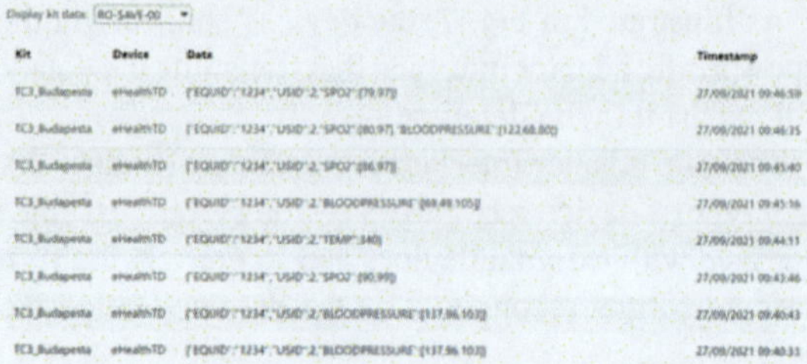

b. Tabular view – unprocessed data

Fig. 9 Data presentation for the eHealth device

A device must be registered in advance; it is associated with a device type (which defines the corresponding adaptation interface). This device will receive a name so that users can easily identify it.

The collected data is stored in tables into the database. To optimize the speed of data access, kits partition them, so that the data from a kit is stored in the same table. The data storage mode is a general one, allowing the saving of any data structure.

6 Adapter Interface for the eHealth Device

The first adaptation interface developed within the SAVE project ensures the connection of eHealth devices to the data collection system.

Figure 9 shows the 2nd iteration for the Dashboard user interface and visualization of raw and unprocessed data, and Fig. 12 presents the processed data.

In addition to this adapter, a user interface has been included in the Admin Centre application, which allows you to view the data, both raw and tabular, processed. A card-based representation component has also been developed for synchronous tracking of data collection from the eHealth device.

7 Admin Centre Web Application

The *SAVE Admin Centre* is a web application developed to efficiently and securely manage data within the SAVE system. It is design mainly for the researchers and the system maintenance staff. The application offers a centralized display, the data being displayed in raw and tabular format, with attached meaning.

Fig. 10 The dashboard module

The Admin Centre was developed using the Angular framework and runs on an Apache Tomcat server.

The access to the application's features is protected using the authentication with the username-password pair and the authorization uses *JSON Web Tokens* (JWT).

The web application consists of several modules: Dashboard, RO Data, Data Kit, Kits, Devices, Device Types, and Users.

The Dashboard module contains a synchronous view (pseudo real-time) of the collected data in the form of cards (see Fig. 10).

The RO Data module allows the visualization of the data coming from the sensors, in raw format, as it is received from the adapter interfaces. It is a component intended only for SAVE system administrators (see Fig. 11).

The Kit Data module displays the data in tabular format, processed to be understood by a human operator (see Fig. 12).

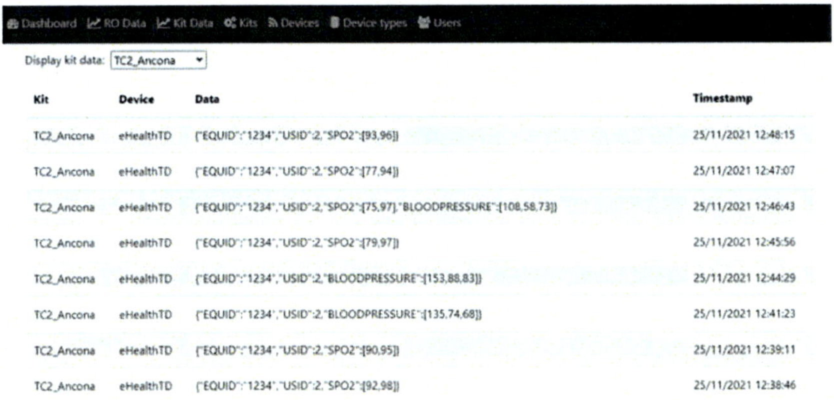

Kit	Device	Data	Timestamp
TC2_Ancona	eHealthTD	{"EQUID":"1234","USID":2,"SPO2":[93,96]}	25/11/2021 12:48:15
TC2_Ancona	eHealthTD	{"EQUID":"1234","USID":2,"SPO2":[77,94]}	25/11/2021 12:47:07
TC2_Ancona	eHealthTD	{"EQUID":"1234","USID":2,"SPO2":[75,97],"BLOODPRESSURE":[108,58,73]}	25/11/2021 12:46:43
TC2_Ancona	eHealthTD	{"EQUID":"1234","USID":2,"SPO2":[79,97]}	25/11/2021 12:45:56
TC2_Ancona	eHealthTD	{"EQUID":"1234","USID":2,"BLOODPRESSURE":[153,88,83]}	25/11/2021 12:44:29
TC2_Ancona	eHealthTD	{"EQUID":"1234","USID":2,"BLOODPRESSURE":[135,74,68]}	25/11/2021 12:41:23
TC2_Ancona	eHealthTD	{"EQUID":"1234","USID":2,"SPO2":[90,95]}	25/11/2021 12:39:11
TC2_Ancona	eHealthTD	{"EQUID":"1234","USID":2,"SPO2":[92,98]}	25/11/2021 12:38:46

Fig. 11 RO data module

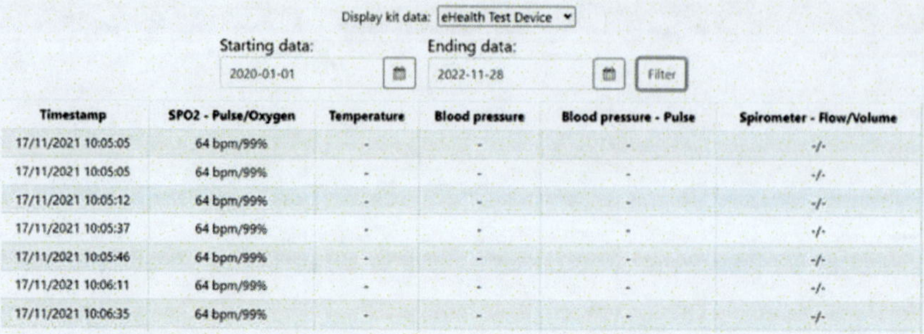

Fig. 12 Kit data component

The Kits module offers the facility to manage the kits recognized by the SAVE system (see Fig. 13).

The Devices module allows the addition, editing and removal of a device and assigning it to a kit (see Fig. 14). These devices have two descriptions—one long and one short—that will be used by the user interfaces to allow them to be easily identified.

The data from unknown devices are ignored by the data collection system.

The Device types module is the tool for the management of the types of devices recognized by the SAVE solution (see Fig. 15).

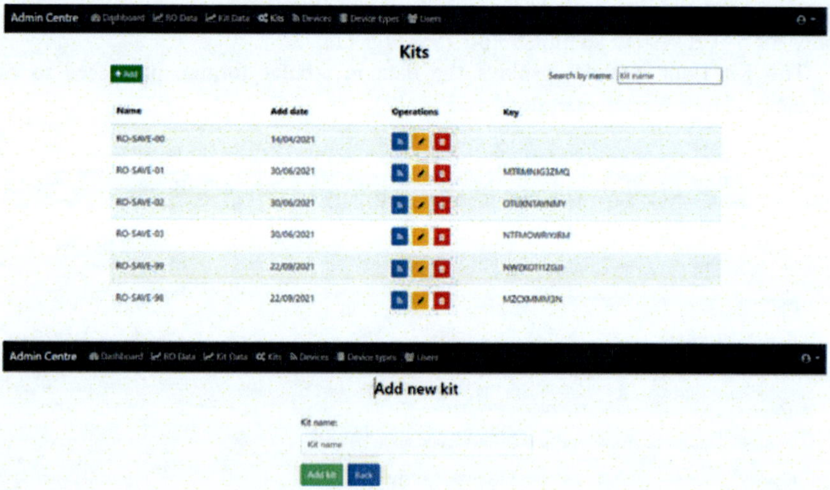

Fig. 13 Kits module

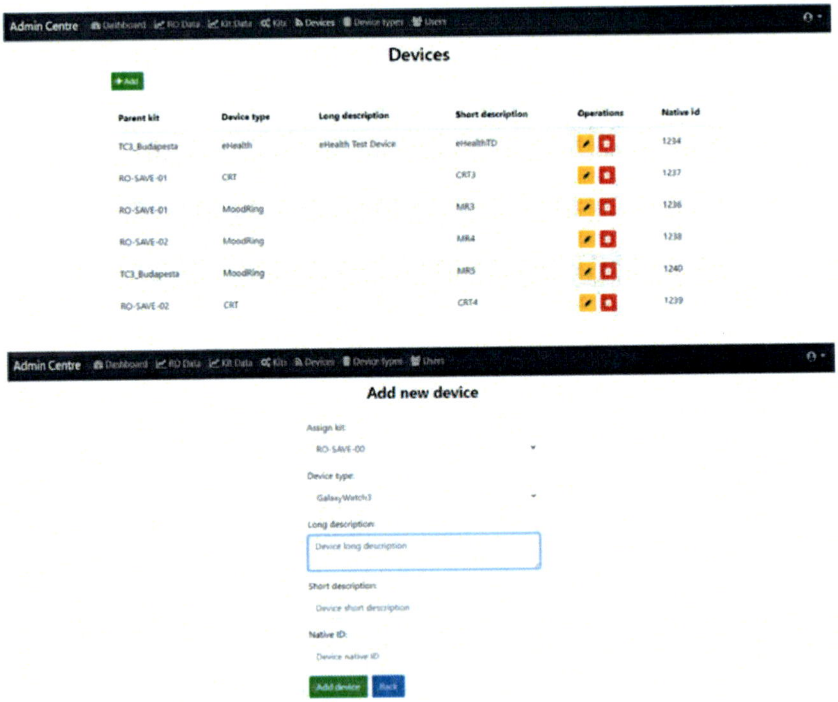

Fig. 14 Device module

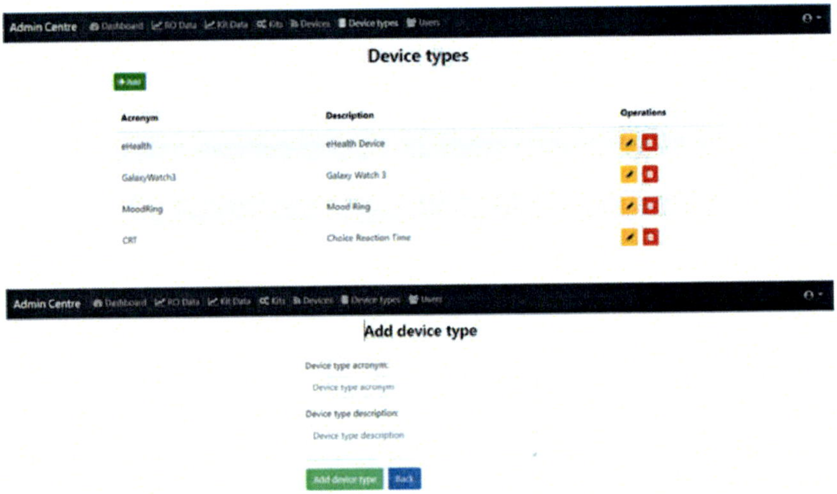

Fig. 15 Device types module

8 Security Aspects

Regarding the security aspects, the SAVE solution uses the current standards to protect the data and the access to its features.

Using HTTPS and the authentication by API key, the communication between the sensors and its adapter or the *Data Collector*, and between the adapter interfaces and the *Data Collector* is protected.

The web applications use authentication by the username-password pair, JWTs for authorization and work on HTTPS.

The passwords are not kept in clear text, but only the SHA-1 hash is stored in the users' table.

The data coming from the sensors are not linked in the database directly to the end-users, but to the kits. Accessing only this data does not reveal anything about the users' identities.

9 Deployment Environment

The SAVE cloud application, being designed on the microservices architecture, and built using the Sprig Boot framework, can be deployed o vast number of infrastructures. The microservices can run independently of a servlet engine/web server (e.g. Tomcat) or can be deployed as a classic Java web application.

In the test environment, the microservices are deployed in a Tomcat server, installed on a virtual machine.

For the production environment, the microservices will be deployed using the Docker containerization engine, Kubernetes being employed for the orchestration.

10 Conclusion

The SAVE system is designed on the model of microservices-based architecture. The frontends of the web applications are developed using the Angular framework and the data is retrieved using REST APIs exposed by the backend, developed using the Spring Boot framework. To ensure data persistence, a relational database management system (Oracle's *MySQL*) and a partitioning scheme are employed. A web application—*Admin Centre*—was created, to efficiently and securely manage data, which consists of several modules: Dashboard, RO Data, Data Kit, Kits, Devices, Device Types, Users. The security of the SAVE system comes from the use of PKI, through HTTPS, whilst using standard protocols to ensure maximum interoperability.

References

1. McKendrick R (2020) Mastering docker, 4th ed. Packt Publishing
2. Stavropoulos TG, Papastergiou A, Mpaltadoros L, Nikolopoulos S, Kompatsiaris I (2020) IoT wearable sensors and devices in elderly care: a literature review. Sensors 20:2826. https://doi.org/10.3390/s20102826
3. Deshpande A, Pal Singh N (2020) Challenges and patterns for modernizing a monolithic application into microservices, IBM, June 2020. https://developer.ibm.com/depmodels/microservices/articles/challenges-and-patterns-for-modernizing-a-monolithic-application-into-microservices/
4. Newman S (2019) Monolith to microservices: evolutionary patterns to transform your monolith. O'Reilly
5. Laszewski T, Arora K, Farr E, Zonooz P (2018) Cloud native architectures: design high-availability and cost-effective applications for the cloud. Packt Publishing
6. Moraru SA, Perniu L, Ungureanu DE, Moşoi AA, Kristaly DM, Sandu F, Manea AC (2018) Home assisted living of elderly people using wireless sensors networks in a cloud system. In: International symposium in sensing and instrumentation in IoT era (ISSI). IEEE, pp 1–8
7. Dey P, Sinha BR, Amin M, Badkoobehi H (2019) Best practices for improving user interface design. Int J Softw Eng Appl 10(5)
8. Microservices Zone. https://dzone.com/articles/what-are-microservices-actually. Last accessed 2021/09/01

Analysis of Indoor Localization Using Beacons for the Visually Impaired: A Systematic Literature Review

Juan Surco-Anacleto and Michael Cabanillas-Carbonell

Abstract Today the most widely used technology is GPS, which does not perform optimally in buildings because it does not have the necessary accuracy. For these cases Bluetooth technology is the most recommended to be used as it provides optimal performance. This paper is focused on comprehensively examining papers published between the years 2016–2020. In the present research, articles were collected from databases such as IEEE Xplore, Scopus, IOPScience, Ebsco and Dialnet systematizing 68 articles between the years 2016–2021.

Keywords Bluetooth · Beacon · Indoor · IOT · Systematic review

1 Introduction

Currently, people with visual impairment (difficulty moving around) around the world have fewer opportunities due to their limited participation, the effects of which result in situations of exclusion. In this context, it is important to investigate the situation of these people, since their development when moving in closed places with obstacles considerably compromises them to accidents, causing them to decide not to carry out their daily activities [1]. Therefore, it is intended to know the use of IOT technologies for indoor location, to facilitate the proper control, monitoring and displacement of visually impaired people improving their activities and preventing accidents when moving in their daily lives [2].

Consequently, the objective of the research is to analyze the best strategies and characteristics to solve the current problems experienced by visually impaired people in their daily lives, determining the location, monitoring and control of their movement [3].

J. Surco-Anacleto (✉)
Universidad Autónoma del Perú, Lima, Peru
e-mail: jsurco@autonoma.edu.pe

M. Cabanillas-Carbonell (✉)
Universidad Privada del Norte, Lima, Peru
e-mail: mcabanillas@ieee.org

X.-S. Yang et al. (eds.), *Proceedings of Seventh International Congress on Information and Communication Technology*, Lecture Notes in Networks and Systems 447,
https://doi.org/10.1007/978-981-19-1607-6_29

2 Methodology

The systematic review of the scientific literature will be used for the elaboration of the article [4]. The article is based on a systematic review of the scientific literature.

2.1 Research Questions

RQ1. What are the best indoor location systems for the visually impaired?

RQ2. What technologies can be used to monitor and control visually impaired people to improve their movement in enclosed spaces?

RQ3. How does a positioning system influence visually impaired people to improve their movement in enclosed spaces?

2.2 Search Strategies

To answer the research questions, a search for published articles was conducted in the main databases EBSCO Host, IEEE Xplore, Dialnet, IOPScience and Scopus. Eighty scientific articles were collected.

The following keywords were considered when applying the search for the research: "Location by beacon", "Application AND interior location", "Applications AND positioning OR beacon", "Diagnostic OR interior systems", "Diagnosis AND big data", "Positioning OR objects by beacons", "Chatbot" and "Importance OR beacons."

2.3 Inclusion and Exclusion Criteria

Of the systematized articles, only articles written in Spanish and English were considered; empirical research articles related to beacons prediction models. The following inclusion and exclusion criteria, presented in Table 1, were applied for the review study.

The process of systematization of articles is shown in Fig. 1, with a total of 80 articles collected, 8 duplicates and 4 that did not meet the inclusion criteria were eliminated, leaving a total of 68 systematized articles.

Table 1 Criteria inclusion and exclusion

	Criterion	
Inclusion	I01	Articles related to systematic reviews of the model in beacon prediction
	I02	Articles related to the development and/or comparison of NCD prediction models (Risk Factors)
	I03	Articles related to the implementation of location models for beacon prediction
	I04	Medical articles related to beacons
Exclusion	E01	Unrelated articles to systematic reviews of the model in beacon prediction
	E02	Articles unrelated to the development of models for beacon prediction
	E03	Articles unrelated to the implementation of models for beacon prediction

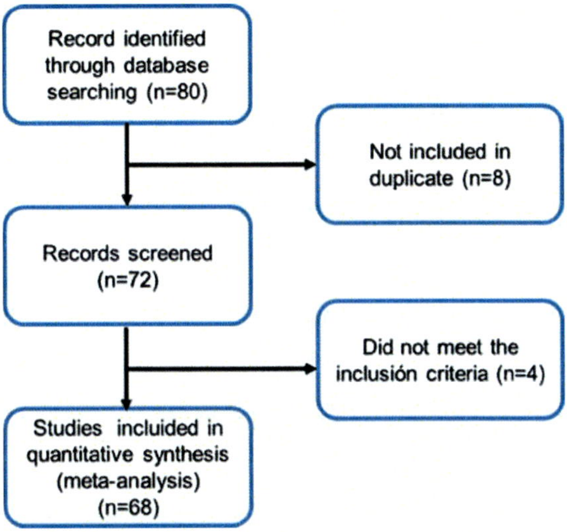

Fig. 1 Inclusion and exclusion stage

3 Results

In Fig. 2, it is observed that a greater number of articles were collected from the IOPSCIENCE database with 27 articles, followed by Scopus with 15 articles complementing IEEE Xplore with 13 articles, EBSCO 12 and Dialnet 1 article.

We selected studies of no more than 6 years, in the period 2016–2021, in Fig. 3, the number of selected articles according to year and continent is shown.

Table 2 shows the scientific articles classified according to their subject matter and selection criteria.

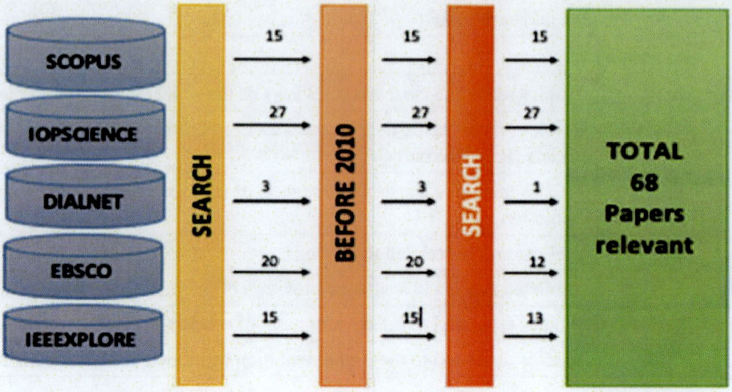

Fig. 2 Collection process of articles from the different databases

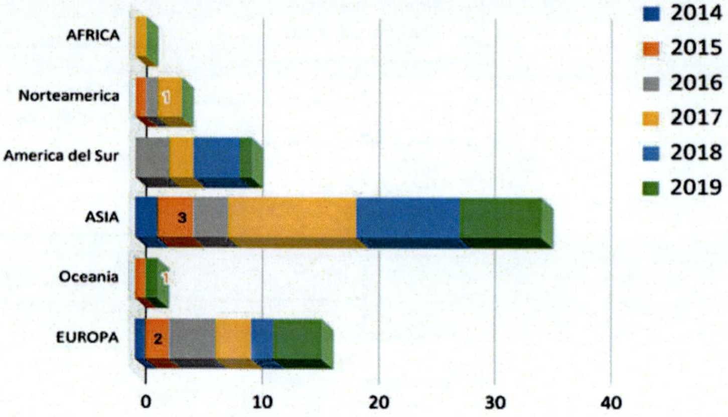

Fig. 3 Total number of articles analyzed per continent and per year

3.1 *Mobile Technologies for Indoor Mobilization of Visually Impaired People*

Regarding mobile application technologies for indoor localization, the benefits of tracking data to speed up the movement of visually impaired people are shown. In addition, the development of the beacon system provides better prediction of objects. In addition, the use of beacon and IoT is achieved through deep learning, prediction from the provided databases.

Table 2 Distribution according to the topics of the articles analyzed

	Criterion	Number of studies
Mobile applications and technologies for indoor mobilization of visually impaired people	Detection system for visually impaired persons	6
	Implementation of chatbot for indoor positioning	12
	Obstacle prediction through beacon positioning	6
	Development of mobile applications	4
	Predictive models for indoor positioning	5
Beacon positioning	Recent challenges and advances in beacon positioning	8
	Impact of beacons on the population	3
	Advances and new findings on beacons	6
	Characteristics of the indoor location	4

3.2 Beacon Positioning

Research on the recent progress and development in indoor positioning, measurements taken in different countries describe studies of systems used to have a better accurate and precise location as opposed to GPS that together with accurate studies fail to determine its great relevance.

The importance of beacons in this research is to reduce incidents by not being able to have an exact location. The studies related to beacons are also related to the transfer of files and their low energy required when using them.

4 Discussion

This systematic investigation of the scientific literature aims to answer the following questions: **What are the best indoor location systems for the visually impaired?**

According to Fig. 3, it can be seen that the articles related to our topic use beacon technology, mobile app, IOT, etc. This result indicates that these categories allow us to have a control and monitoring for the movement of people with visual impairment.

According to Table 2, we can see the categories of articles related to our topic using beacons technology for tracking and positioning. This result indicates that this category is one of the most used to control and track visually impaired people.

The present research systematizes the importance of indoor location technology for the monitoring and control of visually impaired people. Tests showed that the use of this technology increases the effectiveness of their management by maximizing user comfort and acceptance. Also, among several experimental IOT projects that by increasing its sensors responds appropriately with the orientation to be processed by beacons [52].

In addition, it was possible to answer the second research question: **What technologies can be used to monitor and control visually impaired people to improve their movement in enclosed spaces?** According to Fig. 4, we can appreciate the number of articles related to our topic for each continent between the years 2016 and 2021 in mention using technologies for monitoring. This result indicates that these categories of technologies allow us to track and monitor people with visual impairment.

According to Table 3, we can see the categories of articles related to our topic "Mobile applications and indoor technologies for visually impaired people" (The mobile application must be connected to a matrix and an IOT system which allows tracking and control by providing the location of the person). In this result indicates that this category is one of the most used in allowing to control the monitoring of visually impaired people in enclosed spaces as it allows the use of such systems.

Finally, it was possible to answer the third research question: **How does a positioning system influence visually impaired people to improve their movement in enclosed spaces?** According to Table 4, we can see the categories of the articles related to our topic "Studies related to beacon positioning" (indoor location systems must be connected to a matrix and an IOT system which allows tracking and control providing a better displacement for people with visual impairment). In this result indicates that this category is used more beacon technology linked to mobile devices

Table 3 Systematized articles on mobile technologies for indoor mobility of people with visual impairment

Topics	Studies
Detection system for visually impaired persons	[5–7]
Implementation of chatbot for indoor positioning	[8–17]
Obstacle prediction through beacon positioning	[18–23]
Development of mobile applications	[24–29]
Predictive models for indoor positioning	[30–35]
Detection system for visually impaired persons	[5–7]

Table 4 Studies related to beacon positioning

Topics	Studies
Recent advances and challenges in beacon positioning	[36–42]
Impact of beacons on the population	[43–50]
Advances and new findings on beacons	
Characteristics of the indoor location	

as it is one of the most used in allowing to control the tracking of visually impaired people in enclosed spaces as it allows the use of such systems.

5 Conclusions

This paper is a systematic review of the scientific literature on the analysis of indoor localization using beacons for the enhancement of visually impaired people. A total of 68 articles were systematized, between the years 2016–2021. These articles have been distributed in groups by their subject matter.

The topics "Indoor localization systems" and "Beacon-based positioning" have already been implemented in mobile applications and indoor positioning systems, which confirms the importance of a localization system for the displacement of visually impaired people. Likewise, the review allows us to answer the questions posed as they have as a strategy the use of beacons and their implementation to detect the positioning of such people and in turn the importance of IoT to automate applications.

In addition, related projects were discovered according to the results and their importance, being able to demonstrate the positive influence in the detection of people for their respective monitoring and control that they require.

It is recommended for future work on the use of beacons, the development of IoT-based applications and systems with data and features related to indoor positioning.

References

1. Aguilar-Moreno E, Montoliu-Colás R, Torres-Sospedra J (2016) Indoor positioning technologies for academic libraries: towards the smart library. Profesional de La Información 25(2):295–302. https://doi.org/10.3145/epi.2016.mar.17
2. Andrade L, Quintero J, Gamess E, Russoniello A (2018) A proposal for a technological solution to improve user experience in a shopping center based on indoor geolocation services. Int J Adv Comput Sci Appl 9(6):389–401. https://doi.org/10.14569/IJACSA.2018.090653
3. Astafiev A, Zhiznyakov A, Demidov A, Makarov M (2021) Construction automatic positioning system of a moving object based on the readings of Bluetooth beacons and modified multelateration method. J Phys Conf Ser 1828(1):012071. https://doi.org/10.1088/1742-6596/1828/1/012071
4. Baek J, Choi Y, Lee C, Suh J, Lee S (2017) BBUNS: Bluetooth beacon-based underground navigation system to support mine haulage operations. Minerals 7(11):228. https://doi.org/10.3390/min7110228
5. Bai L, Ciravegna F, Bond R, Mulvenna M (2020) A low cost indoor positioning system using Bluetooth low energy. IEEE Access 8:136858–136871. https://doi.org/10.1109/ACCESS.2020.3012342
6. Chien CF, Chen HT, Lin CY (2020) A low-cost on-street parking management system based on Bluetooth beacons. Sensors (Switzerland) 20(16):1–21. https://doi.org/10.3390/s20164559
7. Chiou SY, Liao ZY (2020) Design and implementation of beacon-based positioning. J Inf Sci Eng 36(3):643–658. https://doi.org/10.6688/JISE.202005_36(3).0010

8. Daudov IM, Gavrilova ZL, Kudashkin VA (2021) Liable Bluetooth tracking technology for enhancement of location-based services. IOP Conf Ser Mater Sci Eng 1111(1):012043. https://doi.org/10.1088/1757-899x/1111/1/012043
9. Daudov IM, Orobey MN, Ignatev IV (2021) Bluetooth based technology for industrial personnel local positioning. IOP Conf Ser Mater Sci Eng 1111(1):012029. https://doi.org/10.1088/1757-899x/1111/1/012029
10. Díez-González J, Álvarez R, González-Bárcena D, Sánchez-González L, Castejón-Limas M, Perez H (2019) Genetic algorithm approach to the 3D node localization in TDOA systems. Sensors (Switzerland) 19(18). https://doi.org/10.3390/s19183880
11. Duong NS, Thi, TMD (2021) Smartphone indoor positioning based on enhanced BLE beacon multi-lateration. Telkomnika (Telecommunication Computing Electronics and Control) 19(1):51–62. https://doi.org/10.12928/TELKOMNIKA.V19I1.16275
12. El Ashry AEM, Sheta BI (2019) Wi-Fi based indoor localization using trilateration and fingerprinting methods. IOP Conf Ser Mater Sci Eng 610(1):012072. https://doi.org/10.1088/1757-899X/610/1/012072
13. Fachri M, Khumaidi A (2019) Positioning accuracy of commercial Bluetooth low energy beacon. IOP Conf Ser Mater Sci Eng 662(5). https://doi.org/10.1088/1757-899X/662/5/052018
14. Filippoupolitis A, Oliff W, Takand B, Loukas G (2017) Location-enhanced activity recognition in indoor environments using off the shelf smartwatch technology and BLE beacons. Sensors (Switzerland) 17(6). https://doi.org/10.3390/s17061230
15. Fuentes MPB, Ibarra KLM, Hernandez CM (2020, November 18) Indoor positioning system prototype using low cost technology. In: Proceedings—2020 IEEE Latin-American conference on communications, LATINCOM 2020. https://doi.org/10.1109/LATINCOM50620.2020.9282288
16. Garcete DA, Vazquez Noguera JL, Villalba C (2018) Centralized indoor positioning system using Bluetooth low energy. In: Proceedings—2018 44th Latin American computing conference, CLEI 2018, January 2018, pp 860–869. https://doi.org/10.1109/CLEI.2018.00109
17. Gómez R, Carlos A, Pedraza LF (2018) Locating mobile devices in indoor environments by analysis of WiFi network radiation and magnetic field distortions. Ingeniare 26(2):203–212. https://doi.org/10.4067/S0718-33052018000200203
18. Guaman BP, Cordero J (2020) Indoor positioning system using Beacon technology. In: Iberian conference on information systems and technologies, CISTI, June 2020. https://doi.org/10.23919/CISTI49556.2020.9141009
19. Gulko VL, Mescheryakov AA (2020) Determination of the bearing and roll angle of a moving object using orthogonally elliptically elliptically polarized beacon signals received in a circular polarization basis. J Phys Conf Ser 1499(1):12019. https://doi.org/10.1088/1742-6596/1499/1/012019
20. Horng GJ, Chen KH (2021) The smart fall detection mechanism for healthcare under free-living conditions. Wirel Pers Commun. https://doi.org/10.1007/s11277-020-08040-4
21. Liu J, Xu F, Zhang Q, Jiang L (2017) An underwater acoustic beacons positioning method using single hydrophone. J Phys Conf Ser 887(1). https://doi.org/10.1088/1742-6596/887/1/012033
22. Mahdin H, Omar AH, Yaacob SS, Kasim S, Fudzee MFM (2016) Minimizing heatstroke incidents for young children left inside vehicle. IOP Conf Ser Mater Sci Eng 160(1). https://doi.org/10.1088/1757-899X/160/1/012094
23. McGuire J, Law YW, Chahl J, Doğançay K (2021) Optimal beacon placement for self-localization using three beacon bearings. Symmetry 13(1):1–14. https://doi.org/10.3390/sym13010056
24. Mohanasundar M, Thelly KJ, Raveendran P, Rajalakshmi S, Deborah SA (2020) Student attendance manager using beacons and deep learning. J Phys Conf Ser 1706(1). https://doi.org/10.1088/1742-6596/1706/1/012153
25. Molina B, Olivares E, Palau CE, Esteve M (2018) A multimodal fingerprint-based indoor positioning system for airports. IEEE Access 6:10092–10106. https://doi.org/10.1109/ACCESS.2018.2798918

26. Niño Rondón CV, Castro Casadiego SA, Medina Delgado B, Guevara Ibarra D, Gómez Rojas J (2020) Processing to video images for social distancing verification during the COVID-19 pandemic. Logos Sci Technol J 13(1). https://doi.org/10.22335/rlct.v13i1.1305
27. Ogiso S, Mizutani K, Wakatsuki N, Ebihara T (2019) Robust indoor localization in a reverberant environment using microphone pairs and asynchronous acoustic beacons. IEEE Access 7:123116–123127. https://doi.org/10.1109/ACCESS.2019.2937792
28. Ordonez-Camacho D, Mosquera-Proano L (2019) Increasing accuracy in positioning by RSSI: an analysis with machine learning algorithms. In: Proceedings—2019 international conference on information systems and computer science, INCISCOS 2019, pp 31–35. https://doi.org/10.1109/INCISCOS49368.2019.00014
29. Current Thought. 2015, vol 15(25). Dialnet (n.d.). Retrieved 9 June 2021, from https://dialnet.unirioja.es/ejemplar/450656
30. Perera C, Aghaee S, Faragher R, Harle R, Blackwell AF (2019) Contextual location in the home using Bluetooth beacons. IEEE Syst J 13(3):2720–2723. https://doi.org/10.1109/JSYST.2018.2878837
31. Pérez-Bachiller S, Gualda D, Pérez MC, Villadangos JM, Ureña J, Cervigón R (2018, November 13) Android application for indoor location using sensor fusion: ultrasounds and inertial devices. In: IPIN 2018—9th international conference on indoor positioning and indoor navigation. https://doi.org/10.1109/IPIN.2018.8533785
32. Rios GG, Olaya J (2020) Master's thesis. https://repositoriocrai.ucompensar.edu.co/handle/compensar/101
33. Schmidt E, Inupakutika D, Mundlamuri R, Akopian D (2019) SDR-Fi: deep-learning-based indoor positioning via software-defined radio. IEEE Access 7:145784–145797. https://doi.org/10.1109/ACCESS.2019.2945929
34. Shen Y, Hwang B, Jeong J (2020) Particle filtering-based indoor positioning system for beacon tag tracking. IEEE Access. https://doi.org/10.1109/ACCESS.2020.3045610
35. Shimizu K, Kushida D (2021) Evacuation guidance system using beacon information and Dijkstra's algorithm. In: LifeTech 2021—2021 IEEE 3rd global conference on life sciences and technologies, pp 319–323. https://doi.org/10.1109/LifeTech52111.2021.9391946
36. Sumitra ID, Supatmi S, Hou R (2018) Enhancement of indoor localization algorithms in wireless sensor networks: a survey. IOP Conf Ser Mater Sci Eng 407(1):012068. https://doi.org/10.1088/1757-899X/407/1/012068
37. Sun D, Wei E, Ma Z, Wu C, Xu S (2021) Optimized CNNs to indoor localization through BLE sensors using improved PSO. Sensors 21(6):1–20. https://doi.org/10.3390/s21061995
38. Uttraphan C, Abdul Aziz F, Helmy Abd Wahab M, Zulkarnain Syed Idrus S (2020) Bluetooth based indoor navigation system. IOP Conf Ser Mater Sci Eng 917(1):012055. https://doi.org/10.1088/1757-899X/917/1/012055
39. Wang S, Zhang J, Liu H (2020) Optimization method of node location accuracy based on artificial intelligence technology. IOP Conf Ser Mater Sci Eng 750(1):012091. https://doi.org/10.1088/1757-899X/750/1/012091
40. Wang W, Wang W (2017) A new method of hybrid indoor localization with inertial sensor. IOP Conf Ser Mater Sci Eng 242(1):012112. https://doi.org/10.1088/1757-899X/242/1/012112
41. Wang W, Zang C, Chen Z, Liu S (2021) Mobile node design of indoor positioning system based on Bluetooth and LoRa network. J Phys Conf Ser 1738(1):12092. https://doi.org/10.1088/1742-6596/1738/1/012092
42. Wang Z, Liu M, Zhang Y (2019) Mobile localization in complex indoor environment based on ZigBee wireless network. J Phys Conf Ser 1314(1):012214. https://doi.org/10.1088/1742-6596/1314/1/012214
43. Wye KFP, Zakaria SMMMS, Kamarudin LM, Zakaria A, Ahmad NB (2021) Recent advancements in radio frequency based indoor localization techniques. J Phys Conf Ser 1755(1):12032. https://doi.org/10.1088/1742-6596/1755/1/012032
44. Yang G, Li X, Shi H (2020) An optimized algorithm for broadcast beacon frequency and transmission power in VANETs. IOP Conf Ser Mater Sci Eng 787(1). https://doi.org/10.1088/1757-899X/787/1/012016

45. Yao H, Shu H, Liang X, Yan H, Sun H (2020) Integrity monitoring for Bluetooth low energy beacons RSSI based indoor positioning. IEEE Access 8:215173–215191. https://doi.org/10.1109/ACCESS.2020.3038894
46. You Y, Wu C (2020) Indoor positioning system with cellular network assistance based on received signal strength indication of beacon. IEEE Access 8:6691–6703. https://doi.org/10.1109/ACCESS.2019.2963099
47. Yukhno S, Petrov Y, Yakovlev V, Tanklevskiy L (2019) Evaluation of the probability of correct positioning of the beacon and its motion parameters in passive search and rescue systems. IOP Conf Ser Mater Sci Eng 666(1):012101. https://doi.org/10.1088/1757-899X/666/1/012101
48. Yundra E (2018) Study of adjustment delay scheme on IEEE 802.15.4 networks at beacon enabled mode. IOP Conf Ser Mater Sci Eng 288(1). https://doi.org/10.1088/1757-899X/288/1/012065
49. Zhao Y, Lmg Y, Xia P (2021) Intelligent vehicle navigation system based on visual detection and positioning. J Phys Conf Ser 1861(1):012115. https://doi.org/10.1088/1742-6596/1861/1/012115
50. Zhu H, Xie Y (2019) Research on application of beacon-based location technology in american universities. IOP Conf Ser Earth Environ Sci 252(5). https://doi.org/10.1088/1755-1315/252/5/052002

Information and Communication Technologies for Employability in Times of COVID-19, a Systematic Literature Review

Jesus Palacios-Loayza, Carlos Ayala-Inca, and Michael Cabanillas-Carbonell

Abstract Due to the pandemic that emerged at the end of 2019 (COVID-19), humanity is going through a great crisis and one of the most affected sectors is work due to the large amount of unemployment that it generated, thus resulting in a very low rate of employability. That is why the present study aims to find the determining factors to increase employability. The study conducted is a review of scientific literature, which collects 61 articles from the databases: IEEE Xplore, Scopus, Proquest, Ebsco, and other databases.

Keywords ICTs · Employability · COVID-19 · Pandemic · Systematic review

1 Introduction

At present, humanity is going through a great crisis worldwide due to the pandemic of COVID-19, due to this, several countries opted to take the restriction of a total quarantine, taking strict measures at a general level, so that the most affected are the population because not only fight with this disease, also faces a high rate of unemployment due to the closure of companies, arbitrary layoffs in others [1]. As time went by, many countries opted to reduce their restrictive measures, reactivating their economies, and allowing various sectors to resume their activities, if they complied with the security protocols created because of the pandemic. To adapt to this new reality, mankind placed more emphasis on the use of technology, opting for the implementation of tools offered by ICTs [2]. The review of the scientific literature allows us to analyze the impact of the use of a mobile application on employability to determine to what extent the use of such technology helps to increase the employability of individuals [3].

J. Palacios-Loayza · C. Ayala-Inca
Universidad Autónoma del Perú, Lima, Perú

M. Cabanillas-Carbonell (✉)
Universidad Privada del Norte, Lima, Perú
e-mail: mcabanillas@ieee.org

X.-S. Yang et al. (eds.), *Proceedings of Seventh International Congress on Information and Communication Technology*, Lecture Notes in Networks and Systems 447,
https://doi.org/10.1007/978-981-19-1607-6_30

2 Methodology

The systematic review of the scientific literature will be used for the preparation of the article.

2.1 *Research Questions*

RQ1. What are the ICTs that help increase the employability of people in times of pandemic? **RQ2.** What are the determinants that influence the employability of people in times of pandemic? **RQ3.** Which countries have conducted the most research on employability in times of pandemic?

2.2 *Search Strategies*

To answer the research questions, a search of articles published in the main databases was carried out IEEE Xplore, Scopus, Proquest y Ebsco. Where 78 scientific articles were collected. At the time of applying the search for our research, the following keywords have been considered in our search equation: "TITLE-ABS-KEY (COVID-19) AND TITLE-ABS-KEY (unemployment) AND (LIMIT-TO (PUBYEAR, 2021) OR LIMIT-TO (PUBYEAR, 2020) OR LIMIT-TO (PUBYEAR, 2019) OR LIMIT-TO (PUBYEAR, 2018) OR LIMIT-TO (PUBYEAR, 2017))."

2.3 *Inclusion and Exclusion Criteria*

For the systematic review study, the following inclusion and exclusion criteria were applied in Table 1.

Table 1 Inclusion and exclusion criteria

Criteria		
Inclusion	I01	Related articles on ICT for employability
	I02	Articles related to labor employability in times of pandemic (COVID-19)
	I03	Articles related to labor unemployment in times of pandemic (COVID-19)
Exclusion	E01	Unrelated articles on ICT for employability
	E02	Articles unrelated to labor employability in times of pandemic (COVID-19)
	E03	Unrelated articles on labor unemployment in times of pandemic (COVID-19)

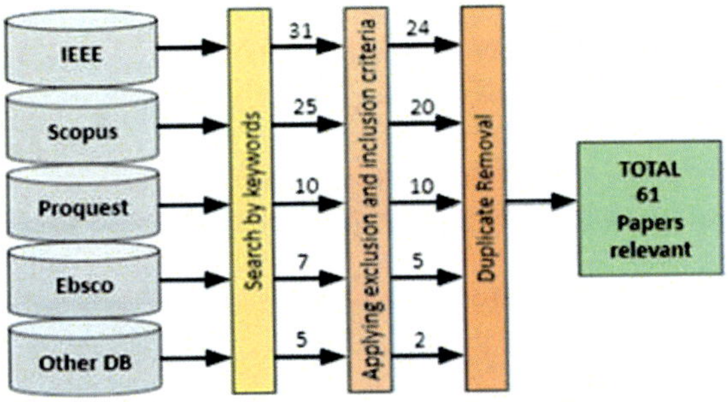

Fig. 1 Document inclusion and exclusion flowchart

Figure 1 shows the total number of research articles according to the databases, going through a filtering process of inclusion and exclusion criteria, resulting in the number of relevant articles.

3 Results

Figure 2 shows a flow chart showing the filtrations performed to reach the selected items.

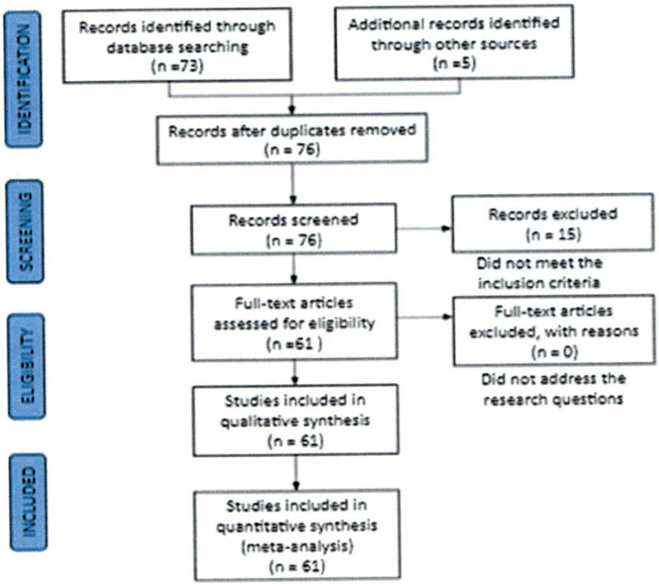

Fig. 2 Prisma diagram methodology

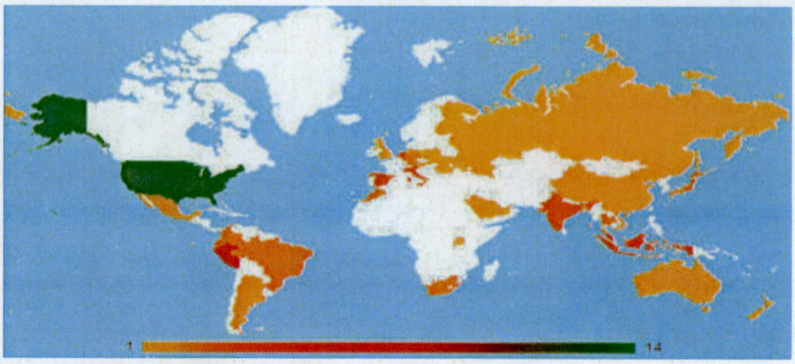

Fig. 3 Map of articles by country

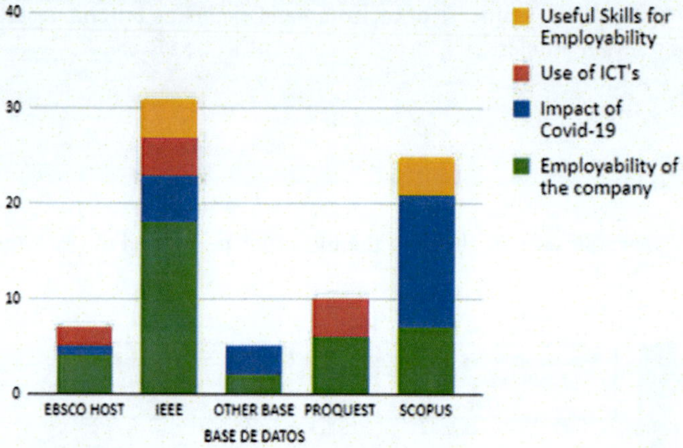

Fig. 4 Articles by database and categories

Figure 3 shows the number of articles published by country, with the United States having the greatest influence.

Figure 4 shows the number of articles published by database and article categories.

In Fig. 5, we can observe the ranking of ICTs that help increase employability during pandemic time, Table 2 show a greater number of investigations in "digital platform" and "data mining".

Figure 6 shows the factors influencing employability in times of pandemic.

Table 3 shows the item classification categories according to the results found.

4 Discussion

This systematic review of the scientific literature aims to answer the proposed questions: **What are the ICTs that help increase the employability of people in times of**

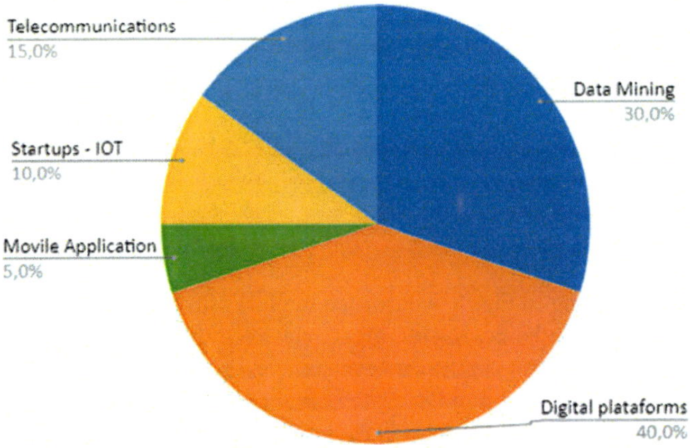

Fig. 5 ICTS in times of pandemics

Table 2 Ranking ICTs that helped employability in times of pandemic

Topics	Studies
Digital platforms	[4–11]
Data mining	[12–16]
Telecommunications	[17–19]
Startups—IOT	[20, 21]
Mobile application	[3, 22]

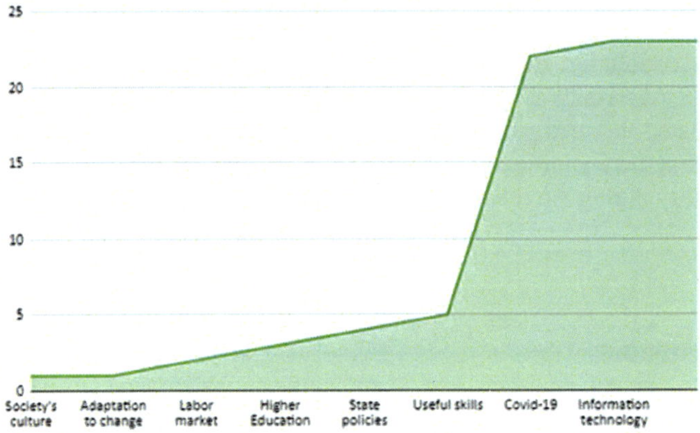

Fig. 6 Influencing factors on employability

Table 3 Classification of articles according to the results obtained

Categories	Articles
Employability of the company	[1, 2, 23–34]
Use of ICT's	[4–22]
Impact of COVID-19	[35–56]
Useful skills for employability	[57–60]

pandemic? According to Fig. 5, we can see the ICTs that help to increase employability in times of pandemic, having great relevance the use of digital platforms, followed using data mining, in the same way we can see a minor use of telecommunications, IOT startups and mobile applications which are influential on a smaller scale. According to Table 2, we can see the categories of articles related to our topic that use information technologies (ICTs). This table gives us greater visibility of which articles support our categories and in turn helped to increase employability in times of pandemic.

In addition, it was possible to answer the second research question: **What are the determinants that influence the employability of people in times of pandemic?** We can observe in Fig. 6, the determinants that influenced employability in times of pandemic, considering COVID-19 and information technologies as the most influential. Additionally, we divided between the factors that positively and negatively influenced employability; based on Fig. 6, the research indicated that information technologies are a positive factor influencing employability, and it also showed that COVID-19 is one of the negative factors influencing employability.

Finally, it was possible to answer the third research question: **Which countries have conducted the most research regarding employability in times of pandemic?** According to the research carried out, we can observe in Fig. 3, the articles related to our topic that come mostly from the United States. This result indicates that there is greater experience in relation to the study that was given to employability during the times of pandemic that has been occurring during the last 3 years.

5 Conclusions

After having carried out a systematic research of the scientific literature of 61 articles related to the topic in question, it is concluded that.

The use of ICT's is playing a fundamental role for employability in times of COVID-19, due to this great importance is being given greater emphasis on the creation and use of these. A clear example is the digital platforms that due to the current restrictions of social distancing increased its use, opening a larger labor market where people and companies benefit.

We can determine that we find positive and negative factors that influence the employability of people, in many cases, it is the way the situation is handled, which will allow us to take advantage of the current circumstances.

We were also able to determine that the United States has a greater concern regarding the impact of employability in times of pandemic due to the large number of scientific articles conducted in this country.

The results of this systematic review may be helpful for future research on the use of ICTs to increase the employability of people in times of pandemic.

References

1. Baranchenko Y, Yizhong X, Lin Z, Lau CK, Ma J (2019) Relationship between employability and turnover intention: the moderating effects of organizational support and career orientation. J Manag Organ 26:1–22. https://doi.org/10.1017/jmo.2019.77
2. Juanpere BA (2020) Medidas fiscales locales para ayudar a la economía y el empleo, también en tiempos de COVID-19. Crónica Tributaria, Instituto de Estudios Fiscales 177(4):11–38. https://ideas.repec.org/a/hpe/crotri/y2020v177i4p11-38.html
3. Alnaghaimshi NI, Alneghaimshi SA (2020) Proposed Arabic mobile application for micro-enterprises: a Saudi Arabian setting. In: 2020 2nd international conference on computer and information sciences (ICCIS), pp 1–6. https://doi.org/10.1109/ICCIS49240.2020.9257719
4. Romero-Saritama JM (2020) Product-based learning adaptation to an online autonomous work strategy in restriction conditions by covid-19. In: 2020 XV Conferencia Latinoamericana de Tecnologias de Aprendizaje (LACLO), pp 1–6. https://doi.org/10.1109/LACLO50806.2020.9381164
5. Sellamy K et al (2018) Web mining techniques and applications: literature review and a proposal approach to improve performance of employment for young graduate in Morocco. In: 2018 international conference on intelligent systems and computer vision (ISCV), pp 1–5. https://doi.org/10.1109/ISACV.2018.8354043
6. Wang T (2020) Intelligent employment rate prediction model based on a neural computing framework and human–computer interaction platform. Neural Comput Appl 32. https://doi.org/10.1007/s00521-019-04019-w
7. Casuat CD, Isira ASM, Festijo ED, Alon AS, Mindoro JN, Susa JAB (2020) A development of fuzzy logic expert-based recommender system for improving students'employability. In: 2020 11th IEEE control and system graduate research colloquium (ICSGRC), pp 59–62. https://doi.org/10.1109/ICSGRC49013.2020.9232543
8. Asfaw AA (2021) The effect of income support programs on job search, workplace mobility and COVID-19: international evidence. Econ Hum Biol 41:100997. https://doi.org/10.1016/j.ehb.2021.100997
9. Yuly AR, Adila RN, Nugrahani F, Waluyo YS, Hammad JA (2020) Working space virtual office prototype in pandemic era. In: 2020 3rd international conference on computer and informatics engineering (IC2IE), pp 388–392. https://doi.org/10.1109/IC2IE50715.2020.9274604
10. Ibadov I, Aksenov A, Iumanova I, Sozykin A (2020) The concept of a dynamic model of competencies for the labor market analysis. In: 2020 Ural symposium on biomedical engineering, radioelectronics and information technology (USBEREIT), pp 511–515. https://doi.org/10.1109/USBEREIT48449.2020.9117691
11. Zhou HA, Gannouni A, Otte T, Odenthal J, Abdelrazeq A, Hees F (2020) Towards a digital process platform for future construction sites. ITU Kaleidoscope: Industry-Driven Digital Transformation (ITU K) 2020:1–7. https://doi.org/10.23919/ITUK50268.2020.9303198

12. Casuat CD, Festijo ED (2019) Predicting students' employability using machine learning approach. In: 2019 IEEE 6th international conference on engineering technologies and applied sciences (ICETAS), pp 1–5. https://doi.org/10.1109/ICETAS48360.2019.9117338
13. Kumar RS, Prakash N, Anbuchelian S (2020) Prediction of job openings in IT sector using long short-term memory model. In: 2020 fourth international conference on I-SMAC (IoT in Social, Mobile, Analytics and Cloud) (I-SMAC), pp 945–953. https://doi.org/10.1109/I-SMAC49090.2020.9243483
14. Mulaudzi R, Ajoodha R (2020) Application of deep learning to forecast the South African unemployment rate: a multivariate approach. In: 2020 IEEE Asia-Pacific conference on computer science and data engineering (CSDE), pp 1–6. https://doi.org/10.1109/CSDE50874.2020.9411581
15. Casuat CD, Castro JC, Evangelista DCP, Merencilla NE, Atal CP (2020) StEPS: a development of students' employability prediction system using logistic regression model based on principal component analysis. In: 2020 IEEE 10th international conference on system engineering and technology (ICSET), pp 17–21. https://doi.org/10.1109/ICSET51301.2020.9265371
16. Montañez CAC, Hurst W (2020) A machine learning approach for detecting unemployment using the smart metering infrastructure. IEEE Access 8:22525–22536. https://doi.org/10.1109/ACCESS.2020.2969468
17. Sathish R, Manikandan R, Priscila SS, Sara BV, Mahaveerakannan R (2020) A report on the impact of information technology and social media on Covid-19. In: 2020 3rd international conference on intelligent sustainable systems (ICISS), pp 224–230. https://doi.org/10.1109/ICISS49785.2020.9316046
18. Siriwardhana Y, De Alwis C, Gür G, Ylianttila M, Liyanage M (2020) The fight against the COVID-19 pandemic with 5G technologies. IEEE Eng Manage Rev 48(3):72–84. https://doi.org/10.1109/EMR.2020.3017451
19. Böhmová L, Pavlíček A (2020) How the modern human resources management can take advantage of information from social media while recruiting. In: 2020 international conference on engineering management of communication and technology (EMCTECH), pp 1–6. https://doi.org/10.1109/EMCTECH49634.2020.9261522
20. Kuckertz A, Brändle L, Gaudig A, Hinderer S, Morales R, Carlos A, Prochotta A, Steinbrink K, Berger E (2020) Startups in times of crisis—a rapid response to the COVID-19 pandemic. J Bus Ventur Insights 13:e00169. https://doi.org/10.1016/j.jbvi.2020.e00169
21. Butgereit L (2020) Using the internet-of-things and clustering algorithms to help allocate temporary work to the unemployed. In: 2020 IST-Africa conference (IST-Africa), pp 1–9
22. Francese R, Gravino C, Risi M, Scanniello G, Tortora G (2017) Mobile app development and management: results from a qualitative investigation. In: 2017 IEEE/ACM 4th international conference on mobile software engineering and systems (MOBILESoft), pp 133–143. https://doi.org/10.1109/MOBILESoft.2017.33
23. Van der Klink J, Bültmann U, Burdorf A, Schaufeli W, Zijlstra F, Abma F, Van der Wilt G (2016) Sustainable employability—definition, conceptualization, and implications: a perspective based on the capability approach. Scand J Work Environ Health 42(1):71–79. Retrieved from http://www.jstor.org/stable/43999197. Accessed 2 June 2021
24. Sato S, Kang T-A, Daigo E, Matsuoka H, Harada M (2021) Graduate employability and higher education's contributions to human resource development in sport business before and after COVID-19. J Hospitality, Leisure, Sport Tourism Educ 28:100306. https://doi.org/10.1016/j.jhlste.2021.100306
25. Hamuľák J, Nevická D (2021) The Slovak v the Danish labor law approach to COVID 19 pandemic. Int Comp Law Rev 20(2):231–238. https://doi.org/10.2478/iclr-2020-0026
26. Maji K, Bera AB (2020) Developing a suitability index algorithm for recruitment. In: 2020 national conference on emerging trends on sustainable technology and engineering applications (NCETSTEA), pp 1–5. https://doi.org/10.1109/NCETSTEA48365.2020.9119944
27. Headd B, Kirchhoff B (2009) The growth, decline and survival of small businesses: an exploratory study of life cycles. J Small Bus Manage:47. https://doi.org/10.1111/j.1540-627X.2009.00282.x

28. Strawn G (2020) IT and 21st century employment in pandemic times. IT Prof 22(3):70–72. https://doi.org/10.1109/MITP.2020.2985823
29. Pryimak V, Melnyk B, Holubnyk O, Kostyshyna T, Brych V (2020) A fuzzy assessment of the development of the national labor market of Ukraine. In: 2020 10th international conference on advanced computer information technologies (ACIT), pp 682–686. https://doi.org/10.1109/ACIT49673.2020.9208915
30. Pauceanu AM, Rabie N, Moustafa A (2020) Employability under the fourth industrial revolution. Econ Sociol 3. https://doi.org/10.14254/2071-789X.2020/13-3/17
31. Gurashi R, Grippo A (2020) How important is culture? Koliko je važna kultura?: Analysis of the most recent data on Italian educational offer and its impact on employment and employabilityAnaliza najsvježijih podataka o talijanskoj obrazovnoj ponudi i njezinom utjecaju na zaposlenost i zapošljivost. Management 25:113–131. https://doi.org/10.30924/mjcmi.25.s.9
32. Araújo MCB, Alencar LH, Mota CMM (2018) Decision criteria for contractor selection in construction industry: a literature review. In: 2018 IEEE international conference on industrial engineering and engineering management (IEEM), pp 637–640. https://doi.org/10.1109/IEEM.2018.8607809
33. Minhas MR, Potdar V, Sianaki OA (2018) A decision support system for selecting sustainable materials in construction projects. In: 2018 32nd international conference on advanced information networking and applications workshops (WAINA), pp 721–726. https://doi.org/10.1109/WAINA.2018.00174
34. Zahmak A, Ghannam O, Nofal O (2020) Comparative study between contractors' and consultants' evaluation of cost overrun factors in building construction projects in UAE. In: 2020 advances in science and engineering technology international conferences (ASET), pp 1–6, https://doi.org/10.1109/ASET48392.2020.9118313
35. Donthu N, Gustafsson A (2020) Effects of COVID-19 on business and research. J Bus Res 117:284–289. https://doi.org/10.1016/j.jbusres.2020.06.008
36. Saito S, Tran H, Qi R, Suzuki K, Takiguchi T, Ishigami K, Noto S, Ohde S, Takahashi O (2021) Psychological impact of the state of emergency over COVID-19 for non-permanent workers: a Nationwide follow-up study in Japan. BMC Public Health 21(1):334. https://doi.org/10.1186/s12889-021-10401-y
37. Twenge JM, Joiner TE (2020) U.S. Census Bureau-assessed prevalence of anxiety and depressive symptoms in 2019 and during the 2020 COVID-19 pandemic. Depression Anxiety 37(10):954–956. https://doi.org/10.1002/da.23077
38. Bonati M, Campi R, Zanetti M, Cartabia M, Scarpellini F, Clavenna A, Segre G (2021) Psychological distress among Italians during the 2019 coronavirus disease (COVID-19) quarantine. BMC Psychiatry 21(1):20. https://doi.org/10.1186/s12888-020-03027-8
39. Hui DS, Azhar EI, Madani TA, Ntoumi F, Kock R, Dar O, Ippolito G, Mchugh TD, Memish ZA, Drosten C, Zumla A, Petersen E (2020) The continuing 2019-nCoV epidemic threat of novel coronaviruses to global health—the latest 2019 novel coronavirus outbreak in Wuhan, China. Int J Infect Dis: IJID: Official Publ Int Soc Infect Dis 91:264–266. https://doi.org/10.1016/j.ijid.2020.01.009
40. Hossain MM, Sultana A, Purohit N (2020) Mental health outcomes of quarantine and isolation for infection prevention: a systematic umbrella review of the global evidence. Epidemiol Health 42:e2020038. https://doi.org/10.4178/epih.e2020038
41. Horesh D, Brown AD (2020) Traumatic stress in the age of COVID-19: a call to close critical gaps and adapt to new realities. Psychol Trauma Theory Res Pract Policy 12(4):331–335. https://doi.org/10.1037/tra0000592
42. Suh J, Horvitz E, White RW, Althoff T (2021) Population-scale study of human needs during the COVID-19 pandemic: analysis and implications. In: Proceedings of the 14th ACM international conference on web search and data mining (WSDM '21), Association for Computing Machinery, New York, NY, USA, pp 4–12. https://doi.org/10.1145/3437963.3441788
43. Cerbara L, Ciancimino G, Crescimbene M, La Longa F, Parsi MR, Tintori A, Palomba R (2020) A nation-wide survey on emotional and psychological impacts of COVID-19 social distancing. Eur Rev Med Pharmacol Sci 24(12):7155–7163. https://doi.org/10.26355/eurrev_202006_21711

44. Grashuis J (2021) Self-employment duration during the COVID-19 pandemic: a competing risk analysis. J Bus Ventur Insights 15:e00241. ISSN 2352-6734. https://doi.org/10.1016/j.jbvi.2021.e00241 (https://www.sciencedirect.com/science/article/pii/S2352673421000196)
45. Brown R, Rocha A (2020) Entrepreneurial uncertainty during the Covid-19 crisis: mapping the temporal dynamics of entrepreneurial finance. J Bus Ventur Insights 14:e00174. ISSN 2352-6734. https://doi.org/10.1016/j.jbvi.2020.e00174
46. Fairlie R (2020) The impact of COVID-19 on small business owners: evidence from the first three months after widespread social-distancing restrictions. J Econ Manage Strat 29:727–740. https://doi.org/10.1111/jems.12400
47. Drake RE, Sederer LI, Becker DR, Bond GR (2021) COVID-19, unemployment, and behavioral health conditions: the need for supported employment. Adm Policy Ment Health 48(3):388–392. https://doi.org/10.1007/s10488-021-01130-w
48. Chodorow-Reich G, Coglianese J (2021) Projecting unemployment durations: a factor-flows simulation approach with application to the COVID-19 recession. J Public Econ 197:104398. ISSN 0047-2727. https://doi.org/10.1016/j.jpubeco.2021.104398
49. Siegfried R, Cecile GP, Haraldur A, Nathalie C, Inggriani L, Lluis TV (2020) Preparing 5.0 engineering students for an unpredictable post-COVID world. In: 2020 IFEES World Engineering Education Forum—Global Engineering Deans Council (WEEF-GEDC), pp 1–5. https://doi.org/10.1109/WEEF-GEDC49885.2020.9293661
50. Mora JJ (2021) Análisis del desempleo y la ocupación después de una política estricta de confinamiento por COVID-19 en Cali. Lecturas De Economía 94:165–193. https://doi.org/10.17533/udea.le.n94a342002
51. Martinez A (2021) Impact of COVID-19 in the production, employment, and digitization of companies in Guanajuato: a first approach. Nova Scientia 13. ISSN 2007-0705. https://doi.org/10.21640/ns.v13ie.2795
52. Barrutia Barreto I, Sánchez Sánchez RM, Silva Marchan HA (2020) Consecuencias económicas y sociales de la inamovilidad humana bajo Covid-19 caso de estudio Perú. Lecturas De Economía 94:285–303. https://doi.org/10.17533/udea.le.n94a344397
53. Fajaryati N, Budiyono, Akhyar M, Wiranto (2021). Instrument development for evaluating students' employability skills. Journal of Physics: Conference Series, vol 1842. In: International conference on science education and technology (ICOSETH), Surakarta, Indonesia (Citation Nuryake Fajaryati et al J Phys: Conf Ser 1842:012035)
54. Hohmeyer K, Lietzmann T (2020) Persistence of welfare receipt and unemployment in Germany: determinants and duration dependence. J Soc Policy 49(2):299–322. https://doi.org/10.1017/S0047279419000242
55. Ahmad M, Khan YA, Jiang C, Kazmi SJH, Abbas SZ (2021) The impact of COVID-19 on unemployment rate: an intelligent based unemployment rate prediction in selected countries of Europe. Int J Fin Econ 1–16. https://doi.org/10.1002/ijfe.2434
56. Arango LE, Flórez LA (2016) Determinants of structural unemployment in Colombia. A search approach. Borradores de Economia 969, Banco de la Republica de Colombia. https://www.proquest.com/docview/2120012959/6F5976C127EF4421PQ/13
57. Pengnate W (2018) Needs of employability skill characteristics based on employers' perception. In: 2018 5th international conference on business and industrial research (ICBIR), pp 598–601. https://doi.org/10.1109/ICBIR.2018.8391268
58. Yoto Y, Marsono, Suyetno A, Tjiptady BC (2020) Teachers internship design to improve students' employability skills in vocational education. In: 2020 4th international conference on vocational education and training (ICOVET), pp 1–4. https://doi.org/10.1109/ICOVET50258.2020.9229902
59. Ali M, Triyono B, Koehler T (2020) Evaluation of Indonesian technical and vocational education in addressing the gap in job skills required by industry. In: 2020 Third international conference on vocational education and electrical engineering (ICVEE), pp 1–6. https://doi.org/10.1109/ICVEE50212.2020.9243222
60. Tavera J, Oré T, Malaga R (2017) La dinámica de la población que no estudia ni trabaja en el Perú: quiénes son, cómo son y cómo han cambiado. Apuntes. Revista De Ciencias Sociales 44(80):5–49. https://doi.org/10.21678/apuntes.80.903

The Acceptance and Challenges of Online Learning over Covid-19 Pandemic

Eang Teng Chan and Mui Joo Tang

Abstract Over the Covid-19 pandemic, physical education has been shifted to online education internationally. Parents, students and teachers faced numerous obstacles from the sudden transition of face-to-face classes to online learning. In this research, three categories of respondents are being studied who are students, teachers and parents, to find out the acceptance and challenges of online learning among them. The research is conducted to determine how online classes have affected students, teachers and parents mentally and physically. Quantitative survey questionnaire has been used to collect all the data from each category. The finding indicates, people still prefer physical classes despites all the conveniences of online learning. The result also indicates the paramount of face-to-face communication in human communication, particularly education. Future research can look more in depth into how the demographic and courses students enrolled affect their acceptance and challenges toward online education, and, what are the improvements to better the interaction via online classes. The study is significant for government and the involved parties to endeavor more challenges in the unknown future.

Keywords Online learning · Computer-mediated communication · Covid-19 pandemic

1 Introduction

The global disruption by the outbreak of Covid-19 has impacted many sectors world widely included higher education sector. As a result of this uncontrollable widespread of virus, the government of Malaysia issued a Movement Control Order (MCO) on 18 March which has affected the educational system nationwide because of the closure of all schools in our country. During the outbreak, the government switched all the face-to-face classes to online learning. Due to the rise of E-learning, the practice of online courses has changed people's way of learning [1]. The Covid-19 pandemic has

E. T. Chan (✉) · M. J. Tang
Tunku Abdul Rahman University College, Kuala Lumpur, Malaysia
e-mail: eangteng@hotmail.com

X.-S. Yang et al. (eds.), *Proceedings of Seventh International Congress on Information and Communication Technology*, Lecture Notes in Networks and Systems 447,
https://doi.org/10.1007/978-981-19-1607-6_31

caused millions of students and teachers to move their connection through online. It is undeniable that online learning comes with great benefits such as lower costs, flexible learning environment. However, the requirement for online learning has become a challenge toward the education systems. During the short period of time, students and teachers face challenges in adapting to the new normal. Students struggle to adapt to online classes because of technical difficulties. Without proper technical support, it is hard to maintain the effectiveness of the online education system [2]. Teaching form has derived digitally and teachers have to stand up and make changes to overcome the challenges. Many schools and colleges were ordered to keep closed to the public during the Covid-19 pandemic. There was no choice but to shift entirely to online education for many educational institutions that were previously reluctant to readjust their traditional pedagogical approach. This research aims to study the acceptance and challenges of online learning based on their experiences over the Covid-19 pandemic in Malaysia. The challenge of this research is to explore the readiness to accept online learning in the future with the growth of technology. It is also to find out the better ways to overcome all the possible challenges faced in online learning where this mode of learning might be inevitable in the near future for all.

2 Literature Review

2.1 Education Systems During Pandemic

According to the reports UNESCO, 87% of the world's understudy populace is influenced by Covid-19 school terminations [3]. UNESCO is dispatching distance learning practices and arriving at understudies who are most in danger [3]. There were more than 1.5 billion understudies in 195 nations that were affected by Covid-19 pandemic school terminations [3]. Covid-19 affects not only students, but also teachers and parents across the world. The education industry suffers in finding a solution to solve the problem of students not being able to attend physical classes. Education industry needs to make sure every student is educated and kept safe at the same time during the Covid-19 pandemic. As a result, distance learning is the only solution to continue the education system [4]. Schools should ensure their lecturers and students are able to adapt to the digital technologies which are used during the E-learning process. Employees and educators should also be familiar with online tools for teaching. Besides, teachers need to ensure students are not cheating during exams also. Before Covid-19, there were inequalities in admission to quality training between understudies in metropolitan and rustic zones and understudies from higher and lower financial status families. The inequalities between understudies could also be expanded by school terminations [4].

Covid-19 affects disadvantaged families because the equipment at home is not available to many students. Closing physical school and introducing distance education lead the student to spend less time studying, stress, and loss of enthusiasm for learning. Some research shows that students retain 25–60% more material on average when learning online, compared to just 8–10% in a classroom. This is mainly due to the ability of students to learn online faster, e-learning takes 40–60% less time to learn than in a traditional classroom setting because students can learn at their own pace, go back and re-read, skip or speed up concepts as they choose [1]. The effectiveness of E-learning is different in each of the age groups. Students should be granted individual consideration with the objective that they will adapt to this learning environment without much of a stretch [2]. For those students who are younger they could not be self-disciplined when they are not under the supervision of the teachers. Therefore, various types of online platforms may be used to communicate with the students.

Guardians from different financial foundations may have distinctive capacity and accessibility to help their kids in their learning cycle at home during the pandemic [5]. Parents are more likely to be able to work from home in more wealthy communities, they are affordable to pay private online tuition [5]. The offspring of single guardians or large families are also liable to experience the adverse effects of the switch, just as understudies with uncommon requirements or incapacities are also liable to experience the adverse effects of the switch, except if assisted advances are quickly set up and adapted to the new learning climate. Parents also need to know the needs of technology such as the computer and the Internet connection. In addition, some of the parents will face the challenges such as they do not understand the process of E-learning. They need to help their children to register accounts and download the E-learning platform if their children are still young.

2.2 *Comparison Between Online and Physical Class*

Online classes have many benefits such as allowing students and teachers to do their work in a flexible and relaxed way. But, online education is actually no large cost savings [6]. In fact, students may have to buy a laptop for online class. An experiment conducted about the impact of more classroom time in both online and traditional face-to-face format. Although the accumulation of test scores throughout the semester of students in face-to-face class with the professors is better than students in online format, they conclude that there is not so much difference in both of these two formats. [6]. It showed that test scores between online and traditional classes are not so much different.

When students have a high degree of self-regulation and put in more cognitive effort, the best perceptions will appear. The students who take an online class are having higher self-regulation than traditional classrooms due to the reason that online students have to put off more effort due to ensure all the works are on track and completed before the deadline [7]. This showed students in online class are more self-regulated compared to students in traditional one.

Relationships within a classroom will lead to a learning community's literacy practices such as social interaction. We like face-to-face communication because it is true, deeper, more authentic, more genuine [8]. In the traditional classroom we can meet each other face-to-face, and discuss works together. Besides, facial expressions allow us to engage in conversation. However online classes are not as immediate as a traditional classroom because it is even more imperative that virtual teachers take the initiative to establish two-way communication with students through synchronous means. Therefore, face-to-face communication is very important for students to have conversation with others.

A person's situation or the environment around may lead to the reason that choosing online or physical class. Therefore, students who chose for online classes may be older and have children. Online classes are useful for them since it is less time consuming and more freedom [9]. Therefore, online class is having more freedom compared to traditional class. Through online, students will do work in a more relaxing way, for instructors, can increase control on students for how they consume class information by personal needs and learning styles [10].

2.3 *Blended Perspectives Toward Online Learning*

According to Bali [11], social presence is needed to increase communication in both traditional and online learning. However, online learning has been assumed as being more cost-effective and flexible than conventional instructional settings, as well as offering resources for more learners to pursue their education. Therefore, most of the people would still prefer practical classes compared to online classes.

Built on the basis of learning philosophy, e-learning can theoretically offer several valuable rewards. For learners, online teaching provides versatility and ease to complete learning units when and when the learner wishes to do so. In addition, online education has been used to minimize costs and offer an effective, structured way to deliver content [12]. The time control of online courses is versatile for both students and teachers. In their spare time, they can do their homework and lessons. In comparison, online courses have saved printing and other transportation expenses. That is the reason why most of the teachers are surprised with the flexibility of the online classes. For students, they are more concerned about their academic performance. But they also cited online learning as a trait that summarized their classroom satisfaction, the teachers teaching and responsiveness and relaxation in the learning environment [13] However, learners face a number of hurdles which include feelings of low trust, lack of sufficient contact with the teacher, lack of adequate support resources and so on. One area of major concern noted in the literature is student experiences of depression and alienation. Bad emotions, such as depression and alienation, are troublesome because they have an effect on cognition, happiness, persistence and completion, negatively affecting student learning environments and outcomes.

2.4 *Theoretical Framework Communication Accommodation Theory*

The theory of Communication Accommodation explains when people adjust or adapt their methods of communication to others [14]. There are two methods of making these improvements in verbal and nonverbal styles: divergence and convergence. Divergence is used to emphasize group identification by encouraging the group distinctions in which they identify. Groups with deep ethnic or racial pride also use divergence. Convergence is used most frequently for social acceptance by powerless people and relies on matching the communication patterns of the person they are talking to. In this scenario, social interaction between teacher and student are deriving into a new form of communication. Teachers and education institutions are using the divergence method, observing and collecting information on which teaching style has more student interaction during online class, improving the effectiveness of online teaching. Such as using a simpler visual and audio during online teaching classes, enabling students to easily illustrate the lessons learnt, allowing students to understand the lesson learnt.

3 Methodology

3.1 *Subjects/Participants*

A total of 95 respondents are included as a sample in this research. 25 of them are parents, 25 are educators and 45 are students. Convenience sampling is employed to gage the random perception of Malaysian regarding the issue of online learning during the Covid-19 pandemic period.

3.2 *Research Design*

Three different sets of survey questionnaires are distributed to the 3 categories of respondents. The research method that is used in this research is the quantitative research method, which is the survey questionnaire. The survey questionnaire will cover **THREE** main areas, which are the demographics, acceptance on online classes and review on online classes experience.

4 Result

4.1 Acceptance of Online Learning

Table 1 shows all the 3 parties involved are at the satisfactory level with the online learning during Covid-19 pandemic. Among all, lecturer indicated the most positive view on this teaching mode due to the flexibility, accessibility and the wide range of available tools. However, parents shared some concern despite the high acceptance of online learning such as the health issues, self-discipline of the children, level of stress during online learning and lacking of interaction between the children and others. Table 1 also sheds the idea that student show comparatively lower level of satisfactory toward online learning due to lacking of physical interactivity with classmates and lecturers which hinder their socialization and discussion in the virtual classrooms. However, all the 3 parties acknowledge the importance of technology and face-to-face communication in online learning despite the fact students generally feel uneasy

Table 1 Satisfactory level on online learning

Parent	
Satisfactory level	Frequency
Very satisfied	6
Satisfied	7
Neutral	6
Dissatisfied	5
Very unsatisfied	1
Total	**25**
Student	
Satisfactory level	
Very satisfied	3
Satisfied	13
Neutral	20
Dissatisfied	8
Very unsatisfied	1
Total	**45**
Lecturer	
Satisfactory level	
Very satisfied	11
Satisfied	10
Neutral	3
Dissatisfied	0
Very unsatisfied	1
Total	**25**

to on the mic or camera in the online classrooms. Students and teachers generally do not have difficulties to possess the means and basic skill in the conduct of online classes due to the support from the institutions and government.

4.2 *Online Learning Experience*

Table 2 shows the preference of delivery mode by parents, students and teachers. Result indicates higher inclination toward physical classes as shown in the Table 2. Parents share the concern that online classes might have burdened their children in terms of the load and stress as they have seen their children spend more time in front of the screen compared to the previous time. This has raised the concern of the parents regarding the health issue such as eyes sore and anxiety due to insufficiency of sleep and increase of stress. At the same time, parents also share the concern if the online learning will deteriorate their children's learning interest which might affect the output of their study. Parents also have the worry that the online learning will affect the performance and learning quality of their children. For students, they generally prefer going back to physical classroom as they have the doubt of the learning quality via the online class. This is due to few factors such as the environment of learning, interactivity and self-discipline. Result shows that they do not

Table 2 The preference of face-to-face leaning or online learning

Parent	
Mode of delivery	Frequency (%)
Online	4 (16%)
Physical	**19 (76%)**
Hybrid	2 (8%)
Total	25 (100%)
Student	
Mode of delivery	
Online	4 (9%)
Physical	**27 (60%)**
Hybrid	14 (31%)
Total	45(100%)
Teacher	
Mode of delivery	
Online	7 (28%)
Physical	**14 (56%)**
Hybrid	4 (16%)
Total	25(100%)

have the confidence of the quality in online learning particularly the practical knowledge which is comparatively harder to be delivered and learned via online classes. The face-to-face interaction with classmates and teachers is also the main concern of the students throughout the learning process. For teachers, they are being more optimistic compared to parents and students where they find online classes as acceptable due to the flexibility in terms of time and the design of the teaching materials. During the online classes, teacher do not face much challenge to have engagements with students. However, the main challenges of teacher in online learning come from the preparation of online teaching materials, time management and the increase of workload and stress. As a result, if all are given the option to choose, they prefer physical classes with the ability of face-to-face communication where mediated online learning somehow has the limitation on it.

5 Conclusion and Discussion

To conclude, each party has a different perception toward online learning. For students, it is luckily as most of them have their own devices that allows them to access to the classes during online learning. The result indicates that the experienced lecturers and the abundance of online materials play an important role in the level of acceptance of online learning among students. Moreover, most of the students feel that online learning failed to provide theoretical and practical knowledge that might jeopardize the quality of the learning via online. For educators, almost all of them are satisfied with the online material and the resources provided by the institutions during online learning. They are embraced by the flexibility of online learning. Yet, the inactive students during online classes make the task more challenging for them too. Based on the result, many teachers find it challenging to engage students via online learning compared to the traditional mode. It is also loaded and stressful for them to prepare the new content for online classes. Hence, most of the teachers also prefer to go back to the system of face-to-face learning when schools are re-opened. For parents, most of them are quite satisfied with their children's online learning structures and confident that teachers can motivate their children in online learning. Besides, some of the parents found out their children have become more disciplined during online learning, and put more effort in their academics during online classes. However, problems such as lack of engagement and interactions with others, and health issues such as sore eyes have been taken into consideration by the parents. Hence, more than half of the interviewed parents prefer to send their children to the physical classes instead of online classes. Communication Accommodation Theory aims to understand and predict why, when and how people change their personal behavior during social interaction, as well as what social implications have been made. Result shows how the 3 parties, parent, student and teacher tries to adapt to the changes of communication mode during the Covid-19 pandemic as that is the only available platform to sustain learning process. The result also indicates clearly that face-to-face communication is still important in both online and physical classes despite the advancement of technology [11]. However, both mediated communication and face-to-face communication somehow complement each other in human communication. To conclude, online learning is acceptable and workable during the Covid-19 pandemic where people are willing to adapt to the changes despites all the challenges. However, this does not deny the fact that face-to-face communication is still paramount as long as it involves human communication. The future research is suggested to look into other variables such as the programs or courses students enrolled and the demographic factor of the samples chosen as they might affect the findings of the research in this area.

Acknowledgements The authors acknowledged the raw materials provided by Franco Lim, Yong Poh Tyng, Yang Char Wen, Kho Phei Kee, and Chuah Tze Ying.

References

1. Li C, Lalani F (2020) The Covid-19 pandemic has changed education forever. This is how. Available at https://www.weforum.org/agenda/2020/04/coronavirus-education-global-covid19-online-digital-learning/. Last accessed 16 Feb 2021
2. Dhawan S (2020) Online learning: a panacea in the time of COVID 19 crisis. J Educ Technol Syst 49(1). https://doi.org/10.1177/0047239520934018
3. UNESCO (2020) Covid 19 educational disruption and response. Available at https://en.unesco.org/node/320920. Last accessed 19 Oct 2021
4. Tadesse S, Muluye W (2020) The impact of COVID-19 pandemic on education system in developing countries: a review. Open J Soc Sci 08(10):159–170. Available at https://www.scirp.org/journal/paperinformation.aspx?paperid=103646. Last accessed 14 Feb 2021
5. Pietro D, Biagi G (2020) The likely impact of COVID-19 on education: reflections based on the existing literature and recent international datasets. JRC Technical report. Available at https://publications.jrc.ec.europa.eu/repository/bitstream/JRC121071/jrc121071.pdf. Last accessed 16 Feb 2021
6. Arias JJ, Swinton J, Anderson K (2018) Online vs. face-to-face: a comparison of student outcomes with random assignment. Available at https://files.eric.ed.gov/fulltext/EJ1193426.pdf. Last accessed 15 Feb 2021
7. Chen C, Jones KT, Morel K (2017) How online learning compares to the traditional classroom. Available at https://www.cpajournal.com/2017/10/09/online-learning-compares-traditional-classroom/. Last accessed 15 Feb 2021
8. Ubell R (2017) Online & blended learning: selections from the field. Available at https://www.routledge.com/rsc/downloads/OLC_FreeBook_Online__Blended_Learning.pdf. Last accessed 16 Feb 2021
9. Stack S (2015) Learning outcomes in an online vs traditional course. Available at https://files.eric.ed.gov/fulltext/EJ1134653.pdf. Last accessed 16 Feb 2021
10. Holmes CM, Reid C (2017) A comparison study of on-campus and online learning outcomes for a research methods course. J Counsellor Prep Supervision 9(2). Available at http://dx.doi.org/10.7729/92.1182. Last accessed 19 Oct 2021
11. Bali ST (2018) Students' perceptions toward online learning and face-to-face learning courses. Available at https://www.researchgate.net/publication/329379022_Students%27_perceptions_toward_online_learning_and_face-to-face_learning_courses. Last accessed 15 Feb 2021

12. Smart KL, Cappel JJ (2015) Students' perceptions of online learning: a comparative study. Available at http://jite.org/documents/Vol5/v5p201-219Smart54.pdf. Last accessed 15 Feb 2021
13. Van Wart M, Ni A, Medina P, Canelon J, Kordrostami M, Zhang J, Liu Y (2020) Integrating students' perspectives about online learning: a hierarchy of factors. Available at https://educationaltechnologyjournal.springeropen.com/articles/10.1186/s41239-020-00229-8. Last accessed 15 Feb 2021
14. Dragojevic M, Grasiorek J, Giles H (2015) Communication accommodation theory. Available at https://doi.org/10.1002/9781118540190.wbeic006. Last accessed 19 Oct 2021

Learn to Ask What You Don't Know

Binay Dahal, Sing Choi, and Kazem Taghva

Abstract Asking questions relates to the cognitive ability of language comprehension and context understanding. For that reason, question generation is a challenging topic in Natural Language Understanding. In this work, we propose a task called "question generation with masked target answer," which emphasizes asking questions from text passages without providing a target answer. Compared to other related question generation tasks, our task demands rigorous language comprehension and closely resembles the question asking ability of humans. We then propose various sequence to sequence-based models leveraging additional information about the text, such as its part of speech and named entity recognition(NER) tags. Results show that the proposed models perform on par with other related question generation tasks, despite lacking the key answer phrase.

Keywords Question generation · Natural language processing · Text modeling

1 Introduction

Question generation has been on the periphery of natural language processing for some time. With the advent of the statistical language modeling techniques, neural question generation has taken off with various neural models based on seq2seq learning [1]. Essentially, the task of question generation is to generate a question based on a given short passage of text. Figure 1 shows few examples of short texts as ***Input text*** and their respective questions as ***Question***. It is clear that although there can be many questions for a given text, the generated question should ask about the target answer of the text.

B. Dahal (✉) · S. Choi · K. Taghva
University of Nevada, Las Vegas, USA
e-mail: binay.dahal@unlv.edu

S. Choi
e-mail: chois1@unlv.nevada.edu

K. Taghva
e-mail: kazem.taghva@unlv.edu

X.-S. Yang et al. (eds.), *Proceedings of Seventh International Congress on Information and Communication Technology*, Lecture Notes in Networks and Systems 447,
https://doi.org/10.1007/978-981-19-1607-6_32

Input text: IBM is headquartered in Armonk, New York.

Masked text: IBM is headquartered in <masked_token>

Question: Where is IBM headquartered?

Input text: Because of the ongoing pandemic, many companies have been forced to operate remotely.

Masked text: Because of the <masked_token>, many companies have been forced to operatoe remotely.

Question: Why are many companies operating remotely?

Fig. 1 Example texts with target answer and their respective questions

Generally regarded as the dual task of question answering, question generation finds its application in various areas. First, it can be used to generate question–answer pairs to serve as a dataset for question answering. Second, the task of question generation can aid in making a chatbot more interactive by learning to ask appropriate questions. These are the applications of classical question generation; however, in our work, we introduce a novel question generation task, which we call "question generation with masked target answer." This task closely resembles the original task of question generation, as shown in Fig. 1; however, the difference is that the target answer in the input text passage is masked while feeding it to the model. The reason behind introducing a novel task in question generation is twofold.

First, asking questions is at the core of human intelligence. It constitutes being able to form syntactically coherent sentences, as well as being able to comprehend the content and context of the text in consideration. Humans are generally able to ask questions even when they cannot hear the key target answer in the sentence. We argue that is the essence of question asking. In Fig. 1, even in the case of ***Masked text***, humans would have enough context to ask a question.

Secondly, "question generation with masked target answer" can be applied to make interactive dialog systems, like Amazon Alexa and Google Assistant more intelligent. We have found that in many cases, while giving instructions to these home assistants, they ask some standard questions without regard to our instructions. Therefore, it makes sense that these systems should ask questions appropriate to our instructions. The contribution of this paper is twofold:

- We propose a novel task, which we call "question generation with masked target answer," the significance of which we have discussed above.
- We propose various models that leverage the information, like parts of speech and named entity recognition, from the text to ask questions based on the input

masked sentences. Through these models, we show that the joint learning approach enhances overall performance, and we also establish a baseline for this novel task.

2 Related Works

The first part of the literature survey discusses works that employ manual rule-based approach for question generation. In the latter part, we briefly present some of the end-to-end neural network-based models.

Pan et al. reported on a broad survey on the recent advances in question generation based on the approaches being taken to tackle the problem [13]. They identify three fundamental aspects of the problem and discuss the variations researchers are adopting in each of these aspects. These aspects are identified to be the learning paradigm, input modality, and cognitive levels.

Most of the works about manual approach involve forming a question template initially and then using such templates to form specific questions. In particular, Heilman et al. created some written rules to perform generic syntactical transformations, which convert declarative sentences into questions [5]. They overgenerate the questions and rank them using some form of classification algorithm. There are some other works that employ somewhat similar approaches [7, 8, 10, 12]. Some research has worked on different modalities than text. Mostafazadeh et al. generated questions from images [11]. Serban et al. used recurrent neural networks to generate factoid questions using a set of triplets (subject, relation, object) [17].

Du et al. performed question generation using end-to-end learning utilizing the seq2seq framework [3]. They employ an attention mechanism to focus on the important aspects of the input while forming the question. Another research study done by Duan et al. approaches the task in two ways. First, they propose a retrieval-based method using a convolutional neural network (CNN); second, they propose a generation-based method using a recurrent neural network (RNN) [4]. Sasazawa et al. tried to improve the results of question generation by utilizing interrogative phrases [16]. They argue that using interrogative phrases that match the target answer is important for generating better questions.

Similarly, Kim et al. proposed a seq2seq framework for question generation using answer separation [6]. They mask the target answers from the sentence but use them separately to generate questions. Few works have been done in visual question generation (VQG). Here, the task is to generate questions based on the input image and target answer. Xu et al. argued that recent trends in VQG treat the problem as a reversed visual question answering, which has higher complexity [21]. Hence, they propose a radial graph convolutional network (radial-GCN) to reduce the complexity. Tuan et al. proposed a model based on bidirectional long short term memory (BiLSTM) to capture greater context looking across multiple sentences [19]. Sun et al. proposed answer-focused and the input context-aware neural question generation model [18]. Chen et al. proposed graph-to-sequence model based on reinforcement learning [2].

3 Approach

3.1 Problem Definition

Given a passage of text $X(x_1, x_2, \ldots, x_n)$, n being the number of tokens, a target answer $A(a_1, a_2, \ldots, a_m)$, m being the number of tokens in the target answer and A being some span of text from the question X, the task of question generation is to generate a question $Q(q_1, q_2, ..., q_k)$, where k can be of arbitrary length; the answer of Q should be given by A based on the passage of text X.

$$\hat{Q} = \arg\max_{Q} P(Q|X, A)$$

However, in our task "question generation with masked target answer," the model tries to generate a question Q, the answer of which should be given by A based on the text $X - A$. Meaning, the input to our model is the passage X, from which we omit the answer A. The motivation behind omitting the answer from the input text is that the model should be able to ask questions even when the answer phrase is not present in the input passage. The model tries to maximize the conditional probability of Q, given $X - A$.

$$\hat{Q} = \arg\max_{Q} P(Q|X - A)$$

3.2 Proposed Models and Joint Objectives

The transformer model for language generation [20] is a seq2seq model developed to mitigate the limitations of recurrent networks, and it has done well in tasks like language generation, question answering, and many others. However, there has not been any notable work on leveraging the model for the task of "question generation." In this research work, we propose models with the underlying transformer architecture. Then, we formulate a couple of joint objectives to learn our task. The underlying model is an encoder–decoder architecture, detailed in [20].

Joint Training with NER tags The input to our model does not contain the target answer phrase, so to ask the right questions, the model should guess the missing answer and generate output accordingly. Our *Joint Objective I* forces the model to learn the name entity recognition tags of the missing target answer, while also making it learn to ask questions. Guessing the NER tags of the target answer enables the model to acquire knowledge about the missing part. Figure 2 shows the model trained with Joint Objective 1.

Given the masked sentence: $X - A$, the model jointly learns to output the questions Q and NER tags sequence of the answer A:A_{ner}

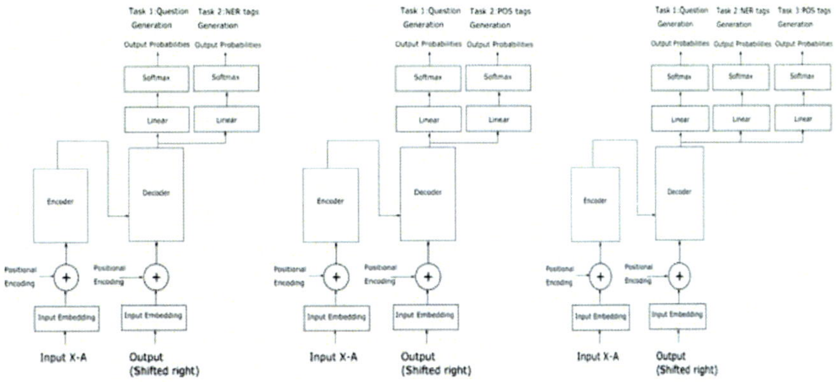

(a) Joint Training I model (b) Joint Training II model (c) Combined Objective(M3) model

Fig. 2 Models with different learning objectives

$$\hat{Q}, \hat{A}_{\text{ner}} = \underset{Q, A_{\text{ner}}}{\arg\max}\, P(Q, A_{\text{ner}} | X - A)$$

Joint Training with POS tags In our task, like most of the other language generation problems, forming grammatically correct sentences is as critical as conveying the correct semantics. Hence, we develop a joint training objective, which we call *Joint Training II* in which the model learns to ask a question given the input text with a masked target answer and also learns to generate the correct Parts of Speech (POS) Tags of the question given the POS tags of the sentence in parallel. Figure 2 is a model trained with this objective.

Given the masked sentence: $X - A$ and POS tag sequence of X: X_{pos}, the model jointly learns to output the questions Q and POS tag sequence of Q:Q_{pos}. Therefore, the Joint Training I objective for our model is given as:

$$\hat{Q}, \hat{Q}_{\text{pos}} = \underset{Q, Q_{\text{pos}}}{\arg\max}\, P(Q, Q_{\text{pos}} | X - A, X_{\text{pos}})$$

The POS sequence generation part of the objective dictates the model to form correct sentences, while learning to ask the right question at the same time.

Combined Objective (M3) In this objective, we combine Joint Training I and Joint Training II, meaning the model learns three tasks at a time. The main task of question generation is aided by NER tags generation and POS tags sequence generation. The combined objective looks like:

$$\hat{Q}, \hat{A}_{\text{ner}}, \hat{Q}_{\text{pos}} = \underset{Q, A_{\text{ner}}, Q_{\text{pos}}}{\arg\max}\, P(Q, A_{\text{ner}}, Q_{\text{pos}} | X - A, X_{\text{pos}})$$

After the model learns each tasks, the evaluation is done only on the main task of question generation.

4 Experimental Setup and Results

In this section, we outline the experimental details to see the performance of the different models and objectives.

4.1 Dataset

SQuAD [15] is a widely used dataset for question answering. Many works have used it to train question generation models. It has over 100k questions from 536 different articles. We use Kim et al. version of the SQuAD dataset [6]. It has been pre-processed and annotated with POS and NER tags using Stanford CoreNLP [9] and has been processed into the train, test, and dev sets. We randomly split the dev set into two splits: ablation study set and eval set as given in Table 1. Then, we mask the target answer phrase from the sentence with a special token "*MASKED_ANSWER*."

4.2 Implementation

We use the pre-trained version of transformer architecture called T5 [14]. It has 12 blocks inside the encoder and decoder. Each block contains the attention layer, feed-forward layer, and normalization layer. For the full details of T5 architecture, refer to the aforementioned paper. T5 has been pre-trained with Colossal Clean Crawled Corpus (C4) data, about 750 GB in size.

We train our models with a single RTX 8000 GPU for a maximum of 10 epochs employing early stopping as a validation strategy. The models are evaluated with an eval set after every 15,000 training steps. Each step has a batch size of 16. The model stops training if it does not improve the eval loss by 0.03 for three subsequent evaluations. The maximum sequence length used is 128. We use Adam optimizer with an initial learning rate of 0.00004 and a greedy search to decode the outputs.

Table 1 Dataset statistics

# Pairs (training)	62,479
# Pairs (test)	9999
# Pairs (eval)	832
# Pairs (ablation)	5790

4.3 Results

We performed an ablation study to see the impact of our joint learning objectives on the model performance. It is shown in Table 2.

Each joint learning objective clearly has a positive impact on the model's performance. This can be seen more distinctly in the test results given in Table 3. Each of the subsequent objectives improves the result in almost every metric. Our combined objective (M3) achieves the best result among all others in the BLEU (1–4) score. In Table 4, we show our model performance on various interrogative question types. The recall metric on almost all question types shows that the model with combined objective performs the best. Figure 3 shows a sample question generated using the different objectives.

Table 2 Ablation study using the ablation set

Models	B-1	B-4	ROUGE
Bare model	24.8	13.55	41.54
Joint I	26.04	14.17	40.60
Joint II	26.68	14.41	41.14
M3	26.46	14.04	39.92

Joint I and joint II is our proposed joint training I and joint training II objective, respectively

Table 3 Evaluation result on the test set

Models	BLEU-1	BLEU-2	BLEU-3	BLEU-4	METEOR	ROUGE
Bare model	34.77	24.10	17.70	13.32	34.82	**40.34**
Joint training I	36.9	25.49	18.58	13.98	**38.4**	39.71
Joint training II	37.9	25.97	18.91	14.15	37.79	39.70
Combined objective (M3)	**39.13**	**27.05**	**19.78**	**14.86**	37.52	39.75

On every metric except ROUGE, our joint learning objectives enhance the performance of the model. **Bare model is the model with single objective of question generation.** For METEOR, we average the individual sentence METEOR score

Table 4 Recall of interrogative words

Question type						
Models	What	Where	When	How	Why	Who
Bare model	0.77	0.51	0.76	0.71	0.17	0.78
Joint I	0.79	0.45	0.68	0.72	0.39	0.75
Joint II	0.79	0.33	0.75	0.74	0.33	0.78
M3	0.8	0.53	0.42	0.8	0.27	0.84

Sentence: in "MASKED_ANSWER" buddhism , the ultimate goal is the attainment of the sublime state of nirvana , achieved by practicing the noble eightfold path also known as the middle way, thus escaping what is seen as a cycle of suffering and rebirth.

Question: what buddhism has a goal of buddhahood or rainbow body?

Bare Model: what type of buddhism aspires to buddhahood or?
Joint I: what buddhism aspires to buddhahood?
Joint II: what religion aspires to buddhahood or rainbow body?
M3: what buddhism aspires to buddhahood or rainbow body?

Fig. 3 Sample question generation using different objectives

5 Conclusion

Through this work, we introduced a task, "question generation with masked target answer," and proposed a series of joint learning objectives to improve our task of question generation. Although our task feeds the model with masked sentences, the evaluation results show it can ask relevant questions, figuring out the missing phrase. These baseline results are in par with other works of question generation despite missing the target answer. However, the results have room for improvement, and one possible extension of this work can look into the recall results and focus on improving particular question types.

References

1. Bahdanau D, Cho K, Bengio Y (2014) Neural machine translation by jointly learning to align and translate. arXiv preprint arXiv:1409.0473
2. Chen Y, Wu L, Zaki MJ (2019) Reinforcement learning based graph-to-sequence model for natural question generation. arXiv preprint arXiv:1908.04942
3. Du X, Shao J, Cardie C (2017) Learning to ask: neural question generation for reading comprehension. arXiv preprint arXiv:1705.00106
4. Duan N, Tang D, Chen P, Zhou M (2017) Question generation for question answering. In: Proceedings of the 2017 conference on empirical methods in natural language processing, pp 866–874
5. Heilman M, Smith NA (2010) Good question! Statistical ranking for question generation. In: Human language technologies: the 2010 annual conference of the North American chapter of the association for computational linguistics, pp 609–617
6. Kim Y, Lee H, Shin J, Jung K (2019) Improving neural question generation using answer separation. In: Proceedings of the AAAI conference on artificial intelligence, vol 33, pp 6602–6609
7. Labutov I, Basu S, Vanderwende L (2015) Deep questions without deep understanding. In: Proceedings of the 53rd annual meeting of the association for computational linguistics and the 7th international joint conference on natural language processing, vol 1: long papers. Asso-

ciation for Computational Linguistics, Beijing, China, July 2015, pp 889–898. https://doi.org/10.3115/v1/P15-1086. https://www.aclweb.org/anthology/P15-1086
8. Lindberg D, Popowich F, Nesbit J, Winne P (2013) Generating natural language questions to support learning on-line. In: Proceedings of the 14th European workshop on natural language generation. Association for Computational Linguistics, Sofia, Bulgaria, Aug 2013, pp 105–114. https://www.aclweb.org/anthology/W13-2114
9. Manning CD, Surdeanu M, Bauer J, Finkel JR, Bethard S, McClosky D (2014) The Stanford coreNLP natural language processing toolkit. In: Proceedings of 52nd annual meeting of the association for computational linguistics: system demonstrations, pp 55–60
10. Mazidi K, Nielsen RD (2014) Linguistic considerations in automatic question generation. In: Proceedings of the 52nd annual meeting of the association for computational linguistics, vol 2: short papers. Association for Computational Linguistics, Baltimore, Maryland, June 2014, pp 321–326. https://doi.org/10.3115/v1/P14-2053. https://www.aclweb.org/anthology/P14-2053
11. Mostafazadeh N, Misra I, Devlin J, Mitchell M, He X, Vanderwende L (2016) Generating natural questions about an image. In: Proceedings of the 54th annual meeting of the association for computational linguistics, vol 1: long papers. Association for Computational Linguistics, Berlin, Germany, Aug 2016, pp 1802–1813. https://doi.org/10.18653/v1/P16-1170. https://www.aclweb.org/anthology/P16-1170
12. Mostow J, Chen W (2009) Generating instruction automatically for the reading strategy of self-questioning. In: AIED, pp 465–472
13. Pan L, Lei W, Chua TS, Kan MY (2019) Recent advances in neural question generation. arXiv preprint arXiv:1905.08949
14. Raffel C, Shazeer N, Roberts A, Lee K, Narang S, Matena M, Zhou Y, Li W, Liu PJ (2019) Exploring the limits of transfer learning with a unified text-to-text transformer. arXiv preprint arXiv:1910.10683
15. Rajpurkar P, Zhang J, Lopyrev K, Liang P (2016) Squad: 100,000+ questions for machine comprehension of text. arXiv preprint arXiv:1606.05250
16. Sasazawa Y, Takase S, Okazaki N (2019) Neural question generation using interrogative phrases. In: Proceedings of the 12th international conference on natural language generation, pp 106–111
17. Serban IV, García-Durán A, Gulcehre C, Ahn S, Chandar S, Courville A, Bengio Y (2016) Generating factoid questions with recurrent neural networks: the 30M factoid question-answer corpus. In: Proceedings of the 54th annual meeting of the association for computational linguistics, vol 1: long papers. Association for Computational Linguistics, Berlin, Germany, Aug 2016, pp 588–598. https://doi.org/10.18653/v1/P16-1056. https://www.aclweb.org/anthology/P16-1056
18. Sun X, Liu J, Lyu Y, He W, Ma Y, Wang S (2018) Answer-focused and position-aware neural question generation. In: Proceedings of the 2018 conference on empirical methods in natural language processing, pp 3930–3939
19. Tuan LA, Shah D, Barzilay R (2020) Capturing greater context for question generation. In: Proceedings of the AAAI conference on artificial intelligence, vol 34, pp 9065–9072
20. Vaswani A, Shazeer N, Parmar N, Uszkoreit J, Jones L, Gomez AN, Kaiser L, Polosukhin I (2017) Attention is all you need. arXiv preprint arXiv:1706.03762
21. Xu X, Wang T, Yang Y, Hanjalic A, Shen HT (2020) Radial graph convolutional network for visual question generation. IEEE Trans Neural Networks Learn Syst

Location-Based Service Discovery for Mobile-Edge Computing Using DNS

Kurt Horvath, Helmut Wöllik, Uran Christoph, and Valentin Egger

Abstract Service discovery combined with security usually plays a role in defining if an application shall be provided in EDGE/FOG or cloud. Most existing solutions focus on the ability of the infrastructure itself to distribute the clients, but it usually raises the question to identify a trustworthy server. We also want to explore the capabilities of mobile devices for service discovery, especially in terms of location awareness, which will aid us to identify the best suitable FOG/EDGE server. We shift the paradigm of searching an edge-instance to addressing an edge based on the location of the mobile device. To do so, we use DNS, and by using sub-domains, we address a location and an app and will use the identified instance for further user processing.

Keywords Fog/Edge computing · Service discovery · Contextual location awareness · DNS · Reverse geocoding

1 Introduction

Cloud computing provides the ability to distribute load on a global scale, but the focus is expanding in dimensions of fog/edge computing (see Kimovski [1], Cao et al. [2] and Hajibaba et al. [3]) The granularity of the segments used splits a country into multiple nodes which are usually based on large scale data centers. Due to the usage of 5G and the urge for low latency, the need arises to increase flexibility in finding the right node to consume a service. In this paper, we want to introduce a method to use regions to partition an area by its administrative districts and beyond. Aim of the proposed solution was also to use existing technology which is already present at most network providers. In the long- term, we want to prove that it is feasible to introduce smaller cells with distinct datacenters to reduce latency and design goals of 5G.

K. Horvath (✉) · H. Wöllik · U. Christoph · V. Egger
Carinthia University of Applied Sciences, Carinthia, Austria
e-mail: horvath@fh-kaernten.at
URL: https://forschung.fh-kaernten.at/roadmap-5g

X.-S. Yang et al. (eds.), *Proceedings of Seventh International Congress on Information and Communication Technology*, Lecture Notes in Networks and Systems 447,
https://doi.org/10.1007/978-981-19-1607-6_33

2 Concept

As described by Sunyaev [4] and Hu [5], we saw the need of a service discovery (SD) approach to enable fog/edge computing capabilities. The key aspect described in this proposal is that a device is usually in some way location-aware, which should be the foundation to identify the correct server to consume a service. In the best case, we face a mobile device, providing a GPS sensor. In the worst case, we might face a desktop machine behind a decentralized firewall with no capabilities to identify its own location. In this case, our concept and also the concepts used by CDNs might show similar results because we have to rely on public IPs to identify a location.

2.1 Naming

Due to the fact that a fully qualified domain can contain an arbitrary amount of sub-domains, also called tags in the chosen nomenclature, we can extend any existing sub-domain by our service. To do so, we append the Service-Name to the existing domain and append geographical qualifiers left of the Service-Name (see Table 1).

In this case, four hierarchical levels are defined to distinguish a higher level granularity from a finer one (see Fig. 1), whereby *B08.* in *B08.Lakeside.9020.kaernten.myApp* defines the finest granularity. The amount of granularity levels is determined by the implementation and the available geographical data.

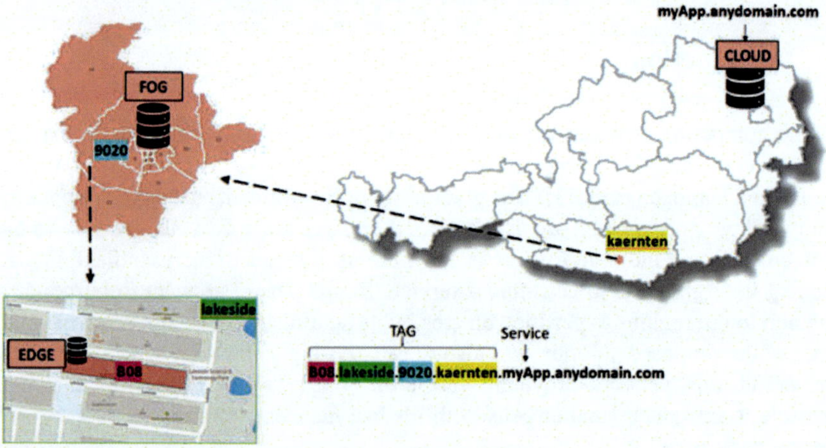

Fig. 1 Concept overview

2.2 Forwarding

The main drawback of our solution is that the client needs to execute JavaScript code to identify the nearest node. This causes additional data transfer but is also indispensable to our aim to transfer the decision making about the node to be used to the client. Hence, it has to be noted that this process has to be repeated, especially when mobile devices shall be supported. Those might alter their position rather frequently and are therefore likely to identify a new better service provider. There are concepts to overcome this obstacle, see Sect. 6

2.3 Workflow

Figure 2 illustrates the workflow of our solution in a sequence diagram.

1. retrieve the JS code.
2. retrieve position using position API of the browser. Remark: depending on the host the position might be retrieved from a GPS module or will be mapped using public IP. Some browsers, such as Chrome, might support location transfer from a mobile device to local device under the assumption that the user might be with the phone.
3. reverse location lookup using AWS Geo-API.[1]
4. aggregation of base URL with possible stage #1 to stage #4 URLs. Stages correspond to a hierarchical level as described in Table 1. The amount of stages depends on the infrastructure available. In general, fewer stages in higher proximity to the client will increase the forwarding performance.
5. resolve aggregated URLs, non-existing URLs can be discarded, resolution itself can be done in parallel.
6. due to the strict hierarchy of the quality of the results, it can defined that stage #4 domain will be preferred over stages #3 to #1 because they will be closer to the client.
7. client will be forwarded to the optimal host using the best URL.

Currently these operations are executed on every refresh of the main page. To enable mobile devices to switch to another fog/edge, the browser should allow the app to trigger this event based on location update (see Sect. 6). This could be enabled by browser plugins/extensions or a independent browser solution.

[1] https://developers.google.com/maps/documentation/geocoding/overview#ReverseGeocoding.

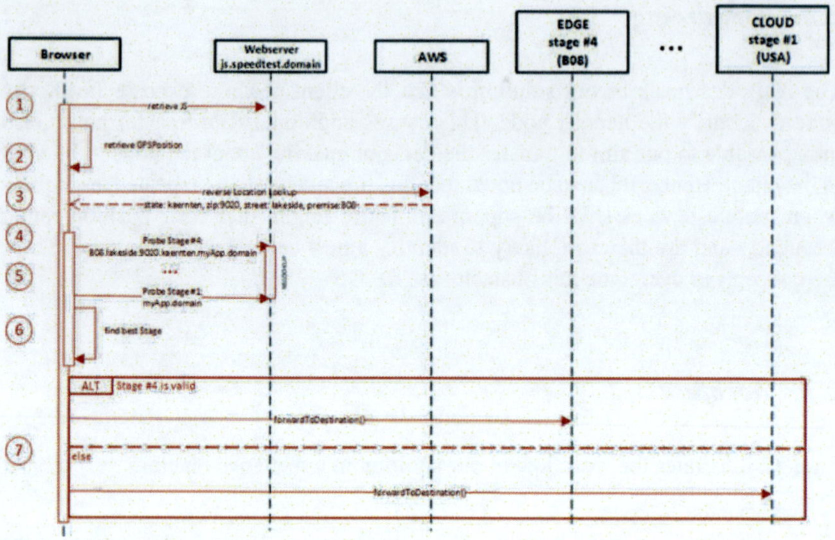

Fig. 2 Location-based processing—sequence diagram

Table 1 Hierarchical levels of domain representations and their corresponding datacenters

Level	Domain	Datacenter
Stage #4	B08.Lakeside.9020.Kaernten.myApp.service	EDGE
Stage #3	Lakeside.9020.Kaernten.myApp.service	EDGE
Stage #2	9020.Kaernten.myApp.service	FOG
Stage #1	Kaernten.myApp.service	CLOUD

3 Reference Implementation and Topology

Our reference implementation is stored in an html file containing the JS code and some CSS. The layout is set for mobile devices so it can be easily included in tests with and without 5G technology. The code itself is publicly available under: https://github.com/kurthorvath/LocationBasedServiceDiscovery.

The reference setup was chosen in the context of the so-called 5G Playground Carinthia to explore capabilities of fog/edge computing utilizing 5G as access network. Table 1 shows four hierarchical levels distinguished by different sub-domains and their corresponding datacenters. To support various 5G operations, the RAN is split into an in-house antenna setup and and an outdoor setup south of the laboratory containing corresponding e/gNB (Fig. 3).

The RAN is directly connected to the core network which is linked to an edge datacenter on the premises. Using an APN gateway, network users can access public Internet. Our investigations use one instance of *openspeedtest* in the edge and a

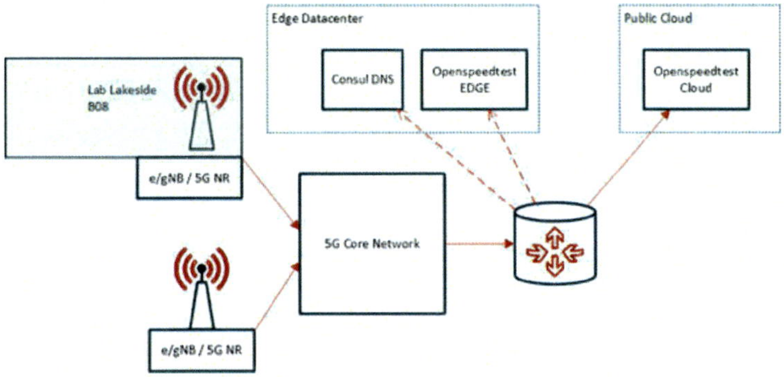

Fig. 3 Infrastructure setup—overview

second one in a public cloud. In terms of simplicity, the option to also add a FOG instance was discarded. In general, ETSI introduces possible 5G network topologies, but also considers concepts by Gupta [6] and Agiwal [7].

Figure 4 shows that our reference implementation where our current location is visible by a marker in the map on the left image. The current position is within building B08, Lakeside premise in postcode 9020 and in the province of Kaernten. Thus, it leads to a valid stage #4 (fine granularity) URL [**B08.Lakeside.9020.Kaernten.** myApp.domain] so stages #3 to #1 are not validated anymore.

Figure 5a shows a reference point outside of the building B08 and outside the Lakeside premise. As visible on the right image in Fig. 5b the validity check executed for *9020.Kaernten.speedtest.domain* on stage #2 is the first one which exists, so the user will be forwarded to this instance.

To validate our forwarding technique, a speedtest is executed[2] to see the performance differences in terms of throughput and latency. Our modified speedtest provides logging output for throughput/latency/jitter. Hence, prolongation of test execution using URL-parameter is provided. The modified speedtest can be pulled from DockerHub (https://hub.docker.com/repository/docker/kurthorvath/5gopenspeedtest).

3.1 Security

As mentioned by Sunyaev [4] and Caprolu [8], one of the key problems in fog/edge computing is security. Especially on edge, it is difficult to identify which server is trustworthy. Under the assumption that the DNS server hosting the main domain is trustworthy, it might be valid to assume that explorable subdomains are trustworthy

2 https://openspeedtest.com/.

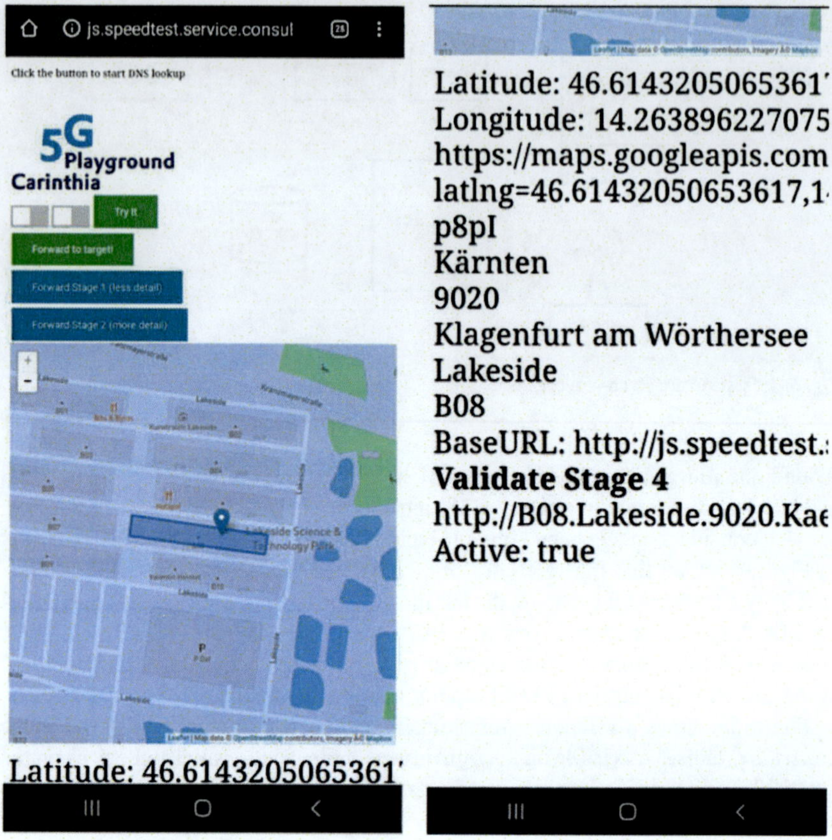

(a) screenshot (b) screenshot (cont.)

Fig. 4 Showcase stage #4—finest granularity

as well, or in other words, if a harmful version of an application is not acknowledged to the main DNS server of a domain, it also will not be discoverable.

4 Performance

Performance measurements have been executed according to the setup described in Sect. 2.3. The setup includes a local 5G network called 5G Playground Carinthia.[3] The edge-instance of our test application is part of the edge nodes near the RAN[4]

[3] https://5gplayground.at/.

[4] Radio access network.

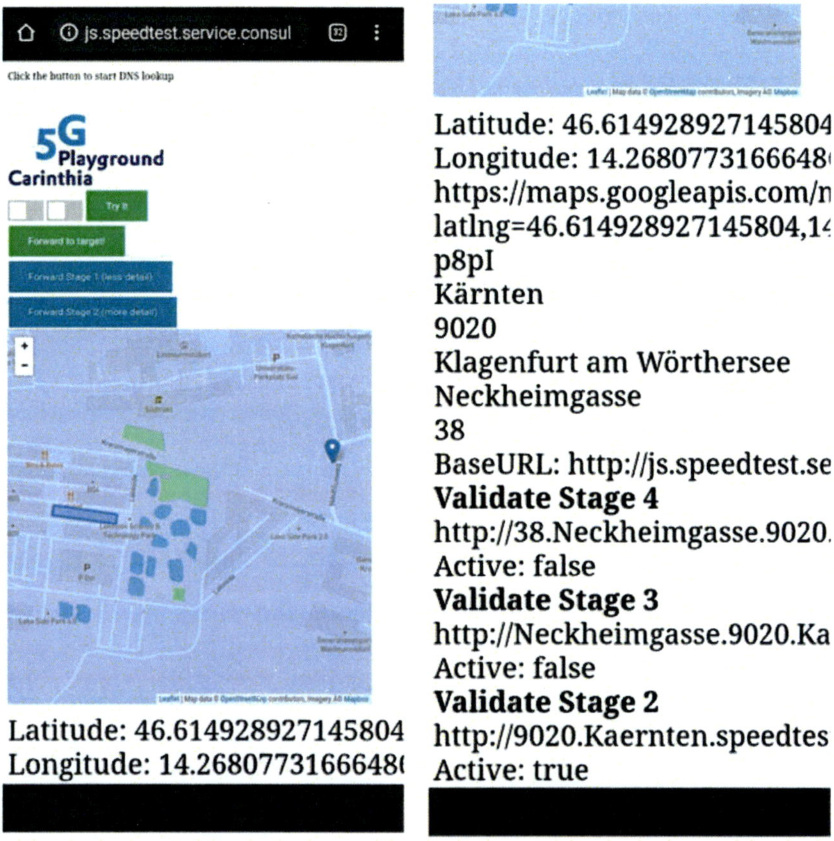

(a) screenshot (b) screenshot (cont.)

Fig. 5 Showcase stage #2 delivering solution for location

itself. The so-called cloud instance is localized in Salt Lake City, Utah, North America.

In Fig. 6a, b, the throughput in general is considerably higher at the edge compared to the provided cloud. Throughput might not be not our key concern but latency does show an even more dramatic difference.

Figure 7 shows that the latency is by average less than 1/4 compared to the cloud instance. This was to be expected, and we are actually looking forward to reducing latency of the edge-instance to less than 20 ms.

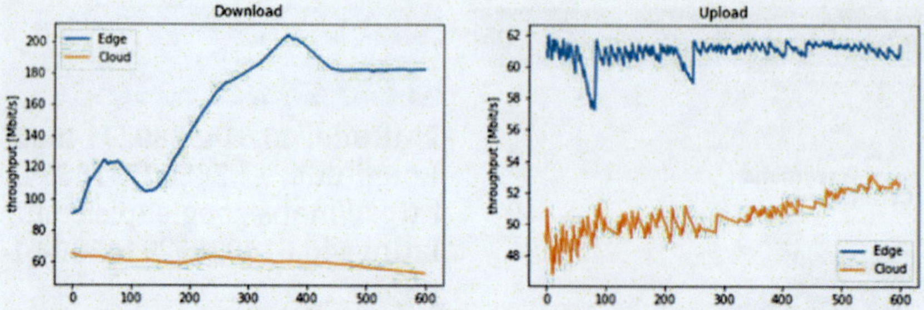

(a) throughput for download [Mbit/s] (b) throughput for upload [Mbit/s]

Fig. 6 Throughput depending on location

Fig. 7 Latency depending on location

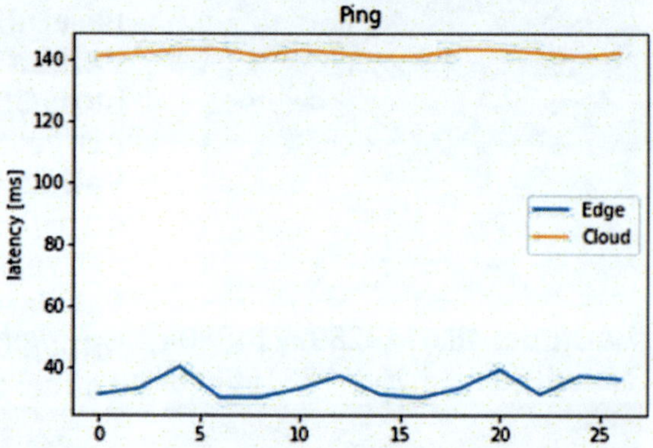

5 Conclusion

Fog/edge computing requires an unobtrusive method to explore its services. In the paradigm of ubiquitous computing and 5G, we see the challenge of constantly moving clients. Distributing fog/edge nodes effected by given or designed topology (Geofence) does work, but suffers from weaknesses by heterogeneity of the available servers. Hence, we see that this approach mainly benefits latency as a measurement. The hierarchical approach proposed here benefits from the predictable execution time because this depends on the amount of hierarchical levels (stages) and the latency to the DNS server. If the geographical partitioning does not fit the load on fog/edge instances, it can be recommended to add more edge instances on the finest granularity level if load does require. In general, we see the main benefit of our solution might be the granularity of services that can be discovered. Further on, we achieve an improvement in overall security due to the fact that the attack surface for an attacker is just a single client. On the other hand, this also shows us the main weak spot. The delivery network per se has no influence on which servers are used by a client. After reaching the finest granularity level, it is difficult for a service provider to introduce new stages on the fly.

6 Future Work

6.1 Geofence-Based Service Discovery

We see that the we can improve granularity of edge instances using geographical context. We propose to follow this step further and want to introduce geofences to allow individual partitioning [9]. One of the biggest drawbacks of the proposed solution might be that we can just partition according to existing topology (state, district, state, street, etc.). Geofences would allow to customize those zones; for example, in the case of 5G, zones can be created according to 5G zones/cells and their datacenters. This would also mean that aggregated domains need not be validated because the definition of a geozone per se defines the validity of those.

6.2 Automatic Forwarding with Browser-Plugin

To overcome the problem to always reload the application code for actual service discovery, there should be an extension/plugin to the most common browser platforms. This plugin can explore edge capabilities for every domain or a subset.

6.3 Mobile-Edge Computing

ETSI published a whitepaper defining the term MEC in the context of radio access networks (RAN). This is very similar to the approach described by Yu [10] and Hu [11] et al. Also, here we see that SD shall be based on DNS. One aspect which needs to be explored is how different operators might share or explicitly not share services in their RAN connected fog/edge servers.

References

1. Kimovski D, Mathá R, Hammer J, Mehran N, Hellwagner H, Prodan R (2021) Cloud, fog, or edge: where to compute? IEEE Internet Comput 25(4):30–36
2. Cao J, Zhang Q, Shi W (2018) Challenges and opportunities in edge computing. In: Edge computing: a primer. Springer, Berlin, pp 59–70

3. Hajibaba M, Gorgin S (2014) A review on modern distributed computing paradigms: cloud computing, jungle computing and fog computing. J Comput Inf Technol 22(2):69–84
4. Sunyaev A (2020) Cloud computing. In: Internet computing. Springer, Berlin
5. Pengfei H, Dhelim S, Ning H, Qiu T (2017) Survey on fog computing: architecture, key technologies, applications and open issues. J Network Comput Appl 98:27–42
6. Gupta A, Jha RK (2015) A survey of 5g network: architecture and emerging technologies. IEEE Access 3:1206–1232
7. Agiwal M, Roy A, Saxena N (2016) Next generation 5g wireless networks: a comprehensive survey. IEEE Commun Surv Tutorials 18(3):1617–1655
8. Caprolu M, Di Pietro R, Lombardi F, Raponi S (2019) Edge computing perspectives: architectures, technologies, and open security issues. In: IEEE International conference on edge computing (EDGE), pp 116–123
9. Bareth U, Kupper A, Ruppel P (2010) geoXmart—a marketplace for geofence-based mobile services. In: 2010 IEEE 34th Annual computer software and applications conference, pp 101–106
10. Yifan Y (2016) Mobile edge computing towards 5g: vision, recent progress, and open challenges. China Commun 13(suppl 2):89–99
11. Sabella D, Sprecher N, Hu YC, Patel M, Young V (2015) Mobile edge computing a key technology towards 5g

Promoting Viable Supply Chain Management (SCM) in the Nigeria Agro-Allied Industry Using Internet of Things

Makinde Oluwafemi Ajayi and Opeyeolu Timothy Laseinde

Abstract Agro-allied supply chain management (ASCM) presents unique issues ranging from dependence on climate, the engagement of many actors, to the bulk of the personnel's lack of literacy, all of which need the use of communication and information technologies (IT). The purpose of the research is to present technologies centered on the Internet of things (IoT) and describe their applicability within the agro-allied industrial supply chain of a developing nation like Nigeria. The study recognized IoT-developed technologies in the framework of ASCM based on literature. In line with the study findings, the application of IoT in the food and agro-allied sector in Nigeria may help boost the growth of the agro-allied supply chain through significant reduction of waste as well as serving users' needs in a long-term manner. In a developing country like Nigeria, IoT-based technology can integrate multiple ASCM tasks in an industrial setting.

Keywords Supply chain management · Nigeria · Agro-allied industry · Internet of things · ASCM · IoT

1 Introduction

A vital driver of economic progress in developing countries is agriculture, and its value in ensuring national food security, reducing import dependency, and creating jobs [1] cannot be over-emphasized. The supply chain must be recognized, mapped, prioritized, and digitized if Nigeria will realize its full agricultural potential [2]. Agriculturalists are expected to generate as much as possible per acre, minimize the risk of crop damage or wastage, reduce operating expenses, and market commodities at their most rewarding prices for Africa to feed its burgeoning population [3]. Furthermore,

M. O. Ajayi
Faculty of Engineering and the Built Environment, University of Johannesburg, Johannesburg, South Africa

O. T. Laseinde (✉)
Mechanical and Industrial Engineering Tech Department, University of Johannesburg, Johannesburg, South Africa
e-mail: otlaseinde@uj.ac.za

X.-S. Yang et al. (eds.), *Proceedings of Seventh International Congress on Information and Communication Technology*, Lecture Notes in Networks and Systems 447,
https://doi.org/10.1007/978-981-19-1607-6_34

it entails digitally monitoring input resources such as manure, seed quality and water as well as limiting the influence of fluctuating factors like the weather and insects. In this, information communication technology (ICT) has played a significant role [4].

ICT can be described as technologies used to connect automated machines and equipment, such as laptops and devices, grids, and allied telecommunication systems [5]. Nigerian agriculture might benefit significantly from the complete economic advantage of using ICT coupled with innovation that cut across the entire agro-allied supply chain. From farming through final retailing and food preparation, agri-food whole supply chain requires contributing to GDP development and being a vital means for job creation in Nigeria [6]. According to the World Bank, agro-allied products impact on real GDP has consistently increased in the recent past, growing from 24.4% in the year 2016 to a projected 25.1% in the year 2017 and a projection of 25.4% in the year 2018. Even though the industry's growth rate has oscillated in latest years, it has continued to be stable, ranging from 4.3% in the year 2014 to 3.7% in the year 2015, it moved to 4.1% in the year 2016, and a projected 3.4% in the year 2017. According to the World Bank, the industry will increase by 3.5% in the following year [7].

Furthermore, in line with the World Bank data, agricultural employment fell from a twenty five year peak of 60.7% in the year 2002 to an unprecedented low rate of 30.6% by 2010, prior it rising to 36.3% in 2016, and 36.6% in the year 2017. Meanwhile, the Nigeria Bureau of Statistics (NBS) states that farm products rose by 10.6% in nominal terms year in year out in the second quarter of 2018. In real terms, the sector grew by 1.2% in the second quartile of year 2018, while its impact on actual GDP stood at 22.9%, according to NBS report [8].

Nigeria recently needed to diversify its economy away from crude oil toward agriculture due to poor prices of oil. Hopefully, agriculture will lift the African continent out of poverty if the technology for mechanized farming is appropriately utilized [9]. However, it is astonishing to learn that most Nigerians and Africans still do agriculture using conventional methods. To lift Nigeria and the continent out of poverty, the government and allied agencies in Nigeria and other African countries must spend extensively on smart agriculture [9].

The engagement of various participants within the supply chain contributes immensely to its sustainability. Their heterogeneity and constantly varying commercial interactions concerning each other during food deliveries is one of the most significant issues that the agro-allied industry has encountered [10]. As a result, the information flow between every member of the supply chain is relatively poor. In recent years, it was clear that inadequate information has become a significant problem in the management of agro-industrial supply chains. To address food crises associated with the Agro-allied Supply Chain (ASC), increased transparency in ASC, trailing, and tracking of food items are required to improve overall performance [11].

One of the essential technologies for handling information in the ASC nowadays is the Internet of things. It is powered by embedded devices such as radars, actuators, as well as system connectivity. For information exchange, these gadgets allow real-time communication and data exchange amongst multiple smart machines/things within the supply chain [12]. Radio-frequency identification (RFID) technology is the most widely utilized IoT technology for tracking and tracing the immediate position of food goods [13, 14]. The information retrieved from RFID readers and tabs, which can log the place and time of occurrence/performance of a products varies, is mainly used to track agricultural products.

On the other side, as mobile network coverage expands into rural regions, farmers may use mobile solutions to improve agro-allied produces and linking them to bigger supply chains, tackling food insecurity, climate change acclimatization, and resistance [15]. AgTech Company, a provider of mobile device, for instance the farmX mobile application, provides farmer summary, intern profiling, transportation services, and agro-allied uses, allowing farmers to establish a digital identity [16]. These contribute to enhancing government-to-farmer access to inputs and extension services. Information from similar network further helps development organizations to reach out to agriculturalists for manure as well as input distribution depending on sex, geography, and farm size. Banks and insurance companies also depend on these mobile platforms to provide necessary information and data. They require efficient information to offer credit and insurance services to female farmers, allowing them to obtain agro-based insurance policy, credit loans, legal services, and rent of equipment at funded or minimal rates, consequently tackling gender inclusiveness issues [17].

Understanding the possibilities digital platforms have provided would assist farm owners to make better real-time, informed, operational and market-based decisions based on fiscal indicators rather than sticking to inherited farming techniques [5]. Furthermore, these services enable farm owners to evaluate facts from past information on related difficulties that could have occurred in the past, allowing them to learn from previous acts and make better operational decisions today [18].

For example, farmers can utilize the trusted agro-sensor drone service to assist them decide the best season of harvest [19]. For example, the server may notify a farm owner that a specific farm land is ripened for harvest sooner than anticipated. It may be beneficial at the inception of a crop cycle because it can generate accurate 3-D maps for initial topsoil examination, which can be helpful to design seed sowing precedents. Farmers might monitor crop wellness as well as detect bacteriological or fungi problems. Drone-borne gadgets scan a farm product with the aid of observable and near-infrared beam to ascertain the plants that reflect different green and NIR radiance [20]. These data are helpful to create multispectral imageries, which trail variations in plants as well as unveil their well-being state.

2 Literature Review

2.1 The Internet of Things (IoT)

The Internet of things signifies a bigger space where the Internet has advanced into a real-life use, including commonplace and everyday devices. Tangible elements for

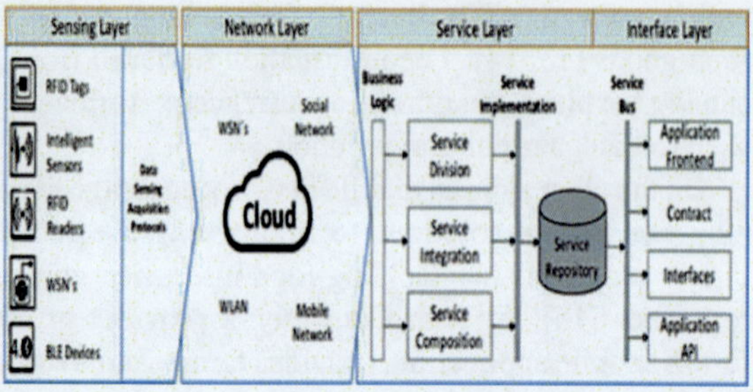

Fig. 1 Internet of things service-oriented architecture [27]

instance equipment, machineries, goods in different stages, and locations in diverse and remote areas are connected virtually [21, 22]. Cyber-systems monitor and operate these devices, which act as physical access points [14].

IoT is a concept that describes a functional network architecture that can reprogram itself with the aid of accepted and interoperable practices [23]. Every unit is recognized as a "thing" with essential natural characteristics such as a virtual reality and smart interfaces that seamlessly incorporate these unit into same operating system through the Internet of things [24]. Because everything is interlinked, the IoT network design must make sure that this cyber-physical interface is well integrated. An array of networks, announcements, organizational models and procedures, as well as security are all part of IoT designs [25].

For an easy change and integration of various unrelated network mechanisms dispersed throughout the whole supply chain, many considerations such as extensibility, scalability, and interoperability must be put in place while developing the IoT architecture [26]. Figure 1 portrays a conventional IoT "service-oriented architecture" (SOA) as opined by Lee et al. [27]. The four levels are described as shown in Fig. 1.

1. The sense layer: This particular layer is being used in conjunction with hardware (such as radio frequency identifications, actuators, and radars). They help to discern and manage tangible systems and collate data.
2. The network layer: This layer provides networking support and data transmission, essentially integrates all entities, and allows them to share information.
3. The service layer: This layer generates and controls service area dependent on the highlighted technology and provides functionality for integrating apps and request within the IoT perfectly.
4. The interface layer: This layer makes it easier for users and other apps to interact.

2.2 *The Role of IoT in Promoting Viable SCM System in the Agro-Allied Industry*

IoT is a relatively new concept that has been around since the 1990s. By the year 2013, it had evolved into a system that incorporated various technologies, ranging from building and home automation to complex wireless networks. It is self-evident that things interacting provide value for both customers and enterprises [28]. The IoT has the ability to provide new opportunities to the agro-allied industry.

IoT-based supply chain management entails providing the right product match with the right size, at the right time, placed in the right location, at the right cost, and in the right state to the right consumer [29]. However, there exist supply–demand discrepancy issues such as understocking, overstocking, as well as delayed supply which have formed common research topics a fore time in the field of business management resulting from the intricacy, insecurity, poor organization, inadequate information among other factors engaged in agro-allied supply chain [5].

Intelligent supply chain management has the advantages of being less expensive, faster, and more effective [30]. However, the ASC is growing more sophisticated, costly, unpredictable, and susceptible to external influences. Supply networks must become much brighter to properly deal with the increasing problems posed by globalization and rising customer demand [31]. The application of IoT for managing supply chain in the agro-allied industry ensures that supply chain processes are more efficient and sustainable by combining numerous domains, such as transmitter computer functions as well as new industry-centered technologies [9].

The future form of the supply system must consider a high-tech setup for handling facts, information, substantial items, produces, and business models [29]. The IoT has become a growing field of study with potentials, transforming traditional ASCM into Smart ASCM. Food processing plants, for example, are furnished with intelligent machines and tools able to accomplish client request with the support of global teams, smart analysis for situation management, and vibrant systems throughout the entire supply chain [5]. Recent changes that each supply chain may encounter to improve international supply chains include worldwide data harmonization and international commercial initiatives [32]. Across supply chain management, IoT has hopes and worries in various tasks spanning from inward bound logistics to outward bound operations. The inward and outward bound logistics are changing due to the mix of mobile computing, real-time data analysis, and cloud technology fashioned through the IoT idea [33]. Third-party logistic (3PL) idea, which can include any organization engaged in outsourcing to transfer products, is one of the most popular techniques for delivering goods. The Internet of things (IoT) facilitates this procedure by saving time and money. Companies can trace goods through their delivery channels using Internet-connected trackers, probably seen as the next generation of RFID technology [33]. As a result, IoT technology can be used to power several aspects of supply chain management. As previously said, such hopes, however, bring with them obstacles [34]. The constraints include

- Excessive costs charged by smart handset companies.
- Absence of collaboration between institutional entities (government and nongovernmental organizations).
- Poor funding and accessibility to capital to grow up the sector.
- Lack of consumer and government understanding of ICT and e-commerce.
- Logistics issues (such as the correctness of provided client addresses for deliveries).
- Challenges of developing business model that support possible growth and development.
- Lack of start-up capital and support for agribusiness and technologies (absence of specialized incubating machines).

2.3 Using the Internet of Things to Evaluate Processes in Agro-Allied Supply Chains (ASC)

RFID, sensor networks, artificial intelligence, cloud computing, and other cutting-edge technologies are all part of the Internet of things [35, 36]. ASCM entails the transportation of food items through many stages, starting from the farmer who supplies, to production, to delivery, and sales, as well as attesting food quality assurance [5]. The following is a basic description of how to evaluate the processes in ASC using IoT.

2.3.1 Production Stage

In the production stage, IoT-based technological applications have a lot of potentials. The raw inputs to work in progress and final products in the agricultural product manufacturing process are separated and trailed with the aid of RFID tags in collaboration with electric product codes (EPC) to ensure consistent output. This RFID code identifies each agricultural product and provides vital information such as the name of product, producer, quality, date of expiration, and product expected useful life. By applying pressure to products before or during packaging, high-pressure processing can help improve product quality [37].

2.3.2 Transportation Stage

Transportation also opens up a lot of possibilities for IoT-dependent technology in the agro-allied industry. Organizations may utilize GPS systems on transportation mediums, such as automobiles, to determine the exact place and state of the motor vehicles throughout the moving phase of agricultural products [38]. This automated transport system may help avoid delays during transit and any accidents or food wastage.

2.3.3 Delivery and Selling Stage

In the agro-allied industrial sector, IoT founded technology has a lot of usefulness in the delivery and selling stages. Data, security, and authentic monitoring systems show how the Internet of things is being used [39]. Devices such as EPC, RFID, and trackers can assist agriculturist and producers in determining product standard, for instance expiration dates, the source of problems, and weaknesses in supply chain management, among other things.

2.3.4 Industrial Internet of Things (IIOTs)

Inside a typical supply chain structure (manufacturer, shipping, distribution center, retailer, and customers) as revealed in Fig. 2, technology undoubtedly plays a generalized function. However, the initiative alters the playing field and increases reliance on information technology. It is not impossible to find a rapid, portable, and relatively inexpensive data that can set in motion future firms. As tracing and tracking activities turn out to be well-organized and dependable, integrating information and technological inventions into ASCM is opening new frontiers for optimizing material design and process flow. The new phase of manufacturing in the nearest future is based on a closed-loop big business model, which will necessitate reliable information technology solutions [40]. The industry 4.0 idea includes the IIoT projects on technology for attaining sustainable industrial development, comprising large data managing structures, cloud processing, RFID app, device computerization, and smart control systems [41]. Industries may further utilize the IIoT by installing radars and close circuit cameras on factory's operational systems for examining, unit reporting, and accident control [42]. The IIoT view suggests that 3D-printing production practices will be highly versatile and efficient. Consequently, the ASCM approach is opening grounds for an automated, viable manufacturing practice.

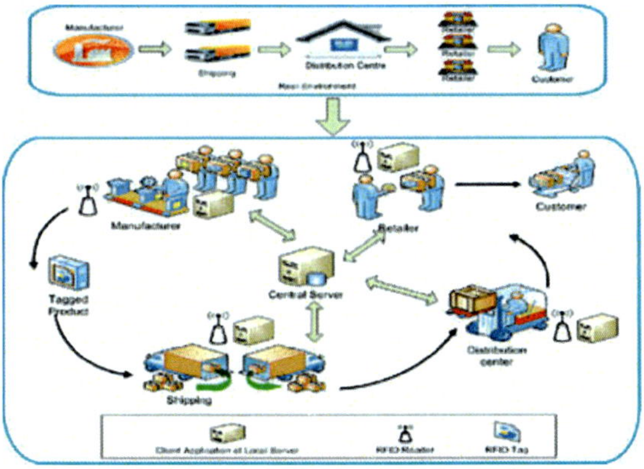

Fig. 2 RFID distributed supply chain system [43]

2.4 IoT-Based Technology Applicability in the Nigerian Agro-Allied Industry Supply Chain

The IoT performs a vital function in agribusiness supply chain administration as experienced in emerging economies. These are listed below [5, 44].

2.4.1 Purification of the Agro-Allied Material Market

The ASC, which uses IoT technologies (RFID, biosensors/wireless sensors/traceability) to meet customer demands for more excellent quality and authenticity, is simply a connection that allows manufacturing, storing, delivery, and trading facilities. Managers can recognize accurate information about items using RFID and radars or wireless sensors for tracing, allowing them to maintain product quality. This also helps to increase food quality while also reducing food waste.

2.4.2 Reduction in the Farmers Burden

The Internet of things will considerably upgrade the intelligibility of each stage throughout the supply chain. The RFID tag can actually perform automated recording of all processes within the agricultural product supply chain. The "bullwhip effect" is expected to be reduced and other costs such as inventory or labor costs, resulting in enhanced performance while the gains go to the farmers.

2.4.3 Potentials of Developing an Efficient Agricultural Product Supply Chain

IoT will ensure that every link in the agro-allied supply chain is served as efficiently as possible [45]. It will adopt AI intense pressure processing machines and AI embedded vibrating spectroscopy technology. Also, tracking methods in distribution will enhance the potentials of agricultural supply chains by ensuring that farm products needs are met. A large portion of the Nigerian population lives in local areas characterized by absence of modern and essential amenities. Healthiness, agriculture job creation, women emancipation, learning, as well as gender equality are all concerns ICT tools and services may address.

3 Conclusion

Although Nigerian innovators continue to confront significant obstacles in scaling, lowering cost of imports, and increasing exports, the agriculture industry shows considerable potential for future digital expansion [46]. With an extensive and rising domestic population, plenty of fertile but uncultivated land, high ICT diffusion and application, as well as various food producing base, the country would meet its mid-term growth ambitions by utilizing ICT and the Internet of things for agricultural productivity. To overcome this challenge, the future of agricultural practices will necessitate fresh and revolutionary partnerships among various players in the agro-allied and food supply chain. Amongst many other partners are Bill & Melinda Gates Foundation, the Fork, Harvestplus, the International Finance Corporation, USAID, FAO, GSMA Agritech, GAFSPfund, CGIAR, and DFID. These establishments actively engage in planning and growing recommended programs and digital policies that help smallholder farmers better their living. These functions are made possible by connecting them with advanced supply chains platforms as well as an avenue to boost agricultural production via better agribusiness practices, entry to funding technology as well as standard raw materials availability.

References

1. Siborurema E (2019) The contribution of urban agriculture to sustainable development: potential role in improving food security and reducing poverty. Stellenbosch University, Stellenbosch
2. Daniels C, Dosso M (2021) Mapping the Potentials for transformative innovation policies in Africa: evidence from Cote d'Ivoire and Nigeria. Entrepreneurship, technology commercialisation, and innovation policy in Africa. Springer, Cham, pp 279–300
3. Adesina AA (2019) Unlocking Africa's agricultural potential. Sustain Glob Food Secur Nexus Sci Policy 446
4. Nair RD, Landani N (2020) Making agricultural value chains more inclusive through technology and innovation. WIDER working paper
5. Luthra S, Mangla SK, Garg D, Kumar A (2018) Internet of things (IoT) in agriculture supply chain management: a developing country perspective. Emerging markets from a multidisciplinary perspective. Springer, Cham, pp 209–220
6. Protopop I, Shanoyan A (2016) Big data and smallholder farmers: big data applications in the agri-food supply chain in developing countries. Int Food Agribus Manag Rev 19(1030-2016-83148):173–190
7. World Bank Group (2018) Nigeria biannual economic update, April 2018: connecting to compete. World Bank
8. Ma L, Musa SO Effect of macroeconomic policy on industrial sector performance in Nigeria
9. Iorliam A, Iorliam IB, Bum S (2021) Internet of things for smart agriculture in Nigeria and Africa: a review. Int J Latest Technol Eng Manag Appl Sci
10. Arunachalam D, Kumar N, Kawalek JP (2018) Understanding big data analytics capabilities in supply chain management: unravelling the issues, challenges and implications for practice. Transp Res Part E Logist Transp Rev 114:416–436
11. Onwude DI, Chen G, Eke-Emezie N, Kabutey A, Khaled AY, Sturm B (2020) Recent advances in reducing food losses in the supply chain of fresh agricultural produce. Processes 8(11):1431
12. Vermesan O, Friess P (2013) Internet of things: converging technologies for smart environments and integrated ecosystems. River Publishers
13. Ben-Daya M, Hassini E, Bahroun Z (2019) Internet of things and supply chain management: a literature review. Int J Prod Res 57(15–16):4719–4742
14. Lee J, Ardakani HD, Yang S, Bagheri B (2015) Industrial big data analytics and cyber-physical systems for future maintenance & service innovation. Procedia cirp 38:3–7

15. Lezoche M, Hernandez JE, Díaz MD, Panetto H, Kacprzyk J (2020) Agri-food 4.0: a survey of the supply chains and technologies for the future agriculture. Comput Ind 117:103187
16. Cavallo A, Ghezzi A, Guzmán BVR (2019) Driving internationalization through business model innovation: evidences from an AgTech company. Multinatl Bus Rev
17. Rutten L, Fanou SL (2015) Innovative and inclusive finance for youth in agriculture. In: Africa Agric. Status Rep. Youth Agric. Sub-Saharan Africa, Alliance a Green Revolut. Africa (AGRA), Nairobi, Kenya
18. Oreszczyn S, Lane A, Carr S (2010) The role of networks of practice and webs of influencers on farmers' engagement with and learning about agricultural innovations. J Rural Stud 26(4):404–417
19. Banik S, Mowla MM, Ahmad I (2019) A strategic routing analysis for agro sensor communications in mobile ad hoc networks. In: 2019 1st international conference on advances in science, engineering and robotics technology (ICASERT), pp 1–6
20. Aasen H, Honkavaara E, Lucieer A, Zarco-Tejada PJ (2018) Quantitative remote sensing at ultra-high resolution with UAV spectroscopy: a review of sensor technology, measurement procedures, and data correction workflows. Remote Sens 10(7):1091
21. Tsiatsis V, Karnouskos S, Holler J, Boyle D, Mulligan C (2014) Internet of things. Academic
22. Tsiatsis V, Karnouskos S, Holler J, Boyle D, Mulligan C (2018) Internet of things: technologies and applications for a new age of intelligence. Academic
23. Kamble SS, Gunasekaran A, Parekh H, Joshi S (2019) Modeling the internet of things adoption barriers in food retail supply chains. J Retail Consum Serv 48:154–168
24. Greer C, Burns M, Wollman D, Griffor E (2019) Cyber-physical systems and internet of things
25. Yao X, Zhou J, Lin Y, Li Y, Yu H, Liu Y (2019) Smart manufacturing based on cyber-physical systems and beyond. J Intell Manuf 30(8):2805–2817
26. Rimal BP, Jukan A, Katsaros D, Goeleven Y (2011) Architectural requirements for cloud computing systems: an enterprise cloud approach. J Grid Comput 9(1):3–26
27. Lan L, Wang B, Zhang L, Shi R, Li F (2015) An event-driven service-oriented architecture for the internet of things service execution. Int J Online Eng 11(2)
28. Kaivo-Oja J, Virtanen P, Jalonen H, Stenvall J (2015) The effects of the internet of things and big data to organizations and their knowledge management practices. In: International conference on knowledge management in organizations, pp 495–513
29. Wu L, Yue X, Jin A, Yen DC (2016) Smart supply chain management: a review and implications for future research. Int J Logist Manag
30. Attaran M (2012) Critical success factors and challenges of implementing RFID in supply chain management. J Supply Chain Oper Manag 10(1):144–167
31. Lal K (2007) Globalization and the adoption of ICTs in Nigerian SMEs. In: Information and communication technologies in the context of globalization, Springer, pp 151–207
32. Attaran M, Attaran S (2007) Collaborative supply chain management: the most promising practice for building efficient and sustainable supply chains. Bus Process Manag J
33. Manners-Bell J, Lyon K (2019) The logistics and supply chain innovation handbook: disruptive technologies and new business models. Kogan Page Publishers
34. Zhao G, Liu S, Lopez C, Lu H, Elgueta S, Chen H, Boshkoska BM (2019) Blockchain technology in agri-food value chain management: a synthesis of applications, challenges and future research directions. Comput Ind 109:83–99
35. Khan ZA, Imran SA, Akre V, Shahzad M, Ahmed S, Khan A, Rajan A (2020) Contemporary cutting edge applications of IoT (Internet of Things) in industries. In: 2020 seventh international conference on information technology trends (ITT), pp 30–35
36. Dopico M, Gómez A, De la Fuente D, García N, Rosillo R, Puche J (2016) A vision of industry 4.0 from an artificial intelligence point of view. In: Proceedings on the international conference on artificial intelligence (ICAI), p 407
37. Chen RS, Chen CC, Yeh KC, Chen YC, Kuo CW (2008) Using RFID technology in food produce traceability. WSEAS Trans Inf Sci Appl 5(11):1551–1560
38. Hassan MU, Rehmani MH, Chen J (2019) Privacy preservation in blockchain based IoT systems: Integration issues, prospects, challenges, and future research directions. Future Gen Comput Syst 97:512–529

39. Tu M, Lim MK, Yang MF (2018) IoT-based production logistics and supply chain system—Part 1: modeling IoT-based manufacturing supply chain. Ind Manag Data Syst
40. Kiritsis D (2011) Closed-loop PLM for intelligent products in the era of the internet of things. Comput Des 43(5):479–501
41. Shahzad Y, Javed H, Farman H, Ahmad J, Jan B, Zubair M (2020) Internet of energy: opportunities, applications, architectures and challenges in smart industries. Comput Electr Eng 86:106739
42. Onu P, Mbohwa C (2018) Green supply chain management and sustainable industrial practices: bridging the gap. In: Proceedings of the international conference on industrial engineering and operations management, pp 786–792
43. Kamaludin H, Mahdin H, Abawajy JH (2018) Clone tag detection in distributed RFID systems. PloS One 13(3)
44. Ramundo L, Taisch M, Terzi S (2016) State of the art of technology in the food sector value chain towards the IoT. In: 2016 IEEE 2nd international forum on research and technologies for society and industry leveraging a better tomorrow (RTSI), pp 1–6
45. Kaloxylos A, Wolfert J, Verwaart T, Terol CM, Brewster C, Robbemond R, Sundmaker H (2013) The use of future internet technologies in the agriculture and food sectors: integrating the supply chain. Procedia Technol 8:51–60
46. Olomola AS, Nwafor M (2018) Nigeria agriculture sector performance review. International institute tropical agriculture, vol 3. Ibadan, Niger

Recovery System of Work Performance by Using Indoor Environmental Changes Based on EEG-Movement Feature Space

Hinata Serizawa and Yoshihisa Fukuhara

Abstract In this paper, we have realized a system to maintain good work performance by conducting the following three experiments. (1) Created a machine learning model with EEG as the objective variable and motion information as the explanatory variable. After that, we established a method for estimating the level of concentration based on movement information. (2) When work performance deteriorates, we intentionally change the indoor environment to verify whether it is effective as a method to recover work performance. (3) While estimating work performance using the model created in No. 1, verify whether work performance recovered from the estimated value by using the changes in the indoor environment verified in No. 2.

Keywords Work performance · Home network · EEG · Biological information processing · Machine learning · Human motion analysis

1 Introduction

In recent years, online classes and telework have become popular worldwide due to the fight against new coronavirus infections. As a result, students have more time to attend classes at home, and working people have more opportunities to do their work at home. In such online classes and telework, there is a limit to the amount of time that each student can concentrate on work, and work efficiency decreases with time. Therefore, there is a growing demand for places and spaces where people can concentrate on their work for a long time. In this paper, we propose a system to realize such a space, which can maintain a good work performance at a desk for a long time. Specifically, the system estimates the work performance during work and automatically changes the room environment when a decline is detected, thereby stimulating the subject's arousal and returning the work performance to a good state.

H. Serizawa (✉) · Y. Fukuhara
Faculty of Data Science, Musashino University, Koutou ku Ariake 3-3-3, Tokyo, Japan
e-mail: s1922065@stu.musashino-u.ac.jp

Y. Fukuhara
Asia AI Institute, Musashino University, Koutou ku Ariake 3-3-3, Tokyo, Japan

X.-S. Yang et al. (eds.), *Proceedings of Seventh International Congress on Information and Communication Technology*, Lecture Notes in Networks and Systems 447,
https://doi.org/10.1007/978-981-19-1607-6_35

This system is expected to create a space where people can devote themselves to their work for a long time.

1.1 What Is Work Performance?

Among the academic publications recorded in IGI global, the word work performance is used in the following sense [1].

1. An accomplishment of the assigned tasks for achieving organization's goal.
2. A kind of evaluation report indicating how well an employee is executing the expected related work activities.

In this paper, it is used in a sense similar to 2. Specifically, it is an indicator of how well the worker is concentrating on the task at hand.

2 Related Work

2.1 Studies to Estimate Work Performance

Many studies have been conducted to estimate work performance, and various biometric data have been used in each study. For example, the number of blinks was measured by a web camera, and it was found that the more blinks, the worse the work performance [2]. In terms of heart rate information, spectral analysis of heart rate variability has been conducted in various studies because heart rate variability is thought to reflect autonomic responses to stress and cognitive tasks [3, 4]. For example, it has been found that the degree of fatigue may be detected by changes in the trend of heart rate [5].

On the other hand, electroencephalogram (EEG) as a biological information is widely used as an evaluation index of the information processing process of the brain. For example, a comparative analysis of the correlations between EEG data and students' thoughts and perceptions during their learning activities has shown that EEG is effective as an index for estimating stress and work performance [6]. This indicates that the frequency characteristics of EEG are very effective as an evaluation index of human work performance. Therefore, in this study, we use the frequency characteristics of EEG as an index to estimate work performance.

2.2 *Relationship Between Room Environment and Work Performance*

In terms of students' thinking ability, there is a certain peak in the room temperature. It has been shown that people's thinking ability starts to decline when the temperature deviates from 26 °C [7]. However, learning efficiency was not just thinking ability, but also concentration and memory. One way to evaluate learning efficiency was to calculate students' performance and found that when the room temperature was lowered from 25 to 20 °C, students' performance increased significantly, contrary to a single change in thinking ability [8]. Therefore, this study aims to promote the arousal of the subjects by changing the room temperature and to recover the work performance.

3 Preliminary Experiments

In this study, as mentioned in the introduction, we will build a system that can estimate the value of work performance during work, and when a decrease in the value is detected, the system will automatically change the room environment to awaken the subject and return the work performance to a good state.

First of all, we have to establish the following two things before conducting the experiment and discussion of this system itself.

I. Estimation of work performance and verification of accuracy.
II. Verification of the changes in room temperature that can stimulate the subjects' arousal.

Therefore, these two preliminary experiments and their results are described in the following. In this study, only I, the author of this paper, will conduct all the experiments.

3.1 *Experiment Environment*

In the experiments described below, we set up a room of 3.4 × 2.6 m^2 room, sit in a chair facing a PC, and wear an electroencephalograph. The room temperature at the beginning of the experiment is about 25 °C, which is relatively comfortable for human beings. It has been found that the color and intensity of the lighting affect the stress and temperature of the subjects [9]. It has also been found that working in a space with high CO_2 concentration and too high or too low humidity has an effect on the subjects' work efficiency [10, 11]. Therefore, during the experiment, the room environment should be ventilated to keep the CO_2 concentration low and to keep the humidity constant.

3.2 EEG

We use EEG as an index for estimating work performance. One of the merits of EEG is that it is highly useful as an index of concentration, as described Sect. 2.1. On the other hand, the disadvantage is that it is worn on the head, and it may be a burden to wear it for a long time to measure the concentration level. To solve this problem, we use EEG values as a reference and build a machine learning model by learning movement information during work, which enables us to estimate the level of concentration using only movement information. This makes it possible to estimate the degree of concentration using only the movement information. We use the EMOTIV EPOC X from EMOTIV as the EEG sensor to measure EEG values. The EPOC X is also equipped with a motion sensor and can measure acceleration. We used the AF3 channel to measure the left frontal lobe because the frontal lobe, where the AF3 channel is located, is responsible for "thinking, coming up with ideas, and making decisions", and we judged it to be the most appropriate measurement site for the experiment. The power spectrum of each frequency can be obtained by Fourier analysis of the original EEG data obtained by the EEG sensor. The frequencies can be classified into five categories according to their magnitudes. The following are the names, frequency ranges, and especially the psychological situations in which the appearance is observed [6].

- δ-wave, 1~4 Hz, During sleep
- θ-wave, 4~8 Hz, During sleep, caution
- α-wave, 8~12 Hz, During relax, close the eyes
- β-wave, 15~20 Hz, During concentrating, exercise
- γ-wave, 30~Hz, During visual processing

In fact, a study using a simple EEG to examine whether EEG can be used as an external indicator of thinking states has shown that the mean value of β/α is higher when thinking is necessary in learning [6]. Therefore, the β/α value as EEG information was used as an index of work performance.

3.3 Preliminary Experiment I

Previously, we constructed a machine learning model with EEG information as the objective variable and motion information as the explanatory variable in order to estimate work performance [12]. In this experiment, we improved the existing method and aimed to improve the accuracy.

Movement information. In this experiment, we used five kinds of motion information as explanatory variables: blink, head acceleration, number of keystrokes, mouse movement distance, and heart rate (see Fig. 1a). All the five data were sampled every one minutes, and the average value was calculated. All these five movements were sampled every 1 min, and the average value was calculated.

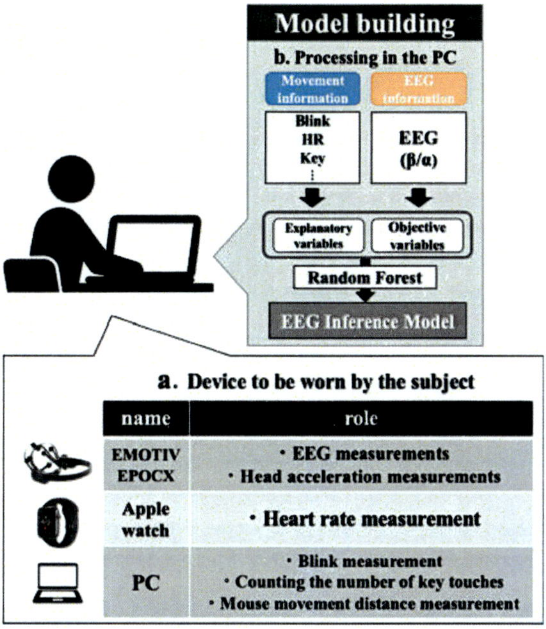

Fig. 1 System diagram of preliminary experiment I

Building machine learning models. From the sampled data with the EEG information (β/α values) as the objective variable and the five motion information as the explanatory variables, we perform a regression analysis using the random forest in scikit-learn to create a machine learning model. This allows us to infer beta/alpha values from movement information and to estimate work performance values without using EEG sensors. In addition, since the random forest is capable of calculating the importance of features from the learned model, we also examine the influence of the motion information on the performance of the task. Six training sessions of 60 min were performed, 5 as training data, and 1 as test data, with the β/α value as the objective variable and the motion information as the explanatory variable. We use the β/α values as the objective variable and the motion information as the explanatory variable (see Fig. 1b).

3.4 Results of Preliminary Experiment I

The experimental results show that the coefficient of determination of the training data is 0.8744 and that of the test data is 0.8025. In the previous study, the score of the training data was 0.8135 and that of the test data was 0.7142. Therefore, we could improve the accuracy of the model compared to the previous study.

Next, we check the importance scores of the features (see Fig. 2). From the graph, we can see that the score of blink is the highest. In other words, it is shown again that

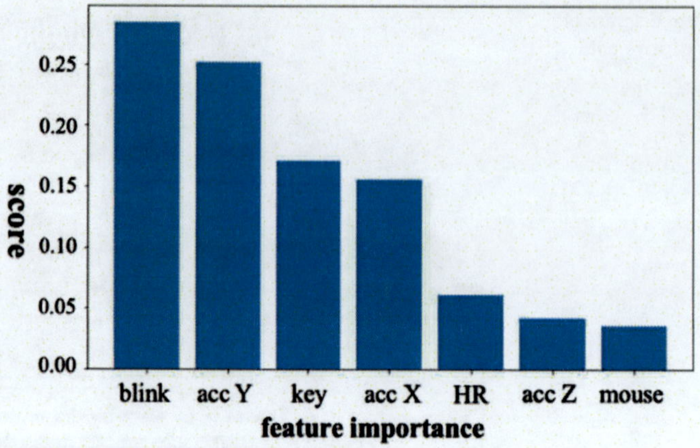

Fig. 2 Feature importance of movement information—present experiment

there is a close relationship between work performance and blinking. The scores of Y-axis acceleration, key, and X-axis acceleration were the highest, while the scores of heart rate, Z-axis acceleration, and mouse were the lowest. In contrast, the scores of heart rate, Z-axis acceleration, and mouse were low. As for acceleration, the scores of X and Y axes were high, and it is considered that back and forth movement is related to work performance. On the other hand, the importance of the mouse was the lowest, probably because, unlike the keyboard, the mouse can be operated easily whether the user is concentrating or not.

By using the trained model, we can estimate the β/α values without measuring the EEG during the work and evaluate the work performance only by the movement information.

3.5 Preliminary Experiment II

In Preliminary Experiment II, we examine the effective way of changing the room temperature to stimulate the subjects' arousal. As in Experiment 1, the evaluation of work performance is based on β/α values. As in Experiment 1, the subjects will write a paper for 70 min while wearing an EEG sensor to record the β/α value, and the EEG and the room environment will be recorded. The experiment was conducted five times in total to confirm the validity.

For recording the indoor environment, we use "NETATMO weather station" by Netatmo. This device is able to measure temperature, humidity, and CO_2 concentration, so we record them.

As mentioned in topic 2.2, it has been found that changes in room temperature have a significant effect on work performance. Therefore, as a hypothesis, we thought that the room temperature can be automatically raised or lowered when the work performance decreases, so as to stimulate the subject's arousal. The following section describes how the room temperature is changed.

Room temperature change. The room temperature that humans can feel relatively comfortable is considered to be around 26 °C [7]. However, even at a comfortable room temperature, work performance inevitably declines during prolonged work. Therefore, we thought that the room temperature could be raised or lowered automatically and deliberately according to the work performance to stimulate the subject's arousal. In order to avoid the subjects to operate the air conditioner directly, we used "Nature Remo", a smart remote control which can be remotely operated from a PC, and set a trigger to operate the air conditioner at a pre-designated time. For the automatic operation of the air conditioner during the experiment, we specifically do the following.

1. start the experiment in a room with a room temperature of around 26 °C and write the paper.
2. at the end of 20 min, the room temperature is intentionally lowered by setting the cooling temperature to 22 °C.
3. at the end of 45 min, the heating temperature is changed to 26 °C to restore the room temperature.
4. 70 min after the start of the experiment, the experiment is terminated.

EEG values will be measured during the experiment in order to see the changes in work performance.

3.6 *Results of Preliminary Experiment II*

The results show the extent to which the room temperature changed every 5 min from the start of the experiment (see Fig. 3) and the average β/α values sampled every 5 min (see Fig. 4).

Fig. 3 Changes in room temperature

minute	exp1	exp2	exp3	exp4	exp5
0	**26.4**	**25.1**	**25.8**	**25.5**	**26.4**
5	0.3	-0.2	0.1	0.1	0.1
10	-0.4	0.1	0.2	0.2	0
15	-0.2	0.4	0.3	0.2	0.2
20	0.2	0.5	0.3	0.1	0
25	-0.4	-0.3	-0.4	-0.7	-0.9
30	-1.2	-1.3	-1.0	-1.7	-1.9
35	-2	-1.8	-1.6	-2.4	-2.9
40	-2.9	-2.3	-2.1	-2.9	-3.6
45	-3.7	-2.9	-2.6	-3.1	-3.9
50	-3.0	-2.6	-2.4	-2.7	-3.4
55	-2.6	-1.7	-2	-1.6	-3
60	-1.5	-0.5	-1.1	-0.9	-2.6
65	-1	0	-0.4	-0.2	-2
70	-0.3	0.4	0.2	0	-1.4

Fig. 4 Changes in β/α temperature

minute	exp1	exp2	exp3	exp4	exp5
0-5	0.633	0.578	0.718	0.664	0.687
5-10	0.650	0.530	0.671	0.676	0.634
10-15	0.513	0.538	0.685	0.689	0.641
15-20	0.582	0.556	0.743	0.665	0.628
20-25	0.600	0.573	0.706	0.729	0.607
25-30	0.647	0.697	0.786	0.719	0.527
30-35	0.796	0.766	0.731	0.764	0.591
35-40	0.705	0.728	0.671	0.714	0.588
40-45	0.619	0.690	0.776	0.706	0.660
45-50	0.575	0.628	0.762	0.743	0.773
50-55	0.690	0.581	0.698	0.677	0.645
55-60	0.615	0.648	0.616	0.597	0.641
60-65	0.576	0.668	0.690	0.593	0.600
65-70	0.559	0.721	0.635	0.577	0.701
AVG	0.629	0.628	0.683	0.656	0.630

Largest value | 2nd largest value | Smallest value | 2nd smallest value

From Fig. 3, we can see that the change in room temperature is well controlled by the cooling and heating operations as expected, although there is some variation in the change in room temperature among the experiments. Figure 4 shows that the β/α values of the four experiments, except for Experiment 5, are larger in the period of 10–20 min after the room temperature is lowered (30–40 min after the start of the experiment). In other words, it is proved that the intentional decrease of the room temperature can awaken the subjects and recover their work performance.

On the other hand, by turning on the heating 45 min after the start of the experiment and returning to the original room temperature, we aimed to continue the high work performance. This may be due to the fact that the subjects were once awakened by lowering the room temperature but were not able to recover their work performance when the room temperature was restored.

However, we considered the possibility that the decrease in work performance was due to the long period of time when the air conditioning was turned on and the room temperature was lowered too much. It was mentioned earlier that the β/α value increased during the first 10–20 min after the air conditioning was turned on, and the change in room temperature during the first 10–20 min after the air conditioning was turned on shows that the room temperature decreased by about −1.5 to −2.0 °C. In other words, when the room temperature dropped by −1.5 to −2.0 °C, the subjects' work performance had recovered.

In this experiment, the air conditioning was turned on for 25 min, so that the work performance had already recovered when the room temperature dropped by −2.0 °C. However, the room temperature was lowered further from that point, which may have encouraged the decline of the work performance. In fact, I myself, the subject, felt the coldness that affected my work performance 20 min after I turned on the air conditioner. Therefore, in the next experiment, we will adjust the room temperature so that it is not too low. Specifically, we set a trigger to turn off the air conditioner when the room temperature drops by 2.0 °C.

4 Experiment

From preliminary experiments, we were able to create a learned model and found that intentionally lowering the room temperature is a factor to recover the work performance. In this experiment, we integrate these two models, estimate the work performance from the motion information, and control the room temperature automatically. When a decrease in β/α is detected, the room temperature is deliberately lowered by automatically turning on the air conditioner using a smart remote control to verify whether it is possible to recover the work performance (see Fig. 5).

As in the preliminary experiment, we will write the paper for 60 min. However, in order to verify the results, it is necessary to decide in detail when and how much the β/α value changes and how much the room temperature is changed. Therefore, we follow the following flow (see Fig. 6).

In Treatment a, the standard value of β/α for checking the decline of work performance was set at 0.565. However, this was because we could not find any previous study on what value of β/α actually indicates bad work performance. Therefore, in Fig. 4, the five values with the smallest β/α values in each experiment were obtained, and the average of these five values (0.565) was used as the criterion for judging poor work performance. In the same way, in Treatment b, the β/α value was set to 0.778 as

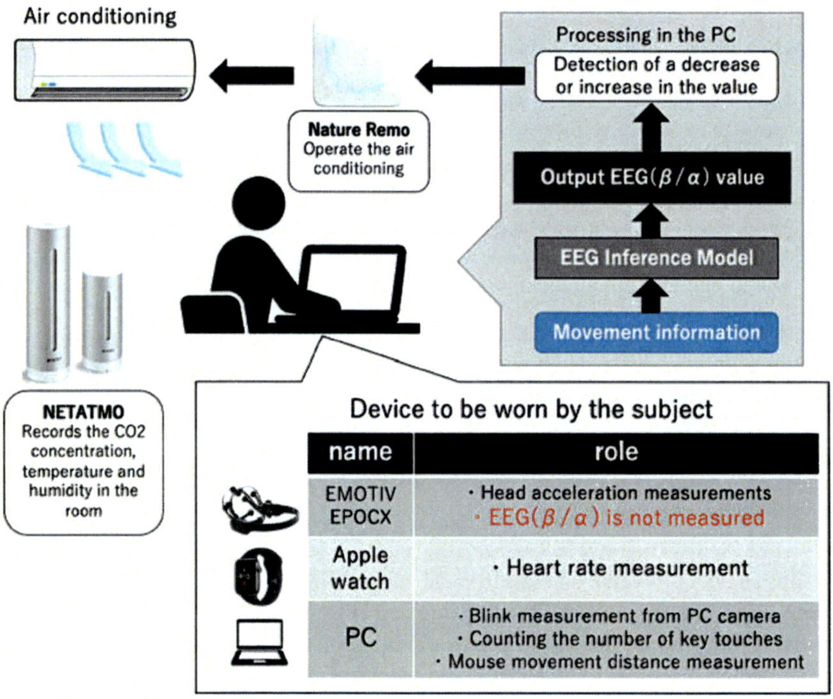

Fig. 5 System diagram of this experiment

Fig. 6 Flowchart in experiment

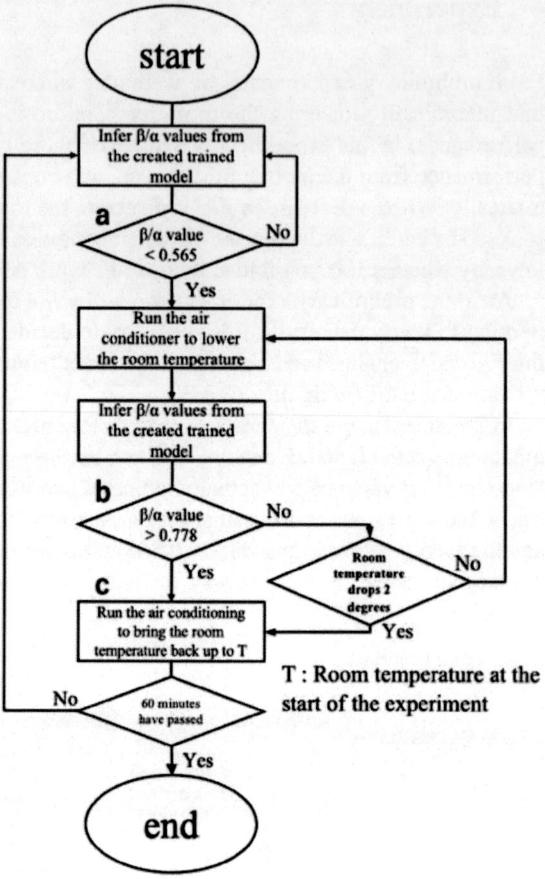

the criterion for judging the recovery of work performance. In Treatment c, the room temperature was returned to that at the beginning of the experiment in order to avoid that the subjects continued to work in the environment after the room temperature was lowered and became accustomed to it.

5 Result

Comparing the β/α value inferred from the model with the correct β/α value obtained from the EEG sensor, we can see that they are close to each other (see Fig. 7).

Particularly important in this study are the points where the β/α value is below 0.565, which is judged to be a decrease in work performance, and the points where the β/α value is above 0.778, which is judged to be a recovery in work performance. Since these values are the triggers for the air conditioner operation, if the inference is

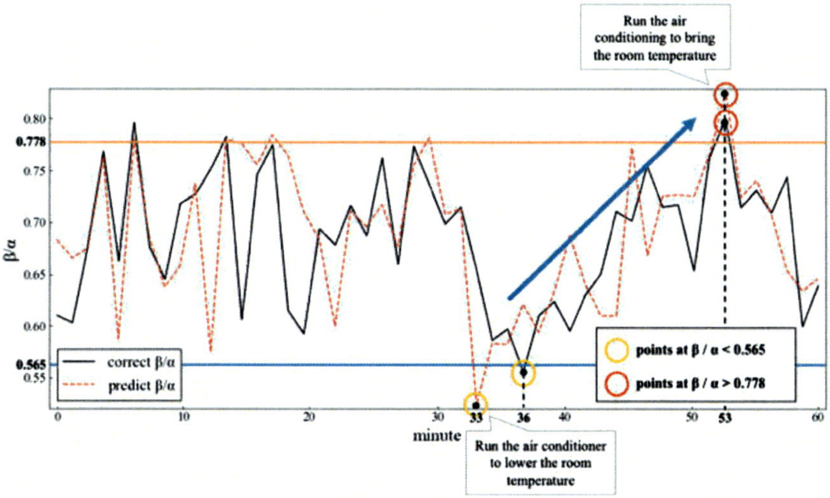

Fig. 7 β/α values obtained from the created model and from the actual EEG sensor

not performed correctly, the room temperature may be lowered because it is judged that the work performance has decreased when it is in a good state, or the room temperature may be stopped when the work performance has not yet recovered sufficiently. For example, the room temperature may be lowered, or the room temperature may be stopped when the subject has not yet fully recovered from the work performance, so that the subject may feel stressed. Therefore, we will focus on these two points.

Result for points at $\boldsymbol{\beta/\alpha}$ < 0.565. The time when β/α was below 0.565 was 33 min, as estimated from the model. The actual value was 36 min. The difference is only 3 min, and we can confirm that the model is able to estimate close to the actual value. When the β/α value fell below 0.565, the work performance was judged to have decreased, and the room temperature was lowered. However, the β/α value increased steadily for about 15–20 min after that, so it was confirmed that the work performance was successfully recovered by lowering the room temperature.

Result for points at $\boldsymbol{\beta/\alpha}$ > 0.778. The time when the β/α value exceeded 0.778 was 53 min, as estimated from the model. The actual value was also 53 min. In this case, there was no difference in the times, indicating that the model was able to guess well.

The results are very good because the overall values are close and there is almost no difference in time at the points where the air conditioner is operated.

6 Conclusion

From the results of these experiments, we confirmed the following two points.

1. to estimate the β/α value from the motion information.
2. to recover the β/α value by automatically changing the room environment based on the estimated β/α value.

From the above results, it is possible to construct a system that recovers and sustains the user's work performance based on information such as blinking and keystroke input without using an EEG sensor.

7 Future Work

In this experiment, we have only one subject, and we have only developed the system with one trained model. It is necessary to consider the use of a single model by several people, because there are always errors in the EEG and movement information that are easily varied by individuals. However, it is not realistic to create and prepare a model for each person who uses the system. Therefore, in the future, we will try to have several people perform the same experiment, compile a certain amount of EEG and movement information data, perform clustering based on the data, and create a machine learning model for each cluster. If we can divide the model into similar user groups instead of individual users, we can save the time and effort to create a model for each user, and we believe that the system can be spread at a lower cost. In this study, we used only the room temperature as the room environment. In the future, we would like to improve the system by adding lighting and sound, so that we can provide changes in the room environment that allow people to recover their work performance.

In order to provide more people with a space where they can devote themselves to work for a long time, we will continue our research to realize these improvements.

References

1. IGI Global Homepage. https://www.igi-global.com/dictionary/exploring-emotional-intelligence-at-work/58875. Last accessed 15 Oct 2021
2. Kabutomori H, Abiko S, Hasegawa D, Sakuta H (2015) Evaluation of concentration by blink detection using a web camera. In: 77th proceedings of the national convention of IPSJ, vol 4. Japan, pp 931–932
3. Jorna P (1992) Spectral analysis of heart rate and psychological state: a review of its validity as a workload index. Biol Psychol 34:237–257. https://doi.org/10.1016/0301-0511(92)90017-O
4. Acharya RU, Joseph KP, Kannathal NC, Lim CM, Suri JS (2006) Heart rate variability:a review. Med Biol Eng Comput 44:1031–1051. https://doi.org/10.1007/s11517-006-0119-0

5. Imai R, Kamiya D, Inoue H, Tanaka S, Sakurai J, Fujii T, Baba A, Ito M (2018) Research on trial of fatigue detection using heart rate data of smart watch. In: 34th proceedings of the symposium on fuzzy systems, vol 34. Japan, pp 407–408. https://doi.org/10.14864/fss.34.0_407
6. Hirai A, Yoshida K, Miyaji I (2013) Comparative analysis of thinking and memory during learning by simple electroencephalography. In: Proceedings of multimedia, DICOMO symposia, Japan, pp 1441–1446
7. Cho H (2017) The effects of summer heat on academic achievement: a cohort analysis. J Environ Econ Manag 83:185–196. https://doi.org/10.1016/j.jeem.2017.03.005
8. Wargocki P, Wyon DP (2007) The effects of moderately raised classroom temperatures and classroom ventilation rate on the performance of schoolwork by children (RP-1257), HVAC&R Research, vol 13. pp 193–220
9. Odahara T, Misu T, Watanabe T, Isshiki M (2017) A study of the effect of illumination color on humans. J Color Sci Assoc Jpn 41(3):145–148
10. Fisk JW, Rosenfeld HA (1997) Estimates of improved productivity and health from better indoor environments. Int J Indoor Environ Health 7:158–172. https://doi.org/10.1111/j.1600-0668.1997.t01-1-00002.x
11. Chikamoto T, Mimura R (2019) Influence of carbon dioxide fluctuation and thermal environment on workability, physiology and psychology. In: E3s Web of conferences. https://doi.org/10.1051/e3sconf/201911102068
12. Serizawa H, Fukuhara Y (2021) Recovery system of work performance by using indoor environmental changes based on EEG-movement feature space. In: 83th proceedings of national convention of IPSJ, vol 4. pp 115–116

Moroccan Sign Language Video Recognition with Deep Learning

Abdelbasset Boukdir, Mohamed Benaddy, Othmane El Meslouhi, Mustapha Kardouchi, and Moulay Akhloufi

Abstract Sign language is the native form of expression used by deaf people in the world. With the recognition techniques applied to sign language, a significant need for developing tools to facilitate the accessibility of information to the deaf public has arisen. Little work deals with recognizing Moroccan sign language (MoSL) for the Moroccan deaf community. In this paper, a deep learning architecture is presented to be used to recognize MoSL signs. The proposed system uses 3D convolution neural networks to describe effectively video sequences containing Moroccan signs. Experiments showed that the system is able reliably to recognize Moroccan word signs, with 99.60% of accuracy.

Keywords Moroccan sign language · Deaf people · Video processing · 3D Convolutional neural networks

A. Boukdir (✉) · M. Benaddy
Laboratory of Engineering Sciences and Energies, FPO,Ibn Zohr University,Ouarzazate, Morocco
e-mail: abdelbasset.boukdir@edu.uiz.ac.ma

M. Benaddy
e-mail: m.benaddy@uiz.ac.ma

O. E. Meslouhi
ENSA–Safi,Cadi Ayyad University,Marrakech, Morocco
e-mail: o.elmeslouhi@uca.ma

M. Kardouchi · M. Akhloufi
Département d'Informatique,Université de Moncton,Moncton, Canada
e-mail: mustapha.kardouchi@umoncton.ca

M. Akhloufi
e-mail: moulay.akhloufi@umoncton.ca

X.-S. Yang et al. (eds.), *Proceedings of Seventh International Congress on Information and Communication Technology*, Lecture Notes in Networks and Systems 447,
https://doi.org/10.1007/978-981-19-1607-6_36

1 Introduction

Sign language was designed to use hand communication, body language, and lip placement instead of sound and spoken language to express the meaning of a deaf person's messages [1]. Understanding the culture surrounding deaf people and recognition of sign language plays an essential role in integrating the deaf population into regular life. Sign languages have a distinct vocabulary and grammar of their own such as Arabic sign language (ArSL) [2], American sign language (ASL) [3], British sign language (BSL) [4], and others.

In this sense, several systems have been created to facilitate the sign languages recognition [5]. These systems are classified into two categories: sensor-based and vision-based systems. Sensor-based techniques employ handheld tools for sign monitoring, such as leap motion sensors [6], Microsoft Kinect [7], and gloves [8]. However, the use of handheld tools can be challenging because of their availability. On the other side, vision-based techniques use only camera-captured image sequences to identify words in signs [9] which makes these techniques very easy to use.

In recent years, many vision-based methods were developed for sign languages recognition. These kinds of methods are based on 2DCNNs or 3DCNNs architectures. In Hayani et al. [10], the authors propose a 2DCNN architecture based on LeNet-5 to recognize isolated sign letters. They achieve an accuracy value of 90.02% with 80% of images from their set. Latif et al. [19] proposed a 2DCNN for Arabic alphabet signs from static images the obtained accuracy is 97.6% using 50,000 Arabic sign images. Boukdir et al. [11] use a combination of 2DCNN model to retrieve features and a recurrent network model to detect the relationship between frame images. Bencherif et al. [14] propose a framework that combines 2D hands and body landmarks from frame sequences using the conjunction of a 3DCNN skeleton network with a 2D landmark convolution network. The obtained score is 89.62% for signer-dependent mode and 88.09% for signer-independent mode. Baccouche et al. [17] use a hybrid model based on an extension of CNN to 3D combined with a recurrent neural network to learn to classify human actions. Huang et al. [18] propose a 3DCNN for sign language recognition, they obtained an average accuracy of 94.2% over multi-channel videos. Boukdir et al. [11] propose a 3DCNN model that learn the spatiotemporal features from small patches. Tests on collection signs give an accuracy of 99%.

In this paper, we introduce a video-based sign language classification system designed for Moroccan sign language. The proposed model is based on a 3D convolutional neural networks (3DCNN) architecture. It is trained on a dataset that contains 56 isolated videos of Moroccan word signs.

The rest of this paper is as follows. Sect. 2 details the proposed 3D CNN model. The experiments of our model and evaluations are discussed in Sect. 3. Finally, the paper achieves in Sect. 4.

2 Proposed Model Architecture

The model proposed in this paper is based on the three-dimensional convolutional neural networks. It takes in its input the video frames of a Moroccan sign language and outputs the corresponding class of the input sign. The proposed model is built with two convolution layers, two pooling layers, and a fully connected layer, along with input–output levels. This architecture is illustrated in Fig. 1, and the details of its structure are given in Table 1.

Convolution is the initial layer that extracts features from the input frame sequences and acts as the core construction unit of the CNN. Within the convolution layer, the kernels are used to extract salient patterns from the input dataset through feed-forward and backward propagation. In this work, we perform this operation with 3 × 3-dimensional filters across the input data.

In a CNN architecture, activation functions are applied to identify which particular element should be activated at a specific location in the network. We used the parametric rectified nonlinear unit (PRenu) [12] activation function which is expressed as:

$$\phi(a) = \begin{cases} 0 & a < 0 \\ a - \lambda \log(1 + a) & a \geq 0 \end{cases} \quad (1)$$

With a is the input parameter and λ is between 0 and 1.

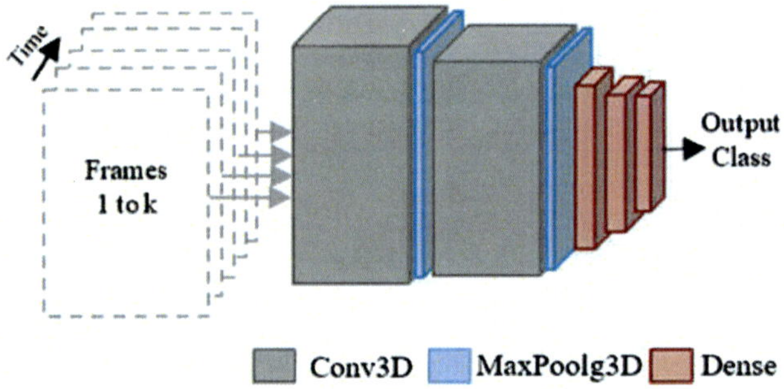

Fig. 1 Proposed 3D CNN architecture

Table 1 Design of proposed 3D CNN architecture model

	Activ.	Output	Stride	# Param
Conv3D	Prenu	16	1 × 1 × 1	1312
MaxPooling3D	–	–	2 × 2 × 2	–
Conv3D	Prenu	8	1 × 1 × 1	3464
MaxPooling3D	–	–	2 × 2 × 2	–
Flatten	–	–	–	–
Dense	Prenu	128	–	50304
Dense	Prenu	64	–	8256
Dense	Softmax	56	–	3575

Pooling is an important technique used in CNN architecture. One of the reasons is that it reduces the memory size of the network by minimizing the links between the convolutional layers and speeds up the training of a CNN. For this purpose, we employ the max-pooling method.

The generated feature maps are fed into the dense layer. Neurons of this layer are fully connected to the neurons of the next layer.

The last layer predicts the class of the input frames of the video sign with a certain probability. Specifically, in the case of a multiclass classification problem, we use the softmax function to return the probabilities of all the classes, where the target class will have the highest probability. The mathematic expression of the softmax function is:

$$f(x_i) = \frac{\exp(x_i)}{\sum_k \exp(x_k)} \tag{2}$$

where x_i are the previous dense layer inputs utilized at each node of the softmax layer and K stands for the number of classes.

3 Experiments

3.1 MoSL Dataset

A Moroccan sign language (MoSL) instructional video [11] was used in the experiments. It is an educational material largely used in Morocco. Each word in the video is repeated a couple of times. We select 56 quotidian vocabularies in Moroccan sign language, with corresponding gesture videos to construct our set of data. A categorical pattern was used to represent each MoSL sign in the videos, which are labeled with an Arabic word. Three samples of frames in the dataset are presented in Fig. 2. Each video was made at 70 frames per second at a resolution of 100 × 100. The model extracts four segments for each video, and each segment has 10 sequences of images.

3.2 Model Training

The training process of the proposed model was run for up to 15 epochs with a batch size of threefold and fourfold cross-validation [13]. To measure the fulfillment of the proposed system, we report the precisions and accuracies over the fourfolds. Hence, we choose the cross-entropy cost function [15] and Adam [16] as the optimization algorithm at 0.001 learning rate. The model is trained on a machine equipped with an Intel Xeon 2.30 GHz CPU and an NVIDIA T4 GPU with 12 GB of memory.

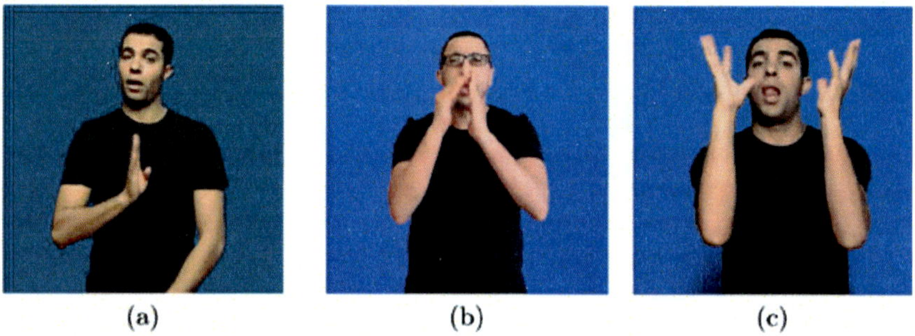

Fig. 2 Three examples of frames in MoSL dataset. **a** "Father," **b** "Address," and **c** "World"

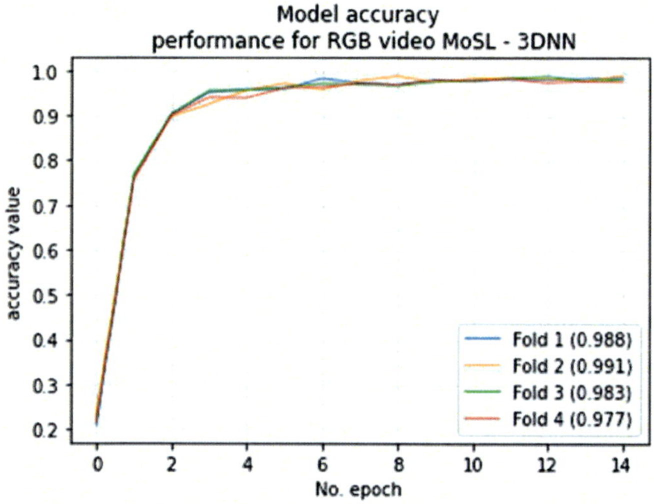

Fig. 3 Accuracy plots of training and validation data

3.3 Experiments

The training and validation accuracy of our model runs four folds on the MoSL dataset as shown in Fig. 3. We can see that the accuracy increases during training for both train and validation datasets without any overfitting phenomenon. Figure 4 shows that the learning loss decreases which indicates that the model performance is growing over time.

Table 2 contains the accuracies and losses values of the classification. The train accuracy grades improved from 99.60%, 97.32%, 96.03%, and 98.90% for the first, second, third, and fourth folds, respectively. However, we found that the lowest performance of 0.011 point loss occurs at fold 1, followed by fold 5 with 0.058, fold 3 with 0.259, and fold 4 with 0.314.

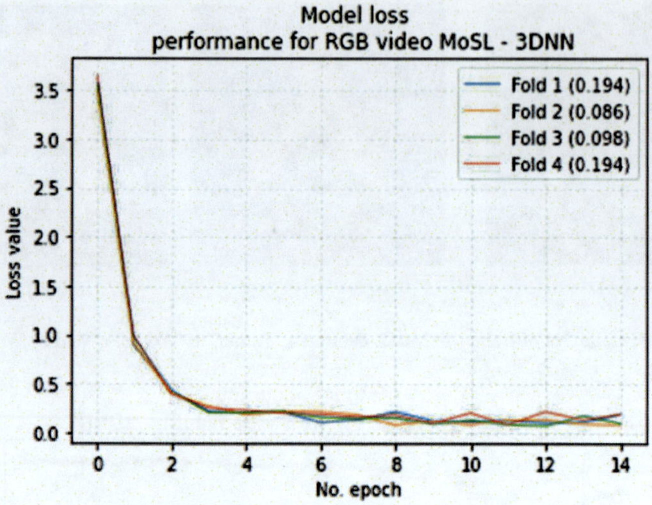

Fig. 4 Loss plots of training and validation data

Table 2 Obtained results of the proposed model over the four folds

Folds	Loss	Accuracy
1	**0.011**	**99.60%**
2	0.314	97.32%
3	0.259	96.03%
4	0.058	98.90%
Average	**0.160**	**97.96% (± 1.38)**

Table 3 Obtained result and comparison with stat of the art

A comparative approach	Accuracy (%)
Baccouche et al. [17]	87.9
Hayani et al. [10]	90.02
Jie et al. [18]	94.20
Latif et al. [19]	97.60
Our model	**99.60**

The performance obtained by our proposed model compared with other systems is given in Table 3. We have chosen for this comparison the fold 1 performance as it is the highest score obtained. It is clear that our proposed model achieves the best accuracy with a variety of classifiers from the state of the art.

4 Conclusion

The Moroccan sign language is the official sign language used by hearing-impaired persons and family members in Morocco. In this work, we proposed a convolutional neural network model for MoSL automatic recognition, and its performance was evaluated on word video datasets. It has significantly improved the MoSL recognition accuracy of some important works already existing in the literature.

References

1. Sandler W, Diane L-M (2006) Sign language and linguistic universals. Cambridge University Press. https://doi.org/10.1017/CBO9781139163910
2. Abdel-Fattah Mahmoud A (2005) Arabic sign language: a perspective. J Deaf Stud Deaf Educ 10(2):212–221. https://doi.org/10.1093/deafed/eni007
3. Clayton V, Ceil L (200) Linguistics of American sign language: an introduction. Gallaudet University Press
4. Rachel S-S, Bencie W (1999) The linguistics of British sign language: an introduction. Cambridge University Press
5. Lucas C (ed) (2001) The sociolinguistics of sign languages
6. Ellen PL, Jake A, Lewis C (2013) The leap motion controller: a view on sign language. In: Proceedings of the 25th Australian computer-human interaction conference: augmentation, application, innovation, collaboration (OzCHI '13). Association for Computing Machinery, New York, USA, pp 175–178. https://doi.org/10.1145/2541016.2541072
7. Aliyu S, Mohandes M, Mohamed D, Badran S (2016) Arabie sign language recognition using the Microsoft Kinect. In: 2016 13th international multi-conference on Systems, Signals & Devices (SSD), pp 301–306. https://doi.org/10.1109/SSD.2016.7473753
8. Abhijith BK, Nair Anoop G, Deepak RK, Krishnan A, Nandi VHR (2016) Smart gloves for hand gesture recognition: sign language to speech conversion system. In: 2016 international conference on Robotics and Automation for Humanitarian Applications (RAHA), pp 1–6, https://doi.org/10.1109/RAHA.2016.7931887
9. Sara B, Rini A, Jimoh ESM, Shafie Amir A (2011) Vision-based hand posture detection and recognition for sign language—a study. In: 2011 4th International Conference on Mechatronics (ICOM), pp 1–6. https://doi.org/10.1109/ICOM.2011.5937178
10. Salma H, Mohamed B, Othmane EM, Mustapha K (2019) Arab sign language recognition with convolutional neural networks. In: International Conference of Computer Science and Renewable Energies (ICCSRE), pp 1–4. https://doi.org/10.1109/ICCSRE.2019.8807586
11. Abdelbasset B, Mohamed B, Ayoub E, Othmane EM, Mustapha K (2021) Isolated video-based Arabic sign language recognition using convolutional and recursive neural networks. Arab J Sci Eng. https://doi.org/10.1007/s13369-021-06167-5
12. El Jaafari I, Ayoub E, Said C (2021) Parametric rectified nonlinear unit (PRenu) for convolution neural networks. SIViP 15:241–246. https://doi.org/10.1007/s11760-020-01746-9
13. Tadayoshi F (2011) Estimation of prediction error by using K-fold cross-validation. Stat Comput 21(2):137–146
14. Bencherif Mohamed A, Mohammed A, Mekhtiche Mohamed A, Mohammed F, Mansour A, Hassan M, Muneer A-H, Hamid G (2021) Arabic sign language recognition system using 2D hands and body skeleton data. IEEE Access 9:59612–59627. https://doi.org/10.1109/ACCESS.2021.3069714
15. Kline Douglas M, Berardi Victor L (2005) Revisiting squared-error and cross-entropy functions for training neural network classifiers. Neural Comput Appl 14:310–318 (2005). https://doi.org/10.1007/s00521-005-0467-y

16. Kingma Diederik P, Jimmy B (2014) Adam: a method for stochastic optimization. arXiv preprint arXiv:1412.6980
17. Moez B, Franck M, Christian W, Christophe G, Atilla B (2011) Sequential deep learning for human action recognition. In: Salah AA, Lepri B (eds) Human behavior understanding. HBU 2011. Lecture Notes in Computer Science, vol 7065. Springer, Heidelberg. https://doi.org/10.1007/978-3-642-25446-8_4
18. Huang J, Zhou W, Li H, Li W (2015) Sign language recognition using 3D convolutional neural networks. IEEE International Conference on Multimedia and Expo (ICME), pp 1–6. https://doi.org/10.1109/ICME.2015.7177428
19. Ghazanfar L, Nazeeruddin M, Roaa A, Rawan A, Jaafar A, Majid K (2020) An automatic Arabic sign language recognition system based on deep CNN: an assistive system for the deaf and hard of hearing. Int J Comput Digital Syst 9(4):715–724. https://doi.org/10.12785/ijcds/090418

Home Automation System and Quality of Life in Low-Income Households: A Systematic Review of the Literature from 2010 to 2021

Jenifer Diana Bustamante-Gonzales, Hugo Eladio Chumpitaz-Caycho, and Franklin Cordova-Buiza

Abstract The aim of this research was to determine what is known about the domotic system and the quality of life in low-income households, through a systematic review between 2010 and 2021. The study made it possible to review important databases such as Scielo, Scopus, EBSCO, and ScienceDirect. The search combination used "domotic system"; "quality of life"; "domotics"; "technology"; "automation". Finally, 30 empirical articles in English and Spanish were systematized. In the analyzed literature, it is evident that the lack of knowledge in the concept of home automation, added to the low investment of assets for research and generation of proprietary technology, has caused a slow inclusion of home automation in different places, which is alarming since it includes important issues such as comfort, security, and energy savings that are applied to homes, buildings, and shopping centers. It was possible to reach the conclusion that the use of home automation in homes would include a better living condition in low-income households, oriented to the reduction of duplicated expenses or the efficient use of resources.

Keywords Quality of life · Home automation system · Low resources

1 Introduction

The systematization of empirical articles allows us to recognize important literature on the variables under study. In this sense, the present study seeks to determine what is known about the domotic system and the quality of life in low-income households. Brush et al. [1] point out that there are many visions of smart homes but that for more

J. D. Bustamante-Gonzales (✉) · H. E. Chumpitaz-Caycho
Engineering Faculty, Universidad Privada del Norte, Lima, Peru
e-mail: n00055970@upn.edu.pe

H. E. Chumpitaz-Caycho
e-mail: hugo.chumpitaz@upn.pe

F. Cordova-Buiza
Research and Innovation Department, Universidad Privada del Norte, Lima, Peru
e-mail: franklin.cordova@upn.edu.pe

X.-S. Yang et al. (eds.), *Proceedings of Seventh International Congress on Information and Communication Technology*, Lecture Notes in Networks and Systems 447,
https://doi.org/10.1007/978-981-19-1607-6_37

than three decades this technology has not been implemented as planned because four main factors were found to impede the success of this new proposal including high cost, inflexibility, manageability, and security conflict. In this regard [2] indicate that around the world every year a great amount of energy is consumed between houses and buildings, which generate a negative effect on the environment; due to the exposure of electromagnetic radiation, it would cause great risks in the health of the human being. The purpose is to design a domotic system with a basic element called Arduino that will decrease electromagnetic radiation. An Android application to serve the user in real time should show a list of devices related to the house so that they can connect with each other and perform the process that the user wants. Given that low-income people do not have enough money to be able to access this home automation system, which in several countries around the world is quoted at a very high price, the aim is to automate the processes that humans perform manually and at no cost to vulnerable people. In this sense [3] indicates that the use of an Android operating system (OS) phone to control home appliances generates time savings and convenience, since the design is composed of an Arduino Mega board and is based on Bluetooth which the phone replaces a remote control for the management of household appliances. For this purpose, Alrumayh and Bhattacharya [4] indicates that home voice assistants are increasingly used to control smart devices through some commands when speaking, mentioning that the Canvas system is a good alternative as context recognition for voice assistants designed for smartphones. Given the positions of the different authors, the objective was to determine what is known about the home automation system and the quality of life in low-income households, through a systematic review between 2010 and 2021. The research question was: What is known about the home automation system and the quality of life in low-income households between 2010 and 2021?

2 Methodology

A systematic review was carried out based on the adaptation of the prism protocol [5]. For the search combination, we used "domotic system", "quality of life", "domotics", "technology", "automation", which allowed a broader search in relation to the topic of study. The criteria for choosing the articles were domotic system and quality of life in low-income households, open access, articles, in English and Spanish language and published between the years 2010 and 2021. Finally, the databases Scopus, Scielo, Ebsco, and ScienceDirect were considered (Fig. 1).

3 Results

A total of 30 empirical articles were obtained. Of which 6 are from the Scopus database, 8 from the Ebsco platform, 7 from Scielo and ScienceDirect, a total of

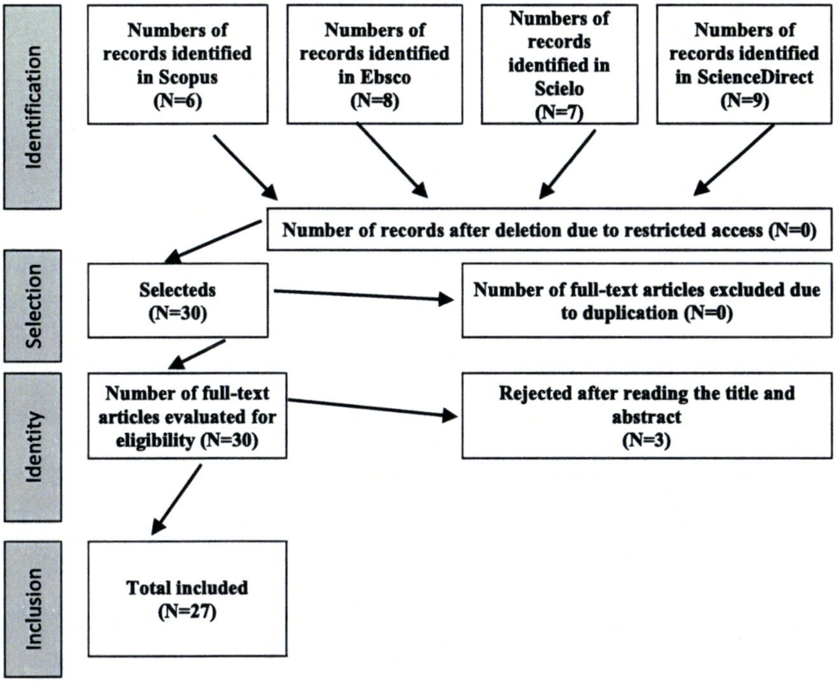

Fig. 1 Flowchart for document search and selection

9 research articles. Applying the exclusion criteria, 27 articles were left, discarding those unrelated to the topic and summary, research articles from other databases were not included, and finally, the 27 articles that best answered the research question were used. On this basis, the results obtained are shown in different tables and figures, in a quantitative and qualitative manner, which allow us to achieve the research objectives (Table 1).

The purpose of the databases is to gather academic production on different areas of knowledge, the indexed scientific journals. Figure 2 shows the databases that were selected; 30% of articles were from ScienceDirect, followed by 27% from Ebsco, 23% from Scielo, and finally Scopus with 20%, respectively.

Table 2 and Fig. 3 show the years of publication, with a total of 11 years, the year that was published the most is 2014 with a total of 6 articles and 22%, in second place the year 2016 with 15%, in third place follows the years 2012, 2018 with 11% each, respectively; in fourth place the year 2010 with 8%, in fifth place the years 2013, 2015, 2020 with 7%; finally, the remaining years with 4% each one, respectively.

On the other hand, answering the research question, what is known about the home automation system and the quality of life in low-income households between 2010 and 2021? It is mentioned that there are a number of people who are afraid of change because the primitive users have mentalized to do things manually, but the adaptation of them to the new technology is sought through minimal changes so that

Table 1 Selected articles in the systematization

Name of article	Journal
Alexa virtual assistant used as an interaction tool for climate monitoring in smart homes using Raspberry Pi and DarkSky APIs	Iberian Journal of Information Systems and Technologies
Implementation of an energy monitoring and domotic control system based on Internet of Things technology	Research and Development
Research of a prototype of a low-cost home automation system	Universidad de Antioquia Engineering School Magazine
Internet of Things using Arduino for home power management	EAN Magazine
Low-cost domotic control system: Support for environmental power generation in Colombia	Tecnura
Mobile device-driven home automation prototype for people with disabilities	International Journal of Combinatorial Optimization Problems and Informatics
Simulation performed on data for the estimation of energy consumption in a smart home	Article
Home Automation: A Design Process for Socially Beneficial Homes	EAN Magazine
Saving electrical energy with a simple home automation system	Academia Journals International Research Congress
Design and implementation of home automation systems based on the Arduino platform for protection, welfare and energy management	Academia Journals International Research Congress
Home Automation System based on ZigBee Technology for Energy Saving for Residential Use	CISTI (Iberian Conference on Information Systems and Technologies/Conferência Ibérica de Sistemas e Tecnologias de Informaçión) Proceedings
The factor to dignify social housing spaces lies in Domotics	Between Science and Engineering
Human voice recognition applied to human voice recognition applied to home automation	Ingenium
Activation of functions in intelligent buildings using voice commands from mobile devices	Engineering Article Research and Technology
Zigbee-based automation systems for homes with smart sensor deployment	Materials Today Magazine
A scalable, cloud-based, Android-compatible home automation system	Computers and Electrical Engineering Article

(continued)

Table 1 (continued)

Name of article	Journal
Design, Specification, and Implementation of a Distributed Home Automation System	Procedia Computer Science
Design of intelligent building control with a view to the needs of senior citizens	IFAC Conference on Programmable Devices and Embedded Systems
Security-based home automation	Procedia Technology
Optimal synthesis and operation of advanced energy supply systems for standard and domotic homes	Energy conversion and management
Design of an IoT home automation system using the MQTT protocol	International Journal of Advanced Trends in Computer Science and Engineering
Domotics in nature: Challenges and opportunities	Conference on Human Factors in Computing Systems—Proceedings
Bluetooth-based home automation system using an Android phone	Jurnal Teknologi
Home Automation: Patent Analysis	Domotic Technology Article
Multiple advantages and security home automation system	European Modelling Symposium on Computer Modelling and Simulation
A domain for ontology-based diagnosis of irregularities in smart homes	Procedia Computer Science
Automatización doméstica inteligente activada por voz (IVA)	MSL Academic Endeavors

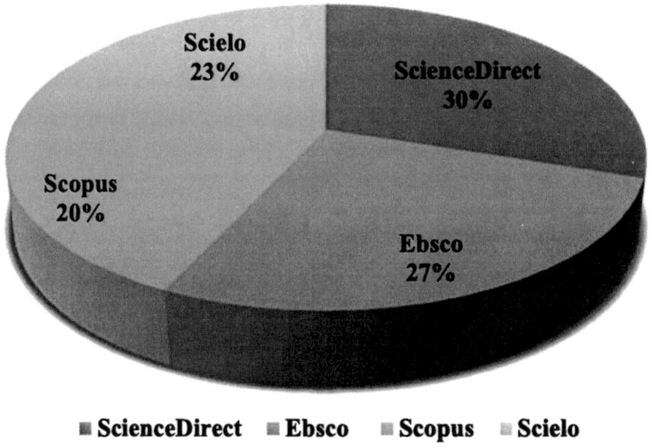

Fig. 2 Articles by database

Table 2 Year of publication

Years	Number	Percentage (%)
2010	2	8
2011	1	4
2012	3	11
2013	2	7
2014	6	22
2015	2	7
2016	4	15
2018	3	11
2019	1	4
2020	2	7
2021	1	4
Total	27	100

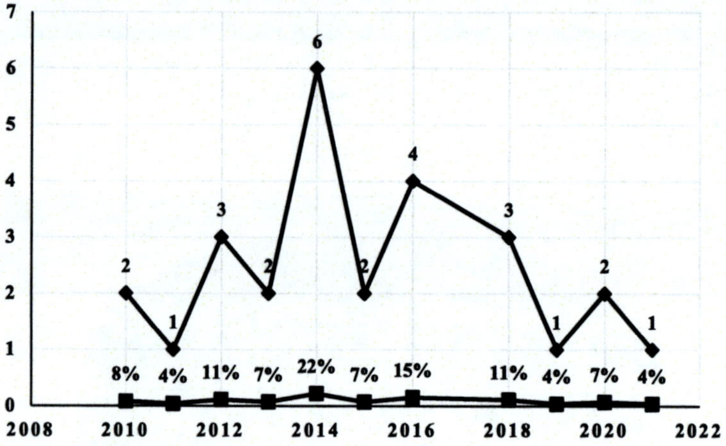

Fig. 3 Year of publication

the user can easily interact with the home automation system [1]. Therefore, in order to [6] indicate that every year software products are more innovative which at the same time the complexity increases, they suggest that Domain Analysis would be of great help for software developers, home automation is a developing domain that for many engineers is difficult to understand the interaction between commonalities and domain variations. The intention is to be able to support the feature modeling technique for the correct domain analysis to ultimately result in efficiency in the implemented system.

4 Discussion

Barrera et al. [7] agree that the lack of knowledge in the concept of home automation, the low investment of assets for research, and generation of proprietary technology has caused a slow inclusion of home automation in different places, which is alarming because it includes important issues such as comfort, security, and energy savings that are applied to homes, buildings, and shopping centers. In this sense, a great number of people believe that the use of home automation in their homes would improve their living conditions, and they see it as a trade of comfort, not as an element of reduction of duplicated expenses or the effective benefit of resources [8]. However, the progress of home automation has increased thanks to the use of Internet of things capabilities that provide emerging technologies that allow automating or tracking various tasks at home [9].

5 Conclusions

It was determined on domotic system and quality of life, through a systematic review between the years 2010 and 2021, that the use of domotics in homes would include a better living condition in low-income households, oriented to the reduction of duplicated expenses or the efficient use of resources. Finally, thanks to this research it was determined that there are few studies in relation to the home automation system and the quality of life in homes, which constitutes a rather slow inclusion of technology, wasting the great benefits it offers as comfort, security, and especially the savings it can generate in homes.

References

1. Brush B, Bongshin L, MahajanR, Agarwal S, Saroiu S, Dixon C (2011) Home automation in the wild: challenges and opportunities, pp 1–2
2. Chioran D, Valean H (2020) Arduino based smart home automation system. Int J Adv Comput Sci Appl (IJACSA) 11(4)
3. Hisham AAB, Ishak M, Teik C, Mohamed Z, Idris N (2014) Bluetooth based home automation system using an android pone. Jurnal Teknologi
4. Alrumayh O, Bhattacharya K (2020) Model predictive control based home energy management system in smart grid. In: Electrical power and energy conference
5. Urrútia G, Bonfill X (2010) Declaración Prisma: Una propuesta para mejorar la publicación de revisiones sistemáticas y metaanálisis. Medicina Clínica
6. Patil P, Aparna R, Chandrasekaran K, Rathnamma M, RamanaV (2020) On feature models of home automation systems towards smart sensing. In: IEEE International conference on communication and signal processing
7. Barrera M, Londoño N, Carvajal J, Fonseca A (2014) Análisis y diseño de un prototipo de sistema domótico de bajo costo. Revista Facultad de Ingeniería Universidad de Antioquia
8. Quintana B, Pereira V, Vega N (2014) El factor para dignificar espacios de vivienda social se encuentra en la Domótica. Entre Ciencia e Ingeniería

9. Calvopiña A, Tapia F, Tello-Oquendo L (2020) Uso del asistente virtual Alexa como herramienta de interacción para el monitoreo de clima en hogares inteligentes por medio de Raspberry Pi y DarkSky API. RISTI—Revista Ibérica de Sistemas y Tecnologías de la Información

Detecting Termites in Wood Structure Using Internet of Things Approach

Nur Zaimah Ahmad, Lutfil Hadi Zaifri, Bazilah A. Talip, and Aznida Abu Bakar Sajak

Abstract Detecting termites in wood structures is complex, and the most available detection methods are potentially damaging to property. The goal of this study is to develop a proof-of-concept termite detection system for an indoor environment. Thermal imaging and microwave radar sensors are used to detect the presence of termites, while a mobile application is used to view the termites' status using a heat map and a wave pattern. Testing is carried out based on the reliability and efficiency of the two methods for detecting termites. The results show that the thermal camera can detect hot and cold spots on the wooden surface up to 15 cm, while the microwave radar sensor can detect termite movement inside the wood up to 3 cm.

Keywords Termites · Thermal imaging · Microwave sensor · Mobile application

1 Introduction

Termites are also known as white ants and are classified as eusocial insects under the taxonomic rank of order, Isoptera. Termites consume a diverse range of organic waste, including dry grass, decaying leaves, animal dung, hummus, and living or dead wood [1]. Termites are abundant in tropical areas, and they cause significant damage [2]. As a result, the condition of residential areas provides termites with an abundant food source. Due to the ecological adaptation of the termites, they have become a threat and create severe problem to plants and building structures.

N. Z. Ahmad (✉) · L. H. Zaifri · A. Abu Bakar Sajak
Computer Engineering Technology Section, Malaysian Institute of Information Technology, Universiti Kuala Lumpur, 50250 Kuala Lumpur, Malaysia
e-mail: nzaimah@unikl.edu.my

A. Abu Bakar Sajak
e-mail: aznida@unikl.edu.my

B. A. Talip
Informatics and Analytics Section, Malaysian Institute of Information Technology, Universiti Kuala Lumpur, 50250 Kuala Lumpur, Malaysia
e-mail: bazilah@unikl.edu.my

X.-S. Yang et al. (eds.), *Proceedings of Seventh International Congress on Information and Communication Technology*, Lecture Notes in Networks and Systems 447,
https://doi.org/10.1007/978-981-19-1607-6_38

One type of termite is the subterranean termite. This termite lives in colonies and enters structures in search of food and moisture. Termites are frequently discovered in structural lumber, furniture, utility poles, boats, and dead tree limbs. Liquid termiticides as chemical barriers in the soil, physical barriers, and termite bait strategies are used to prevent and control subterranean termites [3]. Termites are difficult to detect, and many detection methods are harmful to property or disturb the termites, which may interfere with the remedial treatment. Knocking on wood to hear a hollow sound caused by subterranean termites feeding on the wood is a traditional method. This is an inefficient and time-consuming method.

Several research related to termites' detection have been reviewed. According to Farkhanda [4], thermal imaging can be used as a non-destructive and quick method of detecting termites in trees and buildings when compared to traditional methods such as knocking and drilling in wood. Thermal imaging generates images based on thermal energy rather than light; the resulting image appears as an X-ray with heat regions. While [5] emphasized that termite infestations produce irregular heat patterns that a thermal camera can detect. When termites enter a structure, they expel heat from their digestive system as carbon dioxide. Aside from that, the mud tubes they build have a high moisture content. These cause thermal changes on the surface of a structure's walls, ceilings, floors, and concealed areas, resulting in irregular heat patterns.

Researchers [6] have developed an advanced sensor that can send an alarm when termite activity is detected inside the wood. The alarm is delivered via SMS. The laboratory tests revealed that the microcontroller was capable of detecting individual termites and the movement of the testing colonies under laboratory conditions. Small dust or dirt accumulation in the light emitter or light detector is expected to make the sensor insensitive. Furthermore, termites are known to leave mud trails in the sensors in order to obstruct the LED emitter. As a result, the light emission served as a source of attraction for the termites rather than a deterrent. Microwave technology, according to Taravati [7], is a non-destructive method for detecting dry wood termites in structures. Microwave technology has been used to create a device called Termatrac. When detecting insect movement, the device connects to the mobile app via Bluetooth and displays a line graph. The experiment found that the device correctly identified termite presence or absence in 90% of cases. In wet environments, however, they are not reflected and thus are not detected. Aside from that, dry wood termites can be reliably detected at a maximum depth of 5 cm into the wood.

In this paper, we want to introduce a new method for detecting termites that is reliable, efficient, and non-destructive. We used a new approach that combines thermal imaging with microwave sensor technology because traditional methods rely primarily on knocking on wood to hear a hollow sound caused by termites. The proposed method can detect termite infestations and determine termite presence. In addition, the collected data is sent to a cloud database and is accessible via a mobile application.

2 Materials and Method

The development of the termite detector prototype has been divided into two modules: the termite detector module and the mobile application module. The system architecture of a termite detector system is depicted in Fig. 1. The Teensy is linked to a thermal camera and a microwave radar sensor. The Teensy is used to connect the Raspberry Pi model 3 B+ to the Blynk server, where the microwave sensor data is uploaded, and the heat map is streamed to the Blynk application. This Blynk application is used to view the microwave's heat map and graph.

2.1 Termite Detector Module

The main module of this study is the termite detector module. This module is intended to detect termite infestations using thermal imaging and microwave sensors. The data will be sent to the cloud, and a mobile application will be used as a client. On mobile applications, users can view the status of the termite's existence and location.

As shown in Fig. 2, the process begins by connecting the mobile application to the termites detector. The thermal imaging sensor measures the temperature of the wood's surface. The Raspberry Pi model 3 B+ will then process the temperature and create a heat map of the wood. If a hot or cold spot appears on the heat map, the microwave radar sensor will be deployed. The data will then be uploaded to the Blynk server, and a line graph will be plotted by the Raspberry Pi model 3 B+ . If the line of the graph wave pattern is inconsistent, it indicates the presence of termites. Otherwise, it indicates that termites are not present.

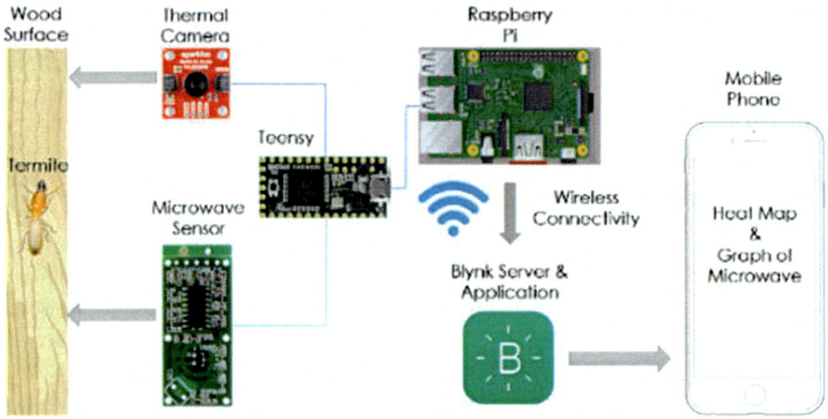

Fig. 1 System architecture of a termite detector

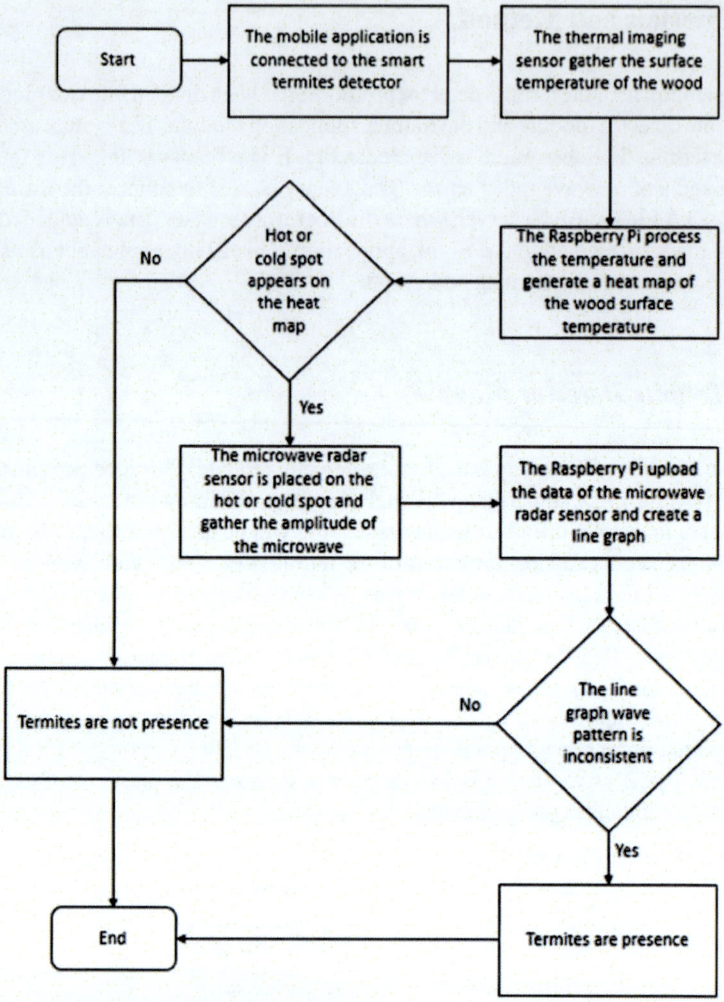

Fig. 2 Termites detector flowchart

The STD circuit design is shown in Fig. 3. The 3.3 V pin of Teensy 3.2 provides power to the sensors, which are the microwave radar sensor and the thermal camera, which are connected to the VIN and 3–6 V pins, respectively. The GND pin is the ground pin, and the AGND pin is the analogue ground pin. To complete the circuit, the ground pins of Teensy 3.2 are connected to the sensors. Furthermore, a serial data (SDA) interface is used to transmit data, and a serial clock (SCL) interface is used to synchronize all data transfers. It is a serial communication protocol that the thermal camera uses to send and receive data from the Teensy. Finally, the analogue pin of

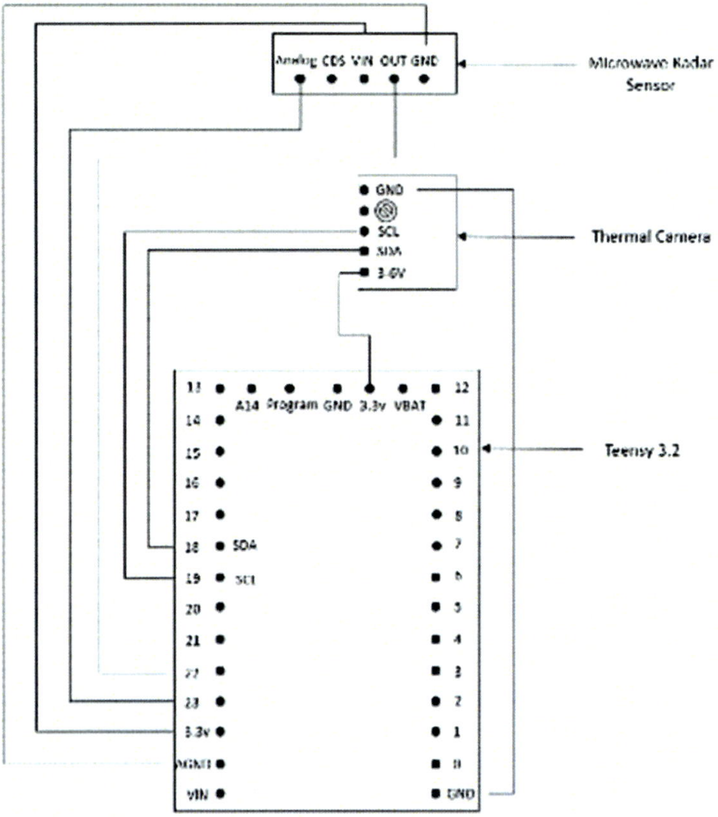

Fig. 3 Termites detector circuit design

the microwave radar sensor sends a signal to Teensy pin 23 with a voltage range of 0–3.3 V. Figure 4 shows a prototype for a termite detector.

2.2 *Mobile Application Module*

The mobile application module serves as a client, allowing users to view the status of termite existence. This module makes use of the Blynk platform, which sends all data to the cloud for data monitoring. The mobile application's layout is divided into three sections. First, the heat map of the thermal image is displayed. Second, it displayed a microwave radar sensor graph. The amplitude of the microwave radar sensor is shown on the Y-axis, and the current time is shown on the X-axis. Third, it displayed the graph of the microwave radar sensor's current amplitude. Figure 5 shows the mobile application for the termite detector.

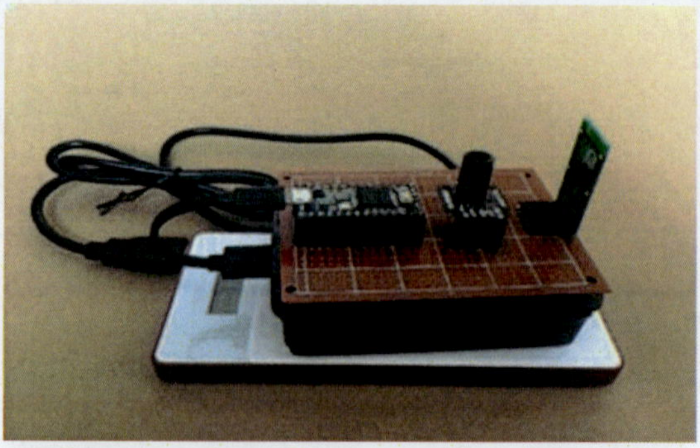

Fig. 4 Prototype of termites detector

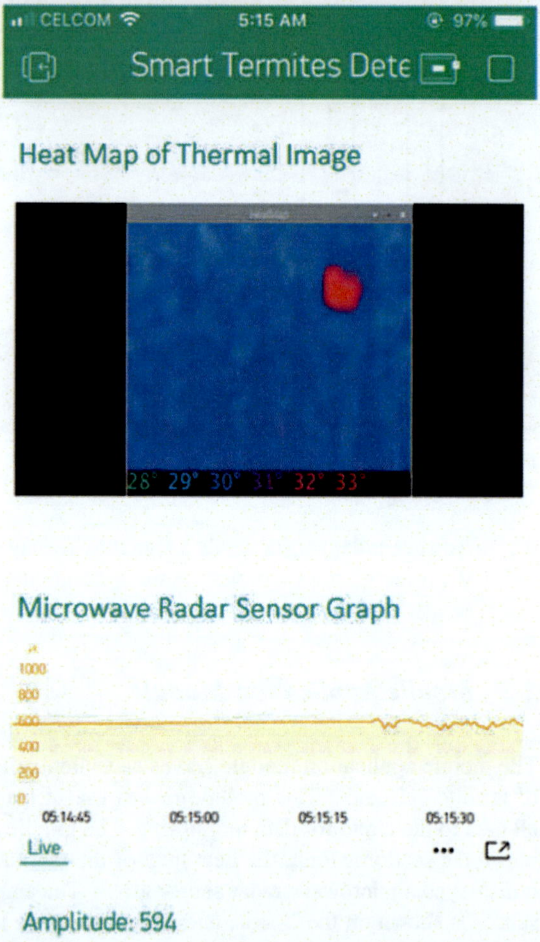

Fig. 5 Termites detector mobile apps

3 Result and Discussion

Reliability and functionality tests have been conducted to test the effectiveness of termites' detector. The tests have been done in two different environments to replicate the termite infestation indoor. A reliability test is adopted to measure the depth of detection using a microwave radar sensor. The depth of detection is based on the thickness of the wood surface. The result has shown that the thickness of the wood surface is increased by 0.5 cm until it reaches its maximum depth of detection. Functionality test is used to test the reliable distance of thermal camera in detecting the termites. Wooden furniture is used as a test object to investigate the existence of the termites. The thermal camera is placed on the surface of the furniture at 5 cm, and it is incremented until no detection occurs.

3.1 Microwave Radar Sensor

The termite is placed in a plastic container measuring 6.0 cm × 4.3 cm. A 0.5 cm thick wood surface with a 5 cm increment is placed on top of the plastic container. For better detection and results, the termites detector prototype is placed directly above the termites and wood surface. Figure 6 depicts the detection frequency, which indicates the presence of termites.

As shown in Table 1, seven different wood thicknesses were tested in two different scenarios: container without termites and container with termites. The absence of termites in the plastic container is consistent across 0.5–2.5 cm of wood thickness. It demonstrates that when termites are not present, there is no movement. As a result, it generates a consistent wave pattern. However, in the case of a plastic container with termite presence, the wave pattern is inconsistent from 0.5 to 3.5 cm. This indicates the presence of termites, and as a result, there is movement in the plastic container, resulting in an inconsistent wave pattern.

Fig. 6 Termites with wood thickness of 1 cm

Table 1 Data gathered by microwave radar

The thickness of the wood (cm)	Plastic container without termite presence		Plastic container with termite presence	
	Graph wave pattern	Peak amplitude	Graph wave pattern	Peak amplitude
0.5	Consistent	800	Inconsistent	900
1.0	Consistent	750	Inconsistent	824
1.5	Consistent	700	Inconsistent	814
2.0	Consistent	666	Inconsistent	697
2.5	Consistent	640	Inconsistent	650
3.0	Inconsistent	640	Inconsistent	640
3.5	Inconsistent	640	Inconsistent	640

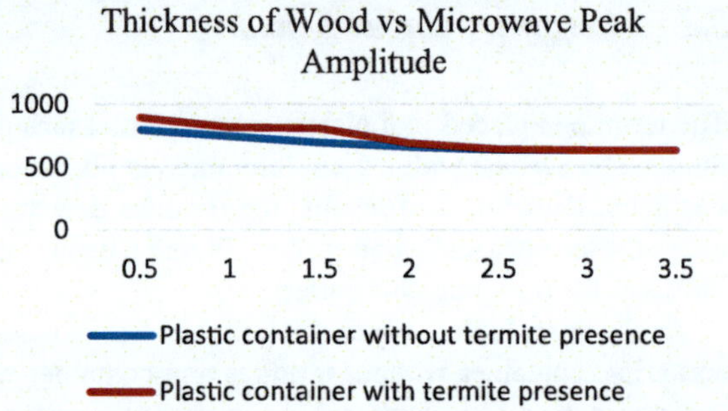

Fig. 7 Microwave radar relationship graph

This study discovered that the thicker the wood, the lower the microwave peak amplitude, as illustrated in Fig. 7. However, at a thickness of 3.0–3.5 cm, the amplitude reading is equivalent and does not distinguish between a plastic container with or without termite presence. This leads to the conclusion that the reliable thickness of a microwave radar sensor for detecting termite presence is up to 2.5 cm.

3.2 *Thermal Imaging Testing*

As shown in Fig. 8, the termite detector was placed 5 cm away from the wooded furniture to test the accuracy of heat images captured by the thermal camera, as shown in Fig. 9. Several tests are also being carried out to assess the thermal camera's ability to capture data on the wooden furniture. Table 2 demonstrates the presence of a cold spot on the wooden furniture at 5–10 cm. Because it is colder than the surrounding temperature, the wooden furniture may have termite mud trails. Furthermore, the presence of a hot spot on the wooden furniture is detected and can be seen at a distance of 10 cm. The hot spot is caused by the termite releasing carbon dioxide

Fig. 8 Existence of termites on the wooden furniture

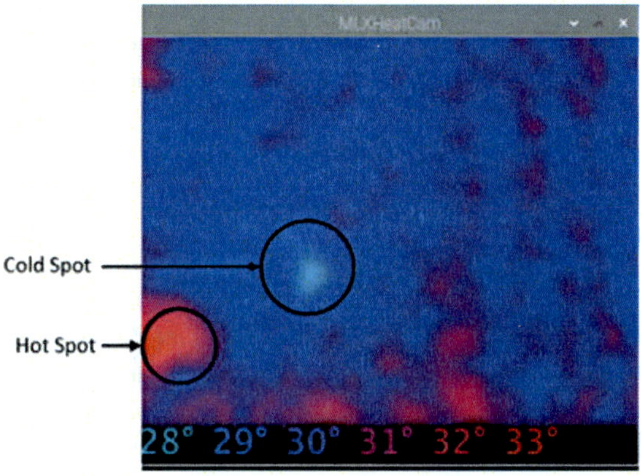

Fig. 9 Heat map shows the presence of termites

Table 2 Heat map reading based on distance of the thermal camera to the wooden furniture

The distance of the thermal camera between the wooden furniture (cm)	Temperature		Presence of	
	Lowest	Highest	Cold spot	Hot spot
5.0	26	31	Yes	No
10.0	28	33	Yes	Yes
15.0	29	29	No	No

gas, which is hotter than the surrounding area. The cold and hot spots, however, are not present at 15 cm.

This study discovered that the distance between the thermal camera and the wooden furniture has a significant influence on thermal camera positioning (see Fig. 10). It demonstrates that the greater the distance between the highest and lowest temperatures, the smaller the difference between them. The thermal camera can detect a temperature difference of 6 °C at a distance of 5–10 cm. As a result, a cold and hot spot can be easily identified with the naked eye. However, the temperature difference at 15 cm is 0 °C, so the only temperature detected is 29 °C. This study

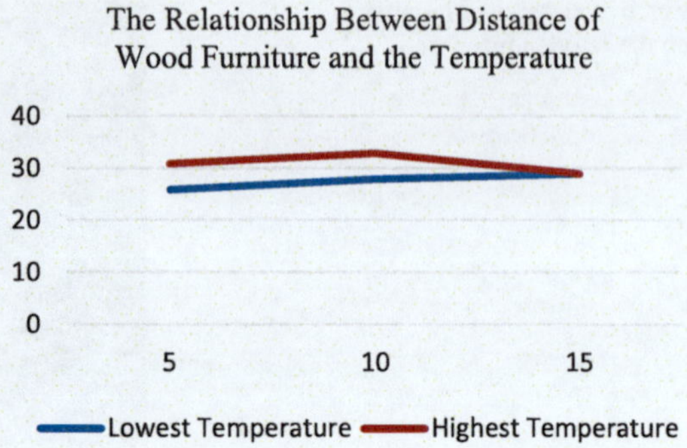

Fig. 10 Graph of the relationship between the distance of wood furniture and the temperature

discovered an intriguing finding: The reliable distance at which a thermal camera can capture data is 10 cm.

4 Conclusion

We highlighted in this paper that the termite detector is designed to detect termite infestations in an indoor environment. This study found that adding two features to the termite detector, a thermal camera and a microwave sensor, can increase detection. A thermal camera is used to generate a heat map, which is then displayed on a mobile application. The presence of termite mud trails is indicated by the cold spot on the heat map, while the hot spot indicates the presence of carbon dioxide. The microwave sensor detects termite movement beneath the wooden surface. The microwave sensor can detect a reliable thickness of wood of up to 2.5 cm. A wave pattern graph is displayed using a mobile application. If it is inconsistent and has a high amplitude, it indicates that termites are moving inside the wooden surface. The limitation of this study is that the Raspberry Pi 3 B+ cannot handle the processing power required to generate a heat map and stream it to the streaming server at the same time. As a result, the video streaming on the mobile application is delayed, and subsequently, it affects the performance of this system.

However, this preliminary research yields an interesting finding on the issues concerning termite detection were not addressed. Prototype development is required for future work to test the system's reliability and feasibility in detecting a larger surface area. Furthermore, using an artificial intelligence approach in detecting termite type may be beneficial in interpreting the data gathered in the production of the heat map and the microwave pattern.

References

1. Brossard M, López-Hernández D, Lepage M, Leprun JC (2007) Nutrient storage in soils and nests of mound-building trinervitermes termites in Central Burkina Faso: consequences for soil fertility. Biol Fertil Soils 43(4)
2. Ghaly A (2011) Termite damage to buildings: nature of attacks and preventive construction methods. Am J Eng Appl Sci 4(2)
3. Kuswanto E, Ahmad I, Dungani R (2015) Threat of subterranean termites attack in the asian countries and their control: a review. Asian J Appl Sci 8(4)
4. Farkhanda M (2013) Biosensors for termite control. IOP Conf Ser: Mater Sci Eng 51(1)
5. Rentoul M (2007) The practice of detecting termites with infrared thermal imaging compared to conventional techniques. In: InfraMation 2007 Proc, no ITC 121A
6. Oliver-Villanueva JV (2012) Advanced wireless sensors for termite detection in wood constructions
7. Taravati S (2018) Evaluation of low-energy microwaves technology (termatrac) for detecting western drywood termite in a simulated drywall system

Performance Evaluation of Boosted 2-Stream TCRNet

Shah Hassan, Md Jibanul Haque Jiban, and Abhijit Mahalanobis

Abstract Target detection in infrared imagery is a particularly challenging problem due to the presence of terrain clutter. The TCRNet-2 CNN architecture was introduced to combat this issue and has been shown to perform better than conventional networks such as faster RCNN and YOLOv3. In this paper, we evaluate the performance of the Boosted 2-Stream TCRNet in detail (including robustness to range variations, performance under day and night conditions) and compare it with that of YOLOv5. A MWIR dataset released by DSIAC is used for training and testing the network. We also propose the MWIR target classifier that recognizes the 10 classes in the NVESD dataset and achieves an accuracy of 65.72% which is state-of-the-art to date.

Keywords TCRNet · Infrared images · NVESD dataset · Target detection · Target classification

1 Introduction

Humans have the ability to detect, locate, and classify objects in standard environments rather easily. The process is fast and accurate, and it allows us to do all sorts of day-to-day tasks from interacting with our surroundings to any other complex task. Although current detection algorithms have shown great results in many vision-related tasks, there is still a long way to go in many other tasks. Detecting targets at distant ranges in a highly cluttered environment in infrared images is one of such tasks. It is even sometimes very difficult for humans to detect targets in such challenging environments. Researchers have been working on solving this task for

S. Hassan (✉) · M. J. H. Jiban · A. Mahalanobis
University of Central Florida,Orlando FL32816, USA
e-mail: shahhassan@knights.ucf.edu

M. J. H. Jiban
e-mail: jibanul@knights.ucf.edu

A. Mahalanobis
e-mail: amahalan@crcv.ucf.edu

X.-S. Yang et al. (eds.), *Proceedings of Seventh International Congress on Information and Communication Technology*, Lecture Notes in Networks and Systems 447,
https://doi.org/10.1007/978-981-19-1607-6_39

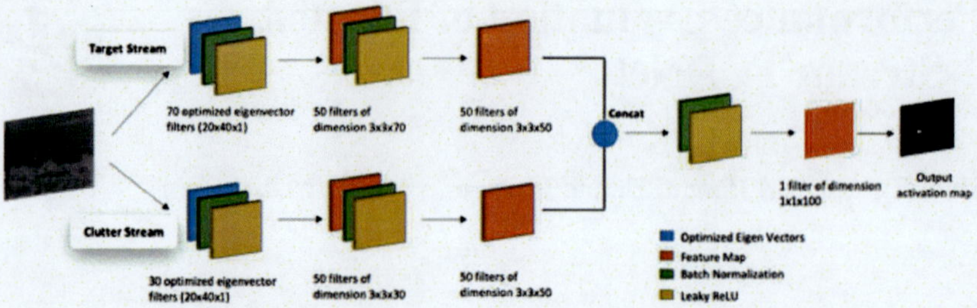

Fig. 1 Architecture of TCRNet-2 network [3]

quite some time [1, 2]; target detection of infrared imagery at low false alarm rates remains a challenging problem.

The TCRNet was introduced to specifically address the problem of finding targets in background clutter. A new loss function referred to as the *target-to-clutter ratio* (TCR) was defined as the ratio between the output energies produced by the network in response to target and clutter. The network also employs analytically derived filters in its first layer that optimally represent target and clutter. Using these fixed filters in the first layers imposes strong priors on the rest of the network, forcing the convolution kernels to be learned such that the TCR metric is optimized. This paper builds on the previous work by rigorously analyzing the performance of the TCRNet using DSIAC MWIR dataset to evaluate its performance at different times of day, its ability to detect targets at different ranges and to compare its performance with the state-of-the-art YOLOv5 object detection network.

2 Two-Stream TCRNet: A Review

Figure 1 exhibits the architecture of two-stream TCRNet [3]. Two-stream TCRNet (TCRNet-2) is similar to the original version of TCRNet [4]; however as shown in Fig. 1, TCRNet-2 has two separate channels to process the target and clutter information, which ensure the maximum discrepancy between the two sub-spaces, whereas the original TCRNet had only one channel. The number of filters (70 for the target stream and 30 for the clutter stream) is determined based on the dominant eigenvalues found using TCR metric. Two more convolution layers with fifty 3×3 filters are added to each stream, and then the output of these two streams is combined. One last convolution layer is added to get the final combined output. Batch normalization [5] and ReLU [6] are used in all layers. The local maxima in the output activation map are used to determine the target locations in the images.

Boosted TCRNet-2 Networks: A Boosted TCRNet model can improve the overall detection rate and reduce false alarm rate. In this process, the primary detector nominates the regions of importance (ROI) that may contain potential targets and

Fig. 2 Example of full-frame images in dataset.

the second network focuses only on the ROIs produced by the primary detector. The final detection score is found by adding the scores produced by the primary and secondary networks. Both the primary and secondary networks have the same architecture; however, the clutter training data is different. The primary detector is trained on target chips that are extracted from full-frame images using ground truth information and clutter chips that are randomly extracted from the same set of full-frame images. The second network uses the same target training chips, but the clutter chip set comprises only of *false positives* produced by the primary network. In essence, the clutter chip set for the second network is produced by applying the primary network to the images at ranges 4000m, 4500m, and 5000m images and extracting the false positive regions (Fig. 2).

3 Experiments and Dataset

TCRNet-2 is trained and tested on automated target recognition (ATR) database provided by DSIAC [7]. This dataset contains both visible and mid-wave infrared (MWIR) imagery of people and ten different vehicular (both civilian and military) targets. However, only MWIR imagery of vehicular targets is used to train and test the TCRNet-2 model. The ATR data were collected during both daytime and nighttime from a distance between 1 and 5 km with 0.5 km increment. The images are 640×514 in size. The dataset contains ground truth information of both target location and target class with much other useful information. This data can be downloaded from [7] which comes with a user reference guide that contains more details about the dataset.

3.1 Result: TCRNet-2 in Different Time Scenarios

The key performance metrics are the percentage of correct detections (P_d) and the number of false detections per square degree (FAR). Figure 3 shows the performance of TCRNet-2 in day, night, and day and night scenarios. It is clear that the nighttime

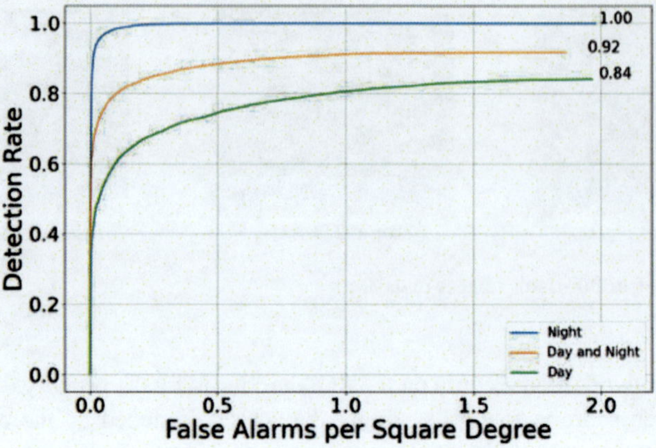

Fig. 3 ROC curves of boosted 2-stream TCRNet comparing day, night, and day and nighttimes

performance is better w.r.t both detection probability and false alarm rate. Specifically, at FAR of 2.0 TCRNet-2 is able to achieve P_d of 0.84 for daytime images and 1.00 for nighttime images. Similarly, at a FAR of 1.0, P_d is 0.8 for the daytime images, and 1.0 for the nighttime images. It can be clearly seen from the nighttime images that TCRNet-2 is able to quickly achieve the maximum P_d very quickly right after the FAS is 0.2. The reason why TCRNet-2 performance is better in the nighttime scenario is that the nighttime images have a lot less challenging clutter than that of the daytime images.

3.2 *Comparison with YOLOv5*

You Only Look Once (YOLO) [8] is a very popular method that does both detection and classification in one step. The latest version is YOLOv5 [9] that we used to compare with Boosted 2-Stream TCRNet for target detection in the cluttered environment.

YOLOv5 is fine-tuned on the NVESD dataset using the same training and testing protocols. The model is trained with batch size 8, image size 640 × 514, 300 epochs, and Adam optimizer with learning rate 0.001. We trained and tested YOLOv5 on Pytorch 1.8.1 using an NVIDIA GeForce RTX 2080 Ti—11 GB GDDR6 GPU.

Figure 4 shows performances of YOLOv5 compared to Boosted TCRNet-2. There are different sizes (s,m,l,x) of YOLOv5 models. We fine-tuned both sizes s and x. Among these two models, size s is found to perform better for this dataset. The maximum Yolov5 P_d is 0.62, whereas at the same false alarm rate the P_d for Boosted TCRNet-2 is around 0.78.

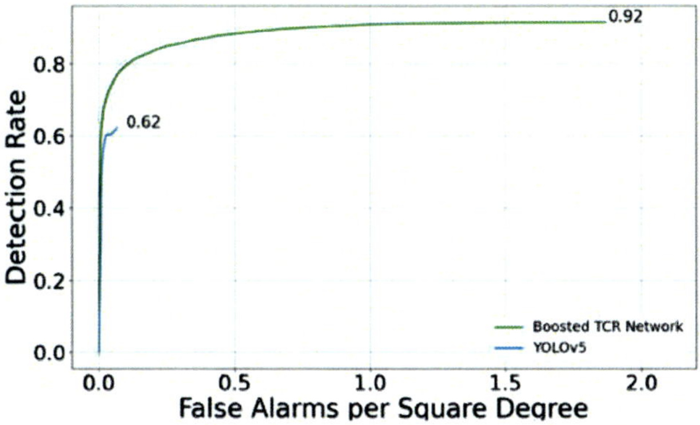

Fig. 4 ROC curves comparing performances of boosted TCRNet-2 and YOLOv5 in both day and night test images. The boosted TCRNet curve shows significantly better performance in terms of detection rate with a margin of 30% over YOLOv5

3.3 *Range Invariance*

Fixed Detection Window Analysis: TCRNet-2 was tested on resized images for 2500 m range with a detection window of 20 pixels radius. However, the manual resizing is not a desirable approach. Therefore, we trained multiple streams TCRNet-2 on different ranges. The first experiment was conducted with three streams: first stream with training chips of range 1000 m, second stream with training chips of range 1500 m, and the third stream for training chips of range 2000 m. The second experiment was conducted on 7 streams, with training chips of ranges 1000, 1250, 1500, 1750, 2000, 2250, 2500 m. The third experiment used three ranges, i.e., 1250, 1750, and 2250 m ranges.

Results for this experiment show that having training images of various ranges can help in achieving almost equal performance on the actual non-resized images. However, defining a detection window for every range is crucial.

Variable Detection Window:

As mentioned earlier, the testing images are manually resized for 2500 m range. However, the manual resizing of the test images is not desired. Therefore, we present a formula to vary the detection radius with respect to the range given in Eq. 1. As seen in Fig. 5, the formula provides a way to estimate the detection window with respect to the size of the target. When the range is 1 km the radius is 50; it decreases to 33.3 as the range increases to 1.5 km, and it keeps on decreasing as the range increases. The radius is 25 for 2000 m range, 20 for 2500 m range, 16.7 for 3000 m range, and 14.3 for 3500 m range. We found that the result without manually resizing the test data is comparable to the scaled images as shown in Fig. 6.

$$DetectionWindow = 2500/(Range) \times 20 \qquad (1)$$

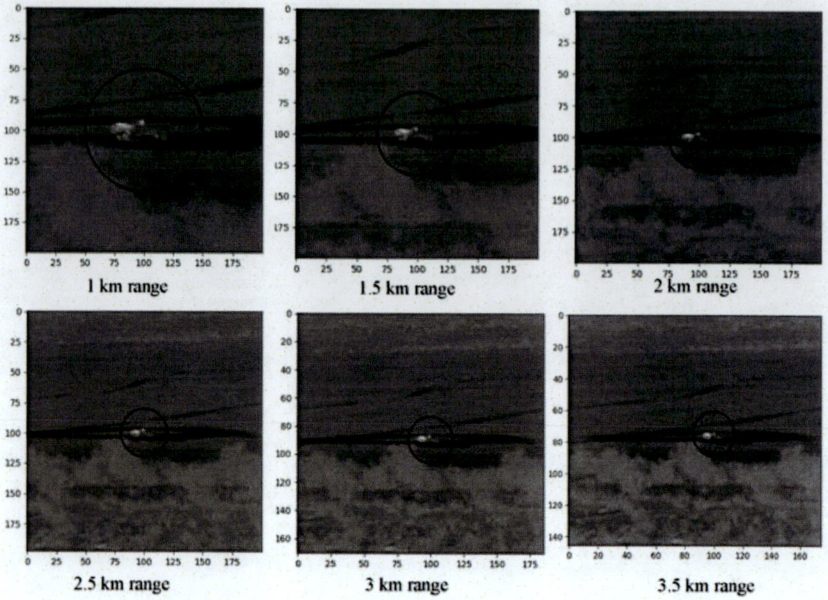

Fig. 5 Variable detection windows

Fig. 6 ROC curves with variable detection window

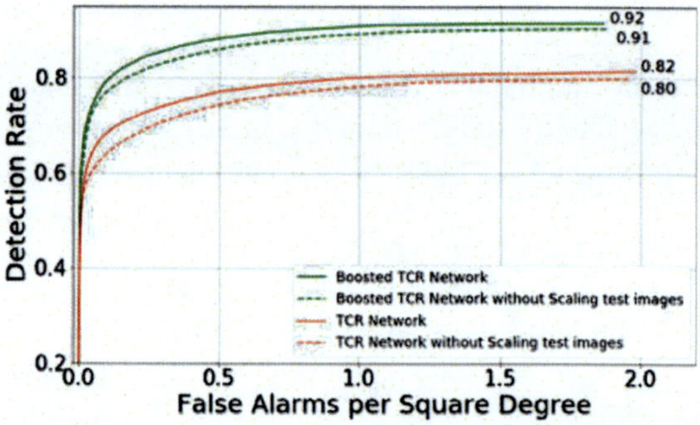

4 MWIR Target Classifier

In order to classify the kind of vehicle in the test images, we also propose a MWIR target classifier. The classifier consists of an input layer, followed by 5 convolution layers, followed by a fully connected layer, and a classification layer. Figure 7 depicts the architecture of the classifier. Each convolutional layer has the filter size of 3×3 and padding of 2. 65 filters are used in the first convolutional layer, 32 in the second layer, 64 in the third, and 32 filters are used in the 4th and 5th convolutional layer. Five convolutional layers are followed by a fully connected layer with ReLU as activation function which is followed by another fully connected layer followed by softmax as the activation function.

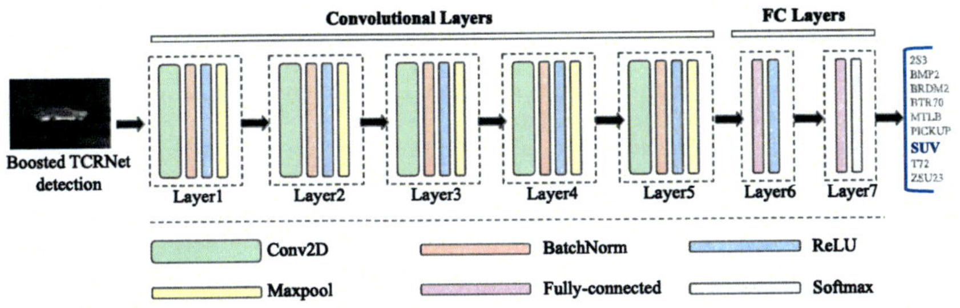

Fig. 7 Architecture of target classifier

We train the MWIR target classifier on 32×64 chips with vehicular target at the center. The chips are extracted from images of the ranges 1, 1.5, and 2 km. We test the classifier on Boosted TCRNet-2 detections in order to complete the pipeline. MWIR Target Classifier achieves 65.72% accuracy which is state-of-the-art on NVESD dataset.

5 Conclusion

This paper evaluates the performance of the Boosted 2-Stream TCRNet for target detection in challenging clutter terrain in MWIR images. First, it significantly outperforms well-known deep learning technique YOLOv5. We showed that the two-stream TCRNet is robust to range variations by testing it on unscaled test images. We also suggest to use a variable detection window ensuring that the targets at variable ranges are correctly detected. Moreover, we showed that a simple CNN classifier can achieve 65.72% accuracy on the NVESD dataset.

References

1. Ratches JA (2011) Review of current aided/automatic target acquisition technology for military target acquisition tasks. Optical Eng 50(7):072001
2. Gundogdu E, Koç A, Alatan AA (2017) Automatic target recognition and detection in infrared imagery under cluttered background. In: Target and background signatures III, vol 10432. International Society for Optics and Photonics, pp 104320J
3. Jiban MJH, Hassan S, Mahalanobis A (2021) Two-stream boosted TCRNet for range-tolerant infra-red target detection. In: IEEE International Conference on Image Processing (ICIP). IEEE, pp 1049–1053
4. McIntosh B, Venkataramanan S, Mahalanobis A (2020) Infrared target detection in cluttered environments by maximization of a target to clutter ratio (TCR) metric using a convolutional neural network. IEEE Trans Aerosp Electronic Syst 57(1):485–496
5. Ioffe S, Szegedy C (2015) Batch normalization: accelerating deep network training by reducing internal covariate shift. In: International conference on machine learning. PMLR, pp 448–456

6. Nair V, Hinton GE (2010) Rectified linear units improve restricted boltzmann machines. In: ICML
7. DSIAC: ATR Algorithm Development Image Database. https://dsiac.org/databases/atr-algorithm-development-image-database/
8. Redmon J, Divvala S, Girshick R, Farhadi A (2016) You only look once: unified, real-time object detection. In: Proceedings of the IEEE conference on computer vision and pattern recognition, pp 779–788
9. YOLOv5-5.0. Ultralytics. https://github.com/ultralytics/yolov5 (2020). Accessed 26 July, 2021

Design of a Cascaded Single-Phase Multilevel Inverter for Photovoltaic Applications

Darío Fernando Yépez Ponce, Héctor Mauricio Yépez Ponce, and William Manuel Montalvo López

Abstract Nowadays, obtaining alternative forms of electrical power is a topic of worldwide interest, since fact that dependence on non-renewable resources such as hydrocarbons, coal, and uranium is becoming more and more costly due to their depletion. An alternative solution to this problem is the generation of electric energy by means of the use of solar energy captured by solar panels and the use of multilevel converters for the conversion to AC, which is how it is normally used in the home and industry. In this paper, we start with the review of the state of the art of the cascade multilevel inverters (CHMLI) and the modulation techniques commonly used for these inverters. Subsequently, the single-phase CHMLI is designed and simulated using MATLAB/Simulink software. To improve the signal quality, Butterworth filters were used for both the input and output of the CHMLI, resulting in an almost pure sinusoid with a very low THD.

Keywords PSPWM modulation for HCMLI · Modular multilevel inverters · DC-AC converters

1 Introduction

Energy generated by unusual energy sources is predicted to satisfy 50% of total energy needs by 2050. Energy demand is increasing too much rapidly, and it is leading to the scarcity of non-renewable energy sources. To satisfy the increasing demand for electric power, renewable energy sources, mainly wind and solar, have become considerably more widely accepted options than the usual ones power generation systems [1, 2].

D. F. Y. Ponce (✉)
Universidad Politécnica Salesiana,Quito170702, Ecuador
e-mail: dfyp1991@gmail.com

D. F. Y. Ponce · H. M. Y. Ponce · W. M. M. López
Instituto Superior Tecnológico,Luis Tello080150, Ecuador
e-mail: wmontalvo@ups.edu.ec

X.-S. Yang et al. (eds.), *Proceedings of Seventh International Congress on Information and Communication Technology*, Lecture Notes in Networks and Systems 447,
https://doi.org/10.1007/978-981-19-1607-6_40

Solar energy is a reliable and clean renewable energy source [3, 4]. Today, PV systems including inverters are connected to the grid to supply electric power to power distribution networks [5, 6] and these modern systems employ some maximum power point tracking (MPPT) technique to make the best use of solar energy [7, 8].

Multilevel inverters (MLIs) reduce a given voltage of some stand-alone DC sources (SDCS), which can be solar cells, batteries, or fuel cells. These inverters improve power quality, generate sinusoidal currents or voltages, reduce common-mode voltages, decrease the number of semiconductor and, due to the low dv/dt, decrease the setback of interruptions with the communication lines [9, 10]. MLI has not only been implemented in single-phase systems, but also in three-phase systems [11].

The most commonly used MLI topology is the three-level full-bridge inverter, because it offers a multilevel output and switches at high frequency. Today, MLIs have a wide range of applications in both domestic and industrial environments [12].

With the development of fast-switching power semiconductor devices, several modulation techniques or methods have been developed for MLIs over the past decades. The fundamental criteria for the choice of any modulation method are sinusoidal output with low THD and low losses. The decrease of THD is mainly the selection criterion mostly used for inverter design. One of the most commonly used modulation strategies today is pulse width modulation (PWM) [13]. Optimal pulse generation allows the objective to be met; deviations in pulse generation cause the semiconductors to err and the system to be suspended for short periods of time [14].

In this chapter, a three-level MLI was carried out and a seven-level single-phase H-bridge topology is proposed together with the PSPWM modulation technique and application for a photovoltaic system. The designed inverter is capable of driving loads with maximum currents of 4 [A], with an output voltage of 110 [VAC], giving a rated power of 440 [W]. To carry out this process, a modular power system was designed with individual galvanically isolated inverters, with their independent ignition control systems and power supplies. To verify the results obtained theoretically, simulations were carried out in MATLAB/Simulink software. The following research question is posed Can the designed CHMLI be standardized and replicated for renewable energy applications that are interconnected to the electrical grid?.

From here on the chapter is presented as follows. In Sect. 2, the state of the art is presented, and the methodology used is given in Sect. 3. The results obtained along with the discussion and future work are shown in Sect. 4, and finally, in Sect. 5, some conclusions are reported.

2 State of the Art

Nowadays, the use of renewable energies is indispensable to mitigate greenhouse gas emissions and contribute to stop global warming.

2.1 *Multilevel Converters*

The term converter is used to define equipment that transforms direct current (DC) or alternating current (AC) into DC–AC, AC–AC, AC–DC, and DC–DC [15].

MLI divides a voltage signal into several stepped DC voltages called levels. There are several topologies; however, they can be encompassed in three major families:

- Diode Clamping Multilevel Inverter (DCMLI).
- Floating Capacitor Multilevel Inverter (FCMLI).
- Cascaded Multilevel Inverter (CMLI).

The main characteristics of CMLI are described below.

- The phase voltage is the sum of the output voltages of the individual full-bridge inverters.
- Great flexibility to be able to increase the number of levels, as only inverters need to be added without having to redesign the power stage.
- As the number of levels increases, the voltage supported by the semiconductor devices decreases.
- It is possible to balance the switching losses, since depending on the number of levels it is possible, that different full-bridge inverter connections provide the same voltage at the output of the MLI.

2.2 *Modulation Techniques*

The control technique used in MLI is shown in Fig. 1, which are basically divided into vector space and voltage level techniques.

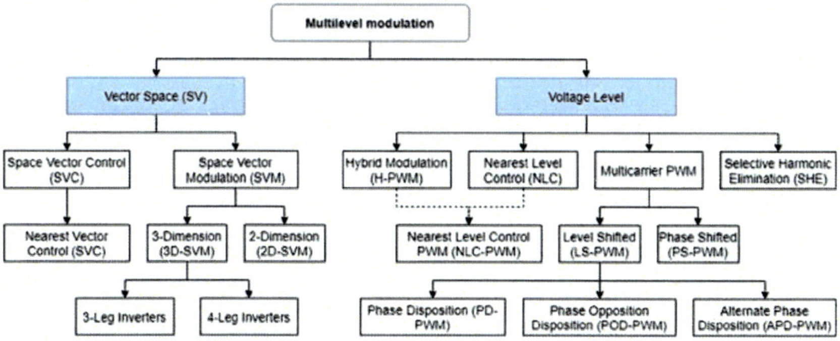

Fig. 1 Multilevel modulation techniques

3 Methodology

For the design of the CHMLI, we started from the design requirements for its subsequent modeling and simulation.

3.1 System Description

Each power cell consists of four switches and an isolated source, and the number of levels in the CHMLI output can be increased by adding more than one cell; however, more switches and sources are required.

The expression of Eq. 1 indicates the relationship that exists between the voltage sources of each cell, where E is the voltage of the first H-bridge, which in the case of CHMLI is the same across all H-bridges.

$$V_{dc1} = V_{dc2} = \cdots = V_{\mathrm{dcs}-1} = V_{\mathrm{dcs}} = E \tag{1}$$

Equation 2 allows to calculate the number of levels that the CHMLI has, where n is the number of levels and s is the number of bridges H.

$$n = 2s + 1 \tag{2}$$

In this type of inverters, the maximum amplitude of the CHMLI must be calculated using Eq. 3.

$$A = sE \tag{3}$$

3.2 PSPWM Modulation

This PWM technique with out-of-phase multicarriers uses n_p carriers of the same amplitude and frequency, but out of phase by an angle $theta$, where this angle is given by Eq. 4.

$$\theta = \frac{360°}{n_p} \tag{4}$$

To apply the PSPWM technique to CHMLI topologies, it is necessary to have two carriers for each full-bridge inverter connected in cascade; therefore, the number of carriers is obtained using Eq. 5.

$$n_p = n - 1 \tag{5}$$

The relationship between the ripple frequency and the switching frequency is given by Eq. 6.

$$f_{\text{ripple}} = n_p f_c \tag{6}$$

3.3 Inverter Modeling

Equations 7 and 8 describe the static model of the inverter.

$$LsI(s) = V_e d(s) - V_o(s) \tag{7}$$

$$CsV_o(s) = I(s) - \frac{V_o}{R}(s) \tag{8}$$

The relationship described in Eq. 9 determines the desired signals to be filtered.

$$\frac{1}{LC} < f_c \tag{9}$$

The natural frequency at which the system should lock is given by Eq. 10, in which it can be determined that it will work a decade earlier than the switching frequency.

$$W_n < \frac{f_c}{10} \tag{10}$$

Equations 11 and 12 establish the relationships that allow determining the values of the capacitor and inductor according to the mathematical model of the system.

$$W_n = \frac{1}{\sqrt{LC}} = \frac{1}{5T} \tag{11}$$

$$\sqrt{\frac{L}{C}} = R \tag{12}$$

3.4 Output Filter Design

Equation 13 describes the relationship between the switching frequency and the cutoff frequency.

$$f_s = 2\pi f_c \tag{13}$$

The calculation of the resistance or load value is made using Eq. 14, for which the design conditions are taken into account.

$$R = \frac{V_{rms}}{I_{rms}} = \frac{60}{4} = 15\Omega. \tag{14}$$

3.5 Input Filter Design

The Fourier analysis performed on the input current of each H-bridge resulted in a component 120 Hz; therefore, the input filter design is performed at this frequency.

Applying Kirchoff's law of nodes, we obtain Eq. 15.

$$I_C = I_o - I_L \tag{15}$$

To control the ripple voltage, Eq. 16 is used.

$$L = \frac{\Delta V_C}{W \Delta I} \tag{16}$$

4 Results and Discussion

In Fig. 2, PSPWM modulation and the design of the CHMLI implemented in Simulink are presented.

Figure 3a presents the carrier waveforms at 20 kHz and the sine waves to generate the modulating wave 60 Hz, and in Fig. 3, you can see the voltage and current at the input of the cascade inverter.

In Fig. 4a, you can see the voltage at the output of each H-bridge and the voltage delivered by the system for connection to the grid. For the Fourier analysis, a frequency 10 times higher than the switching frequency (400 kHz) was used and four cycles were taken for the analysis. In Fig. 4, the THD analysis of the CHMLI output signal is performed to validate the developed design.

The THD value in CHMLI designed is 2.30%, which guarantees the quality of the signal obtained. This value is below the THD obtained in similar works.

For future work, the implementation of CHMLI with solar panels, and the following development boards Arduino, Raspberry, and DSP will be performed to contrast the results obtained so far and compare these results with similar implementations. All this with the idea of making low-cost systems to popularize their use in Ecuador.

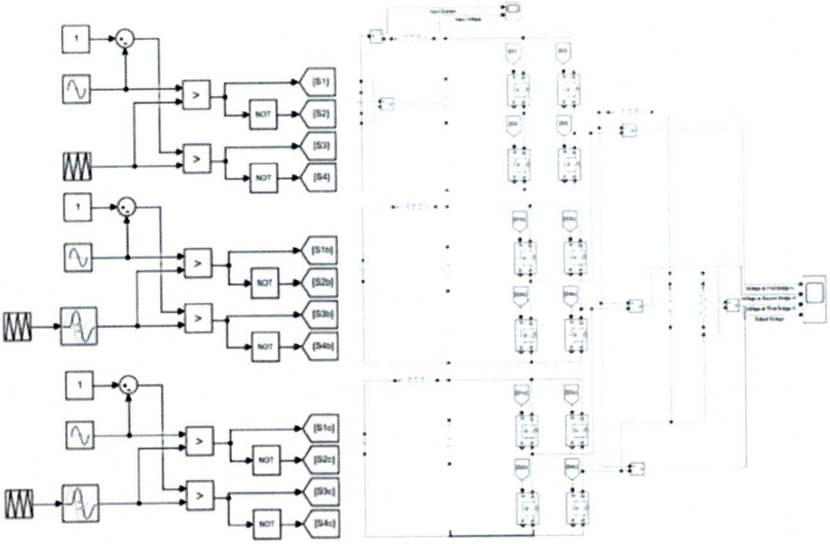

(a) Modulation. (b) CHMLI design.

Fig. 2 Simulation in Simulink

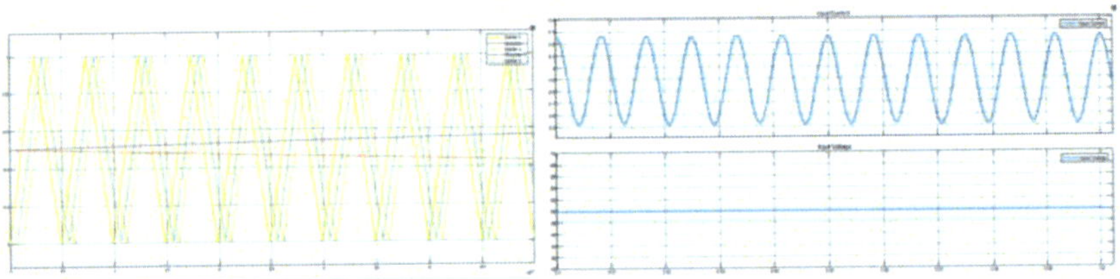

(a) PSPWM Modulation. (b) Input Current and Voltage.

Fig. 3 Simulations

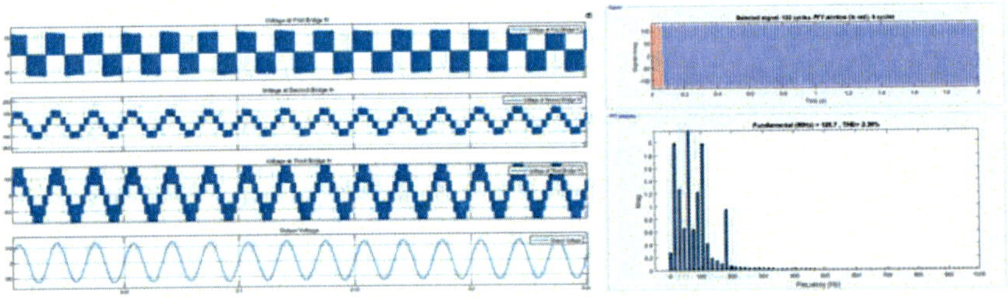

(a) Output Voltages. (b) Fourier analysis at the output.

Fig. 4 Results

5 Conclusions

- The PSPWM modulation technique shifts the frequency of the output voltage ripple to higher frequencies, reducing the size of the output filter and obtaining good performance in the reproduction of distorted signals.
- A seven-level CHMLI was designed, and the Butterworth approximation was used to design the output filter, obtaining a THD value of 2.30%.
- Different simulation tests were performed to observe the behavior of the multilevel inverter under the use of a resistive load, concluding that: The multilevel inverter has an excellent performance.
- Based on the CHMLI design performed and simulated, the CHMLI will be implemented with solar panels and the Arduino, Raspberry, and DSP development boards to contrast the actual performance of the CHMLI.

References

1. Nademi H, Das A, Burgos R, Norum LE (2016) A new circuit performance of modular multilevel inverter suitable for photovoltaic conversion plants. IEEE J Emerg Sel Top Power Electron 4(2):393–404. https://doi.org/10.1109/JESTPE.2015.2509599
2. Stonier AA, Lehman B (2018) An intelligent-based fault-tolerant system for solar-fed cascaded multilevel inverters. IEEE Trans Energy Convers 33(3):1047–1057. https://doi.org/10.1109/TEC.2017.2786299
3. Lazzarin RM, Noro M (2018) Past, present, future of solar cooling: technical and economical considerations. Sol Energy 172:2–13. https://doi.org/10.1016/j.solener.2017.12.055
4. Hasan MM, Abu-Siada A, Dahidah MSA (2018) A three-phase symmetrical DC-Link multilevel inverter with reduced number of DC Sources. IEEE Trans. Power Electron 33(10):8331–8340. https://doi.org/10.1109/TPEL.2017.2780849
5. Al-Shetwi AQ, Sujod MZ, Blaabjerg F, Yang Y (2018) Fault ride-through control of grid-connected photovoltaic power plants: a review. Sol Energy 180:340–350. https://doi.org/10.1016/j.solener.2019.01.032
6. Mirhosseini M (2019) Sensitivity analysis, adaptability improvement and control of grid-connected photovoltaic power plants under grid frequency variations. Sol Energy 184(March):260–272. https://doi.org/10.1016/j.solener.2019.03.072
7. Elmelegi A, Aly M, Ahmed EM, Alharbi AG (2019) A simplified phase-shift PWM-based feedforward distributed MPPT method for grid-connected cascaded PV inverters. Sol Energy 187(May):1–12. https://doi.org/10.1016/j.solener.2019.05.021
8. Amir A, Amir A, Selvaraj J, Abd Rahim N (2019) Grid-connected photovoltaic system employing a single-phase T-type cascaded H-bridge inverter. Sol Energy 199:645–656. https://doi.org/10.1016/j.solener.2020.02.045
9. Sinha A, Chandra Jana K, Kumar Das M (2018) An inclusive review on different multi-level inverter topologies, their modulation and control strategies for a grid connected photo-voltaic system. Sol Energy 170(June):633–657. https://doi.org/10.1016/j.solener.2018.06.001
10. Fernão Pires V, Cordeiro A, Foito D, Fernando Silva J (2018) Three-phase multilevel inverter for grid-connected distributed photovoltaic systems based in three three-phase two-level inverters. Sol Energy 174(April):1026–1034. https://doi.org/10.1016/j.solener.2018.09.083
11. Shi Y, Wang L, Xie R, Shi Y, Li H (2017) A 60-kW 3-kW/kg five-level T-Type SiC PV inverter with 99.2% peak efficiency. IEEE Trans Ind Electron 64(11):9144–9154. https://doi.org/10.1109/TIE.2017.2701762

12. Suresh Y, Panda AK (2012) Research on a cascaded multilevel inverter by employing three-phase transformers. IET Power Electron 5(5):561–570. https://doi.org/10.1049/iet-pel.2011.0150
13. Riyaz A, Iqbal A, Tariq M (2020) Five-phase twenty-seven level inverter using single DC source for photovoltaic application. In: 2020 IEEE international conference computer power communication technology GUCON, pp. 594–598. https://doi.org/10.1109/GUCON48875.2020.9231200
14. Chitra S, Valluvan KR (2020) Design and implementation of cascaded H-Bridge multilevel inverter using FPGA with multiple carrier phase disposition modulation scheme. Microprocess. Microsyst. 76:1–8. https://doi.org/10.1016/j.micpro.2020.103108
15. Jimenez O (2012) Estudio de Tcnicas de Modulaci6n para el Inversor Multinivel en Cascada Hfbrido (Simétrico-Asimétrico)

An IoT Architecture to Enhance Monitoring and Predictive Maintenance for Cultural Heritage Buildings

Mario Casillo, Massimo De Santo, Marco Lombardi, Rosalba Mosca, Domenico Santaniello, and Carmine Valentino

Abstract The significant value related to cultural heritage (CH) is inestimable, and it also represents a resource in economic terms due to tourism. Therefore, CH must be safeguarded to avoid losses that in many cases can no longer be recovered. Due to the variability of elements and conditions, each step in CH protection and conservation is represented by a wide range of variable factors. In this scenario, the ability to intervene seems limited, given the variety of factors to be monitored remotely. However, the deployment of current technology, which retains the ability to provide low-cost sensors, could significantly contribute. The purpose of this paper is to introduce an IoT-based methodology for cultural heritage preservation. Expert users have tested a system prototype to monitor and manage a portion of a building within the Archaeological Park of Pompeii, predict *deterioration*, and schedule conservation interventions and choose the combination of interventions. Preliminary results are promising.

Keywords Cultural heritage · Cultural heritage preservation · Internet of things

M. Casillo · M. De Santo · M. Lombardi · R. Mosca · D. Santaniello (✉) · C. Valentino
DIIn, University of Salerno, Fisciano, SA, Italy
e-mail: dsantaniello@unisa.it

M. Casillo
e-mail: mcasillo@unisa.it

M. De Santo
e-mail: desanto@unisa.it

M. Lombardi
e-mail: malombardi@unisa.it

R. Mosca
e-mail: rmosca@unisa.it

C. Valentino
e-mail: cvalentino@unisa.it

X.-S. Yang et al. (eds.), *Proceedings of Seventh International Congress on Information and Communication Technology*, Lecture Notes in Networks and Systems 447,
https://doi.org/10.1007/978-981-19-1607-6_41

1 Introduction

The significant value of cultural heritage (CH) is inestimable, and it is also a resource in economic terms due to tourism. Therefore, CH must be safeguarded to avoid losses that in many cases can no longer be recovered. However, the conservation capacity of CH has high variability and often, in addition to the natural decay of material properties, depends on external conditions such as weather and human factors [1–3].

One of the main tools for CH preservation is monitoring, and it allows maintenance and any necessary interventions. The phases of CH preservation are related to study, prevention, maintenance, and restoration activities. Prevention is represented by the activities that limit the risk conditions related to the conservation of CH. Maintenance represents activities and interventions aimed at preserving the integrity, identity, and functionality of the CH. Restoration represents the final stage of the process, encompassing the direct intervention procedures aimed at restoring the integrity, identity, and functionality of the CH.

Due to the variability of elements and conditions, each phase of CH protection and conservation is represented by a wide range of variable factors. The factors are related to the elements' physical, mechanical, and chemical characteristics and the microclimatic and climatic conditions of the area of interest [4, 5].

In such a scenario, the ability to intervene seems limited, given the multiplicity of factors to be monitored remotely. However, the diffusion of today's technology, which retains the ability to provide low-cost sensors, could significantly contribute. Nowadays, low-cost devices can communicate with humans and each other by exhibiting intelligence; such a phenomenon is called the Internet of things (IoT) [6]. This paradigm allows having a virtual representation of reality with all the heterogeneous and interconnected physical elements represented in a single digital environment [7, 8]. In fact, IoT helps to take advantage of different heterogeneous sensors able to support monitoring [9–12] in order to obtain data regarding:

- Material alterations;
- Structural alterations;
- Environmental site condition.

One of the methodologies for CH maintenance is to rely on digital twins, i.e. a digital domain representation employing the acquired data [13]. Combined with machine learning detection, such representation support predictive maintenance, representing a form of maintenance performed according to the prediction of significant features and parameters [14, 15]. This intervention aims to identify in advance the natural deterioration of the elements based on the collected data and optimize the work processes for the preservation of the CH. Whilst respecting the predetermined constraints, each CH has different needs; a recent issue is to suggest the proper intervention. This issue is addressed with the method of multicriteria analysis. This methodology can identify the best intervention response by analyzing several characteristic criteria [16].

In such a scenario, it could be helpful to propose a methodology based on the IoT paradigm for monitoring and supporting expert users to preserve cultural heritage.

The following section describes the proposed methodology, and the third section deals with the case study to validate the proposed methodology. Finally, conclusions and future developments are provided.

2 The Proposed Approach

The proposed architecture aims to combine several aspects to preserve cultural heritage buildings, such as monitoring and managing through the Internet of things paradigm. Figure 1 shows the proposed architecture, which is articulated according to different levels.

The first level consists of IoT nodes representing the sensors, i.e. hardware devices for monitoring the CH [17]. These devices can retrieve information about the structural conditions of materials, alterations, and environmental and micro-environmental conditions. In addition, this level also includes actuators, devices capable of performing actions that aim to safeguard the CH. These devices can be

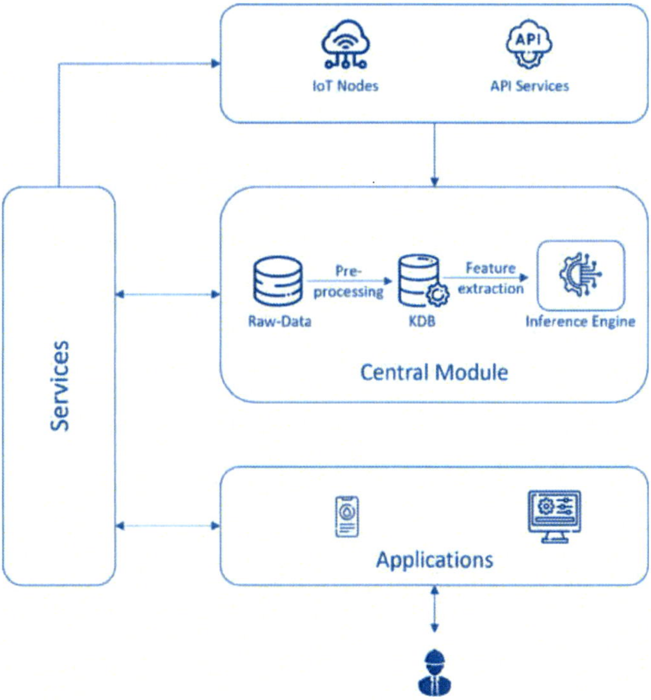

Fig. 1 System architecture

used to activate the air conditioning systems. Furthermore, the system is fed with services and data from outside, such as descriptions of degradation mechanisms and intervention techniques [18].

The second level is represented by the heart of the proposed architecture. This level is responsible for validating and pre-processing the data. This phase is necessary for archiving the data that take place in the knowledge database (KDB). The processed data, which represent the knowledge of the system, are employed through Bayesian approaches to predict future events [19–21]. In fact, the inference engine is in charge of selecting different learning models capable to model reality. These models, fixed some factors, can be queried to predict the occurrence of events in the immediate future.

In summary, the central layer contains the core information flow on which monitoring and decision-making applications are based to enhance CH preservation. The links between the various modules and applications are managed by the services module, which is able to transfer knowledge through the entire architecture and let things happen. The knowledge and models information is exchanged to the application module through the services module, which involves the three critical aspects of cultural heritage conservation.

The applications are devoted to expert and technical users. In particular, the monitoring, which is ensured through the collection, processing, and visualization of data, is done with the open-source platform ThingsBoard [22]. The monitoring provides information about the conditions of the site of interest and is able to provide support for decisions such as humidity control, temperature, and air quality control. Through the dashboards offered by the system, users can access the information needed to monitor the CH in real-time. Moreover, other applications can suggest predictive maintenance and decision support interventions based on learning algorithms present in the inference engine.

3 Case Study

The critical aspects addressed by the proposed architecture cover several aspects related to CH preservation. In particular, the support for expert users to monitoring and management, the prediction of deterioration and planning of maintenance and the identification of the best interventions to be performed.

The chosen case study concerns a portion of a building in the Archaeological Park of Pompeii, "Bakery of Popidio Prisco". For this purpose, a prototype based on the described architecture has been built, integrated with low-cost, open standard, and open-source technologies. The prototype, equipped with the sensors reported in Table 1, has collected data for about a year: from April 2020 to May 2021. The variety of sensors used has allowed collecting information about the architectural and structural components of the building under examination. In particular, chemical and physical corrosion phenomena, humidity formation, and possible kinematic deformations have been monitored through the cameras. The micro-ambient and

Table 1 The sensors of the IoT-based prototype

Control parameters	Sensors
Material and structural alterations	Raspberry Pi high quality HQ camera—12MP (Adafruit Industries, New York, NY, USA)
Peak ground acceleration	ADXL345 (Adafruit Industries, New York, NY, USA)
Indoor air temperature	DHT22 (Adafruit Industries, New York, NY, USA)
Indoor air humidity	DHT22 (Adafruit Industries, New York, NY, USA)
Indoor air quality	PMSA003I air quality breakout—STEMMA QT/Qwiiic (Adafruit Industries, New York, NY, USA)
Outdoor air quality, temperature, pressure, humidity	BME680 breakout (Adafruit Industries, New York, NY, USA)
Weather station	Weather Metres (SparkFun Electronics, Niwot, Colorado, USA)
Temperatures of the bodies analyzed	MLX90640 24 × 32 IR thermal camera breakout—110 degree FoV (Adafruit Industries, New York, NY, USA)
Peak ground acceleration	ADXL345 (Adafruit Industries, New York, NY, USA)

environmental conditions were also generally monitored through relative humidity sensor, dew point, temperature, air quality, etc. The dew point is essential to avoid mould and moisture on vulnerable parts of the cultural heritage building. This parameter is calculated using temperature and relative humidity. The building control loop was created using a Raspberry Pi 4 board. The platform used for the management is based on ThingsBoard, which enables data visualization, processing, and enable remote intervention. The environment for managing and monitoring is represented by a dashboard with a graphic interface accessible from desktop or mobile.

The ability of the system to recognize the maintenance needed is based on an inferential engine module, which is able to exploit machine learning algorithms. The testing phase is focussed on the ability of the system to prevent the walls moisture from preventing possible damage to the building. The phenomenon on which attention was focussed in this experimental phase is that of surface condensation moisture, specifically by monitoring specific conditions, such as indoor and outdoor humidity, air temperature, dew point, wall surface temperature, wind speed and direction, precipitation, and camera images. The system detected the presence of all conditions on the building that could lead to surface condensation moisture damage at multiple points.

Condensation on walls is caused by excess moisture in indoor areas, which can be highly damaging. Warm air is able to hold more moisture suspended than an equal volume of cold air. When a warm mass comes into contact with a cold surface, it drops in temperature and gives up the moisture, and it can no longer hold in suspension. Humidity condenses on the walls in tiny droplets, and its presence encourages the formation of fungal colonies that blacken the walls. In addition, moisture causes the rotting of porous materials. Mould on walls is formed more often in neuralgic areas and can lead to structural damage. Therefore, it is necessary to improve the air quality in the rooms to prevent mould formation through aeration and dehumidification systems.

4 Experimental Results

The experimental phase presented so far has been conducted to test the system's ability to provide monitoring information, reliability, and ability to support expert users.

The experimental campaign was conducted developing a system prototype and involving 35 expert users. The expert users, aged 35–65, interacted with the prototype and answered a questionnaire, which sections are reported below. Section A: ability to view monitoring.

- Section B: system support capability
- Section C: system reliability

The questionnaire was evaluated according to the Likert scale:

- TA"I totally agree"
- A"I agree"
- U"Undecided"
- D"I disagree"
- TD"I totally disagree"

Figure 2 shows the results, which report a high-user satisfaction. Expert users provided meaningful comments on the usability and system availability, demonstrating sufficient satisfaction of the expert users. The monitoring section shows high results. The expert users found the dashboard capable of showing the correct information, leading to preserving cultural heritage buildings. The second and third sections were also rated positively.

The prototype collected data regarding the site's environmental conditions and the building's conditions, such as indoor and outdoor air temperature and humidity, dew point, humidity, wind speed and direction, and precipitation, for about one year, gathering about 2000 examples. The data records were divided into a test set (25%) and a training set (75%).

The database so organized was used by a machine learning algorithm. In particular, an algorithm known as K2 hill climbing [23] was used to perform this case study. The algorithm is able to learn the structure and weights of a Bayesian network that was checked with the test set [24].

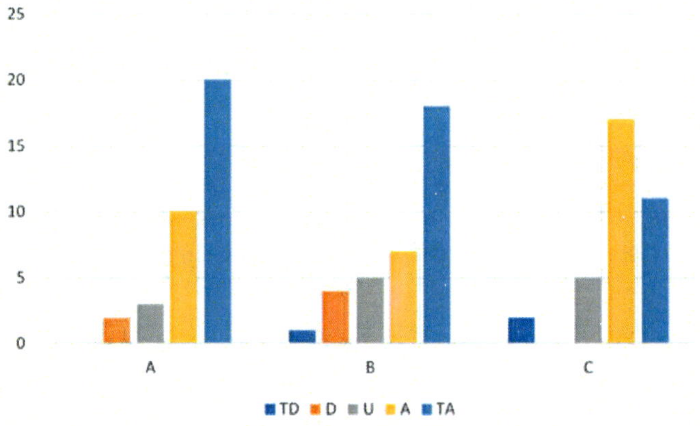

Fig. 2 Questionnaire answers

Table 2 Confusion matrix

	Reference		
Prediction		Yes	No
	Yes	367	107
	No	130	1625
Overall accuracy 89, 63%			

The test was performed to understand the system's ability to prevent degradation and appropriate support to the expert users to suggest maintenance. The parameter controlled is the surface condensation, which was allowed to appear in the first stage only for experimental purposes. The system's prediction data were compiled into a confusion matrix, which is shown in Table 2. Table 2 reports the ability of the system to suggest the use of an air humidifier to prevent condensation on the walls. The confusion matrix shows an overall accuracy of about 89%, a recall of 74%, and a precision of 77%. The showed results are promising even if they could be improved.

5 Conclusions

The present work aimed to present an IoT-based architecture to enhance cultural heritage preservation. The system can exploit data from sensors and process them for active monitoring and lifecycle management of cultural heritage. The proposed architecture has experimented with a prototype, which performed promising results. The results showed the prototype to provide proper support to users in monitoring and to manage an asset by efficiently scheduling maintenance interventions. Future developments will involve more buildings and the introduction of other control parameters.

More in-depth, experiments will then be done on predictive maintenance and decision support applications, expanding the database.

References

1. Bertolin C (2019) Preservation of cultural heritage and resources threatened by climate change. Geosciences (Switzerland) 9(6). https://doi.org/10.3390/geosciences9060250
2. Merello P, García-Diego FJ, Zarzo M (2012) Microclimate monitoring of Ariadne's house (Pompeii, Italy) for preventive conservation of fresco paintings. Chem Cent J 6(1). https://doi.org/10.1186/1752-153X-6-145
3. Casillo M, Colace F, Conte D, Lombardi M, Santaniello D, Valentino C (2021) Context-aware recommender systems and cultural heritage: a survey. J Ambient Intell Humaniz Comput. https://doi.org/10.1007/s12652-021-03438-9
4. Chemeli A, Njoroge JM, Agufana PB (2021) Climate change and immovable cultural heritage in Kenya: impact and response strategies. Handb Clim Change Manage. https://doi.org/10.1007/978-3-030-22759-3_91-1
5. Jara AJ, Sun Y, Song H, Bie R, Genooud D, Bocchi Y (2015) Internet of things for cultural heritage of smart cities and smart regions. https://doi.org/10.1109/WAINA.2015.169
6. Ashton K (2009) That 'Internet of Things' thing. RFiD J. https://doi.org/10.1016/j.amjcard.2013.11.014
7. Carratu M, Ferro M, Pietrosanto A, Sommella P, Paciello V (2019) A smart wireless sensor network for PM10 measurement. In: 2019 IEEE international symposium on measurements and networking, M and N 2019—proceedings, July, pp 1–6. https://doi.org/10.1109/IWMN.2019.8805015
8. Casillo M, Colace F, Fabbri L, Lombardi M, Romano A, Santaniello D (2020) Chatbot in industry 4.0: an approach for training new employees. https://doi.org/10.1109/TALE48869.2020.9368339
9. Maksimovic M, Cosovic M (2019) Preservation of cultural heritage sites using IoT. March. https://doi.org/10.1109/INFOTEH.2019.8717658
10. Lee W, Lee DH (2019) Cultural heritage and the intelligent internet of things. J Comput Cult Heritage 12(3). https://doi.org/10.1145/3316414
11. Zanella A, Bui N, Castellani A, Vangelista L, Zorzi M (2014) Internet of things for smart cities. IEEE Internet Things J 1(1):22–32. https://doi.org/10.1109/JIOT.2014.2306328
12. Sun Y, Song H, Jara AJ, Bie R (2016) Internet of things and big data analytics for smart and connected communities. IEEE Access 4. https://doi.org/10.1109/ACCESS.2016.2529723
13. Jouan P, Hallot P (2020) Digital twin: research framework to support preventive conservation policies. ISPRS Int J Geo-Inf 9(4). https://doi.org/10.3390/ijgi9040228
14. Abdelgawad A, Yelamarthi K (2016) Structural health monitoring: internet of things application. In: Midwest symposium on circuits and systems. https://doi.org/10.1109/MWSCAS.2016.7870118
15. Dong CZ, Catbas FN (2021) A review of computer vision–based structural health monitoring at local and global levels. Struct Health Monit 20(2). https://doi.org/10.1177/1475921720935585
16. Yau Y (2009) Multi-criteria decision making for urban built heritage conservation: application of the analytic hierarchy process. J Build Appraisal 4(3). https://doi.org/10.1057/jba.2008.34
17. Čolaković A, Hadžialić M (2018) Internet of Things (IoT): a review of enabling technologies, challenges, and open research issues. Comput Netw. https://doi.org/10.1016/j.comnet.2018.07.017
18. Formato A, Ianniello D, Pellegrino A, Villecco F (2019) Vibration-based experimental identification of the elastic moduli using plate specimens of the olive tree. Machines 7(2). https://doi.org/10.3390/machines7020046

19. Colace F, de Santo M, Lombardi M, Pascale F, Santaniello D, Tucker A (2020) A Multilevel graph approach for predicting bicycle usage in London area. In: Yang XS, Sherratt S, Dey N, Joshi A (eds) Fourth international congress on information and communication technology. Advance Intelligent Systems Computing, vol 1027. Springer, Singapore, pp 353–362. https://doi.org/10.1007/978-981-32-9343-4_28
20. Clarizia F et al (2020) A multilevel graph approach for rainfall forecasting: a preliminary study case on London area. Concurrency Comput: Pract Experience 32(8). https://doi.org/10.1002/cpe.5289
21. Colace F, Santaniello D, Casillo M, Clarizia F (2017) BeCAMS: a behaviour context aware monitoring system. https://doi.org/10.1109/IWMN.2017.8078374
22. Henschke M, Wei X, Zhang X (2020) Data visualization for wireless sensor networks using thingsboard. https://doi.org/10.1109/WOCC48579.2020.9114929
23. Monti S, Cooper GF (1998) Learning hybrid Bayesian networks from data. In: Learning in graphical models. https://doi.org/10.1007/978-94-011-5014-9_19
24. Colace F, Lombardi M, Pascale F, Santaniello D (2018) A Multilevel graph representation for big data interpretation in real scenarios. In: 2018 3rd international conference on system reliability and safety (ICSRS), Nov 2018, pp 40–47. https://doi.org/10.1109/ICSRS.2018.8688834

A BIM-Based Approach for Decision Support System in Smart Buildings

Francesco Colace, Caterina Gabriella Guida, Brij Gupta, Angelo Lorusso, Francesco Marongiu, and Domenico Santaniello

Abstract Building information model (BIM) is primarily a 3D digital representation of a structure and features such as geometry, spatial relationships, and geographic information to support integrated design. BIM in recent years has evolved from simple 3D to interacting with virtual and augmented reality. Such interaction aims to improve work productivity, home comfort, and entertainment, common Internet of things (IoT) goals. The IoT represents a possible evolution of the use of the Internet in which objects are able to communicate data about themselves and with other devices autonomously. One of the main goals of IoT is to build a digital copy of the real world. Therefore, BIM and IoT can integrate through data acquired in a BIM model; this model can be helpful in predictive analysis as needed. This paper aims to describe a methodology that allows the visualization and representation of data from sensors within the BIM environment to support decisions, sometimes complex, requiring interdisciplinary expertise. The study focusses on a real case study: a scale

F. Colace · A. Lorusso · F. Marongiu · D. Santaniello (✉)
DIIn, University of Salerno, Fisciano, Italy
e-mail: dsantaniello@unisa.it

F. Colace
e-mail: fcolace@unisa.it

A. Lorusso
e-mail: alorusso@unisa.it

F. Marongiu
e-mail: fmarongiu@unisa.it

C. G. Guida
DICiv, University of Salerno, Fisciano, Italy
e-mail: cguida@unisa.it

B. Gupta
Department of Computer Science and Information Engineering, Asia University, Taichung 413, Taiwan
e-mail: bbgupta@asia.edu.tw

Research and Innovation Department, Skyline University College, P.O. Box 1797, Sharjah, United Arab Emirates

Staffordshire University, Stoke-on-Trent ST4 2DE, UK

X.-S. Yang et al. (eds.), *Proceedings of Seventh International Congress on Information and Communication Technology*, Lecture Notes in Networks and Systems 447,
https://doi.org/10.1007/978-981-19-1607-6_42

prototype of a single-family house that includes several sensors capable of producing data that feed a database based on the predictive/decision-making phase developed through machine learning techniques. The proposed methodology integrates an IoT-based platform that allows communication between sensors and Dynamo software to access sensor data, automatically updating the information contained in the BIM model.

Keywords Building information model—BIM · Internet of things · Digital twin

1 Introduction

In recent years, the technology of BIM has taken a position of relevance within the building and industrial design becoming an integral part of digital innovation, and this design is focussed in particular on the systems and parts of the artefacts. All operations in the BIM environment concerning design have been internationally standardized by ISO 19650-3, which is a well-structured guideline about information management and its use. The latest innovation regarding systems concerns the integration of static data with dynamic data extrapolated from a network of sensors put in the field thanks to the advent of the Internet of things (IoT) in order to manage applications inherent to buildings [1–3]. Basic BIM is mainly used to represent elements and objects that comply with the Industry Foundation Classes (IFC) standard, and a universal model created to make all types of modelling operations required in design interact by allowing direct and intuitive visualization of 3D models created by other users [4]. Thus, this format facilitates design interoperability by enabling immediate communication and modification without the need for additional specialized representation codes, which can effectively reduce discrepancies and allow for the rapid resolution of inconsistencies and errors amongst professionals. Collecting and evaluating the correct data to acquire could allow for a more accurate assessment and design changes that result in more comfortable environments. With the advancement of technology, data have grown exponentially in recent years. This growth is partly due to the spread of low-cost technology, which has enabled the development of a paradigm in which devices are always connected to the Internet and are defined as smart for their ability to exchange information autonomously: Internet of things (IoT) [5–7]. Due to enabling technologies, the IoT has emerged from its initial phase as a revolutionary technology to support a fully integrated Internet projected to manage different application scenarios such as smart industries, smart cities, and smart buildings. A crucial role of the IoT, in fact, is to create a digital copy of reality, generating digital scenarios for the management of different sectors. In this scenario, particular attention is paid to data visualization, which allows data to be analyzed, monitored, and processed using computer graphics and interactive technologies. The data representing the digital twin can be derived from a wide variety of sources: from sensors transmitting various aspects of its operating conditions; historical data relating to past conditions; data provided by human experts, such as engineers, technicians, physicians, with specific and relevant knowledge; data collected from other [8] similar machines, or from the systems, and environment of which it may be a part. Information retrieved from any database accessible through the Internet. The digital twin may also use machine learning and artificial intelligence systems, to process data and produce new knowledge or predict operational scenarios through analysis of collected data. Digital twins can integrate the Internet of things, artificial intelligence, machine learning, and data analytics with spatial network graphs [8–10]. They can create digital simulation models that update and change when their physical counterparts change. A digital twin can continuously learn and update from multiple sources. It can also represent, in real-time, the state, working condition, or location of objects or the physical system. In several projects have tried the integration between the information collected from sensors and the information intrinsic to the BIM model to visualize the data and manage it for a building [11] and it was concluded that the data obtained mainly from the diagnosis of the structural health of buildings are difficult to integrate with other data from other monitoring methods in order to support the management and visualization of data, so a dynamic approach to BIM parameterization was proposed by integrating and visualizing the data from sensors in real-time. In fact, this workflow provides a dynamic inspection of critical points of structural performance and makes an immediate update of the conditions, and this approach will give more value to the data; in fact, the sensor data will constantly update the model in IFC format and will be useful for future design and maintenance decisions. Lather [12], on the other hand, has developed a framework that exploits the inherent characteristics of the spatial model and the location of the sensors by incorporating the sensor data into a monitored building management system (BMS) to improve the management part of the facilities. On the other hand, Mccaffrey [13] developed a graphical user interface based on a user-friendly visualization that integrates the peculiarities of BIM models with data in the service of managers and users to perform an even immersive visualization of spatial data and facilitate the understanding of building performance information, greatly increasing the capacity of energy management by supporting all decisions in the operational phase. So this kind of approach will also have positive feedback on consumption and indirectly also on the environment. Another approach has proposed a platform to manipulate the collected data of the sensors into visualizable data according to the needs and the visualization easier as a visualization with a chromatic variation of the results on the same BIM model that represents the building. So, this type of visualization would allow a manager to see the distribution of data based on the indicated point within the model and then be able to make corrective factors in a targeted way acting from the outside. This study seeks to integrate data visualization from IoT-based sensors to a BIM-based model, which can be extended to support and revolutionize the smart home concept, where all sensors are represented in the BIM model and can interact and exchange information with each other. This paper aims to describe a methodology that allows the visualization and representation of sensor data within the BIM environment to support decisions, sometimes complex, that require interdisciplinary skills. The study is based on a real case study: a scale prototype of a single-family house that includes several sensors capable of producing data that feed a database at the basis of the predictive/decision-making phase developed through machine learning techniques. The proposed methodology integrates an IoT-based platform that allows communication between sensors and Dynamo software to access data, automatically updating the information contained in the BIM model [14]. This study develops on the capability of intelligent management and monitoring in a BIM environment through the following steps: modelling technology that focusses on creating a visualized three-dimensional model of the rooms of the single-family house and managing the information of the spatial elements; collecting data on an IoT-based platform, which allows data storage and management; data integration is achieved by developing an automated sensor data exchange system between the sensor data collection platform, parametric design software (such as Dynamo), and BIM modelling platform (such as Autodesk Revit); use of machine learning techniques for autonomous management of the environment for comfort enhancement.

2 Proposed Approach

The proposed methodology can be approached in the final steps of the design phase with the aim of evaluating the most suitable solution amongst the various possible design proposals. To start with, design rules are established and then a design proposal is made, and only then are the various components characterized through parameters by the designer according to the desired improvements. For this reason, parametric design has been enthusiastically adopted in various BIM applications as an engine for managing and displaying the possible design variables; this has led to the development of effective design software, even though it still has some shortcomings for a complete design of all the construction phases. Dynamo is an open-source visual programming application from Autodesk that aims to be accessible to both programmers and newcomers to computer languages such as Python. It offers users the ability to visually describe behaviour, define custom logical elements, and execute code segments using various textual programming languages. Dynamo can be used as a stand-alone product (Sandbox) or embedded in other Revit programmes (Dynamo for Revit). There is a difference between visual programming and textual programming; in fact, the former is a type of coding that, unlike the former, does not require skills in compiling line code or familiarity with a textual programming language. So visual programming uses a visual interface where the designer has to connect nodes or blocks with specific functionalities. The combination of these nodes forms a large network of different functionalities that can achieve complex design scenarios. This approach gives the possibility to explore the design path better than text-based programming and opens up the world to new players such as civil engineers or architects, which was previously reserved for experienced programmers. Furthermore, visual programming software such as Dynamo allows designers to set up automation or calculation tables through visual ease of node-based computation so that designers can process input data and proceed with geometric and structural parameter checks. For these reasons, the civil and architectural design sector has undergone a significant evolution towards visual programming, and therefore, many design studios are adopting and equipping themselves for the use of these types of innovative design, progressively abandoning the old 2D vector design techniques such as CAD. So visual programming allows a new design paradigm, changing the concept of design from a series of static modelling operations to a dynamic and customizable flow of design operations.

2.1 Workflow

The proposed workflow involves collecting sensor data, their integration, a parametric control mechanism, visualization modules for easy management and monitoring, and the possibility of triggering autonomous choice mechanisms to improve environmental comfort, as shown in Fig. 1. This workflow is developed along two main phases used in architectural design: visual programming and parametric design. Within the visual programming environment, the steps for data management and visualization are connection with the data collection platform; setting values and data display methods; connection with the BIM environment. Parametric design to achieve the objective of visualizing the data obtained to improve comfort, the steps to follow are scale-up of the real building with sensors; implementation of the digital twin (BIM model): establishment of a 3D virtual space, positioning of the sensors detected by an IoT-based platform; parameterization: according to the decision objectives linked to the different scenarios, the operating rules of the custom nodes are implemented within the visual programming environment by automatically updating the digital twin; visualization of results and implementation of machine learning-based rules to allow the model to make autonomous decisions to improve the environmental comfort [15–17].

3 Case Study

The case study focusses on the reproduction of a digital twin of an independent house scale model. Firstly, the house reproduced in the laboratory is connected to several sensors that monitor different parameters of the real state and are interconnected to a microcontroller. The acquired data are stored on a cloud platform, ThingsBoard [18, 19] which allows the data to be read and displayed correctly, filtering the type of data according to the selected sensor. The next step is the real-time visualization of the data coming from the sensors in a BIM environment; this is possible thanks to the use of Dynamo. In fact, using the dedicated API, it automatically connects to the ThingsBoard platform for real-data acquisition, and then, employing a display, directly to the digital twin in BIM. The ability to use the data in real-time for a

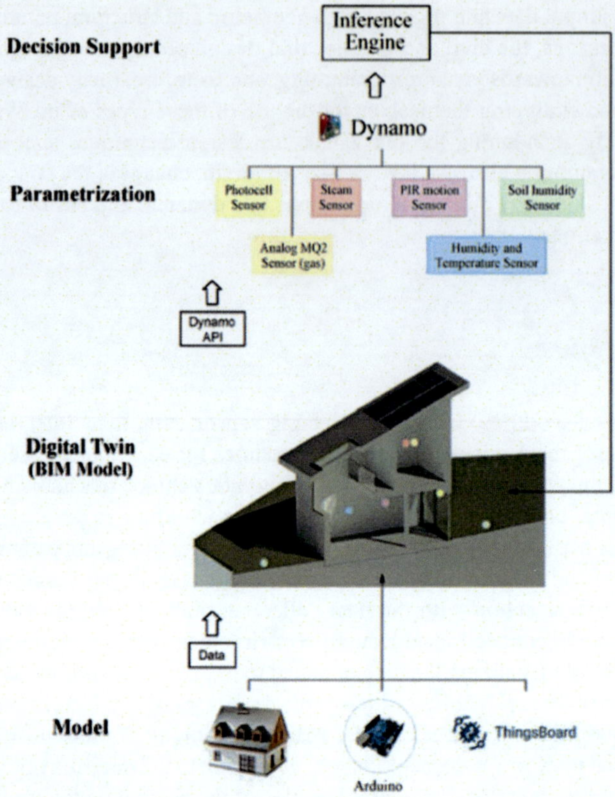

Fig. 1 Case study workflow

predictive study of living comfort, becoming a decision support tool, allows changes to be made in real-time based on needs (Fig. 1).

3.1 *Arduino-Based Prototype*

The heart of the project is the Arduino board, a microcontroller-based on a new generation microchip very performant. This board has 14 pins between inputs and outputs of which 6 are analogue inputs and the others digital, as well as being equipped with a ceramic resonator, a USB connection, a power jack, and a reset button. This technology gives the possibility to connect the microcontroller to a computer through a simple USB cable or it can be powered directly by electricity or a battery. So the Arduino board acts as a physical hub between all the sensors fielded for the desired monitoring. The data collected by the position of the sensor in the reconstructed model in the lab are

- The *photocell sensor* is a resistor made by the photoelectric effect of a semiconductor. It turns out to be sensitive to light and its variations, so depending on the intensity of light that captures varies the value of resistance. This sensor works that as the intensity increases the resistance decreases and consequently, the analogue port voltage increases, so the analogue reference value of the microcontroller also increases. On the contrary, when the light intensity decreases, the resistance tends to increase and inversely the voltage of the port decreases, and as expected the analogue value of the microcontroller decreases. In summary, this technology can use it to display the corresponding analogue value and thus perceive the intensity of light within an environment. This sensor has been used to monitor and measure the intensity of light and thus control and manage the ambient lighting system.
- The *vapour sensor* is one of the most common sensors used in environmental comfort monitoring. Its physical principle means that it can quantify the percentage of water deposited on parallel strips printed on the circuit. The greater the presence of water, the greater will be the connection between the wires, so as the contact area increases, its conductivity will also increase, and therefore, the output voltage will increase constantly. This sensor can also detect the percentage of water vapour, humidity, in the air. So the vapour sensor can be used as both an outdoor rainwater detector and an indoor humidity detector, with different applications and purposes. The first case can be assumed for garden irrigation management and the second for room ventilation.
- The *motion sensor* is a sensor that uses pyroelectric infrared technology and can capture the infrared signals of the passage of a moving person or animal and produce signals to be converted. Therefore, this type of sensor can be applied to different occasions of monitoring movement of people. Usually, conventional pyroelectric infrared sensors are very large, complex, and not very reliable. Instead the sensor in question integrates a digital infrared pyroelectric sensor and connection pins, this makes it more reliable, consuming less electricity, and a circuit easier to understand and apply.
- The *smoke sensor* can also be used as a multi-gas detector; in fact, it turns out to be a very sensitive sensor, fast in response, stable, durable and has a simple start-up system. It has the ability to detect both the presence of flammable gas and smoke in the range of 300–10000 ppm. In addition, its peculiarity leads it to have good sensitivity to the presence of natural gas, liquefied petroleum gas, and other types of smoke, especially the smoke of alkanes.
- The *soil moisture sensor* sees the presence of water within the soil. If the soil needs water, then the analogue value output from the sensor will be low; otherwise, the value will be higher. This sensor can be useful for managing an automatic watering device; in fact, it allows you to detect if there is a need for water for the plants and thus prevent them from wilting. The sensor consists of two probes that are inserted into the soil, and electric current will be passed through it, and then, the sensor will be able to read the degree of resistance by reading the current variations between the two parts of the probe and convert that resistance value into moisture content.

- *Humidity and temperature* sensor consist of two elements, such as a capacitive humidity sensor and a thermistor. This sensor also has a very simple chip inside that converts the data from analogue to digital giving the humidity and temperature values. This type of signal is easy to display and use any microcontroller.

3.2 *Platform: ThingsBoard*

ThingsBoard is an open-source cloud platform dedicated to the management of IoT devices. It can communicate with the devices in real-time through different protocols, including the MQTT used to connect the sensors for structural monitoring to the platform. ThingsBoard also enables data visualization via dashboards, which can be customized and extended according to specific needs. They were then leveraged to implement the visualization layer of the architecture. In particular, time sequences, alarms, and a widget developed ad hoc for the visualization of data on a three-dimensional model. The platform makes it possible to securely forecast, monitor, and control data elements from the Internet of things using external APIs and thus define the relationship between devices, resources, customers, or any other entity and also collects and stores data from telemetry. The platform also collects and stores telemetry data. Thanks to the presence of integrated widgets, the platform allows an intuitive visualization of the data and customization of the interface display mode. In fact, the dashboards can be shared with other users and thus expand functionality, also allowing devices to be controlled remotely and device data to be sent to other systems.

3.3 *Data Visualization*

This study used Dynamo to establish an automatic visualization of the data from the sensors. The transmission interface is carried out through the ThingsBoard platform, which processed the data and automatically transmitted it into the BIM model. The data can be viewed and catalogued within the specific dashboards within the ThingsBoard environment, and then, the data will be automatically projected within the digital twin, and this is possible thanks to the creation of dedicated APIs for direct visualization between ThingsBoard and Dynamo (Fig. 2).

4 Conclusions

The purpose of this work was to introduce a methodology that allows the visualization and representation of data from sensors within the BIM environment to support decisions, sometimes complex, that require interdisciplinary skills. The case

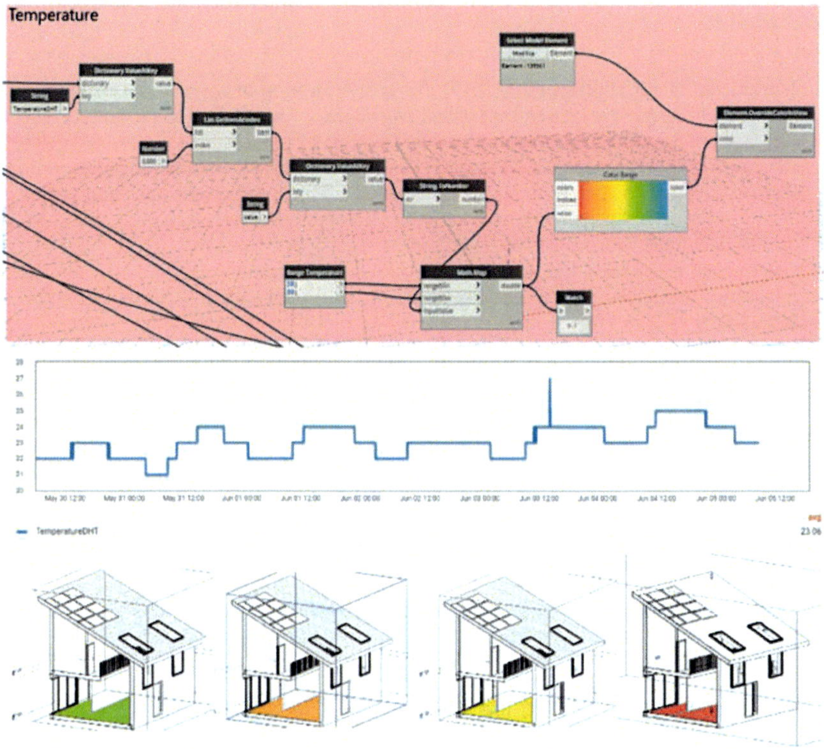

Fig. 2 Temperature data visualization

study was a small-scale prototype of a single-family house that includes several sensors. A preliminary experimental phase was conducted for this case study using Bayesian networks. The proposed methodology integrates the use of an IoT-based platform, ThingsBoard, which allows communication between sensors and Dynamo software to access sensor data, automatically updating the information contained in the BIM model [20]. The experimental results, although preliminary, gathered promising results. They showed that the system is able to learn and manage some actions autonomously, supporting users. Future developments include expanding the database, which could improve the system, introducing contextual parameters that could improve the system's performance in terms of reliability and the development of an application that could help users manage the building.

References

1. Jourdan M, Meyer F, Bacher JP (2019) Towards an integrated approach of building-data management through the convergence of building information modelling and internet of things. J Phys: Conf Ser 1343. https://doi.org/10.1088/1742-6596/1343/1/012135
2. Clarizia F, Colace F, Lombardi M, Pascale F, Santaniello D (2018) A multilevel graph approach for road accidents data interpretation. In: Lecture notes in computer science (including subseries lecture notes in artificial intelligence and lecture notes in bioinformatics), vol 11161. LNCS, pp 303–316. https://doi.org/10.1007/978-3-030-01689-0_24
3. Colace F, Elia C, Guida CG, Lorusso A, Marongiu F, Santaniello D (2021) An IoT-based framework to protect cultural heritage buildings. https://doi.org/10.1109/smartcomp52413.2021.00076
4. Venugopal M, Eastman CM, Sacks R, Teizer J (2012) Semantics of model views for information exchanges using the industry foundation class schema. Adv Eng Inform 26(2). https://doi.org/10.1016/j.aei.2012.01.005
5. Ashton K (2009) That 'internet of things' thing. RFID J 22(7):97–114
6. D'aniello G, Gaeta M, Reformat MZ (2017) Collective perception in smart tourism destinations with rough sets. In: 2017 3rd IEEE international conference on cybernetics, CYBCONF 2017—proceedings, June 2017, pp 1–6. https://doi.org/10.1109/CYBConf.2017.7985765
7. Formato A, Ianniello D, Pellegrino A, Villecco F (2019) Vibration-based experimental identification of the elastic moduli using plate specimens of the olive tree. Machines 7(2):46. https://doi.org/10.3390/machines7020046
8. Casillo M, Colace F, de Santo M, Lorusso A, Mosca R, Santaniello D (2021) VIOT_Lab: a virtual remote laboratory for internet of things based on ThingsBoard platform. In: 2021 IEEE frontiers in education conference (FIE), Oct 2021, pp 1–6. https://doi.org/10.1109/FIE49875.2021.9637317
9. el Saddik A (2018) Digital twins: the convergence of multimedia technologies. IEEE Multimedia 25(2). https://doi.org/10.1109/MMUL.2018.023121167
10. Rasheed A, San O, Kvamsdal T (2020) Digital twin: values, challenges and enablers from a modeling perspective. IEEE Access 8. https://doi.org/10.1109/ACCESS.2020.2970143
11. Delgado JMD, Butler LJ, Brilakis I, Elshafie MZEB, Middleton CR (2018) Structural performance monitoring using a dynamic data-driven BIM environment. J Comput Civil Eng 32(3). https://doi.org/10.1061/(ASCE)CP.1943-5487.0000749
12. Lather JI, Amor R, Messner JI (2017) A case study in data visualization for linked building information model and building management system data. June 2017. https://doi.org/10.1061/9780784480823.028
13. McCaffrey R, Coakley D, Keane M, Melvin H (2015) Development of a web-based BMS data visualisation platform using building information models. In: CIBSE technical symposium, April. https://doi.org/10.13140/RG.2.1.4169.5840
14. Tang F, Ma T, Zhang J, Guan Y, Chen L (2020) Integrating three-dimensional road design and pavement structure analysis based on BIM. Autom Constr 113. https://doi.org/10.1016/j.autcon.2020.103152
15. di Filippo A, Lombardi M, Marongiu F, Lorusso A, Santaniello D (2021) Generative design for project optimization. https://doi.org/10.18293/dmsviva21-014
16. de Simone MC, Guida D (2020) experimental investigation on structural vibrations by a new shaking table. pp 819–831. https://doi.org/10.1007/978-3-030-41057-5_66
17. Matos R, Rodrigues H, Costa A, Rodrigues F (2022) Building condition indicators analysis for BIM-FM integration. Arch Comput Methods Eng. https://doi.org/10.1007/s11831-022-09719-6
18. Henschke M, Wei X, Zhang X (2020) Data visualization for wireless sensor networks using ThingsBoard. https://doi.org/10.1109/WOCC48579.2020.9114929
19. Guida CG, Gupta BB, Lorusso A, Marongiu F, Santaniello D, Troiano A (2021) An integrated BIM-IoT approach to support energy monitoring. In: CEUR workshop proceedings, vol 3080.

20. Matos R, Rodrigues F, Rodrigues H, Costa A (2021) Building condition assessment supported by building information modelling. J Build Eng 38:102186. https://doi.org/10.1016/j.jobe.2021.102186

Deficiencies of Computational Image Recognition in Comparison to Human Counterpart

Vladimir Vinnikov and Ekaterina Pshehotskaya

Abstract The paper is concerned with the cases where machine-based image recognition fails to succeed and becomes inferior to human visual cognition. We consider the computational experiments on the set of specific images and speculate on the nature of these images that is perceivable only by natural intelligence. We deduce that image recognition and computer vision both based on machine learning or even more sophisticated AI models are unable to represent features of human vision due to the lack of tight coupling with the respective physiology.

Keywords Image recognition · Computer vision · Machine learning · Neural networks · Hidden images · Human vision

1 Introduction

Recent advances in cloud computing and neural networking resulted in the wide use of recognition software in a variety of applications. Such software rapidly became an integral part of social networking, banking, and surveillance systems as well as vehicle autopiloting. The state-of-the-art recognition is capable of nearly real-time processing of vast volumes of visual information with a high success ratio. Such high-performing technologies provide users with very convenient services, facilitating mundane operations like authorization and acquisition of metadata from physical media. However, at the same time, these technologies feed the said services with the user data and essentially establish control over users' actions. Such control usually

WWW home page: https://www.researchgate.net/profile/Vladimir_Vinnikov.

V. Vinnikov (✉)
Department of Computer Sciences, Higher School of Economics, Pokrovsky Blvd 11, Moscow, Russian Federation 109028
e-mail: vvinnikov@list.ru

E. Pshehotskaya
Department of Information Security, Moscow Polytechnic University, Avtozavodskaya st., 16 building 4, Moscow, Russian Federation 115280

X.-S. Yang et al. (eds.), *Proceedings of Seventh International Congress on Information and Communication Technology*, Lecture Notes in Networks and Systems 447,
https://doi.org/10.1007/978-981-19-1607-6_43

manifests itself as acts of automatic censorship by the means of social networks. There are much more severe cases of user control, namely tracking of a person of interest via a heterogeneous network of cameras. Therefore, for certain applications, it is worth rendering the ethically ambiguous technologies of visual recognition ineffective, while in other applications, it is desirable to make recognition robust and fail-safe.

The paper is structured as follows. Section 2 outlines the common properties of the challenging cases that demonstrate the deficiencies of computational image recognition. Section 3 introduces the set of images serving as examples of said deficiencies. Section 4 describes the results of respective image recognition and draws a connection from abstract tests to real-world scenarios. Section 5 discusses some hints on how to overcome such deficiencies. Finally, Sect. 6 presents our conclusions.

2 General Considerations on the Challenging Cases for Recognition

As the paper title suggests, we would consider cases, where computer image recognition is unable to see what humans can without a significant effort. The cases should be as simple as possible, implying a singular object to detect. The objects themselves should exhibit as free of ambiguity as possible. It is preferable to have objects representing geometrical shapes.

To fit into the majority of recognition algorithms, the input image should be of moderate dimensions in pixels. We assume that the image size of about a half of the screen resolution should suffice. It is preferable to have source images in the vector format to avoid lossy pixel-sampling while resizing. For the sake of simplicity, it is desirable to have grayscale images for the cases, where colors are not the source of the challenge.

The above-mentioned properties of the cases can be justified by the following reasoning. First, every material object can be coarsely approximated via a set of geometrical 3D primitives like spheres, cones, cylinders, as well as various low vertex polyhedrons. Such an approach to object representation was widely used in old computer games, especially in simulators before the onset of graphic accelerators. It means that geometrical primitives are sufficient to constitute the scene recognizable by human perception.

Second, the whole concept of color is context-dependent, since color perception varies with environment light and the optical properties of the object itself. For example, the object perceived as red-colored under white light would appear black under red light. Such circumstances add excessive complexity to the initial stage of image or scene recognition and are to be accounted for at the later sophisticated stages. It is also worth mentioning that many vision enhancers like night vision devices deal only with light intensity and essentially provide grayscale images as an output.

3 The Set of Challenging Images

The challenges to computer image recognition can be of different nature, namely:

- heavily dithered images;
- images with contrast-based optical illusions.

The challenging images of the first category usually contain a set of rarefied and condensed dots colored from a shallow palette. Such images are used in computer games of the indie genre. The second category is based upon the innate capability of the human visual perception that augments the image with its parts assumed "missing" (e.g., see variants of Kanizsa illusions). This capability is known as a reification being one of the basic principles of Gestalt Theory. There is a corpus of works concerned with illusory and impracticable contours (e.g., see [1–9]).

In the current study, we consider only the second category of challenging images as being the more demonstrative. The selected images are as follows:

- a white triangle overlapping three black circles on a white background (Kanizsa triangle illusion. See Fig. 1, courtesy of Isaac Farias);
- a white square overlapping four black circles on a white background (Kanizsa square illusion. See Fig. 2);
- a white five-pointed star overlapping five black circles on a white background (See Fig. 3, courtesy of Katarzyna Malinowska);
- a white cubic wire-frame projection overlapping eight black circles on a white background (see Fig. 4, courtesy of Wikimedia Commons);
- a projection of a white sphere with conical additions on a white background (See Fig. 5, courtesy of Craig Ward);
- a projection of a black and white soccer ball on a white background (See Fig. 6, courtesy of Adam Wiskerchen);
- a pattern of hatched vertical stripes hiding a panda image (see Fig. 7, authorship by Ilja Klemencov).

Fig. 1 Hidden triangle

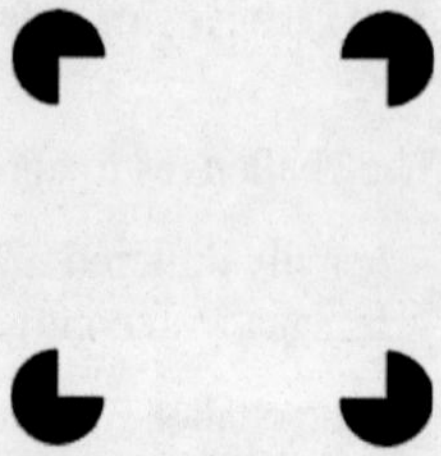

Fig. 2 Hidden square

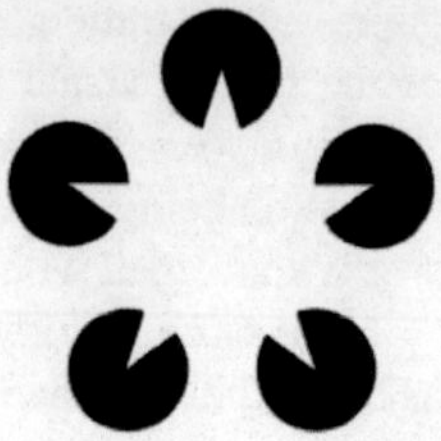

Fig. 3 Hidden star

Fig. 4 Hidden cubic frame

Fig. 5 Hidden sphere

Fig. 6 Hidden soccer ball

Fig. 7 Hidden panda

4 Recognition Results for Challenging Images

We used several online recognition services (e.g., Watson Visual Recognition https://visual-recognition-code-pattern.ng.bluemix.net/). No service was able to detect the hidden shape. We provide the results returned by the mentioned service for the challenging images. The results are represented as tables, containing top recognized classes for each of the images (Tables 1, 2, 3, 4, 5, 6, and 7).

One can see that even the highly sophisticated algorithms of Watson Visual Recognition service carry out only the "literal" detection and recognition and remove white background from consideration. It seems that once the background color is established, any image area with the color same or close to the background is considered

Table 1 Top recognized classes for the hidden triangle

Class	Score
Coal black color	1
Crescent shaped	0.818
Photograph with lines	0.786
Rescue equipment	0.786
Mouthpiece	0.5

Table 2 Top recognized classes for the hidden square

Class	Score
Coal black color	1
Crescent shaped	0.838
Photograph with lines	0.801
Tool	0.795
Lamp	0.527
Electric lamp	0.502
Lamp/electric lamp/arc lamp	0.5

Table 3 Top recognized classes for the hidden star

Class	Score
Coal black color	0.993
Cell	0.6
Food/food product/food ingredient/food seasoning/sassafras	0.571
Microorganism	0.57
Microorganism/paramecium (organism)	0.528
Pick (for string instrument)	0.502

Table 4 Top recognized classes for the hidden cubic frame

Class	Score
Coal black color	0.903
Nature	0.795
Black color	0.667

Table 5 Top recognized classes for the hidden sphere

Class	Score
Figure/star	0.955
Coal black color	0.885
Bird	0.793
Black color	0.518

Table 6 Top recognized classes for the hidden soccer ball

Class	Score
Coal black color	0.953
Polygon	0.942
Figure	0.942
Figure/polygon/concave polygon	0.802
Figure/polygon/convex polygon	0.676

Table 7 Top recognized classes for the hidden panda

Class	Score
Fabric	0.801
Fabric/net	0.796
Ash gray color	0.725
Black color	0.607
Supporting structure/grating	0.59
Framework	0.59
Supporting structure	0.59

Table 8 Top recognized classes for the gray triangle

Class	Score
Conical	0.639
Suction cup	0.608
Mechanism	0.598
Coal black color	0.5

Table 9 Top recognized classes for the red triangle

Class	Score
Dark red color	0.962
Conical	0.93
Round shape	0.798
Figure	0.796

background. For example, the soccer ball case is recognized as belonging to the polygon class.

On the other hand, the benchmark images of the Kanizsa triangle highlighted with gray or red color are recognized much better, yielding a "conical" class among the results (Tables 8 and 9).

5 Discussion

The results showed the inability of computer image recognition algorithms to detect "human-visible" parts of the challenging images. This deficiency is explicitly confirmed in [10]. One can argue that the said impracticable parts do not exist in reality, and therefore, the correct result should be their non-recognition.

We disagree with the validity of the non-recognition since the considered images have a physical basis and very important real-world counterparts. Every challenging image essentially represents a 2D projection of a 3D scene, where the contrast between the object and the background converges to zero. At the same time, the contrast circles represent marker lights outlining the dimensions of the object. The real-world examples range from bridges, cranes, and skyscrapers to port terminals and airport runways, as well as other infrastructure, machinery, and landmarks exposed to the risks of collision. For example, the fatal crash incidents of Tesla autopilots were caused exactly by the failure to distinguish trailer shape from background sky.

The nature of such deficiencies is a difference in the topology of artificial networks designed for recognition and natural networks with the latter having a greater number of neurons and more dense interconnections. To overcome the challenges, one can introduce a preliminary step to image recognition. At this step, some physics-based filters can be applied to the input image. For example, an unsteady process of area color-filling with simulated surface tension and viscosity can be used. Next, the image-by-image recognition of the resulting sequence of images would return the sequence of scored class vectors. Then, at the ending step, the classes with the most stable scores within a sequence can be retrieved and used as a final output.

We assume that the most appropriate approach to image and scene recognition would rely on the inverse problem of recognition using the library of 3D objects ranging from geometrical primitives to real-life models. Such inverse problem builds a scene from 3D objects, renders a projection of the scene to the plane, compares the obtained projection with an input image, and minimizes the differences by iteratively adjusting object positions and sizes. This problem is computationally expensive like all inverse problems. Also, it can suffer from non-uniqueness of the solution due to ambiguities of real-world object sizes. To overcome these complexities, the recognition can also use auxiliary information from range sensors like lidars and radars.

The rendering part of the inverse problem can be regarded as some kind of prediction based on feedback from a comparative part of the iterative algorithm. Using the idea of prediction, it is possible to build an approach to the recognition of illusory shapes that relies on neural networks with feedback (e.g., see [11]). This technique is a trade-off to computationally expensive 3D scene inverse problems and provides solutions that detect illusory shapes with varying accuracy.

6 Conclusion

In the presented study, we experimented on computer recognition of images with hidden objects. Employing Watson Visual Recognition as well as other online services, we revealed that the recognition algorithms fail to detect illusory contours. We also noted the connection between challenging images and the real-world scenarios of objects poorly contrasted against a background. Considering results, we outlined the possible techniques to overcome the deficiencies of currently used approaches to computer image recognition.

The study had the following limitations. We only consider static images but not sequences of frames, representing either the varying layout of the scene or the varying light and exposure conditions. We also use the recognition software trained on datasets of common images. We believe that no special sets with hidden shapes were employed. However, the hidden shapes come in great variety, which is challenging to represent as a training set. Therefore, we assume that these limitations are mild and only insignificantly affect the study.

References

1. Kanizsa G (1955) Quasi-perceptional margins in homogenously stimulated fields. Rivista di Psicologia 49:7–30
2. Simmons S (1996) About the triangle Princeton.edu Website, Retrieved on 1 May 2012
3. Gregory RL (1997) Eye and brain. Princeton University Press, Princeton
4. Homan DD (2000) Visual intelligence: how we create what we see. W.W. Norton & Company
5. Koch C (2004) The quest for consciousness: a neurobiological approach. Roberts & Company Publishers
6. Norretranders T (1999) The user illusion: cutting consciousness down to size. Penguin
7. Mendola J, Dale A, Fischl B, Liu A, Tootell R (1999) The representation of illusory and real contours in human cortical visual areas revealed by functional magnetic resonance imaging. J Neurosci 19(19):8560–8572
8. Knebel J, Murrah M (2012) Towards a resolution of conflicting models of illusory contour processing in humans. Neuroimage 59(3):2808–2817
9. Sary Gy, Koteles K, Kaposvari P, Lenti L, Csifsak G, Franko E, Benedek G, Tompa T (2008) The representation of Kanizsa illusory contours in the monkey inferior temporal cortex. Eur J Neurosci 28(10):2137–2146
10. Baker N, Erlikhman G, Kellman PJ, Lu H (2018) Deep convolutional networks do not perceive illusory vontours. In CogSci
11. Pang Z, O'May CB, Choksi B, VanRullen R (2021) Predictive coding feedback results in perceived illusory contours in a recurrent neural network. arXiv preprint arXiv:2102.01955

Electronic Health Record's Security and Access Control Using Blockchain and IPFS

Md. Yeasin Ali, Suhaib Ahmed, Muhammad Iqbal Hossain, A. B. M. Alim Al Islam, and Jannatun Noor

Abstract An electronic health record (EHR) typically contains sensitive medical records, personal information, doctors' provided prescription, and other physical histories of a patient. This digital approach remodeled the health sector while increasing privacy concerns and possibility of security breaches. This paper proposes an EHR system based on blockchain, interplanetary file system (IPFS), and cryptographic functions and includes features like secure access control having accountability, transparency, immutability of data in a cost-efficient patient-centered architecture which is free from third-party interruption. Here, we divided data into two categories and three types of participants who are verified with digital certificates are granted permission by the patients and then they can access data. Finally, we build and investigate a simple implementation to analyze the cost of the system and propose some approaches to optimize it.

Keywords EHR · Blockchain · IPFS · Access control · Data privacy · Data integrity · Medical record · Health record

1 Introduction

It is the era of rapid technological growth where healthcare technology has seen enormous evolution. Electronic health records are the electronic health information of patients stored in digital format. Currently, this digital way of storing data increases the opportunity to provide more efficient treatment to patients. Also, it helps to coordinate the treatment process for different health service providers. These data help to track the health information of a patient boosting the treatment process

Md. Y. Ali · S. Ahmed (✉) · M. I. Hossain · J. Noor
Department of Computer Science and Engineering, BRAC University, Dhaka, Bangladesh
e-mail: suhaib.ahmed@g.bracu.ac.bd

A. B. M. Alim Al Islam
Department of Computer Science and Engineering, Bangladesh University of Engineering and Technology, Dhaka, Bangladesh

X.-S. Yang et al. (eds.), *Proceedings of Seventh International Congress on Information and Communication Technology*, Lecture Notes in Networks and Systems 447,
https://doi.org/10.1007/978-981-19-1607-6_44

and minimizing the cost. However, opportunity creates challenges that are indifferent in healthcare. Interestingly, these new challenges open doors to new solutions.

1.1 Why Security and Privacy Is a Concern in Healthcare Technology?

In recent years, EHR systems have modernized healthcare sectors but increased huge concern to data privacy and security for both patients and the country. Other than cyber attack, internal unauthorized access or disclosures [1] and fault of system [2] are now a serious concern similar to cyber attack [3]. Even these can cause the death of patients too [4]. It also causes loss of billions of dollars. Currently, research is increasing in this area. However, still it is in high risk of vulnerability.

1.2 Assessment and Use of Blockchain and IPFS in Healthcare Ecosystem

Storing data in a centralized database is more exposed to security attacks like single-point-of-failure, arbitrary modification, and even transparency issues. Some EHR systems use cloud services to store data but cannot ensure immutability. Blockchain [5] is good for security but can get costly for limitations such as lack of scalability, interoperability, excessive energy usage, and so on. Using only cloud services have flaws like the risk of data confidentiality, insecure interface. However, blockchain combined with IPFS [6] in a single ecosystem can help us redesign an architecture to overcome these issues of a traditional EHR system. Together it is decentralized, immutable, transparent, and cost-efficient while maintaining interoperability, faster, and removing third-party dependency.

1.3 Research Contribution

Based on our study, contributions that we make in this paper are as follows-

- We propose an architectural model of an electronic health record system using IPFS and Blockchain that is patient-centered, transparent, and immutable, having secure access control and accountability. Additionally, we make it cost-efficient by categorizing data into two parts called payment data and health records, where health records are the large size data which are stored inside IPFS.
- We build a simple personal Ethereum-based prototype of our architectural model. Also, we propose a systematic file structure for IPFS. Then we analyze some char-

Table 1 Comparison among some EHR system

Author	$C1$	$C2$	$C3$	$C4$	$C5$	$C6$	$C7$	$C8$	$C9$	$C10$	$C11$
Chen et al. [9]	×	✓	✓	Partial	×	×	✓	✓	✓	✓	×
Omar et al. [10]	✓	×	×	×	Not Specified	✓	✓	×	✓	✓	✓
Radhakrishnan et al. [11]	×	×	✓	✓	×	✓	✓	×	✓	✓	×
Harshini et al. [12]	×	×	×	✓	×	×	✓	×	✓	✓	×
Chinnasamy et al. [13]	✓	×	✓	×	Not Specified	×	×	✓	✓	✓	✓
Li et al. [14]	×	✓	✓	×	×	✓	✓	×	✓	✓	✓
Oliveira et al. [15]	×	✓	×	✓	×	×	✓	×	✓	✓	✓
Proposed Architecture (2021)	✓	✓	✓	✓	✓	✓	✓	✓	✓	✓	✓

$C1$—In every situation patient's permission is needed; $C2$—The person who stores data has the accountability for storing it; $C3$—User categorized access control; $C4$—Data stored directly by specialist/Institution; $C5$—Trace both health and payment data; $C6$—Multiple authentication/verification; $C7$— Transperent; $C8$—Cost efficient; $C9$—Immutable; $C10$—Data integriety; $C11$—Confidentiality

acteristics of blockchain transactions and the cost for storing data in blockchain compared to our proposed architecture to evaluate the cost efficiency of the proposed model.

- Finally, we propose some approaches to optimize data size to reduce cost of our model.

2 Related Work

In recent years, making medical data more secure, patient-centered, cost-efficient, and easy for insurance claims gathered much attention. Recent approaches for managing electronic medical records incorporate merely storing in a Blockchain [7]. Different categories to store data into the blockchain while offering methods to limit the access control are proposed. A research [8] proposes a hybrid architecture allowing EHR data access control using blockchain and edge nodes. They assessed some aspects of the model using the hyperledger composer fabric.

In this research [16], a two-layer storage model is proposed that keeps medical data off-blockchain and shares on a blockchain using a certificate-less aggregate signature technique based on the multi-trapdoor hash function to minimize costs.

Moreover, we review other work and show a comparison in Table 2. These papers have a good combination of blockchain and cloud storage but do not cover all the necessary features that we provide.

3 Proposed Architecture

Our system divides users into three participants. First, owners of health records, whose data are stored in blockchain and IPFS categorized as patients. Second, medical specialists, surgeons, and others who produce health records or treat patients are categorized as physicians. Finally, data questers include insurance companies, researchers, or others who search health data for ethical, legal, or research purposes. Data questers can access data but restricted to store it. Here, each participant has a public key and a private key. Only patients have two extra symmetric keys called access key and session key. The access key controls the access of both IPFS storage and blockchain with the help of the session key. Moreover, digital certificates are provided by an official certification authority (CA) to ensure the identity of the user who wants to send and receive the data. All public keys are given by X.509 certificates issued by a trusted CA to verify the user's public key by the globally approved X.509 PKI standard [17].

All patients data are categorized as payment data and health records. Payment data are text-based and stored in blockchain since it cannot be erased or changed. Moreover, it helps both the patient and the insurance company to claim insurance easily. These payment data include bills for doctor's appointments, operations, treatments, hospitals, medicines, and deductibles. It increases transparency and prevents medical facilities from overcharging patients for treatment. Besides health records are both image and text-based data which takes more storage in IPFS. These data cover every health-relevant event such as surgical history, obstetric history, medical encounters, immunization records, medications, family history, social history, habits, development history, and demographics. Finally, to achieve patient-centered access control, our proposed model of EHR system can be described in three segments which are:

3.1 Key Generation and Store

First, the patient registers in the system and generates an access key and a session key. Then encrypts the access key using his public key and stores that encrypted access key in the blockchain. It is the first transaction where the access key is stored into the block to ensure integrity. Figure 2 shows a general approach of the access key generation and storing process. Noticeably, patients only store and maintain the access key and the session key. Therefore, if the key revocation is necessary, the patient can generate a new access key and use it to encrypt all data again. Then this

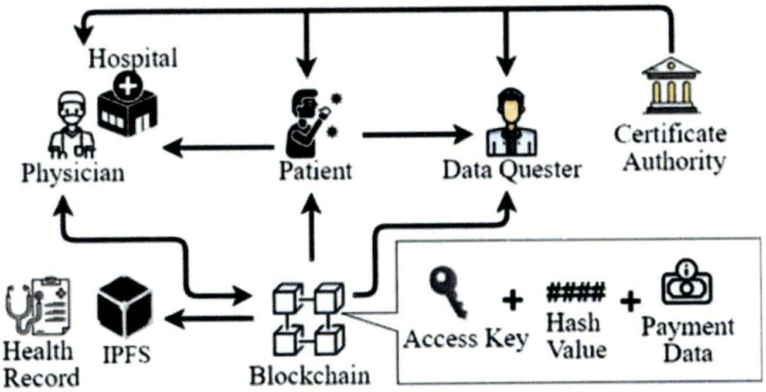

Fig. 1 Overview of the proposed system

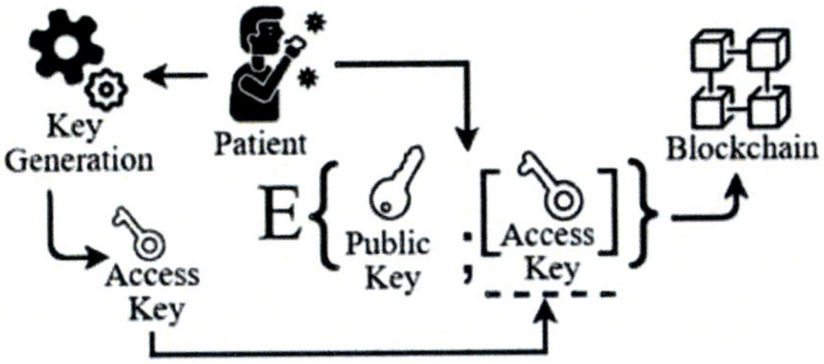

Fig. 2 Access key generation and storing

access key is also encrypted with the public key and stored inside the blockchain. The process is similar for a private key also. It ensures the continuity of the flow of the architecture.

3.2 *Data Access*

To access data, at first, data receivers that are data questers or physicians send an access request to the patient with the digital certificate attached with the public key. Now, the patient retrieves his previously encrypted access key from the blockchain. The payment data and previous hash of IPFS are encrypted using cascade encryption by the physician. Therefore, the patient can decrypt it using the old session key followed by the access key. Now, the patient encrypts payment data and the hash of IPFS using the access key and then the encrypted data is again encrypted using the newly generated session key. These multiple encryptions are known as cascade encryption or cascade cipher. If payment data is not needed, only the hash of IPFS is encrypted. Then the access key is encrypted using the receiver's public key. Finally, all this encrypted data is stored in the blockchain. Figure 3 shows the sharing of the access key through blockchain (Fig. 1).

In addition, while these tasks are being operated, the patient generates an one-time password (OTP) and encrypts this along with the data address of IPFS, and the session key with the receiver's public key and sends it to that particular receiver through a secure communication channel. Even if someone breaks into this communication channel, the data is still secure as it is encrypted with the receiver's public key. Therefore, their private key is the only way to decrypt it. Moreover, the access key is encrypted using the receiver's public key after verifying them by their digital certificate. The OTP is to access the data stored in IPFS. Noticeably, sending multiple data through a communication channel contradicts using the features of blockchain. Fortunately, even by using only OTP, one cannot access IPFS data as it also needs the previously stored health record's hash of IPFS from the blockchain. Therefore, this contradiction is necessary for our model because this ensures that the receiver can access data in IPFS. This ability is gathered from blockchain by getting the access key with the direct consent of the patient where OTP and session key represent the direct consent. It also reduces the possibility of sending the data to the wrong person as double communication is maintained by blockchain and a direct channel. Finally, the receivers have the access key retrieved from blockchain and session key, OTP, and the address of data received from the patient through the communication channel. Therefore, the receiver can decrypt the encrypted access key using the private key and by using the access key and session key all other encrypted data retrieved from the blockchain can be decrypted. But, before getting access to the encrypted data of IPFS, receivers need to provide the previous hash of the IPFS data, which is retrieved from the blockchain and the OTP received from the patient directly by the communication channel.

3.3 Data Store

Finally, after getting data access, only physicians can generate new data of the patient. Thereafter, the physician uses the access key to encrypt the medical records that can be image-based and text-based and store them in IPFS. It returns a new hash of all the health records existing in IPFS. Then, the payment data and the new hash value are encrypted using cascade encryption also known as cascade cipher. It increases the operating time but maintains the patient-centered access control. It also ensures that physicians cannot access these data in the future because this session key is valid for a short period until the patient generates a new session key. While storing data in IPFS and blockchain, the physician digitally signs data with his private key guaranteeing the non-repudiation of storing data. Figure 4 shows a general idea of the data storing process. In the future, to approve access to data for any receiver, the patient decrypts this cascade encrypted data using the session key and generates a new session key and by using this new key along with other key, data are encrypted. The rest approaches are similar and sequential to Sects. 3.1, 3.2, and 3.3. Our proposed model is patient-centered. But the patient cannot store any data into IPFS or blockchain because if

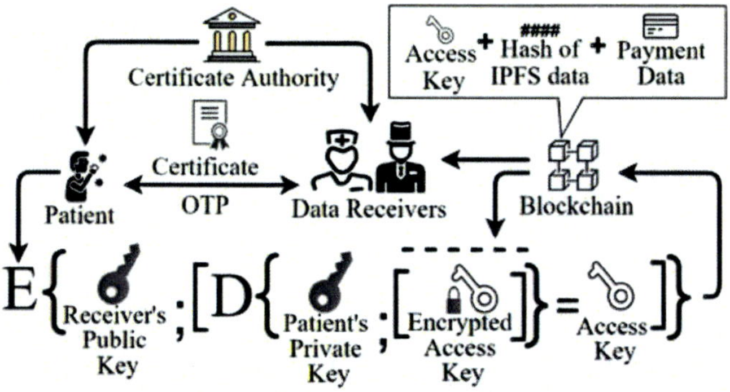

Fig. 3 Access key sharing via blockchain

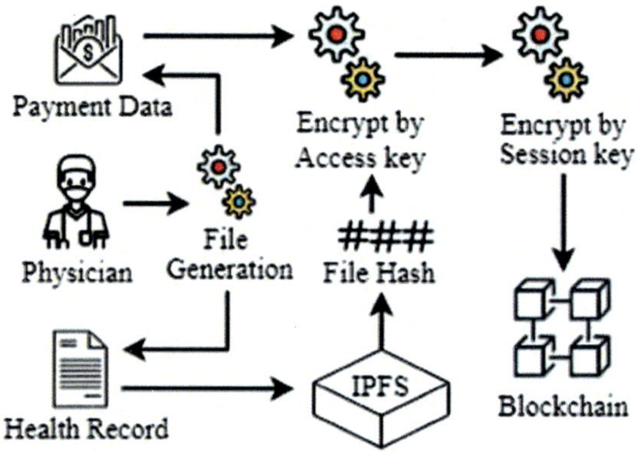

Fig. 4 Generate and store data

the patient is allowed to store data, they can alter the data before storing them, which increases the possibility of insurance fraud.

3.4 Systematic File Structure of IPFS

File management of stored data is essential to make the system faster and user-friendly. That is why we organize data by creating a separate parent folder for each patient under the root folder of IPFS. The parent folder contains two child folders. One is for storing image-based data and another one is for storing text-based data. These data are arranged with the dates to keep track. It helps physicians to track and treat patients more efficiently.

4 Experimental Analysis

We use remix IDE[1] to create smart contract and analyze the cost for storing data in Ethereum blockchain. We then build a prototype shown in Fig. 7 to analyze prospects of blockchain as per our model in Ethereum platform. The prototype is built and tested in a computer of Intel(R)Core(TM)i3-7100 CPU, 4 GB Ram, and the operating system is Windows10-Version20H2(OS Build 19042.867).

4.1 Cost Characteristics and Analysis to Store Data in Blockchain

We use a simple smart contract to analyze the cost for storing data in blockchain using Remix-IDE. We store 1 kb to 70 kb of data. First, we store some data of equal length three times consecutively. Next, data size is increased and stored in blockchain again for three times consecutively where data are different but equal in length. In this manner, Figs. 5 and 6 show the gas used for storing data started from 1 kb to 70 kb. In Ethereum, cost is calculated by the amount of gas which is used for transaction and execution. Here transaction costs are the amount of gas used for the base transaction. To store each byte in blockchain, some fixed amount of gas is used as transaction cost that can be found in Ethereum yellow paper.[2] Besides, execution cost varies from system to system depending on the smart contract. We identify some special characteristics of execution cost and transaction cost that is if we store any data in the blockchain and next if we again store larger data in size compared to the most recent stored data size, the execution cost and transaction cost increases. But, if the stored data length is equal to the previous one, then the execution cost reduces significantly but transaction cost remains same. If we again store data of same size, then both of the cost will remain same. It is same for the rest of the consecutive transaction if their data lengths are equal. From Figs. 5 and 6, we see storing 1 kb data in blockchain needs 91288 gas or 7.20 USD as transaction cost and 649614 gas or 51.21 USD as execution cost. Similarly after a transaction of 1 kb, if the next transaction is also of 1 kb, then the execution cost is reduced to 165400 gas or 13.04 USD and the transaction cost remain the same as previous and so on for the rest of the consecutive transaction if the data are equal in size. This is too costly compared to the data size we are storing.

We calculate all the costs based on January 17, 2021, when the average price for 1 gas was 63.89 Gwei and price of 1 Ether(eth) was 1233.73 USD.[3]

---

[1] Remix- Ethereum IDE. https://remix.ethereum.org.

[2] Ethereum Yellow Paper- https://ethereum.github.io/yellowpaper/paper.pdf.

[3] Price chart of ether- https://etherscan.io/chart/etherprice.

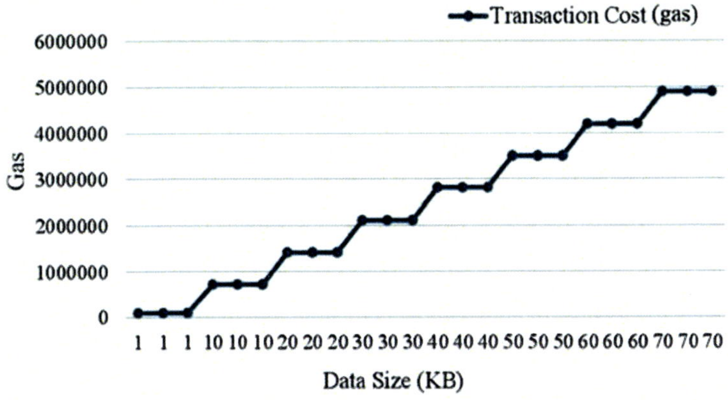

Fig. 5 Transaction cost in gas

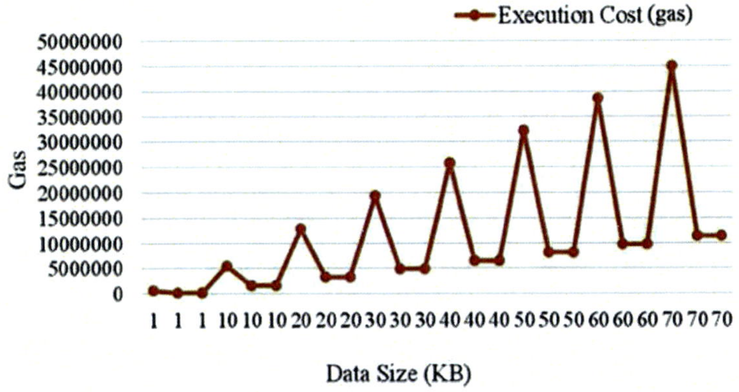

Fig. 6 Execution cost in gas

4.2 *Cost Assessment for Blockchain Using Proposed Architecture*

To evaluate the model using the prototype, we store the same data used in Sect. 4.1. However, now as per our architecture, we store these data in IPFS through an Application Programming Interface (API)[4] instead of blockchain. This returns a hash of 46-byte of the stored data which is automatically stored inside the blockchain. The workflow of the prototype is shown in Fig. 7.

As we store the health records in IPFS and the hash in blockchain, we determine the cost for storing this hash. This does not include the payment data and keys that we propose in the model. Figure 8 shows the cost for storing the 46-byte hash, in the

[4] API of IPFS that were used in this paper from https://infura.io/.

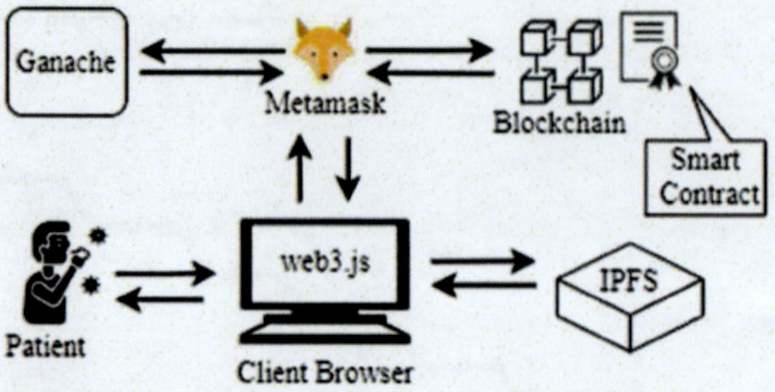

Fig. 7 Prototype workflow

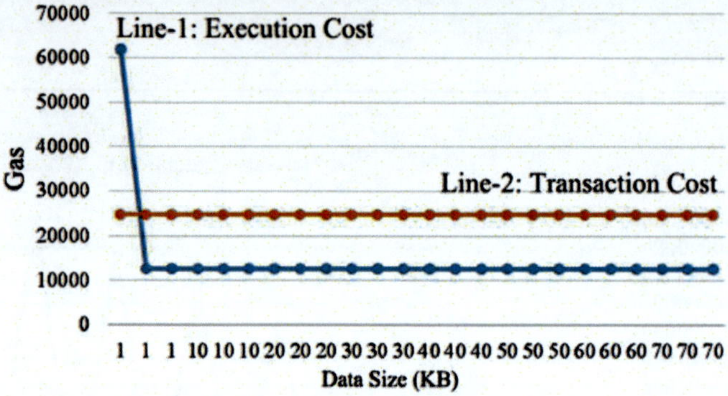

Fig. 8 Gas usage for 46 bytes of hash of different size of data

amount of gas. As the hash size is fixed, each of the transaction cost remains the same. Line-2 of Fig. 8 illustrates this transaction cost. Line-1 is the execution cost represents a similar characteristic for the execution cost of blockchain as we mentioned in Sect. 4.1. Comparing the cost for the same data from Figs. 5, 6, and 8, we see a notable difference which is, for example, for 1 kb data, in our approach, the total transaction cost and execution cost stand for the first transaction at 86715 gas worth of 6.82 USD and second transaction takes a total of 37515 gas worth of 2.96 USD and same for rest transaction. However, in Sect. 4.1 we see, storing this same data directly into the blockchain, in first transaction takes 740,902 gas worth of 58.41 USD and second transaction takes 256,688 gas worth of 20.24 USD. So the proposed approach can reduce cost while preserving the security measurement. Though execution cost may vary in other situations as it depends on the smart contract, the transaction cost is approximately similar for other use cases.

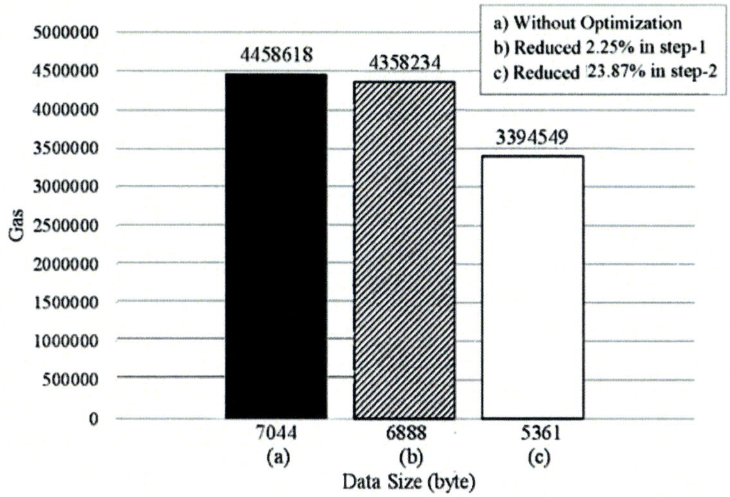

Fig. 9 Execution cost optimization

5 Cost Optimization

To analyze the prototype, we store health records inside IPFS and the corresponding hash into the blockchain. However, as per our proposed architecture, we also need to store payment data in the blockchain. We notice title of the diagnostic test may take a remarkable space in payment data. Therefore, we analyze a total of 401 titles of different diagnostic tests of 7044 bytes total. After two steps of optimization, 21.59% data is reduced. Figures 10 and 9 show optimization of transaction and execution cost in gas of two steps. After first step, total gas costs were reduced by a total of 8.73 USD and after second step reduced by 91.87 USD which is 23.75% less than the initial cost.

Steps that we follow to optimize the title of diagnostic tests are-

- Remove parenthesis from the title. Instead, dash(-) can be used. For example, instead of writing "X-ray (KUB)" write "X-ray-KUB."
- Use acronyms whenever it is possible. For example, instead of writing "Magnetic Resonance Imaging," write "MRI."

We also recommend using acronyms of the test title if possible. Moreover, instead of using full title of the test, a specified numeric code between 4 and 6 digits can be used for the specific test title. As of our analysis, we find a title takes 16 characters on average. Therefore, representing them using numeric numbers can reduce the data size lot more than our result.

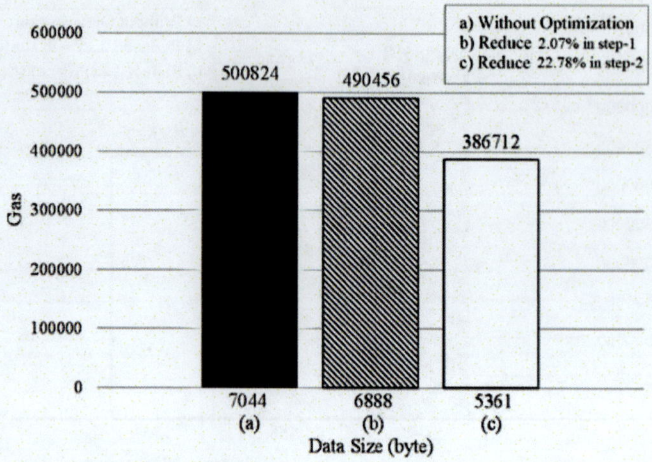

Fig. 10 Transaction cost optimization

6 Conclusion and Future Work

One of our key goal is safeguarding data access through patient's permission. But, situation may arise when patient is unconscious and unable to grant access. We must find a way to access data without compromising security and privacy. Besides, price of cryptocurrencies always fluctuates which may raise a situation when per transaction fee becomes too high. So, way to limit the cost is yet to figure out. Finally, till now blockchain has created enormous impact to build protected systems. Our patient-centered architecture is another upgraded addition to those where IPFS, cryptographic functions together with blockchain assures cost efficiency, secure access control having accountability, transparency, immutability of data without third-party interruption.

Acknowledgements This work was funded by *Department of Computer Science and Engineering, BRAC University, Dhaka, Bangladesh.*

References

1. "Ten Defendants Charged": Ten Defendants Charged in $1.4 Billion Rural Hospital Pass-Through Billing Scheme. https://www.justice.gov/opa/pr/ten-defendants-charged-14-billion-rural-hospital-pass-through-billing-scheme (2020). Accessed 26 Apr 2021
2. Schulte F, Erika F (2020) Electronic health records creating a 'New Era' of health care fraud. Kaiser Health News. https://khn.org/news/electronic-health-records-creating-a-new-era-of-health-care-fraud-officials-say. Accessed 10 May 2021
3. Huet N (2021) Several French hospitals crippled by cyberattacks. Euronews. https://www.euronews.com/2021/02/16/several-french-hospitals-crippled-by-cyberattacks. Accessed 10 Apr 2021

4. Eddy M, Perlroth N (2020) Cyber attack suspected in German woman's death. New York Times. https://www.nytimes.com/2020/09/18/world/europe/cyber-attack-germany-ransomeware-death.html. Accessed 10 May 2021
5. Yaga D, Mell P, Roby N, Scarfone K (2018) Blockchain technology overview. https://doi.org/10.6028/nist.ir.8202
6. "IPFS": How IPFS works. IPFS. https://docs.ipfs.io/concepts/ (2021). Accessed 6 Oct 2021
7. Magyar G (2017) Blockchain: solving the privacy and research availability tradeoff for EHR data: a new disruptive technology in health data management. in: 2017 IEEE 30th Neumann Colloquium (NC). https://doi.org/10.1109/nc.2017.8263269
8. Guo H, Li W, Nejad M, Shen CC (2019) Access control for electronic health records with hybrid blockchain-edge architecture. In: 2019 IEEE international conference on blockchain (blockchain). https://doi.org/10.1109/blockchain.2019.00015
9. Chen Y et al (2018) Blockchain-based medical records secure storage and medical service framework. J Med Syst 43(1):1–9. https://doi.org/10.1007/s10916-018-1121-4
10. Omar AA, Bhuiyan MZ, Basu A, Kiyomoto S, Rahman MS (2019) Privacy-friendly platform for healthcare data in cloud based on blockchain environment. Future Generation Comput Syst 95:511–521. https://doi.org/10.1109/CITS.2018.8440164
11. Radhakrishnan BL, Joseph AS, Sudhakar S (2019) Securing blockchain based electronic health record using multilevel authentication. In: 2019 5th International Conference on Advanced Computing & Communication Systems (ICACCS). https://doi.org/10.1109/icaccs.2019.8728483
12. Harshini VM, Danai S, Usha HR, Kounte MR (2019) Health record management through blockchain technology. In: 2019 3rd International Conference on Trends in Electronics and Informatics (ICOEI). https://doi.org/10.1109/icoei.2019.8862594
13. Chinnasamy P, Deepalakshmi P (2018) Design of secure storage for health-care cloud using hybrid cryptography. In: 2018 second International Conference on Inventive Communication and Computational Technologies (ICICCT). https://doi.org/10.1109/ICICCT.2018.8473107
14. Li H, Zhu L, Shen M et al (2018) Blockchain-based data preservation system for medical data. J Med Syst 42:141. https://doi.org/10.1007/s10916-018-0997-3
15. de Oliveira MT, Reis LH, Carrano RC, Seixas FL, Saade DC, Albuquerque CV, Fernandes NC, Olabarriaga SD, Medeiros DS, Mattos DM et al (2019) Towards a blockchain-based secure electronic medical record for healthcare applications. In: ICC 2019—2019 IEEE International Conference on Communications (ICC). pp 1–6. https://doi.org/10.1109/ICC.2019.8761307
16. Shu H, Qi P, Huang Y, Chen F, Xie D, Sun L (2020) An efficient certificateless aggregate signature scheme for blockchain-based medical cyber physical systems. Sensors 20(5):1521. https://doi.org/10.3390/s20051521
17. Gerck E (1999) Overview of certification systems: X.509, CA, PGP and SKIP. Meta-Certificate Group(MCG)

Implementing Butterfly Key Expansion Using Post-Quantum Algorithms

Ahmad Salman and Zachary Blankinship

Abstract Vehicular ad-hoc networks are important components in intelligent transportation systems which provide a reliable means of communication. Given that a security breach in such networks can result in fatal situation, the security of these networks is as important as their reliability. However, implementing security is not always a straightforward task especially when there are different devices with different capabilities. Additionally, current public-key infrastructure makes use of public-key algorithms that are susceptible to quantum-computation attacks making them not usable if quantum computers become a practical reality. In this paper, we present a public-key infrastructure implementation which makes use of a post-quantum algorithm providing an extra layer of protection to it. We compare the results of our implementation to implement which make the use of classical algorithms and show that they are very comparable in terms of speed, power, and energy consumption.

Keywords Post-quantum cryptography · Autonomous vehicles · Unmanned aerial vehicles · VANET · Public-key cryptography · V2X

1 Introduction

Since their introduction in the beginning of the millennia, vehicular ad-hoc network (VANET) has been an important building block of intelligent transportation systems (ITS) [1]. VANET consists of groups of moving or stationary vehicles connected by a wireless network. These vehicles may communicate with remote endpoints, such as manufacturer towers known as vehicle-to-infrastructure (V2I) communication or with other vehicles in transit that come within the range known as vehicle-to-vehicle (V2V) communication.

A. Salman (✉) · Z. Blankinship
James Madison University, Harrisonburg, VA 22801, USA
e-mail: salmanaa@jmu.edu

Z. Blankinship
e-mail: blankiza@dukes.jmu.edu

X.-S. Yang et al. (eds.), *Proceedings of Seventh International Congress on Information and Communication Technology*, Lecture Notes in Networks and Systems 447,
https://doi.org/10.1007/978-981-19-1607-6_45

VANETs have numerous legitimate use-cases, such as collision prevention, safety and blind-spot monitoring, dynamic route-scheduling, and real-time traffic condition monitoring [2]. This type of network is unique in that it undergoes topology changes in quick succession, given that the vehicles are typically moving in and out of range from one other.

The traditional use-case for VANETs is interconnectivity between motor vehicles on roads. However, emerging cases of ITS focus on VANET architectures which include other devices such as electric vehicles (EVs), autonomous vehicles (AVs), and unmanned ariel vehicles (UAVs) [3]. These devices are expected to run all VANET operations as well as other applications while relying on batteries as power source. This requirement means that any future customization applied to the VANET configuration must be optimized for low-power consumption, so as to not overload existing applications and device components.

Unlike traditional networks, security of VANETs is at an utmost importance as human lives are involved. Security services such as confidentiality, integrity, authentication, non-repudiation, and availability are not only necessary for the privacy and protection of data but also for the safety of human beings. Currently, there is no one standard for security in VANETs although there has been several proposed frameworks to provide standardization for VANETs [4, 5].Nevertheless, current VANETs rely on cryptographic functions such as the advanced encryption standard (AES) and elliptic curve cryptography (ECC) to provide secret-key (SK) services and public-key infrastructure (PKI), respectively. However, current public-key (PK) algorithms will not survive the eventual introduction of quantum computing making the need for replacement PK algorithms that are immune to quantum computational attacks a necessity.

In this paper, we introduce a new potential implementation for securing VANETs and address two main elements in which a modern VANET should deliver. The first is having a method to provide PKI that is reliable, fast, and secure against current attacks as well as quantum-computation attacks. The second is to measure the power consumption of the implementation to test its suitability for battery-powered devices in VANETs.

The remaining of the paper is organized as follows. Section 2 reviews the algorithms and protocols used to provide PKI in VANETs. Section 3 discusses the previous work performed in this area of research. In Sect. 4, we discuss the components used in our experiments and how we conduct them. The results are presented and analyzed in Sect. 5. Finally, we conclude the paper in Sect. 6.

2 Background

There are a number of established PK algorithms and protocols that are being used in several applications ranging from resource-limited and IoT devices to high-speed implementations [6].

2.1 Elliptic Curve Cryptography

ECC is a widely used public-key cryptosystems (PKC) which was introduced in mid-80s by Miller [7] and Koblitz [8] independently. It has become part of the National Institute of Standards and Technology (NIST) standard for public keys since the late 90s [9]. ECC is based on the algebraic structure of elliptic curves over finite fields. ECC allows smaller keys compared to other PKC, such as RSA, to provide equivalent security which makes it a preferable choice when it comes to implementing PKI. The main operation in ECC is the scalar multiplication operation which is the multiplication of a generator point G in the field by a scalar integer.

2.2 N-th Degree Truncated Polynomial Ring Units (NTRU)

The N-th degree truncated polynomial ring units (NTRU) is a public-key cryptosystem that uses lattice-based cryptography to encrypt and decrypt data [10]. It consists of two algorithms, NTRUEncrypt and NTRUSign, which are used for data encryption and digital signatures, respectively. Unlike ECC, it is resistant to quantum-computation attacks such as Shor's algorithm.

NTRU is a valid replacement candidate for ECC, as it is one of the Finalists in the NIST post-quantum cryptography (PQC) standardization competition [11].

2.3 Butterfly Keys

Butterfly key is a cryptographic construction that allows a device to request an arbitrary number of certificates, each with different keys and each encrypted with a different encryption key, using a request that contains only one verification public key "seed" and one encryption public key "seed" and two "expansion functions" [12]. The expansion functions allow a second party to calculate an arbitrarily long sequence of statistically uncorrelated (from the point of view of an outside third party) public keys such that only the original device knows the corresponding private keys. Without Butterfly keys, the device would have to send a unique verification key and a unique encryption key for each certificate. Butterfly keys reduce the upload size, allowing requests to be made when there is only spotty connectivity, reduce the work to be done by the requester to calculate the keys and smooth peak requests [13].

3 Previous Work

Several research works focused on security of vehicular network and VANET. Goyal et al. [5] focus on describing different VANETs communication architectures with focus on authentication schemes and how they can be achieved. Kaur et al. [14] discuss the various challenges associated with security in VANETs, different attacks that can be performed on VANET applications and networks, and the security requirements for these networks. Bittle et al. [15] tackle the issue of the overhead posed on VANETs by the addition of security protocols and propose the usage of cross-layer coordinated message assembling mechanisms that can reduce design weaknesses. Karimireddy and Bakshi [16] proposed a hybrid security framework which combines RSA public-key algorithm with the advanced encryption standard (AES) secret-key algorithm to increase the accuracy of packet delivery as well as the security level. Bariah et al. [4] survey different attacks on VANETs and propose PKI and identity-based (ID-based) technique as a solution to protecting against most of these attacks through security services such as non-repudiation and digital signatures. The usage of Butterfly keys in PKI for VANETs has shown promising results. In [17], Hammi et al. introduce a PKI implementation which uses the Butterfly scheme and shows that it has a better performance than traditional key management algorithms. Another implementation of Butterfly keys in VANETs is devised by the Crash Avoidance Metrics Partners (CAMP) LLC, which is made up of partners including Honda, Hyundai-Kia Motors, Nissan, Ford, General Motors, Mazda, and Volkswagen, and in conjunction with the United Status Department of Transportation (USDOT). In this implementation are three main parties: the end-device, the Registration Authority (RA), and the Certificate Authority (CA). The device (e.g., a vehicle) submits a single Certificate Signing Request (CSR) with "seed" keys, which are collated and organized by a Registration Authority (RA), and forwarded to the CA for signing. The device then receives the CA's signed response, and derives many certificates at once. A full description of CAMP's implementation may be found at [13].

4 Methodology

4.1 Vehicle Ad-Hoc Networks (VANETs)

VANETs have multiple use-cases and features. Each of these features introduced by VANETs require V2V and V2I communications to produce the relevant information. These communications occur in quick succession and may only last a matter of seconds or minutes. With these newly introduced features, they also introduce certain concerns regarding the network itself with the three main ones being as follows:

Data Security and Authenticity This concern is spawned by the number of peers required for the device to interact within the VANET. Each endpoint requires a compatible security implementation and proof-of-authenticity from a commonly agreed authority. The solution to this concern is the introduction of Certificate Authorities (CAs), which may fulfill and sign certificates on the device's behalf. This provides both confidentialities through encryption and integrity though certificate signing.

Data Dissemination This is another concern related to the previous, in that data must be sent quickly and successively between VANET endpoints. This requires low-bandwidth security implementations that can adapt to rapid changes in network topology and status. A VANET endpoint must be able to securely communicate with multiple other devices without compromising or leaking communications with a previous device. The solution to this concern is the ability to create multiple keys and certificates quickly while in the vicinity of a CA.

Power Consumption Many devices within the VANET are battery-powered making power consumption a concern. These devices typically have small power supplies, compared to an enterprise tower installation. Physical space must be conserved for other critical components aboard the device. Therefore, it is required that all security components consume the least amount of power as possible and do so efficiently without compromising the rest of the device's internal components or suffering a long time for computationally intensive cryptographic functions.

4.2 Butterfly Key Expansion

Butterfly keys are a credential management system specifically designed for V2V and V2I communications within a VANET and are being developed by the US Department of Transportation in conjunction with the CAMP LLC. as described previously. The general workflow of CAMP is illustrated in Fig. 1.

Butterfly keys have four use-cases, device bootstrapping, certificate provisioning, (mis)behavior reporting, and certificate revocation. These features are fulfilled by the concept of pseudonym certificates, and the involvement of multiple authority entities that help collate and assign these certificates. This works by a VANET device interacting with a Registration Authority (RA) that acts as a middle-entity between an end-device and a CA. The RA maintains privacy by having no knowledge of the secret keys involved with the Certificate Signing Requests (CSRs) and the certificates assigned by the CA. Instead, the RA merely aggregates CSRs and certificates between end-devices and CAs and keeps track of which device submitted which request. This allows for one VANET device to submit one request and receive multiple certificates in return. However, this process has some challenges which are described as follows

Certificate Revocation Privacy Certificate revocation can happen as a result of possible insider attacks among the triad of entities: the device, RA, and CA. This could theoretically be achieved by nature of the Security Credential Management System (SCMS), which is different from a typical PKI system due to two characteristics: the size of the SCMS and the number of certificates the SCMS will generate. If it becomes an official standard for communications between VANET members, the SCMS is expected to generate over 300 billion certificates per year [18]. As such, insider security is a major concern among the members of the SCMS. The solution to

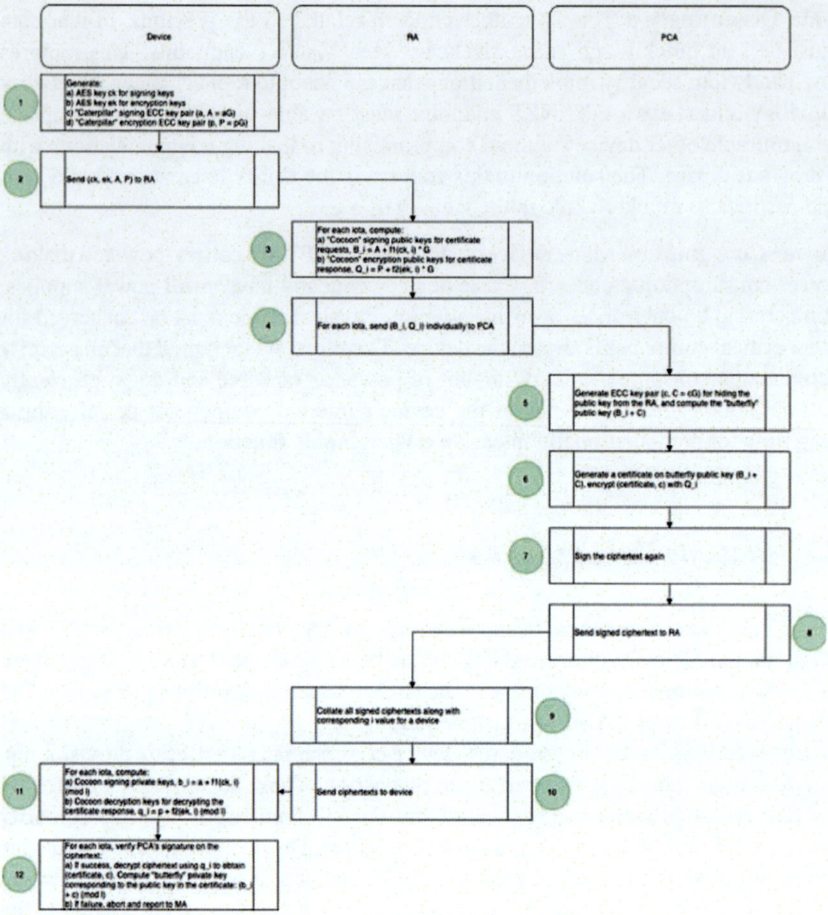

Fig. 1 CAMP workflow

this is frequent certifications and revocations among the VANET devices (e.g., every 5 min). This makes it difficult for any one entity to have the ability to decrypt communications from a device for an extended period of time. Butterfly key expansion provides such ability.

Certificate Revocation Efficiency This challenge spawns as a result from the solution to the previous concern. That is, if certificate revocations are occurring rapidly, certificate generation becomes an expensive operation for the device. The solution to this is the CAR-2-CAR Communication Consortium (C2C-CC) model [19], which issues multiple certificates to a device at the same time in batches, and is valid for the same duration of time. Limitations are placed on the certificates that they are used only for specific functions and thus are separately used to communicate with

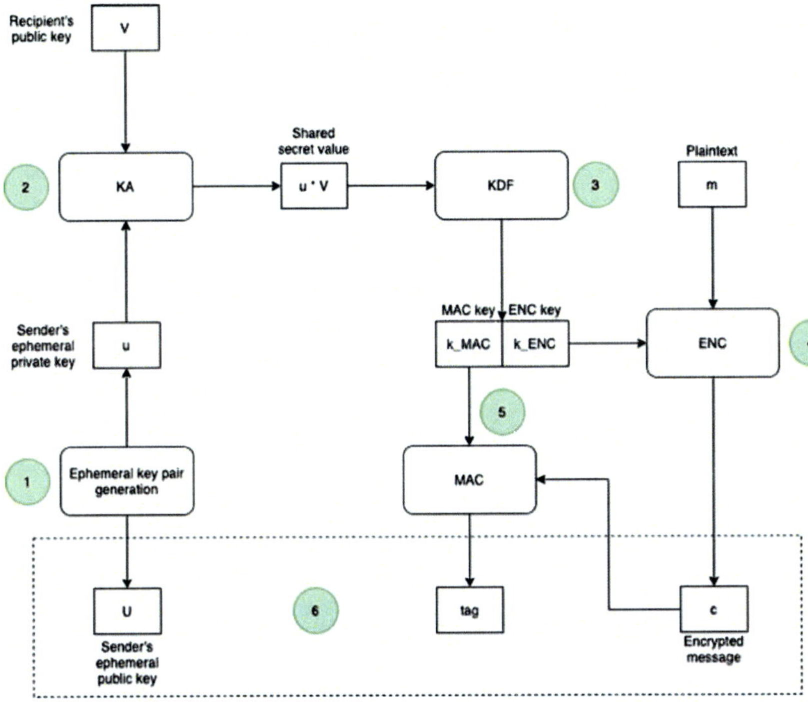

Fig. 2 Elliptic curve integrated encryption scheme workflow

different devices within the VANET. For example, a device may use CertA for bootstrapping and startup, then use CertB to obtain travel information, and then use CertC to communicate with another device.

In the CAMP implementation, end-entity certificates contain a linkage value (LV), which is derived from linkage seed material to support revocation. Once published, the seed revokes all certificates belonging to the device it was published to. However, not knowing the seed is sufficient to protect against an eavesdropper who would not be able to tell which certificates belong to which device.

ECC key pairs are used primarily for encryption of both certificate signing requests (CSRs) by an end-device and certificate responses from a CA. The CAMP implementation of Butterfly keys uses the Elliptic Curve Integrated Encryption Scheme (ECIES) as shown in Fig. 2, which provides semantic security against an adversary who is allowed to use chosen-plaintext and chosen-ciphertext attacks.

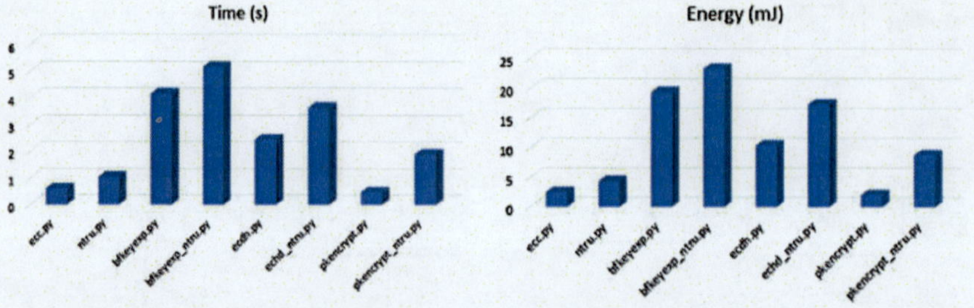

(a) Time Required to Run Each Script (b) Energy consumption for Each Script

Fig. 3 Time and energy consumption for the CAMP scripts

4.3 Implementation

To protect against quantum-computation attacks, we replaced the implementations of the CAMP core scripts (ecc.py, bfkey.py, ecdh.py, and pkenc.py), which made use of ECC, with new implementations (ntru.py, bfkey_n.py, ecdh_n.py, and pkenc_n.py) which make use of NTRU as the core PKC. Modifications to the high-level protocols used in CAMP were kept to the minimum to maintain a fair comparison between the original implementation and our design which is immune to quantum-computation attacks. We chose to make our implementation on Raspberry Pi 4 with a 4-core CPU and 4 GB RAM running Raspbian GNU/Linux 10 (buster). We performed time and power consumption measurements on each of the core scripts for the original (using ECC) and modified (using NTRU) ones.

5 Results

The results presented in Table 1 represent the average time, memory usage, and power and energy consumption over 10 runs for each of the scripts. It can be shown that the running time of the core NTRU requires about 58% more time than ECC. Consequently, the higher protocols which make use of these core algorithms have also increased when using NTRU as shown in Fig. 3a. However, the increase in time is acceptable given the protection against quantum-computation attacks NTRU provides. Additionally, the power consumption for the NTRU implementations has shown slight increase compared to those from the original implementations based on ECC. Finally, the increase in both time and power consumption resulted in increase in the energy consumption of each of the NTRU scripts as visualized in Fig. 3b. Overall the increase is acceptable for the extra protection provided by using NTRU as the core PKC.

Table 1 Summary of time required to run each script, memory usage, and power and energy consumption for each of the original and modified scripts

Script	Description	Time (s)	Mem (kb)	Power (mW)	Energy (mJ)
ecc.py	Performs the scalar multiplication operation	0.618	15408	4.06	2.50908
ntru.py	Implements the core PKI with NTRU	1.054	20808	4.16	4.38464
bfkey.py	Performs Butterfly key Expansion	4.154	30712	4.65	19.3161
bfkey_n.py	implements bfkey.py using NTRU	5.156	15648	4.71	24.28476
ecdh.py	Performs the Diffie-Hellman key-exchange	2.438	15308	4.26	10.38588
echd_n.py	implements ecdh.py using NTRU	3.627	20692	4.76	17.26452
pkenc.py	Encrypts data using ECC	0.471	15408	4.31	2.03001
pkenc_n.py	implements pkenc.py using NTRU	1.895	20492	4.62	8.7549

6 Conclusion and Future Work

In this paper, we presented the implementation of PKI for VANETs using a post-quantum algorithm. It was observed that the NTRU implementation of the CAMP Butterfly keys took more time to execute compared to the ECC counterparts and consuming slightly more power overall. As a future work, we would like to optimize the implementation of the NTRU for time usage, power consumption, and optimize the Butterfly key implementation for Python 3.

Acknowledgements This work was funded by the Commonwealth Cyber Initiative (CCI), an investment in the advancement of cyber R&D, innovation and workforce development (Grant no. E2235831).

References

1. Toh C-K (2001) Ad Hoc mobile wireless networks: protocols and systems
2. Bhoi S, Khilar P (2014) Vehicular communication: a survey. Networks IET
3. Raza A, Bukhari SHR, Aadil F, Iqbal Z (2021) An UAV-assisted VANET architecture for intelligent transportation system in smart cities. Int J Distributed Sensor Netw
4. Bariah L, Shehada D, Salahat E, Yeun CY (2015) Recent advances in VANET security: a survey. In: Vehicular Technology Conference (VTC Fall)
5. Goyal AK, Kumar Tripathi A, Agarwal G (2019) Security attacks, requirements and authentication schemes in VANET. In: 2019 international conference on Issues and Challenges in Intelligent Computing Techniques (ICICT), pp 1–5
6. Salman A, Ferozpuri A, Homsirikamol E, Yalla P, Kaps J, Gaj K (2017) A scalable ECC processor implementation for high-speed and lightweight with side-channel countermeasures. In: 2017 international conference on ReConFigurable Computing and FPGAs (ReConFig)
7. Miller VS (1985) Use of elliptic curves in cryptography. Lecture Notes in CS
8. Koblitz N (1987) Elliptic curve cryptosystems. In: Mathematics of computation
9. Digital Signature Standard (DSS) (2019) National Institute of Standards and Technology (NIST) FIPS Publication, pp 186–185
10. IEEE Draft Standard Specification for Public-Key Cryptographic Techniques Based on Hard Problems Over Lattices. In: IEEE Unapproved Draft Std P1363.1/D12, Oct 2008
11. Post-Quantum Cryptography Standardization. https://csrc.nist.gov/Projects/post-quantum-cryptography/post-quantum-cryptography-standardization
12. Kumar V (2017) Special cryptographic primitives in SCMS. SCP1: Butterfly Keys
13. Crash Avoidance Metrics Partners LLC. https://wiki.campllc.org/display/SCP/SCP1%3A+Buttery+Keys
14. Kaur R, Singh TP, Khajuria V (2018) Security issues in vehicular Ad-Hoc network. In: 2nd international conference on trends in electronics and informatics
15. Bittl S, Roscher K, Gonzalez AA (2015) Security overhead and its impact in VANETs. In: 8th IFIP Wireless and Mobile Networking Conference (WMNC)
16. Karimireddy T, Bakshi AGA (2016) A hybrid security framework for the vehicular communications in VANET. In: 2016 international conference on Wireless Communications, Signal Processing and Networking (WiSPNET)
17. Hammi B, Monteuuis JP, Labiod H, Khatoun R, Serhrouchni A (2017) Using butterfly keys: a performance study of pseudonym certificates requests in C-ITS. In: 2017 1st Cyber Security in Networking Conference (CSNet), pp 1–6
18. U.S. Department of Transportation—Research and Innovative Technology Administration. Vehicle-to-Vehicle (V2V) Communications for Safety
19. BiBmeyer N, Stiibing H, Schoch E, Gotz S, Stolz JP, Lonc B (2011) A generic public key infrastructure for securing car-to-X communication. In: 18th world congress on intelligent transport systems

Perceived Readiness of Information and Communication Technology Policy in Supporting Mobile Learning in Times of COVID-19 at South African Schools

Baldreck Chipangura

Abstract This study answers the research question, "how do teachers perceive the readiness of information and communication technology (ICT) school policies in supporting the provision of remote m-learning as a strategy for mitigating learning disruptions caused by COVID-19?" To investigate this question, the study adopted and adapted the UNESCO policy guideline for m-learning as an evaluation framework. Data were collected through semi-structured interviews from five schools in South Africa, and ten teachers participated in the study. The sampled teachers perceived the ICT policies at the selected schools as outdated, not recognizing mobile devices as tools for learning, and silent on providing teachers with resources that facilitate m-learning. The results indicate that the ICT policies at the sampled schools were perceived as not supporting the UNESCO policy guideline for m-learning. Perceptions of teachers were investigated in this study because teachers are the custodians of m-learning at schools.

Keywords M-learning · ICT school policy · COVID-19

1 Introduction

The uncertainty brought by the COVID-19 pandemic in school learning brought anxiety to learners, parents, teachers, and governments. In responding to the effects of COVID-19, some South African schools turned to online learning technologies to save learning from a total collapse. One technology adopted as a learning tool for alleviating COVID-19 disruptions was mobile technology. A literature search revealed that mobile technologies are not new in learning and have been utilized for m-learning in the last decade [8, 11, 12]. In South Africa, mobile technologies are the preferred tools for learning because most citizens primarily depend on mobile technologies as their main information and communication tools [10].

B. Chipangura (✉)
University of South Africa, Florida Campus, Johannesburg, South Africa
e-mail: chipab@unisa.ac.za

X.-S. Yang et al. (eds.), *Proceedings of Seventh International Congress on Information and Communication Technology*, Lecture Notes in Networks and Systems 447,
https://doi.org/10.1007/978-981-19-1607-6_46

Whilst m-learning interventions guarantee learning continuity at schools in times of COVID-19 disruption, they are confronted by constraints that include lack of policies that support m-learning [5, 7]. Earlier studies carried in South Africa found school ICT policies outdated and not recognizing mobile technologies as tools for learning [1, 15, 18]. As such, the policies failed to encourage, allocate resources, support, or force the schools to implement m-learning. Accordingly, this study anticipates that remote m-learning will mitigate learning disruptions caused by COVID-19 if ICT school policies support teachers when delivering m-learning. Therefore, this study investigated how teachers perceived the readiness of ICT school policies in supporting the provision of remote m-learning as a strategy for mitigating COVID-19 disruptions. Perceptions of teachers were investigated in this study because they are the custodians of m-learning at schools.

2 Theoretical Framework: UNESCO Policy Guideline for Mobile Learning

The premise of the UNESCO policy guideline for m-learning is that mobile technologies enhance learning by providing an interface for accessing learning resources and streamlining educational management [18]. The policy guideline provides a set of ten recommendations that policy makers should adopt in creating their m-learning policies. The ten guidelines are focussed on creating or updating ICT policies, teacher training, supporting teachers, creating m-learning content, gender equality with respect to m-learning, equal access for all learners, cyber safety and providing mobile connectivity. This study adopted and adapted the UNESCO policy guideline for m-learning [18] as an evaluation framework. Only four recommendations deemed to have a direct impact on supporting teachers when providing m-learning at schools were selected and they are create or update policies related to m-learning, train and support teachers to provide m-learning, create and optimize educational resources for use on mobile devices, and use mobile technology to improve communication.

The selected guidelines were used to lead literature analysis in this study, which focussed on policy, training, and m-learning creating resources.

2.1 Policy

The South African department of education regulates technology use at schools through the white paper on education [6]. The white paper guides schools in integrating electronic devices into the curriculum, providing infrastructure, and training teachers, just to list some important objectives. Even though the white paper provides guidelines, the South African department of education has been criticized for not having a fully-fledged policy on m-learning [1, 15].

At school level, there is a dearth of literature on m-learning policies in South Africa even though numerous case studies were carried at some schools. Learning from published South African studies [7, 13], it was inferred that the studies encountered implementation challenges, and the blame was put on the lack of policies that support m-learning. The vacuum on m-learning policies at schools has been recognized by UNESCO, which proposed a policy guideline for m-learning [18].

2.2 *Training*

Earlier studies identified lack of m-learning teaching skills amongst teachers as a challenge constraining the implementation of m-learning at schools [1, 11]. The studies concurred that lack of m-learning skills amongst teachers was due to lack of training. As a result of the COVID-19 pandemic, UNESCO reiterated that for teachers to be in a better position to facilitate online learning, they need at least a week to prepare for the transition [17]. Hence, for South Africa and many other developing countries, the teachers would not only need to prepare for online learning, but they urgently need to prepare for m-learning. The reason being that mobile devices are practical tools that can provide equal access to learning resources in South Africa in times of COVID-19 disruptions.

2.3 *M-learning Resources*

Blau and Hameiri [3] proposed that both learners and their parents or caregivers require access to mobile learning resources to facilitate learning at home. In that respect, Bano et al. [2] observed that learners require access to resources that enable them to engage in peer tutoring, virtual learning, as well as accessing learning content that include lesson slides, video clips, study manuals, and assignments. Moreover, parents require access to resources that facilitate communication with the school, access to their children's academic profiles such as attendance registers, exam marks, assignment submission, and marks, as well as fees payments [9]. Missen et al. [14] observed that the design of mobile learning resources or applications must provide a positive user experience to all the stakeholders, which are teachers, parents, caregivers, and learners.

3 Methodology

The study purposefully collected and analyzed data from ten teachers at five private schools in South Africa. Private schools were sampled because they implemented online learning as a strategy for mitigating COVID-19 disruptions. From each school, two teachers were interviewed. There were 7 female and 3 male teachers. The five schools were given pseudo names: A, B, C, D, and E, and the participants were given names that reflected the school, for example, Participant A1 or Participant D3.

The interview protocol consisted of four opened ended questions formulated from the four components of the UNESCO policy guideline for m-learning [18] adopted in this study. The interview questions were as follows: Does your school have an ICT policy? How does the ICT school policy ensure that teachers are trained to provide m-learning? How do the ICT school policies support teachers when creating m-learning content? and How do ICT policies support mobile communication in learning?

Data analysis followed Braun and Clarke's [4] deductive thematic analysis, and the UNESCO policy guideline for m-learning [18] was used as an evaluation framework and to structure deductive coding.

4 Results

This section presents the results of data analysis. The components of the UNESCO policy guideline for m-learning (UNESCO 2013) were used as the evaluation framework.

4.1 Status of ICT Policies with Respect to Supporting Mobile Learning

Three schools (B, D, and E) had ICT policies and two schools did not have. The schools with ICT policies belonged to a group of schools, whilst schools with no ICT policies were standalone and privately owned. At school (E), the policy was perceived outdated and having been published in 2009 and at school (B), and the policy was said to be currently under review. Participant E9 said, "*the only two policies that I am aware of are the computer policy and the responsible use of technology policy. .... supporting m-learning, it is something we have not paid attention to in the formal policy.*"

The consensus amongst the eight participants from four schools (A, B, C, and E) was that school ICT policies were not designed to support m-learning but to enforce ethical use of mobile devices. Because of that, school (A) banned learners from bringing cell phones to school but allowed them to bring tablets with no Internet connection. Participant A1 said, "*we have a social media policy for the children's safety but not an ICT policy yet. Personal cell phones, no! tablets, they can bring to school, but no Internet connection.*"

Hence, it can be concluded that the teachers perceived the policies as not aligned with the UNESCO policy guideline for m-learning and not ready to support m-learning.

4.2 Train and Support Teachers to Provide M-learning

All the schools provided teachers with refresher computer training, but three schools (B, D, and E) supported m-learning training. Refresher courses enable teachers to catch-up and gain competence with new teaching technologies. Participant A2 said, *"there is training for the stuff regarding general computing, how to use your regular office programmes to teach. However, we do not have training that is specific towards mobile devices."*

Teachers from schools (B, D, and E) indicated that their schools were migrating from Google classroom to Microsoft Teams, and they had received training. Microsoft teams was recognized for its ability to create content that is accessible on multiple devices that include mobile devices. Teachers from schools (A and B) indicated that they were trained to teach through virtual classrooms, where they can share content, interact, and communicate with students. Participant B4 said, *"......not the policy itself but the company(school) supports us with training, right now, we are rolling out Microsoft teams, previously we used Google classrooms and so forth, so it is been out there for many years at our group of schools."*

School policies supported different modes of training, which included instructor led training by an internal or external technology champion, online self-training (schools C, D, and E) and seminars (school D). School (D) belonged to a group of schools, and they held seminars to share educational technologies used within their schools. Participant D8 said, *"two months ago, we had a session where teachers from different schools came to us and had a seminar on teaching technologies. They showed us how they use technologies at their school, and we also did showcase ours."*

It can be concluded that policies at the schools were perceived as supporting teacher training with respect to using computer technologies.

4.3 Create and Optimize Educational Content for Use on Mobile Devices

Teachers at the five schools perceived the ICT policies as supporting the provision of software and hardware for creating online content. However, the teachers perceived the policies as silent on supporting creation of content that is designed specifically for m-learning. Software packages supported for creating content at some of the schools (B, C, D, and E) included MS Word, Excel, PowerPoint, and Adobe. The content was uploaded on virtual online platforms, for example, Google classroom, MS Teams, or e-learning platforms. The teachers perceived content uploaded on such platforms as accessible on mobile devices. Participant B4 said, *"we are not using any specific software for formal content creation. Teachers do their work, whether it is in PowerPoint presentation or word document or whatever and they can upload on e-learning platform."*

4.4 Use Mobile Technology to Improve Communication

The teachers perceived some schools (A, B, C, and E) as not having formal mobile device communication policies. However, the schools were perceived to have mechanisms for regulating mobile communication, for example, code of ethical behaviour when using WhatsApp. Even though schools had no formal mobile device communication policies, mobile devices were used for communication. Participant E10 said, *"but there is not a formal policy, it is a very informal structure. So, most subjects have WhatsApp groups. The teachers use WhatsApp to send announcements and learners use it to check for update."*

The teachers perceived schools as supporting several communication tools, such as D6 communicator, SMS, Email, and WhatsApp. Whilst the codes of conduct for using these tools were similar across the schools, the codes of conduct for WhatsApp were different. At schools (A and C), WhatsApp communication was not allowed on tuition matters, whilst at schools (B, D, and E), it was allowed. However, all the schools discouraged WhatsApp personal conversations with learners or parents. Participant B4 said, "*we use our phones a lot, all our learners are on WhatsApp groups. We send message out early in the morning to make announcements.*"

At most schools, policies supported mobile communication for administrative communication amongst teachers as well as between the school and the parents.

5 Discussion

The study uncovered that the teachers perceived the status of m-learning policies at the sampled schools as not ready to fully support m-learning in times of COVID-19 pandemic. The policies were perceived as obstructing the deployment of m-learning at schools. The findings conform with prior studies that found ICT policies restrictive and an obstruction to m-learning [16]. On the other hand, the findings contradict the UNESCO policy guideline for mobile learning [18] and the South Africa white paper on e-learning [6], which recommended schools to adopt and integrate technologies in teaching and learning. Even though the policies were perceived as restrictive, the schools were taking some initiatives to introduce remote m-learning as a strategy for mitigating COVID-19 disruptions. The findings confirm that the rate at which technologies are permeating schools is ahead of the rate at which schools are updating their policies to embrace new technologies [5].

The teachers perceived the school policies as supporting continuous professional development. The findings show that the schools are complying with the requirements of the UNESCO policy guideline on m-learning [18] and the South Africa white paper on e-Learning of 2004 [6], which require teachers to receive training. However, only three out of five schools were perceived as supporting teachers with m-learning training whilst the other two schools were supporting general ICT training. Even though policies at some schools were perceived as not supporting m-learning, the teachers perceived the technologies provided by schools as supporting m-learning.

The South Africa e-learning white paper [6] together with the UNESCO policy guideline on m-learning [18], stipulate that school curriculums must be supported through the creation, optimization, and distribution of electronic content to facilitate e-learning or m-learning. In this study, the policies were perceived as complying with the guidelines because at all the schools, it was compulsory to create e-learning content and to use e-learning platforms. The e-learning platforms have underlying technological intelligence for adapting content for mobile device access, and a teacher does not need any skill to create mobile content. The results of this study show a deviation from the results of past studies [7], which expected teachers to have technical skills to create mobile learning content.

This study revealed some antagonistic perspectives on the use of WhatsApp in learning. Some schools prohibited teachers from communicating with learners through WhatsApp, whilst at other schools, it was permitted. The schools that discouraged WhatsApp found it unethical to communicate with minors on social media. However, WhatsApp was allowed for administrative communication with parents at all the schools. The implication on policy is that schools should recognize the advantages and ramifications that come with WhatsApp.

This study was limited by that the sampled schools were private urban schools in Pretoria, South Africa. Given that only teachers at the private schools were interviewed, other stakeholders such as school principals, learners, and parents could have different perspectives on the readiness of school policies in supporting m-learning in times of disaster. Therefore, the results of this study could have been limited by the characteristics of the sample.

6 Conclusion

The study revealed that school ICT policies were generally perceived as designed to restrict the use of mobile devices and social media at the schools. However, the teachers perceived their schools as encouraging mobile technology integration in teaching even though the policies were not supporting them. The findings reflect that the rate at which mobile technologies have permeated teaching and learning at the schools in South Africa is ahead of the rate at which schools are updating their policies to take advantage of the technologies in teaching. The contribution of the paper lies in identifying how the teachers perceive the support that is provided by policy when providing remote m-learning during periods of disaster such as the COVID-19. The perceived support is valuable as it determines if teachers will have the confidence to initiate the provision of remote m-learning as a mitigating strategy when disasters strike teaching and learning at schools. The findings of this study are valuable for they inform the formulation of ICT policies that inform remote learning in periods of disasters such as COVID-19 disruptions in developing countries and the world over.

References

1. Aluko R (2017) Applying UNESCO guidelines on mobile learning in the South African context: creating an enabling environment through policy. Int Rev Res Open Distrib Learn 18(7). https://doi.org/10.19173/irrodl.v18i7.2702
2. Bano M, Zowghi D, Kearney M, Schuck S, Aubusson P (2018) Mobile learning for science and mathematics school education: a systematic review of empirical evidence. Comput Educ 121:30–58. https://doi.org/10.1016/j.compedu.2018.02.006
3. Blau I, Hameiri M (2017) Ubiquitous mobile educational data management by teachers, students and parents: Does technology change school-family communication and parental involvement? Educ Inf Technol 22(3):1231–1247. https://doi.org/10.1007/s10639-015-9456-7
4. Braun V, Clarke V (2013) Successful qualitative research: a practical guide for beginners. Sage, New York
5. Chipangura B, Van Biljon J, Botha A (2015) Evaluating mobile-centric readiness of higher education institutions: the case of institutional policies and information systems students: issues in educational informatics: renewing our human resources for the digital economy. African J Inform Commun 2015(15):4–13. https://journals.co.za/doi/10.10520/EJC189258
6. Department of Education (2004) White paper on e-education: transforming learning and teaching through information and communication technologies. Government Gazette 470(26734). Cape Town, South Africa
7. Herselman M, Botha A, Mayindi D, Reid E (2018) Influences of the ecological systems theory influencing technological use in rural schools in south Africa: a case study. In: International conference on advances in big data, computing and data communication systems (icABCD). IEEE, USA, pp 1–8. https://doi.org/10.1109/ICABCD.2018.8465432
8. Herselman M, Botha A (2014) Designing and implementing an information communication technology for rural education development (ICT4RED) initiative in a resource constraint environment: Nciba school district, Eastern Cape. CSIR, Pretoria, South Africa
9. Hershkovitz A, Elhija MA, Zedan D (2019) WhatsApp is the message: out-of-class communication, student-teacher relationship, and classroom environment. J Inform Technol Educ 18:63–95. https://doi.org/10.28945/4183
10. ICASA (2020) State of the ICT sector in South Africa—2020 report. https://www.icasa.org.za/uploads/files/State-of-the-ICT-Sector-Report-March-2020.pdf. Last Accessed 21 Oct 2021
11. Isaacs S, Roberts N, Spencer-Smith G (2019) Learning with mobile devices: a comparison of four mobile learning pilots in Africa. South African J Educ 39(3). https://doi.org/10.15700/saje.v39n3a1656
12. Jack C, Higgins S (2019) Embedding educational technologies in early years education. Res Learn Technol 27. https://doi.org/10.25304/rlt.v27.2033
13. Jantjies M, Joy M (2015) Mobile enhanced learning in a South African context. J Educ Technol Soc 18(1):308–320
14. Missen MMS, Javed A, Asmat H, Nosheen M, Coustaty M, Salamat N, Prasath VS (2019) Systematic review and usability evaluation of writing mobile apps for children. New Rev Hypermedia Multimedia 25(3):137–160. https://doi.org/10.1080/13614568.2019.1677787
15. Ng'ambi D, Brown C, Bozalek V, Gachago D, Wood D (2016) Technology enhanced teaching and learning in South African higher education–a rearview of a 20-year journey. Br J Edu Technol 47(5):843–858. https://doi.org/10.1111/bjet.12485
16. Ott T, Magnusson AG, Weilenmann A, af Segerstad YH (2018) "It must not disturb, it's as simple as that": students' voices on mobile phones in the infrastructure for learning in Swedish upper secondary school. Educ Inform Technol 23(1):517–536. https://doi.org/10.1007/s10639-017-9615-0
17. UNESCO (2021) Back to school: Preparing and managing the reopening of schools. https://en.unesco.org/news/back-school-preparing-and-managing-reopening-schools. Last Accessed 21 Oct 2021
18. UNESCO (2013) UNESCO policy guidelines for mobile learning. UNESCO. Paris, France

Application of Random Forest Model in the Prediction of River Water Quality

Turuganti Venkateswarlu and Jagadeesh Anmala

Abstract Excessive runoffs from various non-point source land uses and other point sources are rapidly contaminating the river water quality in the Upper Green River watershed, Kentucky, USA. It is essential to maintain the stream water quality as the river basin is one of the significant freshwater sources in this province. It is also important to understand the water quality parameters (WQPs) quantitatively and qualitatively along with their important features as stream water is sensitive to climatic events and land use practices. In this study, a model was developed for predicting one of the significant WQPs, fecal coliform (FC) from precipitation, temperature, forest land use factor (FLUF), agricultural land use factor (ALUF), and urban land use factor (ULUF) using random forest (RF) algorithm. The RF model, a novel ensemble learning algorithm, can even find advanced feature importance characteristics from the given model inputs for different combinations. This model's outcomes showed a good correlation between FC and climate events and land use factors ($R^2 = 0.94$), and precipitation and temperature are the primary influencing factors for FC.

Keywords Water quality · Land use factors · Random forest · Fecal coliform

1 Introduction

Large rivers are particularly good indicators of cumulative impacts. Most streams and rivers function as integrators of terrestrial landscape characteristics and receivers of contaminants from both the atmosphere and the landscape. The well-being of lotic biotas has probably been affected by various hydrologic changes, viz., habitant alteration and landscape transformations to rivers and streams [1]. Intensified fertilizer

T. Venkateswarlu (✉) · J. Anmala
Birla Institute of Technology and Science, Pilani, Hyderabad Campus, Hyderabad, Telangana 500078, India
e-mail: p20170426@hyderabad.bits-pilani.ac.in

J. Anmala
e-mail: jagadeesh@hyderabad.bits-pilani.ac.in

X.-S. Yang et al. (eds.), *Proceedings of Seventh International Congress on Information and Communication Technology*, Lecture Notes in Networks and Systems 447,
https://doi.org/10.1007/978-981-19-1607-6_47

use for agriculture in rural areas of developing countries and various anthropological deeds have substantially influenced the physical and chemical parameters of stream water quality at regional and global scales [16]. In recent study of Park et al. [21] discovered that extents of land use types in riparian zones, including urban, forest, and agricultural areas have a larger effect on biotic societies in rivulets than land cover spatial varieties. The percentage of forest in riparian regions has the most significant impact on the stream environment. In separate studies, Bu et al. [5], Chen, and Lu [8] found an interrelation between land use patterns and stream WQPs using various correlation and factor analysis methods. As a result, the current study focuses on building a model that defines the land use features or factors (LUFs) that impact the stream water quality in the Upper Green watershed, Kentucky, USA. The objectives of the present study are as follows: (i) to predict FC concentrations from temperature, precipitation, and LUFs using RF model and ANNs, (ii) to identify the important features for FC, and (iii) to recommend the suitable model for predicting the stream water quality accurately. The following section highlights the literature of a few research papers that show the influences of land uses on water quality.

2 Literature Review

Several research works have been conducted worldwide to examine the impacts of land use on water quality in a watershed applying different statistical and spatial approaches. Few study results showed that agricultural, urban, and forested, land uses affected water quality and aquatic biome. According to the statistics, urban land uses have a lower supply of nutrients than agricultural land uses. However, a more significant source of suspended sediment and agricultural, and urban streams has modest variations in water quality when compared to forested streams [9]. In another study, Bolstad and Swank [7] discovered continuous, persistent, downstream variations in WQPs along Coweeta Creek linked with artificial changes in land use. Ren et al. [22] explored the relation between urbanization and water quality in Shanghai. The data demonstrate that greater urbanization correlates with a higher degradation of water quality. Although there are some links between land use and water quality, the relationship is complicated. The relationships in different watersheds are likely to be site or geographically specific [20]. They also vary with temperature and seasonal variations in streamflow [17]. Advanced statistical and spatial methodologies and integrated models might be used to assist in comprehending the complicated association between water quality and land use.

It is also necessary to precisely predict WQPs from land uses because several studies have identified connections between public health and stream WQPs such as FC. In recent years, Mokondoko et al. [19] found zones of high- and low-cholera frequency owing to E. coli prevalence in rivers and streams at varied sizes in central Veracruz, Mexico. The regression models coupled with geographical information system (GIS) models were used to find the relations between fecal effluence and land use in Murrells Inlet, a South Carolina estuary located between Georgetown and Myrtle Beach [14]. The regression model was used solely to estimate nutrient and FC concentrations in the lower Piedmont of West Georgia watersheds from land uses under groundwater flow and storm flow conditions [23]. At 24 Eastern Canadian catchments, linear mixed effect models were used to estimate FC and turbidity concentrations based on land use, climate, pedology, and morphology [10]. Anmala et al. [2] created a model based on ANN-GIS for predicting FC from climate parameters such as precipitation, temperature, and innovative land use factors in Upper Green River Basin, Kentucky, USA. According to the findings of [18], the rise in pastures and artificial surfaces is largely related to increased soluble salts and FC concentration in the Zêzere watershed (Portugal). Similarly, [27] created a one-of-a-kind spatial regression model of FC contamination across North Carolina (USA), using urban development characteristics and land cover. As per the above literature research, water quality models are necessary to preserve watershed health, riverine aquatic biome, and water quality from environmental deterioration. A thorough understanding of environmental informatics, particularly the hydrologic impacts of land use changes, is essential. Furthermore, an honest assessment model for prediction of WQPs under diverse conditions is necessary [24]. According to Zampella et al. [28], the developed models must adequately define the relation between watershed disturbance and water quality conditions and highlight the relationships between distinct WQPs and land use patterns. The RFs, a decision tree model, may be applied to extract quantifiable results and draw meaningful conclusions as they are referred to as "white box" models since the accumulated knowledge can be represented clearly. Bui et al. [6] used an RF model to enhance the water quality indices prediction in northern Iran's Talar watershed. Similarly, the RF model performance was superior to the M5 model tree in the prediction of biochemical oxygen demand (BOD) in Karun River at Ahvaz location, Iran [11]. Likewise, Victoriano et al. [26] coupled RF and GIS models to predict the degree of pollution from phosphate, coliform, pH, nitrate, dissolved oxygen (DO), BOD, and total suspended solids (TSSs), in the Marilao-Meycauayan-Obando River System (MMORS) of Philippines. In another study, Green et al. [12] predicted the concentrations of solutes at the Hubbard Brook Experimental Forest using RF models effectively than support vector machine. Similarly, concentrations of nutrient load from various non-point sources are predicted using this robust and flexible machine learning method [13, 15]. In the context of the previously stated research, this study makes a unique approach by building RF models utilizing existing data to forecast the FC from precipitation, temperature, and land use variables in the Upper Green River watershed. The RF model is examined in depth in this study and is used to determine critical characteristics for FC from supplied causative factors. Section 3 contains information on the study area, and gives the modeling approach. Section 4 presents and examines the findings. Finally, in Sect. 5, the inferences are derived from the results.

3 Study Area and Methodology

The current investigation was carried out in Kentucky, USA, on the Green River watershed. The Green River's main branch courses east to west across many provinces, including Taylor, Green, Hart, and Edmonson. This is one of the most essential freshwater riverine resources in the US. The Green River watershed has a nutrient-rich river basin and a eutrophic system with Karst characteristics. Agriculture, urban, rural, undeveloped woods, and industrial sectors are all included in the Green River watershed [3]. The sampling sites of the watershed are as shown in Fig. 1. More important topographical and hydrological details can be found at Venkateswarlu et al. [25].

In small test tubes, water samples were collected from 42 sampling locations that spread the entire watershed which are located on the Green River and its tributaries. The data were collected for six months (May 2002 to October 2002) [3]. The concentrations of WQPs were determined in the laboratory mainly using titrations. The data of temperature and daily precipitation were collected from the Kentucky Climate Center. To estimate the daily precipitation data at all of the sample locations within the Green River basin, the inverse distance-weighted technique was employed with ArcGIS. The two-day cumulative daily precipitation data are then computed for each sample location. Anmala et al. [2] provide the following land use variables for the primary land use in the Green River watershed:

$$\text{ALUF} = \sin^{-1}\sqrt{\text{agricultural area/catchment area}}$$

$$\text{ULUF} = \sin^{-1}\sqrt{\text{urban area/catchment area}}$$

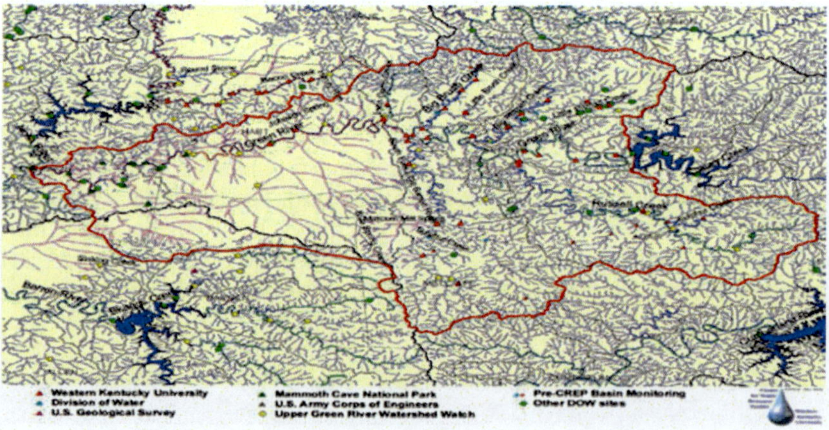

Fig. 1 Sampling sites of the Upper Green River watershed ("with permission from ASCE")

$$\text{FLUF} = \sin^{-1}\sqrt{\text{forest area}/\text{catchment area}}$$

The RF model was employed using the aforementioned land use characteristics and climatic variables to predict FC. The RF model was created with the help of Google Colaboratory in Jupyter notebook environment and Python libraries such as Matplotlib, Scikit-learn, Pandas, and NumPy. The datasets X variable, which contains the five input variables, and y variable, which contains the target variable, are imported first into the Jupyter notebook. After that the dataset is split into two parts: training and testing (70% for training and 30% for testing). To predict FC from temperature, precipitation, and land use variables, regression tree-based models were built using standard Python tools. The details of the RF algorithm are briefly discussed in the next section.

3.1 *Random Forest*

Breiman suggested random forests in 2001. The law of large numbers aids in avoiding the problem of overfitting. The RF model becomes accurate regressors and classifiers with inherent randomness [4]. The trees are produced randomly by selecting all instances from a training set with restoration from the original data. A few variables that are fewer than the greatest number of input variables are picked at random, and the finest split is utilized to split the node. The cutting down of trees is avoided for the trees to reach their maximum potential. After growing many trees, choosing is held to identify the most prominent class in this model. The RF approach delivers equivalent results to bagging and boosting without altering the training set. It is an additive model that employs a bagging method with ensemble learning representatives. To decrease bias, the RF method uses an arbitrary sample predictor before separation at each node.

In many cases, it is just as essential to have a precise model as it is to have an interpretable model. The RF models can help in various ways knowing the feature importance. For example, by the superior appreciation of the model's reasoning, one may validate that it is true and enhance the model by concentrating solely on the key parameters. Various methodologies, viz., (i) default Scikit-learn's feature importance, (ii) permutation feature importance, and (iii) drop column feature importance were used for interpreting feature importance using the RF model. In default Scikit-learn's method, every node in a decision tree is a provision for splitting values in a single feature so that alike dependent variable values end up in the same set after the division, while in permutation feature importance method is based on monitoring how haphazard reshuffling of each predictor impacts model performance. Drop column feature importance is totally intuitive as it compares the feature significance by matching a model with all features versus a model with this feature removed for training. The first method is simple to retrieve and fast in computations, and second method can be applied to any reasonably efficient model, and there is no requirement to retrain the model after each change to the dataset. The last method is the most accurate model as it removes all negative feature importance's from given inputs.

The methods described above were used to find the most essential characteristics for FC from the given climatic variables and land use factors. It should be noted that the more precise our model, the more we may rely on feature significance steps and other explanations. The outcomes of the RF model predictions and other significant aspects are presented in the next section.

4 Results and Discussions

The RF model was simulated for all parameters using the Python program in Google Colaboratory. The model was developed for given WQPs by dividing 30% of the data for testing and 70% of data for training out of 225 samples. Firstly, FC was predicted individually from precipitation, temperature, ALUF, ULUF, and FLUF. The corresponding R^2 value of FC for training is 0.93 and for testing 0.66, and the overall R^2 value is 0.94, respectively. The accuracy of predictions of the developed model was further assessed by computing root mean square error (RMSE) and mean arithmetic error (MAE) where RMSE—gives the information on how the residuals are dispersed about the best-fit line, and MAE—indicates the mean error in the prediction or average arithmetic error.

The training, testing, and overall R^2 values of FC indicate a strong correlation between the WQPs mentioned above and temperature, precipitation, ULUF, FLUF, and ALUF's. Further, the RF model results are compared with feedforward network results. The ANN model was developed using MATLAB 2017 by giving temperature, precipitation, ULUF, FLUF, and ALUF inputs to predict target WQPs separately. The respective training, testing, and overall R^2 of the feedforward network model of FC are 0.817, 0.903, 0.846. The corresponding RMSE and MAE values are 1540.487, 829.456, respectively. The RF model outperformed the ANN model in training and overall R^2 values for FC. The results also show that WQPs could be predicted using the RF model with higher accuracy. The corresponding R^2 value plots of WQPs are shown in Fig. 2.

Then, the model was run to find out the essential features for each variable by running the codes of default Scikit-learn's feature importance and drop column feature importance in Jupyter notebook. The results of default feature importance values of the RF model for precipitation = 0.59, temperature = 0.22, ULUF = 0.05, FLUF = 0.06, and ALUF = 0.09 and respective plots are shown in Fig. 3. The drop column feature importance plots of FC indicate that precipitation and temperature are primary influencing factors for the WQP in the Upper Green River watershed. Though forest, urban, and agricultural land use factors also impact the WQPs, maximum distribution leanings of parameters could be better predicted using field data of precipitation and temperature at the catchment scale [25]. This is in consistent agreement with the investigation of Liang et al. [15]. In the later stages, the correlations between WQP with each causal parameter were found. For example,

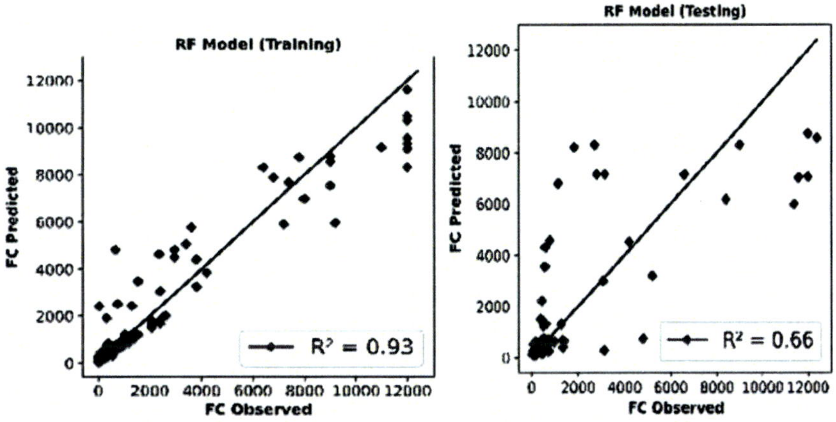

Fig. 2 Training and testing R^2 values of FC of RF model

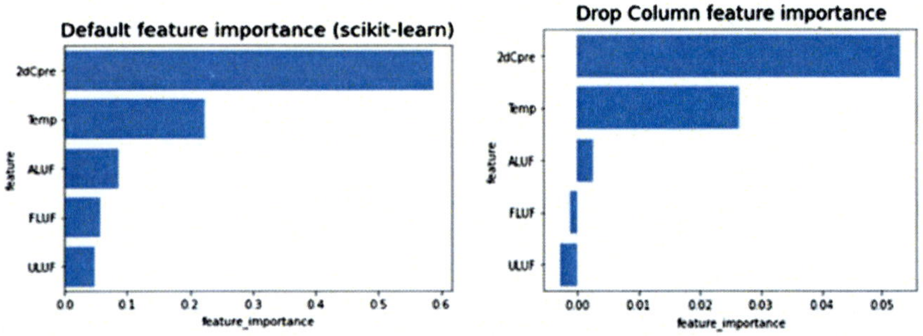

Fig. 3 Default feature importance and drop column feature importance values for FC

the correlations between FC versus precipitation, FC versus temperature, FC versus ULUF, FC versus FLUF, and FC versus ALUF were computed. For the remaining parameters also, the correlations have been found similarly. Afterward, the RF model was further investigated to determine the relation between FC with different combinational groups. In Group 1, various combinations of causal parameters versus WQPs were predicted without including precipitation, and for Group 2, the temperature was excluded, and both were not included in Group 3 combinations. Finally, WQPs were predicted using different combinations, including both temperature and precipitation. The respective computed R^2, RMSE, and AME values are presented in Table 1 for the combinations mentioned above. After interpreting the training, testing, and overall R^2 values of various combinations from Table 1, a poor correlation has been observed between land use factors and WQPs when applied individually. It can be observed that the accuracy of predictions gets better when any of the climatic parameters (either temperature or precipitation or both) is combined with the land use

Table 1 R^2, RMSE, and AME values of FC for various input combinations using RF models

Inputs*	Training			Testing			Overall		
	R^2	RMSE	MAE	R^2	RMSE	MAE	R^2	RMSE	MAE
T, P, U, A, F	0.93	865.974	458.117	0.66	2205.924	1166.460	0.94	835.430	435.731
P	0.62	2021.395	1119.344	0.68	2116.114	1232.665	0.65	2031.898	1154.870
T	0.75	1626.508	909.963	0.08	3609.241	2175.524	0.76	1679.497	966.477
U	0.09	3130.060	2118.074	−0.17	4076.120	2701.690	0.09	3130.06	2118.07
F	0.15	3019.880	2031.839	−0.21	4133.842	2712.260	0.08	3310.048	2339.923
A	0.23	2871.880	1926.360	−0.50	4601.964	3000.250	0.09	3288.363	2329.461
P, T	0.92	944.813	486.288	0.79	1727.306	1028.987	0.93	893.108	452.519
T, U, F, A	0.87	1188.817	711.606	0.15	3462.329	2309.384	0.88	1197.023	745.030
T, U	0.84	1316.550	786.077	0.25	3253.955	2139.370	0.86	1301.210	797.244
T, F	0.85	1276.451	769.036	0.11	3541.467	2291.414	0.84	1366.370	802.484
T, A	0.87	1172.408	709.675	0.09	3576.697	2261.626	0.88	1197.056	753.748
T, U, F	0.85	1254.070	747.778	0.14	3485.577	2340.247	0.87	1247.453	771.970
T, U, A	0.87	1171.598	704.641	0.18	3404.256	2291.567	0.88	1180.830	745.886
T, F, A	0.87	1176.280	701.173	0.12	3518.937	2268.930	0.88	1192.773	737.683
P, U, F, A	0.91	970.399	512.570	0.60	2373.348	1306.417	0.91	1045.101	570.445
P, U	0.84	1298.343	686.123	0.69	2095.911	1187.975	0.86	1302.106	711.520
P, F	0.89	1095.709	545.370	0.56	2500.990	1344.369	0.88	1180.243	631.386
P, A	0.88	1151.973	578.486	0.59	2418.133	1363.394	0.88	1204.670	633.449
P, U, F	0.91	964.003	516.234	0.59	2416.471	1306.288	0.91	1048.415	580.074
P, U, A	0.90	1014.974	543.186	0.59	2391.040	1319.177	0.91	1045.460	569.992
P, F, A	0.90	1034.255	513.021	0.60	2376.495	1311.260	0.90	1103.921	586.149
U, F, A	0.27	2803.393	1850.990	−0.52	4627.077	2991.400	0.11	3252.860	2280.165
U, F	0.27	2806.030	1856.765	−0.50	4606.935	2991.731	0.27	2806.030	1856.765
U, A	0.27	2802.880	1848.829	−0.53	4639.553	2998.265	0.11	3252.868	2280.238
F, A	0.27	2813.650	1859.209	−0.51	4618.219	2982.330	0.11	3257.610	2284.131
T, P, U	0.91	966.229	492.235	0.72	1986.110	1143.604	0.93	927.527	486.533
T, P, F	0.94	804.421	431.737	0.68	2122.654	1131.758	0.94	821.962	423.420
T, P, A	0.93	853.813	450.876	0.71	2019.106	1135.964	0.95	800.281	411.811
T, P, U, F	0.93	857.417	460.762	0.66	2198.161	1171.646	0.94	844.823	443.626
T, P, U, A	0.93	892.396	473.192	0.66	2204.037	1175.402	0.94	828.063	431.414
T, P, F, A	0.94	825.761	432.554	0.69	2087.984	1117.510	0.94	822.492	419.678

* T = Temperature P = Precipitation U = ULUF F = FLUF A = ALUF

factors for WQPs. Hence, we can conclude that land uses are important parameters for comprehensively describing river water quality in any watershed.

5 Conclusions

From the above section, the conclusions below could be drawn.

- The concentration of FC is predicted with lower RMSE and MAE values and with higher R^2 values from precipitation, temperature, ULUF, ALUF, and FLUF as inputs into the RF model in comparison with the feedforward neural network model.
- Although the feature importance of the RF model shows the lower values for land use factors, the regression values of various combinations of land use factors and climatic variables indicate that the accuracy of predictions of WQPs gets better with the inclusion of land use factors.
- Hence, agricultural, urban, and forest land uses and precipitation and temperature are essential to comprehensively describe and understand the water quality of the Upper Green River watershed.
- In general, RF models showed persistent predictions and outperformed the feedforward neural network models for the considered watershed in the training and overall R^2 value. The model can provide important features out of the given data so that the authorities can take necessary actions to improve the water quality of rivers.

Acknowledgements The second author would like to thank CSIR, India grant No. 24(0356)/19/EMR-II of the project titled "Experimental and Computational studies of Surface Water Quality parameters from Morphometry and Spectral Characteristics." The authors would like to express their gratitude toward Dr. Ouida Meier, Dr. Stuart Foster, Prof. Albert Meier, Ms. Jenna Harbaugh, and Mr. Tim Rink of Western Kentucky University, Bowling Green, KY, the USA, for helping us with the dataset.

References

1. Allan JD, Flecker AS (1993) Biodiversity conservation in running waters. Bioscience 43(1):32–43
2. Anmala J, Meier OW, Meier AJ, Grubbs S (2015) GIS and artificial neural network-based water quality model for a stream network in the Upper Green River basin, Kentucky, USA. J Environ Eng 141(5):04014082
3. Anmala J, Venkateshwarlu T (2019) Statistical assessment and neural network modeling of stream water quality observations of Green River watershed, KY, USA. Water Supply 19(6):1831–1840
4. Breiman L (2001) Random forests. Mach Learn 45(1):5–32

5. Bu H, Meng W, Zhang Y, Wan J (2014) Relationships between land use patterns and water quality in the Taizi River basin, China. Ecolog Ind 41:187–197. https://doi.org/10.1016/j.ecolind.2014.02.003
6. Bui DT, Khosravi K, Tiefenbacher J, Nguyen H, Kazakis N (2020) Improving prediction of water quality indices using novel hybrid machine-learning algorithms. Sci Total Environ 721:137612
7. Bolstad PV, Swank WT (1997) Cumulative impacts of landuse on water quality in a southern Appalachian watershed. J Am Water Resour Assoc 33(3):519–533. https://doi.org/10.1111/j.1752-1688.1997.tb03529.x
8. Chen J, Lu J (2014) Effects of land use, topography and socio-economic factors on river water quality in a mountainous watershed with intensive agricultural production in East China. PLoS ONE 9(8):1–12. https://doi.org/10.1371/journal.pone.0102714
9. Coulter CB, Kolka RK, Thompson JA (2004) Water quality in agricultural, urban, and mixed land use watersheds. J Am Water Resour Assoc 40(6):1593–1601. https://doi.org/10.1111/j.1752-1688.2004.tb01608.x
10. Delpla I, Rodriguez MJ (2014) Effects of future climate and land use scenarios on riverine source water quality. Sci Total Environ 493:1014–1024. https://doi.org/10.1016/j.scitotenv.2014.06.087
11. Golabi MR, Farzi S, Khodabakhshi F, Geshnigani FS, Nazdane F, Radmanesh F (2020) Biochemical oxygen demand prediction: development of hybrid wavelet-random forest and M5 model tree approach using feature selection algorithms. Environ Sci Pollut Res 27(27):34322–34336
12. Green MB, Pardo LH, Bailey SW, Campbell JL, McDowell WH, Bernhardt ES, Rosi EJ (2020) Predicting high-frequency variation in stream solute concentrations with water quality sensors and machine learning. Hydrol Process. https://doi.org/10.1002/hyp.14000
13. Harrison JW, Lucius MA, Farrell JL (2020) Prediction of stream nitrogen and phophorus concentrations from high-frequency sensors using random forests regression. Sci Total Environ. https://doi.org/10.1016/j.scitotenv.2020.143005
14. Kelsey H, Porter DE, Scott G, Neet M, White D (2004) Using geographic information systems and regression analysis to evaluate relationships between land use and fecal coliform bacterial pollution. J Exp Mar Biol Ecol 298(2):197–209. https://doi.org/10.1016/S0022-0981(03)00359-9
15. Liang K, Jiang Y, Qi J, Fuller K, Nyiraneza J, Meng F-R (2020) Characterizing the impacts of land use on nitrate load and water yield in an agricultural watershed in Atlantic Canada. Sci Total Environ 729:138793
16. Mattikalli NM, Richards KS (1996) Estimation of surface water quality changes in response to land use change: application of the export coefficient model using remote sensing and geographical information system. J Environ Manage 48(3):263–282. https://doi.org/10.1006/jema.1996.0077
17. de Mello K, Valente RA, Randhir TO, dos Santos ACA, Vettorazzi CA (2018) Effects of land use and land cover on water quality of low-order streams in Southeastern Brazil: watershed versus riparian zone. Catena 167(September 2017):130–138. https://doi.org/10.1016/j.catena.2018.04.027
18. Meneses BM, Reis R, Vale MJ, Saraiva R (2015) Land use and land cover changes in Zêzere watershed (Portugal)—water quality implications. Sci Total Environ 527–528:439–447. https://doi.org/10.1016/j.scitotenv.2015.04.092
19. Mokondoko P, Manson RH, Pérez-Maqueo O (2016) Assessing the service of water quality regulation by quantifying the effects of land use on water quality and public health in central Veracruz, Mexico. Ecosyst Serv 22:161–173. https://doi.org/10.1016/j.ecoser.2016.09.001
20. Namugize JN, Jewitt G, Graham M (2017) Effects of land use and land cover changes on water quality in the uMngeni river catchment, South Africa. Phys Chem Earth 105(April 2017):247–264. https://doi.org/10.1016/j.pce.2018.03.013
21. Park S-R, Kim S, Lee S-W (2021) Evaluating the relationships between riparian land cover characteristics and biological integrity of streams using Random Forest algorithms. Int J Environ Res Public Health 18:3182. https://doi.org/10.3390/ijerph18063182

22. Ren W, Zhong Y, Meligrana J, Anderson B, Watt WE, Chen J, Leung HL (2003) Urbanization, land use, and water quality in Shanghai 1947–1996. Environ Int 29(5):649–659. https://doi.org/10.1016/S0160-4120(03)00051-5
23. Schoonover JE, Lockaby BG (2006) Land cover impacts on stream nutrients and fecal coliform in the lower Piedmont of West Georgia. J Hydrol 331(3–4):371–382. https://doi.org/10.1016/j.jhydrol.2006.05.031
24. Tong STY, Liu AJ, Goodrich JA (2009) Assessing the water quality impacts of future land-use changes in an urbanising watershed. Civ Eng Environ Syst 26(1):3–18. https://doi.org/10.1080/10286600802003393
25. Venkateswarlu T, Anmala J, Dharwa M (2020) PCA, CCA, and ANN modeling of climate and land-use effects on stream water quality of karst watershed in Upper Green River, Kentucky. J Hydrol Eng 25(6):05020008. https://doi.org/10.1061/(asce)he.1943-5584.0001921
26. Victoriano JM, Lacatan LL, Vinluan AA (2020) Predicting river pollution using random forest decision tree with GIS model: a case study of MMORS, Philippines. Int J Environ Sci Dev 11(1):36–42
27. Vitro KA, BenDor TK, Jordanova TV, Miles B (2017) A geospatial analysis of land use and stormwater management on fecal coliform contamination in North Carolina streams. Sci Total Environ 603–604:709–727. https://doi.org/10.1016/j.scitotenv.2017.02.093
28. Zampella RA, Procopio NA, Lathrop RG, Dow CL (2007) Relationship of land-use/land-cover patterns and surface-water quality in the Mullica River basin 1. JAWRA J Am Water Resour Assoc 43(3):594–604

Supervised Learning-Based PV Output Current Modeling: A South Africa Case Study

Ely Ondo Ekogha and Pius A. Owolawi

Abstract Photovoltaic (PV) plants utilization for green solar energy is growing exponentially in demand as industries committed to move away from carbon energy sources such as coals, oil, or gas. However, for efficient green solar energy utilization, a precise prediction method is required to minimize design composition wastage. The measured output current determined by empirical method will be compared with the predicted current obtained from the proposed neural network (ANN) and random forest (RF) methods. The comparative analysis of the measured and the proposed models is evaluated by using the minimum root means square error (RMSE), mean absolute percentage error (MAPE), and mean bias error (MBE). The obtained results suggest the superiority of RF over the ANN with improvement performance metrics values of 173% for RMSE, 39% for MAPE, and 188% for MBE.

Keywords Forecasting PV current · Random forest · Artificial neural network

1 Introduction

Global warming has taken another level of importance during United Nation Climate Change Conference (COP21) in 2015 where 196 countries decided to reduce greenhouse gas (GHG) emission to nearly zero by the end of twenty-first century [1]. According to [2], South Africa is ranked 14th worldwide for its greenhouse gases emission, and the country is committed to alleviate the impact of greenhouse gases by 42% by 2025. Green energy utilization appears to bring the solution to reduce GHG and leverage long-term expenses [3]. As a matter of fact, photovoltaic produces energy from solar irradiations, and its massive production has drastically reduced the cost of PV systems [4]. However, harnessing solar energy has been affected by weather uncertainties such as rain, snow, and clouds. As industrials are moving

E. O. Ekogha (✉) · P. A. Owolawi
Tshwane University of Technology, Pretoria 0001, South Africa
e-mail: ondoekoghae@tut.ac.za

P. A. Owolawi
e-mail: owolawipa@tut.ac.za

X.-S. Yang et al. (eds.), *Proceedings of Seventh International Congress on Information and Communication Technology*, Lecture Notes in Networks and Systems 447,
https://doi.org/10.1007/978-981-19-1607-6_48

toward PV systems energy integration, forecasting photovoltaic (PV) output power has become the interests of many researchers. The importance of good energy prediction ensures robust and sustainable design, adequate planning, and reasonable forecast, which leads to cost-effectiveness of deploying solar energy system [5]. Furthermore, the services delivered by infrastructure planners and economists find their importance in the level of precision provided by the predicted PV output current [6].

Therefore, this paper will be structured as follows: Sect. 2 discusses about different techniques and approaches from previous researchers for predicting current from PV panel. Then, on Sect. 3, the theory of the models is presented, followed by evaluation metrics and obtained results, while discussions are presented in Sect. 4, and lastly, Sect. 5 presents the conclusion of the paper.

2 Literature Review

The impact of PV panel aging as a de-rating factor in predicting PV output power shows that the aging factor is rated at 1% per year when analyzing the accuracy from the relation between the predicted power at standard conditions and the measured one using RMSE, correlation coefficient, and standard deviation formulas on a grid-connected PV system [6]. A study claims on accuracy improvement on single-diode PV model output power to proceed by eliminations of variables with less effect on the output power using principal component analysis algorithm and then optimize the important variables using hybrid charge system search thus improving the prediction error by 25.59% when comparing to differential evolution or particle swarming optimization methods [7]. Predicting output current from grid-connected PV system using RF model has been investigated in [8]; the paper demonstrates performances of predicting output current using six months of hourly meteorological data against output current from a 3 KWh PV system. The accuracy of the model is claimed to be 2.7482 for RMSE, 8.7151 for MAPE, and −2.5772% for MBE. Deterministic point prediction and prediction interval of PV output power were developed in [9] using in a combination of ANN techniques and nonparametric kernel density estimation. Similarly, a hybrid method combining deep convolutional neural network constructs a two-dimensional data form with both daily and hourly timescale data through variation mode decomposition in [10] to achieve better forecasting performances over commonly used method. Two back to back irradiance mapping models for sky images over global irradiance are proposed on [11] using supervised machine learning (CNN) combines with long to short-term memory neural network technique. Comparing the proposed model with ANN model based on probability and deterministic methods, a better performance of CNN is observed particularly. Theocharides et al. [12] implemented a supervised ensemble method on gradient boosting machine learning for predicting PV output power; the results claimed RMSE prediction was 0.80% while some days approached 0.50%. A model merging two machine learning architectures for predicting output power of a PV plant is presented in [13] which includes seasonal auto-regressive integrated moving average (SARIMA) model and

ANN model. The results claims that the error of prediction can be reduced for up to 10% when the methods are merged than used individually, meanwhile the robustness and redundancy are ensured thanks to parallel architecture of the two methods.

3 Model Presentation

3.1 Solar Panel

A PV panel is built on array of cells, whereby each cell made with polycrystalline has a voltage of 0.5 V. Cells are connected in series to increase the voltage and in parallel to increase the output current. Figure 1 [14] depicts an equivalent circuit of a solar cell, whereby current is generated from a source which is connected with a single diode in parallel along with a resistor also mounted in parallel, a resistance connected in series is added to the circuit.

The generated current is proportional to the incident of light *G*; a diode current is generated when light falls on the cell the series and parallel resistances R_S and R_{SH} complete the characteristics of the PV cell under different load. The output current of the solar cell is the result of the current generated by the incident of light on the photocell I_L subtracted by the current flowing on the diode I_o and the shunt current I_{sh} [14] which gives the expression

$$I = I_L - I_o\left(e^{\left(\frac{q(V+I*R_S)}{a*k*T}\right)} - 1\right) - \frac{V - I * R_S}{R_{SH}} \tag{1}$$

I_o is the diode saturation current; q, a, and k are constants for electron charge, diode factor, and Boltzmann; V and T are the PV voltage and temperature [14]. However, the empirical model for PV panel considers only solar radiation and temperature, whereby the power from the PV is linear to the amount of irradiance and inversely proportional to the increase of temperature [15], the output current is given by the relation:

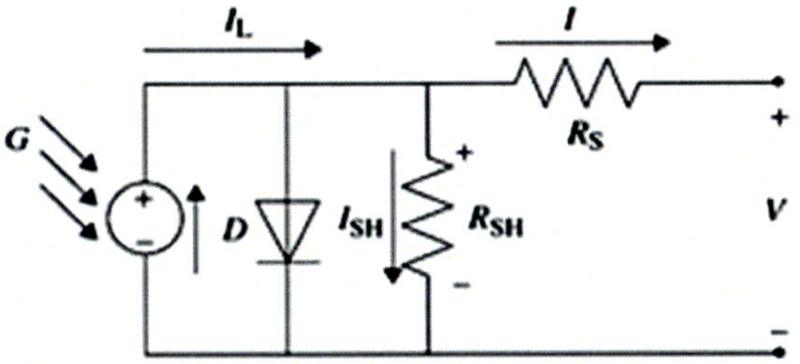

Fig. 1 Equivalent circuit of solar cell

$$I_{\mathrm{PV}}(t) = \frac{P_m\left(\frac{G(t)}{G_{\mathrm{STD}}}\right) - \alpha_T (T(t) - T_{\mathrm{STD}})}{V_{\mathrm{PV}}(t)} \quad (2)$$

where the maximal power of the photovoltaic panel is P_m, the irradiance at standard test conditions is G_{STD}, while the temperature is T_{STD}, G is the instantaneous irradiance, α_T is the temperature coefficient of PV power [15].

3.2 Random Forest

The RF model used is built on MATLAB function "*treebagger*" which creates an ensemble of trees from initials sample data by the process of bootstrap aggregation with replacement to overcome overfitting and have a better generalization [16]. Each tree is then combined by randomization; then, the results are determined by classification or regression technique [17]. Practically, the model begins by forming numbers of trees *ntree* of the size of the initial dataset by bootstrapping. Then, the bootstrap samples randomly split to form unpruned trees called decision trees [14]. Bootstrap aggregation is an ensemble technique which includes various machine learning methods using bagging with replacement to reduce variances, determining variable importance, while improving accuracy. Majority of votes from the decision trees or the averages value is then determined by classification or regression by the process of bootstrap aggregation [18]. Error rates are defined as the data which are not found in the bootstrap samples after each bootstrap iteration, they are called "out-of-bag" data (OOB) [14]; according to [18], estimation error may represent up to 36% of initial dataset.

Random forest determines its important variables from the rate of change observed in OOB samples. Important variables are given by how much OOB error occurs when permuting a single variable from all samples trees, while others remained constant [18]; the mean square error (MSE) is then averaged before and after permutation for all bootstrap samples [17]. A formula to determine variable importance (VI) in [14] can be described as:

$$\mathrm{VI}^{(t)}(f) = \frac{\sum_T \left(\frac{\sum_{x_i \beta^{c(t)}} I\left(L_j = c_i^{(t)}\right)}{\left|\beta^{c(t)}\right|} - \frac{\sum_{x_i \beta^{c(t)}} I\left(L_j = c_{i,\pi f}^{(t)}\right)}{\left|\beta^{c(t)}\right|} \right)}{T} \quad (3)$$

where $\beta^c(t)$ is OOB data for a single tree, the tree number is given by t, T the sum of all trees, meanwhile $c_i{}^{(t)}$ and $c_{i,\pi f}{}^{(t)}$ represent the predicted values in a single sample of tree for each permutation of the sample variable x_i before and after. The true label is represented by L_j.

Optimization criterion is made at each decision tree by looking at the minimum mean square error (MSE). Using classification technique, the criterion is defined

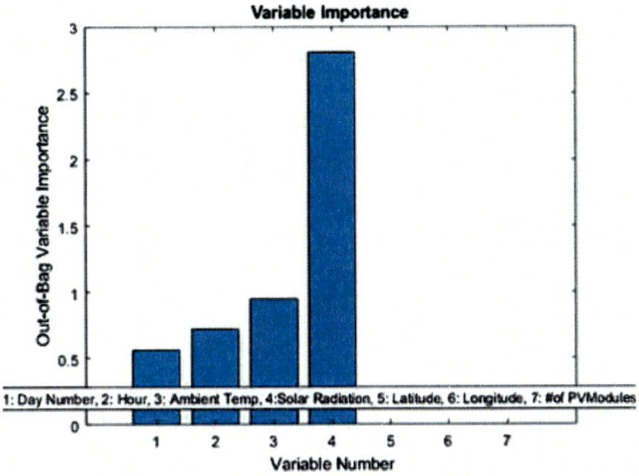

Fig. 2 Variables importance metrics

by Gini algorithm, deviance (entropy), or twoing algorithm. On the other hand, regression technique defines the RF predictors by the average of all sample trees [14] as displayed in Fig. 2. The formula for Gini criterion technique is given by Eq. 4, meanwhile the deviance or entropy is given by Eq. 5 [19].

$$\text{Gini impurity} = \sum_{i=1}^{C} f_i(1 - f_i) \tag{4}$$

$$\text{Entropy} = \sum_{i=1}^{C} -f_i \log(f_i) \tag{5}$$

where f_i is given by the number of alterations of variable i at node C.

3.3 *ANN Technique*

Artificial neural network is a supervised machine learning technique that can be used to solve nonlinear problems adapted to climate uncertainties. It is constructed with separated layers of neurons that are connected by weighted links, where information is processed from one neuron to the others. The layers are labeled as inputs, hidden, and output neurons; information is transferred from one neuron to the other following the sigmoid algorithms given in Eq. (6).

$$f(z_i) = \frac{1}{1 + e^{-z_i}} \tag{6}$$

$$z_i = \sum_{j=1}^{4} w_{ij} + \beta_i \quad (7)$$

z_i is the weighted summation of inputs given by the incoming signal x_j and its weighted link w_{ij} plus bias error β_i. Our inputs are defined as irradiance, temperature, day length, latitude, longitude, and the number of PV modules, while we chose four neurons at the hidden layers based on the fact that this number must be ¾ of the inputs neurons [14].

3.4 Statistical Tools

The root means square error (RMSE) determines the efficiency for the prediction of the output current. The lower the value the better the efficiency, the equation is given by

$$\text{RMSE} = \sqrt{\frac{1}{n}\sum_{i=1}^{n}(P_i - M_i)^2} \quad (8)$$

The mean bias error (MBE) determines the correlation between the predicted values and the measured ones. A negative value determines the level of under forecasting, meanwhile a positive value results in the level of over forecasting.

$$\text{MBE} = \frac{1}{n}\sum_{i=1}^{n}(P_i - M_i) \quad (9)$$

The mean absolute percentage error (MAPE) provides the overall accuracy of the technique used. The equation is given by:

$$\text{MAPE} = \frac{1}{n}\sum_{i=1}^{n}\left|\frac{M - P}{M}\right| \quad (10)$$

M represents the actual criterion, while P represents the result predicted.

4 Evaluation Metrics and Results

In this paper, the ground-based data collected are from the Southern African Universities Radiometric Network (SAURAN); the chosen location is the station GIZ University of Pretoria at latitude −25.75308 and longitude 28.22859, whereby four years daily data (2016–2019) have been captured.

Table 1 depicts the average radiation per month from the daily irradiance. Throughout the year 2016–2019, the lowest irradiances happen to be in the months of May, June, and July; these months will mark the unfavorable seasons. On the other hand, the months of October, November, and December show some relatively high irradiances which will be depicted as the most favorable season of the year meanwhile.

RF model can be divided into three stages. The first stage is to determine the important variables, and the selected number of trees of the model should be half of the training sample [14]. The input data given along with the solar radiation are the day number of the year, the day length of irradiance, the latitude, longitude, the number of panels. The graph in Fig. 2 describes the important variables where the irradiance has a value of 2.803 followed by the ambient temperature with a rate of 0.9523 over 3, and the daylight in hours with a rate of 0.724 over 3, then finally the day number with a rate of 0.563 over 3. The input variables such as latitude, longitude, and PV panel number have no influence on the output current, and therefore will be discarded on the second stage.

Cluster analysis is performed to detect outliers or noise from the training sample set; this method allows similar dataset to be grouped according to the density criteria

Table 1 Meteorological evaluation metrics

	Irradiances				Temperature			
Years	2016	2017	2018	2019	2016	2017	2018	2019
Months								
Jan	274.36	253.64	309.70	288.17	22.66	22.47	22.60	23.54
Feb	282.29	214.55	239.06	245.47	24.24	21.35	21.65	23.13
Mar	229.59	248.92	216.69	245.19	20.68	21.68	20.38	21.64
Apr	215.55	199.84	186.05	183.07	19.51	17.91	18.54	18.75
May	170.09	173.43	178.08	188.52	15.18	15.25	15.29	16.99
Jun	159.95	169.60	165.61	174.38	13.28	14.55	13.67	14.16
Jul	175.08	171.71	174.18	189.32	13.79	15.50	14.55	16.38
Aug	223.12	212.36	198.33	201.39	18.75	16.88	17.82	18.14
Sep	237.48	240.21	248.78	257.42	19.75	20.15	20.91	21.31
Oct	283.25	263.08	285.30	272.01	23.05	21.14	20.14	23.22
Nov	269.69	313.39	289.28	296.37	21.84	20.30	23.38	21.74
Dec	272.46	268.15	311.10	247.82	22.04	23.24	22.88	23.11

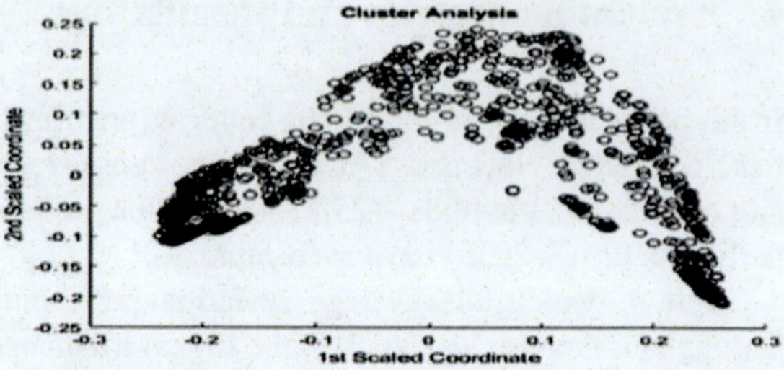

Fig. 3 Cluster analysis in training stage

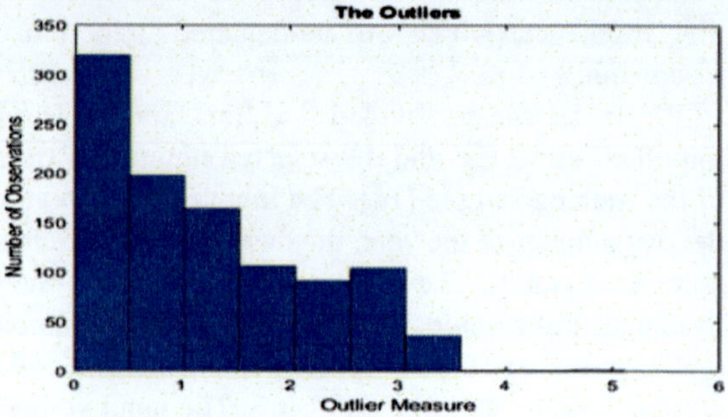

Fig. 4 Outliers measured in training stage

[8], see Fig. 3. Outliers represent cases that are removed from the main training dataset based on the fact that their proximities to all other cases in the dataset are generally small [20]. The graph from the outliers in Fig. 4 depicts 319 observations on the first pattern, 197 on the second pattern, 164 on the third pattern, 106 on the fourth pattern, 91 on the fifth pattern, 104 on the sixth pattern, 36 on the seven pattern, 0 on the eighth pattern, and 1 on the ninth pattern. These outliers will be replaced in the training set for a more accurate prediction [8].

On the second stage, we used the importance variable as inputs of the model to find the optimum number of trees then leaves per tree using statistical metrics from the minimum value for RMSE, MAPE, MBE, and elapse time [14]. For this model, the minimum value of RMSE is 0.003694 at tree number 29 and leave 1; the minimum value of elapse time is 0.012037 s at tree number 3501 and leaves.

On the third stage, the optimum tree number is 29 on leaves number 1 following the minimum RMSE indicator. As a result, the statistical measurements present the superiority of choosing the RF model over ANN model as depicted on Table 2,

Table 2 Statistical measurements

	Accuracy measurements indicators		
	RMSE	MAPE	MBE
RF model	0.0670	0.1335	−0.0028
ANN model	0.9312	0.0897	−0.0990

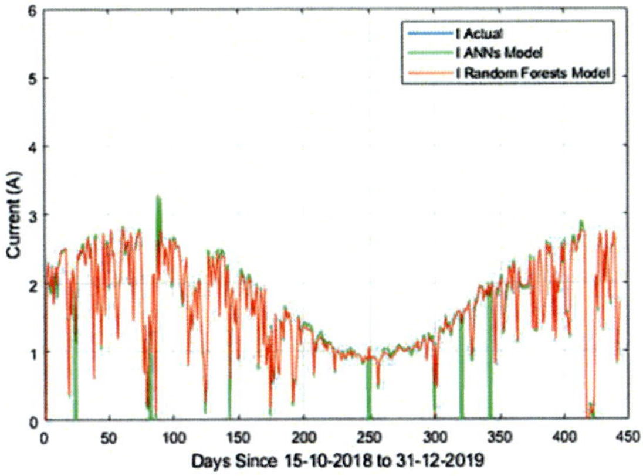

Fig. 5 Daily output current prediction between RF model and ANN model from 15-10-2018 to 31-12-2019

Table 3 Models evaluation statistics on different seasons

	From May to July (unfavorable)			From October to December (favorable)		
	RMSE %	MAPE %	MBE %	RMSE %	MAPE %	MBE %
RF model	3.85	2.65	−1.41	3.73	9.51	0.46
ANN model	86.94	9.84	−9.07	0.46	0.77	−0.032

whereas Fig. 5 presents the graph of performances between ANN model and RF model for over one year.

Taking into consideration that unfavorable solar radiation season has been identified in the months of May, June, and July, and favorable solar radiation season happens during the months of October, November, and December; the evaluation statistics demonstrates that RF model is suitable for unfavorable season, whereas ANN model is ideal for favorable season; this is depicted on Table 3.

5 Conclusion

This paper showcases predictions of a PV output current using RF model under different types of seasons in Pretoria and compares them to ANN model. The statistical tools for measuring accuracy of the model such as RMSE, MAPE, and MBE reveal the superiority of using RF model over ANN model in global scale with performance indicators of 0.0670 for RMSE, 0.1335 for MAPE, and −0.0028 for MBE.

However, ANN-based model has presented better performances on good weather conditions, whereas RF model is more efficient in bad weather conditions. This discrepancy could help for a change of strategies in harnessing solar radiation for a better rendering of PV output current.

References

1. Antonanzas J, Osorio N, Escobar R, Urraca R, Martinez-de-Pison FJ, Antonanzas-Torres F (2016) Review of photovoltaic power forecasting. Sol Energ 136:78–111
2. Borel-Saladin JM, Turok IN (2013) The impact of the green economy on jobs in South Africa. S Afr J Sci 109:01–04
3. Emmanuel BO, Owolawi PA, Srivastava VM (2017) Hybrid power systems for GSM and 4G base stations in South Africa. In: 2017 IEEE AFRICON 2017, pp 1003–1008
4. Fernandez-Jimenez LA, Muñoz-Jimenez A, Falces A, Mendoza-Villena M, Garcia-Garrido E, Lara-Santillan PM et al (2012) Short-term power forecasting system for photovoltaic plants. Renew Energ 44:311–317
5. Zeng J, Qiao W (2013) Short-term solar power prediction using a support vector machine. Renew Energ 52:118–127
6. Zainuddin H, Shaari S, Omar AM, Sulaiman SI (2011) Power prediction for grid-connected photovoltaic system in Malaysia. In: 2011 3rd international symposium and exhibition in sustainable energy and environment (ISESEE), pp 110–113
7. Huang Y-C, Huang C-M, Chen S-J, Yang S-P (2019) Optimization of module parameters for PV power estimation using a hybrid algorithm. IEEE Trans Sustain Energ
8. Ibrahim IA, Khatib T, Mohamed A, Elmenreich W (2018) Modeling of the output current of a photovoltaic grid-connected system using random forests technique. Energ Explor Exploit 36:132–148
9. Liu L, Zhao Y, Chang D, Xie J, Ma Z, Sun Q et al (2018) Prediction of short-term PV power output and uncertainty analysis. Appl Energ 228:700–711
10. Zang H, Cheng L, Ding T, Cheung KW, Liang Z, Wei Z et al (2018) Hybrid method for short-term photovoltaic power forecasting based on deep convolutional neural network. IET Gener Transm Distrib 12:4557–4567
11. Wang F, Zhang Z, Chai H, Yu Y, Lu X, Wang T et al (2019) Deep learning based irradiance mapping model for solar PV power forecasting using sky image. In: 2019 IEEE industry applications society annual meeting, pp 1–9
12. Theocharides S, Venizelou V, Makrides G, Georghiou GE (2018) Day-ahead forecasting of solar power output from photovoltaic systems utilising gradient boosting machines. In: 2018 IEEE 7th world conference on photovoltaic energy conversion (WCPEC) (A Joint Conference of 45th IEEE PVSC, 28th PVSEC and 34th EU PVSEC), pp 2371–2375
13. Vrettos E, Gehbauer C (2019) A Hybrid approach for short-term PV power forecasting in predictive control applications. In: 2019 IEEE Milan PowerTech, pp 1–6
14. Khatib T, Elmenreich W (2016) Modeling of photovoltaic systems using Matlab: simplified green codes. Wiley
15. Khatib T, Mohamed A, Sopian K, Mahmoud M (2012) An iterative method for calculating the optimum size of inverter in PV systems for Malaysia. Electr Rev 88:281–284
16. Guide GS (1984) MATLAB® 7
17. Pham LT, Luo L, Finley A (2021) Evaluation of random forests for short-term daily streamflow forecasting in rainfall-and snowmelt-driven watersheds. Hydrol Earth Syst Sci 25:2997–3015
18. Liaw A, Wiener M (2002) Classification and regression by random forest. R News 2:18–22
19. Ronaghan S (2018) The mathematics of decision trees, random forest and feature importance in Scikit—learn and spark
20. Breiman L (2015) Random forests. In: Breiman L, Cutler A (eds) Random forests-classification description

Use of Social Networks by Russian Politicians

Olga Gris and Anna Sosnovskaya

Abstract This article presents the results of a study of social media accounts of the heads of the regions of the Russian Federation for the period January-May 2021 and a survey of 54 government officials for the period August-November 2021. The study shows what communication tasks are addressed through social networks, how government officials perceive and assess them and what problems they face. The data assessment techniques derive from actor-network theory.

Keywords E-government · Social network · Communication · Russia · ANT

1 Introduction

In the context of a low level of trust in public authorities, a decrease in electoral activity, as well as in the conditions of communication restrictions due to the pandemic, it becomes even more necessary to ensure the transparency of communication and the visibility of heads of government as guarantors of governance. Using social media, it can help achieve this goal. But the possibility of realizing visibility and achieving transparency will be due to the specific characteristics of social networks, such as their instant interactivity and ability to build horizontal networks.

The introduction of social media into the space of public administration through communication between governors and the population in the Russian Federation is a new and little-studied phenomenon. This is due to the fact that their use began only 3–4 years ago [14]. We did not find published articles on E-government and actor-network theory (ANT) [10, 11] and E-government in Russia and Western countries. There are, however, a number of relevant contemporary articles addressing the interaction of government and social networks. For example, a study by Canadian authors (2019) examined how Justin Trudeau's personal life is used on his Instagram

O. Gris (✉) · A. Sosnovskaya
RANEPA, Moscow, Russia
e-mail: gris-oa@ranepa.ru

A. Sosnovskaya
e-mail: sosnovskaya-am@ranepa.ru

X.-S. Yang et al. (eds.), *Proceedings of Seventh International Congress on Information and Communication Technology*, Lecture Notes in Networks and Systems 447,
https://doi.org/10.1007/978-981-19-1607-6_49

account to support the values and ideas of the Liberal Party of Canada, and how celebrity cultural codes are mobilized to discuss political issues such as the environment and youth [9]. Researchers from the United States (2019) studied how the driving forces of voter behavior are reflected on Twitter [5]. Johnson et al. [8] investigated the ecology of hate and Internet fraud [8]. It has been shown that the benefits of government social network use include enhanced citizen satisfaction with transparency [2]. However, current practice does not yet live up to the promise of social media to better engage citizens and facilitate collaborative dialog with stakeholders, as has been shown of local authorities in Europe [2], local authorities in Egypt [1], and local governments in China [12].

In this article, we will consider the case of the Russian Federation, describing the features of the functioning of accounts in social networks of authorities. In the course of the research, we asked the following questions: How does this network function? What tasks may be solved by means of social networks by authorities? What are the interests and communication practices, the strategies of the acting agents? What are the implications of using social media for government?

To identify practices and their conditions for social networks as an e-government tool, we analyzed the use of social networks by the heads of the regions of the Russian Federation. The choice of the accounts of the heads of regions for analysis is due to the fact that they are the central media person representing the image of power among the population. We analyzed the qualitative and quantitative characteristics of the social networks of the heads of the regions to gage how effectively they are being used. An analysis of all social networks used showed that the heads of regions use Instagram accounts as the main ones. Therefore, following actor-network theory, we take into account how this network functions and affects the communication of authorities. In addition, we will consider how Russian politicians position their communication strategies within the requirements of network agents. According to the reconstructed functions of communication, the most mastered functions are described, behind which the structure of communication is visible. The description of this framework contributes to the scientific knowledge base of how governments, voters, and the media interact, an area of growing activity that requires scientific study. In addition, in part because it is an ever-updating process, political actors also fail to rationally explain their communication strategies except through the performance of communicative functions.

Whether researchers acknowledge the fact or not, whether for good or ill, government officials are becoming Instagram's new influencers, attracting the attention of large audiences, which is bound to lead to a restructuring of power relations on the network. Using ANT in our study of political network communication, we analyzed the network to identify and understand the parts and relationships that make it up. This analysis provides some answers to our research questions as to how the introduction of social media into management contributes to the formation of network relationships. ANT is advantageous in that it allows assembly to be considered as a continuous process and to apply social criticism to the various stages of this process.

Our research was prompted by this shift in the use of social media by policymakers, which are understood as an active representation tool and perceived as a necessary tool, although the rules for using and verifying information on these platforms are still in their infancy. Flexible electronic communication systems, the nature of which is not defined, are used as reliable tools. The system of practices is dictated not by the norms of the use of these media in the official context, but by their actual procedures. The stages of development these media undergo follow from the specifics of the functioning of new media, where their own growth and coverage occur.

Latour's theory [10], which explains the self-agency of non-human agents in network interactions, complements traditional social theories of communication, providing a quick response to challenges concerning how communication is constructed. The criteria for communication effectiveness should not only derive from the particularities of traditional media, such as audience coverage, but also new ones coming from the actor's interaction with the network in a situation of multiple social problems. ANT explains the specifics of new media in all their diversity, where content analysis would otherwise have to be carried out piecemeal on the specifics of each platform, with the platforms in a situation of constant change. Political network communication as a transformation is shown, ways of using the resources of the platform are revealed, and the consequences in real life of such a device of the communicative sphere are shown.

2 Methodology

To achieve the objectives of this study, that is, a qualitative description of communication and assembly of a network of participating actors, we carried out the following research tasks:

(1) Conducted a literature review using scientometric analysis and identifying keywords that guide the performative network. Identified functional expectations regarding the use of social networks by authorities. For this purpose, the Snyder methodology was used [6], with the technical programs VOSviewer_1.6.16_exe and CitNetExplorer_1.0.0_exe., carrying out the selection of keywords for the selection of the latest literature based on the title and research objectives.

(2) Conducted discursive analysis of the accounts of the heads of regions of the Russian Federation in social networks (85/-2 heads of regions of the Russian Federation, a total of 141,495 records published in January-May 2021 were analyzed) identified communication blocks and their strategies. Conducted content analysis of communication blocks in accordance with the message strategy. Compared communication strategies with type of communication and functional value.

(3) Conducted targeted interviews with 54 government officials to identify strategies and restrictions in government accounts, expectations, and challenges they face. Transcription of the audio text, codification, clustering, and interpretation of the data obtained.

3 Results

Analysis of the identified works using the Snyder methodology allowed us to create our own classification of the functions of social networks, the implementation of which, according to the authors [7, 8], will ensure the achievement of a system of public administration focused on the interests of citizens. Our analysis reveals nine such functions (more details can be found in the talk by Sosnovskaya, Gris, II International Scientific Forum on Computer and Energy Sciences (WFCES II 2021), November 11–12, 2021, Almaty, Kazakhstan, 2021). Next, we analyzed the messages themselves, carrying out their codification. We identified the communication strategies used by the heads of government and their relative frequency. We also compared these strategies with communication type (unilateral, bilateral, interactive). The data obtained by this analysis are presented in Table 1.

As a result of content analysis of messages, the dominance of the one-sided type of communication and communication strategy was revealed. The least developed type is the interactive type of communication. Discussions between heads and stakeholders were not found. "Mobilization for communication" and "productive discussion", that is, the 7th and 12th strategies represent gaps and zones for development.

At the next stage, a survey was conducted to identify how government officials perceive and evaluate social networks as a tool for solving problems facing the state, what problems they face, and what they see as pros and cons. All government officials

Table 1 The results of the analysis of messages within the type of communication. The table shows in the first column the percentage of messages that implement the strategy [S], and in the second—the percentage of accounts of heads of regions using the strategy [H]

No	Message strategies	Communication type					
		unilateral		bilateral		interactive	
		% of S	% of H	% of S	% of H	% of S	% of H
1	Informing	25.8	97				
2	Information support of citizens	10.9	97				
3	Business personalization	32.8	97				
4	Personalization personal	1.23	56				
5	Managing experiences	15.4	97				
6	Promotion of the region	5	97				
7	Mobilization for communication			0,1	5		
8	Mobilizing for action			0,4	97		
9	Feedback			8.3	97		
10	Open dialog					0.02	8.2
11	Supporting initiatives					0.05	43.5
12	Productive discussion					0	0

or their delegated authority have social media accounts. In 6 cases, comments are closed in accounts, which is 11%. In one account, comments are open, but there is no work with comments, and there are no feedback tools at all. Clustering and codification of statements made it possible to distinguish 403 semantic units on 7 topics. The results of codification and interpretation of the data obtained are presented in Table 2.

All respondents noted that social networks are primarily a tool for informing and feedback. For the most part, respondents stated that it is controversial at the moment to introduce social networks into the formation of a positive image of a leader or government body. Only 15% of respondents described social networks as a tool for forming a positive image of the authority, the leader. According to respondents, citizens use social networks as a channel for conveying certain problems to the authorities. However, according to respondents, they are used to an even greater extent to express negative emotions about the activities of the authorities. Thanks to the research a high level of dysfunction in the implementation of vertical communication was revealed when using horizontal communication of social networks. Many statements indicate the unwillingness of the authorities themselves to cope with the flow of information. These statements are confirmed by the results of the authors' research on the need to form information and communication competencies among civil servants [4].

Thus, there is a contradiction between the use of open network technologies, democratic attitudes on the one hand, and the administrative system of management inherent in the Russian Federation, on the other; Representatives of the authorities who enter into communication.

4 Discussion

Our study of the accounts of social networks of the heads of the regions of the Russian Federation shows the specifics of contemporary Russian state communication in social networks. These technical agents provide different opportunities and make different requirements of users, neither of which are fully exploited by the heads of regions. Further research could help to explain this pattern of under-exploitation, perhaps using the concept of affordance. Researchers have investigated the conditions for the emergence of certain strategies using the concept of affordance, used in field theory, and actor-network theory [11]. The concept of «affordance» arose from study of the interaction between actors and their environment [3]. This point of view implies that several possibilities of the same object may arise in connection with different points of view of the actor. Thus, for example, it is possible not to be aware of all Instagram's possibilities. For example, the main omission in all accounts is the non-use of geotags, since Instagram maps the interests and movements of users. Such metadata could be used in urban planning processes.

Three aspects of communication suggest themselves as targets for future study in terms of affordance by social media platform actors: (1) There are no interactions in

Table 2 Results of codification and interpretation of attitudinal data

Codification of semantic units /*examples*	% *	Interpretation
Informing citizens *"Information is posted", "we convey to citizens", "inform", "report", "explain", "post reports", "documents", "broadcast"*, etc.	100	Implementation of tasks (functions) of information and information support. One-way communication Following the principles of vertical communication. **
Involvement in events *"We were able to attract", "invited", "mobilized", "organized citizens"*, etc.	5	Implementation of the task (function) of mobilization to action. One-way communication Following the principles of vertical communication
Work with appeals, comments of citizens *"We answer questions", "we initiate an inspection based on appeals", "we receive information from citizens", "we work with comments", "we respond through personal messages", "we redirect requests to the necessary structural units"*, etc.	87	Implementation of feedback task (function). Two-way communication Adherence to the principles of vertical communication
Negative assessment of the use of social networks. Emotionalization *"A lot of hate leads to a deterioration in the image", "the use of obscene expressions", "unconstructive statements", "few constructive comments, appeals", "about the impersonality of appeals"*, etc.	76	On the part of citizens, social networks are perceived as a channel for expressing attitudes to the activities of the authority Use of horizontal communication principles
Negative assessment of the use of social networks from the standpoint of organizing work with feedback from citizens *"They ask questions that are not related to the activities of the authority", "it is impossible to process the entire flow of appeals", "and so many tasks, but you have to be distracted by social networks", "the deadlines set is not realistic for a constructive answer", "often you simply do not understand how to answer", "it is not clear how to work with the negative", most often the answers to citizens are of a formal nature"*, etc.	70	Unwillingness of the authorities to cope with the flow of information, structural, and organizational problems Dysfunction of the implementation of the principles of vertical communication when using the horizontal in essence communication of social networks
Positive evaluation of communication through social networks *"Instant receipt of information about the problems that have arisen", "the awareness of the authorities is increasing", "reducing the time for delivering information to the authorities", "accelerating the decision-making process", "we have entered into a constructive dialog with citizens"*, etc.	27	Implementation of feedback tasks (functions). Qualitative impact of the use of social networks as a feedback channel on the system of authorities Adaptation of vertical communication characteristic of the system of state power to horizontal communication of social networks

(continued)

Table 2 (continued)

Codification of semantic units /*examples*	% *	Interpretation
Overcoming the problems of using social networks in communication with authorities *"Setting strict rules allowed to restore order in accounts", "constantly moderate the account", "remove negative comments"*, etc.	48	Implementation of strategies for subordination of horizontal communication of social networks to vertical communication characteristic of the system of state power

* % of respondents who expressed an opinion
** By vertical communication, we mean communication used in the system of power relations, where one subject of communication has certain resources for decision-making

the accounts. Interactive communication would only be achieved if the government and the leader see themselves as part of the network and recognize that many actors on the network have valuable information that can contribute to governance. (2) Officials who listen to citizens on social media need to have a higher degree of openness, respect for difference, and reflexivity, which translates into personalized images. Heads of regions can claim to be influencers among citizens, which, although they are instruments of a different ideological platform, can lead to civic participation in governance. (3) All researched accounts were subject to criticism, attacks, and negativity. Each leader chose one or a few strategies for working with the negative (either comment, ignore, or delete). Researchers have noted that social media is a private business that inconsistently enforces its own rules on offensive language, confusing the problem, and making it difficult to solve [8]. American researchers [13] have suggested centralized control over the Internet space and social networks—either the establishment of government control or the regulation of social media as public enterprises. One option would be to proactively moderate social media posts, where people are essentially earning the right to good behavior, rather than the current practice of pre-moderation. Another idea is random moderation, in which people are encouraged to behave because they do not know when their comments can be viewed, like a virtual Panopticon, as described by M. Foucault. A third option would be the creation of levels of participation with different obligations and different standards of removal (including a permanent ban). Regardless of the method used to combat harassment of politicians on social media, it is clearly important that platforms act quickly and consistently to enforce their own rules against offensive language.

5 Conclusion

Social networks of accounts of heads of regions, as actors of network relations, are horizontal and informal in nature. They suppose the equality of participants in communication. We see that when entering this space, citizens follow this setting, trying to enter into a dialog, expressing an opinion, commenting on events, and making demands of representatives of the authorities. Representatives of the authorities use one-way communication to their advantage and only use two-way communication to a small extent, leaving many requests unanswered and almost completely ignoring the possibilities of interaction. Using social networks, political actors were not ready to obey the "rules". Put another way, the existing network of actors' interaction remains unwilling to allow the relevant power structures to abandon their usual vertical communication practices and continue to adhere to the previously universally normative principles of a hierarchical structure. They regard social networks as a communication resource that supports the verticality of power relations.

However, the presence of such strong actors as social networks, active citizens, and ideological guidelines for the development of the institution of electronic participation cannot leave this situation unchanged. An ANT approach to studying communicative networks assumes that agents acting in communication are able to change the communicative landscape, and by extension the behavior of all participants. It seems likely that the above actors, including the technologies of social networks and their use for communication between authorities and citizens, not only led to the identified contradictions, but will also contribute to further changes. One of the most likely scenarios for further development is the adoption of social networks as a tool that ensures the openness of government bodies, involving citizens in discussion and decision-making. In this case, actors from power structures who are not ready to rebuild their communication strategies from vertical to horizontal, not ready to engage in interaction, can be pushed out by the network to the periphery of communication.

The presented results are just the beginning of a large study, focusing on related research questions: Can social media generally be viewed as an actor contributing to the development of democracy? What is the impact of social media on citizens? Will their level of activity in political discussion and involvement in the process of making political and managerial decisions increase? Can social networks become a platform for the manifestation of group interests and interdepartmental communication? These and other questions will form the basis for future research.

References

1. Abdelsalam HM, Reddick CG, Gamal S, Al-shaar A (2013) Social media in Egyptian government websites: presence, usage, and effectiveness. Gov Inf Q 30(4):406–416. https://doi.org/10.1016/j.giq.2013.05.020
2. Bonsón E, Torres L, Royo S, Flores F (2012) Local e-government 2.0: social media and corporate transparency in municipalities. Gov Inform Q 29(2):123–132
3. Dini AA, Wahid F (2017) Four strategies of social media use among Indonesian politicians. In: Choudrie J, Islam M, Wahid F, Bass J, Priyatma J (eds) Information and communication technologies for development. ICT4D 2017. IFIP advances in information and communication technology, vol 504. Springer, Cham
4. Gris OA, Lebedeva LG (2021) Actual competences for the Russian civil service system. In: Ashmarina SI, Mantulenko VV (eds) Digital economy and the new labor market: jobs, competences and innovative HR technologies. IPM 2020. Lecture notes in networks and systems, vol 161. Springer, Cham. https://doi.org/10.1007/978-3-030-60926-9_49

5. Grover P, Kar AK, Dwivedi YK, Janssen M (2019) Polarization and acculturation in US Election 2016 outcomes "Can twitter analytics predict changes in voting preferences. Technol Forecast Soc Change 145:438–460. https://doi.org/10.1016/j.techfore.2018.09.009
6. Hannah S (2019) Literature review as a research methodology: an overview and guidelines. J Bus Res 104:333–339. https://doi.org/10.1016/j.jbusres.2019.07.039
7. Hofmann S, Beverungen D, Räckers M, Becker J (2013) What makes local governments' online communications successful? Insight from a multi-method analysis of facebook. Gov Inf Q 30(4):387–396. https://doi.org/10.1016/j.techfore.2018.09.009
8. Johnson NF, Leahy R, Restrepo NJ et al (2019) Hidden resilience and adaptive dynamics of the global online hate ecology. Nature 573:261–265. https://doi.org/10.1038/s41586-019-1494-7
9. Lalancette M, Raynauld V (2019) The power of political image: Justin Trudeau, Instagram, and celebrity politics. Am Behav Sci 63(7):888–924
10. Latour B (2005) Reassembling the social. Oxford University Press, Oxford
11. Low J (2015) After Method: Disorder and Social Science [Text]/trans. from English. In: Gavrilenko S, Pisarev A, Khanova P Sci. ed. translated by Gavrilenko S, Gaidar M Institute Publishing House
12. Ma L (2018) The post-adoption behaviors of government social media in China. In: Sobacı M, Hatipoğlu İ (eds) Sub-national democracy and politics through social media. Public administration and information technology, vol 29. Springer, Cham
13. Sobaci MZ (2016) Social media and local governments: an overview. In: Sobaci MZ (ed) Social media and local government. Springer, New York, pp 3–21
14. Sosnovskaya AM (2021) Society's digital transformation: the impact of social media on social practices. Digital transformation in the development of economy and society, materials of the XV international scientific and practical conference. Science-Unipress, Voronezh, pp 122–130

Design and Implementation of Verifiable Blockchain-Based e-voting System

Seiwoong Choi, HeeSeok Choi, and Kwang Sik Chung

Abstract Voting is the most representative way to express individual decision making in a democratic society. The classic voting method requires a lot of confirmation procedures, and requires a lot of time and money. In order to solve these kinds of problems, there have been efforts to introduce e-voting with IT services advantages, such as reduced election costs and shorter tally hours, compared to the existing voting methods. Despite many advantages, e-voting is not widely used for various technical concerns, including the risk of data manipulation. There has been much effort recently to apply the blockchain for voting data integrity to e-voting as a way to reduce the risk of voting data security and integrity. Blockchain could guarantee the integrity of voting data, but it has problems to be directly applied to e-voting system. In this paper, by applying cryptographic algorithms to the blockchain, we propose a verifiable blockchain-based e-voting system that allows voters to verify their votes while separating voters and voting results. Proposed the verifiable blockchain-based e-voting system satisfies the requirements of voting, including completeness, soundness, privacy, un-reusability, eligibility, fairness, verifiability.

Keywords e-voting system · Blockchain · e-voting system requirements · Verifiable voting system

S. Choi
Graduate School, Department of Information Science, Korea National Open University, Seoul, Korea
e-mail: woodyc@naver.com

H. Choi
ATGLab R&D Center, Seoul, Korea
e-mail: hs.choi@atglab.co.kr

K. S. Chung (✉)
Department of Computer Science, Korea National Open University, Seoul, Korea
e-mail: kchung0825@knou.ac.kr

X.-S. Yang et al. (eds.), *Proceedings of Seventh International Congress on Information and Communication Technology*, Lecture Notes in Networks and Systems 447,
https://doi.org/10.1007/978-981-19-1607-6_50

1 Introduction

Research is actively being conducted to use blockchain based on Distributed Ledger technology in e-voting as a way to solve the risk of vote manipulation and difficulty in monitoring. Blockchain technology provides irrevocability by sharing and managing the same ledger on distributed network servers through agreement. This irrevocability may provide the integrity and verifiability of the vote. Using various algorithms and blockchain, the proposed e-voting system meets the requirements of electronic voting.

The remainder of the paper is structured as follows: Sect. 2 reviews related e-voting system requirements and the blockchain-based e-voting system, cryptography algorithm to be applied to e-voting systems. Section 3 presents the process and implement of verifiable blockchain-based e-voting System. Section 4 concludes the paper by explaining that the proposed e-voting system meets the seven requirements of e-voting, and mentions the limitations of the study and future research directions.

2 Related Works

Won-geun et al. [1] summarizes the requirements for e-voting into completeness, soundness, privacy, unreuability, eligibility, fairness, verifiability. The encryption of e-voting content is a key technology that maintains the confidentiality of e-voting. E-voting using various methods of encryption algorithms was proposed. The e-voting results in [2] are homomorphically encrypted and stored, but when the secret key to verify the result is exposed, the voting results of voters are likely to be exposed because the contents of each vote are disclosed. Therefore, further discussion is needed on how to separate voting rights from voting rights. Cramer et al. [3] proposed a method of using the public bulletin board and ElGamal cryptographic algorithm to support confidentiality, verifiability and completeness. The voting method is based on a vote of pros and cons on a particular matter, and the vote of pros and cons shall be counted without any decoding using the homomorphic characteristics of the ElGamal. Gòdia [4] uses MixNet to disassociate voters from their votes and adds an Elliptic Curve-ElGamal algorithm to aggregate them. In [5], critical signature algorithms using various elliptical curves were compared and analyzed.

3 Implementation of Verifiable Blockchain-Based e-voting System

3.1 Development Environment

In the proposed e-voting system, we implement a blockchain using Ethereum version 1.9.24. The Ethereum source has been modified to apply to the proposed e-voting system. The smart contract uses solidity of 0.8.1, and uses web3j, which is a JSON-RCP implementation, for communication with external modules.

We implement the e-voting system using Java 14 and use the Bouncy Castle library and the Rings library as libraries related to encryption. The Flutter is used as a framework for developing a UI of the proposed e-voting system for the user.

Figure 1 shows the prototype system configuration. The websocket server is implemented by spring boot. The app is implemented so that the voter registration terminal, the voting terminal, and the voter verification terminal are each independently operated on the Android tablet. The evidence of voting content and ballot are stored using MongoDB. The voting server is implemented as a Spring Boot REST server. The ballot counting program is an independent application implemented in Java 14. The multiplex encryption system for the election auditing team encryption is not implemented as a separate system, but as an encryption module included in the voting server.

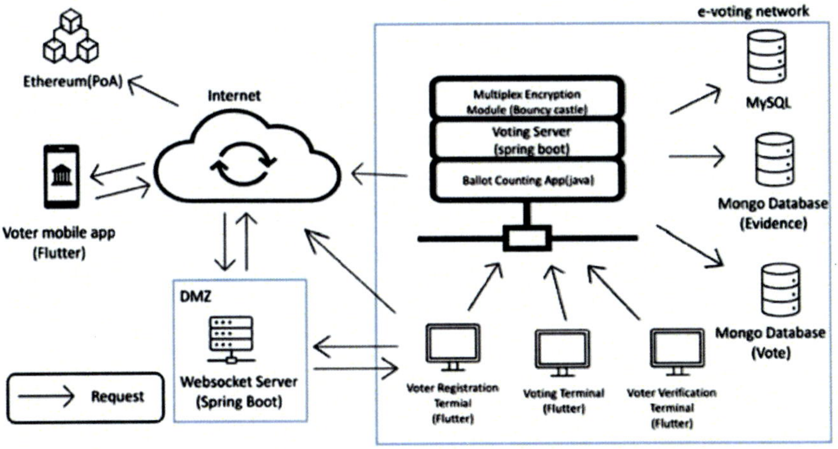

Fig. 1 Prototype system configuration

3.2 Component Implementation

The extract number component uses a TensorFlow handwritten image recognition model to recognize number from handwritten candidate number image.

The proposed e-voting system uses asymmetric keys of various algorithms.

- Signature of handwritten candidate number image (The voter Input Candidate Number Image Signature): the signature on the handwritten candidate number image is submitted to the ECDSA signature algorithm that can verify the signer's public key from the signature.
- Voting Evidence: the primary encryption is performed using the voter's encryption key. The primary encrypted evidence of voting content is re-encrypted with the PgpElgaml algorithm by the encryption key of the evidence of voting.
- Candidate Number Partial Homomorphic Encryption: this uses the ElGamal algorithm with multiplicative homomorphic properties.
- Multiplex Encryption: election audit team encryption is sequentially performed using the secp256r1 elliptic curve encryption algorithm.

3.3 Prototype System Implementation

Voting preparation stage

The smart contract is created by the Election Commission during the voting preparation stage. The voting system generates an asymmetric key pair for the encryption of voter voting content evidence and an asymmetric key pair for the partially homomorphic encryption of voting information. Among the generated asymmetric key pairs, the encryption key is stored in the voting terminal, while the decryption key is stored in a separate offline storage.

Voting right registration stage

Figure 2 shows the process of registering voter voting rights.

(1) The voter creates a blockchain address and an asymmetric key to encrypt the voting content evidence in the voter mobile application.
(2) The voters are confirmed their voting rights at the polling place.
(3) and (4) The voting right registration terminal presents the constituency information and the QR code containing the UUID to the voter mobile application.
(5)–(8) The voter mobile application registers voting right in the smart contract using the web socket server and voting right registration terminal.
(9)–(11) The voter mobile application receives the hash of the result of storing the voting right registration transaction.

Voter voting stage

Figure 3 shows the voter voting process.

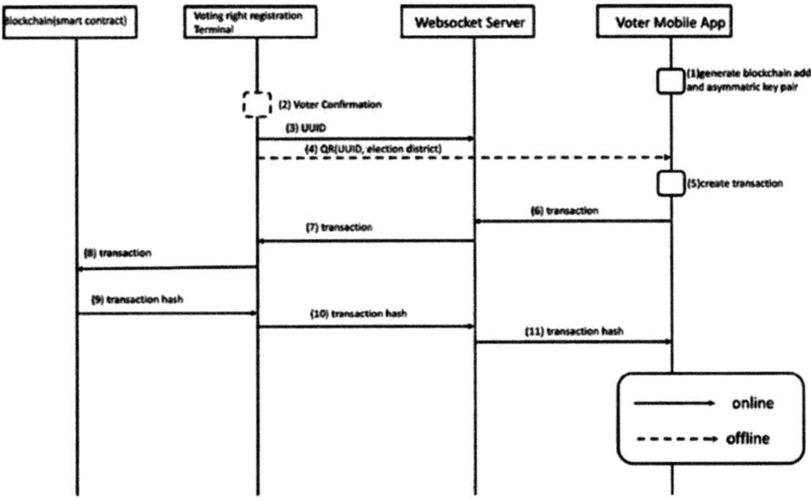

Fig. 2 Voting right registration stage

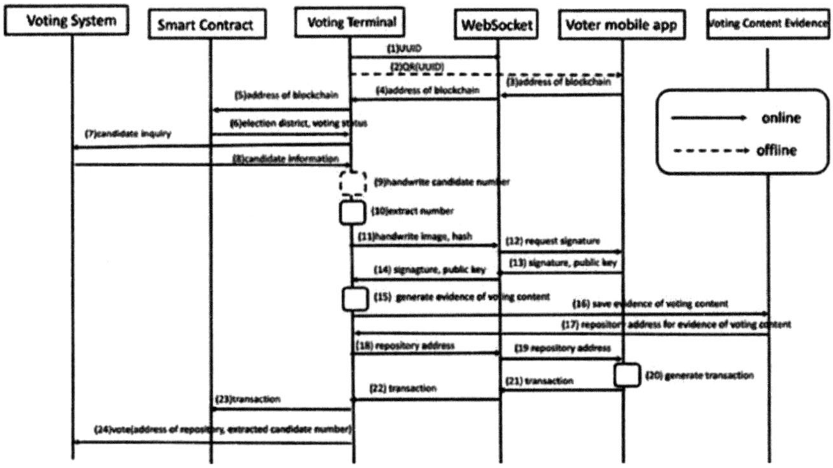

Fig. 3 Voter voting stage

(1)–(2) The voting terminal displays a QR code with UUID to voters.
(3)–(6) The voting terminal uses the blockchain address in step (4) to check the right of the voters, and whether they voted.
(7)–(9) The voting terminal displays a list of candidates to voters. Voters manually input the candidate number into the voting terminal.
(10) The voting terminal recognizes the handwritten candidate number and receives confirmation from the voters.

(11)–(14) The voter mobile application transmits the signature and encryption key for the hash of the handwritten candidate number to the voting terminal.
(15) The voting terminal generates evidence of voting content using the encryption key of step (14), and re-encrypts the generated evidence of voting content using the encryption key of the voting system created in the voting preparation stage.
(16) and (17) The voting terminal stores the encrypted voting content evidence in the storage and receives the storage address.
(18) and (19) The voting terminal transmits the storage address of the voting content evidence to the voter mobile application.
(20)–(23) The voter mobile application creates a transaction to indicate voting completion on the smart contract, and sends it to the blockchain for saving.
(24) The voting terminal transmits the recognized candidate number and voting content proof data storage address to the voting system.

Voting system voting process stage

Figure 4 shows how the voting system that receives the vote from the voting terminal processes the vote.

(1) The voting system converts the candidate number received from the voting terminal into a candidate identification number.
(2) The voting system uses the ElGamal algorithm for partial homomorphism encryption of the candidate's unique number, and the encryption key used at this time is generated in the voting preparation stage.
(3) The voting system obtains a hash of the partially homomorphically encrypted candidate identification number and the voter blockchain address, and stores it in the blockchain.

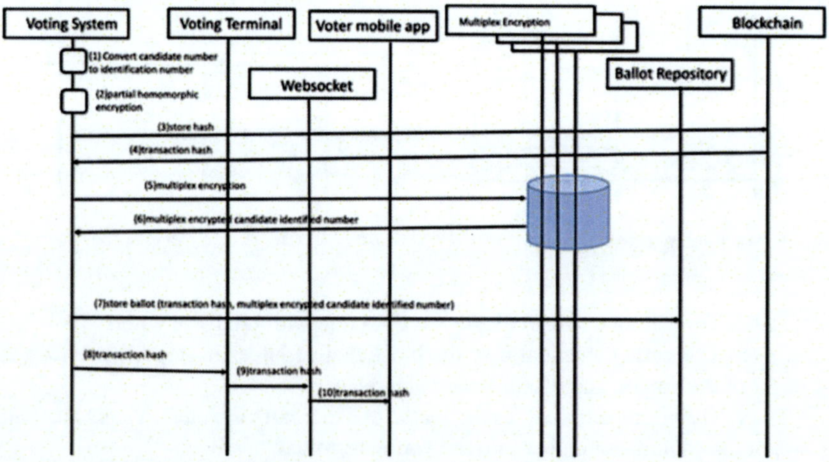

Fig. 4 Voting system voting process

(4) The voting system receives the hash of the blockchain-stored transaction.
(5) The voting system requests the election monitoring team for multiplex encryption of the partially homogeneous encrypted candidate identification number.
(6) The voting system receives the multiplex-encrypted candidate identification number from the election monitoring group.
(7) The voting system stores the multiplex-encrypted candidate identification number of step (6) and the blockchain-stored transaction hash of step (4) in the ballot repository.
(8)–(10) The voting system delivers the blockchain-stored transaction hash to the voter mobile app through the voting terminal and websocket server.

Ballot counting stage

Figure 5 shows how the ballot counting process proceeds.

(1) The ballot counting system requests the encrypted ballot from the ballot repository.
(2) and (3) The ballot counting system decrypts the requested ballot using a multiplex encryption system.
(4) The voting system compares whether the voter blockchain address included in the ballot retrieved from the ballot repository and the voter blockchain address included in the decrypted ballot are the same.
(5) The voting system generates a SHA256 hash from the decrypted ballot.
(6) and (7) The voting system requests and receives the hash of the ballot from the blockchain.
(8) The voting system compares the hashes of steps (5) and (7).

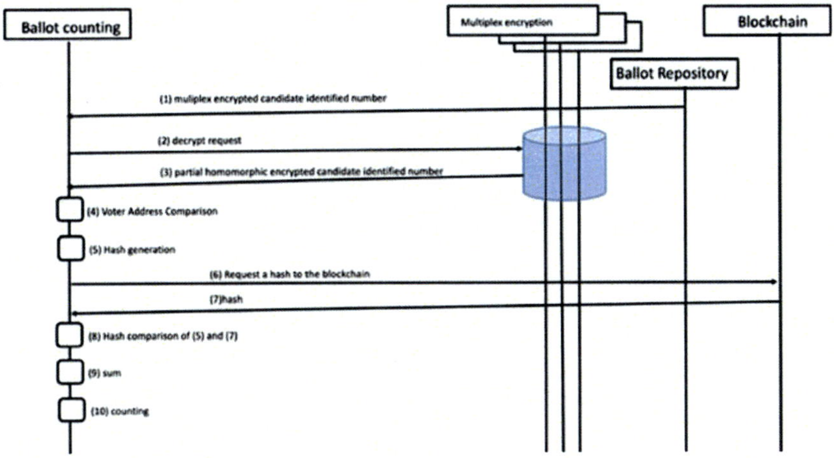

Fig. 5 Ballot counting stage

(9) The ballot counting system adds the partially homomorphic encrypted candidate identification numbers. The counting system uses a calculation unit for the efficiency of demarcation.
(10) The ballot counting system decrypts the sum using a decryption key for partial homomorphic encryption, and converts the decoded sum into a polynomial expression using the candidate identification number to obtain the number of votes for each candidate.

Voter verification stage

After the ballot counting is completed, voters can check whether their voting results are reflected or not by using the voter verification system. Figure 6 shows the voter verification process.

(1) The voter verification system receives the voting evidence decryption key and the partial homomorphic decryption key from the voting system.
(2) and (3) The voter sends the voting contents evidence storage address, voting contents evidence decryption key, and voter blockchain address to the voter verification system.
(4)–(6) The voter verification system uses the decryption keys of steps (1) and (3) to decrypt the voting content evidence.
(7) The hash of the handwritten input candidate number included in the decoded voting content evidence is calculated.
(8) The voter verification system verifies whether the blockchain address of the signature included in the evidence of voting content is the same as the block chain address of step (3).

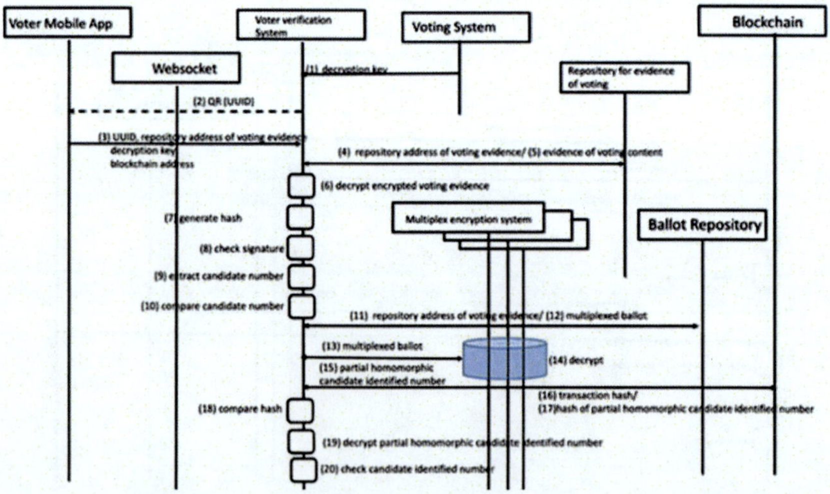

Fig. 6 Voter verification stage

(9)–(10) It is confirmed whether the candidate number included in the voting content certification data and the candidate number recognized in step (9) are the same.
(11) and (12) The voter verification system requests the multiplex-encrypted voting contents using the voting contents evidence repository address.
(13)–(15) The voter verification system requests the multiplex encryption system for decryption to obtain the partially homomorphic encrypted voting contents.
(16) and (17) The voter verification system uses the blockchain transaction hash to request the hash of the partially homomorphic encrypted candidate identification number.
(18) The hash of the partially homomorphic encrypted voting content in step (15) is compared with the hash of step (17).
(19) The voter verification system uses the decryption key of step (1) to decrypt the partially homomorphic encrypted voting content and confirms the candidate's unique number.
(20) The voter verification system compares the candidate identification number in step (19) with the candidate number in step (9), to confirm whether the voter's vote is reflected in the voting result.

4 Conclusion

In spite of e-voting with various advantages such as cost and time compared to the existing voting method, many countries have not officially introduced and implemented national e-voting systems. And they have not yet secured enough voter credibility to officially introduce it. Election must comply with the seven e-voting requirements of completeness, soundness, privacy, unreuability, eligibility, fairness, verifiability. Previous researches did not satisfy all the e-voting system requirements by targeting only some of the requirements of the e-voting system.

In this paper, we designed and implemented the blockchain-based verifiable e-voting system that includes the entire voting process including the encryption of the voting results and the accompanying verification mechanism. We believe that this study will contribute to the realization of e-government by increasing voter confidence in e-voting. In the blockchain-based verifiable e-voting system, the polling place voting method was adopted to secure the soundness of e-voting. But the need for remote e-voting is increasing due to the development of communication technology. In the case of remote e-voting, there is a possibility that voters may disclose their votes for specific purposes. In addition, the security is weak compared to the polling place voting. Accordingly, additional research is needed on a method that can satisfy the requirements of e-voting in remote Internet voting.

References

1. Won-geun H, Heesun K, Kwangjo K (2000) Requirements of electronic election protocol. J Inform Sec 10(1):63—69. Korea Information Security Association
2. Sang-woo H, Min-soo B, Kyung-ho H (2019) Development of blockchain based electronic voting system using homomorphic cryptography. J Korea Telecommun Assoc 44(1):171–174
3. Cramer R, Gennaro R, Schoenmakers B (1997) A secure and optimally efficient multi-authority election scheme. In: International conference on theory and applications of cryptographic techniqes, Eurocrypt 1997: Advances in Cryptology—Eurocrypt '97, pp 103–118
4. Gòdia SS (2011) An electronic voting platform with elliptic curve cryptography. Enginyeria Tècnica en Informàtica de Sistemes Escola Politècnica Superior Universitat de Lleida
5. Ertaul L, Lu W (2005) ECC based threshold cryptography for secure data forwarding and secure key exchange in MANET(I). In: NETWORKING 2005. Networking technologies, services, and protocols; performance of computer and communication networks; mobile and wireless communications systems, 4th International IFIP-TC6 networking conference, Waterloo, Canada, 2–6 May 2005

DevOps Best Practices in Highly Regulated Industry

Ruth G. Lennon

Abstract DevOps has an important role in supporting critical decisions in software and systems development in highly regulated industry. To determine best practice, I have reviewed industry papers, standards, and comments of implementers. A systematic review of key reports on DevOps was conducted. Surveys conducted by a number of organizations with over 277,000 respondents are considered here. Key questions posed are how is DevOps perceived by industry, what are the key benefits from an industrial perspective, and what practices contribute to success. Results of this analysis show that while commonly accepted commercial reports make interesting reading; they are not sufficient to base critical decision in highly regulated industry. Furthermore, standards are required to establish confidence in process and product. This paper provides insights and guidance for software and systems development with DevOps practices in a highly regulated environment. I present the need for consistent quality to be encapsulated through industry standards.

Keywords DevOps · Best practice · Regulations · Standards · Highly regulated industry · Meta-study · Industry perspective · Risk appetite

1 The Need for Guidance

1.1 Introduction

DevOps has been accepted by many as a positive concept. As with all trending terms, this leads a drive to implement it in industry. However, the fast rise of the concept, passed from company to company, without solid guidance has left many confused. It has resulted in many versions of the concept through the design of new terms as per Table 1. The problem with these variations is that they show a profound lack of understanding of the concept DevOps. At its core DevOps focuses on promoting change through techniques including enhanced communications, to improve quality

R. G. Lennon (✉)
Atlantic Technological University, Letterkenny, Ireland
e-mail: ruth.lennon@lyit.ie

X.-S. Yang et al. (eds.), *Proceedings of Seventh International Congress on Information and Communication Technology*, Lecture Notes in Networks and Systems 447,
https://doi.org/10.1007/978-981-19-1607-6_51

Table 1 DevOps derivations with focused application areas

Name	Focus	Name	Focus
AIOps	Artificial intelligence for operations	AgileOps	Agility and business value are core to ops and hybrid cloud solutions
BizOps	Decision support systems	BlockchainOps	Blockchain
ChatOps	Integrating chat into workflows	CloudOps	Managing cloud infrastructure
ComplianceOps	Support end-to-end continuous compliance to policies and procedures	DataOps	Data analytics
DevSecOps	Security and accountability	FinOps	Managing financial cost of transactions in cloud-based projects
GitOps	Infrastructure as code (IaC)	MLOps	Machine learning
ModelOps	Machine learning models	NetOps	Network Operations
PrivacyOps	Integration of privacy requirements and regulations	QualityOps	Integrating QA into CI/CD
RiskOps	Frameworks for IT risk	SRE	Applies software engineering practices to IT and Ops
StorageOps	Applying DevOps to storage	TestOps	Test automation
UXOps	User experience	WebOps	Web application development

throughout. Moreover, DevOps includes concepts that are merely highlighted via many of the new terms in Table 1. In other cases, terms in Table 1 narrow the scope further to create new silos which had not previously existed. Other cases still simply show how DevOps principles can be applied to a specific domain. The most widely accepted generic terms that I have found over the past 5 years are listed here. Terms coined by and primarily used by research organizations are not listed. The core focus of each of the terms in the table may be arguable which of course further proof of misunderstood terms is.

While the table could be expanded with many more examples, the principles of DevOps can be applied to many domains. Regardless, the choice of name as "DevOps" can be considered a poor one as each domain area vies for stronger visibility as per Table 1. I would be remise if I did not mention the related term of NoOps, where all tasks can be automated such that there is little need for an in-house or dedicated operations team, but that requires discussion in greater detail than can be provided here. More obscure examples were not included here, of which HappinessOps was one of my favorites. What would it look like if I could automate continuous delivery of high-quality secure happiness and apply it to the human life cycle?

Table 2 Variety of focus in DevOps definitions

Company	Definition
Azure	"A compound of development (Dev) and operations (Ops), DevOps is the union of people, process, and technology to continually provide value to customers" [1]
PhoenixNAP	"… a set of practices that the development (Dev) and operations (Ops) teams implement to build, test, and deploy software faster and easier" [2]
Medium	"DevOps is a software development strategy which bridges the gap between the developers and the IT staff" [3]
IntelliPaaT	"DevOps (development and operations) is one of the widely used tools that help in the field of development and operations" [4]

What is DevOps? There are many conflicting definitions of DevOps in addition to the extrapolations shown. Some contain core commonality but focus on different elements as key to the principles of DevOps. Some focus on value for the customer but not on continuous improvement. Others focus on improving deployment times but not on collaboration and communication. In many of the cases, surveyed key characteristics of DevOps were missing from the core definition. Refer to Table 2, while some are later extrapolated into significantly more detailed descriptions including the missing concepts, they were often so long or obtuse that the key concepts remained hidden or undervalued.

From this, we can see that not only are the core concepts under disagreement, but it is also not clear to some whether DevOps is a set of practices, a strategy, a tool(s), or a compound of items to provide value.

1.2 The Need for Market Analysis Meta-Study

The purpose of the meta-study was to (a) to determine the desire for stability in DevOps principles and practices, (b) review and evaluate literature regarding key gains to be obtained from DevOps implementation, and (c) synthesis knowledge of key factors and processes that from an industry perspective, contribute to these gains. While much work has been presented by industry on the broad gains of DevOps, detailed work has also been presented by academia on narrow aspects of improvement. The broad qualitative nature of discussions on DevOps and, indeed, the human aspects of the processes and practices make a meta-analysis approach ideal to provide both quantitative and qualitative support for research work undertaken here. This provides controls and rigor to the more qualitative aspects of the research.

A systematic search of commercial whitepapers, academic journals, and databases led to a list of papers to be reviewed. After the identification process, works were screened for duplicate work presented across media. Criteria for eligibility began with filtering for full text articles written in English to aid comparison. Criteria for inclusion involved commercial acceptance determined by number of downloads, sponsoring companies, citations, and feedback from experts in the field. Early evaluations indicated common themes. It quickly became evident that there is a need for a common understanding of the core principles of DevOps. It is also clear that support for regulatory bodies through a structured approach to DevOps for building reliable and secure systems is needed.

Regulated Industry. The contributions of this paper are as follows:

1. A systematic review of industry perspectives on the stability of DevOps concepts is presented.
2. Industry-specified key gains obtainable from DevOps implementation are identified.
3. The need for standardization in DevOps is highlighted.

Organization. The rest of the article is organized as follows: Sect. 2 provides information on the methodology applied and the selection of key works in the meta-study. Section 3 presents the challenges identified. Key findings are presented in Sect. 4. In Sect. 5, existing guidance is examined outlining the case for a new DevOps standard which is presented in Sect. 6. Best practice for highly regulated industry is considered in Sect. 7 which leads to the conclusions on the research are provided in Sect. 8.

2 Meta-study Methodology

2.1 Introduction

Numerous studies exist focused on a variety of concepts core to DevOps in media such as research papers [5] to surveys of the state of the DevOps in industry [6–14]. Despite the longevity of DevOps, much confusion remains as to the definition, and indeed how to implement, it has been noted by Leite et al. [32]. Leite, after systematic literature review (SLR), does propose a definition of DevOps but fails to consider aspects of quality such as safety and security. This paper describes an exploratory study on the state of practice of DevOps.

2.2 Review Protocol

To conduct this systematic literature review of the state of practice, a three-phase approach was taken. In the first phase, a survey of available literature was carried out as previously described. In phase two, the most relevant literature was selected based on predefined common criteria. In phase three, the contents of each item were individually analyzed to assess the quality of the publication. The selected papers were chosen to reflect the overall DevOps landscape. Each survey defined its own approach to data collection, reporting and rigor of process. From the selected papers, common aspects can be identified to provide a high-level view of trends emerging as a result of analysis across multiple years and across a variety of surveys in any given year.

2.3 Review Data

Data from some surveys have been collected since 2011 even though the published surveys, presented here, began to appear in 2013. Some of the studies are carried out yearly although sponsorship may vary year on year. As a result, distribution channels for the study varied somewhat. This meta-study takes the largest, commercially accepted studies, and provides an analysis of their findings. This study is a longitudinal study of ten years' worth of survey data from several organizations, with over 277,000 responses. Reports considered in the meta-study are briefly described in Table 3.

Figures vary across the reports, but where data are available, respondents run across the world with an average of 51% from South America, 29% across Russia and Europe, 10% from Asia, 3% from South America, and only 1% from Africa. It is noted that this may have been due to the low level of translation to other languages. The State of the Developer Ecosystem report from 2021 [19] has been translated into nine languages to aid distribution. From the data obtained to date, there is an indication of increasing interest in such DevOps surveys' internationally with greater numbers of respondents from Asia in later years of the surveys. It is anticipated that over the next few years, several industry surveys will have greater parity across continents.

Demographics from the surveys presented here show a broad range of technologies and management. However, an increasing number of C-suite managers are filling in these surveys which may indicate that they see such reports as import. Refer to Fig. 1, the reports surveyed include a wide variety of industry. However, this paper focuses particularly on those which are highly regulated. This includes: finance, insurance, health care and pharmacy, telecommunications, and energy sectors. Responses from highly regulated industry represent approximately 33% of the market as noted the surveys.

2.4 Research Questions

To focus the research more clearly, I pose three questions. The questions move from perception, to potential benefits to a path to achieve the benefits.

- RQ1: How is "DevOps" perceived in industrial surveys of DevOps implementation? The focus is to determine how the DevOps is globally perceived across

Table 3 Reports included in the meta-study

Years	Name	Report focus areas	Key sponsor
2013–2021 [6–14]	State of DevOps	DevOps adoption and benefits	Puppet
2017–2021 [15–19]	State of the Developer Ecosystem	Environments, languages, general welfare, and culture questions	JetBrains
2019 [20]	2019 Global Developer Survey	Developer processes and feedback loops	GitLab
2020 [21]	Mapping the DevOps Landscape	Responsibilities across functional domains	Gitlab
2021 [22]	A maturing DevSecOps landscape	Moving from culture shift to what the next steps are	GitLab
2017 [23]	Interop ITX and InformationWeek 2017 State of DevOps	DevOps adoption	Interop
2019 [24]	2019 State of DevOps	The need for DevOps and its advantages	Interop
2020 [25]	2020 State of DevOps	Challenges for DevOps adoption	Information Week (Interop)
2021 [26]	2021 State of DevOps	DevOps challenges and the pandemic	Information Week (Interop)
2017–2020 [27–30]	DevSecOps Community Surveys	Security practices with DevOps	Sonatype
2020–2021 [31]	Azure Annual DevOps Report: Enterprise DevOps Report 2020–2021	Implementing DevOps at enterprise scale	Microsoft

industry including highly regulated industry. This will help determine the need for stability.

- RQ2: What positive outcomes can be gained from implementing DevOps practices from the perspective of industry? Benefits determined from an industry perspective may have different priority than those discussed in academic literature.
- RQ3: What practices and processes from an industry perspective contribute to successful implementation of DevOps? To obtain the greatest value from implementing DevOps, what supporting processes and practices should be put in place.

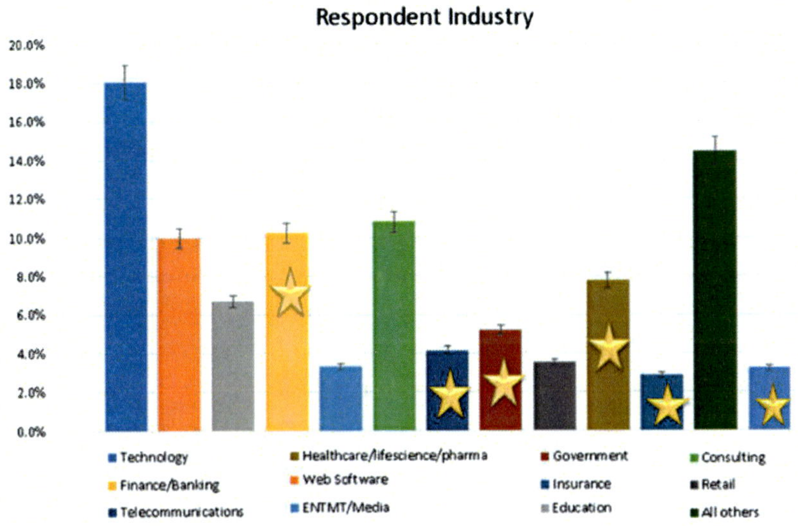

Fig. 1 Respondent breakdown by industry

3 The Need for Guidance

3.1 Introduction

Despite significant progress in methodologies for software development, processes to support quality in the software development life cycle and a general awareness of the need for security, it is worth noting that 72% of respondents to a survey [22] is still not in cross-functional teams. Approximately, 50% of respondents to a survey [30] state that they consider security important, but they do not have time to commit to it. Without clear guidance on how to proceed, some companies may find it difficult to implement appropriate practices.

Regulated Industry. Responses to surveys reviewed came from a wide range of sectors. Most interestingly, 50% of participants on the DevSecOps Community Surveys [27–30] came from the financial services, banking, and technology sectors. Participation from the government sector also increased approximately twofold each year. This is also reflected in other surveys. Data from the 2011–2013 Dora State of DevOps survey and reports [6] indicate respondents roles ranged from C-suite (17%), consultants (8%) and administrators/engineers (75%). Within industry, these are the roles where organizational strategy is set and implemented. From the combined surveys reviewed thus far, there are an increasing number of managers responding to the surveys indicating an interest in results at all organizational levels. Management roles are highlighted with the suitcase symbol in Fig. 2. Notably, the roles of DevOps

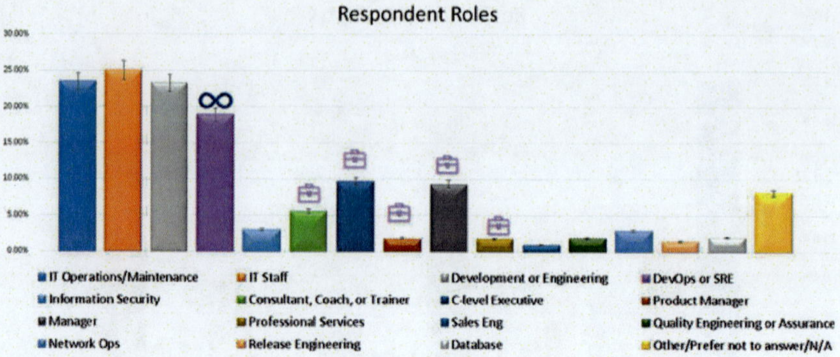

Fig. 2 Range of respondent roles

and SRE are also increasingly indicated as job titles or roles. This is highlighted using the infinity symbol in Fig. 2.

Key Factors Cited. Some of the surveys provided ask respondents to rank their organization regarding DevOps practices. Rankings have changed over the ten years of surveys and vary somewhat between surveys. For the most part [11–13, 27–30], align in providing categories of elite, high, medium, and low performers. However, the rapid change of capabilities has resulted in changing criteria to meet these rankings. This results in changes to the steps or categories as shown in Fig. 3. Metrics

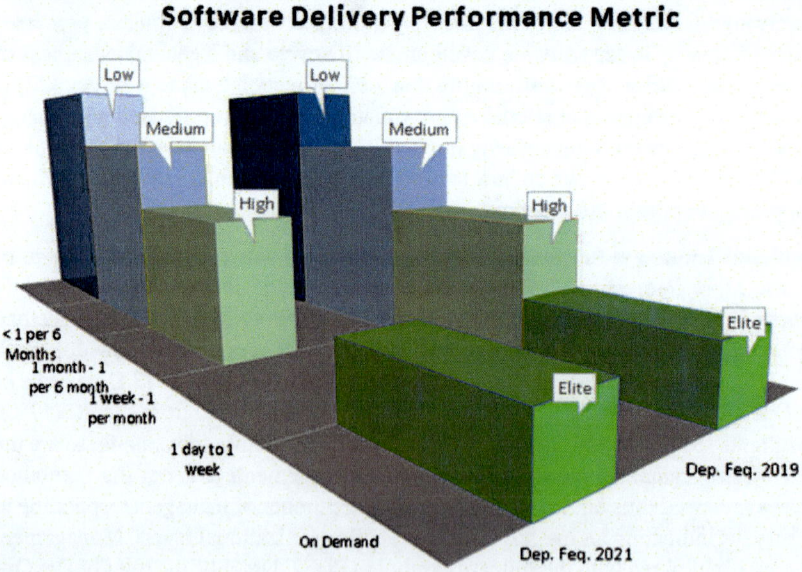

Fig. 3 Software delivery performance metrics 2019 [12] and 2021 [14]

used to measure the categories include deployments, change lead time, mean time to recover, and change failure rates. These metrics are widely used and can be found in a variety of the surveys analyzed.

Deployments. Most of the documents reviewed refer to deployments, whether regarding the tools and practices to aid deployment [20–22] or frequency of deployments [6–13]. The DevSecOps survey of 2020 [30] indicates that 55% of respondents deployed at least once a week, while the Interop State of DevOps report 2020 [25] focuses on the time to move to deployment after completing development. In 2020, it indicated that 32% deployed in 1 day or less. Themes of seeking practices supporting high frequency deployments are common across surveys.

Change Lead Time. One of the core concepts of agile development is the ability to adapt to change. Taking that change from requirement to deployment is referred to as change lead time. The application of automation to support the change process can enhance confidence in the change management process [21]. Many of the challenges cited in comments from [30] arise from poor change management processes. Providing clear structure for such processes can provide solutions to a number of problems.

MTTR. It is unrealistic to assume that failures don't happen. Metrics to consider failures, or more appropriately, recovery from failures is important. While mean time to repair (MTTR) is very broad and does not consider the root of the problem, it is still used as a general measure of stability. Most surveys consider stability in some form. In addition, DevOps provides high levels of observability supported by enhanced communications enable greater feedback loops and trackability when failures occur.

Failure Rate. The speed at which code is pushed to production is rapidly increasing. Even with automated test procedures, code smells and poor practices can lead to failures in production. Identifying the level of failures that escape quality gates can help establish if supporting practices need improvement. An Infoweek survey [24] identified that in 2019, 14% of participants was experiencing 11–20 failures per month which dropped to only 6% in 2020. The change failure rate identified by puppet for medium maturity organizations in 2016 [9] was 31–45%, 2019 [12] was 0–15%, and 2020 [13] was 16–30%. The increase in 2020 is due to a widening gap between maturity levels in organizations.

4 DevOps Meta-study Findings

The broad topic of DevOps spans many themes and sub-topics depending on the interpretation selected. In this paper, analysis begins with broad findings narrowing to focus on those with specific impact on highly regulated industry. Consideration of DevOps maturity, governance, regulatory need for change to security and automation is provided here with respect to outcomes achievable from an industry perspective (RQ2).

4.1 Maturity

Currently, there is no standard which provides a definition of DevOps maturity. Thus, surveys [6–13] have asked respondents to self-identify as elite, high, medium, or low with regard to maturity. In a similar structure, respondents in other surveys have been asked if the use DevOps exclusively right down to no use of DevOps [21]. In the meta-study carried out, various levels of performers are identified. Although there is some variation in criteria over the years, the term elite performers is consistent. It is expected that "Elite" level will shift in criteria as technologies improve our process and capabilities to perform. In keeping with this, descriptors such as "Deploy Frequently" vary significantly from report to report. If the levels are considered with the capabilities available at the time, a more meaningful evaluation can be considered. This is also true for the other characteristics of lead time for changes, mean time to repair (MTTR), and change failure rate. With these in mind, the state of practice surveys shows that elite performers indicated reduced costs, reduced time to market, minimal inventory, value for money, and better return on investment where they had implemented DevOps practices.

Where organizations have self-identified as elite performers, they are noted [9] to spend 22% less time on unplanned work and 50% less time on fixing vulnerabilities. The same report noted that to achieve elite level should focus on the implementation of standards to accelerate adoption of automation. As an additional benefit to increased automation is the automation of security practices. Nevertheless, the subjective nature of their self-identification as either elite or not must be taken into account. As previously mentioned, the reports selected for this meta-study are well accepted in industry hence their presentation here. Without access to the raw data, more in-depth review of the suitability of their self-selection cannot be made. Common themes reported from the elite performers include deployment on demand, low lead times, MTTR of less than 1 h, and change failure rate of 15% or lower. Further, review of the impact of such metrics, again self-reported, is less time spent on unplanned work or rework due to improved process and less time remediating security issues due to the automation of security practices. It is questionable if the reported time gained to bring in additional work opportunities is correct due to potential issues with the figures reported. This cannot be confirmed as many of the reports, most notably, the state of DevOps reports [6–13] does not release their raw data. The DevSecOps Community Reports [15–19, 27–30] have made available their raw data for this analysis. As such, other aspects of metanalysis can be made in more detail.

4.2 DevOps and Governance

The prominence of such reports, even with issues of self-reporting and the lack of availability of raw data, will remain. However, I would suggest that companies take more time to review and analyze what information is presented and why it is presented. Reports which make available their raw data go a long way to dispelling concerns with analysis of data. Regardless, making a strategic decision on process based solely on commercially funded reports, from the maker of tools to be used in the process, is questionable.

Scaling can be hard and expensive. Making incorrect decisions can be more costly still. Governance at scale can be harder again due to separate knowledge stacks which reduces knowledge sharing. Teams don't often have the breadth of skills or depth of expertise to create and operate all aspects of a full infrastructure and application stack. Measurement, information, skills, policy, and process can all be defined, put in place, and measured for conformance to aid benchmarking of organizational status. This can help provide a more accurate reflection of the status of the organization rather than simply self-reporting. This can take some of the risk out of scaling.

Transparency into controls inspires confidence of meeting required regulatory and governance needs. This supports better policy adoption enabling the governance champions with even broader impact and faster uptake integration. This in turn reaches higher levels of alignment across the organization through iteration and again enabling more standardization forming a virtuous circle with automation.

4.3 Regulatory Need for Change

Disaster drives change in many areas. There are many new laws across continents that require best practice to be followed. We don't just need to comply with the laws; we need to prove that we did. Many companies lack an up-to-date register of legal and regulatory obligations. Where companies have an awareness of their obligations, they may lack the management systems to carry out risk assessment on their ability to comply with any given requirement. Obligations vary across countries but can include internal, national, and international requirements. Many companies lack an up-to-date register of compliance obligations—covering internal, local, state, federal, and international requirements. Others need a methodology to rank the risk related to each obligation.

"Organizations are facing continual changes to their regulatory requirements, so building a strategy to manage the pace and complexity is essential to avoid costly fines, reputational damage, or even a loss of the right to operate" [32]. The State of the Developer Ecosystem report from 2021 [5] indicates a long list of items that developers would like to spend more time on. The list ranges from shift left security and automated testing and code review, right through to integrating AI and ML for writing code. Many of these topics provide benefits in not only time saving but added value to the SDLC. In highly regulated environments, risk benefit analysis is carried out for all changes including that of the SDLC and supporting components.

4.4 Security and Automation

One of the key findings of the DevSecOps Community surveys [30] was that developers, despite an understanding of its importance, do not have sufficient time to spend on security.

DevOps prioritizes automation as an enabler for processes. A Forbes article [36] from 2017 states that intelligent automation brings cost savings of between 40 and 75%. This can be in part from freeing up valuable people to focus on the interesting problems. Automating consistent processes aids data collection for more informed decision-making.

From the DevSecOps Community reports [27–30], it has been found that companies that are more elite in their DevOps practices integrate automated security practices. In highly regulated environments, this provides an opportunity to meet regulatory requirements with minimum disruption. This also provides an opportunity to demonstrate compliance more easily during audits.

5 Existing Guidance

In this section, existing guidance is discussed which can mitigate some of the challenges. While the nature of what constitutes best practice [33] has been debated, for the purpose of this work, we can consider it as a process that is considered near optimal for a given application in a specific scenario. Thus, best practices should not be considered so broadly generic as to be applicable to all scenarios. As an alternative to the almost anecdotal structure of best practice, standards can be considered to provide a level of rigor when defining processes for improvement. Standards to improve process and to document practice can guide the implementation of practices such as DevOps. In the case of highly regulated industry, the use of standards can support improvements to process as established by reviewing best practice. Continuous improvement on process is often required by regulatory authorities and auditors. Standards can themselves be considered a form of best practices in that they provide a near optimal solution for a specific scenario. Consider the ISO/IEC/IEEE 12207 standard for systems and software engineering specifically for software life cycle processes. This standard is adapted to more specific scenario to provide guidance for the adopter of a given standard. ISO 29110 for example aligns with 12207 but maps directly to very small entities. Despite variations across business domain and country, many key standards are repeated throughout due to their core focus on improvement of process and practice.

5.1 Regulated Industry

Highly regulated industries follow many standards coming from the ISO, IEE, or other international standardization bodies in addition to their own area specific standardization bodies. Samples of standards applicable to a variety of highly regulated industry areas are shown in Table 4. In addition, numerous standards exist focused on a variety of areas core to DevOps. Each standard can be combined with others to provide a virtuous circle of standards. Relevant elements from each can be taken as part of the tailoring process to ensure that the most appropriate standards for a given area at a given point of time can be selected. Despite variations across business domain and country, many key standards are repeated throughout due to their core focus on improvement of process and practice. An example can be seen in Table 4.

In addition to very specific standards and regulations, some broad regulations exist. For example, GDPR and SOX can be applied to many areas and may business domains. Recent breaches in security such as SolarWinds [33] have caused greater requirements for rigor beyond those previously stated for the given domain. With a rise in notable attacks, the requirement for rigor is likely to increase with compliance to standards a mere minimum requirement.

Table 4 Sample standards to industry area mapping

Industry area	Standard number	Focus area
National Security	IEEE 15288.1-2014	IEEE Standard for Application of Systems Engineering on Defense Programs
	ISO 28000	Supply chain security management systems
	ISO 31000	Risk management
	IEEE 1633-2016	IEEE Recommended Practice on Software Reliability
	ISO/IEC 27000	Information security management
Airline	ISO/IEC 27001	Information security management system
	ISO/IEC 27002	ISMS controls details; dependent on 27001 standard
	ISO/IEC 27018	Protection of PII in public clouds
	ISO/IEC 27701	Privacy information management system (PIMS)
	AS9100D	Quality management systems—requirements for aviation, space, and defense organizations
Banking	ISO 5116	Improving transparency in financial and business reporting
	ISO 3531	Financial services—financial information eXchange
	ISO/IEC 27001	Information security management system
	ISO 20022	Financial services—universal financial industry message scheme

6 IEEE 2675 DevOps Standard

To establish best practices, with an agreed level of rigor, organizations may look to standards provided by the IEEE, ISO, and others. The IEEE 2675 Standard for DevOps: Building reliable and secure systems including application build, package, and deployment provide guidance on outcomes, process, activities, and tasks. By standardizing DevOps, we can establish pivot points through checkpoints of factors needed for quality and scalability. Defining clear process to follow can enable not only uptake of continuous improvement processes but benchmarking against others. The application of benchmarking can be achieved through the application of activities and tasks with suitable measures for conformance. This also leads to greater accuracy in benchmarking the maturity of the organization as elite.

Key to this standard is an understanding of why DevOps was created. Software complexity, risk, scalability, security, and many more issues are often not addressed at the highest levels as many processes focus solely on moving to the next stage of development without appropriate communications. The standard notes that: "DevOps was created to provide solutions to constantly changing complex problems, where reducing organizational risk and improving security and reliability are critical requirements" [34]. To this end, DevOps is a full lifecycle endeavor established to reduce risk and add value to the software development lifecycle. Understanding the need for continuous improvement, which is "baked-in" to the standard, will also help accelerate successful adoption of the processes, activities, and tasks. Further discussion on this standard is provided in relation to the specific topics in Sect. 7.

7 Best Practice and Highly Regulated Industry

Long established and widely accepted best practice are few, but they do exist. Highly regulated industry requires significantly greater rigor often turning to standards to establish guidance on how to achieve best practice, for example, guidance on processes for continuous improvement throughout the lifecycle, strong governance, and enhanced security. Discussion on activities to achieve best practice and standardization is provided here (RQ3).

7.1 Continuous Improvement and Risk Management

Continuous improvement is needed to stay ahead of the competition. In highly regulated industry, where risk appetite is low, there is little room for innovative improvement. Many questions arise as a result. How can risk be managed in an environment where developers can continuously pushing code to production? What controls are in place to ensure that code is reliable, safe, secure, and well governed? We need to manage risk. We need to ensure policies are applied as well as written. Governance needs to become part of the company strategy. It must be supported by a community of people who understand the positive and negative risks. Without standards, improving governance and auditing becomes significantly more challenging. In [8], it is stated that: "…there's no secret to achieving both speed and reliability and delivering higher-quality products and services at lower cost. Our research shows this can be achieved with the right practices in place."

Continuous improvement reduces overall risk and is core to standardizing reliability, quality, and security in processes. Adding security, reliability, and safety to the SDLC to improve risk appetite can support informed decision-making in highly regulated industries. Continuous automation aids improvement through governance, compliance, and consistency across standardized processes. Continuous communications between all stakeholders drive change through confidence in quality processes. The DevOps standard outlines activities to support this. "DevOps demands higher levels of integration, collaboration, automation, and established feedback systems for continuous improvement into the life cycle process model. Higher-level process capabilities specifically bring security into prominence in the activities and tasks." IEEE 2675 DevOps [35].

7.2 *Governance*

Belief in strong practices can be hard to show. Manual recording proof process and practice can be not only time-consuming but often results in out-of-date information. DevOps encourages automation of data collection and documentation to aid transparency. This establishes the validity of practice at any given point in time. The standardization of DevOps encourages transparent and automated compliance practices throughout the organization. It provides a mechanism to establish the validity of practices. It defines practices to automate the collection of process data, documenting processes resulting in reduced cost. Such standardization improves visibility across the organization enabling informed decision-making. Transparency into controls inspires confidence of meeting required regulatory and governance needs. The Enterprise DevOps Report [31] reported that Tichaona Zororo, a director at ISACA, stated "Governance should be principles and outcome based not rule based." This is an important distinction as it places greater importance on values. Thus, compliance for highly regulated industry would include flexibility in demonstrating how they have met the requirement. In [35], there is a recommendation on expediting "continuous governance over risk-based decisions that aligns with the environment, the people, and their skills, as well as the capability to evaluate and manage dynamic risks."

7.3 Security

Security is a concern. We cannot hide from it. We need to highlight security practices as critical to all standards. After all, what is a standard but an accepted measure of authority, value, and excellence. We need to standardize and integrate security across all aspects of software and systems development so that we can promote building reliable, safe, and secure systems. This leads to greater integration of automated security practices which could have stopped that type of risk found with Equifax [37] among others. In highly regulated industries, evidence of having applied appropriate security controls is as essential as having carried out the task itself. We need to encourage action and evidence by baking it into our standards. The 2675 DevOps standard [35] states that "there is no DevOps without a continuous focus on security."

7.4 DevOps Culture

Cultural impacts are often observed as DevOps practices require and are rewarded by increased collaboration through feedback loops. As a direct result of this, a common language of understanding can be determined supporting process management and governance. Efficiencies in communication enable continuous improvement often a mandated requirement in highly regulated industry. DevOps culture supports fast reaction to emerging market opportunities. This culture of communication and coordination across stakeholders helps refine the value chain proposition. In regulated environments, controls, both internal and external, are required to reduce risk. Controls such as standardized processes enable more detailed modeling and forecasting which in turn enables organizations to assess their risk appetite more accurately. In this way, they can become market leaders by reacting quickly to new opportunities in a strategic way.

7.5 Open Problems

In Table 4, existing standards and guidance have been presented which promote solutions to a part or parts of the problems currently experienced during the SDLC. Unfortunately, no single standard covers the myriad of problems experienced with modern large-scale systems in complex environments in highly regulated industry. Scalability, safety, and security need to be considered in small-scale systems as they may rapidly expand to large scale in a short period of time. To address this need, there is a necessity for further guidance on tailoring standards. Tailoring is the process by which small to medium-sized organizations, or those in an early stage of adoption, can indicate partial conformance while in the process of obtaining full conformance.

Many standards offer such some guidance on this process including the DevOps standard [35]. However, stepped levels of conforms would prove useful.

8 Conclusions

The first question considered here was to establish how DevOps is perceived in industry. It has been established that much confusion still exists. Poor interpretation of the term combined with domains vying for visibility in the process leads to greater confusion. It is hard to establish best practice when confusion exists. The creation of the DevOps standard should help resolve this situation.

Research [38] has explored the challenges encountered with DevOps implementation proposing solutions to mitigate the challenges. Each snowflake solution simply adds to the problem. It could be suggested that due to the early stage of DevOps adoption, some challenges are not yet defined or not properly understood. By analyzing both academic and commercial work, a clearer understanding of the needs of industry and paths for optimization can be established. This paper described the perspective of industry as seen from commercial surveys; academic work is presented in another paper. From the surveys analyzed, we determine positive outcomes of DevOps practices. Only, a few are listed here.

As a structured mechanism to discover and apply best practice, the use of standards is strongly recommended. Each time an organization applies a standard, they improve the quality of their processes, reducing risk, and promoting greater productivity. Each of the standards work together to reduce negative risk and promote continuous improvement. The DevOps standard [35] discussed here promotes an open culture of perpetual improvement for reliable, safe, secure systems. Standardizing DevOps practices enable highly regulated companies to be both agile and reliable increasing their risk appetite in judicious manner. This research underscores the need for agility, safety, security, and continuous improvement in highly regulated industry underpinned by internal best practices for their own domain and supported by standards.

9 Limitations

Limitations of this research spawn in the most part, from lack of access to underlying data. In the case of commercially sponsored surveys, I would question the impact of the sponsoring company on the findings. Other limitations such as geographical scope and access to raw data should also be considered. The narrow geographical scope of some surveys is counterbalanced by the wider scope of others with an increased number of respondents. Some surveys [15–19, 27–30] included access to the cleaned raw data, while others did not [6–13]. Figures that when individual yearly figures were combined did not correspond with the quoted combined figures lead to questions as to the rigor of process applied by a commercial company. Further work in this research will include expanding the number of sources to obtain further raw data to validate the findings. Additionally, the second part of this work it to cross-compare with findings from academic sources.

Acknowledgements This paper and the research behind it would not have been possible without the exceptional support of the IEEE 2675 DevOps working group members. Special thanks to Hasan Yasar for providing access to raw data and Sarah Baker for discussions on Summa de Summae.

References

1. What is DevOps, Microsoft. https://azure.microsoft.com/en-us/overview/what-is-devops/#devops-overview. Last accessed: 2021/10/23
2. Danicic D What is DevOps pipeline and how to build one, PhoenixNAP. https://phoenixnap.com/blog/devops-pipeline. Last accessed: 2021/10/23
3. Kulshrestha S Who is a DevOps engineer?—DevOps engineer roles and responsibilities, Edureka. https://www.edureka.co/blog/devops-engineer-role. Last accessed: 2020/08/23
4. What is DevOps? https://intellipaat.com/blog/what-is-devops/. Last accessed: 2020/08/23
5. Jabbari R, bin Ali N, Petersen K, Tanveer B (2016) What is DevOps? A systematic mapping study on definitions and practices. In: Proceedings of the scientific workshop proceedings of XP2016 (XP '16 Workshops). ACM, New York, USA, pp 1–11. Article 12
6. 2013 State of DevOps Report, Puppet (2013)
7. Forsgren N, Kim G, Kersten N, Humble J (2014) 2014 State of DevOps Report, Puppet
8. 2015 State of DevOps Report, Puppet Labs (2015)
9. Brown A, Forsgren N, Humble J, Kerseten N, Kim G (2016) 2016 State of DevOps Report, Puppet & DORA
10. Forsgren N, Humble, Kim G, Brown AJ, Kerseten N (2017) 2017 State of DevOps Report, Puppet & DORA
11. Mann A, Brown A, Stahnke M, Kersten N (2018) 2018 State of DevOps Report, Puppet & Splunk
12. Forsgren N, Smith D, Humble J, Frazelle J (2019) 2019 Accelerate State of DevOps, Dora & Google Cloud
13. Brown A, Stahnke M, Kersten N (2020) 2020 State of DevOps, Dora & Circleci
14. Smith D, Villalba D, Irvine M, Stanke D, Harvey N (2021) 2021 Accelerate State of DevOps, Google Cloud
15. The state of developer ecosystem in 2017. https://www.jetbrains.com/research/devecosystem-2017/. Last accessed: 2021/10/23
16. The state of developer ecosystem in 2018. https://www.jetbrains.com/research/devecosystem-2018/. Last accessed: 2021/10/23
17. The state of developer ecosystem in 2019. https://www.jetbrains.com/research/devecosystem-2019/. Last accessed: 2021/10/23
18. The state of developer ecosystem in 2020. https://www.jetbrains.com/research/devecosystem-2020/. Last accessed: 2021/10/23
19. The state of developer ecosystem in 2021. https://www.jetbrains.com/research/devecosystem-2021/. Last accessed: 2021/10/23
20. 2019 Global Developer Survey, GitLab (2019)
21. 2020 Mapping the DevSecOps Landscape, GitLab (2020)
22. 2021 A Maturing DevSecOps Landscape, GitLab (2021)
23. Bruno E (2017) Interop ITX and information week 2017 state of DevOps, Interop

24. Salamone S (2019) 2019 State of DevOps Research Report, Interop
25. Morgan L (2020) 2020 State of DevOps Research Report, Interop
26. Harvey C (2021) 2021 State of DevOps Research Report, Interop
27. DevSecOps Community Survey, Sonatype (2017)
28. DevSecOps Community Survey, Sonatype (2018)
29. DevSecOps Community Survey, Sonatype (2019)
30. DevSecOps Community Survey, Sonatype (2020)
31. Jhaveri S, Reijnen C, Thair S Enterprise DevOps Report 2020–2021, Microsoft
32. Munro J, Dougall M, Robinson N (2017) Regulatory compliance—staying ahead of change, KPMG. https://home.kpmg/au/en/home/insights/2017/02/regulatory-compliance-ahead-of-change.html. Last accessed 2021/10/20
33. Maire JL, Bronet V, Pillet M (2005) A typology of "best practices" for a benchmarking process. Benchmarking Int J
34. Badhwar R (2021) Defensive measures in the wake of the SolarWinds fallout. In: The CISO's transformation 2021. Springer
35. IEEE 2675-2021 (2021) IEEE standard for DevOps: building reliable and secure systems including application build, package, and deployment. IEEE
36. Kirk D (2017) How much is intelligent automation saving you? Forbes. https://www.forbes.com/sites/kpmg/2017/09/21/how-much-is-intelligent-automation-saving-you/. Last accessed 2021/10/19
37. Zou Y, Schaub F (2018) Concern but no action: consumers' reactions to the equifax data breach. In: Extended abstracts of the 2018 CHI conference on human factors in computing systems. ACM
38. Toh Z, Sahibuddin S, Naz'ri Mahrin M (2019) Adoption issues in DevOps from the perspective of continuous delivery pipeline. In: Proceedings of the 2019 8th international conference on software and computer applications (ICSCA '19). ACM

Citizens' Use of Social Media: A Thematic Analysis on Digital Co-Production in Disaster Management

Vicente A. Pitogo and Jesterlyn Q. Timosan

Abstract Disaster Management is a method for saving lives and property that has been built over many years at great expense and effort. Forecasting demand, determining needs, procuring, storing, and handling inventory and coordination, and distributing relief to reduce losses before, during, and after any disaster are all important aspects of disaster management. Since natural disasters strike without warning, disaster management requires well-organized and coordinated planning before, during, and after the event. The use of ICT-based co-production can significantly improve disaster management by effectively and conveniently performing different operations to include diverse stakeholders. This study focuses on citizens utilizing social media in Disaster Management anchored on the co-production theory, as the framework in analyzing the interaction of citizens with the government through social media platform. Co-production theory considers other stakeholders outside an organization by their input toward producing products or services. Using qualitative case study design, this research was able to identify the factors that lead citizens to use social media during disaster and crisis situations using reflexive thematic analysis technique, for locating, analyzing, and interpreting meaning patterns (themes) in qualitative data. Findings indicated that using social media it will strengthen the communication between citizens and government or with the constituents, the ability to generate situational awareness about the crisis and disaster situations. The factors that lead citizens to used social media are the "Socialization" and "Altruism" themes. Those themes come from the interview conducted and transcribed the audio recordings into text conversation, collecting all the common among codes and produces themes. How social media facilitate citizens in co-producing valuable information includes "Collective Action," "Collaboration," and "Venue for Engagement" themes. The researchers further urge that this research be expanded to include not only ordinary citizens who use social media, but also social media leaders and community

V. A. Pitogo (✉) · J. Q. Timosan
College of Computing and Information Sciences, Caraga State University, Butuan City, Philippines
e-mail: vapitogo@carsu.edu.ph

J. Q. Timosan
e-mail: jqtimosan@carsu.edu.ph

X.-S. Yang et al. (eds.), *Proceedings of Seventh International Congress on Information and Communication Technology*, Lecture Notes in Networks and Systems 447,
https://doi.org/10.1007/978-981-19-1607-6_52

organizations, as they play an important role in society during disasters. The government may have all the means and equipment needed to mitigate disaster risks, but citizens, particularly community leaders, have the knowledge and expertise that the government needs to act and respond in those unfortunate situations.

Keywords Disaster management · Co-production · Reflexive thematic analysis · Social media

1 Introduction

Disaster Management (DM) is a method for saving lives and property that has been built over many years at great expense and effort [1]. It also refers to the defense of a large number of lives and properties, when a catastrophe occurs, whether man-made or caused by natural phenomena [2]. Based on the records of the Emergency Management Database (EM-DAT), one of the most comprehensive international catastrophe databases, has recorded 7348 catastrophe incidents globally between 2000 and 2019. According to the United Nations Economic and Social Commission for Asia and the Pacific (ESCAP), Asian and Pacific countries are more vulnerable to catastrophes than other parts of the world. The high frequency and impact of catastrophes in Asia is partly due to the continent's size and ecosystems, such as river basins, flood plains, and seismic fault lines, which together provide a high risk of natural catastrophes [3]. The imminent growing number and broader scale of disasters indicate that there is an urgent need to strengthen the capability of every nation to prevent wider spread and mitigate impending untoward events [4]. In the Philippines, the government is also moving into the same direction, and in fact, formulated various DM plans and programs. The Philippine Disaster Risk Reduction Management Act of 2010, also known as Republic Act No. 10121, attempts to encompass disaster preparedness, disaster response, disaster prevention and mitigation, restoration, and recovery down to local government units. In Caraga region, located at the southern part of the Philippines, always experiencing disaster and calamity, mostly hit by typhoons and its effect on flooding and landslide. In recent years, it has been struck by many typhoons. Agaton, on the other hand, devastated Butuan City and the Caraga region in 2014, resulting in 29 deaths, 4 injuries, and 38 people missing [5].

Social media technology has created thriving opportunities and challenges that have an impact on society by using it in a variety of ways, such as information dissemination [6], participatory budgeting [7], calamities and disaster phenomena [8], and, events like protests and social movements [9]. The novelty of this platform even influences government entities to capitalize the social media for better public services, thus, forging interaction between the public and the government [10]. This leads in many government initiatives to consider social media in the delivery of services to the public and digital inclusiveness to the citizens [11]. In addition, the emergence of information and communications technology (ICT) makes it easy for public workers to communicate with the people, giving rise to a nascent possibility of co-producing public service both collective and individual collaboration [12]. Also, ICT-enabled public services shaped the way government and people interact with each other. It also aids in building dynamic interaction among public entities and the general public that are the potential resource of co-production [13]. However, different issues, such as people becoming passive or even unconscious co-producers, have emerged as a result of the rising usage of ICT platforms in co-production [14]. Technological progress may have a positive influence on people's ability to co-produce, but also on the likelihood that existing disparities would be increased. With this, the goal of this research is to explore the dynamics of citizens in Butuan City for utilizing social media in co-producing insightful information with the government in mitigating disaster risks and managing its effect using a qualitative case study technique anchored on a reflexive thematic analysis [15]. To further investigate this phenomenon, specific research questions are framed:

RQ1:What factors that lead citizens to use social media during disaster and crisis situations?
RQ2:How social media facilitate citizens in co-producing valuable information for disaster management?

2 Review of Related Literature

2.1 Emergency and ICT-Based Disaster Management

Emergency management is "a discipline that deals with risk and risk avoidance" [16]. Risk compasses a wide variety of concerns as well as a diverse cast of characters. The number of cases in which the scope of emergency management, or the emergency management system, is wide. This backs up the argument that emergency management is critical to everyone's everyday safety and that it should be included into daily decisions rather than being used solely in times of catastrophe. Disaster management seeks to reduce the disaster's overall impact and necessitates complete preparedness in terms of organizational preparation, communication, and collaboration among all stakeholders, resource availability and professional engagement [17]. ICTs can aid in the gathering of data and the making of smarter decisions for a more successful disaster management plan. Cities are vulnerable to those suffering of lack advanced technologies, human settlements, education and by many factors including social media and economic inequity. Today's ICTs are employed in a variety of ways by the governmental, business, and civil society sectors, engaged them to processes of decentralized decision-making [18]. Social media apps like Facebook, Twitter, Flickr, and YouTube have the potential to help with disaster recovery [19]. During normal times and during crises, the need for knowledge is a major motivator for using social media [20]. Social media gives real-time disaster details that no other medium can deliver [21]. One of the key reasons the public utilizes social media after a tragedy is to gather significant information and to keep in connected with family and friends [22].

2.2 *Co-production Theory*

In 1970s and 1980s, the term "co-production" entered the public administration lexicon. The phrase was created by Elinor Ostrom and companions at Indiana University's Workshop in Political Theory and Policy Analysis, which can be described as "the process by which individuals who are not in the same organization contribute inputs to provide a good or service." The "co" side of co-production initially included two actors: "regular producers" and "citizen producers" or "co-producers," according to its original definition [23]. Co-production research involves wide history, traverse fields in sociological, research on the voluntary sector, and public administration [24]. Recent concept of co-production since it specifically describes the relationship between professionals and people while still leaving open the type of contribution and function of the organization in question. Citizens needed to become active customers, and citizen happiness became a drive for improved government service delivery [25, 26].

3 Materials and Methods

Research is defined as systematic creative processes attempt to gain knowledge and using it to develop or verify proof, resolve challenges, build new theories, and give workable ideas [27]. The model in qualitative research design [28] is shown in Fig. 1.

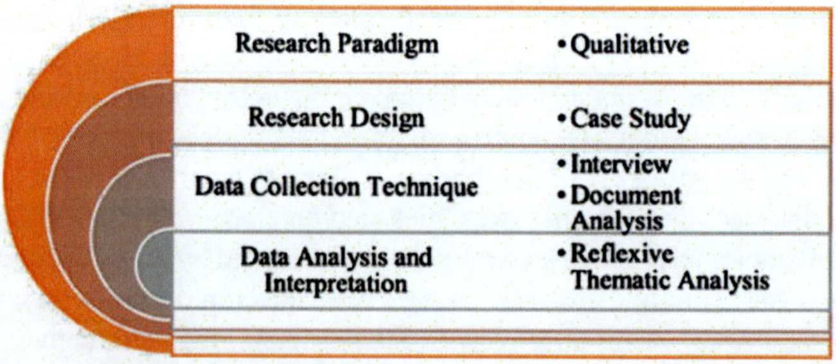

Fig. 1 Qualitative research model

3.1 Research Paradigm, Design, Collection, and Interpretation

Qualitative research used to collect data that is non-numerical, such as emotions, thoughts, and experiences [29]. The aim of qualitative research is to find solutions to the "how," "when," and "what" questions about a phenomenon [30]. Qualitative analysis aims to expose the viewpoints of the subjects or patients that are the focus of the study [31]. In which, this study is anchoring on the qualitative paradigm.

Case study research, in particular qualitative case study research, maybe a useful technique for solving complicated, real-world problems. When compared to other social research approaches, case studies have a reputation for lacking objectivity and consistency. This is one of the key reasons for well-articulated research design and development. Despite this caution, case studies are frequently used since they can reveal information that other methods cannot. In this study, the case site is Butuan City, a highly urbanized city in the Caraga Region, which also experienced extreme weather disturbances and calamities.

For data collection, researchers conducted fifteen (15) semi-structured interviews in the citizens of Butuan City. This includes interview conducted in person or over the phone. The conversations were recorded via phone voice recorder, and then during the discussion, note-taking equipment was utilized to allow for later transcription of the interview. Thematic analysis (TA) is a qualitative data analysis approach for discovering, evaluating, and interpreting meaning patterns (themes). TA is unique among qualitative analytic approaches in that it offers a method, a tool, or a technique that is not restricted by theoretical commitments rather than a methodology [32]. Based on the data, codes, and themes were analyzed and interpreted and presented in the next section.

4 Results and Discussions

By applying the principles of Reflexive Thematic Analysis, the following codes/themes are presented (Table 1).

Socialization and Altruism

Citizens tends to use social media because of socialization—the entertainment, communication, to be updated to information in what's happening in the world and

Table 1 Code matrix

RQ	Themes
RQ1	Socialization and altruism
RQ2	Collective action, collaboration and venue for engagement

trends, being watchful, and learning to enhanced skill. People use social media to interact and socialized other citizens during disaster. Altruism is also one of the factors that citizens tend to use social media during disaster and crisis situations. Act of altruism includes helping others or helping those in needs. Network contagion effects allow social causes to reach a huge number of interconnected individuals quickly, effectively, and at a cheap cost in the age of online social media. Understanding viral altruism and its basic behavioral features can aid in the long-term sustainability of good social transformation [33].

Collective Action, Collaboration, and Venue for Engagement

Social media facilitate citizens in co-producing valuable information for disaster management by collective actions include asking help, learning, and sharing information and dissemination of information. People help so as to get help in return as well, and together, act for the betterment of the community [34].

Collaboration is also one of the reasons how social media facilitate citizens in co-producing valuable information in disaster management. Collaboration or co-production and partnership as well as asking and informing the government. Social media allows for a new paradigm of knowledge management that includes both official and informal communication, as well as cooperation through a number of apps [35].

Citizens create valuable information on using social media in disaster management because it is a "venue for engagement" where citizens spread valuable information, a powerful platform for safety precautions, posting awareness to community and where they can share their own ideas.

5 Conclusions and Future Works

This paper concluded that socialization and altruism are the factors that lead citizens to use social media during disasters and crises situation. Citizens utilize social media to learn and to behave in a socially acceptable manner. Citizens have these disinterested and unselfish habits of concern for the well-being of others. We also found out that social media facilitate citizens in co-producing valuable information for disaster management by having a collective actions, collaboration, and venue for engagement. Collective action in a way that they spread awareness, learning, and information sharing or dissemination, as well as providing a warning during a crisis or tragedy. Those acts taken together to improve their condition and attain mutual objectives. Collaboration and information sharing among citizens to government organizations is becoming increasingly important for promoting efficiency and productivity as well as for enhancing citizen services. Likewise, social media also provide an avenue for engagement [36], a platform for citizen involvement, where they may freely share their own learnings and ideas, as well as an effective platform for the provision of safety precautions and security measures on certain inevitable event.

By investigating the dynamics of citizens in Butuan City using social media for disaster management, the theory of co-production was able to provide insightful information, such as collaboration, act of altruism, and active engagement during unforeseeable disaster phenomena, which is a vital way of co-producing valuable knowledge that the government should aspire to. Citizens, like ordinary people, who can use technology, particularly social media, can be a valuable partner and cohort in reducing disaster risks and hazards, according to the study.

5.1 *Future Works*

The researchers make the following recommendations for using social media in disaster management.

To reduce disaster risks and repercussions, disaster management can use social media ads as another method for information flow. It enhances collaborative efforts for immediate aid and increases the speed with which information is disseminated during emergencies. Second, although the study used a single case study design on Butuan City, it would be worthwhile to investigate other case sites, such as LGUs in other areas and provinces, as they may provide valuable and relevant perspectives not found in the current study. Third, the researchers further urge that this research be expanded to include not only ordinary citizens who use social media, but also citizen leaders and community organizations, as they play an important role in society during disasters.

References

1. Hossain MS, Gadagamma CK, Bhattacharya Y, Numada M, Morimura N, Meguro K (2020) Integration of smart watch and Geographic Information System (GIS) to identify post-earthquake critical rescue area part. I. Development of the system. Prog Disaster Sci 7:100116
2. Mohan P, Mittal H (2020) Review of ICT usage in disaster management. Int J Inf Technol 12(3):955–962
3. UNESCAP (2019) Summary of the Asia-Pacific Disaster Report 2019
4. Saarikko T, Nuldén U, Meiling P, Pessi K (2020) Framing crisis information systems: the case of WIS. In: Proceedings of the 53rd Hawaii international conference on system sciences, vol 3, pp 2167–2176
5. Espina P (2016) Butuan disaster official: information key to risk reduction. Rappler, July 2016
6. Tai K-T, Porumbescu G, Shon J (2020) Can e-participation stimulate offline citizen participation: an empirical test with practical implications. Public Manag Rev 22(2):278–296
7. Kassen M (2018) E-participation actors: understanding roles, connections, partnerships. Knowledge management research and practice. Taylor and Francis Ltd., Political Sciences, Eurasian Humanities Institute, Astana, Kazakhstan
8. Li L, Zhang Q, Tian J, Wang H (2018) Characterizing information propagation patterns in emergencies: a case study with Yiliang Earthquake. Int J Inf Manage 38(1):34–41

9. Roberts JM, Ibrahim J (2019) Contemporary left-wing activism: democracy, participation and dissent in a global context, vol 1. Routledge studies in radical history and politics. Routledge, Taylor & Francis Group, Abingdon, Oxon, New York, NY, p 218
10. Khan A, Krishnan S (2018) Social media enabled e-participation: review and agenda for future research. e-Serv J 10(2):45
11. Chu P-Y, Huang T-Y, Hung Y-T, Lee C-P, Tseng H-L, Huang W-L (2017) A longitudinal research of public value and electronic governance development in Taiwan. In: International conference on theory and practice of electronic governance, pp 459–464
12. Lember V (2018) The increasing role of digital technologies in co-production and co-creation. Public Procure. Innov. Int. Perspect. March 2018, pp 115–127
13. Cordella A, Paletti A (2018) ICTs and value creation in public sector: manufacturing logic vs service logic. Inf Polity 23(2):125–141
14. Lember V, Brandsen T, Tonurist P (2019) The potential impacts of digital technologies on co-production and co-creation. Public Manag Rev 21(11):1665–1686
15. Braun V, Clarke V (2006) Using thematic analysis in psychology. Qual Res Psychol 3(2):77–101
16. Haddow GD, Bullock JA, Coppola P (2017) Introduction to emergency management
17. Khorram-Manesh A (2017) Handbook of disaster and emergency management. 34(2)
18. Hanna NK (2010) Implications of the ICT revolution. In: Transforming government and building the information society
19. Ahmed A (2011) Use of social media in disaster management. In: International conference on information systems 2011, ICIS 2011, vol 5
20. Merrifield M, Palenchar MJ (2012) Uncertainty reduction strategies via Twitter: the 2011 wildfire threat to Los Alamos National Laboratory. In: Proceedings of the annual meeting of the association for education in journalism and mass communication
21. Zou Q (2017) Research on cloud computing for disaster monitoring using massive remote sensing data. In: IEEE 2nd international conference on cloud computing and big data analysis, vol 5(1), pp 1–8
22. Coche J, Romera-Rodriguez G, Montarnal A (2021) Social media processing in crisis response: an attempt to shift from data to information exploitation. In: Proceedings of the 54th Hawaii international conference on system sciences, p 2285
23. Ostrom E (1996) Crossing the great divide: synergy, and development. World Dev 24(6):1073–1087
24. Verschuere B, Brandsen T, Pestoff V (2012) Co-production: the state of the art in research and the future agenda. Voluntas 23(4):1083–1101
25. Pestoff V (2012) Co-production and third sector social services in Europe: some concepts and evidence. Int J Volunt Nonprofit Organ 23(4):1102–1118
26. Pacot MPB, Marcos N (2019) Feature-based stitching algorithm of multiple overlapping images from unmanned aerial vehicle system, vol 1, January 2018, pp 17–24
27. Mahoney J, Goerts G (2012) A tale of two cultures: contrasting quantitative and qualitative research. Polit Anal 14(3)
28. Myers MD (2019) Qualitative research in business and management. Thousand Oaks, CA: Sage Publications Limited
29. Clark K, Vealé B, Zaleski F (2018) Caring for the transgender patient. Radiol Technol 90(1)
30. Green J, Thorogood N (eds) (2014) Qualitative methods for health research (introducing qualitative methods series) (9781446253090). SAGE
31. Haven T, Van Grootel L (2019) Preregistering qualitative research. Accountability Res 26(3)
32. Guest G, MacQueen K, Namey E (2014) Validity and reliability (credibility and dependability) in qualitative research and data analysis. In: Applied thematic analysis
33. Van Der Linden S (2017) The nature of viral altruism and how to make it stick. Nat Human Behav 1(3)
34. Mancur O (1971) The logic of collective actions: public goods and the theory of groups. Harvard University Press

35. Razmerita L (2013) Collaboration using social media: the case of Podio in a voluntary organization. In: Lecture notes in computer science (including subseries lecture notes in artificial intelligence and lecture notes in bioinformatics), vol 8224 LNCS
36. Pitogo VA, Ramos CDL (2020) Social media enabled e-participation: a lexicon-based sentiment analysis using unsupervised machine learning. In: ACM international conference proceeding series, pp 518–528

Robot Welding Path Planning and Application Based on Graphical Computing

Jingjing Lou, Xujiang Yu, Yongfei Chen, Zhubing Sun, and Pengfei Zheng

Abstract As a classic problem in artificial intelligence research, robot welding path planning has been extensively studied. Related scholars have also proposed many solutions, such as heuristic algorithm, neural network, genetic simulated annealing algorithm, improved genetic algorithm. But there are still deficiencies in welding torch posture, welding position, and robot motion stability. Because of the characteristics of the welding seam have a vital influence on the path planning of the welding robot, it is also the basis for ensuring the welding accuracy. From the perspective of graphic calculation, the graphical computing method of precise welds is analyzed, the point cloud graphics of the welded parts are used to calculate the overlap of primitives, and the accurate and rapid extraction of weld features are realized by changing the graphic representation of the welded parts model. According to the connection characteristics of the weldment, the characteristics of the weldment are collected, and a simple, fast, and more versatile method for extracting and calculating weld features is designed, and the weld features are discretized. Discrete weld feature points are used as the basis for path planning of the welding robot to carry out reasonable welding path planning, which reduces the manual teaching process and workload. Finally, a robot welding path planning method based on graphical computing is proposed, and corresponding simulation experiments are carried out.

Keywords Graphical computing · Point cloud · Welding robot · Path planning

1 Introduction

Welding is an indispensable surface forming technique in modern industrial high quality and efficient manufacturing technology. It is a method of bonding workpieces by heat or pressure or both, and with or without filler material. In essence, it is a

J. Lou (✉) · X. Yu · P. Zheng
Yiwu Industrial and Commercial College, Yiwu 322000, China
e-mail: chinglou@ywicc.edu.cn

Y. Chen · Z. Sun
Zhejiang Linix Motor Co., Ltd, Dongyang 322118, China

X.-S. Yang et al. (eds.), *Proceedings of Seventh International Congress on Information and Communication Technology*, Lecture Notes in Networks and Systems 447,
https://doi.org/10.1007/978-981-19-1607-6_53

processing method in which metal atoms on two separated solid surfaces are brought close to the lattice distance (0.3–0.5 mm) by a suitable physicochemical process to form a metal bond, thus achieving a permanent connection. Its advantages include easy forming, low production cost, adaptability, material saving, high structural strength, good sealing, and easy automation, especially in auto body processing, the use of industrial robots for automatic welding of body is the most typical application.

In welding production, there are many types of components, variable types of welds and complex spatial distribution, and these characteristics make it much more difficult to automate welding using robots. In robotic welding technology, the proper planning of the welding path is crucial. Welding path planning, as a classical problem in artificial intelligence research, has been extensively studied by many scholars and corresponding planning algorithms have been proposed, such as artificial potential field method, genetic algorithm, simulated annealing algorithm [1–8]. In addition, the use of manual instruction for welding path planning is also more common, but the manual instruction method is time-consuming and the accuracy is difficult to guarantee. Therefore, a graphical computing-based robot welding path planning method is proposed in this paper from a graphical perspective and simulated in the RobotStudio environment. Finally, based on the simulation, the OTC FD-B6 welding robot is used for the corresponding example welding test, and the test verifies the effectiveness of the method for robot welding path planning.

2 Graphical Representation of Welding Problems

In general, the welding process of any component can be abstracted as a number of three-dimensional graphics connection problem, the weld is the spatial intersection of these three-dimensional graphics. As shown in Fig. 1, where (a) is the welding model of a tee pipe; (b) is the welding assembly diagram of a complex box. It is obvious that the welds are all intersecting curves of each component in the assembly

(a) Welding diagram of tee pipe (b) Welding assembly diagram of complex box

Fig. 1 Weld seam characteristics in welded assemblies

attitude. Therefore, the problem of calculating the weld curve is essentially a graphical problem, and the graphical calculation can be used to find the exact weld curve. In welding robot path planning, the teaching method is based on curve fitting of several manually collected feature points to obtain the welding trajectory curve. Obviously, the more feature points collected, the more accurate and time-consuming the welding trajectory curve is. Based on this need, a discrete representation of the weld profile is used to plan the weld path.

3 Graphical Calculation Method of the Weld Path

3.1 *Parametric Calculation Method of the Weld Curve*

3.1.1 Establishment of the Geometric Model of the Weld Curve

The process of building the geometric model is illustrated by taking the example of a cylindrical tube docked with a flat plate. As shown in Fig. 2, the projection of the cylindrical tube and the flat panel buttress is shown, and the full section of the cylindrical tube is performed. As can be seen from the figure, the left half of the cylindrical tube has the inner skin in contact with the surface of the flat plate, and the right half has the outer skin in contact with the surface of the flat plate. At the plain lines aa_1 and bb_1 of the interface surface, the inner and outer skins of the cylindrical tube are in contact with the flat surface at the same time.

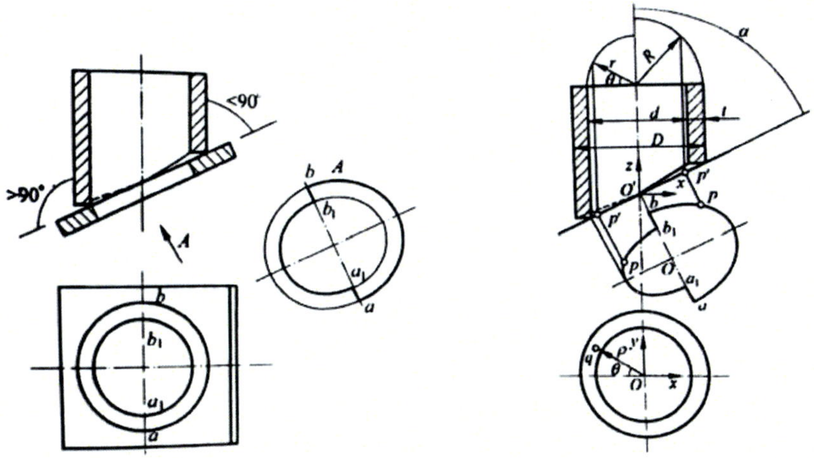

(a) Cylindrical tube butted against a flat plate (b) Geometric intersection diagram

Fig. 2 Mathematical model of the docking of cylindrical tube and flat plate

3.1.2 Establishment of Mathematical Model of Weld Curve

The mathematical model for solving the docking of a cylindrical tube to a flat plate is shown in Fig. 2. The dimensions of the structure are known to be D, t and a, $d = D - 2t, r = \frac{d}{2}, R = \frac{D}{2} = r + t$. Taking any point p, on the interface curve, it can be seen from Fig. 2. That $x_p = -r\cos\theta$, $y_p = r\sin\theta$. Because, $\frac{r\cos\theta}{-z_p} = \tan\alpha$, therefore, $z_p = -\frac{r\cos\theta}{\tan\alpha}$.

Thus, the coordinates of the point on the interface curve are:

$$\begin{cases} x_p = -r\cos\theta \\ y_p = r\sin\theta \\ z_p = -\frac{r\cos\theta}{\tan\alpha} \end{cases} \quad (1)$$

where the values of θ are: $0 \le \theta \le \frac{\pi}{2}$ and $\frac{3\pi}{2} \le \theta \le 2\pi$.

According to Eq. (1), the coordinates of the point p on the epithelial interface curve can be obtained as:

$$\begin{cases} x_p = -R\cos\theta \\ y_p = R\sin\theta \\ z_p = -\frac{R\cos\theta}{\tan\alpha} \end{cases} \quad (2)$$

where the values of θ is: $\frac{\pi}{2} \le \theta \le \frac{3\pi}{2}$.

If Q is any point on the interface surface and ρ is the distance from that point to the axis, for the left half of the interface surface, it have:

$$\begin{cases} x_q = -\rho\cos\theta \\ y_q = \rho\sin\theta \\ z_q = -\frac{\rho\cos\theta}{\tan\alpha} \end{cases} \quad (3)$$

where the values of ρ is: $r \le \rho \le R$, the values of θ is: $0 \le \theta \le \frac{\pi}{2}$ and $\frac{3\pi}{2} \le \theta \le 2\pi$.

Similarly, for the right half of the interface surface, it have:

$$\begin{cases} x_q = -\rho\cos\theta \\ y_q = \rho\sin\theta \\ z_q = -\frac{R\cos\theta}{\tan\alpha} \end{cases} \quad (4)$$

where the values of ρ is: $r \le \rho \le R$, the values of θ is: $\frac{\pi}{2} \le \theta \le \frac{3\pi}{2}$.

3.2 General Calculation Method of Weld Curve

In order to further reduce the computational complexity of the weld curve, a reduced dimensional intersection calculation method of the geometry is proposed. As shown in Fig. 3, the point cloud model of the welded part is collected using 3-D digital scanning technology, and the point cloud data of the intersection area is approximated to find the mean value.

If point P_i of weld 1 and point P_j of weld 2 satisfy $dis(P_i, P_j) \leq \delta$, where δ is the pre-set accuracy value, then the point $P_{c(x,y)} = (P_{i(x,y)} + P_{j(x,y)})/2$ on their intersection curve. These discrete points are the feature points on the weld curve, similar to the manually taught path points. Since these weld curve characteristic points are obtained by graphical calculation method, the number of points is high and the calculation process is simple. In contrast, the graphical calculation method for path planning of the welding robot saves a lot of manual demonstration time and also improves the accuracy of the welding path. A detailed description of this generic calculation method for weld profiles can be found in the literature [9].

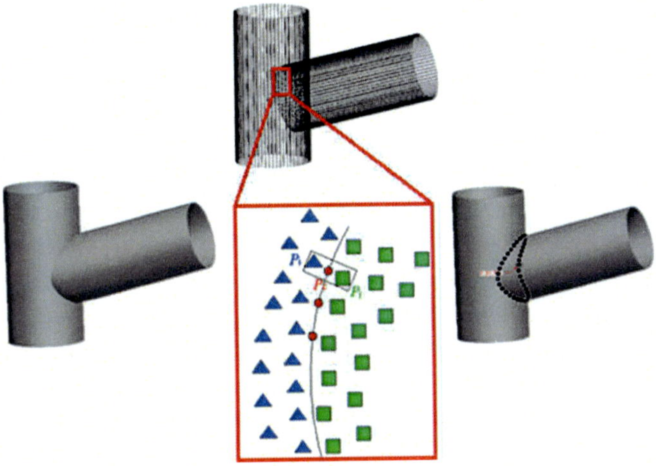

Fig. 3 General calculation model of weld curve

4 Example Verification

4.1 *Calculation of the Welding Path for Unequal Diameter Pipes*

The following is an example of the unfolding correction calculation for two circular pipes with a spatially curved weld. As shown in Fig. 4, a welding case of two round tubes of unequal diameters, intersecting at a 55° angle. Where, (a) is the 3-D model of the intersection of the fittings; (b) is the actual interface surface of the two fittings obtained by the surface intersection method; (c) is the calculation of the neutral layer curve by the interface surface, and the red bolded curve in the figure is the neutral layer curve.

In order to further verify the effect of the unfolding correction method on the promotion of welding quality, the industrial robot offline programming software RobotStudio is used to simulate and verify the welding path, and the collision detection function in the software is used to monitor the collision of the entire welding process of the robot to ensure that no collision occurs during the entire welding process. Therefore, the above-mentioned welding path can generate the corresponding rapid program, which can be imported into the welding robot control cabinet to be programmed on site.

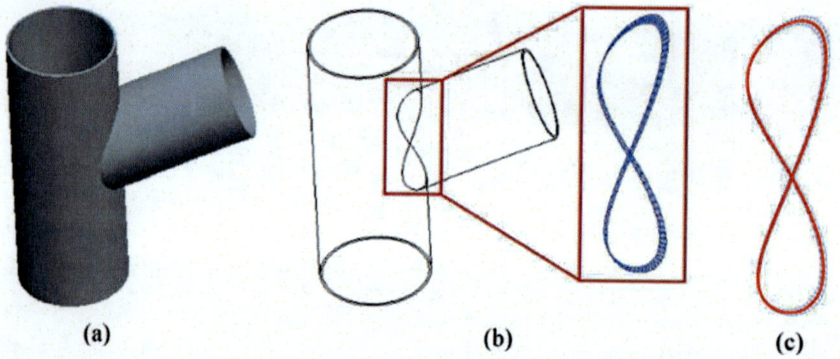

Fig. 4 Arbitrary fillet butt welding of unequal diameter circular pipe parts. **a** Three-dimensional model of the intersection of the tubes. **b** Theoretical intersection curve. **c** Neutral layer intersection curve

(a) Welding robot. (b) Tooling turntable.

Fig. 5 Welding robot welding example

4.2 Weld Test of Unequal Diameter Pipe

In order to more intuitively verify the rationality of the welding path calculation method proposed in this paper, the OTC FD-B6 welding robot is now used to physically weld the unequal diameter pipe weldments in Fig. 4. As shown in Fig. 5, the size of welded parts: the diameter of pipe 1 is 50 mm, the wall thickness is 4 mm; the diameter of pipe 2 is 40 mm, the wall thickness is 4 mm; two pipes fittings are 55° intersecting connection, the joint surface using V-bevel design. The relevant welding parameters are as follows: welding current is 150 A; voltage is 17 V; welding speed is 25 mm/s; welding wire type is ER50–6.

In order to facilitate comparative testing, this example uses two sets of round pipe parts of the same material and size combination, choose the same welding parameters for welding and forming. As shown in Fig. 6, the automatic welding results obtained by using the manual teach-in method for welding path planning can be seen from the two different poses of the tube forming (a) and (b): the residual height of the weld is obvious and the quality of the weld is poor. As shown in Fig. 7, the automatic welding results obtained by using the graphical calculation method for weld path planning can be found from its two different poses (a) and (b): the weld residual height is better and the weld quality is also significantly better than that of the teach-in method.

5 Conclusion

A graphical computational method of the robot welding path planning problem is studied, which abstracts the welding path planning as a surface intersection problem

(a) Attitude 1 (b) Attitude 2

Fig. 6 Example of welding by manual demonstration method

(a) Attitude 1 (b) Attitude 2

Fig. 7 Example of welding by this method

of the welded parts. Using the advantages of the three-dimensional point cloud model, the high-dimensional calculation of complex three-dimensional graphics is simplified to the one-dimensional calculation of geometric elements, which greatly reduces the complexity of the calculation, reduces the time for manual demonstration of the planning path, avoids the chance of random errors easily generated by the manual demonstration method, and improves the accuracy of the welding path calculation. Combined with the industrial robot offline programming software RobotStudio for welding path planning, and the welding of unequal diameter pressure pipes as an example was experimentally verified. The test results show that: the welding process is smooth, no collision phenomenon, good quality weld forming, no weld collapse, weld through, edge biting and other phenomena occur, the proposed welding path calculation method can effectively improve the accuracy of welding forming.

References

1. Wang X, Yan Y, Gu X (2019) Spot welding robot path planning using intelligent algorithm. J Manuf Process (42):1–10
2. Zhang J, Wang Q, Xiao G et al (2021) Filling path planning and polygon operations for wire arc additive manufacturing process. Math Prob Eng (2021):6683319
3. Zhou P, Peng R, Xu M et al (2021) Path planning with automatic seam extraction over point cloud models for robotic arc welding. IEEE Robot Autom Lett 3(6):5002–5009
4. Wang T, Xue Z, Dong X et al (2012) Autonomous intelligent planning method for welding path of complex ship components. Robotica 39:428–437
5. Fang HC, Ong SK, Nee AYC (2017) Robot path planning optimization for welding complex joints. Int J Adv Manuf Technol (90):3829–3839
6. Liu Y, Tian X (2019) Robot path planning with two-axis positioner for non-ideal sphere-pipe joint welding based on laser scanning. Int J Adv Manuf Technol (105):1295–1310
7. Ghariblu H, Shahabi M (2019) Path planning of complex pipe joints welding with redundant robotic systems. Robotica 37:1020–1032
8. Zhou X, Zhao Q, Zhang D (2019) Discrete fireworks algorithm for welding robot path planning. J Phys (1267):012003
9. Zheng P, Liu Q, Lin D et al (2019) An algorithm for computing intersection of complex surface parts and its application. IEEE 304–310

The Interpolation-Vandermonde Method for Numerical Solutions of Weakly Singular Volterra Integral Equations of the Second Kind

E. S. Shoukralla, B. M. Ahmed, Ahmed Saeed, and M. Sayed

Abstract The solution of the second kind weakly singular Volterra integral equations (VIEs) was solved applying an interpolation technique based on the Vandermonde matrix. We devised optimal rules for the node distributions of the two kernel variables, ensuring that the singularity in the kernel is eliminated completely. The unknown solution is interpolated based on the Vandermonde matrix through three matrices in total, one of which is a monomial matrix. Five matrices, two of which are monomial basis, are used to convert the interpolated singular kernel into a double-interpolated non-singular polynomial. A linear system is produced by substituting the interpolated unknown function into the integral equation. The unknown coefficients are obtained from the direct solution of this system, and then, the interpolated solution is obtained. Using the lowest degree of interpolation, the results for the two solved examples strongly converge to the exact ones. This demonstrates the method's originality and accuracy.

Keywords Vandermonde matrix · Singular integral equation · Lagrange interpolation · Weakly singular kernels · Computational methods

1 Introduction

When Dirichlet or Niemann conditions are usually met, the application of the integral equation method to solve problems with initial, boundary, or mixed values of

E. S. Shoukralla (✉) · M. Sayed
Faculty of Electronic Engineering, Menoufia University, Shibin Al Kawm, Egypt
e-mail: shoukralla@el-eng.menofia.edu.eg

M. Sayed
e-mail: Mohamed.abdelkader@el-eng.menofia.edu.eg

B. M. Ahmed · A. Saeed
Faculty of Engineering and Technology, Future University in Egypt, Cairo, Egypt
e-mail: Basma.magdy@fue.edu.eg

A. Saeed
e-mail: asaeed@fue.edu.eg

X.-S. Yang et al. (eds.), *Proceedings of Seventh International Congress on Information and Communication Technology*, Lecture Notes in Networks and Systems 447,
https://doi.org/10.1007/978-981-19-1607-6_54

Laplace equations, electromagnetic wave equations, or Helmholtz equations, and others yields equivalent singular boundary integral equations [1–5]. Singularities in integral equations are caused by the unknown function at the integration domain's endpoints, or when one of the kernel's variables approaches the other variable, or by both singularities. Shoukralla et al. [6–9] created a number of analytical approaches for resolving the first-class Fredholm integral equations with logarithmic weakly kernel, where the unknown functions and kernel singularities are considered. The proposed method applying Lagrange interpolation based on the Vandermonde matrix, together with a treatment of the kernel singularity analytically. The kernel singularity was completely isolated using one rule we devised to govern the approach of selecting the variable distribution nodes in such a way that no negative or zero values appear under the square root sign. Ramadan et al. [10] obtained exact solution on the class of all kinds of Volterra integral equations, and the homotopy perturbation method has been modified in this paper by combining it with a formulation of a power series that Taylor formulated first. Pourgholi [11] provided a Tau method depends on Legendre Wavelet basis for giving approximate solution to Volterra integral equations with weakly singular. Shoukralla et al. [12–15] used matrices to modify the barycentric Lagrange interpolation formula and solved non-singular second kind Volterra equations for the first time, with remarkable success, employing various techniques, and obtaining exact numerical solutions or strongly convergent solutions relying on the smoothness of the given data functions and the kernel provided. These methods may be adequate for solving singular Volterra equations, but the most crucial question remains how to eliminate the kernel singularity using the Lagrange interpolating method via the Vandermonde matrix. The major purpose of this study is to use the Vandermonde matrix to apply the Lagrange interpolation formula to interpolate both the unknown and given data functions using polynomials of the same degree. As a result, each of these functions is represented by three matrices: the functional values, the known Vandermonde matrix, and the monomial basis functions matrix of the primary argument. We developed the approach of [9] for interpolating the weakly singular kernel and devised an innovative rule for optimum node distribution for the two sets of nodes corresponding to the two kernel variables. Thus, for any value of the nodes, we ensure that the denominator of the kernel never approaches zero or becomes imaginary, and we obtain a double interpolant non-singular kernel in the form of the product of five matrices, where two of which are monomial basis for the kernel variables, two Vandermonde matrices subjected to the two kernel variables, and the fifth matrix is a square known coefficient matrix. By using a substitute, the single unknown interpolant polynomial on both sides of the integral equation and taking into account the substitution of the double interpolates kernel and the single interpolate data function on the right side of that equation, we were able to transform the required solution into a linear algebraic system. The unknown coefficient matrix is found by solving this system, and therefore, the unknown function is obtained. We solved different examples for different values of the upper integration limit, all for. The obtained results, including absolute errors, are compared to the exact solutions and found strongly converge to them, as shown in the tables and figures. This highlights the method's uniqueness as well as its ability to deliver accurate results with the lowest degree of interpolation.

2 Vandermonde-Interpolation Method

Assume the weakly singular VIEs of the second kind

$$\tau(x) = \beta(x) + \int_a^x \kappa(x,t)\tau(t)\mathrm{d}t\,;\, x \in I = [a,b] \tag{1}$$

where $\beta(x)$ is a given real continuous function defined on I, $\tau(x)$ is the unknown function to be found, and $\kappa(x,t)$ is a given continuous kernel on the domain $\Omega = \{(x,t) : a \le x \le t \le b\}$ where b is a positive real number. For the weakly singular kernel $\kappa(x,t)$, it is usually presented by: $\kappa(x,t) = \frac{g(x,t)}{(x-t)^\alpha}$ where $0 < \alpha < 1$ with the assumption that $\|\kappa(x,t)\|_2 \le M < \infty$ where $x, t \in \Omega$. It is assumed that the Volterra operator which is defined by $V\tau = \int_a^x \kappa(x,t)\tau(t)\mathrm{d}t$ acting in $L_2[a,b]$. We put $\tau(x)$ in the tabulated form function $\tau(x_i) = \tau_i$ for $\{x_i\}_{i=0}^n \subset [a,b]$ where $x_i = a + ih$ with size of the step $h = \frac{b-a}{n}$. Let $\tau_n(x)$ be the Lagrange interpolating polynomial of nth degree that interpolates $\tau(x)$ at the $(n+1)$ equal distance distinct nodes $\{x_i\}_{i=0}^n$. such that $\tau_n(x_i) = \tau(x_i) = \tau_i\ \forall i = \overline{0,n}$. Then,

$$\tau_n(x) = UX(x) \tag{2}$$

Here, $U = \left[a_j\right]_{j=0}^n$ of order $1 \times (n+1)$ of unknown coefficients, $X(x) = \left[x^j\right]_{j=0}^n$ is the $(n+1) \times 1$ monomial basis functions a matrix of column. The unknown coefficients $\left\{a_j\right\}_{j=0}^n$ are determined by resolving the equation of linear algebraic system

$$\widetilde{X}U = A \tag{3}$$

where A is the column matrix $A = [\tau(x_i)]_{i=0}^n$, where $\{x_i, \tau(x_i)\}_{i=0}^n$ are the interpolation knots that interpolate $\tau(x)$, whereas the Vandermonde matrix $\widetilde{X}$ is given by

$$\widetilde{X} = \left[x_{ij}\right]_{i,j=0}^n;\ x_{ij} = x_i^j,\ x_{ij} = 1\,\forall i = \overline{0:n},\ j = 0 \tag{4}$$

By substituting the matrix U which is obtained by solving (3), we get $\tau_n(x)$

$$\tau_n(x) = A^T\left(\widetilde{X}^{-1}\right)^T X(x) \tag{5}$$

By the same way, the given data function $\beta(x)$ can be interpolated like $\tau(x)$ to get

$$\beta_n(x) = F^T \left(\widetilde{X}^{-1}\right)^T X(x) \tag{6}$$

where $F^T = [\beta(x_i)]_{i=0}^n$ is the row matrix of the order $1 \times (n+1)$. The kernel $\kappa(x,t)$ will be interpolated with regard to both variables x and t. To interpolate the kernel, we have to ensure the conservation of the denominator of the kernel to never become zero or imaginary, that is, $x > t$. We set one innovative rule that enable to overcomes any kernel's singularity when $x \to t$ and when $x \to 0$. We interpolate $\kappa(x,t)$ with respect to x on the right-half interval and with respect to t on the left-half interval. Then, we choose the sets of nodes $\{\tilde{x}_i\}_{i=0}^n$ and $\{\tilde{t}_i\}_{i=0}^n$ as follows

$$\tilde{x}_i = \left(\frac{b-a}{2} + \delta\right) + i \times h;\ \tilde{t}_j = (a+\delta) + j \times h;\ h = \left(\frac{b+a}{2} - 2\delta\right)/(n-1) \tag{7}$$

where $\delta = \frac{b}{\alpha n}$; $\alpha \geq 0$. Analogical to (6) and (7), we get the matrix–vector single interpolate kernel $\kappa_n(x,t)$ of degree n that corresponding the set of nodes $\{\tilde{x}_i\}_{i=0}^n$ in the form

$$\kappa_n(x,t) = X^T(x)\tilde{X}^{-1}K(\tilde{x}_i,t) \tag{8}$$

where $K(\tilde{x}_i,t)$ is the $(n+1) \times 1$ column matrix

$$K(\tilde{x}_i,t) = \left[\kappa(\tilde{x}_i,t)\right]_{i=0}^n \tag{9}$$

Now, likewise (8), each entry of the set $\{\kappa(\tilde{x}_i,t)\}_{i=0}^n$ will be interpolated corresponding to the variable t according to the set of nodes $\{\tilde{t}_i\}_{i=0}^n$ given by (7). As a result, we have the matrix–vector through double interpolate kernel $\kappa_{n,n}(x,t)$ in the following form

$$\kappa_{n,n}(x,t) = X^T(x)\tilde{X}^{-1}K\left(\widetilde{T}^{-1}\right)^T X(t) \tag{10}$$

where $\widetilde{T}$ is Vandermonde matrix corresponding to the variable t such that

$$\widetilde{T} = \left[t_{ij}\right]_{i,j=0}^n;\ t_{ij} = t_i^j,\ t_{ij} = 1\,\forall i = \overline{0:n},\ j = 0 \tag{11}$$

From (6) and (10), we get

$$\kappa_{n,n}(x,t)\tau_n(t) = X^T(x)\tilde{X}^{-1}K\left(\widetilde{T}^{-1}\right)^T P(t)\tilde{X}^{-1}A \tag{12}$$

where $K = \left[\kappa_{ij}\right]_{i,j=0}^{n}$ is the square known matrix $(n+1) \times (n+1)$ whose items are determined by $\kappa_{ij} = \kappa\left(\tilde{x}_i, \tilde{t}_j\right) \forall\, i, j = \overline{0:n}$ $P(t) = X(t)X^T(t) = \left[t^{i+j}\right]_{i,j=0}^{n}$ of order $(n+1) \times (n+1)$.

By replacing $\tau_n(x)$ given by (6) with $\tau(t)$ in the right hand side of Eq. (1), we find that

$$\tau_n(x) = \varphi(x) + X^T(x)\tilde{X}^{-1}K\left(\tilde{T}^{-1}\right)^T\tilde{P}(x)\tilde{X}^{-1}A \tag{13}$$

where $\tilde{P}(x) = \int_0^x P(t)\mathrm{d}t$. Now, we investigate an innovative approach to transform the solution of integral Eq. (1) into an algebraic system of equation without need to apply the collocation method. The implication of this concept starts by replacing $\tau(x)$ in the left side of (1) and in the right side with $\tau_n(x)$ that was given by (13), replacing the kernel $\kappa(x, t)$ in the right side with $\kappa_{n,n}(x, t)$ that was provided by (10), and finally, replacing the given data function $\beta(t)$ in (1) with $\beta_n(t)$ based on (6). Hence, we get the system shown below

$$\begin{aligned} &X^T(x)\tilde{X}^{-1}K\left(\tilde{T}^{-1}\right)^T\tilde{P}(x)\tilde{X}^{-1}A - X^T(x)\tilde{X}^{-1}K\left(\tilde{T}^{-1}\right)^T\Psi(x)\tilde{X}^{-1}A \\ &= X^T(x)\tilde{X}^{-1}K\left(\tilde{T}^{-1}\right)^T\tilde{P}(x)\tilde{X}^{-1}F \end{aligned} \tag{14}$$

where

$$\Psi(x) = \int_0^x \left(P(t)\tilde{X}^{-1}K\left(\tilde{T}^{-1}\right)^T\tilde{P}(t)\right)\mathrm{d}t, \quad \tilde{P}(x) = \int_0^x P(t)\mathrm{d}t \tag{15}$$

Depending on some matrix-algebra abbreviations, we get the system (15) in the following form

$$\left(\tilde{P}(x)\tilde{X}^{-1} - \Psi(x)\tilde{X}^{-1}\right)A = \tilde{P}(x)\tilde{X}^{-1}F \tag{16}$$

By solving (17), we find the coefficient unknown matrix A, and then, we substitute it into (5) to find the interpolate solution

$$\tau_n(x) = X^T(x)\tilde{X}^{-1}\left(\tilde{P}(x)\tilde{X}^{-1} - \Psi(x)\tilde{X}^{-1}\right)^{-1}\tilde{P}(x)\tilde{X}^{-1}F \tag{17}$$

3 Computational Results

We designed a MATLAB code to find the interpolant solution $\tau_n(x)$ and for any value of the parameter b. For different values of the parameter b, we find numerical solutions for two weakly singular (VEIs) of the second kind. Upon using (7), we set $\alpha = 3$ for the two examples. We denote the absolute error by $R_2^b(x_i) = |\tau(x_i) - \tau_2^b(x_i)|$, where $\tau(x_i)$ denotes the exact solution values and $\tau_2^b(x_i)$ denotes the interpolate numerical solution values of degree 2 at the integration interval $[b/10, b]$.

Example 1

$$\tau(x) = -1 + 4x - 3x^2 + x^5 + \frac{2\sqrt{x}(-3465 + 8x(1155 - 693x + 160x^4))}{3465}$$
$$- \int_0^x \frac{\tau(t)}{\sqrt{x-t}} dt \, ; \; x \in [0, 1] \tag{18}$$

where the exact solution is $\tau(x) = x^5 - 3x^2 + 4x - 1$ [11]. In Table 1, shown the numerical solutions $\tau_2^{0.8}(x_i)$. In Fig. 1, shown the graphs of $\tau_2^{0.8}(x_i)$ compared with the exact ones.

Example 2

Table 1 The exact solutions $\tau(x_i)$, the interpolate solutions $\tau_2^{0.8}(x_i)$, and the absolute errors $R_2^{0.8}(x_i)$

x_i	$\tau(x_i)$	$\tau_2^{0.8}(x_i)$	$R_2^{0.8}(x_i)$
0.08	−0.699196723200000	−0.634795599168448	0.0644011240315525
0.16	−0.436695142400000	−0.393607671077958	0.0430874713220418
0.24	−0.212003737600000	−0.201784907416050	0.0102188301839500
0.32	−0.0238445568000000	−0.037482642326264	0.0136380855262644
0.4	0.130240000000000	0.0972583786440818	0.0329816213559182
0.48	0.254280396800000	0.205428080781716	0.0488523160182844
0.56	0.354273177600000	0.300042060143000	0.0542311174569997
0.64	0.438574182400000	0.398249482871648	0.0403246995283516
0.72	0.518291763200000	0.511703194008525	0.00658856919147499
0.80	0.607680000000000	0.631227743357679	0.0235477433576791

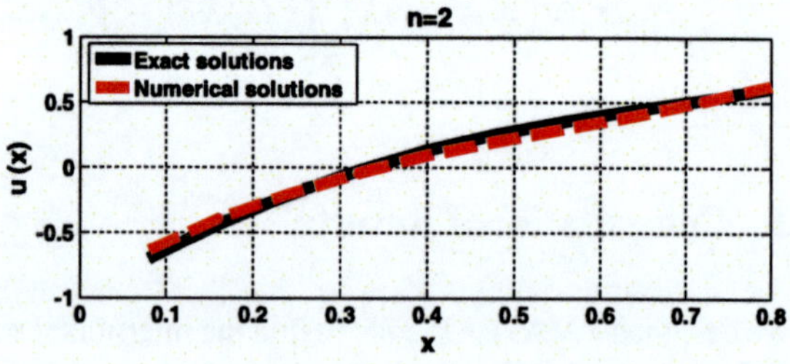

Fig. 1 Exact solutions $\tau(x_i)$ and numerical solutions $\tau_2^{0.8}(x_i)$

Table 2 The exact solutions $\tau(x_i)$, the interpolate solutions $\tau_2^{0.2}(x_i)$, and the absolute errors $R_2^{0.2}(x_i)$

x_i	$\tau(x_i)$	$\tau_2^{0.2}(x_i)$	$R_2^{0.2}(x_i)$
0.02	0.230145812847464	0.326874166761865	0.0967283539144010
0.04	0.301356159017631	0.367940297487299	0.0665841384696676
0.06	0.348931137694249	0.406489320887903	0.0575581831936541
0.08	0.384988673287066	0.434261114022825	0.0492724407357592
0.10	0.414059169439569	0.451507942706269	0.0374487732666997
0.12	0.438393491935992	0.464444537873983	0.0260510459379914
0.14	0.459289868775332	0.481149417807427	0.0218595490320949
0.16	0.477570933038754	0.507545851071353	0.0299749180325988
0.18	0.493793144720939	0.543506488074266	0.0497133433533264
0.20	0.508351517944768	0.577632355563481	0.0692808376187130

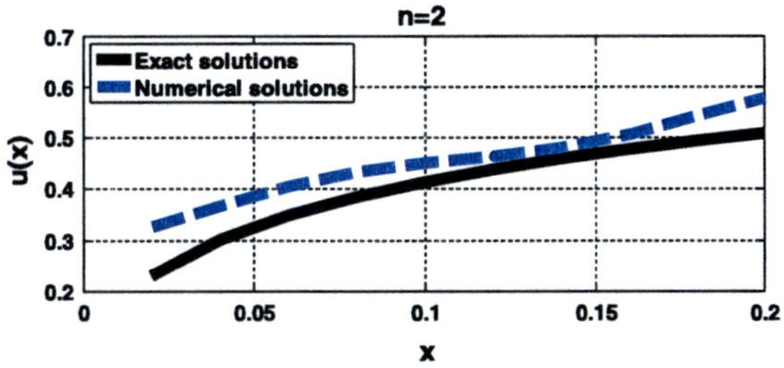

Fig. 2 Exact solutions $\tau(x_i)$ and numerical solutions $\tau_2^{0.2}(x_i)$

$$\tau(x) = 2\sqrt{x} - \int_0^x \tau(t)(x-t)^{-\frac{1}{2}} dt \,;\, x \in [0, 1] \tag{19}$$

where the exact solution is $\tau(x) = 1 - e^{\pi x} erfc(\sqrt{\pi x})$ [11]. In Table 2, shown the numerical solutions $\tau_2^{0.2}(x_i)$. In Fig. 2, shown the graphs of $\tau_2^{0.2}(x_i)$ compared with the exact ones.

4 Conclusion

An interpolation method based on the Vandermonde matrix is provided for solving the second kind weakly singular Volterra integral equations (VIEs). By designing the best node distributions for the two kernel's variables and guaranteeing that the denominator never becomes imaginary or zero, the technique presents new principles for isolating the kernel singularity. The substitution of the interpolated solution into the integral equation yields an equivalent algebraic linear system without employing the collocation approach. The numerical solutions of the two solved examples converge faster to the exact ones while using the lowest interpolant degree. Thus, the suggested method is superior and capable to provide high-accuracy results.

References

1. Atkinson KE (2010) The numerical solution of integral equations of the second kind. Cambridge University Press
2. Wazwaz AM (2015) A first course in integral equations-solutions manual, 2nd ed. World Scientific Publishing Co. Pte. Ltd
3. Kumar P, Dubey GC (2015) An application of Volterra integral equation by expansion of Taylor's series in the problem of heat transfer and electrostatics. IOSR J Math (IOSR-JM) 11(5):59–62
4. Keaveny EE, Shelley MJ (2011) Applying a second-kind boundary integral equation for surface tractions in Stokes flow. J Comput Phys 230(5):2141–2159
5. Hatamzadeh S, Naser-Moghadasi M (2008) An integral equation modelling of electromagnetic scattering from the surfaces of arbitrary resistance distribution. Prog Electromagn Res B 3:157–172
6. Shoukralla ES (2020) A numerical method for solving Fredholm integral equations of the first kind with logarithmic kernels and singular unknown functions. Int J Appl Comput Math 6(6):1–14
7. Shoukralla ES (2021) Application of Chebyshev polynomials of the second kind to the numerical solution of weakly singular Fredholm integral equations of the first kind. IAENG Int J Appl Math 51(1):1–16
8. Shoukralla ES, Markos MA (2018) The economized monic Chebyshev polynomials for solving weakly singular Fredholm integral equations of the first kind. Asian-Eur J Math 12(1):2050030-1–2050030-10
9. Shoukralla ES (2021) Interpolation method for solving weakly singular integral equations of the second kind. Appl Comput Math 10(3):76–85
10. Ramadan MA, Moatimid GM, Taha MH (2020) A powerful method for obtaining exact solutions of Volterra integral equation's types. Glob J Pure Appl Math 16(2):325–339
11. Pourgholi R, Tahmasebi A, Azimi R (2017) Tau approximate solution of weakly singular Volterra integral equations with Legendre wavelet basis. Int J Comput Math 94(7):1337–1348
12. Shoukralla ES, Ahmed BM (2020) The Barycentric Lagrange interpolation via Maclaurin polynomials for solving the second kind Volterra integral equations. In: 2020 15th international conference on computer engineering and systems, pp 1–6
13. Shoukralla ES, Ahmed BM (2019) Multi-techniques method for solving Volterra integral equations of the second kind. In: 2019 14th international conference on computer engineering and systems (ICCES), pp 209–213
14. Shoukralla ES, Elgohary H, Ahmed BM (2020) Barycentric Lagrange interpolation for solving Volterra integral equations of the second kind. J Phys Conf Ser 1447(1):012002. IOP Publishing.
15. Shoukralla ES, Ahmed BM (2020) Numerical solutions of Volterra integral equations of the second kind using Lagrange interpolation via the vandermonde matrix. J Phys Conf Ser 1447(1):012003). IOP Publishing

Adoption of Cloud-Based Communicable Disease Surveillance in Taiwan: Chief Information Officers' Perspectives of Hospitals

Pi-Jung Hsieh and Hui-Min Lai

Abstract Cloud-based communicable disease surveillance (CCDS) allows the infection control staff of Taiwanese hospitals to simplify many of their reporting procedures and improve patient safety more efficiently and cost-effectively. Despite its great potential, there are gaps in health information technology researchers' understanding of how hospitals decide to adopt CCDS. Therefore, our study proposes a research model that incorporates technological, organizational, environmental, transaction cost, and sociological perspectives to determine the factors that influence hospitals' intention to adopt CCDS. A series of surveys was used to empirically test the organizational adoption model through the chief information officers (CIOs) of Taiwanese hospitals. The 206 valid questionnaires were selected for data analysis. The results showed hospitals' decision to adopt CCDS as based on relative advantages, hospital size, top management support, government policies, uncertainty, and trust. These findings provide valuable insights and implications for medical informatics practices to facilitate the likelihood of cloud technology adoption.

Keywords Cloud-based communicable disease surveillance · Innovation adoption · Transaction costs · Social exchange

1 Introduction

During the current coronavirus disease (COVID-19) pandemic, the surveillance and control of communicable diseases are critical for managing public health worldwide. To effectively collect, analyze, and monitor epidemic information among hospitals, the Taiwan Centers for Disease Control (CDC) has applied cloud computing

P.-J. Hsieh
Department of Hospital and Health Care Administration, Chia Nan University of Pharmacy and Science, Tainan, Taiwan, ROC
e-mail: beerun@seed.net.tw

H.-M. Lai (✉)
Department of Business Administration, National Taichung University of Science and Technology, Taichung, Taiwan, ROC
e-mail: hmin.mis@msa.hinet.net

X.-S. Yang et al. (eds.), *Proceedings of Seventh International Congress on Information and Communication Technology*, Lecture Notes in Networks and Systems 447,
https://doi.org/10.1007/978-981-19-1607-6_55

applications, building the cloud-based communicable disease surveillance (CCDS) platform by using hospitals' information systems, and connecting with the virtual private network (VPN) to help hospitals report infection cases more effectively and conveniently. The CCDS mission allows hospitals to simplify the reporting process and thus ensure the timeliness of surveillance reports and improve patient safety efficiently and cost-effectively [1]. It also enables hospitals to automatically receive laboratory test results and reduces the need for manual entry. Hospitals can participate in the project on a voluntary basis. However, the CCDS has some challenges and limitations, such as the system requires a higher level of information infrastructure and information specialists to support the implementation of CCDS. Thus, hospital scale is a determinant of CCDS adoption. Further, the low adoption rate of CCDS is another challenge since system adoption is voluntary. As of 2019, of the 427 hospitals in Taiwan, 60 and 66 hospitals joined CCDS for automatic infectious disease notification and laboratory data submission, respectively [2]. Thus, to gain CCDS-enabled benefits (e.g., to collect, analyze, and share CCDS information regarding epidemics among hospitals), hospitals must first adopt CCDS.

CCDS adoption integrates information technology (IT) adoption with healthcare factors and thus needs distinct theorization within technology adoption. Although prior studies have identified several contextual factors influencing health IT adoption in healthcare organizations [3–5], the understanding of organizational CCDS adoption is limited and fragmented. CCDS is somewhat distinct due to its highly complicated characteristics. It needs heavy IT resource investment and must be combined with the hospital information system (HIS) and the laboratory information system (LIS) to increase the timeliness of reports and to analyze the trends and prevalence of specific pathogens. Furthermore, communicable disease surveillance reports must be shared between hospitals and the Taiwan CDC. Hospital managers have had to consider social contexts to overcome any information or power threats (such as government policies and competitive pressure). Thus, these unique features make CCDS adoption a challenging task for hospital chief information officers (CIOs).

Prior studies have investigated the factors influencing technology adoption, primarily based on the innovation diffusion theory (IDT) [6], and on the technology-organization-environment (TOE) framework [7]. These studies have demonstrated that innovation characteristics (relative advantage, complexity, and compatibility) may influence decisions regarding health IT adoption [8, 9]. However, the TOE framework does not consider key determinants, such as economic and social issues, which are important for technology adoption. Prior studies have identified this limitation and have argued for integrative approaches that combine two or more theoretical perspectives to explain technology adoption [10]. These have primarily investigated technology adoption, yet they have not tested economic and social factors. To the best of our knowledge, few empirical studies have been conducted to examine the technological, organizational, environmental, economic, and social contexts that influence CCDS adoption within and outside health organizations. We therefore seek to combine economic and social theories to identify the factors affecting hospitals' willingness to adopt CCDS. From a practical standpoint, explaining why hospitals should adopt CCDS can assist government agencies and system managers in developing appropriate strategies that will increase hospital adoption and its positive effects on medical care. Our present study fills this gap, with the following research purposes:

(1) Provide a comprehensive set of key factors driving the adoption of CCDS via the TOE framework.
(2) Investigate the impacts of technological, organizational, environmental, transaction cost, and social exchange contexts on CCDS adoption.
(3) Offer policymakers and system managers' valuable guidelines for implementing CCDS.

2 Literature Review

2.1 Cloud-Based Communicable Disease Surveillance (CCDS)

Taiwan CDC has developed and established the national communicable disease surveillance system to provide hospitals across the country with a Web site for reporting cases and predominant information related to communicable disease occurrences. However, the communicable disease reports still rely on manual operations. To quickly collect and monitor epidemic disease surveillance data among hospitals and increase the surveillance efficiency and efficacy, the Taiwan CDC launched the CCDS project. The two cloud applications and aims of CCDS are (1) to increase the timeliness of reports from the notifiable disease surveillance system via hospital electronic medical records (EMRs) and (2) to establish a laboratory automated reporting system that will adopt examination results from LISs in order to analyze the trends and anomalies of pathogens. CCDS not only integrates infectious disease-related databases in real time and effectively, but it also allows the Taiwan CDC to quickly analyze and monitor epidemic information. By using CCDS, hospitals can also make early and informed decisions to undertake disease prevention and stop contagion. Thus, the Taiwan CDC can quickly formulate effective epidemic prevention policies to control the COVID-19 disease in the current pandemic.

2.2 Technology-Organization-Environment (TOE) Framework

The TOE framework's purpose is to provide an understanding of the adoption of innovation technology in organizational contexts [7]. It posits that the technology adoption depends on technological, organizational, and environmental contexts. The technological context refers to technology portfolios, technology expertise, and technological characteristics, which have been claimed to influence decisions about adopting technological innovations [8, 10]. The organizational context comprises the organizational scope, size, strategies, structure, and top management championship [7]. The environmental context encompasses the limitations of and opportunities for technology adoption, which involve the various environment actors who may affect the decision-making process, competitors, and interactions with the government [7]. Bose and Luo [11] suggest that the TOE framework does not offer a specific model for describing the factors that affect an organization's innovation adoption; instead, it offers a taxonomy for classifying innovation adoption factors in their respective contexts. Bose and Luo [11] further argue that researchers should consider the specific context of technical innovation. The TOE framework has been utilized in prior research to explain the adoption of health IT, such as in open-source software [3], the picture archiving and communication system (PACIS) [12], and cloud computing [9]. Thus, the TOE framework offers a set of useful theoretical insights for organizational technology adoption.

2.3 Combining TOE Framework and Innovation Diffusion Theory (IDT)

The IDT states that five innovation attributes influence an organization's technology adoption decisions [6]. *Relative advantage* refers to the advancement that the new technology may bring to an organization. *Compatibility* involves whether the new system is perceived as compatible with prior experiences, practices, and needs. *Complexity* denotes the perceived difficulty of using the new system. *Trialability* describes how easily a new system may be experimented with. *Observability* is the extent to which the outcomes of using the new technology are evident to others. The IDT has been widely adopted to understand the adoption of new technology [13, 14]. Among the IDT's five cited attributes, many technology adoption studies rule out trialability and observability since they are not related to cloud-computing technology [13, 15]. Thus, following the general guidance of IT research, relative advantage, compatibility, and complexity are included in our study.

To enhance the understanding of innovation adoption, researchers have therefore combined the TOE framework with the IDT [9, 10, 16]. The TOE framework perspectives overlap with the IDT's innovation characteristics. For example, Yang et al. [10] suggest that the health IT context in the TOE framework includes the cognition of innovation characteristics from the IDT. Furthermore, the IDT's internal and external organizational features include the same concepts as those in the TOE framework's organizational context [8]. However, there are also major differences between the two theories. The IDT suggests including individual characteristics (top management support) in the organizational context, while the TOE framework does not specify the role of individual characteristics. Moreover, the IDT does not concern the influence of the environmental context, while the TOE framework provides an overall perspective for explaining innovation adoption by including the technological, organizational, and environmental contexts. Thus, in this study, we combine the IDT and the TOE framework to understand hospitals' CCDS adoption decisions.

2.4 *Transaction Cost Theory (TCT)*

Originally proposed by Williamson [17, 18], the transaction cost theory (TCT) attempts to define organizational boundaries by justifying activities that should be completed within the organization or those that should be completed by suppliers. This decision-making process should focus squarely on the risk of opportunism. If the risk is too high, then the activity should be done within the organization. Furthermore, the TCT offers a framework of three determinants that lead to opportunism—asset specificity, uncertainty, and frequency [18]. *Asset specificity* refers to the degree to which robust investments are utilized to support transactions. *Uncertainty* pertains to the degree to which organizations are confronted with unforeseen actions; these raise the requirement for the proper manner to enforce contracts between cooperative parties and the requirement for greater flexibility in coordinating responses to unforeseen problems. *Frequency* denotes the volume of transactions in a given time [19]. The TCT has been employed in past studies to explain technology adoption, such as EMR exchange [4], organizational virtualization [19], and electronic data interchange [20]. In an EMR exchange study, transaction cost factors have been confirmed as constituting a major context in cultivating IT executives'/directors' positive attitudes toward adopting health IT [4]. Therefore, in our study, the above-mentioned three transaction cost factors are used as critical determinants to offer a supplementary, theoretical foundation for understanding hospitals' decisions to adopt CCDS.

2.5 *Social Exchange Theory (SET)*

The SET was initially developed to analyze individual behavior [21] and was later extended to the organizational context, with a focus on the importance of social norms. The SEC states that people and organizations interact to maximize their rewards and minimize their costs [22]. The application of the SET has spread across different areas, such as technology adoption decisions, cross-hospital EMR exchange [4], contact tracing mobile applications [23], and social media drivers [24]. The cited researchers have proposed several determinants, based on the concept that technology adoption can bring benefits—positing that technology adoption depends on exercising power, reciprocal investments, reputation, and trust. In an electronic data interchange (EDI) usage study, these relational factors contribute significantly to EDI usage [20]. This study mainly uses relational factors (namely reciprocal investments and reputation) to identify the factors that determine intention toward

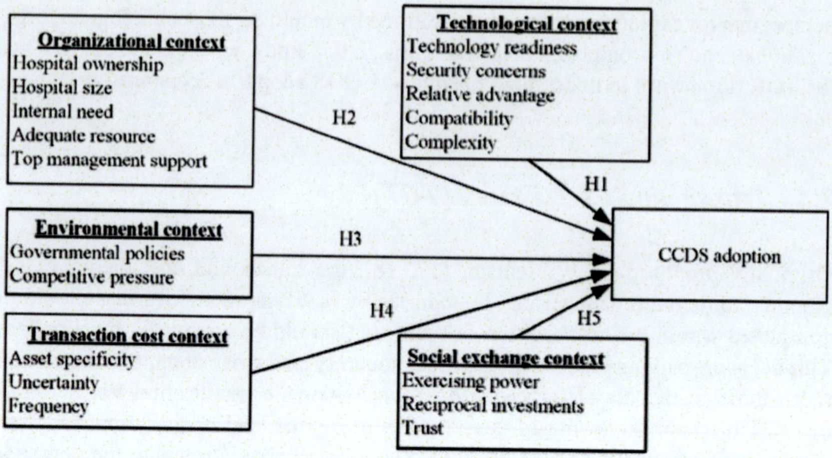

Fig. 1 Research framework

hospital—the Taiwan CDC relationships. In an EMR exchange study, trust has been proven to affect EMR exchange among hospitals [4]. Therefore, we adopt the SET to explain the concept of technology adoption at the organizational level.

3 Research Methodology

3.1 Research Model

To explain the organizational decisions related to technology adoption, the model proposes that technological, organizational, environmental, transaction cost, and social exchange contexts influence CCDS adoption. Figure 1 shows the research model of this study.

3.2 Research Hypotheses

The technological context may influence organizational decisions to adopt technology, as identified by two variables—technology readiness and security concerns. Technology readiness comprises the technical resources available for technology adoption in an organization. It includes the structural aspects (e.g., the technological infrastructure) and human resources [8]. When hospitals have experience with developing and building health IT, they can use higher levels of technological readiness to support the adoption of innovations [10]. Hence, hospitals with a higher level of technological readiness are more suitable for adopting CCDS. Furthermore, CCDS implementation not only integrates HISs but also empowers the Taiwan CDC to effectively collect hospital disease databases. However, the Internet-based connections from hospitals to the Taiwan CDC may be viewed as potential security concerns, which may diminish hospitals' readiness to adopt CCDS. Chang et al. [25] confirm that security protection is a critical predictor of technology adoption. Therefore, we formulate the following hypotheses:

H1a. Technology readiness will positively influence CCDS adoption.
H1b. Security concerns will negatively influence CCDS adoption.

In this study, if CCDS is perceived as offering more benefits when compared with existing practices and systems (such as the communicable disease reporting system), IT adoption will be encouraged. For example, CCDS allows hospitals to automatically receive test results and reduces the double entry of data. Moreover, since compatibility refers to the extent to which CCDS may be integrated with a hospital's operation process and existing information systems, the hospital's CIO may perceive CCDS as compatible if it fits well with the hospital's current environment. Compatibility can lead to a greater intent to adopt CCDS. Finally, the adoption of a new technology is inhibited or discouraged if the adopter perceives it as complex [8]. Conversely, the easier it is to incorporate CCDS into organizational operations, the greater the opportunity of its adoption. Martins et al. [13] have found relative advantage and complexity to be important predictors of technology adoption, and Zhu et al. [26] have found compatibility to be a key factor influencing technology adoption. Thus, we propose the following hypotheses:

H1c. Relative advantage will positively influence CCDS adoption.
H1d. Compatibility will positively influence CCDS adoption.
H1e. Complexity will negatively influence CCDS adoption.

The organizational context includes hospital ownership, hospital size, internal needs, adequate resources, and top management support. Hospital ownership may guide organizational strategies, based on hospital missions and values [5]. Thus, hospital ownership can influence CCDS adoption. Hospital size can also influence the adoption of cloud computing [5] since larger hospitals associated with more financial capability and adequate human resources [5]. Larger hospitals are often considered adopters of new technology as they have rich resources [16]. Thus, hospital size is a determinant of CCDS adoption. An organization's internal needs can also serve as project triggers [10]. By using CCDS, workflows of disease reporting can be more simplified than those of existing systems. In particular, the time required to input data can be significantly reduced, thereby promoting and encouraging CCDS adoption. Additionally, adequate resources are critical to adoption success [25]. The adoption of cloud computing technology is usually a large project and budget undertaking for hospitals. If a given hospital has a sufficient budget, adequate human resource support, and ample time, then CCDS adoption will be met positively. Top management support is the commitment offered to promote the desired environment for new technology adoption [13]. Chang et al. [12] also indicate that top management support generally bodes well for health IT adoption in Taiwan's hospitals. When top hospital managers understand the importance of CCDS, they tend to play a critical role in influencing other staff members to use it. Without top management support, it is unlikely that the decision to adopt CCDS will be effective. Previous researchers have found that these organizational contexts—hospital ownership, hospital size, internal needs, adequate resources, and top management support—are important predictors of technology adoption [10, 25]. Therefore, we propose the following hypotheses:

H2a. Hospital ownership will positively influence CCDS adoption.
H2b. Hospital size will positively influence CCDS adoption.
H2c. Internal needs will positively influence CCDS adoption.
H2d. Adequate resources will positively influence CCDS adoption.
H2e. Top management support will positively influence CCDS adoption.

The environmental context may influence organizational decisions to adopt technology, as identified by two factors—government policies and competitive pressure. Government policies have often driven the adoption of new technologies [8]. The technology support that a government can provide includes IT infrastructure, training policies, and enough labor [27]. It has been suggested that in Taiwan, government support is needed to promote hospitals' adoption of health IT [25] since the Taiwan CDC receives incomplete disease reports from hospitals. To combat this, the Taiwan CDC can offer IT infrastructure, financial incentives, project promotion, and counseling. Then, hospitals might choose to adopt CCDS to reduce the time required for reporting. Further, pressure from healthcare partners and competitors may trigger hospitals' use of new technologies. Vest [28] has found that competitive pressure from the healthcare industry forces many hospitals to adopt health information exchange. By adopting CCDS, hospitals can gain advantage from greater work efficiency and higher speed in receiving patient test results. Therefore, CCDS adoption is expected to be influenced by the proportion of surrounding CCDS adopters. Hence, we propose the following hypotheses:

H3a. Supportive government policies will positively influence CCDS adoption.
H3b. Competitive pressure will positively influence CCDS adoption.

The transaction cost context includes asset specificity, uncertainty, and frequency. *Asset specificity* can be described as the degree to which the value of an organization's capital is specific to its relationships with other organizations [20]. The asset specificities in CCDS adoption are human, physical, and temporal. Human assets refer to healthcare professionals' knowledge of communicable disease surveillance. Physical assets address specialized equipment and related technology infrastructure. Temporal assets refer to the degree to which timely performance by a healthcare professional is critical. Asset specificity can strengthen the cooperative relationship between hospitals and the Taiwan CDC. Chang et al. [4] also indicate both associations in the EMR exchange context. *Uncertainty* occurs when there is insufficient information that reduces confidence in health IT adoption decisions Chang et al. [4] Uncertainty increases the need for appropriate ways to negotiate contracts between hospitals and the Taiwan CDC, as well as the requirement for greater flexibility in coordinating responses to unforeseen problems. Thus, uncertainty will result in lower adoption intentions. Chang et al. [4] confirm uncertainty as a significant predictor of resistance to technology. Additionally, Chan and Chong [27] state that a high *frequency* of interaction shows the significance of a seller's service for a buyer's operations. As a result, high frequencies will facilitate higher incentives for hospitals to increase their coordination. Chan and Chong [27] have provided support for the positive effect of technology adoption. Thus, we propose the following hypotheses:

H4a. Asset specificity will positively influence CCDS adoption.
H4b. Uncertainty will negatively influence CCDS adoption.
H4c. Frequency will positively influence CCDS adoption.

The social exchange context involves exercising power, reciprocal investments, and trust. *Power* is often exercised with the intention to immediately influence the other party's actions [20]. The Taiwan CDC, taking the CCDS project, often exercises its power to immediately influence hospitals' actions, for instance, by offering rewards and assistance. Iacovou et al. [29] show that power exercised by large trading partners positively affects initial adoption. *Reciprocal investments* imply that business partners intend to guarantee stable, cooperative relationships [4]—such as the reciprocal investments between hospitals and the Taiwan CDC in training, workshops, and facilitation sharing. Reciprocal investments affect a hospital's willingness to adopt CCDS. Prior studies have presented reciprocal investments as critical factors that significantly influence organizational innovation adoption [4]. Additionally, the SET advocates *trust* as influencing cooperation between organizations and states that trust has been proven to influence interorganizational relations [30]. For example, trust facilitates information exchange and improves the effectiveness of joint problem solving. Chang et al. [4] confirm that trust is a significant predictor of technology adoption. Thus, we suggest the following hypotheses:

H5a. Exercising power will positively influence CCDS adoption.
H5b. Reciprocal investments will positively influence CCDS adoption.
H5c. Trust will positively influence CCDS adoption.

3.3 *Questionnaire Development*

Our questionnaire included two parts. The first part was used to collect respondent characteristics, including gender, age, number of employees, and hospital ownership. The second part was used based on the constructs of technology readiness, security concerns, relative advantage, compatibility, complexity, internal needs, adequate resources, top management support, government policies, competitive pressure, asset specificity, uncertainty, frequency, exercising power, reciprocal investments, trust, and CCDS. They were measured using a seven-point Likert scale (from 1 = strongly disagree to 7 = strongly agree). The survey items were adapted from those of previous studies [5, 8, 9, 19, 20, 25] and modified to suit the CCDS context. For the data analysis, we adopted partial least squares estimations for the statistical analysis since they impose fewer restrictions on sample size, with unbiased estimates for model validation [31].

3.4 Data Collection

The study participants were CIOs from Taiwanese hospitals that had not adopted CCDS. In total, 427 hospitals are listed by the Taiwan Joint Commission on Hospital Accreditation. In total, 241 hospitals were successfully contacted to secure their collaboration, and a questionnaire was sent to the CIO of each hospital.

4 Analysis and Results

A total of 218 responses were returned, with 12 incomplete responses. The resulting 206 valid responses constituted a response rate of 85.48%. Most of the questionnaire respondents were males (72.3%) between the ages of 41 and 50 years (42.2%). The majority (38.3%) was number of employees between 101 and 500. The hospital ownership of 45.6% of the respondents was private hospital. Reliability was examined according to a composite reliability (CR) greater than 0.6 [32]. Convergent validity was measured according to the following three criteria: (a) CR greater than 0.6, (b) average variance extracted (AVE) greater than 0.5, and (c) item loading (λ) greater than 0.5 [33]. For the discriminant validity, the square root of each factor's AVE should exceed its correlations with any other factor [32]. Table 1 presents the outcomes of the reliability and the validity assessments. In the present study, the CR values were all greater than 0.8; the item loadings ranged from 0.54 to 0.97, and the AVEs ranged from 0.60 to 0.93. Furthermore, the square root of the AVE for a factor was greater than its corresponding off-diagonal factors. These outcomes presented our scales as having reliability and validity.

The test outcomes from the structural model are shown in Fig. 2. The data analysis results partially supported this study model. CCDS adoption was jointly predicted by relative advantage ($\beta = -0.16$, standardized path coefficient, $p < 0.05$), hospital size ($\beta = 0.31$, $p < 0.001$), top management support ($\beta = 0.18$, $p < 0.01$), government policies ($\beta = 0.21$, $p < 0.001$), competitive pressure ($\beta = 0.14$, $p < 0.01$), uncertainty ($\beta = 0.18$, $p < 0.05$), and trust ($\beta = 0.13$, $p < 0.05$). Together, these variables explained 44.1% of the variance of CCDS adoption. As a result, hypotheses 1c, 2b, 2e, 3a, 3b, 4b, and 5c were all supported. Furthermore, technology readiness, security concerns, compatibility, hospital ownership, internal needs, adequate resources, asset specificity, frequency, exercising power, and reciprocal investments did not significantly affect CCDS adoption. Thus, hypotheses 1a, 1b, 1d, 1e, 2a, 2c, 2d, 4a, 4c, 5a, and 5b were not confirmed.

Table 1 Reliability and validity of the scale

Construct	Item loading	CR	AVE
CCDS adoption (CA)	0.86–0.89	0.91	0.77
Relative advantage (RA)	0.82–0.94	0.93	0.81
Asset specificity (AS)	0.86–0.89	0.93	0.76
Reciprocal investment (RI)	0.87–0.90	0.95	0.79
Top management support (TM)	0.91–0.94	0.96	0.84
Compatibility (CO)	0.89–0.94	0.94	0.84
Complexity (CP)	0.83–0.92	0.94	0.74
Frequency (FR)	0.94–0.97	0.96	0.90
Government policies (GP)	0.85–0.95	0.94	0.83
Internal need (IN)	0.85–0.96	0.95	0.85
Exercising power (EP)	0.93–0.95	0.94	0.88
Competitive pressure (CP)	0.86–0.92	0.93	0.80
Adequate resource (AR)	0.78–0.93	0.95	0.79
Security concern (SC)	0.95–0.98	0.98	0.93
Technology readiness (TR)	0.54–0.89	0.81	0.60
Trust (TU)	0.84–0.89	0.95	0.75
Uncertainty (UN)	0.85–0.93	0.98	0.82

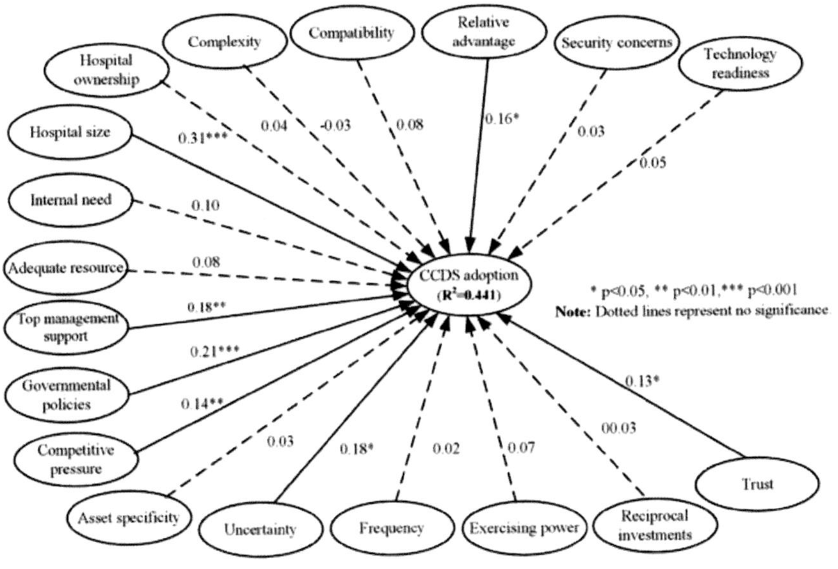

Fig. 2 Results of the structural model analysis

5 Discussion

We explored how the discussed factors affected hospitals' intentions to adopt CCDS. In the proposed model, the explained variance ($R^2 = 0.44$) proved better than the findings of previous studies [8, 34] in explaining organizational intentions to adopt technological innovation. This implied that the TOE framework, integrated with the IDT, the TCT, and the SET, might be a robust research model for explaining organizational intentions to adopt cloud service. Drawing from the IDT and the TOE framework, in this study, we posited relative advantage, hospital size, top management support, government policies, and competitive pressure as critical factors influencing hospitals' adoption intentions. Of the three innovation factors, relative advantage was confirmed as positively influencing technology adoption. This finding is consistent with the results of prior studies on technology adoption [8, 15]. The results suggest that hospitals consider it worthwhile to evaluate whether CCDS adoption may leverage greater organizational advantages. The data showed compatibility and complexity as insignificant in CCDS adoption. With respect to compatibility, the findings perhaps indicate that hospital executives do not believe CCDS to be consonant with their existing information systems and practices and may believe that CCDS does not satisfy their specific needs. With respect to complexity, the Taiwan CDC usually provides promotion strategies, such as on-site consulting or training, to connect the emerging system to hospitals. Therefore, hospital executives perceive the effects of CCDS complexity as not obvious. The data also showed technology readiness as insignificant in CCDS adoption. This result is consistent with the findings of prior studies on technological adoption, which have indicated that technology readiness may not affect technology acceptance [11]. A plausible explanation for this is that the Taiwan CDC purchases and maintains the necessary hardware and software in its own facility. Thus, hospitals avoid the capital expenditures of acquiring hardware, such as servers and data storage equipment. Consequently, technology readiness may not affect CCDS adoption. Our findings also indicated that the indirect effects of security concerns did not inhibit CCDS adoption. Again, this finding is consistent with those of past studies [8, 28]. A possible reason is that the recent advances in privacy-enhancing and encryption techniques ensure data security in the cloud-based environment. These security technologies build organizational trust and control over data when using cloud services. This may explain the lack of attention to security and privacy when adopting a cloud service.

The relevance of the organizational context specifically highlighted the importance of hospital size and top management support. Similar to previous technology adoption studies [8, 26], hospital size was found to be important in determining CCDS adoption. Larger hospitals have more resources to support the expenditure and investment risk of a new system. Top management support was also found to be critical for technology adoption, which is consistent with the results reported by Chan and Chong [27]. The levels of CCDS adoption were higher when management commitment was obtained. However, the data showed hospital ownership as insignificant in CCDS adoption. This result confirms the findings obtained by Marques [5].

A possible explanation is that hospitals in Taiwan are cornerstones of the national epidemic prevention system and are the dominant forces in controlling and managing communicable diseases. Hospitals are so important to begin with that it does not matter who owns them. Similarly, the data revealed internal needs as insignificant in CCDS adoption. This finding is consistent with those of a prior study on technology adoption [25]. Although CCDS can promote the timeliness of communicable disease reports, for a hospital, such communicable disease reports are few. Therefore, internal needs may not necessarily influence CCDS adoption. Additionally, adequate resources did not significantly affect CCDS adoption, but this result is inconsistent with those of Lian et al. [9]. To encourage hospitals to adopt CCDS, the Taiwan CDC has provided IT infrastructure, consulting services, training, and adequate resources to hospitals. Thus, hospitals' internal organizational resources are less important for CCDS adoption decisions.

In the environmental context, our study showed competitive pressure as a key factor for the likelihood of CCDS adoption. This finding coincides with the results of prior studies on technology adoption [27]. This implies that hospitals tend to execute changes more aggressively when they face fierce competition. Furthermore, this study, consistent with the study of Hsu et al. [34], showed government policies as driving CCDS adoption. The CCDS program has provided infrastructure and incentives, such as funds to promote CCDS adoption. Thus, government policies have played a key role in promoting CCDS adoption.

The results suggest that uncertainty negatively and directly affects CCDS adoption—meaning that higher uncertainty leads to greater resistance to CCDS adoption. Our finding is consistent with the results obtained by Liu et al. [19]. However, asset specificity did not significantly affect CCDS adoption. There are two possible reasons why our study varies from previous studies [4, 20]. First, CCDS uses hospitals' EMRs and connects with the VPN to transfer data. Second, hospitals in Taiwan have accumulated a wealth of information and communication technology application experience and have obtained extensive domain knowledge in medical care. Therefore, asset specificity is less significant in health care than in other sectors. This study also showed that frequency did not drive CCDS adoption, which is inconsistent with the finding of Chan and Chong [27]. One possible explanation is that the target participants were executives of hospital information departments, which had not adopted CCDS. Their communicable disease reports were still largely paper based. Thus, frequency of contact was less important for their CCDS adoption decisions.

Our results indicated a significant and positive relation between trust and CCDS adoption. This finding is consistent with that of Chan and Chong [27]. The major reason is that CCDS implementation requires integrating a hospital's information system. Hospital executives must confirm that they can trust the Taiwan CDC before sharing epidemic information with it. As such, higher trust will facilitate hospitals' intention to adopt CCDS. However, exercising power and reciprocal investments did not significantly affect CCDS adoption. There are two possible reasons for the difference between this study's results and those of previous studies [4, 27]. First, hospitals in Taiwan are cornerstones of the national epidemic prevention system and are the dominant forces in collecting and managing epidemic information. Hospitals are claimed to perform the related task of implementing a communicable disease reporting system. Thus, reciprocal investments are less important for CCDS adoption decisions. Second, although the Taiwan CDC, with its higher power, can persuade or even force hospitals to report epidemic cases, CCDS is a national project with a limited budget. In the context of voluntary adoption, this may explain the lack of attention to the exercise of power when considering an adoption decision.

5.1 Research and Managerial Implications

The present study provides several implications and contributions. First, we have integrated four theoretical perspectives to build the research model at the organizational level. The model integrates the TOE framework with the IDT, the TCT, and the SET—all of which underlie CCDS adoption. This is distinct from past research on cloud technology, which falls short of holistically assessing the integrated influences of economic and social factors. Thus, we offer new perspectives for future researchers that may help in encouraging hospitals to adopt cloud technology. Second, economic and social factors of technology adoption have not been clearly identified or examined in prior studies. Based on the TCT and the SET, this study contributes to technology adoption research by explicitly conceptualizing and measuring economic and social factors at the organizational level. Our study has confirmed that uncertainty is a critical economic factor inhibiting technology adoption. Concerning the role of social factors, the driving forces positively affect CCDS adoption. Third, our data were collected from healthcare organizations that use cloud services at the organizational level. Thus, it would be useful to duplicate our research model across other cloud services and in different industries to strengthen the robustness of our study's results.

The present study also has several implications for practical action. First, hospitals tend to view CCDS as a tool for gaining competitive advantage. Thus, the Taiwan CDC should focus more on promoting and validating CCDS benefits when marketing its healthcare services to hospitals for CCDS adoption. Second, this study suggests that when hospital managers are familiar with the relative advantages of CCDS, this eliminates their uncertainty about CCDS use and enhances their willingness to adopt the system. Thus, the government, working closely with hospitals, must develop long-term CCDS implementation plans, intervention strategies, and workshops to drive CCDS adoption. Third, smaller hospitals often lack the financial capacity to support the enormous expenditure of CCDS. The Taiwan CDC could offer training programs to help small hospitals in upgrading related technological capabilities, which would, in turn, increase their willingness to adopt CCDS. Fourth, we suggest that uncertainty is a critical economic factor inhibiting CCDS adoption. Thus, system managers' preparation for training and consulting services for hospital IT staff can decrease technological barriers and uncertainties at the initial adoption stage. Fifth, the Taiwan CDC should take a stronger position about providing financial aid and IT training for non-adopters, and it should build related policies to promote CCDS to hospitals.

5.2 *Limitations and Future Research Directions*

The limitations of this study must be acknowledged. First, our study was conducted in one country; thus, the results are limited to a specific national culture, as well as the culture of the medical community. Second, although the TCT and the SET perspectives represent a comprehensive theoretical explanation for economic and social factors, not all factors exist in a specific context. There may be additional economic and social factors influencing technology adoption—beyond asset specificity, frequency, exercising power, and reciprocal investments—which should be assessed in future research. Finally, the results drawn from our study cannot readily be generalized to other industries. A study in another industry—which might have different technology needs and technological capabilities—could offer different outcomes. We suggest that future research be extended to different industries to overcome this study's limitations.

References

1. Hsieh PJ, Lin WS (2019) Understanding the performance impact of the epidemic prevention cloud: an integrative model of the task-technology fit and status quo bias. Behav Inf Technol 39(8):899–916. https://doi.org/10.1080/0144929X.2019.1624826
2. Taiwan Centers for Disease Control (2020) Centers for Disease Control Annual Report. Taiwan Centers for Disease Control, Taiwan. https://www.cdc.gov.tw/File/Get/VzxSlitRN9UWzV7N9z6bNA. Last accessed 2021/10/13
3. Marsan J, Pare G (2013) Antecedents of open source software adoption in health organizations: a qualitative survey of experts in Canada. Int J Inf Manage 82:731–741
4. Chang IC, Hwang HG, Hung MC, Kuo KM, Yen DC (2009) Factors affecting cross-hospital exchange of electronic medical records. Inform Manag 46:109–115. https://doi.org/10.1016/j.im.2008.12.004
5. Marques A, Oliveira T, Dias SS, Martins MFO (2011) Medical records system adoption in European hospitals. Electron J Inf Syst Eval 14(1):89–99. https://doi.org/10.1002/9781118093467.ch1
6. Rogers EM (1995) Diffusion of innovations. Free Press, New York
7. Tornatzky LG, Fleischer M (1990) The processes of technological innovation. Lexington Books, Lexington
8. Oliveira T, Thomas MA, Espadanal M (2014) Assessing the determinants of cloud computing adoption: an analysis of the manufacturing and services sector. Inform Manag 51:497–510. https://doi.org/10.1016/j.im.2014.03.006
9. Lian JW, Yen DC, Wang YT (2014) An exploratory study to understand the critical factors affecting the decision to adopt cloud computing in Taiwan hospital. Int J Inf Manage 34(1):28–36. https://doi.org/10.1016/j.ijinfomgt.2013.09.004
10. Yang Z, Kankanhalli A, Ng BY, Lim JTY (2013) Analyzing the enabling factors for the organizational decision to adopt healthcare information systems. Decis Support Syst 55:764–776. https://doi.org/10.1016/j.dss.2013.03.002
11. Bose R, Luo X (2011) Integrative framework for assessing firms' potential to undertake green IT initiatives via virtual organizational theoretical perspective. J Strat Inf Syst 20(1):38–54. https://doi.org/10.1016/j.jsis.2011.01.003

12. Chang IC, Hwang HG, Yen DC, Lian JW (2006) Critical factors for adopting PACS in Taiwan: views of radiology department directors. Decis Support Syst 42(2):1042–1053. https://doi.org/10.1016/j.dss.2005.08.007
13. Martins R, Oliveira T, Thomas MA (2016) An empirical analysis to assess the determinants of SaaS diffusion in firms. Comput Human Behav 62:19–33. https://doi.org/10.1016/j.chb.2016.03.049
14. Kwon WS, Woo H, Sadachar A, Huang X (2021) External pressure or internal culture? An innovation diffusion theory account of small retail businesses' social media use. J Retail Consum Serv 62:e102616. https://doi.org/10.1016/j.jretconser.2021.102616
15. Ifinedo P (2001) An empirical analysis of factors influencing internet/e-business technologies adoption by SMEs in Canada. Int J Inf Technol Decis Mak 10(4):731–766. https://doi.org/10.1142/S0219622011004543
16. Hung SY, Hung WH, Tsai CA, Jiang SC (2010) Critical factors of hospital adoption on CRM system: organizational and information system perspectives. Decis Support Syst 48(4):592–603. https://doi.org/10.1016/j.dss.2009.11.009
17. Williamson OE (1979) Transaction cost economics: the governance of contractual relations. J Law Econ 22(2):233–263
18. Williamson OE (1985) The economic institutions of capitalism. Free Press, New York
19. Liu C, Sia CL, Wei KK (2008) Adopting organizational virtualization in B2B firms: an empirical study in Singapore. Inform Manag 45(7):429–437. https://doi.org/10.1016/j.im.2008.06.005
20. Son JY, Narasimhan S, Riggins FJ (2005) Effects of relational factors and channel climate on EDI usage in the customer–supplier relationship. J Manag Inf Syst 22(1):321–353. https://doi.org/10.1080/07421222.2003.11045839
21. Homans G (1958) Social behavior as exchange. Am J Sociol 63(6):597–606
22. Obeng E, Nakhata C, Kuo HC (2019) Paying it forward: the reciprocal effect of superior service on charity at checkout. J Bus Res 98:250–206. https://doi.org/10.1016/j.jbusres.2019.02.003
23. Fox G, Clohessy T, Werff LVD, Rosati P, Lynn T (2021) Exploring the competing influences of privacy concerns and positive beliefs on citizen acceptance of contact tracing mobile applications. Comput Human Behav 121:e106806. https://doi.org/10.1016/j.chb.2021.106806
24. Ferm LEC, Thaichon P (2021) Customer pre-participatory social media drivers and their influence on attitudinal loyalty within the retail banking industry: a multi-group analysis utilizing social exchange theory. J Retail Consum Serv 61:e102584. https://doi.org/10.1016/j.jretconser.2021.102584
25. Chang IC, Hwang HG, Hung MC, Lin M, Yen DC (2007) Factors affecting the adoption of electronic signature: executives' perspective of hospital information department. Decis Support Syst 44(1):350–359. https://doi.org/10.1016/j.dss.2007.04.006
26. Zhu K, Dong S, Xu SX, Kraemer KL (2006) Innovation diffusion in global contexts: determinants of post-adoption digital transformation of European companies. Eur J Inf Syst 15:601–616. https://doi.org/10.1057/palgrave.ejis.3000650
27. Chan FTS, Chong AYL (2012) A SEM-neural network approach for understanding determinants of interorganizational system standard adoption and performances. Decis Support Syst 54:621–630. https://doi.org/10.1016/j.dss.2012.08.009
28. Vest JR (2010) More than just a question of technology: factors related to hospitals' adoption and implementation of health information exchange. Int J Med Inform 79(12):797–806. https://doi.org/10.1016/j.ijmedinf.2010.09.00
29. Iacovou CL, Benbasat I, Dexter AS (1995) Electronic data interchange and small organizations: adoption and impact of technology. MIS Q 19(4):465–485. https://doi.org/10.2307/249629
30. Liu YHS, Deligonul S, Cavusgil E, Chioud JS (2018) Always trust in old friends? Effects of reciprocity in bilateral asset specificity on trust in international B2B partnerships. J Bus Res 90:171–185. https://doi.org/10.1016/j.jbusres.2018.05.012
31. Chin WW, Marcolin BL, Newsted PR (2003) A partial least squares latent variable modeling approach for measuring interaction effects: results from a Monte Carlo simulation study and an electronic-mail emotion/adoption study. Inf Syst Res 14(2):189–217. https://doi.org/10.1287/isre.14.2.189.16018

32. Chin WW (1998) Issues and opinion on structural equation modelling. MIS Q 22(1):7–16
33. Hair J, Black WC, Babin BJ, Anderson RE, Tatham RL (2006) Multivariate data analysis, 6th edn. Pearson Education, New Jersey
34. Hsu PF, Ray S, Hsieh YYL (2014) Examining cloud computing adoption intention, pricing mechanism, and deployment model. Int J Inf Manage 34:474–488. https://doi.org/10.1016/j.ijinfomgt.2014.04.006

Design and Evaluation of a Novel and Modular Educational Robot Platform Based on Technology Acceptance Model

Avraam Chatzopoulos, Konstantinos Kalovrektis, Apostolis Xenakis, Elefterios Chondrogiannis, Michail Papoutsidakis, Michail Kalogiannakis, and Sarantos Psycharis

Abstract In this research, we design an open, easy-to-use robotics platform for education applications, focused on primary education. Our platform is statistically evaluated and is modular, expandable, and scalable in terms of supporting the development of new modules. Our proposed platform, in contrast to other commercial ones, is easy to use, cheap, and modular. Additionally, we present initial results regarding the evaluation of the usage of the proposed robotic structure under the technology acceptance model (TAM) in terms of easiness of usage. According to the results, the proposed open educational robotic platform shows a positive effect toward its usage by active teachers.

Keywords Educational robotics · STEM · Open platforms · TAM

A. Chatzopoulos (✉) · M. Papoutsidakis
University of West Attica, Aigaleo Attiki, Greece
e-mail: xatzopoulos@uniwa.gr

M. Papoutsidakis
e-mail: mipapou@uniwa.gr

K. Kalovrektis · A. Xenakis
University of Thessaly, Galaneika, Lamia, Greece
e-mail: kkalovr@uth.gr

A. Xenakis
e-mail: axenakis@uth.gr

E. Chondrogiannis
Agricultural University of Athens, Athens Attiki, Greece
e-mail: elhon@aua.gr

M. Kalogiannakis
University of Crete, Rethymnon Crete, Greece
e-mail: mkalogian@edc.uoc.gr

S. Psycharis
ASPETE, Athens Attiki, Greece
e-mail: spsycharis@aspete.gr

X.-S. Yang et al. (eds.), *Proceedings of Seventh International Congress on Information and Communication Technology*, Lecture Notes in Networks and Systems 447,
https://doi.org/10.1007/978-981-19-1607-6_56

1 Introduction

Educational robotics is considered a new trend in education introduced into the classrooms, which enriches the teaching environment and promotes building knowledge activities [1]. Its playful nature creates an environment helpful for learning that increases students' interest in STEM activities and programming [2, 3]. The term educational robotics is an approach to science technology engineering and mathematics education (known as STEM) [4] that defines a broad knowledge area [5] and refers to a collection of robotic (technology) platforms, educational programs, activities, resources, and learning theories [4, 6]. Educational robotics is a useful and innovative tool [7] offering students a hands-on, practical understanding of the daily-life things that do not fully understand, such as actuators (motors, LEDs, buzzers, lights, etc.), sensors (temperature, proximity, motion, light, etc.), as well as failures and problems related to software and hardware bugs [5]. Educational robotics requires students to design and program a robot and creates its parts and especially helps young students to develop motor skills, coordination, communication, social and mathematical skills, as well as learn to code and programming [5, 8]. Moreover, educational robotics is used in many teaching environments as an innovative learning and teaching tool that helps students: (i) to develop high-level skills, (ii) to understand objects through creating multiple representations of them, (iii) to develop communication and increase collaboration between them, (iv) to develop and improve students' knowledge by solving complex problems (mostly authentic), (v) to increase students' abstract design conception, and (vi) to support students' learning through research and experimentation and boost their knowledge's development in the fields of the STEM areas [2]. Aside from the fact that educational robotics may use "unplugged" activities, the first move toward it is to select and use an appropriate robotic platform, robotic device, or educational robot [4]. In the market, there is a big variety of such tools [9]. Choosing the "appropriate" one depends on the available supported educational activities to help the students to fulfill specific pedagogical objectives. These objectives can be part into two basic categories: (i) building a robot and (ii) handling the robot. Consequently, we may distinguish two basic types of robotic tools: (i) programmable robots (e.g.,Thymio, Edison, Bee-Bot, Ozobot, mBot, etc.) and (ii) robotics construction kits (e.g., Lego® WeDo, VEX V5, Makeblock, Robotis Dream kits, Hexbug VEX, Lego® Mindstorms, etc.) [1].

2 Educational Robot Platforms

On the international online market, there are many commercial robots or robotic construction kits for education classified into the two above categories. The *programmable robots* (Edison, Thymio, Bee-Bot, Blue-Bot, etc.) are usually mobile wheeled robots based on the original Papert's turtle and his logo programming language concept [10], while the *robotics construction kits* (Lego® WeDo 2.0, Mindstorms, Edison, Thymio, VEX EDR, etc.) are building blocks used to make a robot. In addition, several robotic kits work and are programmed by popular microcontrollers such as the BBC's micro: bit—a programmable device introduced for purely educational purposes [11]—, the Arduino's boards [12–16] and the Raspberry Pi are, a cheap computer, with a credit-card size, use to learn to program and to practice with projects [17, 18]. The above robot's list is limited, however representative, and there are plenty of other commercial products such as MouseBot, Kid First Coding, Evo, Tinkerbots, KUBO, ProBot, etc. The aim of us mentioning all the above, is to primarily record the main limitations of the well-known robotic platforms and to propose a low-cost improved one that eliminates the aforementioned limitations. To summarize, such limitations are as follows: (i) the robot is costly, thus is not affordable to all, (ii) can't be expanded with more sensors and/or actuators, (iii) is not open source, and it can't be expanded in terms of software and hardware by its community, (iv) can't be customized to promote the "A" factor ("Art" in STEAM term), (v) can't be used without specific software (it needs program installation) or hardware (demand modern hardware specifications), and vi) use the Internet for its programming.

3 Proposed Robot Platform

The robot's design and hardware architecture focused on quick development, expandability, and ease of use. Several specifications had to be considered to fulfill its educational purpose. These are: (i) robot's hardware cost should be low cost, (ii) robot's hardware is free from difficult-to-find electronic parts, (iii) robot uses as much open-source hardware and software as possible and is also open source, (iv) robot can be easily programmed by any home device: personal computer, smartphone, tablet, (v) its main advantage over other educational commercial robots is that its user does not need to download and install any software, (vi) a block-based language (VPL or visual programming language) is embedded into the robot too, (vii) the educational community (students, teachers) and stakeholders (parents, open laboratory, maker spaces, etc.) may easily assemble the robot and expand it with future sensors and actuators.

3.1 Architecture and Hardware

Robot's block diagram is shown in Fig. 1. Espessif system's ESP32 is used as the main microcontroller of the robot.

ESP32 is a cheap, low-power System on a Chip (SoC) that integrates dual-mode Bluetooth and Wi-Fi, and it can perform as a complete stand-alone system. Although the primary idea was to use the popular and low-cost Arduino UNO [19], the ESP32

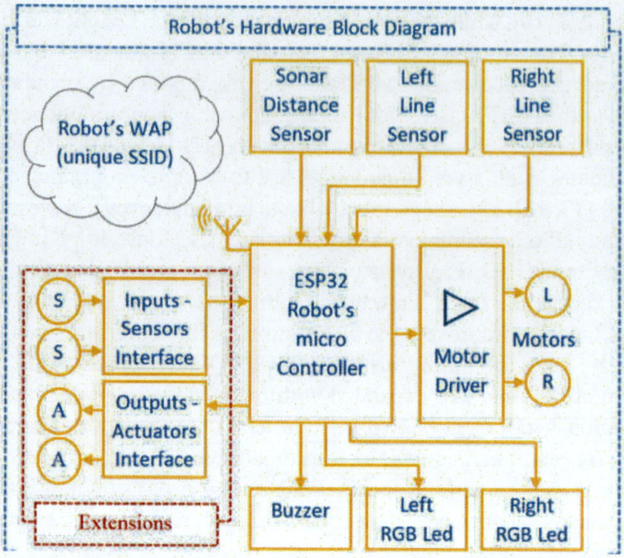

Fig. 1 Robot's architecture block diagram

(a successor to the ESP8266 [20]) is a better choice because it has the same low cost and it comes with enhanced features and peripherals. It is equipped with Xtensa dual-core (or single-core) 32-bit LX6 microprocessor, working at 160 MHz (or 240 MHz), 448 KB ROM, 320 KB RAM, Bluetooth: v4.2 BR/EDR, BLE, and Wi-Fi: 802.11 b/g/n. Its peripheral interfaces include 34 GPIOs, 12-bit ADC (18 channels), 2 × 8-bit DACs, 10 touch (capacitive sensing GPIOs) sensors, 4 × SPI, 2 × I^2S and 2 × I^2C interfaces, 3 × UART, CAN bus 2.0, infrared remote controller, motor and LED PWM (up to 16 channels), Hall effect sensor, SD/SDIO/CE-ATA/MMC/eMMC host controller, and an SDIO/SPI slave controller, among others. In addition, it is also programmed (like Arduino UNO) in C/C++ through the Arduino IDE [14]. ESP32 is responsible: (i) to providing the robot's wireless access point (WAP with a unique SSID), (ii) to act as an intermediate web server serving the clients' (user's devices, e.g., PC, tablet, smartphone) requests, (iii) to provide the robot's user interface (UI) to the users' devices, (iv) to read users' visual programming language (VPL) commands, and convert them to robot directions, and finally, (v) to incorporate all the necessary functions for smooth cooperation of the above operations.

As shown in Fig. 2, many users—clients—may have access to a single robot; however, this can be easily configured and only one user can control a single robot. The "multi-user" specification is chosen in case there is only one robot available to a student class, so to give access to all students. However, this specification will be a little confusing when applied in practice, so it may be configured via the robot's firmware. The robot's controller, ESP32, is connected to the robot's motors through a motor driver [13] it drives through electronic circuit interfaces [16], the other robot's actuators (RBG LEDs and buzzer) and reads the signals of the robots' sensors

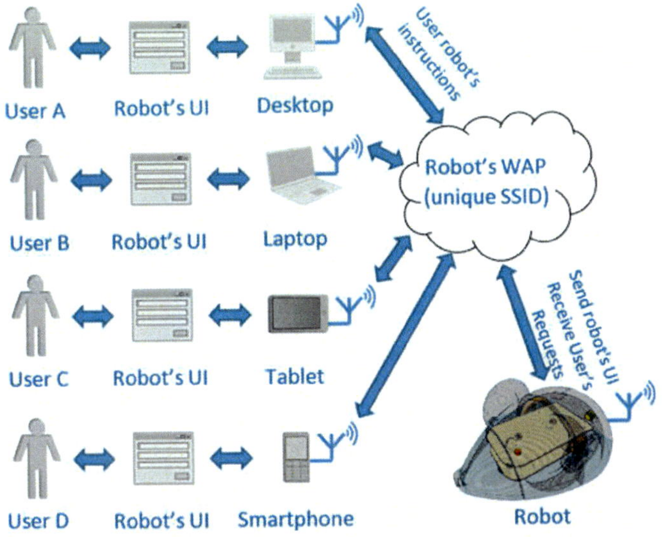

Fig. 2 Robot's multi-users operation block diagram

(supersonic sonar, line sensors). ESP32 has a multi-role that operates under multi-tasking software which is used for every robot's task.

3.2 *Robot Design*

Robot's shell (form and shape) was designed using Autodesk's Tinkercad© an online cloud-based and free 3D software (and more) application used by the educational community and all over the world to think, create, and make. Previous versions of the robot's shell were based on other freely designs under the same license [21, 22]; however, the authors of this paper found the concept of an open freely provided robot's shell as a starting point, to be more creative for the students that boost the "A," (Art term on the STEAM education), while this version of the robot shell -a 3D mouse- was chosen as a "starting shell", because it is based on an open-source 3D printed design [23], easy to assemble, and according to users' feedback [21], it is quite cute. In the future, there will be a collection of new robot designs and shells (based on animal or other figures) that can be easily replaced according to the community's needs. The robot's *perspective view* is shown in Fig. 3, and the robot's *transparent* and *bottom view* is shown in Fig. 4 (left and right accordingly).

Figure 4 *transparent view* shows the robot with a *shell face* revealing its base with its hardware parts (mechanical parts, electronic circuit, battery, etc.).

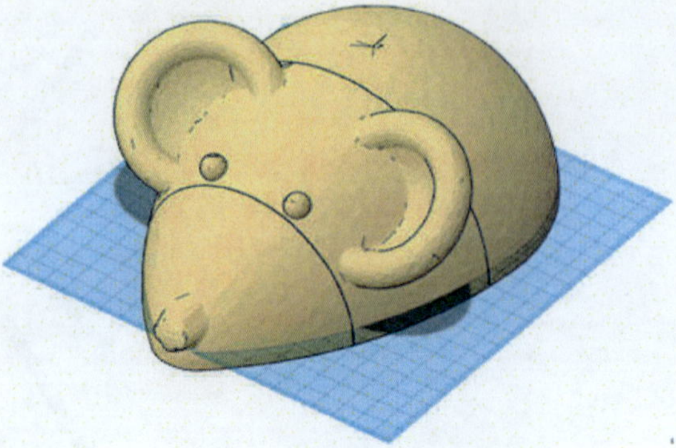

Fig. 3 Robot's perspective view

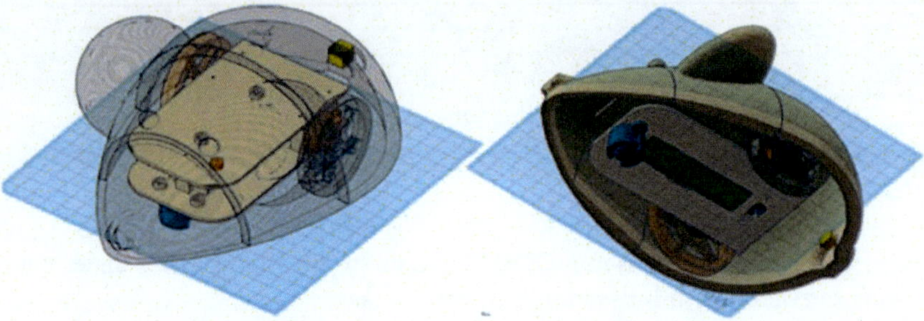

Fig. 4 Robot's transparent shell view (left) and bottom view (right)

4 Robot Software

Robot's software consists of two parts: the *robot's visual programming language (VPL) and* the *robot's firmware.* The robot's firmware is responsible for (i) setting up the robot's WAP, (ii) hosting a web server, (iii) serving clients' requests, (iv) driving the robot's actuators, (v) reading the robot's sensors data, and (vi) interfacing all these for a smooth operation. Robot's VPL is an integrated custom language and is based on Dethe Elza's Blockcode [21, 24].

Its main advantage over other famous VPLs (e.g., Google's Blockly) is due to its small memory footprint (500 lines of code) [25]. In addition, it is simple enough and is capable of use in an ESP32, leaving more memory for the students' programs and robot's activities. Robot's VPL is written using HTML code, cascade style sheets (CSS), and JavaScript (JS and JS libraries), so the robot can be programmed using only the user's device browser. The user does not download or install any software or application. This VPL builds a slick *user interface*, easy to use, where the user can explore using his device's (PC, tablet, and smartphone) browser. In Fig 5, left image shows the robot's introductory screen that is loaded to the user's device browser when its device is connected to the robot's access point (WAP). There are three operation modes, the user has to choose between *easy*, *medium,* and *hard*.

Fig. 5 Robot's intro screen (left image) and *easy* operation mode (right image)

Easy mode (Fig. 5 right image) is for controlling the robot's area movement by pressing the appropriate buttons–keys. In this mode, the user becomes familiar with the robot's buttons, keys, icons, and by playing—have fun—with the robot he understands its movements. *Medium mode (*Fig. 6) is for programming the robot through its block-based language. In this mode, the user programs the robot according to his needs using VPL's blocks. VPL's blocks are non-text representations that use icons and input numbers that can be dragged around the screen, attached to other blocks, and chained together and represent the robot's code to be executed (Fig. 6). Finally, the user must press the *Start Block* to run his program and execute his commands on the robot. *Hard mode*, is currently under development and according to the educational community feedback, provides the users with more advanced blocks, and a simulation area, where the robot moves in the virtual world as its program is executed.

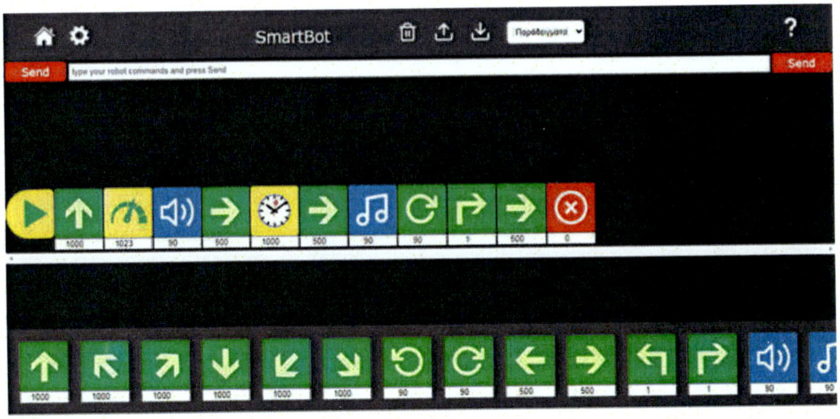

Fig. 6 Robot's *medium* operation mode (robot's VPL with a block-based program example)

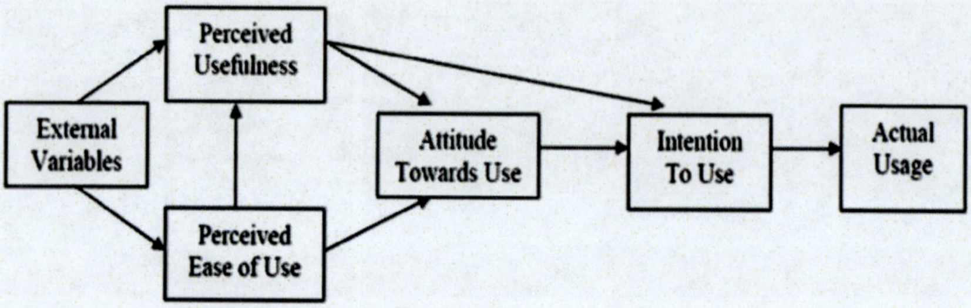

Fig. 7 Technology acceptance model (TAM) [27]

5 Research Methodology

5.1 The Technology Acceptance Model (TAM)

The technology acceptance model (TAM) was developed by Davis [26, 27]. TAM is the application of the most influential extensions of Ajzen and Fishbein's theory of reasoned action (TRA) in the literature. TAM tries to explain and predict the acceptance of information technology by people. The TAM model has four basic factors which explains the technology acceptance; (i) perceived ease of use, (ii) perceived usefulness, (iii) attitude toward use, and (iv) the behavioral intention to use, which leads to (v) actual usage. Figure 7 shows the relationship between the factors of the model.

In the paper, we present the initial part of the results for the factors of perceived ease of use.

Perceived Ease of Use Handling Evaluation Scenario

The robotic mechanism was used by 120 ASPAITE students in a script to evaluate its handling at an initial level of perceived ease of use factors. Students were given a series of ten moves that the robot should perform on a map to evaluate the robot's handling by students of different gender and different field of study (humanitarian vs science studies). In the factors of perceived ease of use, we were focused on the below hypotheses.

Hypothesis 1: There is no difference between gender and the use/learn of the robot (H0).

Hypothesis 2: There are no differences between the field of study (humanitarian and science studies) and the use of the robot (H0).

Upon completion of the tests, to evaluate the use at initial level factors of perceived ease of use, we shared the TAM questionnaire to 120 students in a department of ASPAITE (www.aspete.gr), a school of pedagogical and technological education in the Greece, where its graduates will be the future teachers in primary and secondary education. We received 116 answered questionnaires. We were processed the answered questionnaires using the open software Jamovi.

6 Results and Discussion

From the received 116 answered questionnaires, 56 males (48.3%) and 60 females (51.7%) were participated. The 65 teachers (51.7%) of the sample had a background in science studies, and 51 teachers (44.0%) had a background in humanitarian studies. Initially, we evaluate the difference between gender and the use/learn of the robot (H0). The result of t-test (Mann–Whitney U) is shown in below Table 1. The results indicate that there was no significant difference ($p < 0.05$) in performance between gender and the use/learn of the robot. We are accepting the H0. There is no difference between gender (male/female) and the use/learn of the robot.

Secondly, we evaluate the difference between the field of study and the use/learn of the robot (H0). The results in Table 2 of t-test (Mann–Whitney U) indicate that there was a significant difference ($p < 0.05$) in performance between the field of

Table 1 Hypothesis 1: T-test (Mann–Whitney U) independent samples

		Statistic	p
A1. It would be easy for me to learn to use the robot	Mann–Whitney U	1554	0.444
A2. It would be easy for me to use the robot the way I want	Mann–Whitney U	1451	0.173
A3. The interaction with the robot was clear/understandable	Mann–Whitney U	1399	0.095
A4. I would consider the robot flexible to interact with	Mann–Whitney U	1515	0.334
A5. It would be easy for me to become proficient in using the robot	Mann–Whitney U	1494	0.262
A6. I would consider the robot easy to use	Mann–Whitney U	1531	0.376

Table 2 Hypothesis 2: T-test (Mann–Whitney U) independent samples

		Statistic	p
A1. It would be easy for me to learn to use the robot	Mann–Whitney U	1116	< 0.001
A2. It would be easy for me to use the robot the way I want	Mann–Whitney U	998	< 0.001
A3. The interaction with the robot was clear/understandable	Mann–Whitney U	1106	< 0.001
A4. I would consider the robot flexible to interact with	Mann–Whitney U	1302	0.036
A5. It would be easy for me to become proficient in using the robot	Mann–Whitney U	1107	< 0.001
A6. I would consider the robot easy to use	Mann–Whitney U	1253	0.016

study and the use/learn of the robot. We are not accepting the H0. The use/learn of the robot depends on the field of study between humanitarian studies and science studies.

7 Conclusions

The result of the primary development are the following: (i) the robot's cost is low, thus it is affordable to all, (ii) it can be expanded with more sensors and/or actuators, (iii) it is open source and expandable in terms of software and hardware, (iv) it can be customized to be used in STEAM educational activities promoting the "A" dimension (i.e. the Arts in STEAM), (v) it can be used without specific software (no need for program/app installation) or hardware (i.e. no demand for modern hardware specifications), and (vi) it does not require Internet access for its programming. According to our initial statistics regarding robots' testing and evaluation, gender does not play a role in properly using and controlling the robot. However, the field of study shows that scientific background facilitates programming skills, necessary to control the robot. In future work, we intend to test our robot's usage according to the other TAM model factors.

Acknowledgements The authors would like to thank all the research participants and the University of West Attica for the financial support provided to them to undertake this research project.

References

1. Papadakis S (2020) Robots and robotics kits for early childhood and first school age. Int J Interact Mob Technol 14:34–56. https://doi.org/10.3991/ijim.v14i18.16631
2. Chatzopoulos A, Papoutsidakis M, Kalogiannakis M, Psycharis S, Papachristos D (2020) Measuring the impact on student's computational thinking skills through STEM and educational robotics projects implementation. In: Kalogiannakis M, Papadakis SJ (eds) Handbook of research on tools for teaching computational thinking in P-12 education. IGI Global, pp 234–284
3. Papadakis S (2018) The use of computer games in classroom environment. Int J Teach Case Stud 9:1. https://doi.org/10.1504/ijtcs.2018.10011113
4. Chatzopoulos A, Papoutsidakis M, Kalogiannakis M, Psycharis S (2019) Action research implementation in developing an open source and low cost robotic platform for STEM education. Int J Comput Appl 178:33–46
5. Papadakis S, Vaiopoulou J, Sifaki E, Stamovlasis D, Kalogiannakis M, Vassilakis K (2021) Factors that hinder in-service teachers from incorporating educational robotics into their daily or future teaching practice. In: Csapó B, Uhomoibhi J (eds) CSEDU 2021—13th international conference on computer supported education computational. SCITEPRESS, pp 12–26
6. Daniela L, Lytras MD (2018) Educational robotics for inclusive education. Technol Knowl Learn. https://doi.org/10.1007/s10758-018-9397-5
7. Tzagkaraki E, Papadakis S, Kalogiannakis M (2021) Exploring the use of educational robotics in primary school and its possible place in the curricula. Springer International Publishing

8. Kalogiannakis M, Papadakis S, Dorouka P (2020) Tablets and apps for promoting robotics, mathematics, STEM education and literacy in early childhood education. Int J Mob Learn Organ 14:255. https://doi.org/10.1504/ijmlo.2020.10026334
9. Papadakis S, Vaiopoulou J, Sifaki E, Stamovlasis D, Kalogiannakis M (2021) Attitudes towards the use of educational robotics: exploring pre-service and in-service early childhood teacher profiles. Educ Sci 11:1–5. https://doi.org/10.3390/educsci11050204
10. Papert S (1980) Mindstorms: children, computers, and powerful ideas. Basic Books Inc., New York
11. Kalogiannakis M, Tzagaraki E, Papadakis S (2021) A systematic review of the use of BBC micro: bit in primary school. In: Pixel (ed.) 10th international conference new perspectives in science education. Filodiritto Publisher, Bologna
12. Melkonian S, Chatzopoulos A, Papoutsidakis M, Piromalis D (2018) Remote control via android for a small vehicle ' s 2-wheels balancing. J Multidiscip Eng Sci Technol 5:8964–8967
13. Mavrovounioti V, Chatzopoulos A, Papoutsidakis M, Piromalis D (2018) Implementation of an 2-wheel educational platform for STEM applications. J Multidiscip Eng Sci Technol 5:8944–8948
14. Xatzopoulos A, Papoutsidakis M, Chamilothoris G (2013) Mobile robotic platforms as educational tools in mechatronics engineering. In: International scientific conference eRA–8. Pireaus, pp 41–51
15. Vordos G, Chatzopoulos A, Papoutsidakis M, Piromalis D (2018) Balance control of a small scale sphere with an innovative android application. J Multidiscip Eng Sci Technol 5:8957–8963
16. Papoutsidakis M, Chatzopoulos A, Drosos C, Kalovrektis K (2018) An arduino family controller and its interactions via an intelligent interface. Int J Comput Appl 179:5–8
17. Papoutsidakis M, Chatzopoulos A, Kalovrektis K, Drosos C (2017) A brief guide for the continuously evolving μ controller raspberry PI mod. B Int J Comput Appl 176:30–33
18. Papoutsidakis M, Kalovrektis K, Drosos C, Stamoulis G (2017) Design of an autonomous robotic vehicle for area mapping and remote monitoring. Int J Comput Appl 167:36–41. https://doi.org/10.5120/ijca2017914496
19. Papoutsidakis M, Tanwar R, Chatzopoulos A, Tseles D (2017) Custom made embedded automation systems for smart homes—part 2: the implementation. Int J Eng Appl Sci Technol 2:16–19
20. Papoutsidakis M, Chatzopoulos A, Piromalis D (2019) Distance control of water temperature via android devices. J Multidiscip Eng Sci Technol 6:11240–11244
21. Chatzopoulos A, Papoutsidakis M, Elza D, Papadakis S, Kalogiannakis M, Psycharis S (2021) DuBot: an open-source, low-cost robot for STEM and educational robotics. In: Papadakis S, Kalogiannakis M (eds) Handbook of research on using educational robotics to facilitate student learning. IGI Global, Hershey PA, USA, pp 441–465
22. Chatzopoulos A, Papoutsidakis M, Kalogiannakis M, Psycharis S (2020) Innovative robot for educational robotics and STEM. In: Kumar V, Troussas C (eds) 16th international conference on intelligent tutoring systems, ITS 2020. Springer, Cham, Athens, Greece, pp 95–104
23. Don Starkey: 3D Mouse by DDStarkey—Thingiverse, https://www.thingiverse.com/thing:61909
24. Elza D (2022) GitHub—dethe_bloc A fork of my blocklib code from architecture of open source applications 500 lines or less. https://github.com/dethe/bloc/
25. Elza D (2022) 500 Lines or less blockcode a visual programming toolkit. http://www.aosabook.org/en/500L/blockcode-a-visual-programming-toolkit.html
26. Davis FD (1989) Perceived usefulness, perceived ease of use, and user acceptance of information technology. MIS Q 13:319–340. https://doi.org/10.5962/bhl.title.33621
27. Davis FD, Bagozzi RP, Warshaw PR (1989) User acceptance of computer technology: a comparison of two theoretical models. Manage Sci 35:982–1003. https://doi.org/10.1287/mnsc.35.8.982

Assessing the Effects of Landmarks and Routes on Neuro-Cognitive Load Using Virtual Environment

Usman Alhaji Abdurrahman, Lirong Zheng, and Usman Haruna

Abstract The study aims to determine whether landmarks and routes influence navigational efficiency. In this study, 79 subjects participated in the experiments, and we evaluated their cognitive loads based on the generated psychophysiological measures and performance features from the driving system. The virtual reality system recorded the participant's heart rate, eye gaze, pupil size, as well as the driving performance metrics. The participants were presented with different landmarks (sufficient and insufficient landmarks) and routes (easy and difficult routes) to help them reach their respective destinations. An analytic strategy method was employed to measure neuro-cognitive load for user classifications. The participants were divided into two groups, each group having two sessions. Each session had either sufficient landmarks or insufficient landmarks. The results showed that insufficient landmarks and difficult routes elicited an increase in heart rate and pupil size, which caused the participants to commit more mistakes. It also showed that easy routes with sufficient landmarks achieved higher-navigation efficiency. These results would help improve the use of landmarks and the design of the driving routes. It could also be used to analyze traffic safety by utilizing the driver's cognition and performance.

Keywords Cognitive load · Driving simulator · Physiological measures · Navigation efficiency · Virtual reality

U. A. Abdurrahman (✉) · L. Zheng
School of Information Science and Technology, Fudan University, Shanghai 200433, China
e-mail: aausman18@fudan.edu.cn

L. Zheng
e-mail: lrzheng@fudan.edu.cn

U. Haruna
Department of Computer Science, University of Terengganu, Terengganu, Malaysia
e-mail: usmancyz01@gmail.com

U. A. Abdurrahman · U. Haruna
Department of Computer Science, Yusuf Maitama Sule University, Kano, Nigeria

X.-S. Yang et al. (eds.), *Proceedings of Seventh International Congress on Information and Communication Technology*, Lecture Notes in Networks and Systems 447,
https://doi.org/10.1007/978-981-19-1607-6_57

1 Introduction

One of people's daily routines is navigating from one place to another. This navigation can be from work to home, from friends to particular shops, and so on [1]. While navigating, people tend to make sequences of decisions, like going straight, turning left, or turning right; this is called route knowledge [2]. Apart from route knowledge, other essential information such as metrics and features are needed for effective navigation. Metrics information such as turn angles and distances is acquired from sensory sources, while features such as landmarks are mostly perceived visually [3]. It was discovered that, in the course of route finding, people tend to make on average one error per week, with 49% of those errors occurring when people turn in the wrong direction [4]. Metric and feature information are needed to reduce the number of mistakes during navigation, as they play a vital role in enriching people's knowledge.

Feature information, such as landmarks, affects our ability to navigate effectively, as they provide positional and orientation information. Landmarks show spatial information of natural objects and, at the same time, reduce user's cognitive loads and promote user's navigational efficiency [5]. For successful wayfinding, accurate integration and memory of landmarks are also required [6]. Research shows that environmental landmarks reduce the number of mistakes participants commit when traversing a route. The landmarks help participants apply the knowledge acquired while traversing a route from one direction when returning to the other [7].

In the current study, participants would use landmarks in locating their respective destinations even though landmarks can be any specific point used for navigation, like intersections, buildings, streets, and so on [6, 8, 9]. In this work, landmarks such as a McDonald's, a convenience store, a gas station, a basketball court, a post office, a church, a Walmart were used as the primary source of landmarks. We will refer to these locations as landmarks in the subsequent sections. Therefore, we hypothesized that sufficient landmarks would reduce the number of mistakes participants commit during the navigational exercise.

One of the targets in this study is the measurement of navigational efficiency. Navigational efficiency is the time needed by participants to complete the assigned task. In this study, participants were given the task of recognizing routes based on the landmarks to reach the destination. The navigational efficiency was determined by analyzing the recorded data of the participants under the driving performance metrics. Consequently, we hypothesized that the navigational efficiency of routes with sufficient landmarks would be significantly higher than that of routes with insufficient landmarks.

Moreover, complex routes affect drivers as they have to process more information, increasing their cognitive driving workload. According to senders [10], the workload is a measure of effort dissipated by a human operator while performing a task, regardless of the performance of the task itself. When workload levels are low, performance is also low because of inattention missed information. As the workload increases, the level of performance increases as well up to a maximum level. This maximum performance represents the optimal workload level for a given task. An additional mental workload leads to an abrupt decrease in performance because of the extra amount of information to be processed, resulting in a high cognitive workload [11]. According to Sweller [12], the cognitive workload is the total amount of cognitive resources needed for processing information in cognitive activities [12]. The cognitive workload is characterized by psychophysiological changes such as alterations in heart rate, skin conductance, behavioral approach, or avoidance. It involves several subcomponents occurring in frontal subcortical circuits [13, 14]. We, therefore, hypothesized that easy and difficult routes used in our study would elicit different cognitive workloads. Also, the high cognitive load would negatively affect the participants applying the knowledge acquired at the beginning of the exercise.

Several studies have investigated the measurement of driving behavior using different psychophysiological parameters [15–18]. Pupil size is well known to respond rapidly to changes in brightness in the visual field and has been used to measure the cognitive load while performing an assigned task [15, 18]. Research shows that heart rate increases when a participant is subjected to more challenging conditions [19–22]. An increase in respiratory rate has been constantly related to the rise in cognitive demand [22–24]. An increase in skin conductance leads to an increase in cognitive workload [17]. Likewise, EEG signals are highly sensitive and reliable for cognitive load measurements [25, 26]. Thus, in this study, pupil size and heart rate were used to measure the cognitive workload of the participants during the navigation exercise.

Virtual reality (VR) technology was used to achieve the stated objectives. The VR technology can create immersive and realistic interactive environments for behavioral learning. Besides, VR technology provides individualized treatment, accurate control of complex stimuli and a structured and safe learning environment [27, 28]. For that reason, we employed a VR-based driving system to investigate the effects of landmarks and routes on navigational efficiency.

2 Methods and Experiments

2.1 Subjects

A total of seventy-nine (79) undergraduates, 36 males and 43 females, participated in the experiment. They were Fudan University students with no real-life driving experience. The participants were grouped into two (group X and group Y); each group had easy and difficult routes, sufficient and insufficient landmarks. Group X had 39 participants (18 males and 21 females), and group Y had 40 participants (18 males and 22 females). All the subjects filled out a consent form, and the university granted approval. The generated data were evaluated and reported in this manuscript.

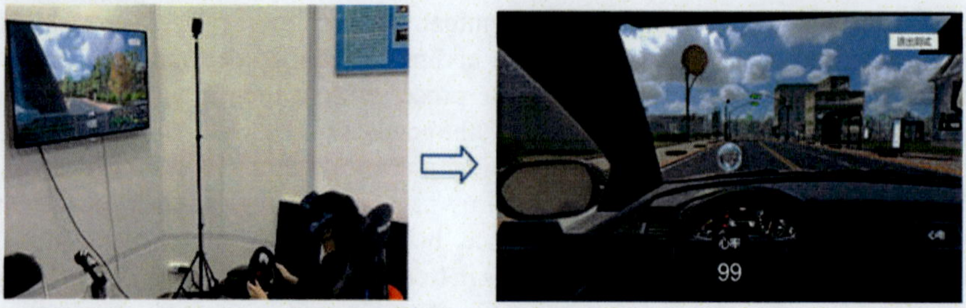

Fig. 1 VR driving system and driving simulator

2.2 Experimental Apparatus

The apparatus used in this study include VIVE Pro Eye for tracking eye data [29] and Logitech G27 steering-wheel controller for controlling the virtual agent vehicle in the driving environment. An Autodesk Maya [30] and Esri CityEngine [31] were used in designing landmarks, intersections, buildings, traffic lights, routes, and cars, while a Unity3D [32] was employed to develop the game platform. The driving platform consisted of city roadways that comprised of straightaways, intersections, several turns, and landmarks (such as a convenience store, a gas station, a basketball court, a McDonald's, a post office, a Walmart, and a church). The driving system (as shown in Fig. 1) kept track of the different psychophysiological measures as well as the driving performance features of the participants. The eye gaze data were recorded at 50 Hz. A heartbeat recording device designed in our lab was used to track the participant's heart rate at 500 Hz while driving throughout the virtual environment.

2.3 Experimental Design and Procedure

The study's main goal is to assess the effects of different landmarks and routes on cognitive workload. There were two (2) groups (group X and group Y) in the study, with each group having two (2) sessions: group X1 (sufficient landmarks and easy routes), group X2 (insufficient landmarks and easy routes), group Y1 (sufficient landmarks and difficult routes), and group Y2 (insufficient landmarks and difficult routes). The two sessions of each group were of the same difficulty levels; however, the number of landmarks varies. Group X1 had seven landmarks, group X2 had three landmarks, group Y1 had seven landmarks, and group Y2 had five landmarks. Moreover, the easy routes had three turns and intersections and three traffic lights, while the difficult routes had five turns and intersections and five traffic lights. Each participant completed two (2) sessions of the assigned group (either X1 and X2 or Y1 and Y2).

Before the commencement of the experiments, the baseline data were collected. Then, a video tutorial was given to the participants to introduce them to the routes they would follow and the landmarks, intersections, and traffic lights they would see to help them complete the given task. The video was played only once, and the participants were asked to memorize all the landmarks along the routes. The instructors then set up the eye-tracking and heartbeat sensors on the participant's bodies and recorded them correctly. After that, each participant was then asked to carry out the assigned task based on the video tutorial they had just watched. During the driving, participants would be using the landmarks to determine the right turn to take to reach their destinations. Thus, whenever a subject made a wrong turn, they were dragged to the starting point to re-watch the video tutorial and start driving again. The performance of the tutorial sessions was not considered in the data analysis; they were just included to help improve the subject's driving performance.

2.4 Data Collection

The recorded data by the eye tracker were preprocessed to remove unwanted data and reduce noise using the median value method [33]. Different data were also extracted for the driving performance features. The performance features used in this study and their meaning are given in Table 1.

3 Results

This study aims to determine the effects of landmarks and routes on navigational efficiency using a VR-based driving system. There were 39 data for group X1 (easy routes and sufficient landmarks) and X2 (easy routes and insufficient landmarks), and 40 data for group Y1 (difficult routes and sufficient landmarks) and Y2 (difficult routes and insufficient landmarks), making the total number of 79 data. A statistical method was employed in analyzing the data.

Table 1 Performance features and their meaning

Performance feature	Meaning
Completion time	Task completion time (in seconds)
Wrong turns	Total number of wrong turns
Barricade counts	Total number of collisions on the edge of the road

3.1 Data Analysis

A t-test method was employed to determine the influence of landmarks and routes on cognitive load. This analysis was carried out as follows:

Assessment of the Spatial Cognitive Load. The cognitive workload of the subjects was assessed using the generated psychophysiological responses and the driving performance data. The responses obtained are presented here.

Psychophysiological Measures versus Cognitive Load. A t-test method was employed for comparison. The heart rate results (in bpm) show that group X using sufficient landmarks was significantly different from group Y using insufficient landmarks ($t = -230.10$, $p = 1.89\text{E}{-26}$). The heart rate of group Y ($M = 76.2$, SD $= 10.1$) was significantly higher than that of group X. Additionally, the heart rate of group X1 was significantly different from that of group X2 ($t = -147.65$, $p = 1.34\text{E}{-14}$), and the heart rate of group X2 ($M = 74.3$, SD $= 8.1$) was significantly higher than that of group X1. Likewise, the heart rate of group Y1 was significantly different from that of Y2 ($t = -158.32$, $p = 1.45\text{E}{-16}$), and the heart rate of group Y2 ($M = 75.9$, SD $= 9.4$) was significantly higher than that of group Y1.

Furthermore, the results of the pupil size (in mm) show that group X using sufficient landmarks was significantly different from group Y using insufficient landmarks ($t = -19.49$, $p = 4.56\text{E}{-13}$). The pupil size of group Y ($M = 5.4$, SD $= 0.71$) was significantly larger than that of group X. Additionally, the pupil size of group X1 was significantly different from that of group X2 ($t = -17.20$, $p = 3.7\text{E}{-12}$), and the pupil size of group X2 (M $= 4.5$, SD $= 0.3$) was significantly larger than that of group X1. Likewise, the pupil size of group Y1 was significantly different from that of group Y2 ($t = -17.53$, $p = 4.8\text{E}{-13}$), and the pupil size of Y2 (M $= 5.4$, SD $= 0.71$) was significantly larger than that of group Y1.

The results obtained from all the groups were associated with the nature of the routes and insufficient use of landmarks.

Driving Performance Features versus Cognitive Load. Similar to psychophysiological measures, the results of driving performance measures (Table 2) show a significant difference between group X and group Y using the t-test method ($t = 9.63$, $p = 2.3\text{E}{-21}$). Likewise, the performance measures of group X1 were significantly different from that of group X2 ($t = 6.41$, $p = 1.9\text{E}{-18}$), and the performance measures of group Y1 were significantly different from that of group Y2 ($t = 7.32$, $p = 2.0\text{E}{-19}$). As shown in Table 2, the performance features of participants who

Table 2 Driving performance features

Performance feature	Group			
	X1	X2	Y1	Y2
Completion time	203.13	250.34	289.37	347.85
Wrong turns	0.24	0.32	0.54	0.98
Barricade counts	0.25	0.36	0.58	0.95

Table 3 Completion time

Level	Group	M	SD	t	p
Sufficient landmarks	X1	203.13	22.48	−90.83	5.16E−12
Insufficient landmarks	X2	250.34	42.05		
Sufficient landmarks	Y1	289.37	32.54	−97.47	7.10E−13
Insufficient landmarks	Y2	347.85	35.13		

Note M = mean; SD = standard deviation; t = inferential statistic; p is the probability of obtaining test results

drove the difficult routes with insufficient landmarks were significantly higher than those who drove easy routes and sufficient landmarks.

Moreover, as hypothesized, participants who drove the difficult routes with insufficient landmarks committed more mistakes than their easy routes counterparts. The cognitive workload significantly differed in responding between difficult and easy routes regarding performance features ($t = 5.54, p = 0.001$), leading to a significant interaction between the two groups.

Navigational Efficiency versus Cognitive Load. This study has focused on completion time regarding navigational efficiency and cognitive workload. Completion time is an essential indicator for the cognitive workload and plays a vital role in practical applications. A higher completion time means that a participant has to use more mental capacity to identify the right way to follow, resulting in a more significant workload.

The navigational efficiency assessment can be obtained by analyzing the completion time (as given in Table 3) and observing whether intergroup differences exist due to different landmarks and routes. The two-sample t-test method was applied for the comparison. The results show that the completion time of group X1 was significantly different from group X2 ($t = -90.83, p = 5.16\text{E}{-}12$), and the completion time of group X2 using insufficient landmarks ($M = 250.34$, SD = 42.05) was significantly higher than that of group X1 using sufficient landmarks ($M = 203.13$, SD = 22.48). Likewise, the completion time of group Y1 was significantly different from group Y2 ($t = -97.47, p = 7.10\text{E}{-}13$). The completion time of group Y2 using insufficient landmarks ($M = 347.85$, SD = 35.13) was significantly higher than that of group Y1 using sufficient landmarks ($M = 289.37$, SD = 32.54).

Influence of Landmarks and Routes on Gender. The use of insufficient landmarks and difficult routes influenced gender. The experimental findings show that female participants generally drove slower than their male counterparts. This resulted in spending much time completing the exercise due to cognitive load. Moreover, the mean change applied on cognitive load, and the completion time has revealed the degree of these effects. As given in Table 4, the mean change of landmarks for females (ΔM = 142.70) was more significant than that of males (ΔM = 112.29). This shows that insufficient landmarks have more influence on females than males.

Additionally, the mean change of difficult routes for males (ΔM = 135.19) was scarcely different from that of females (ΔM = 133.22). This shows the effect

Table 4 Task completion mean

Source	Gender	Mean sufficient landmarks	Mean insufficient landmarks	ΔM	Mean easy routes	Mean difficult routes	ΔM
Completion time	Male	194.13	306.42	112.29	189.34	324.53	135.19
	Female	207.25	349.95	142.70	210.32	343.54	133.22

of landmarks on navigational efficiency is the same for both genders (males and females).

4 Discussion

The results show that landmarks (sufficient landmarks, insufficient landmarks) and routes (easy routes, difficult routes) affect navigational efficiency. This section discusses the results obtained in this study.

4.1 Effect of Landmarks on Driving Workloads

We hypothesized that the use of insufficient landmarks would elicit an increase in psychophysiological activation, such as increased pupil size, eye gaze, and heart rate. This can be seen from the results (in Table 2) where the participants had an increase in their psychophysiological activations which resulted in participants committing more driving errors. Sufficient landmarks have reduced the number of mistakes participants commit during the navigational exercise. Previous studies have supported these findings [7, 34]. Landmarks primarily facilitate traveling between specific places instead of assisting the overall layout of a space. These landmarks provided positional information, which helped address the participant's most common mistake of going straight instead of turning left or right [7].

Furthermore, we hypothesized that the navigational efficiency of routes with sufficient landmarks would be significantly higher than that of routes with insufficient landmarks. This has also been achieved considering the results obtained regarding all the group's completion time. The previous studies supported the results [5, 7, 35]. The results show that the completion time of groups X2 and Y2 is significantly higher than that of groups X1 and Y1, respectively. This is because the participants required more cognitive resources to search for the proper routes to follow to reach the destination.

4.2 Effect of Workloads on Psychophysiological Metrics

As expected, navigations using insufficient landmarks increased in psychophysiological activation, such as heart rate and pupil size. The increased workload was related to the conditions of the cognitively challenging task (searching for the right turn due to insufficient landmarks and difficult routes), which led to an increase in the response level.

The results obtained for the pupil size have been supported by several findings as indicated in [17, 36, 37]. According to Kahneman et al. [38], pupillary diameter increases as the amount of information loaded into working memory increases. In addition, several articles have supported our findings on heart rate [39, 40]. Verway et al. [41] tested participants to cognitive tasks compared to a control task where no additional cognitive task was given while driving. The results obtained showed decreased heart rate variability and increased heart rate (decreased IBIs) when performing the cognitive tasks.

4.3 Effect of Workloads on Driving Performance

In this study, driving features like completion time, wrong turns, and barricade counts were considered and investigated for the driving performance. These parameters were quantified and measured under the different cognitive workloads and assessed the participant's driving performance. Group X2 and Y2 participants spent more time reaching the destination than their group X1 and Y1 counterparts for the completion time. This happened due to the additional time needed by the participants in searching for the right turn/direction to reach the destinations. Our results are consistent with Fan H. et al. [5]. The navigational task completion time of the group using the full-landmark map was significantly higher than that of the group using the key-landmark map.

The wrong turns and barricade counts were also higher in groups X2 and Y2 than in groups X1 and Y1 for the same reasons. The cognitively challenging task of searching for the right turn using insufficient landmarks was tedious enough, especially on the difficult routes. This added more information into the participant's working memory, which required more cognitive resources to process them. This result is in line with previous research conducted by Lyu et al. [42], in which the speed maintenance and lane deviations were significantly different under different levels of cognitive workload.

4.4 *Effect of Psychophysiological Metrics and Driving Performance on Gender*

According to Underwood et al. [43], the cognitive workload is higher for novice drivers than skilled drivers, as novice drivers need to pay much attention while driving. However, only novice drivers were considered for the experiments in our case. As shown from the results, insufficient landmarks and difficult routes influenced gender. The female participants paid more attention to the landmarks and drove slower than their male counterparts. Thus, they made more mistakes (higher wrong turns and barricade counts) during the driving. Moreover, female participants had higher completion times than male counterparts [15, 42].

5 Conclusions

This study investigated the effects of landmarks and routes on navigational efficiency using a novel VR-based driving system. All the stated objectives of the study have been achieved by analyzing both the psychophysiological and performance features of the subjects. The statistical method was used for this analysis. Insufficient landmarks and difficult routes were used to trigger the cognitive workload in the participants. The participants were expected to use these landmarks to locate the right directions to reach the pre-assigned destinations based on the knowledge acquired at the beginning of the exercise.

These results would help improve the use of landmarks and the design of the driving routes. Additionally, it could be used to analyze traffic safety using driver's cognition and performance.

References

1. Stokes G, Karen L (2011) National travel survey analysis. Transport Studies Unit, School of Geography and the Environment
2. Siegel AW, White SH (1975) The development of spatial representations of large-scale environments. Adv Child Developm Behav 10:9–55(10)
3. Montello DR (1998) A new framework for understanding the acquisition of spatial knowledge in large-scale environments. Spatial Temporal Reasoning Geographic Inform Syst 143–154
4. Williamson J, Barrow C (1994) Errors in everyday route-finding: a classification of types and possible causes. Appl Cogn Psychol 8(5):513–524
5. Fang H, Xin S, Zhang Y, Wang Z, Zhu J (2020) Assessing the influence of landmarks and paths on the navigational efficiency and the cognitive load of indoor maps. ISPRS Int J Geo-Inform 9(2):82
6. Lee PU, Barbara T (2005) Interplay between visual and spatial: the effect of landmark descriptions on comprehension of route/survey spatial descriptions. Spat Cogn Comput 5(2–3):163–185

7. Waller D, Yvonne L (2007) Landmarks as beacons and associative cues: their role in route learning. Memory Cognition 35(5):910–924
8. Ruddle RA et al (2011) The effect of landmark and body-based sensory information on route knowledge. Mem Cognit 39(4):686–699
9. Denis M (1997) The description of routes: a cognitive approach to the production of spatial discourse. Current Psycol Cogn 16:409–458
10. Raubal M, Winter S (2002) Enriching wayfinding instructions with local landmarks. In: International conference on geographic information science, Springer, Berlin, Heidelberg
11. Senders JW (1970) The estimation of operator workload in complex systems. Syst Psychol 207–216
12. Cafiso S, La Cava G (2009) Driving performance, alignment consistency, and road safety: real-world experiment. Transport Res Record 2102(1):1–8
13. Sweller J (1999) Instructional design. Australian educational review
14. Bonelli RM, Cummings JL (2007) Frontal-subcortical circuitry and behavior. Dialogues Clin Neurosci 9(2):141
15. Ray RD, Zald DH (2012) Anatomical insights into the interaction of emotion and cognition in the prefrontal cortex. Neurosci Biobehav Rev 36(1):479–501
16. Abdurrahman UA, Yeh SC, Wong Y, Wei L (2021) Effects of neuro-cognitive load on learning transfer using a virtual reality-based driving system. Big Data and Cognitive Comput 5(4):54
17. Zhang L et al (2017) Cognitive load measurement in a virtual reality-based driving system for autism intervention. IEEE Trans Affect Comput 8(2):176–189
18. Parsons TD, Courtney CG (2016) Interactions between threat and executive control in a virtual reality stroop task. IEEE Trans Affect Comput 9(1):66–75
19. Pomplun M, Sunkara S (2003) Pupil dilation as an indicator of cognitive workload in human-computer interaction. In: Proceedings of the international conference on HCI
20. Jorna and Peter GAM (1992) Spectral analysis of heart rate and psychological state: a review of its validity as a workload index. Biol Psychol 34(2–3):237–257
21. Charlton SG, O'Brien TG (2019) In: Handbook of human factors testing and evaluation. CRC Press
22. Lenneman JK, Shelly JR, Backs RW (2005) Deciphering psychological-physiological mappings while driving and performing a secondary memory task
23. Back RW, Seljos KA (1994) Metabolic and cardiorespiratory measures of mental effort: the effects of level of difficulty in a working memory task. Int J Psychophysiol 16(1):57–68
24. Brookhuis KA, De Waard D (2001) Assessment of drivers'workload: performance and subjective and physiological indexes. Stress Workload Fatigue
25. Mehler B, Reimer B, Coughlin JF, Dusek JA(2009) Impact of incremental increases in cognitive workload on physiological arousal and performance in young adult drivers. Transp Res Record 2138(1):6–12
26. Brookings JB, Wilson GF, Swain CR (1996) Psychophysiological responses to changes in workload during simulated air traffic control. Biol Psychol 42(3):361–377
27. Gevins A, Smith ME (2000) Neurophysiological measures of working memory and individual differences in cognitive ability and cognitive style. Cerebral Corex 10(9):829–839
28. Gevins A et al (1998) Monitoring working memory load during computer-based tasks with EEG pattern recognition methods. Human Factor 40(1):79–91
29. Strickland D (1997) Virtual reality for the treatment of autism. Stud Health Technol Inform 44:81–86
30. Armougum A et al (2019) Virtual reality: a new method to investigate cognitive load during navigation. J Environ Psychol 65:101338
31. VIVE VIVE (2021) https://www.vive.com/uk/product/vive-pro-eye/overview/. Last Accessed 12 Nov 2021
32. Autodesk, Autodesk (2021) https://www.autodesk.com/. Last Accessed 12 Nov 2021
33. Esri ArcGIS CityEngine (2021) https://www.esri.com/en-us/arcgis/products/arcgis-cityengine/. Last Accessed 12 Nov 2021
34. Unity3D Unity. www.unity3d.com. Last Accessed 12 Nov 2021

35. Komogortsev OV et al (2010) Standardization of automated analyses of oculomotor fixation and saccadic behaviors. IEEE Trans Biomed Eng 57(11):2635–2645
36. Jansen-Osmann P, Fuchs P (2006) Wayfinding behavior and spatial knowledge of adults and children in a virtual environment: the role of landmarks. Exp Psychol 53(3):171–181
37. Ishikawa T, Montello DR (2006) Spatial knowledge acquisition from direct experience in the environment: Individual differences in the development of metric knowledge and the integration of separately learned places. Cogn Psychol 52(2):93–129
38. Querino E et al (2015) Cognitive effort and pupil dilation in controlled and automatic processes. Translational Neurosci 6(1):168–173
39. van der Wel P, Steenbergen HV (2018) Pupil dilation as an index of effort in cognitive control tasks: a review. Psychon Bull Rev 25(6):2005–2015
40. Kahneman D, Beatty J (1966) Pupil diameter and load on memory. Science 154(3756):1583–1585
41. Goldstein DS et al (2011) LF power of heart rate variability is not a measure of cardiac sympathetic tone but maybe a measure of modulation of cardiac autonomic outflows by baroreflexes. Experim Physiol 96:1255–1261
42. Solhjoo S et al (2019) Heart rate and heart rate variability correlate with clinical reasoning performance and self-reported measures of cognitive load. Sci Reports 9(1):1–9
43. Verwey WB, Veltman HA (1996) Detecting short periods of elevated workload: a comparison of nine workload assessment techniques. J Exp Psychol Appl 2(3):270

On the Transposition of Translator Functions in a Digital Communicative Environment

Lyudmila Kushnina, Elena Alikina, Irina Perlova, Kristina Permiakova, and Marina Khudaiberdina

Abstract The article is based on the idea that in the conditions of a digital communicative environment, there are some changes in the structure of the human-translator activity and their functions, some of which are transferred to machine translation. It allows us to talk about the transposition of the translator linguistic persona's functions. The research aims at finding ways to integrate translation activities into a new communicative environment associated with digital platforms, with the development of new linguistic technologies in the process of transcultural communication. The methodological basis of the research is the concept of translation space based on a synergetic approach to translation, which assumes the harmonization of the source and target meanings, i.e., their coordination and proportionality. The result of the study is understanding the correlation between the subject-centric and text-centric functions of a human-translator developing their reflexive, empathic, and other cognitive abilities, which cannot be delegated to machine translation. Therefore, we observe the transposition process of the translator language persona's functions when the translator retains all subject-centric functions, and part of the text-centric functions are performed by the translation software. At this time, the social and cultural responsibility of the translator increases since the translation is assigned by the human and performed for the human.

Keywords Transposition of translator functions · Harmonious translation · Digital communicative environment

1 Introduction

Modern translation studies, like many other humanities and social sciences, are entering a new scientific paradigm, responding to the challenges of scientific and technological progress—the emergence and intensive development of the digital communication environment and digital translation [1–4].

L. Kushnina · E. Alikina · I. Perlova (✉) · K. Permiakova · M. Khudaiberdina
Perm National Research Polytechnic University, Komsomolsky Prospekt, 29, 614990 Perm, Russia
e-mail: perl_telecom@mail.ru

X.-S. Yang et al. (eds.), *Proceedings of Seventh International Congress on Information and Communication Technology*, Lecture Notes in Networks and Systems 447,
https://doi.org/10.1007/978-981-19-1607-6_58

According to Garbovsky, digital translation science is a new kind of translation as a system of interaction between human-translator cognitive-communicative activity and digital information and communication means [5, p. 10]. As a result of a new translation reality introduction, the translation paradigm of the digital communicative medium is formed which is characterized by reformatting the relationship between the linguistic persona of the translator and translation software. It leads to the shift in some traditional translator's functions on the one hand, and on the other hand, to the emergence of new ones.

In our paper, we hypothesize that the translation process in the era of digital translation science acquires new functions, in other words, there are transposition transformations. The ideas of comparing the translation with transposition were put forward by scholars before. It was possible to talk about transposition at the language level. Translation as transposition is now at a new level, where, along with language and the natural intelligence of the man, artificial intelligence is emerging, which, on the one hand, is created by the man himself, on the other hand, to which the man must adapt. Researchers continue to explore translation procedures in the view of the functions the human-translator will continue to perform [6], but they also need to know how existing functions will change, what new ones may emerge, what cognitive properties of the persona, and what cognitive processes will develop.

In the framework of this article, we will focus our attention on the problems of the human-translator, assuming that the dialogue "human-translator" and "machine translation" has started, and every new replica in this dialogue will become a new stage to understand the process.

Based on the idea of text centrism and subjectivity, which determine our understanding of the translation process, we believe that in the new communicative-digital reality, text-centric functions are more likely to pass to the machine, and subject-centric functions continue to be performed by a person, but their content is transformed. The essence of the transformation is that machine translation mastery will constantly improve, reducing the share of human translation skills. At the same time, we are convinced that this share will never disappear completely, and the problem of the translator's linguistic persona will also be relevant. The extent of these changes remains to be explored. For example, there are already studies showing that the combined activities of a human and a machine require extra effort from the human to pre-edit the source text and post-edit the translation.

Our study focuses on the anthropocentric and sociocultural aspects of this interaction. We believe that the sociocultural responsibility of the translator's linguistic persona extends to all aspects of their activity: visible and invisible, conscious and unconscious, stereotypical and creative, digital and non-digital. In other words, the activity of a human-translator and machine translation is ultimately judged by a human and performed for a human. It confirms the principle of anthropocentrism in translation science and the social responsibility of the translator's linguistic persona, which is aimed at obtaining a high-quality product and successful and harmonious interaction of individuals, languages, and cultures. In this interaction, the priority of a human-translator is based on their sociocultural function.

The emergence of modern linguo-technologies does not mean that the cognitive and cultural mission of the translator, aimed at the transfer of knowledge in the intercultural discursive digital space, is significantly changing or disappearing. We should obviously expect changes in the content and structure of the translator's linguistic persona activity, in the emergence of a new type of specialists, such as a digital linguist, digital translator, translator-editor, translator-technical writer. It only means that translation science is moving to a new level of its development, and we are at its starting point [7].

2 Problem Statement and Research Goals

In a broad sense, the goal of the research is to integrate translation activity into a digital communicative environment represented by a network of translators. It involves coordinating the translation discourse with existing digital platforms and developing new linguo-technologies. In a narrow sense, our goal is to identify the human-translator functions associated with transposition change that involves the cognitive abilities of the linguistic persona, such as reflection and empathy.

In the new communicative and social reality, with the inevitable expansion of the machine translation capabilities, there is a reorganization of the human-translator activity.

3 Literature Review

Let us consider the results of the research of domestic linguists and translation theorists, who lay the theoretical foundations of the new translation paradigm. We will call it the information-cognitive paradigm, where the information component implies the activity of artificial intelligence or machine translation, and the cognitive component implies the activity of natural intelligence or a human-translator.

The linguistic basis of our research is the works of Kotyurova and Sokolova on scientific communication and Chernyavskaya in the field of text theory and discourse [8, 9]. The problems of artificial intelligence are closely related to the problems of natural intelligence and natural human language, stated by Ryabtseva [10]. The search for ways to study artificial intelligence is also connected with the study of the persona linguistic consciousness and the translator language consciousness, presented in the works of Ufimtseva [11].

Linguo-personology focused on the study of a persona in the language, explores the linguistic persona of the translator following the models of linguistic personality presented by Karaulov, Bogin, Bushev, Krasnykh, Tarnaeva, etc. [12–16]. As Bushev emphasizes, a modern translator is influenced by the mutual influence of cultural identity and foreign culture, which requires the following competencies: analytical, creative, emotional, and the ability to recognize and produce texts in two contact languages [14]. Tarnaeva refers the translator to an elitist type of speech culture because they can create written or spoken text of any functional speech style [16]. According to our concept of translation space, the translator must have a harmonious translation world view. By generating a harmonious target text, they create a quality text, which is as unique and inimitable as the source.

What happens to the translator's linguistic persona in the new conditions?

Firstly, let us refer to the statements of Alekseeva and Mishlanova, who are studying the problem of knowledge transfer from the standpoint of cognitive translatology. They state that translation as a language activity is not a simple manipulation with the source text, not the replacement of one text by another, but a complex mental process based on such parameters as intertextuality and iconicity of a language sign. The target text, thus always turns out to be the process of creating a new text rather than recreating the original as it looks in terms of translation technique (text engineering) [17, p. 79].

The ideas of cognitive translatology were developed by European scientists who published a collective monograph "Cognitivisme et traductologie" edited by Gue Achard-Bayle et Christine Durieux in 2020. The authors of the monograph emphasize the following idea: "La traduction renvoie, au-delà des considérations linguistiques du passage entre langues-cultures, à une reflexion cruciale sur la nature de cet acte même, sur les fondements ontologiques et la nature de la réalité perçue et representée par la conscience" [18, p. 7] (The translation refers, beyond the linguistic considerations of the passage between languages and cultures, to a crucial reflection on the nature of this act itself, on the ontological foundations and the nature of reality perceived and represented by consciousness). We also base our translation reflections on the works of the modern French scientist Lederer [19].

4 Material and Research Methods

The concept of the translation space put forward by one of the article authors is the cornerstone of our research [20]. Its essence is as follows.

The translation is regarded as an interaction process of texts and discourses belonging to different linguo-cultures. The key notions of the translation process are meaning, meaning synergy, and meaning harmonization. This process can be represented as a field containing the core and the periphery [21]. The core of the translation space, its content is an invariant for the translator, there factual meaning is formed, and it is the only explicit meaning. Heterogeneous implicit meanings are formed in peripheral text fields (energy and phatic ones), as well as in the fields of translation communication subjects. But the translator seeks not the sum of the meanings of all its fields but their synergy, implying the augment of new meanings acceptable to recipients of the host culture. The synergetic effect for each linguo-culture will be different. Synergy effect is the production of a harmonious target text, i.e., a text, which meanings are proportional to those of the source text.

As a material for analysis, we used poetic texts presented in the bilingual Russian French edition of "Anthologie de la poésie Russe pour enfants" translated by Abril [22].

5 Research Results

As a result of the linguo-translation analysis of poetic texts, which translation was performed by a human, we have identified harmonious variants of translations that represent the emotive-empathic and reflexive abilities of the translator's linguistic persona and which are not peculiar to machine translation We assume that machine translation of texts is done quickly and efficiently, achieving an adequate or equivalent ratio between the source and its translation. But a harmonious translation that conveys not only the proportionality of meanings but their reinterpretation, increment, and unique "feeling" can only be expressed by a human-translator. Here is an example:

Кисточка Гордился рыжий львенок, Гордился рыжий львенок Своей пушистой кисточкой На кончике хвоста: - Наверно, я художник Конечно, я художник Раз у меня есть кисточка На кончике хвоста…	L'artiste Un lionceau était fier d'avoir en fait la queue au pinceau aux longs poils "C'est sans doute que Je suis un artiste Un peintre sans égal si j'en ai guise de queue ce pinceau aux longs poils [22], pp. 120–121]	Brush The red lion cub was proud, The red lion cub was proud, With his fluffy brush At the tip of the tail: - I guess I'm an artist Of course, I'm an artist Since I have a brush At the tip of the tail

We recognize this translation from Russian into French as harmonious and note the presence of dictionary inconsistencies when the translator achieves semantic correspondences. So, the title "Кисточка" ("Brush") is translated as "L'artiste" ("Artist"), which is quite reasonable, since it reflects the key idea of the text: having a fluffy brush on the tip of the tail, the lion cub imagines himself an artist. "Un peintre sans égal" ("выдающийся художник"). The target text does not contain the combination "рыжий львенок", ("red lion cub"), while the combination "Un peintre sans égal" ("outstanding artist") is added. But the translator manages to recapture the atmosphere, joyful mood, pride, and confidence of the lion cub imaging to be an artist. It can be expected that a small French-speaking reader will like and appreciate this text as if it was created in their native language.

If to assume that there can be machine translation of this poetic text, the content of the text or its factual explicit meaning would likely be expressed adequately, but the implicit meanings would remain non-verbalized. They are assets of a human-translator who can interact with machine translation.

As our numerous observations have shown, the functions of a human-translator will not disappear, and their reasonable transposition will occur during the interaction with machine translation.

Considering the following factors:

- Main provisions of linguo-personology related to analytism, creativity;
- Emotionality of the translator's linguistic persona;
- The ideas of cognitive translation studies about the mental component of the translation process focused on creating the target text, but not just on recreating the source text.

The concept of translation space about the harmonious worldview of the exclusive translating persona it is possible to draw general conclusions concerning the transposition of translation functions in light of the new paradigm:

1. A human-translator, as a unique linguistic persona, who needs to coordinate their actions with other network translators and with machine translation, retains functions corresponding to certain competencies, which cannot be delegated to other subjects of translation.
2. A human-translator forms, in their mind, a translation picture of the world, a harmonious translation outlook oriented toward the interaction of heterogeneous cultures and the understanding of recipients belonging to those cultures.
3. A human-translator has the properties of reflection, empathy, emotionality, creativity, analytism, etc., which allow them to create harmonious texts enriching other cultures.

6 Conclusion

To sum up, this paper shows our attempt to analyze the transposition of the human-translator functions, who should coordinate their actions with other network translators and with the machine translation. We did not study the issue of machine translation in the framework of this article. We were interested in the creation of translation space to achieve a harmonious translation discourse that can become an integral part of the target culture. A human-translator retains and develops cognitive abilities that cannot be delegated to machine translation. These functions reflect the subject-centricity of the translation process in its cognitive dimension, related to natural intelligence. As far as text-centricity of translation is concerned, this process tends toward artificial intelligence. It is the perspective of further research.

References

1. Picard RW (2010) Affective computing: from laughter to IEEE. IEEE Trans Affect Comput 1(1):11–17. https://doi.org/10.1109/T-AFFC.2010.10,lastaccessed2021/10/20
2. Broek EL (2012) Affective computing: a reverence for a century of research. In: Esposito A, Esposito AM, Vinciarelli A, Hoffmann R, Müller VC (eds) Cognitive behavioural systems. Lecture notes in computer science, vol. 7403. Springer, Berlin, Heidelberg https://doi.org/10.1007/978-3-642-34584-5_39. Accessed 18 Oct 2021

3. Alekseeva LM, Mishlanova SL (2020) Kommunikatsiya i menedzhment znaniya v gumanitarnykh naukakh [Communication and knowledge management in the humanities]. Publishing house of Perm national research polytechnic university, Perm (in Russian)
4. Yv Gambier (2016) Perevod i perevodovedeniye na perekrestke tsifrovykh tekhnologiy [Translation and translation science at the crossroads of digital technologies]. Philology oriental studies journalism bulletin of St. Petersburg state university 9:56–74 (in Russian)
5. Garbovsky NK, Kostikova OI (2019) Intellekt dlya perevoda: iskusnyy ili iskusstvenniy [Intellect for translation: skilled or artificial]. Bulletin of Moscow University Translation Theory 22(4):3–25 (in Russian)
6. Pym A (2018) A typology of translation solutions. In: The journal of specialised translation https://www.researchgate.net/publication/327619732. Accessed 21 Oct 2021
7. Kushnina LV, Zubkova EV, Pogorelaya NG (2020) Refleksiya perevodchika vs metaperevodcheskaya deyatelnost [Reflexion of the translator vs metatranslation activity]. World of Science Sociology Philology Culture 2(11) (in Russian)
8. Kotyurova MP, Sokolova NB (2017) Sovremennyy nauchnyy tekst skvoz prizmu diskursivnykh izmeneniy [Modern scientific text through the prism of discursive changes]. Publishing house of Perm national research university, Perm (in Russian)
9. Cherniavskaya, V.E.: Lingvistika teksta. Lingvistika diskursa [Linguistics of text. Linguistics of discourse]. Lenand, Moscow (2014). (in Russian)
10. Ryabtseva NK (2005) Yazyk i estestvennyy intellekt [Language and natural intelligence. Academia, Moscow (in Russian)
11. Ufimtseva NV (2017) Etnopsikholingvistika kak razdel teorii rechevoy deyatelnosti [Ethnopsycholingvistics as a section of the speech activity theory]. In: Bubnova IA, Zykova IV, Krasnyh VV, Ufimtseva NV (eds) Neopsycholinguistics and psycholinguculturology: new sciences about a speaking human. Gnosis, Moscow (in Russian), pp 21–96
12. Karaulov YN (2007) Russkiy yazyk i yazykovaya lichnost [Russian language and language persona]. LKI, Moscow (in Russian)
13. Bogin GI (2004) Tipologiya ponimaniya teksta // Obshchaya psikholingvistika [Typology of text understanding // General psycholinguistics]. Anthology. Labirint, Moscow (in Russian)
14. Bushev AB (2010) Russkaya yazykovaya lichnost professionalnogo perevodchika [Russian linguistic persona of the professional translator]. Moscow. (in Russian)
15. Krasnyh VV (2003) "Svoy" sredi "chuzhikh": mif ili realnost ["At home among strangers": myth or reality]. Gnosis, Moscow (in Russian)
16. Tarnaeva LP (2008) Kontseptsiya yazykovoy lichnosti v kontekste problem perevodovedeniya [Concept of linguistic persona in the context of translation theory problems]. Bulletin of Leningrad University 2(13):55–68 (in Russian)
17. Alekseeva LM, Mishlanova SL (2019) Transfer znaniia v gumanitarnykh naukakh [Transfer of knowledge in the humanities]. Publishing House of Perm national research polytechnic university, Perm (in Russian)
18. Cognitivisme et traductologie (2020) Approches sémantiques et psychologiques. Sous la direction de Guy Achard-Bayle et Christine Durieux. Classiques Garnier, Paris
19. Lederer M (1994) La traduction aujourd'hui. Hachette, Paris
20. Kushnina LV (2009) Teoriya garmonizatsii: opyt kognitivnogo analiza perevodcheskogo prostranstva [Harmonization theory: experience of cognitive analysis of translation space]. Publishing house of Perm national research polytechnic university, Perm (in Russian)
21. Kushnina L, Perlova I, Permiakova K (2021) Knowledge transfer in intercultural technical communication in view of translation synergetic paradigm. In: Yang XS, Sherratt S, Dey N, Joshi A (eds) Proceedings of sixth international congress on information and communication technology. Lecture notes in networks and systems, vol 235. Springer, Singapore (2021). https://doi.org/10.1007/978-981-16-2377-6_23
22. Anthologie de la poésie russe pour enfants. Traduction et choix de Henri Abril. Editions Circe (2006).

Factors Affecting Intelligent Enterprise Resource Planning System Migrations: The South African Customer's Perspective

Precious Mushayi and Thembekile Mayayise

Abstract Enterprise Resource Planning (ERP) systems are a strategic component of most organisations' information systems and have evolved to become intelligent ERPs. ERP migrations are often marked by huge costs which is a challenge for developing economies. It remains unclear what the determinants for ERP migrations by organisations are for developing economies in the era of digital transformation. The aim of this study was to identify the factors that influence the decisions of ERP customers in developing countries to adopt intelligent ERPs. The Technological-Organisational-Environmental (TOE) and the institutional theory frameworks were used as the foundation upon which these factors were studied. The study focused on adopters and non-adopters of SAP S/4 HANA within a South African context. Guided by the positivist paradigm, a sample of 95 South African based companies was selected where four employees were targeted per company to respond to an electronic survey. Data was collected using an online data collection tool called, QuestionPro. The data was analysed through the IBM SPSS data analysis tool. The findings revealed that ICT infrastructure, availability of cyber-security systems, mimetic forces, and normative pressures are the influencing factors for intelligent ERPs adoption. The contributions of this study are discussed in this paper.

Keywords ERP · Institutional theory · Technological-organisational-environmental framework

1 Introduction

The concepts of innovation and customer-centricity have pushed ERP vendors to reimagine and redesign how ERPs store, process, and extract data [1]. New and improved Intelligent ERP solutions have been launched, and customers are being

---

P. Mushayi (✉) · T. Mayayise
University of the Witwatersrand, 1 Jan Smuts Avenue, Johannesburg, South Africa
e-mail: prmushayi@gmail.com

T. Mayayise
e-mail: Thembekile.Mayayise@wits.ac.za

X.-S. Yang et al. (eds.), *Proceedings of Seventh International Congress on Information and Communication Technology*, Lecture Notes in Networks and Systems 447,
https://doi.org/10.1007/978-981-19-1607-6_59

encouraged to adopt these next-generation ERP solutions by ERP vendors [2]. Cloud-based intelligent ERPs offer added benefits such as seamless access to systems through various means and devices. Given the difficult history of failed, delayed, and costly ERP implementation projects [3], ERP customers might be hesitant to invest in Intelligent ERP systems as the implementation of Intelligent ERPs might not yield the perceived benefits that ERP vendors and their partners portray. Customers are now faced with the difficult decision of whether to adopt the new and improved Intelligent ERP solutions or not. This decision to adopt Intelligent ERPs is even more complex for customers in developing nations such as South Africa, where factors such as poor infrastructure and lack of relevant skills might inhibit the adoption of advanced technologies such as Intelligent ERPs.

The aim of this study is to investigate the factors that influence the migration decision of adopters and non-adopters to a new and intelligent ERP platform. The TOE framework and institutional theories are used in this study to determine the influence of technology, organisational and environmental influences on migration decisions.

1.1 Problem Statement

The main research question that this study seeks to answer is, "What are the factors affecting ERP migrations in South Africa?" Specifically, the study seeks to address the following sub-question:

- What are the factors that determine whether organisations in developing countries such as South Africa migrate to the new ERP versions that have been introduced by ERP vendors?

2 Literature Review

The literature review summarises the key findings from existing literature on intelligent ERP systems and factors affecting migration to intelligent ERP systems in developing nations are also discussed.

2.1 Overview of Existing Literature on Intelligent ERP Systems

Intelligent ERPs are designed to deliver an improved user experience as they automate repetitive tasks and augment how less frequent tasks are performed. These systems are highly efficient as they can analyse large volumes of data using in-memory computing. Examples of intelligent ERPs include SAP S/4 HANA, Salesforce Einstein, Oracle's Adaptive Intelligent Applications for ERP, and Microsoft's Dynamics CRM with Azure [4]. In 2015, SAP launched SAP S/4 HANA, a product that runs solely on the SAP HANA database and incorporates a revolutionary user experience through the Fiori Launchpad and UI5 applications. These applications make it easy for users to transact on any mobile device, which improves business user satisfaction and productivity [5].

2.2 *Factors Impacting on Intelligent ERP Systems Adoption*

The introduction of intelligent ERP solutions has the potential to enhance the productivity and competitiveness of most organisations. However, there are numerous factors to be considered by customers when it comes to adopting such technologies.

(a) Technological context

Technological capability and the ability to manage cyber-security and data breaches is key in promoting adoption of technologies such as intelligent ERP by organisations [6]. Organisations confident in cyber-security systems are often willing to adopt online technologies [7].

(b) Organisational context

Large and profitable organisations have been shown to more likely adopt new technologies [8] compared to smaller and struggling institutions. Moreover, top management's support has been found to be key to overcoming the barriers to ERP technology adoption.

(c) Environmental context

Coercive forces in a firm's operating environment can be seen when stakeholders put pressure on the organisation to make decisions such as when SAP gave support deadlines to their customers for new releases of their ERP [2]. Coercive forces have an impact on an organisation's technology decisions. Mimetic or competitive pressure has been shown to be a main driver of technology adoption. Normative pressures exerted by different parties have been shown to influence ERP adoption [9]. Regulations such as data privacy and control of carbon emissions and accounting standards have an influence on adoption decisions.

3 Theoretical Background and Research Model

This section identifies and validates the use of existing theoretical frameworks, upon which the research model for this study was established. A set of hypotheses statements were formulated based on the proposed model.

3.1 The Technology–Organisation–Environment Framework

The TOE framework identifies the large influence that the context of an organisation has on the adoption and implementation of innovative technologies. The context is split into technological, organisational, and environmental. The technological context relates to the technologies that are already in existence in the firm, as well as the existing technologies that have not been acquired and used by the firm. The organisational context refers to the descriptive characteristics of the organisation, such as firm size, internal communication processes, management structures, and the firm's resources. Lastly, the environmental context relates to the arena, e.g. service providers, competitors, and regulations in which a firm operates. The TOE framework has a solid theoretical basis and has been used in adoption studies because of its adaptability [10] hence chosen for this study.

3.2 The Institutional Theory

The institutional theory posits that the environment of an institution is crucial in shaping the organisation's structure and action. The theory claims that over time, organisations will become more and more alike as they succumb to isomorphic pressure and the need for legitimacy within their industry [11]. Institutional isomorphism is a result of organisations in a similar industry conforming to these pressures and eventually becoming similar to the leading firms in the industry [12].

The research model for this study was based on a combination of the TOE framework and the institutional theory. Prior technology adoption models have combined the TOE framework and the institutional theory [13].

3.3 Other Technology Adoption Frameworks

Other frameworks such as the Technology acceptance model, Unified theory of acceptance and the Diffusion of innovation theory were considered in this study however these were discarded because they have been applied predominantly in studies that focused on technology adoption and the direct impact they had on individual users and not on the organisations as such.

3.4 Research Model

The research model is illustrated in Fig. 1. Nine variables were positioned within the TOE framework as predictors of intelligent ERP adoption. Three of these variables were derived from the institutional theory and positioned within the environmental context of the TOE model. These predictors are: Technological (ICT infrastructure, technical skills, and cyber-security); Organisational (organisational size, top management support); Environmental (government regulations and three factors from the institutional theory, which are coercive forces, normative pressure, and mimetic pressure). The dependent variable is a dichotomous measure to determine whether an organisation will adopt intelligent ERP or not.

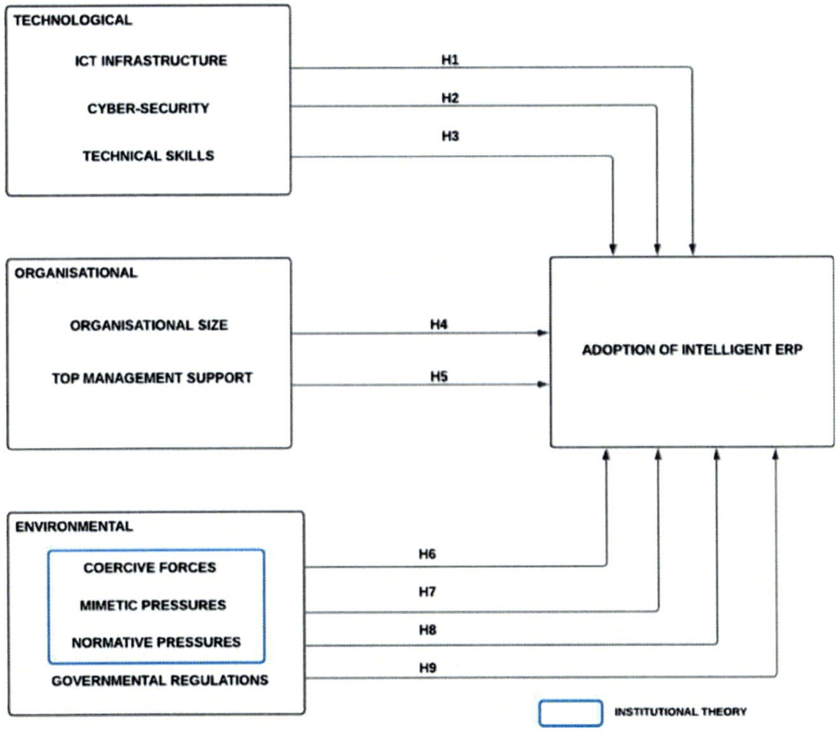

Fig. 1 Research model

Table 1 Summary of proposed hypotheses

TOE context	Hypothesis
Technological	H1-Access to ICT infrastructure is positively related to intelligent ERP adoption H2-The availability of cyber-security systems is positively related to intelligent ERP adoption H3-The availability of technical skills is positively related to intelligent ERP adoption
Organisational	H4-The size of an organisation is positively related to intelligent ERP adoption H5-Top management support is positively related to intelligent ERP adoption
Environmental	H6-Coercive forces are positively related to intelligent ERP adoption H7-Mimetic pressure is positively related to intelligent ERP adoption H8-Normative pressure is positively related to intelligent ERP adoption H9-Increased regulations are positively related to intelligent ERP adoption

Based on the research model in Fig. 1 and the findings from the literature review, nine hypotheses were proposed in the technological, organisational, and environmental contexts. The summary of hypothesis is outlined in Table 1.

4 Research Methodology

This study seeks to answer the research question posed in Sect. 1.1 using a research model depicted in Fig. 1. The research methodology that guided this study is discussed in this section. A positivist approach was deemed appropriate as similar studies have been conducted within this paradigm [14]. The survey research strategy was employed in this study and South African organisations were sampled in this study. An online questionnaire was used as the data collection instrument for the survey.

4.1 Data Collection

Non-probabilistic purposive sampling was used. The sample was made up of 95 South African potential SAP customer organisations. The link to the survey was sent electronically to four employees from each of these 95 organisations where 380 participants were targeted. Email addresses were found on LinkedIn social networking website and respective company's websites for relevant participants to be contacted. The data collected from the first section of the questionnaire relates to the demographic information whereas the second section determined if the respondents' organisation had adopted SAP S/4 HANA or not. Those that had adopted SAP S/4 HANA were required to confirm the deployment option that their organisation had chosen for the intelligent ERP. The non-adopters were required to confirm if their company had plans to migrate to SAP S/4 HANA. Several follow-up questions were posed to the respondents, depending on whether or not they had adopted SAP S/4 HANA. A 7 point Likert scale was used on the questionnaire.

5 Results

An external validity test was conducted to ascertain the existence of sufficient intercorrelation and enough overlaps in the variables to conduct a factor analysis [15]. The Bartlett test and the Kaiser–Meyer–Olkin (KMO) measures were used to test external validity. The reliability of the scale was conducted using the Cronbach's alpha test, which is a measure of internal consistency that depicts, how closely related a set of items are as a group [16]. Finally, Structural Equation Modelling (SEM) was used to test the hypotheses proposed for the research model. Given that there were adopters and non-adopters in the sample, the study broke down the SEM into two, the first one being for adopters and the second one for non-adopters. The SEM estimation was linked to the hypothesis formulated in the methodology section.

5.1 *Sample Demographics*

The data of 84 respondents was used as the final dataset for analysis. Thirty-three per cent of the respondents were on middle management to senior management level and 67% of the respondents were IT specialists from different domains. Figure 2 depicts the career profile of the participants.

Based on the information depicted in Fig. 2 it is evident that the majority of the participants were on specialist roles, with very few participants occupying senior management roles.

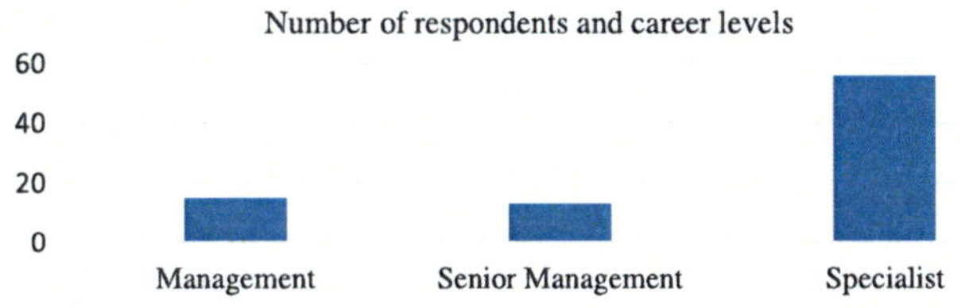

Fig. 2 Number of respondents and career levels

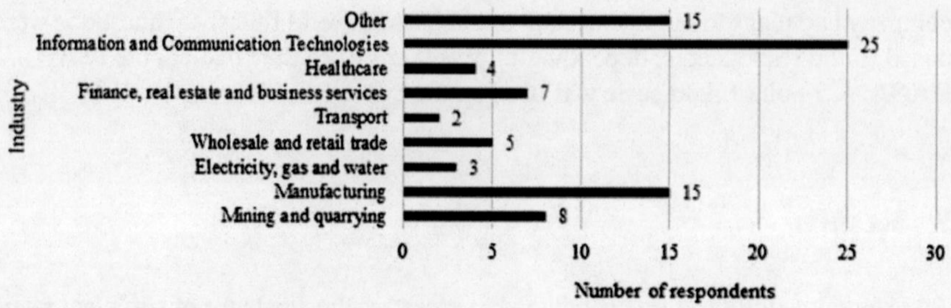

Fig. 3 Industrial sector distribution

5.2 *Industrial Sector Distribution*

Most respondents were employees of organisations in the ICT industry, followed by those from manufacturing as shown in Fig. 3. Transport and electricity, gas and water were the least represented industries, contributing only two and three respondents, respectively. Some of the sectors that fell in the "Others" section were the public sector, logistics, as well as oil and gas sectors.

Most of the respondents in the data sample were employees of large enterprises. Thirty-three respondents represented firms with more than 10,000 employees and 25 came from firms with an employee base of between 1001 and 5000 people. Only three respondents were employed by firms that had between 1 and 50 employees.

5.3 *Findings*

Reliability

Cronbach's alpha is a measure of internal consistency; that is, how closely related a set of items are as a group. It is a measure of scale reliability [16]. A Cronbach value greater than or equal to 0.9 is regarded as excellent, a value between 0.9 and 0.8 is good, between 0.8 and 0.7 is acceptable and below 0.7 is not acceptable.

According to Table 2, Factors 1, 2, and 5 have a Cronbach value of at least 0.8, which is acceptable. The other models have low Cronbach values and were thus not considered for the Structural equation modelling (SEM).

Table 2 Cronbach's alpha test

	All	Factor 1	Factor 2	Factor 3	Factor 4	Factor 5	Factor 6	Factor 7
Scale reliability coefficient	0.892	0.859	0.804	0.462	0.692	0.923	0.483	0.291
Number of items in the scale	25	6	5	4	4	2	2	2

Table 3 Hypothesis outcomes-adopter's model and non-adopters model

Adopters				Non-adopters		
Hypothesis	Result (p-value)	Result t-statistic	Outcome	Result (p-value)	Result t-statistic	Outcome
H1	0	6.05	Accepted	0.007	3.24	Accepted
H2	0.001	4.13	Accepted	0	6.13	Accepted
H3	0.002	2.16	Accepted	0.53	0.37	Rejected
H4	0.069	1.98	Rejected	0.001	3.5	Accepted
H5	0.155	1.54	Rejected	0.281	1.09	Rejected
H6	0.129	1.62	Rejected	0.992	0.01	Rejected
H7	0.026	2.28	Accepted	0.032	2.2	Accepted
H8	0.001	3.54	Accepted	0.03	2.21	Accepted
H9	0.292	1.1	Rejected	0.338	0.97	Rejected

Hypothesis Testing

Considering the tests and the analysis done, the hypotheses were tested using SEM. Given that there were adopters and non-adopters in the sample, the study broke down the SEM into two, the first one being for adopters and the second one for non-adopters. The SEM estimation was linked to the hypothesis formulated in the methodology section. Table 3 depict the proposed hypotheses for the adopters and non-adopters models and the result of testing the hypotheses for each model. The p-values and the t-statistics for each construct in both the adopters and non-adopters models are presented. The p-value should be less than 0.05 for the relationship to be statistically significant at the 5% significance level. The t-statistics should also be greater than or equal to 2 for the relationship to be statistically significant [17].

5.4 *Discussion*

This study was split into two models for adopters and non-adopters of intelligent ERP. Hypotheses were proposed for the relationship between the dependent variable "adoption of intelligent ERP" and nine independent variables. Based on the outcome of this study, access to ICT infrastructure, availability of cyber-security systems and availability of technical skills are positively related to intelligent ERP adoption for both adopters and non-adopters. From an organisational context perspective the size of an organisation is positively related to intelligent ERP adoption only for the non-adopters. It is only top management support which did not have positive influence on ERP adoption for both groups. Mimetic and normative pressures are positively related to intelligent ERP adoption for both groups.

6 Conclusion

In the context of a developing country such as South Africa, four factors were found to have a significant impact on the migration to new ERP versions within the digital economy, namely, access to ICT infrastructure, the availability of cyber-security systems, mimetic forces, and normative pressures. These factors were confirmed as being significant by both adopters and non-adopters of the intelligent ERP, SAP S/4 HANA as depicted in Tables 3. The findings of this study can be beneficial to ERP vendors, ERP customers as well as government institutions.

References

1. Piccarozzi M, Aquilani B, Gatti C (2018) Industry 4.0 in management studies: a systematic literature review. Sustainability 10(10)
2. ASUG (2018) Maintenance 2040 https://support.sap.com/en/release-upgrade-maintenance/maintenance-information/maintenance-strategy/s4hana-business-suite7.html (Accessed 29 Mar 2021)
3. Mahmood F, Khan AZ, Bokhari RH (2019) ERP issues and challenges: a research synthesis. Kybernetes, vol ahead-of-print
4. i-SCOOP, From ERP to intelligent ERP in the smart factory and supply chain. https://www.i-scoop.eu/industry-4-0/erp-intelligent-erp/ (Accessed 12 Feb 2021)
5. Deloitte (2017) Deloitte_ERP_Industrie-4–0_Whitepaper.pdf—Industry 4.0 Is your ERP system ready for the digital era Investing in Germany | A guide for Chinese | Course hero, 2017. https://www.coursehero.com/file/35526389/Deloitte-ERP-Industrie-4-0-Whitepaperpdf/ (Accessed 04 Feb 2021)
6. Bayode A (2019) 4th industrial revolution: challenges and opportunities in the South African context. Waste Manage, p 7
7. Belkhamza Z, Wafa SA (2009) The effect of perceived risk on the intention to use e-commerce: the case of algeria. J Internet Bank Commer 14
8. Zhu K, Kraemer KL (2005) Post-adoption variations in usage and value of E-business by organizations: cross-country evidence from the retail industry. Inf Syst Res 16(1):61–84
9. Awa HO, Ukoha O, Igwe SR (2017) Revisiting technology-organization-environment (T-O-E) theory for enriched applicability. The Bottom Line; Bradford 30(1):2–22. https://doi.org/innopac.wits.ac.za/10.1108/BL-12-2016-0044
10. Ramdani B, Kawalek P, Lorenzo O (2009) Predicting SMEs' adoption of enterprise systems. J Enterp Inf Manage 22(1/2):10–24
11. Martins MF, Oliveira T, Dias SS (2011) Medical records system adoption in european hospitals
12. Liang H, Saraf N, Hu Q, Xue Y (2007) Assimilation of enterprise systems: the effect of institutional pressures and the mediating role of top management. MIS Quarterly 31(1):59. https://doi.org/10.2307/25148781
13. Soares A, Palma-dos-Reis A (2008) ResearchGate, https://www.researchgate.net/publication/3077057_Why_Do_Firms_Adopt_E-Procurent_systems_using_logistic_regression_to_empirically_test_a_conceptual_model (Accessed 28 Apr 2021)
14. Awa HO, Ukoha O, Emecheta BC (2016) Using T-O-E theoretical framework to study the adoption of ERP solution. Cogent Bus Manage 3(1):1196571

15. Stapleton CD (1997) Basic concepts in exploratory factor analysis (EFA) as a tool to evaluate score validity: a right-brained approach
16. Streiner DL (2003) Starting at the beginning: an introduction to coefficient alpha and internal consistency 80(1):pp 99–103
17. Bartholomew DJ, Knott M, Moustaki I (2011) Latent variable models and factor analysis: a unified approach. John Wiley & Sons, London, UK
18. Awa HO, Ukoha O, Igwe SR (2017) Revisiting technology-organization-environment (T-O-E) theory for enriched applicability 30(1), pp 2–22
19. SAP (2020) The intelligent enterprise | Key Components and Solutions, SAP, https://www.sap.com/africa/products/intelligent-enterprise.html (Accessed 07 Dec 2020)
20. Awa HO, Ukoha O, Igwe SR (2017) Revisiting technology-organization-environment (T-O-E) theory for enriched applicability. The Bottom Line 30(01):2–22. https://doi.org/10.1108/BL-12-2016-0044

Analysis of an Efficient ZnO/GeTe Solar Cell Using SCAPS-1D

Mostafa M. Salah, A. Zekry, Mohamed Abouelatta, Ahmed Shaker, Mohamed Mousa, and Ahmed Saeed

Abstract Most of the solar cells that dominate the market are single-junction solar cells. These solar cells use mainly silicon, and some of them use relatively new materials like copper, indium, gallium, selenide (CIGS) and perovskite. These materials show a good performance, but they have a limitation of performance and are also costly or unstable. The strategy for improving the performance of single-junction solar cells in this paper is based on the use of promising material. The proposed single-junction solar cell uses germanium telluride (GeTe) as an absorber layer and zinc oxide (ZnO) as an electron transport layer. Germanium telluride has main advantages compared to many materials. It has a high electrical conductivity and a small energy gap, allowing it to absorb a larger portion of the input spectrum. The cell shows a 21.58% power conversion efficiency at room temperature using input spectrum air mass (AM 1.5). The effects of the doping profiles and thickness of the used materials are studied and improved to find the highest possible performance, and this improves short-circuit current density, open-circuit voltage and fill factor resulting in increasing the efficiency to be 33.18%.

Keywords SCAPS-1D · Single-junction solar cell · GeTe · High efficiency · ZnO

1 Introduction

Nowadays, the demand for energy is rapidly increasing. Clean energy, especially from solar cells, shows a promising solution [1]. The move from conventional to green energy is related to a set of principles. One of these principles is to increase device reliability [2]. Currently, the photovoltaic market is dominated by crystallized silicon solar cells in multi-crystalline and monocrystalline forms. So far, the efficiency of crystalline silicon cells has surpassed 25% [3]. But these cells have a comparatively

M. M. Salah · M. Mousa (✉) · A. Saeed
Electrical Engineering Department, Future University in Egypt, Cairo, Egypt
e-mail: mohamed.mossa@fue.edu.eg

M. M. Salah · A. Zekry · M. Abouelatta · A. Shaker
Faculty of Engineering, Ain Shams University, Cairo, Egypt

X.-S. Yang et al. (eds.), *Proceedings of Seventh International Congress on Information and Communication Technology*, Lecture Notes in Networks and Systems 447,
https://doi.org/10.1007/978-981-19-1607-6_60

high production cost. CIGS thin-film cells are very competitive cells, which have achieved a power conversion efficiency (PCE) of over 23% [4]. Perovskite solar cells (PSCs) have also recently emerged, displaying quick progress and opening up new avenues in photovoltaics. PSC's devices currently have an efficiency of over 22% [5]. However, PSCs continue to confront substantial development problems, especially long-term reliability versus environmental stressors like temperatures [6]. In this regard, the demand for new absorber materials is still existing. Cadmium telluride (CdTe) solar cells have achieved a PCE of over 21% [7]. It should be noted that cadmium telluride is a toxic material. GeTe as an absorber material is proposed to be a good candidate for the conventional costly, unstable and toxic materials. In the 1960s, there was a lot of interest in GeTe. It has been studied continuously from both a technical and fundamental standpoint, as well as from a phase-change, thermoelectric and ferroelectric material [8]. Recently, it was used as absorber material in simulated solar cells [9]. GeTe has promising optical and electrical properties. Also, it has a high absorption coefficient and bandgap energy of 0.8 eV [10], which allows it to have a high and extended quantum efficiency. GeTe has a resistivity of 2×10^{-4} Ω-cm [11]. This material enables high doping concentration exceeding 10^{21} cm^{-3} [10]. According to its bandgap, GeTe is a very effective material for the bottom subcells of the multi-junction solar cells [12, 13]. GeTe has superior thermoelectric characteristics, which allow it to have good stability at high temperatures. Besides the unique set of electronic and optical properties, GeTe is a nontoxic material. GeTe thin films may be deposited using a variety of low-cost processes, including thermal evaporation, pulsed laser deposition and magnetron sputtering. SCAPS-1D is widely regarded as a useful tool for simulating a wide variety of solar cells. It has been used and validated with experimental cells in this manner [14]. SCAPS-1D can mimic solar cells with up to seven layers [15]. Mostly, all material properties, such as permittivities, affinities and doping concentrations, can be valued [16].

This work presents a proposed single-junction solar cell that uses GeTe as an absorber layer and zinc oxide as an electron transport layer. The rest of this work is divided into the following sections: The suggested GeTe solar cell and how it has been studied as well as the impact of temperature on performance measures are illustrated in Sect. 2; Sect. 3 shows the optimization of the designed cell, the optimization of the thicknesses of GeTe and ZnO, and the optimization of the doping of both GeTe and ZnO; Sect. 4 provides a comparison between the initial designed GeTe cell and the optimized one; Finally, Sect. 5 concludes our work.

2 GeTe with ZnO

The structure of the proposed GeTe solar cell is shown in Fig. 1a. A glass substrate, fluorine-doped tin oxide (FTO) as front contact (4.6 eV), ZnO as n-material and GeTe as p-material are the layers of the proposed Gete solar cell design. For the interface between the GeTe layer and the back contact metal, a flat band model is utilized. Figure 1b illustrates the energy band diagram of GeTe and ZnO materials. Table 1

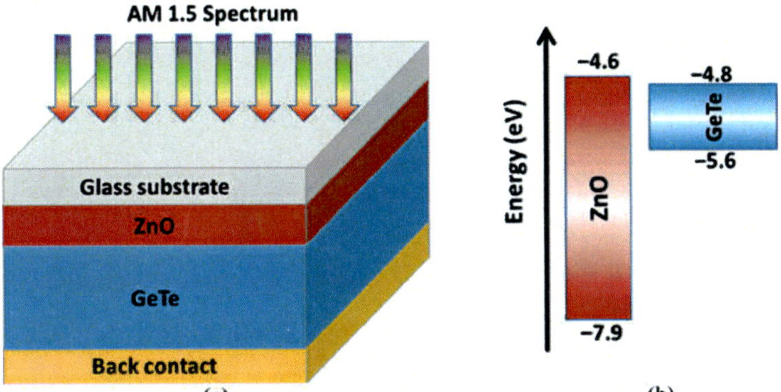

Fig. 1 GeTe solar cell **a** structure and **b** the constituting layers energy diagrams

Table 1 Simulation parameters of GeTe and ZnO layers

Parameters	GeTe	ZnO
E_g(eV)	0.8	3.30 [21]
Thickness (nm)	500	300
Relative permittivity	36	9.0 [9]
χ(eV)	4.8	4.6
N_c(1/cm^3)	10^{16}	2.2×10^{18}
μ_e(1/(cm^2 V s))	100	100 [22]
N_D(1/cm^3)	0	10^{18}
N_v(1/cm^3)	10^{17}	1.8×10^{19}
μ_p(1/(cm^2 V s))	20	25 [23, 24]
N_A(1/cm^3)	2×10^{16}	0
N_t(1/cm^3)	10^{14}	10^{15} [25, 26]

shows the parameters of the materials that were used. To be close to the practical devices, a neutral defect with Gaussian distribution is used. For all analysis, the incident spectrum is AM 1.5 at room temperature (300 K) is used. The absorption coefficients (α) for used materials are estimated by (1) [17]:

$$\alpha(E) = A_\alpha \sqrt{h\nu - E_g} \tag{1}$$

where A_α is pre-factor of 10^5 (1/(cm^1 eV$^{1/2}$)), h is the Planck's constant (eV s), ν is the spectrum frequency (Hz), and E_g is the material energy gap (eV). Lowering the recombination at GeTe/ZnO interface is an effective approach to extract carriers efficiently. To achieve that, ZnO's conduction band offset (CBO) should be 0 to 0.3 eV lower than GeTe [18, 19]. CBO is given by (2) [20] as

$$CBO = (\chi_G - \chi_Z) \tag{2}$$

where χ_G and χ_Z are the material affinities of the GeTe and ZnO, respectively. According to (2) and the GeTe affinity of 4.8 eV, the optimum CBO is 4.5–4.8 eV which is achieved by ZnO.

The simulated J/V curve for the GeTe cell is given in Fig. 2a. The modeled GeTe solar cell has a PCE of 21.58%, FF 78.48%, V_{OC} 0.55 V and J_{SC} of 50.02 mA/cm^2. Figure 2b illustrates the quantum efficiency (QE) of the modeled cell. According to the absorber material energy gap, the cut-off wavelength is 1550 nm. The temperature stability of the designed GeTe cell is investigated from 280 to 360 K. The results show that the performance parameters of this cell are stable and immune to temperature

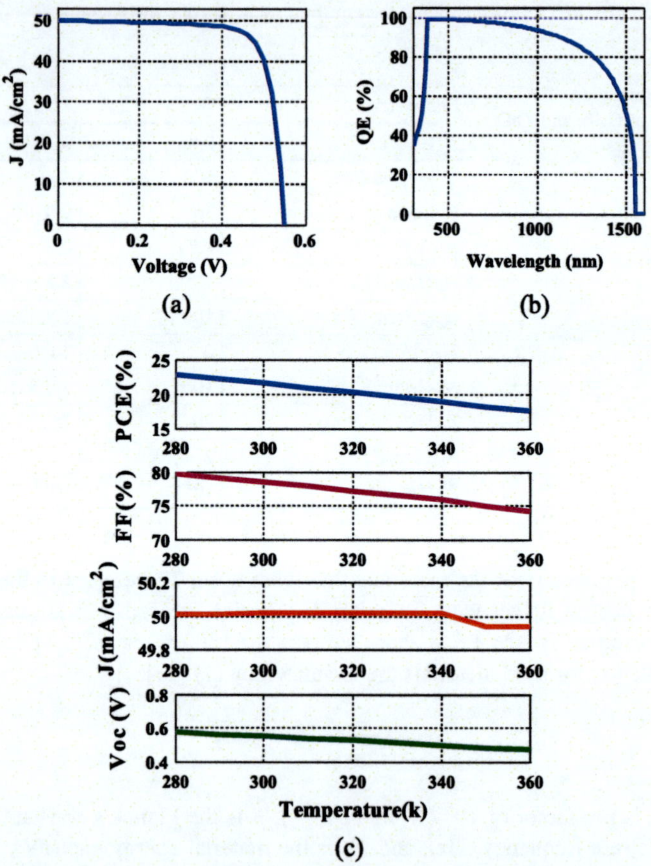

Fig. 2 Performance curves of the suggested GeTe cell **a** J/V, **b** QE and **c** performance metrics with temperature variations

variations. Figure 2c illustrates the variation of main parameters of performance of designed GeTe cell with temperature.

3 Optimizing the GeTe Solar Cell

The thickness of the solar cell layers can be modified to improve performance metrics significantly. Furthermore, the conductivity of used materials in solar cells significantly impacts the performance parameters. The doping, either n-type or p-type, can be used to regulate the conductivity of the materials. To improve the effect of thickness, doping concentrations on GeTe and ZnO are explored in the following subsections.

3.1 Thickness Optimization

The effects of GeTe and ZnO thicknesses have been studied in this subsection. The influence of GeTe thickness on performance metrics is shown in Fig. 3a. It can be shown that, as the thickness of the GeTe increases, the performance metrics are enhanced, but no significant changes after 3.5 μm. At this thickness, the cell has a PCE of 26.99%, FF of 82.37%, V_{OC} equals to 0.62 V and a J_{SC} 53.21 mA/cm^2. Figure 3b shows how the thickness of ZnO affects the performance characteristics. It can be shown that as the thickness of ZnO decreases the performance metrics

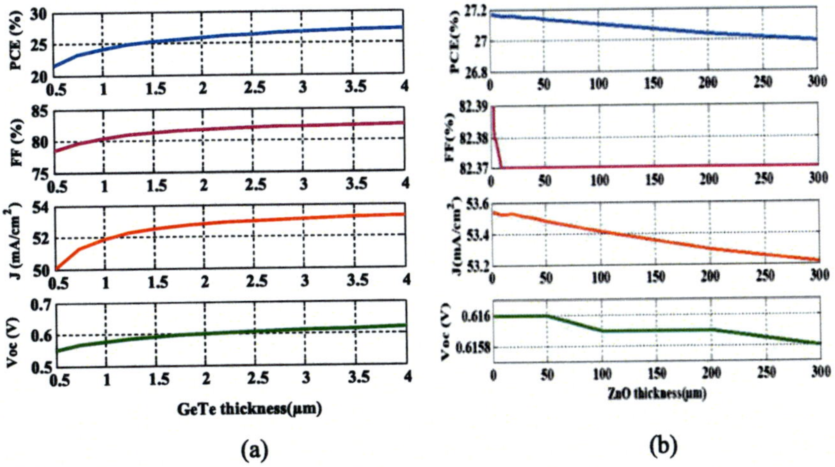

Fig. 3 Performance metrics variations depending on the thickness of **a** GeTe and **b** ZnO

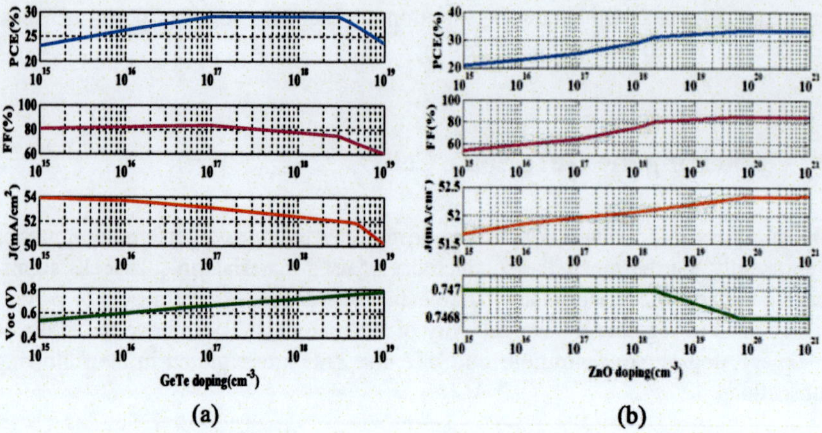

Fig. 4 Performance metrics variations depending on the doping concentration of **a** GeTe and **b** ZnO

enhances, but no significant change below 10 nm. At this thickness, the cell has a PCE of 27.16%, no change in FF, V_{OC} equals to 0.62 V and J_{SC} 53.53 mA/cm^2.

3.2 Doping Optimization

The effect of GeTe and ZnO doping concentrations has been studied in this subsection. Figure 4a shows how the concentration of GeTe doping affects performance measures. As the doping concentration of GeTe increases, the performance metrics are enhanced till 3×10^{18} cm^{-3}, and then, the performance metrics decrease. At this doping concentration, the cell has a PCE of 29.11%, a FF of 74.83%, V_{OC} equals to 0.75 V and J_{SC} 52.1 mA/cm^2. Figure 4b shows how the concentration of ZnO doping affects the performance characteristics. It can be shown that as the doping concentration of ZnO increases the performance metrics are enhanced and the PCE is constant after 6×10^{19} cm^{-3}. At this doping concentration, the cell has a PCE of 33.18%, a FF of 84.95%, a V_{OC} equals to 0.75 V and J_{SC} 52.31 mA/cm^2.

4 Comparison of the Initial and Optimized Cell Designs

The J/V curves of the initial designed GeTe cell and the optimized thickness and doping concentrations are shown in Fig. 5a. For more physical explanation, Fig.5b and 5c shows the energy band diagrams of the initial and optimized cells. As shown in Fig. 5c, the spike CBO in Fig. 5b is solved after optimization.

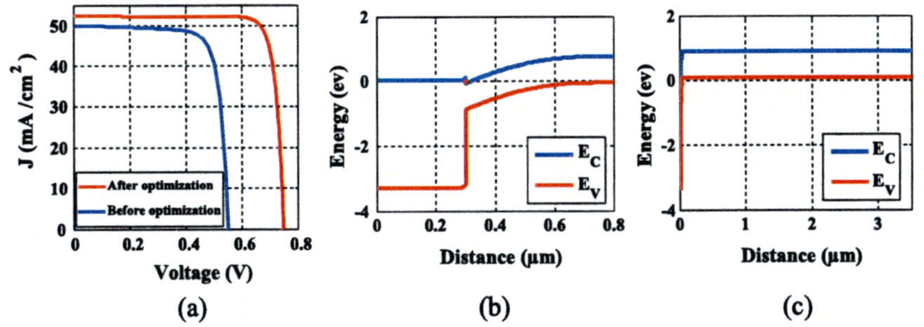

Fig. 5 Performance of the designed GeTe cell **a** J/V, **b** Energy diagram before optimization and **c** Energy diagram after optimization

5 Conclusions

In this study, the results indicate that GeTe is a promising material due to its high short-circuit current density and open-circuit voltage, both of which are essential to improving the performance of a single-junction solar cell. Besides of its electrical conductivity, another factor contributing to GeTe's high current density is its low energy gap, which allows it to absorb more of the incident spectrum. The PCE increases from 21.58 to 26.99% when the absorber layer thickness is increased from 0.5 to 3.5 μm. Decreasing the thickness of the ZnO layer to 10 nm increased the PCE to 27.16%. Optimizing the doping of GeTe and ZnO layers increases the PCE to 33.18%, due to increasing of the open-circuit voltage. The weakness of this cell is the dependence of the PCE on increasing the thickness of GeTe, which is not preferred in some photovoltaics. The simulated design should be applied practically to verify the output performance parameters. The future work of this research can be extended to study the possibility of using this cell as a part of a tandem cell especially it shows an excellent performance in current density, which is one of tandem cells main limitations.

References

1. Fouda S, Salem MS, Saeed A, Shaker A, Abouelatta M (2020) Thirteen-level modified packed u-cell multilevel inverter for renewable-energy applications. In: 2020 2nd international conference on smart power and internet energy systems (SPIES), IEEE, pp 431–435
2. Abdelhalim Z (2020) A road map for transformation from conventional to photovoltaic energy generation and its challenges, pp 407–410
3. Zekry A, Shaker A, Salem M (2018) Solar cells and arrays: principles, analysis, and design. In: Advances in renewable energies and power technologies, pp. 3–56. Elsevier
4. Nakamura M, Yamaguchi K, Kimoto Y, Yasaki Y, Kato T, Sugimoto H (2019) Cd-free Cu (In, Ga) (Se, S) 2 thin-film solar cell with record efficiency of 23.35%. IEEE J Photovoltaics 9(6):1863–1867

5. Yang WS, Park BW, Jung EH, Jeon NJ, Kim YC, Lee DU, Shin SS, Seo J et al (2017) Iodide management in formamidinium-lead-halide–based perovskite layers for efficient solar cells. Science 356(6345):1376–1379
6. Zekry A, Yahyaoui I, Tadeo F (2019) Generic analytical models for organic and perovskite solar cells. In: 2019 10th international renewable energy congress (IREC), IEEE, pp 1–6
7. Green M, Dunlop E, Hohl-Ebinger J, Yoshita M, Kopidakis N, Hao X (2021) Solar cell efficiency tables (version 57). Prog Photovoltaics Res Appl 29(1):3–15
8. Gervacio-Arciniega JJ, Prokhorov E, Espinoza-Beltran FJ, Trapaga G (2012) Characterization of local piezoelectric behavior of ferroelectric GeTe and Ge2Sb2Te5 thin films. J Appl Phys 112(5):052018
9. Mousa M, Amer FZ, Mubarak RI, Saeed A (2021) Simulation of optimized high-current tandem solar-cells with efficiency beyond 41%. IEEE Access 9:49724–49737
10. Bahl SK, Chopra KL (1969) Amorphous versus crystalline GeTe films. II. optical properties. J Appl Phys 40(12):4940–4947
11. Madelung O, Rossler U, Schulz M (1999) Collaboration: authors and editors of the volumes III. 17B-22A-41B, zinc oxide (ZnO) crystal structure, lattice parameters semiconductors of landolt-börnstein-group iii condensed matter
12. Naby MA, Zekry A, El Akkad F, Ragaie HF (1993) Dependence of dark current on zinc concentration in ZnxCd1 − xS/ZnTe heterojunctions. Solar Energy Mater Solar Cells 29(2):97–108
13. Salem MS, Zekry A, Shaker A, Abouelatta M, Abdolkader TM (2019) Performance enhancement of a proposed solar cell microstructure based on heavily doped silicon wafers. Semicond Sci Technol 34(3):035012
14. Abdelaziz W, Shaker A, Abouelatta M, Zekry A (2019) Possible efficiency boosting of non-fullerene acceptor solar cell using device simulation. Optical Mater 91:239–245
15. Burgelman M, Decock K, Khelifi S, Abass A (2013) Advanced electrical simulation of thin film solar cells. Thin Solid Films 535:296–301
16. Niemegeers A, Burgelman M, Degrave S, Verschraegen J, Decock K (2018) SCAPS (Version: 3.3. 07) manual. Nov 7
17. Mousa M, Amer FZ, Mubarak RI, Saeed A (2021) High-efficiency modified tandem solar cell: simulation of two-absorbers bottom subcell. Optik 168458
18. Basyoni MS, Salem MM, Salah MM, Shaker A, Zekry A, Abouelatta M, Alshammari MT, Al-Dhlan KA, Gontrand C (2021) On the investigation of interface defects of solar cells: lead-based versus lead-free perovskite. IEEE Access 9:130221–130232
19. Abdelaziz S, Zekry A, Shaker A, Abouelatta M (2020) Investigating the performance of formamidinium tin-based perovskite solar cell by SCAPS device simulation. Optical Mater 101:109738
20. Salem M, Shaker A, Zekry A, Abouelatta M, Alanazi A, Alshammari MT, Gontand C (2021) Analysis of hybrid hetero-homo junction lead-free perovskite solar cells by SCAPS simulator. Energies 14(18):5741
21. Mousa M, Amer FZ, Saeed A, Mubarak RI (2021) Simulation of high-efficiency perovskite-based tandem solar cells. In: 2020 6th international symposium on new and renewable energy (SIENR), IEEE, pp 1–5
22. Mousa M, Salah MM, Amer FZ, Saeed A, Mubarak RI (2020) High efficiency tandem perovskite/CIGS solar cell. In: 2020 2nd international conference on smart power and internet energy systems (SPIES), pp 224–227, IEEE
23. Salah MM, Hassan KM, Abouelatta M, Shaker A (2019) A comparative study of different ETMs in perovskite solar cell with inorganic copper iodide as HTM. Optik 178:958–963
24. Mousa M, Amer FZ, Saeed A, Mubarak RI (2021) Two-terminal perovskite/silicon solar cell: simulation and analysis. In: 2021 3rd novel intelligent and leading emerging sciences conference (NILES), IEEE, pp 409–412

25. Salem MS, Salah MM, Mousa M, Abouelatta M, Shaker A, Alzahrani AJ, Alanazi A, Ramadan R EFFICIENT PEROVSKITE SOLAR CELL WITHOUT ELECTRON TRANSPORT LAYER
26. Salah MM, Abouelatta M, Shaker A, Hassan KM, Saeed A (2019) A comprehensive simulation study of hybrid halide perovskite solar cell with copper oxide as HTM. Semicond Sci Technol 34(11):115009

Looping Through Color Space: A Simple Augmentation Method to Improve Biased Object Detection

Pascal Lampert, Janis Jung, Andreas Hubert, and Konrad Doll

Abstract In this work, we address the challenging problem of color-dependent and imbalanced datasets. For many use cases, the training of models based on such data will not generalize well enough and fail even on slight domain variations. This issue is usually addressed by artificially extending the data by manipulating input data or using synthetic data. In this context, we introduce a new augmentation method for extended color mapping from single-channel depth images that reduce color dependency and decrease the amount of annotated data needed for comparable model performance. We found that this method improves the generalization of models for depth-based hand detection on our dataset captured at a manual assembly workspace. Additionally, we validated our results on a publicly available dataset.

Keywords Data augmentation · Object detection · Biased detection · Feature engineering · Computer vision · Depth encoding · Colormap

P. Lampert (✉) · J. Jung · A. Hubert · K. Doll
University of Applied Sciences Aschaffenburg, Würzburger Str. 45, 63743 Aschaffenburg, Germany
e-mail: pascal.lampert@th-ab.de

J. Jung
e-mail: janis.jung@th-ab.de

A. Hubert
e-mail: andreas.hubert@th-ab.de

K. Doll
e-mail: konrad.doll@th-ab.de

X.-S. Yang et al. (eds.), *Proceedings of Seventh International Congress on Information and Communication Technology*, Lecture Notes in Networks and Systems 447,
https://doi.org/10.1007/978-981-19-1607-6_61

1 Introduction

1.1 Motivation

In the field of computer vision, deep learning-based algorithms constitute the state-of-the-art for a significant number of problems. One drawback of these algorithms is that they require many annotated training data to obtain good performance. Another one is that most algorithms cannot easily be adapted to slightly different domains for real-life applications. Heavily domain-specific style changes present a challenging task to quickly and feasible handle rapid domain changes, with the need to annotate data and retrain the used models. This fact makes impressive methods impractical for many actual use cases. In this paper, we present a new augmentation method for depth-based object detection. As a use case, we apply our method for intelligent assistance in manual assembly processes. We came across two interesting observations. First, we get a higher degree of generalization from training only on depth data. Second, we introduce a method for data augmentation that addresses training issues with small and imbalanced datasets.

1.2 Related Work

In general, image datasets tend to be biased and are unlikely to generalize well on underrepresented or unknown domains [1]. This is a well-known problem in computer vision, with research spanning across several subcategories. Here we want to emphasize both Domain Adaptation and Augmentation to deal with the generalization problem of small datasets.

Domain Adaptation Unlearning methods show promising results for biased datasets where a set of known but unwanted target features can be described. For the training procedure, a loss is formulated, such that a certain set of undesirable features are learned to be no longer distinguishable [2, 3]. For available but unlabeled target domain data, a variety of well-studied Unsupervised Domain Adaptation methods show an improvement in the model generalization in the target domain. The common practice is to reduce the domain shift by directly aligning the source and target marginal distributions, thus reducing the influence of domain-specific information [4, 5].

Augmentation: With more variation in training data, the model does improve its ability to generalize. Generating large datasets to increase the amount of rich training data is a challenging and expensive task. Therefore, many training methods use augmentation techniques to extend the training data and increase the detection performance. Some augmentation techniques can directly operate on input images such as cropping, flipping, changing contrast [6], whereas others fabricate data by synthesis or simulation. As an example [7] exploit the MANO model [8] to remap synthetic hands into source depth images. Other [9] utilize a realistic game engine to generate precise depth data in order to make their model generalize better with less real-world data. Further, methods [10] combine 3D and 2D data sources in order to achieve better results with fewer data.

1.3 Main Contributions and Outline of This Paper

The main contribution of this paper is to provide an augmentation approach to deal with small amounts of data and the imbalances that often accompany it. In our use case, the object's color and surface pattern are subject to substantial changes that cannot be anticipated well. To compensate for these uncertainties, we use depth instead of RGB images. Unfortunately, one of the rising problems with static depth datasets is that they tend to be biased to specific depth ranges. Thus, we introduce an augmentation method using color-encoded depth images that normalize the object's appearance and mitigate this problem. To prove this, we evaluate our approach using a biased example dataset.

The remainder of this paper is structured as follows: In Sect. 2, we address the underlying methodology. This section focuses on preprocessing the input data and, in particular, on augmentation through colormap rotation. Section 3 presents our use case, followed by the results of the proposed method on our dataset, and confirms the outcome on a public dataset. Finally, we conclude and give a short outlook in Sect. 4.

2 Method

The proposed augmentation method introduces a rotation of the colormap. For this purpose, the color mapping of the depth images that serve as the basis for the augmentation is described first, followed by the explanation of the augmentation method and the need for a continuous colormap.

2.1 Depth to Color Mapping

The input for the depth to color mapping is single-channel depth values $D(u, v)$, with u and v being the pixel position. The depth values $D(u, v)$ are normalized between 0 and 1, resulting in $D_{\mathrm{norm}}(u, v)$, which are used as base for color mapping and in the further course of the paper referred to as grayscale. The encoding of the depths through HSV colormap is done by a linear mapping, resulting in a three-channel depth image $D_{\mathrm{hsv}}(u, v, c)$ with c as the color channel:

$$D_{\mathrm{hsv}}(u, v, c) = F_{\mathrm{hsv}}\left(1529 \cdot D_{\mathrm{norm}}(u, v)\right) \tag{1}$$

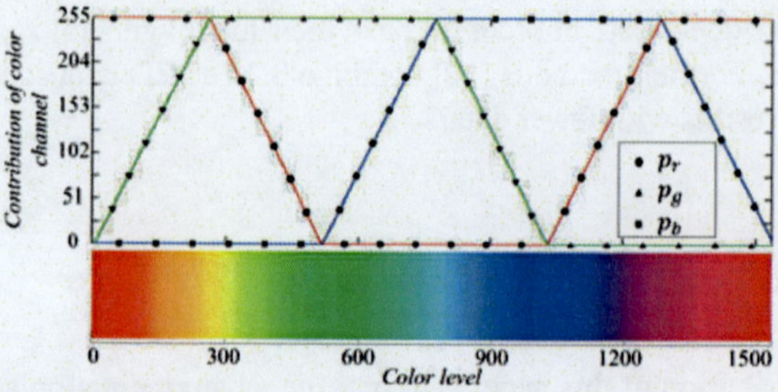

Fig. 1 HSV colormap with reference function of RGB channel contributions

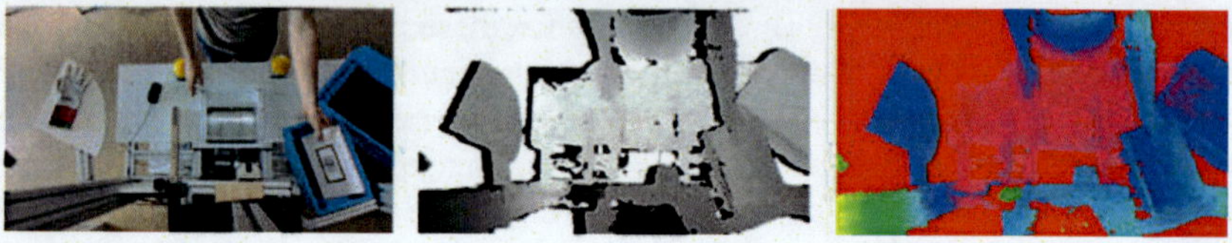

Fig. 2 Input samples. Left: RGB image. Center: Grayscale depth image ($D_{\text{norm}}(u, v)$). Right: Depth image encoded with HSV colormap ($D_{\text{hsv}}(u, v, c)$)

with

$$F_{\text{hsv}}(d): \quad d \mapsto [p_r, p_g, p_b]. \tag{2}$$

The input of the reference function $F_{\text{hsv}}(d)$ is $d = 1529 \cdot D_{\text{norm}}(u, v)$. 1529 corresponds to the distinct values of the HSV colormap. Figure 1 shows the full HSV colormap at the bottom with the reference function for the contribution of RGB channels (p_r, p_g, and p_b). A sample input collection of the RGB image on the left, the depth image $D_{\text{norm}}(u, v)$ represented as grayscale in the center, and the color-encoded depth image $D_{\text{hsv}}(u, v, c)$ on the right is shown in Fig. 2.

2.2 *Colormap Rotation*

To reduce imbalances in datasets, we introduce a method to extend the data by rotating the colormap artificially. For example, a dataset that covers only a particular depth range for the appearance of objects, leading to an undesired depth sensitivity. As a result, models might not generalize well to detect the same objects in other depth ranges. Furthermore, the problem is not solved by mapping from depth to HSV colormap because the imbalance is just shifted from depth to color space.

Within the scope of this work, a method for augmenting the depth images was developed where the depth information is distributed over the entire encoding range of the colormap by rotating the colormap and thus eliminating existing imbalances. Figure 3 represents the relative color frequency of objects—left and right hands—

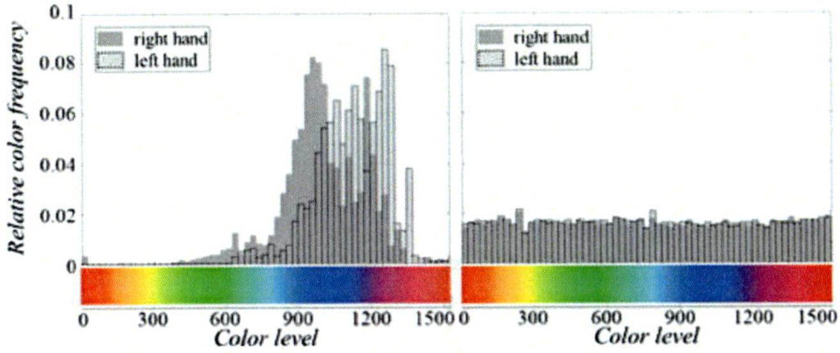

Fig. 3 Relative frequency of color levels corresponding to depth values. Left: High imbalances, preferably in the blue color spectrum (between 800 and 1300). Right: Approximate uniform distribution by $N = 5$ colormap rotations

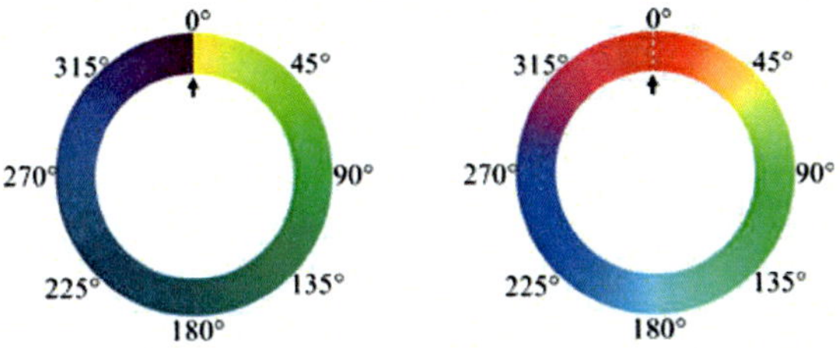

Fig. 4 Annular arrangement of two colormaps with angle indications and marked boundary point (arrow). Left: viridis colormap with color jump on boundary edge. Right: HSV colormap with soft overflowing boundary edge

occurring in our dataset, visualizing the imbalance before (left) and the result after colormap rotation (right). Visible are the strong imbalances of the objects in the blue color spectrum between color levels 800 and 1300. Due to the augmentation method, a more uniform distribution is achieved.

In principle, all colormaps can be rotated, but the choice is crucial as some will create undesired side effects such as artifacts in the image. Artifacts are mainly edges in the middle of objects due to indistinguishable transitions at colormap borders, e.g., grayscale or viridis colormap. To illustrate this, Fig. 4 shows two colormaps in circular representation. On the left side, the viridis colormap reveals a jump from purple to yellow at 0°. In contrast, the HSV colormap on the right side has a differentiable transition.

The augmentation is performed by random rotations of the HSV colormap so that the encoding of the depth values is performed with different color levels in each case. The same rotation must be applied for each pixel to convert an entire image, but it is possible to rotate the colormap several times for the same image. To distribute the objects approximately equally over the entire colormap, the rotation range is split into a fixed number N of equal subranges $s = \{s_0, s_1, \ldots, s_{N-1}\}$, where a single random angle $\phi(s_i)$, with $0 \leq i \leq N-1$, is picked for each subrange (cf. Eqs. 3 and 4).

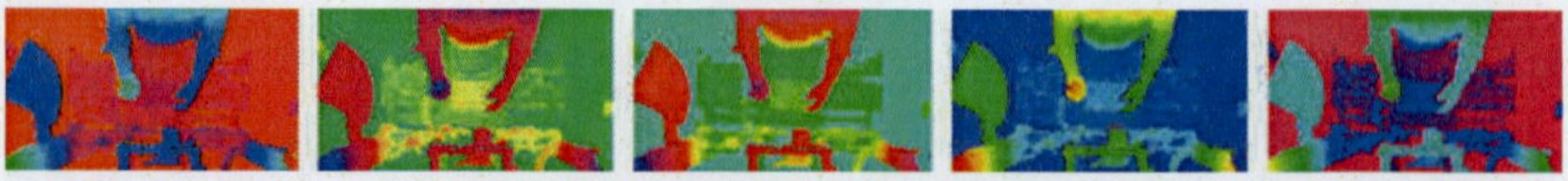

Fig. 5 Augmentation of a single depth image ($D_{\text{norm}}(u, v)$) using $N = 5$ colormap rotations resulting in five augmented images ($D_{\text{aug},\phi(s_i)}(u, v, c)$) with randomly determined rotations ($\phi(s_i)$) from left to right: 21 °, 136 °, 162 °, 263 °, 325 °

Thus, for each depth image ($D_{\text{norm}}(u, v)$), N color images ($D_{\text{aug},\phi(s_i)}(u, v, c)$) are generated from different colormap rotations. Figure 5 shows the resulting augmented images with $N = 5$ randomly rotated colormaps.

$$s_i = \left[\frac{360°}{N} i \; ; \; \frac{360°}{N}(i + 1) \right) \tag{3}$$

$$\phi(s_1) = \text{random}(s_i) \tag{4}$$

3 Experimental Results

The following section will introduce the use case on which we evaluate our method. We show that color mapping using HSV colormap leads to better results than simple grayscale and apply the proposed augmentation method of colormap rotation as further improvement.

3.1 *Use Case*

Our goal is to understand human–environment interaction in manual assembly processes by using camera-based detection. We want to recognize differences in the flow of work processes to assist workers intelligently. During assembly, hands play a central role by manipulating the environment. Therefore, it is crucial to have precise and robust hand detection from images. Depth images are used as model input to ensure color independence of hand detection, even over the high color variation of skin and gloves. An Intel® RealSense™ D435 depth camera is mounted overhead, pointing downwards to the working area. Each scene is recorded in 30 frames per second (FPS) with a resolution of 1920 × 1080 pixels (px) for RGB and 1280 × 720 px in 16-bit resolution for depth images ($D(u, v)$). The dataset contains 1858 RGB-D images with 3577 bounding box annotations of hands from several scenes recorded during assembly. We randomly split the dataset for training, validation, and evaluation. Thus, 70 % of the data is used for training (T), 10 % for validation during training (V), and 20 % for evaluating the resulting models (E).

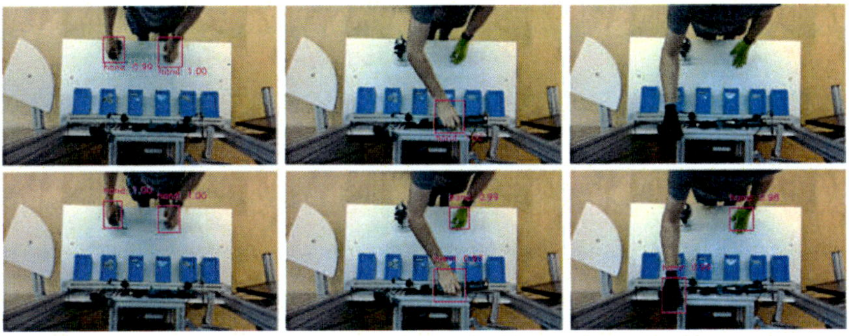

Fig. 6 Representative images are showing a varying usage of gloves which weren't part of the training process. The top row shows the hand detections for a model trained on RGB images, and the bottom row displays the detections for a model trained on the corresponding depth images. The detections are marked as magenta-colored bounding boxes with their corresponding model score. An example of wearing no gloves is shown on the left, where both models achieve similar results. However, when switching to another domain, using gloves, as in the images in the center and on the right, the RGB model (top) cannot resolve the color change, while the depth model (bottom) can

In general, the proposed augmentation method is independent of the chosen model. In our case, an SSD300 [11] model pre-trained on the COCO dataset [12] is used to evaluate our methods. The model consumes 300×300 px images as input and is chosen due to real-time constraints at the manual assembly workstation. The depth images are resized and padded to match the expected input size and avoid image distortion. For all different considerations, the whole network is trained from the same starting point. For small datasets, it is hard to train well generalizing models. For example, our dataset contains only images of one kind of gloves or images without gloves. To show that our method generalizes very well for gloves of varying colors that are not available for training, we have also trained a model on the RGB color images for comparison. Figure 6 shows the detection results for the color and depth model, respectively. For an imbalanced dataset, the color model seems to focus on the actual color of the hands. In contrast, the depth model has no problems detecting hands wearing gloves of different colors due to missing color variance in depth data.

3.2 *Depth Encoding*

For the proposed augmentation method, it is essential to map the depth images into a continuous colormap, to be able to rotate without creating side effects as mentioned in Sect. 2. Here, we resort to the described HSV colormap. By evaluating two models, one trained with grayscale (m_{gray}) and another with HSV color encoding (m_{hsv}), we show that the simple HSV encoding of the depth images already leads to an improvement. The training showed, that the loss for m_{gray} rises again after 100,000 steps, which is an indicator for overfitting. This means that the model cannot

Table 1 Validation results for the HSV (m_{hsv}) and grayscale model (m_{gray})

Model	mAP@0.75IoU (%)
m_{gray}	86.6
m_{hsv}	**93.2**

generalize well with the available data. In the case of m_{hsv}, a continuous decline in loss occurred, which stops falling at around 400,000 steps and also does not begin to increase. Furthermore, m_{hsv} performs better in terms of the maximum achieved mAP and training speed. At 100,000 steps, it has already reached a value more than 5 % higher than m_{gray}.

From both encoding methods, the overall training results are presented in Table 1. m_{hsv} achieves an mAP of over 93 %. The performance is an indication for better-distributed depth information in the HSV colormap. The resulting HSV mapped depth images have a higher similarity to RGB images, making object detection easier for the model, pre-trained with RGB images.

It has been hypothesized that using depth information would reduce color independence, but it cannot be implied in a general sense. It is indeed the case that the depths do not provide any information about the skin color and the wearing of gloves, but because of the defined colormap, depth is inherently represented in the mapped colors. Due to given depth imbalances in the dataset, as shown in Fig. 3, the imbalance is shifted to the color spectrum. Therefore, indirect color dependencies might appear in the model representation when training with HSV mapped depth images.

3.3 *Augmentation Through Colormap Rotation*

The m_{hsv} trained in the previous section serves as the baseline of the investigations in this section. The difference between the models is solely in the input data. The used dataset for m_{hsv} with a total number of 3,577 annotated hands is multiplied by the colormap rotation w.r.t. the number of rotations. To investigate the need for full circle rotation of the colormap for achieving color independence, we apply two versions of augmentation, which differ in the possible values of the rotation angle ($\phi(s_i)$). First, with $\phi(s_i) \in [0°, 360°]$, the whole spectrum of the colormap is used ($m_{\text{aug\_full}}$). Second, with $\phi(s_i) \in [-70°, 70°]$, the colormap rotation is applied partially with 70° clockwise and counterclockwise ($m_{\text{aug\_partial}}$). The use case defines the partial rotation angle to cover the range where hands can appear. For both options, $N = 5$ rotations were selected, and the resulting augmented datasets count to a total number of 17,885 annotated hands.

The achieved performance from the training of the models with augmented input data (cf. Table 2) resulted in consistently better mAPs than the baseline (m_{hsv}). Furthermore, there are minimal differences for $m_{\text{aug\_full}}$ and $m_{\text{aug\_partial}}$ – the latter is

Table 2 Validation results for the models with augmented input data trough colormap rotation

Model	mAP@0.75IoU (%)
m_{hsv}	93.2
$m_{\text{aug\_full}}$	95.58
$m_{\text{aug\_partial}}$	**95.63**

Table 3 Evaluation results of the detectors—comparing evaluation with the different test datasets E_{hsv} and E_{ext}

Model	F_{1_hsv} (%)	F_{1_ext} (%)	$F_{1\_\text{ext}} - F_{1\_\text{hsv}}$ (%)
m_{hsv}	83.5	33.2	−50.3
$m_{\text{aug\_full}}$	**87.9**	**88.1**	+0.2
$m_{\text{aug\_partial}}$	86.2	70.9	−15.3 %

slightly but not significantly better in terms of mAP. The different rotation ranges do not seem to have any influence during training. Compared to m_{hsv}, the augmented models are characterized by a decrease in loss even beyond the 400,000 steps. Regarding the mAP, a slower increase is recorded, which exceeds m_{hsv} after approximately 250,000 steps. In contrast, no significant difference could be seen between $m_{\text{aug\_full}}$ and $m_{\text{aug\_partial}}$ in either curve, which is attributable to the similar complexity of the supplied data.

To evaluate the trained models (m_{hsv}, $m_{\text{aug\_full}}$ and $m_{\text{aug\_partial}}$) w.r.t. color independence, two different evaluation datasets are introduced. Therefore, the evaluation dataset (E_{hsv}) is encoded using HSV colormap. To determine a model's color independence, we artificially extend the test data by colormap rotation to cover the whole colormap range (E_{ext}). The model performances in both evaluations—naming F_{1_hsv} and F_{1_ext}—must show comparable results to achieve color independence. On the other hand, if there are significant differences in the evaluation metrics, there is likely an expression of color sensitivity.

Table 3 summarizes the evaluation results. When comparing the differences of F_{1_hsv} and F_{1_ext} for each model, the m_{hsv} recorded the most significant performance spread with −50.3 % for the two evaluation sets. Thus, the model seems to react sensitively to color-mapped depth ranges, which are predominantly present in the training dataset. The model $m_{\text{aug\_partial}}$ shows similar but less distinctive behavior. The constrained rotation resulted in better detection performance of F_{1_ext}. The $m_{\text{aug\_full}}$ shows no significant difference in F_1 scores, and therefore, color-independent detection can be assumed.

In the following, the frequencies of detections and annotations are compared w.r.t. the different color levels. The results stating color independence are shown more clearly in Fig. 7. First, the annotations (lighter gray) are displayed in a histogram as a discrete distribution in the different color levels corresponding to the depth values. Subsequently, the indicated models' true positive (TP) detections (darker gray) were

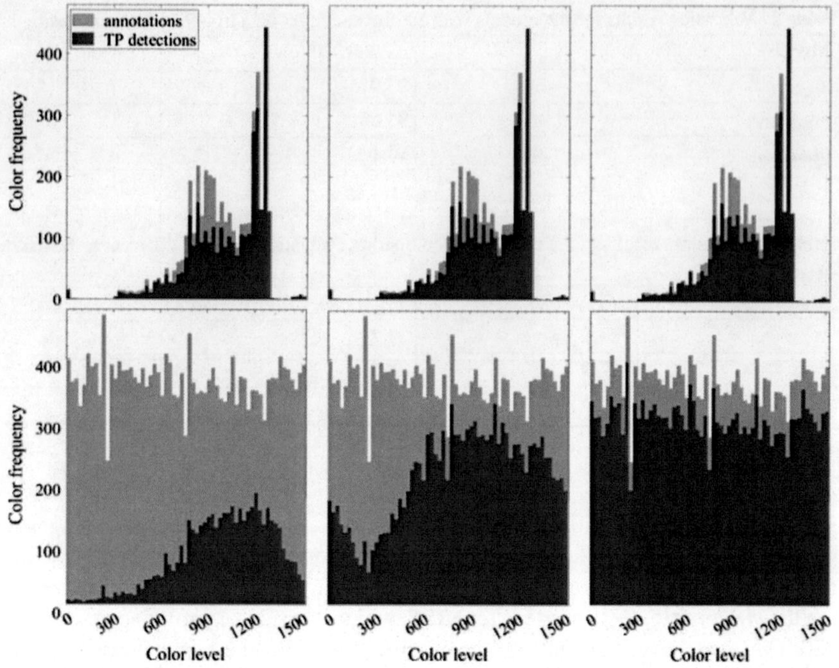

Fig. 7 Evaluation results of the models m_{hsv} (left), $m_{\text{aug\_partial}}$ (middle) and $m_{\text{aug\_full}}$ (right) showing color frequencies of detections and annotations w.r.t. the different color levels, using E_{hsv} (top) and E_{ext} (bottom)

plotted against the annotations. The conformity of these two distributions can be interpreted as recall—the aim is to achieve the maximum overlap and thus a recall of almost 100 %. For each model, the distributions for the evaluation with E_{hsv} (top) and E_{ext} (bottom) are compared.

Only moderate differences can be seen in a direct comparison of the top row images of all models. The distributions on the bottom visualize color-sensitive behavior for the models m_{hsv} (left) and $m_{\text{aug\_partial}}$ (middle), because TPs are not approximately equally present in all color levels. In contrast, the detections of the $m_{\text{aug\_full}}$ model (right) are evenly distributed over the entire evaluation set, and no color level preferences are apparent. Additionally, for model $m_{\text{aug\_full}}$ with better-balanced input data, the TP rate per color level is significantly increasing for the dominating colors of model m_{hsv} (between 800 and 1300). As a result, the assumption of color independence from Table 3 is confirmed.

Table 4 Evaluation results of the hand detectors trained with HANDS dataset—comparing evaluation with the different test datasets

Model	F_{1_hsv} (%)	F_{1_ext} (%)	$F_{1\_ext} - F_{1\_hsv}$ (%)
m_{hands_hsv}	96.4	1.65	−94.75
m_{hands_aug}	94.5	93.4	−1.1

3.4 Public Dataset: HANDS

To confirm our results, we applied the presented method to a publicly available dataset called HANDS [13]. The ground truth consists of 12,000 depth images of individuals captured from the front performing poses with their hands. For our purposes, only the bounding boxes of the hands are of interest.

In total, two models were trained, which differed only in terms of input data. The input data preparation was carried out as follows: Mapping the depth data using HSV colormap (m_{hands_hsv}), and mapping the depth data using HSV colormap including augmentation with $N = 5$ rotations (m_{hands_aug}).

As in the previous section, the evaluation is performed with two sets—one with HSV color mapping and the other artificially extended with colormap rotation. The obtained results (Table 4) reflect our expectations and show a high qualitative similarity to the evaluation of the dataset before.

The m_{hands_hsv} model shows better result of F_{1_hsv}, which comes due to the high color sensitivity caused by strong imbalances in the HANDS dataset. This can be seen in the very extreme drop of the F_{1_ext}. Additionally, it seems that the model is not generalizing to a broader spectrum of colors. The model m_{hands_aug} shows comparable results, as in the previous section. There is about a 1 % difference in the performance of the two evaluation sets, confirming the improvement in color-independent detection.

4 Conclusions and Future Work

In this chapter, we presented an extended color mapping for depth images to overcome the challenging problem of color dependency. We showed the different degrees of generalization between RGB and depth image trained models. We found that a depth encoding with HSV colormap further improves the performance compared to a simple grayscale encoding from 86.6 % to 93.2 % in $mAP@0.75IoU$. However, especially in small datasets dealing with high imbalances is a challenging problem—as they exist in the two datasets we used. Neglecting these will lead to an undesired sensitivity w.r.t. the object's color appearance and therefore to bad generalization results of the trained model. Our augmentation method reduces the color dependency in the depth domain by rotating the underlying colormap. The augmentation eliminates the imbalances and simultaneously increases the data volume. As a result, the trained models showed a much better degree of generalization in the evaluation due to the higher amount of color variation. A limitation of the method is the pure applicability to depth images, and therefore, limitations from depth-based models occur here as well, like lower detail level compared to color images and higher chance of false positives for similarly shaped objects.

The model resulting from the proposed method is used at a workstation to assist workers in manual assembly. Furthermore, with color-independent hand detection, we can adapt to a diversity of assembly tasks. Robust hand detection is necessary for this context to achieve better results in activity recognition which again leads to more possibilities in intelligent assistance.

Acknowledgements This work results from the projects GesInA40, within the funding initiative: "KMU-innovativ" and DiProLeA, both supported by the Federal Ministry of Education and Research (BMBF), grant numbers 01IS19043C and 02J19B123.

References

1. Tommasi T, Patricia N, Caputo B, Tuytelaars T (2015) A deeper look at dataset bias. CoRR http://arxiv.org/abs/1505.01257
2. Alvi MS, Zisserman A, Nellåker C (2018) Turning a blind eye: explicit removal of biases and variation from deep neural network embeddings. CoRR http://arxiv.org/abs/1809.02169
3. Kim B, Kim H, Kim K, Kim S, Kim J (2018) Learning not to learn: Training deep neural networks with biased data. CoRR http://arxiv.org/abs/1812.10352
4. Morerio P, Cavazza J, Murino V (2017) Minimal-entropy correlation alignment for unsupervised deep domain adaptation. CoRR http://arxiv.org/abs/1711.10288
5. Sun B, Saenko K (2016) Deep CORAL: correlation alignment for deep domain adaptation. CoRR http://arxiv.org/abs/1607.01719
6. Howard AG (2014) Some improvements on deep convolutional neural network based image classification. CoRR http://arxiv.org/abs/1312.5402
7. Zhang Z, Xie S, Chen M, Zhu H (2020) Handaugment: a simple data augmentation method for depth-based 3d hand pose estimation
8. Romero J, Tzionas D, Black MJ (2017) Embodied hands: modeling and capturing hands and bodies together. ACM Trans Graph (Proc SIGGRAPH Asia) 36(6):245:1–245:17 https://doi.org/10.1145/3130800.3130883
9. Haji-Esmaeili MM, Montazer G (2018) Playing for depth (2018). http://arxiv.org/abs/1810.06268
10. Iqbal U, Doering A, Yasin H, Krüger B, Weber A, Gall J (2017) A dual-source approach for 3d human pose estimation from a single image. http://arxiv.org/abs/1705.02883
11. Liu W, Anguelov D, Erhan D, Szegedy C, Reed SE, Fu C, Berg AC (2015) SSD: single shot multibox detector. http://arxiv.org/abs/1512.02325
12. Lin T, Maire M, Belongie SJ, Bourdev LD, Girshick RB, Hays J, Perona P, Ramanan D, Dollár P, Zitnick CL (2014) Microsoft COCO: common objects in context http://arxiv.org/abs/1405.0312
13. Nuzzi C, Pasinetti S, Pagani R, Coffetti G, Sansoni G (2021) Hands: an rgb-d dataset of static hand-gestures for human-robot interaction. Data in Brief 35:106,791. https://www.sciencedirect.com/science/article/pii/S2352340921000755%7D

Detection of Retinopathy of Prematurity Stages Utilizing Deep Neural Networks

Nazar Salih, Mohamed Ksantini, Nebras Hussein, Donia Ben Halima, Ali Abdul Razzaq, and Sohaib A. Mahmood

Abstract Retinopathy of prematurity is the leading cause of blindness in children around the world. This paper exhibited ten deep convolutional neural networks (DCNNs) models to detect ROP stages in fundus images using deep neural networks. A dataset of 3720 fundus images was collected from the private clinic Al-Amal eye centre, which consisted of 3 classes of ROP stages. A training dataset and a test dataset were created from the images. VGG16, ResNet50, ResNet101, ResNet152, SqueezNet1_0, SqueezNet1_1, DenseNet121, DenseNet169, AlexNet169, and Inception_v3 were trained to make differential diagnoses and then tested. The classification accuracies for the highest three DCNN (ResNet152, DenseNet169, Inception_v3) were 73.95, 77.14, and 99.50%, respectively.To conclude, after training with an extensive dataset, the Inception v3 DCNN model presented large potential in facilitating the diagnosis of ROP stages utilizing fundus images.

Keywords Artificial intelligence · Deep learning · Deep convolutional neural network · Retinopathy of prematurity

1 Introduction

In 1940, Terry was the first scientist to describe and identify the retinopathy of prematurity (ROP) as retrolental fibroplasia by a complete retinal detachment behind the lens [1]. Since that time, it has been considered as the main reason for what is called

N. Salih (✉) · M. Ksantini · D. B. Halima
CEMLab, ENIS, University of Sfax, Sfax, Tunisia
e-mail: nazar.s2009@yahoo.com

M. Ksantini
e-mail: mohamed.ksantini@ipeis.usf.tn

N. Hussein
Biomedical Engineering Department, Al-Khwarizmi College of Engineering, Baghdad University, Baghdad, Iraq

A. A. Razzaq · S. A. Mahmood
Ibn AL Haitham Teaching Eye Hospital, Baghdad, Iraq

X.-S. Yang et al. (eds.), *Proceedings of Seventh International Congress on Information and Communication Technology*, Lecture Notes in Networks and Systems 447,
https://doi.org/10.1007/978-981-19-1607-6_62

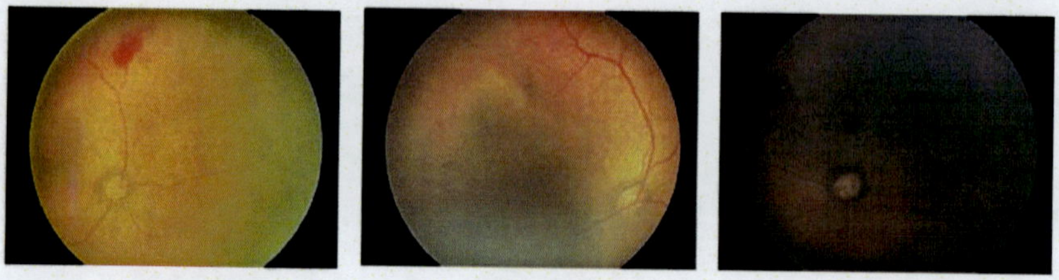

Fig. 1 From left to right, retina images represent the stages (2–4) of ROP

childhood blindness all around the world [2–4]. Thus, the percentage of survived premature babies born before they finish the gestational age less than 37 weeks has been increased after neonatal intensive care units [5]. Each year, about 10% of babies are premature babies, about 15 million babies according to global estimation [6].

Nowadays, ROP is regarded as a significant public health problem [7]. The lack of qualified ophthalmologists to test and treat ROP and the absence or delay in screening are significant contributors to ROP blindness [7]. Early detection and treatment of ROP can significantly improve the visual acuity of high-risk patients. As a result, early detection of ROP is critical for avoiding vision loss [8]. Premature retinopathy is classified according to the International Classification of Retinopathy of Prematurity (ICROP) guidelines published in [9–11]; ROP is classified based on the severity of the disease into stages (1–5), described in Fig. 1. Stage1 is the precaution for retinal detachment, which does not significantly differ from the healthy stage. Stage5 was a complete retinal detachment; therefore, it does not have a retina image. (a) Stage2 was an addition of depth and width to the demarcation line (Ridge); (b) stage3 was a presence of extraretinal fibrovascular proliferation; (c) stage4 was a partial retinal detachment.

For many reasons, this study is going to investigate the diagnosis of only stages "2–4" ROP:

1. As mentioned above, stage1 is not significantly different from the healthy condition.
2. Stage5 is a completely detached retina that is irreversible and leads to total blindness.
3. Other stages are essential to be classified as they could be managed by either laser or injection or surgical treatment.

Table 1 ROP datasets

	Stage2	Stage3	Stage4
Train set (80%)	1005	979	992
Validation set	126	123	124
Test set (20%)	251	245	248
Total	1382	1347	1364

2 Methodology

2.1 Dataset and Implementation

1. **Data**
   A total of (3720) fundus images from the ROP screening (from 2015 to 2019) were collected from the Private Clinic Al-Amal Eye centre in Baghdad, Iraq. All images were obtained using a RetCam3 imaging system. The resolutions of the fundus images were 640 × 480 pixels, though they are resized to 224 × 224 when inputting to our deep learning models.
2. **Image Labelling**
   Two senior ophthalmologists who had over 15 years of experience working with patients with ROP have been involved in the study. These experts labelled the fundus images and classified them as stage2, stage3, and stage4.
3. **Data Partition**
   As shown in Table 1, the dataset used for training, validation, and testing the model is split at random.
4. **Implementation**
   Intel Core™ i7-5500 CPU At 2.40, 16 GB RAM was used for the entire method. Installing Python 3.9.5 was done on a Windows 10 environment, with everything done in that environment.

2.2 Network Architecture

On the ImageNet dataset, Inception v3 achieved a 78.1% accuracy rate, making it a frequently used image recognition model. Multiple researchers have contributed to the model, which is the sum of their efforts. "Rethinking the Inception Architecture for Computer Vision" by Szegedy and colleagues was published in 2015 [12].

The DCNN model we tested was Inception v3. Inception v3 has three types of inception modules: inception A, inception B, and inception C, as shown in Fig. 2. Convolution modules of this type can generate discriminatory features whilst reducing the number of parameters. Each inception module has multiple convolutional and pooling layers running in parallel. It is possible to reduce the number of parameters in the inception modules by using small convolutional layers, such as 3

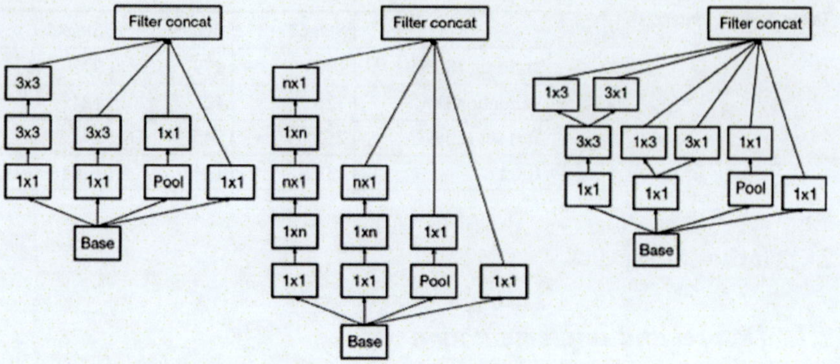

Fig. 2 Inception v3 modules: A, B, and C, respectively (left to right)

× 3, 1 × 3, 3 × 1, and 1 × 1. It is possible to stack three inception A modules, five inception B modules, and two inception C modules. Image size in the dataset was 224 x 224, but Inception v3 expects 224 x 299 as an input image. We did not scale the images to 299 × 299 pixels when training and testing Inception v3. This simply changed the size of the feature maps generated during the process, not the number of channels, so the results were acceptable. Inception modules and convolutional layers resulted in a feature map with 2048 channels in a 5 × 5 format.

Softmax classifiers were added to the model, and the class with the highest probability was selected as the predicted one. Several convolutional layers and pooling layers make up each inception module. The dimensions of convolutions denote an n m layer, and pooling is indicated by pooling. Original Inception v3 has 1000 classes, but we had just three: stage1, stage2, and stage3 Consequently, we reduced the number of output channels from 1000 to 3 in the final layer.

3 Results and Discussion

In present study, we covered a large variety of models by selecting them from different classification models, with each having a different number of layers, such as VGG16 [13], ResNet50 [14], ResNet101 [14], ResNet152 [14], SqueezNet1_0 [15], SqueezNet1_1 [15], DenseNet121 [16], DenseNet169 [16], AlexNet169 [17], and inception_v3 [12]. All these ten DNN models were selected to achieve our primary aim of detecting ROP stages. From these outcomes, it was selected the highest accuracy models which are ResNet152, DenseNet169, and Inception_v3.

The ability of our model system Inception v3 to distinguish between the three stages of ROP classification from fundus images was assessed. The results showed that the system was able to achieve 99.5% accuracy, a sensitivity of 1.0, and a specificity of 1.0 for each stage of the process. In addition, the F1-score of 1.0 is grading the ROP stages as stage2, stage3, stage4 and AUC of 0.1, ROC of 1.0 (Table

Table 2 Performances evaluation of the three deep neural network models.

Architecture	ResNet152		DenseNet169		Inception_v3	
	Sensitivity	Specificity	Sensitivity	Specificity	Sensitivity	Specificity
Stage2	0.65	0.88	0.84	0.88	1.0	1.0
Stage3	0.69	0.81	0.66	0.91	1.0	1.0
Stage4	0.76	0.86	0.86	0.88	1.0	1.0
Accuracy	7395		7714		995	
Precision	0.70		0.78		1.0	
Recall	0.70		0.79		1.0	
F1-score	0.70		0.78		1.0	
AUC	0.77		0.86		1.0	
ROC	0.81		0.87		1.0	

2). It was also compared the system's performance with the results obtained by another two models (ResNet152, DenseNet169).

ResNet152: Achieved an accuracy of 73.95%, F1-score of 0.70, AUC of 0.70, and ROC of 0.81. DenseNet169: Achieved an accuracy of 77.14%, F1-score of 0.78, AUC of 0.86, and ROC of 0.87.

For each image, three confusion matrixes are shown in Fig. 3 to show how different predictions have been assigned to each one. There are two sets of data here: a predicted label for each sample and a true label for each sample. On the heat map, each diagonal line represents the percentage of images that were correctly classified into the corresponding category. Elements that are not diagonal show how many images were incorrectly classified and by how much.

Figure 4 shows the ROC curve for the algorithms that detect stages of ROP. The ROC for DenseNet169, ResNet152, and Inception v3 algorithms was 0.81, 0.87, and 1.0, respectively.

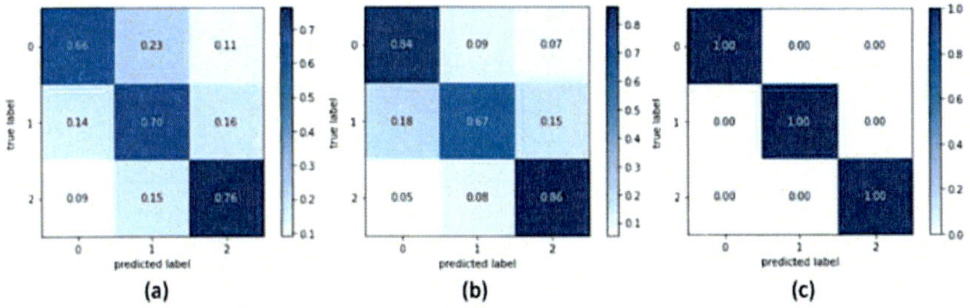

Fig. 3 Three confusion matrixes for the three deep neural network models. **a** Confusion matrix of DenseNet169. **b** Confusion matrix of ResNet152. **c** Confusion matrix of Inception v3

Fig. 4 Receiver operating characteristics (ROCs) curves for algorithms detecting stages of ROP. **a** DenseNet169. **b** ResNet152. **c** Inception v3

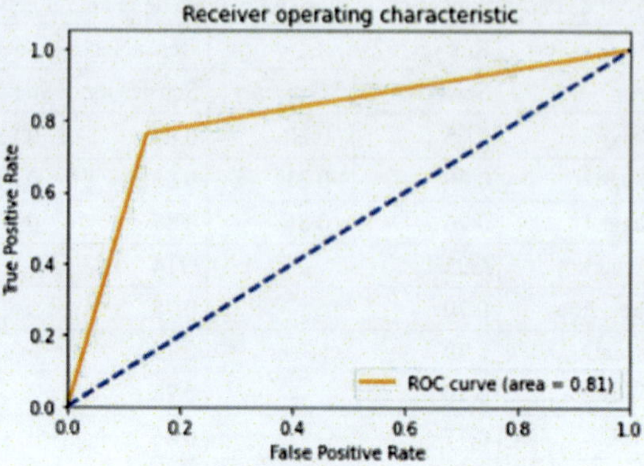

(a)ROC of DenseNet169 algorithm (0.81)

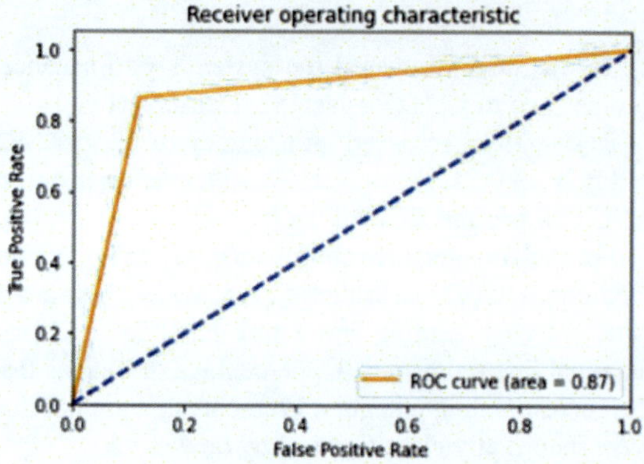

(b)ROC of ResNet152 algorithm (0.87)

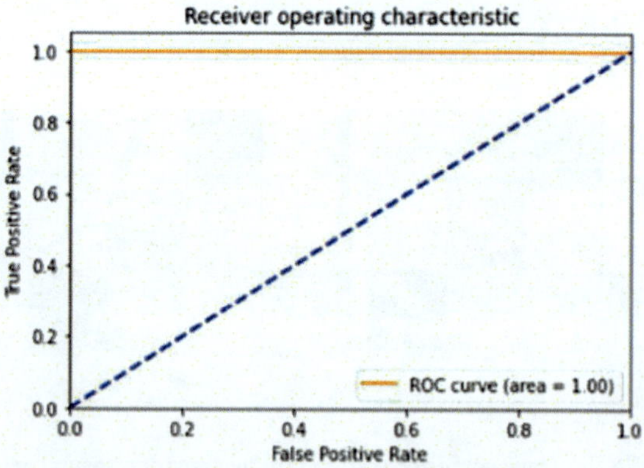

(c)ROC of inceptionv3 algorithm (1.0)

4 Conclusion

Infants born prematurely and with low birth, weights are more likely to suffer from retinopathy of prematurity (ROP). It causes retinal vascular multiplication, leading to visual loss and eventually, retinal detachment, resulting in blindness.

Convolutional neural networks (CNNs) are a fast-developing approach for automated image processing in several medical disciplines.

The Inception v3 DCNN model showed considerable promise in facilitating the identification of ROP phases based on fundus pictures after training on a large dataset. In this paper, it was applied (10) deep neural architectures for detecting the stages of ROP in preterm infants. From these outcomes, the highest accuracy models were selected (ResNet152, DenseNet169, Inception_v3). Our results showed that the Inception_v3 achieved an accuracy of 99.50% for detecting the stage of ROP. To help advance medical reform in the new environment, we will be focussing our efforts on improving algorithms and building a larger training dataset.

References

1. Terry TL (1942) Extreme prematurity and fibroblastic overgrowth of persistent vascular sheath behind each crystalline lens* *from the massachusetts eye and ear infirmary. This investigation is made possible through the special fund for research for pathology laboratory. Am J Ophthalmol 25(2):203–204. https://doi.org/10.1016/S0002-9394(42)92088-9
2. Early Treatment for Retinopathy of Prematurity Cooperative Group (2005) The incidence and course of retinopathy of prematurity: findings from the early treatment for retinopathy of prematurity study. Pediatrics 116(1):15–23. https://doi.org/10.1542/peds.2004-1413
3. Zin A, Gole GA (2013) Retinopathy of prematurity-incidence today. Clin Perinatol 40(2):185–200. https://doi.org/10.1016/j.clp.2013.02.001
4. Solebo AL, Teoh L, Rahi J (2017) Epidemiology of blindness in children. Arch Dis Child 102(9):853–857. https://doi.org/10.1136/archdischild-2016-310532
5. Goldenberg RL, Culhane JF, Iams JD, Romero R (2008) Epidemiology and causes of preterm birth. The Lancet 371(9606):75–84. https://doi.org/10.1016/S0140-6736(08)60074-4
6. Martin JA, Kochanek KD, Strobino DM, Guyer B, MacDorman MF (2005) Annual summary of vital statistics–2003. Pediatrics 115(3):619–634. https://doi.org/10.1542/peds.2004-2695
7. Rashaed SA (2019) Retinopathy of prematurity—a brief review. Dr Sulaiman Al Habib Med. J. 1(3–4):58–64. https://doi.org/10.2991/dsahmj.k.191214.001
8. Huang Y-P et al (2020) Deep learning models for automated diagnosis of retinopathy of prematurity in preterm infants. Electronics 9(9). https://doi.org/10.3390/electronics9091444
9. Patz A (1984) An international classification of retinopathy of prematurity. Pediatrics 74(1):127–133
10. An international classification of retinopathy of prematurity. II. The classification of retinal detachment. The international committee for the classification of the late stages of retinopathy of prematurity. Arch Ophthalmol Chic Ill 1960 105(7):906–912, Jul. 1987
11. International Committee for the Classification of Retinopathy of Prematurity (2005) The international classification of retinopathy of prematurity revisited. Arch Ophthalmol Chic Ill 1960, 123(7):991–999. https://doi.org/10.1001/archopht.123.7.991
12. Szegedy C, Vanhoucke V, Ioffe S, Shlens J, Wojna Z (2016) Rethinking the inception architecture for computer vision. In: 2016 IEEE conference on computer vision and pattern recognition (CVPR), pp 2818–2826. https://doi.org/10.1109/CVPR.2016.308

13. Simonyan K, Zisserman A (2015) Very deep convolutional networks for large-scale image recognition. ArXiv14091556 Cs. [Online]. Available: http://arxiv.org/abs/1409.1556. Accessed 23 Jun 2021
14. He K, Zhang X, Ren S, Sun J (2015) Deep residual learning for image recognition. ArXiv151203385 Cs. [Online]. Available: http://arxiv.org/abs/1512.03385. Accessed 23 Jun 2021
15. Iandola FN, Han S, Moskewicz MW, Ashraf K, Dally WJ, Keutzer K (2016) SqueezeNet: AlexNet-level accuracy with 50x fewer parameters and <0.5MB model size. ArXiv160207360 Cs. [Online]. Available: http://arxiv.org/abs/1602.07360. Accessed 23 Jun 2021
16. Huang G, Liu Z, van der Maaten L, Weinberger KQ (2018) Densely connected convolutional networks. ArXiv160806993 Cs. [Online]. Available: http://arxiv.org/abs/1608.06993. Accessed 23 Jun 2021
17. Krizhevsky A, Sutskever I, Hinton GE (2017) ImageNet classification with deep convolutional neural networks. Commun ACM 60(6):84–90. https://doi.org/10.1145/3065386

Iterative Approach for Reduction of Index-2 Periodic Models Using Generalized Inverse Procedure

Atia Afroz, Mohammad-Sahadet Hossain, Musannan Hossain, and Mashrur Wasek

Abstract This chapter studies the structure-preserving iterative approach for reduction of index-2 periodic models descriptor systems using generalized inverses of periodic matrix pairs. This work fulfills two objectives. The first part of our research is concerned with the discrete-time generalized a system which reformulate from the discrete-time descriptor system by changing the structure of the system. Then the periodic matrix pairs are computed from the generalized inverse matrices, and the reformulated system is represented by a cyclic lifted structure. Smith method is exploited to find the iterative solutions of the associated Lyapunov equations of the cyclic lifted system. The original periodic system is contained in the solutions of the periodic Lyapunov equations. The periodic system is then reduced by using projectors computed from those periodic solutions. The above procedures are applied to reduce an artificial problem of the index-2 periodic structure. To verify the accuracy and performance of the algorithm, we have demonstrated the results obtained from numerical simulations.

Keywords Balance truncation · Smith iterative method · Index-2 periodic descriptor systems · Cyclic iteration approach

---

This project was supported by the North South University Research Grant: CTRG-20/SEPS/07.

A. Afroz · M.-S. Hossain (✉)
Department of Mathematics and Physics, North South University, Dhaka, Bangladesh
e-mail: mohammad.hossain@northsouth.edu

A. Afroz
e-mail: atia.afroz@northsouth.edu

M. Hossain · M. Wasek
Department of Electrical and Computer Engineering, North South University, Dhaka, Bangladesh
e-mail: musannan.hossain@northsouth.edu

M. Wasek
e-mail: mashrur.wasek@northsouth.edu

X.-S. Yang et al. (eds.), *Proceedings of Seventh International Congress on Information and Communication Technology*, Lecture Notes in Networks and Systems 447,
https://doi.org/10.1007/978-981-19-1607-6_63

1 Introduction

The discrete-time system of index-2 that we study in this research has a very usual state-space form

$$\underbrace{\begin{bmatrix} E_{1,p} & 0 \\ 0 & 0 \end{bmatrix}}_{E_p} \underbrace{\begin{bmatrix} x_{1,p+1} \\ x_{2,p+1} \end{bmatrix}}_{x_{p+1}} = \underbrace{\begin{bmatrix} A_{1,p} & A_{2,p} \\ A_{2,p}^T & 0 \end{bmatrix}}_{A_p} \underbrace{\begin{bmatrix} x_{1,p} \\ x_{2,p} \end{bmatrix}}_{x_p} + \underbrace{\begin{bmatrix} B_{1,p} \\ 0 \end{bmatrix}}_{B_p} u_p,$$
$$y_p = \underbrace{\begin{bmatrix} C_{1,p} & 0 \end{bmatrix}}_{C_p} \begin{bmatrix} x_{1,p} \\ x_{2,p} \end{bmatrix}, \quad p = 0, 1, \ldots, K-1. \tag{1}$$

The system is K-periodic, and $p = 0, 1, \ldots, K-1$. The matrices $E_{1,p}, A_{1,p} \in \mathbb{R}^{n_{1,p} \times n_{1,p}}$ have full rank and $A_{2,p} \in \mathbb{R}^{n_{1,p} \times n_{2,p}}$, $B_{1,p} \in \mathbb{R}^{n_{1,p} \times m_p}$, and $C_{1,p} \in \mathbb{R}^{p_p \times n_{1,p}}$. Here $x_{1,p} \in \mathbb{R}^{n_{1,p}}$, $x_{2,p} \in \mathbb{R}^{n_{2,p}} (n_{1,p} > n_{2,p})$ are the states and $n_{1,p} + n_{2,p} = n_p$. The dimension of the system is $\mathbf{n} = (n_0, n_1, \ldots, n_{K-1})$.

The periodic Lyapunov equations that represent the energy quotients of system (1) from different input–output are given by

$$E_p M_{p+1} E_p^T - A_p M_p A_p^T = B_p B_p^T, \tag{2a}$$
$$E_{p-1}^T N_p E_{p-1} - A_p^T N_{p+1} A_p = C_p^T C_p, \tag{2b}$$

where $p = 0, 1, \ldots, K-1$, $\{M_p\}_{p=0}^{K-1}$ and $\{N_p\}_{p=0}^{K-1}$ are the unique solutions of (2a) and (2b), respectively. Here M_p represents the controllability and while N_p represents the observability Gramian, respectively, which are also periodic. We assume this periodic system is stable in the reference that all finite eigenvalues $\{E_p, A_p\}_{p=0}^{K-1}$ lies inside the unit disk [1].

Following the work of [2], system (1) can be represented as a generalized periodic system by adopting the algebraic part of (1) into its finite difference part as

$$\tilde{E}_p \tilde{x}_{1,p+1} = \tilde{A}_p \tilde{x_p} + \tilde{B}_p u_p, \quad \tilde{y}_p = \tilde{C}_p \tilde{x_p} \tag{3}$$

where

$$\tilde{E}_p = \Pi_p^T E_{1,p} \Pi_{p+1}, \; \tilde{A}_p = \Pi_p^T A_{1,p} \Pi_p, \; \tilde{B}_p = \Pi_p^T B_{1,p}, \; \tilde{C}_p = C_{1,p} \Pi_p, \tag{4}$$

are the transformed matrices, and $\Pi_p = I - A_{2,p} \left(A_{2,p+1}^T E_{1,p}^{-1} A_{2,p} \right)^{-1} A_{2,p+1}^T E_{1,p}^{-1}$, for $p = 0, 1, \ldots, K-1$, is an oblique periodic projector. Details of this transformation and the formulation of the periodic projector Π_p with its significant properties can be found in [3]. However, the null space is still present in system (3). Hence, an explicit generalized system of dimension $(n_{1,p} - n_{2,p})$ can be formulated by a transformation $\Pi_p = \Omega_p \Psi_p^T$, where $\Omega_p, \Psi_p^T \in \mathbb{R}^{n_{1,p} \times (n_{1,p} - n_{2,p})}$ such that $\Omega_p \Psi_p^T = I_p$.

Using the notation $\hat{x}_p = \Omega_p^T \tilde{x}_{1,p}$, we can rewrite (3) as a generalized system in the explicitly form:

$$\hat{E}_p \hat{x}_{1,p+1} = \hat{A}_p \hat{x}_p + \hat{B}_p u_p, \qquad \hat{y}_p = \hat{C}_p \hat{x}_p, \tag{5}$$

with the system matrices, $\hat{E}_p = \Psi_p^T E_{1,p} \Psi_{p+1}$, $\hat{A}_p = \Psi_p^T A_{1,p} \Psi_p$, $\hat{B}_p = \Psi_p^T B_{1,p}$, $\hat{C}_k = C_{1,p} \Psi_p$.

Model reduction for such system has been proposed in [3], and also in the LTI form [4] of continuous system. However, the model proposed in [3] encounters two different critics:

- The sparsity of the system matrices is destroyed in (3) due to the pre- and post-multiplication with dense projectors.
- Explicit computation of the projectors is time-consuming.

We exploit the approach of generalized inverse matrices of periodic matrix pairs corresponding to the periodic system (1) that exhibits the reflexivity. Therefore, explicit computation of the projectors can be averted and the model reduction process is then applied to the reformed periodic system.

The residual of the paper is outlined in the following manner. In Sect. 2, we discuss the generalized inverses of periodic matrix pairs associated with the eigenstructures of the periodic matrix pairs. We present the cyclic lifted representation of (2a) and (2b) in Sect. 3 and illustrate the iteratively solutions of the lifted discrete-time algebraic Lyapunov equations (LDALEs). We develop a cyclic computation of the LDALEs here, which functions directly with the periodic matrices and avoid explicit lifted construction. A Balanced Truncation Model Order Reduction (BT-MOR) is considered in Sect. 4, and numerical results with graphical illustrations from an index-2 model problem are shown in Sect. 5. Finally, this paper ends with concluding remarks that are exposed in Sect. 6.

2 Periodic Generalized Inverse Matrices

The idea of the periodic system of the reflexive generalized inverse matrices of periodic matrix pairs (1) is used. As a result, explicit projector calculation can be omitted. The reformed periodic system of generalized inverses is next subjected to the model reduction procedure. For periodic systems, the generalized inverses have been considered in [5] and [6].

We develop the idea of the reflexive generalized inverses from the periodic system (1), and the periodic matrix pairs $\{(\tilde{E}_p, \tilde{A}_p)\}_{p=0}^{K-1}$ can be used for this separation for $p = 0, 1, \ldots, K-1$, which considered from the periodic Kronecker canonical form

$$U_p \tilde{E}_p V_{p+1} = \begin{bmatrix} I_{n_{p+1}^f} & 0 \\ 0 & I_{n_p^\infty} \end{bmatrix}, U_p \tilde{A}_p V_p = \begin{bmatrix} A_p^f & 0 \\ 0 & I_{n_p^\infty} \end{bmatrix}, \tag{6}$$

where U_k, V_k are nonsingular transformation matrices [7]. Then, we obtain for $\bar{E}_p$ the *reflexive generalized inverse matrices* as

$$E_p^{\dagger} = V_{p+1} \begin{bmatrix} I_{n_{p+1}^f} & 0 \\ 0 & 0 \end{bmatrix} U_k, \tag{7}$$

for $p = 0, 1, \ldots, K-1$. Additionally, the reflexive generalized inverse matrices preserve the expressions

$$E_p^{\dagger} E_p E_p^{\dagger} = E_p^{\dagger}, \quad E_p E_p^{\dagger} = \Pi_p, \quad E_p^{\dagger} E_p = \Pi_{p+1}, \tag{8}$$

for $p = 0, 1, \ldots, K-1$.

3 Iterative Solutions of Lifted Lyapunov Equations

The solutions to Lyapunov equations (2a) and (2b) are used in a wide range applications for model reduction of larger models. Several approaches for solving Lyapunov equations used in so many literature including Smith method, alternating direction implicit method (see [8], [9] and [10]). A very popular method to solve (2a) and (2b) is the time-invariant reconstruction of (2a) and (2b), known as lifted representation. Following [1], periodic equations (2a) and (2b) are analogous to the lifted periodic discrete-time algebraic Lyapunov equations

$$\mathcal{E}\mathcal{M}\mathcal{E}^T - \mathcal{A}\mathcal{M}\mathcal{A}^T = \mathcal{B}\mathcal{B}^T, \tag{9a}$$
$$\mathcal{E}^T\mathcal{N}\mathcal{E} - \mathcal{A}^T\mathcal{N}\mathcal{A} = \check{\mathcal{C}}^T\check{\mathcal{C}}, \tag{9b}$$

here $\mathcal{E}$, $\mathcal{A}$, $\mathcal{B}$, and $\mathcal{C}$ are in the lifted forms, $\mathcal{M} = \text{diag}(M_1, \ldots, M_{K-1}, M_0)$ and $\mathcal{N} = \text{diag}(N_1, \ldots, N_{K-1}, N_0)$ (see [1] and [11]).

The transfer function of (1) in its lifted form is represented as below

$$\mathcal{H}(z) = \mathcal{C}(z\mathcal{E} - \mathcal{A})^{-1}\mathcal{B}. \tag{10}$$

The iterative solution of (9a) requires to find the inverse of the matrix $\mathcal{E}$. Since $\tilde{E}_p$ are singular for $p = 0, 1, \ldots, K-1$, it is obvious that $\mathcal{E}$ is singular. Directly finding the inverse of $\mathcal{E}$ is not possible. In that case, we consider the reflexive periodic inverses $E_p^{\dagger}$ for $p = 0, 1, \ldots, K-1$. Then, the reconstruction of $\mathcal{E}$ as follows

$$\mathcal{E}^{\dagger} = \text{diag}\left(E_0^{\dagger}, E_1^{\dagger}, \ldots, E_{K-1}^{\dagger}\right).$$

Therefore, the $\mathcal{E}^{-1}$ and $\mathcal{E}^{-T}$ are replaced by the $\mathcal{E}^{\dagger}$ and $\mathcal{E}^{\dagger^T}$ in construction of iterative solutions of (9a). Multiplying by $\mathcal{E}^{\dagger}$ and $\mathcal{E}^{\dagger T}$ on the left-hand side and right-hand side, respectively, then we obtain,

$$\mathcal{M} - (\mathcal{E}^{\dagger}\mathcal{A})\mathcal{M}(\mathcal{E}^{\dagger}\mathcal{A})^{T} = \mathcal{E}^{\dagger}\mathcal{B}(\mathcal{E}^{\dagger}\mathcal{B})^{T}. \tag{11}$$

We can use the following Smith iterations (see [8] and [9]) for determining an approximation of the solution $\mathcal{M}$ in (11),

$$\mathcal{M}_i = \sum_{l=0}^{i-1}(\mathcal{E}^{\dagger}\mathcal{A})^{l}\mathcal{E}^{\dagger}\mathcal{B}(\mathcal{E}^{\dagger}\mathcal{B})^{T}\left((\mathcal{E}^{\dagger}\mathcal{A})^{T}\right)^{l}. \tag{12}$$

Using the Cholesky factorization $\mathcal{M}_i = \mathcal{R}_i\mathcal{R}_i^{T}$, where

$$\mathcal{R}_i = \left[\mathcal{E}^{\dagger}\mathcal{B}, (\mathcal{E}^{\dagger}\mathcal{A})\mathcal{E}^{\dagger}\mathcal{B}, \ldots, (\mathcal{E}^{\dagger}\mathcal{A})^{i-1}\mathcal{E}^{\dagger}\mathcal{B}\right] \tag{13}$$

is the Cholesky factor. In every iteration (13) [11], it is not possible to preserve the estimated block diagonal structure of the solution which is the main disadvantage of the Smith method for this system. We present a cyclic permutation matrix to resolve this problem in each of the iterations of (13). We assume the following permutation matrix

$$\Omega = \operatorname{diag}(I_{\nu_0}, I_{\nu_1}, \ldots, I_{\nu_{K-1}}), \tag{14}$$

where I_{ν_p} signifies a square identity matrix with the size ν_p which states the number of columns of B_p, for $p = 0, 1, \ldots, K-1$. We present a cyclic permutation $\sigma^{i}(\Omega)$ at each i iteration step in (13), where $\sigma^{i}(\Omega)$ changes at each iteration step by a forward block shift [11].

Therefore, for $p = 0, 1, \ldots, K-1$ and for $i = 1$ we get,

$$\sigma^{1}\Omega = \sigma(\Omega) = \begin{bmatrix} 0 & I_{\nu_0} & 0 & \ldots & 0 \\ 0 & 0 & I_{\nu_1} & \ldots & 0 \\ \vdots & \vdots & \vdots & \ddots & \vdots \\ 0 & \ldots & \ldots & 0 & I_{\nu_{K-2}} \\ I_{\nu_{K-1}} & \ldots & \ldots & 0 & 0 \end{bmatrix}, \tag{15}$$

here Ω is just the last block column of forward shift in (16).

We want to put a remark that the periodicity property [11] is well-preserved for this permutation matrix $\sigma^{K}(\Omega) = \sigma^{0}(\Omega)$. Therefore, (13) takes the new form

$$\mathcal{R}_i = \left[\mathcal{E}^{\dagger}\mathcal{B}\,\Omega, (\mathcal{E}^{\dagger}\mathcal{A})\mathcal{E}^{\dagger}\mathcal{B}(\sigma\Omega), \ldots, (\mathcal{E}^{\dagger}\mathcal{A})^{\dagger}\mathcal{E}^{\dagger}\mathcal{B}(\sigma^{i-1}\Omega)\right]. \tag{16}$$

The iterations (16) allow us to determine periodic Cholesky factor R_p directly just avoiding the lifted constructions. The following illustrates some of the periodic iterative computations of $R_{i,p}$ at the ith iteration step.

For $i = 1$, we get

$$R_{1,0} = E_{K-1}^{\dagger}\tilde{B}_{K-1},$$
$$R_{1,1} = E_0^{\dagger}\tilde{B}_0,$$
$$\vdots$$
$$R_{1,K-1} = E_{K-2}^{\dagger}\tilde{B}_{K-2}.$$

For $i = 2$, we get

$$R_{2,0} = E_{K-1}^{\dagger}A_{K-1}^{-1}R_{1,K-1},$$
$$R_{2,1} = E_0^{\dagger}A_0^{-1}R_{1,0},$$
$$\vdots$$
$$R_{2,K-1} = E_{K-2}^{\dagger}A_{K-2}^{-1}R_{1,K-1}.$$

and so forth. Algorithm 1 summarizes the complete computation. In step 9, it is worth noting that we eliminate the superfluous columns of R_p under the tolerance τ by the QR decomposition method.

Algorithm 1: For controllability Gramian Cyclic composition of LDALEs with the Smith method.

Input: $\tilde{A}_p$, $E_p^{\dagger}$, and $\tilde{B}_p$.
Output: Periodic low rank of R_p such that $M_p = R_p\,(R_p)^T$.
1: **for** $p = 0, 1, ..., K-1$ **do**
2: $R_{1,p} = E_{p-1}^{\dagger}\tilde{B}_{p-1}$
3: $R_{1,p}^{B} = R_{1,p}$
4: **end for**
5: **for** $i = 2, 3, ...,$ **do**
6: **for** $p = 0, 1, ..., K-1$ **do**
7: $R_{i,p} = E_{p-1}^{\dagger}A_{p-1}R_{i-1,p-1}$
8: $R_{i,p}^{b} = \left[R_{i-1,p}^{b} \;\; R_{i,p}\right]$
9: $R_p = QR(R_{i,p}^{b}, \tau_p)$
10: **end for**
11: **end for**

4 Reduction of the Model Under Balanced Truncation

We estimate a reformulated reduced-order model from the original model of order $\bar{r} = (\bar{r}_0, \bar{r}_1, \bar{r}_2, \ldots, \bar{r}_{K-1})$ for (1) as

$$\begin{aligned}\bar{E}_p\bar{x}_{p+1} &= \bar{A}_p\bar{x}_p + \bar{B}_p u_p,\\ \bar{y}_p &= \bar{C}_p\bar{x}_p,\end{aligned} \tag{17}$$

where the system matrices are of compatible orders and $\bar{r}_p \leq n_p$ for $p = 0, 1, \ldots, K-1$. Also, $\sum_{p=0}^{K-1} r_p = \mathbf{r}$ and $\mathbf{r} \ll \mathbf{n}$. For the model reduction method (see [12] and [13]), we directly follow the strategy of [14] and compute the projection matrices using the Cholesky factors of the periodic Gramian $P_p = R_p R_p^T$, $Q_p = L_p^T L_p$.

We also construct the reduced lifted transfer function $\hat{\mathcal{H}}$ for (17) and compute the H_∞ infinite-norm by checking the error bound to understand the reduced-order model's transfer function which is acceptable estimation of the transfer function from the original system

$$||\mathcal{H} - \hat{\mathcal{H}}||_{H_\infty} \leq 2\sum_{p=0}^{K-1} \text{trace}\left(\Sigma_{2,p}\right). \tag{18}$$

Here $\Sigma_{2,p}$ includes all the available Hankel singular values (HSVs) which have already eliminated (see [14] and [15]).

5 Numerical Findings

Our suggested algorithm's accuracy and performance are evaluated incorporating suitable data that is specific to index-2 descriptor systems. The model example is taken from [3]. The sparsity structures of the periodic matrices are same as of (1). We consider $K = 3$ and $p = 0, 1, 2$. The dimension of the original system is $\mathbf{n} = (10, 10, 10)$. The dimensions of $A_{1,p}$ and $E_{1,p}$ are 8×8, whereas the dimensions of $B_{1,p}$ and $C_{1,p}$ are 8×2 and 3×8, respectively, for $p = 0, 1, 2$. We obtained the reduced-order system of dimension $\bar{r} = (6, 6, 6)$ with reduction tolerance 10^{-1}.

Using Algorithm 1, we determine the periodic low-rank Cholesky factor R_p. In order to obtain an acceptable R_p, the algorithm is run 14 times. To obtain the observability Gramian, we determined the low-rank Cholesky factor L_p.

Finite eigenvalues are plotted from the original system and reformulated reduced lifted systems, in Fig. 1a. We see a nice match between them. Figure 1b shows the decreasing patterns of Hankel singular values (HSVs) of the periodic subsystems of $p = 0, 1, 2$.

To determine the correctness in their lifted form, we compare the transfer function of a reduced-order model to the original model. In Fig. 2a, we observe the original system, equivalent reduced system, and the projected system of the transfer functions. We observe that they have a strong resemblance. Despite the notable overlapping of the original and reduced form to the transfer functions, we still need to compute the reduced-order transfer function aimed at checking the error. In Fig. 2b, we can notice from the reduced-order system that the absolute error is fairly insignificant.

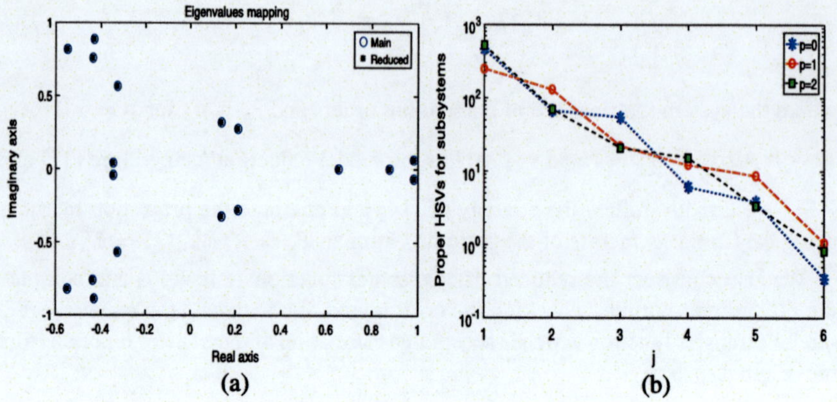

Fig. 1 **a** Eigenstructures, and **b** Proper HSVs of original (main) system

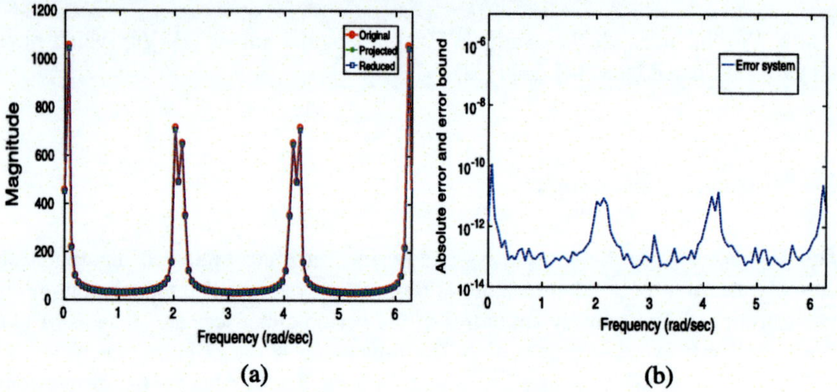

Fig. 2 **a** Norm of transfer functions, **b** Error in transfer functions

In order to demonstrate the efficiency with the exactness of the reduced model, we have plotted the linear simulation for a periodic input $u = \cos t$ both for the original system and reduced systems in Fig. 3a and the body diagram in Fig. 3b of those systems are also plotted. In all the above cases, we see very satisfactory results. Thus, we observe the reduced-order model is effective with the actual system.

6 Conclusion

In this paper, we exploited the generalized inverses of periodic matrices and from the periodic Lyapunov equations found the iterative solutions. These solutions are then applied to approximate reduced-order models from their corresponding original model which is based on balanced truncation. Our numerical findings show that our

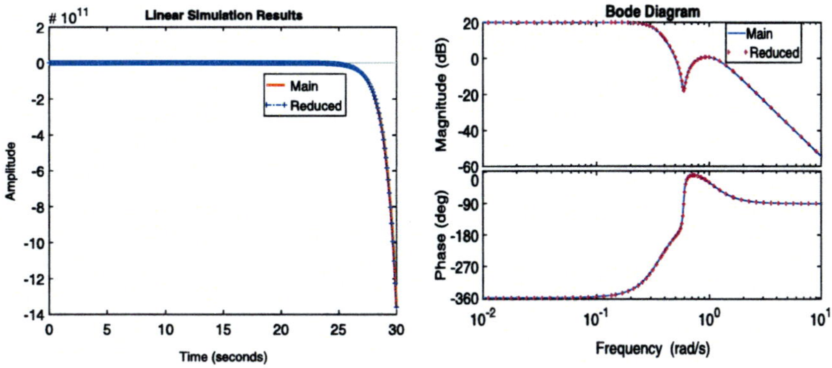

Fig. 3 **a** Linear simulations, and **b** Bode diagrams for main and reduced systems (to 1st output from 1st input)

proposed approach has essentially the same response characteristics as the original system, whereas our proposed approach developed from the reduced-order model. Our proposed technique process works on the reformulated generalized system but not in the original sparse system which is the key drawback of our article. Developing a more robust iterative technique for the MOR, where the sparsity patterns of the original system are restored, will be our next forwarding of this present MOR method.

References

1. Benner P, Hossain M-S, Stykel T (2011) Model reduction of periodic descriptor systems using balanced truncation. Lec Notes Elec Eng 193–206
2. Heinkenschloss M, Sorensen DC, Sun K (2008) Balanced truncation model reduction for a class of descriptor systems with application to the Oseen equations. SIAM J Sci Comp 30(2):1038–1063
3. Hossain Khan E, Hossain M, Galib Omar S, Tahsin A, Monir Uddin M (2018) K-cyclic Smith iterative method for model reduction of index-2 periodic control systems. In: 2018 International conference on innovations in science, engineering and technology (ICISET), pp 151–156
4. Hossain M-S, Hossain Khan E, Monir Uddin M, Galib Omad S (2020) An efficient model reduction strategy for discrete-time index descriptor control system
5. Varga A (2004) Computation of generalized inverses of periodic systems. In: 2004 43rd IEEE conference on decision and control (CDC) (IEEE Cat. No.04CH37601), vol 5, pp 5397–5402
6. Chu EK-W, Fan H-Y, Lin W-W (2007) Projected generalized discrete-time periodic Lyapunov equations and balanced realization of periodic descriptor systems. SIAM J Mat Anal Appl 29(3):982–1006
7. Stykel T, Simoncini V (2012) Krylov subspace methods for projected Lyapunov equations. Appl Numer Math 62(1):35–50. ISSN 0168-9274
8. Smith R (1968) Matrix equation XA + BX = C. SIAM J Appl Math 16(1):198–201
9. Zhou B, Lam J, Duan G-R (2009) On smith-type iterative algorithms for the stein matrix equation. Appl Math Lett 22(7):1038–1044
10. Benner P, Saak J, Uddin MM (2016) Balancing based model reduction for structured index-2 unstable descriptor systems with application to flow control. Num Alg Control Opt 6(1):1–20

11. Benner P, Hossain M-S (2017) Structure preserving iterative methods for periodic projected Lyapunov equations and their application in model reduction of periodic descriptor systems. Numer Algorithms 76(4):881–904
12. Mehrmann V, Stykel T (2005) Balanced truncation model reduction for large-scale systems in descriptor form. In: Dimension reduction of large-scale systems, Springer, pp 83–115
13. Varga A (2000) Balanced truncation model reduction of periodic systems. In: Proceeding of CDC'2000, Sydney, Australia, vol 3, pp 2379–2384
14. Hossain M-S (2011) Numerical methods for model reduction of time-varying descriptor systems. PhD thesis
15. Benner P, Hossain M-S, Stykel T (2014) Low-rank iterative methods for periodic projected Lyapunov equations and their application in model reduction of periodic descriptor systems. Numer Algorithms 67(3):669–690

Smart Village Crop Planning: Enhancing Farmer's Decision-Making Culture with Data-Driven Predictive Model

Ariza Nordin and Faizah Ahmad Faizar

Abstract Crop planning prevents inappropriate crop selection and rotation by farmers which can cause economic loss and biodiversity issues. Technology solution for crop planning is a digital innovation in conserving traditional crop varieties and achieving optimal yield. Solutions range from database-oriented crop decision support systems to big data analytics platforms. However, technological capability must match the situated stage of farmers' digital fluency and skill as traits of their digital culture for decision-making. This research presents a work to explore two aspects of the smart village crop planning application model, which are the pragmatic use of the prediction model and rural community as-is digital culture. As a result, a smart village crop planning application model recommends three components, crop predictor, collaborative tool, and ask expert application, to enhance farmer's crop planning decision-making culture.

Keywords Crop planning · Digital culture · Data-driven · Predictive model · Decision-making · Situational analysis · Smart village

1 Introduction

The concept of a smart village promotes sustainable agriculture with rural community traditional crop variety farming [1, 2]. Crop planning is crucial for conserving traditional crop varieties and achieving optimal yield [2]. Crop planning prevents inappropriate crop selection and rotation by farmers which can cause economic loss and biodiversity issues. Over the decades, crop planning in villages is a community decision-making activity underpinned by farmers' collective indigenous and experiential knowledge with risk perceptions toward disaster and economic demand [3, 4]. With the lack of ICT infrastructure in the past, farmers' constraints are related complexity of data collection and processing problems.

A. Nordin (✉) · F. A. Faizar
Faculty of Computer and Mathematical Sciences, Universiti Teknologi MARA, Shah Alam, Malaysia
e-mail: ariza@uitm.edu.my

X.-S. Yang et al. (eds.), *Proceedings of Seventh International Congress on Information and Communication Technology*, Lecture Notes in Networks and Systems 447,
https://doi.org/10.1007/978-981-19-1607-6_64

Today, smart village digital infrastructure has enabled biophysical and environmental information such as soil and rainfall data collection using a technological solution known as the Internet of Things [5–9]. Numerous studies have proposed from crop decision support systems to big data analytic platform solutions [5–9] to realize the potential of technology-enabled crop planning tools to achieve the vision of smart village [9] and Sustainable Development Goals (SDGs) 2030 [10]. Besides the technology capability and functional aspect of a solution, from a sociotechnical perspective, situational analysis of community and culture aspect of digital informational activity and decision-making are crucial to design the non-functional components of the solution. This research presents a work to explore two aspects of the crop planning application model for the smart village, the pragmatic use of a predictive model with consideration of community as-is decision-making behavior and the digital culture maturity stage.

2 Conceptual Framework

2.1 Crop Planning

Crop planning is a pre-cultivation activity involving decision-making to select and rotate crops for optimal yield [11]. Crop cultivation with an effective plan increases crop revenues due to improved soil structure and broken reproductive cycles of pests and decreases farming costs due to reduced need for fertilizers (due to improved soil structure) and pesticides (because of lower pest populations) [11, 12]. Crop planning based on multiple crops with rotation benefits is an integral part of sustainable agriculture as it increases the biodiversity in the farm [11–13].

2.2 Farmer's Decision-Making Framework

Crop planning requires farmers' decision-making process. Waldman et al. [14] informed that the decision-making of smallholder farmers in developing countries has been researched from anthropology, cognitive science, economics, geography, political science, psychology, and rural sociology domains. Based on the work by Klein et al. published in 1993 [15], more recent research [16, 17] placed the decision-making process in agriculture within a naturalistic decision-making framework. The framework described a decision-maker, in the context of agriculture and naturalistic theory; a farmer is comfortable acting as street-smart and a hands-on practitioner who relies on experiences and heuristic knowledge to solve problems [18]. The farmer makes assessments of situations, classifies and interprets problems based on knowledge and experience, and decides for the best option to be agreed upon collectively within the local community [18, 19]. Klein et al. [15] contributed a recognition-primed decision (RPD) model of rapid decision-making. In this model, information availability is crucial for experts to generate options for decisions.

2.3 *Digital Culture*

The emergence of the smart village concept in Europe [20] has been phenomenal and followed by many nations worldwide with the emphasis of preserving community values and conserving biodiversity for a sustainable ecosystem [21]. The concept attempts to empower poor rural communities to find solutions in solving problems they faced today and, in the future, to retain talents and minimize the rural–urban migration [21]. The new social system of empowerment embraces digital culture which determines experiences and opportunities with digital technologies and the networked environment new practices, opportunities, and threats [22, 23].

2.4 *Crop–Soil Prediction Models*

Numerous research outputs were published for crop prediction data analytics and cropping decision systems [24–30]. Crop–soil prediction models are outputs of data analytics for the agriculture domain aimed at decisions on crop cultivation using data mining techniques. Research outputs were reported consistently discussing supervised machine learning techniques with feature selection as the method to provide fit crop planning and yield prediction models. Feature selection is a method to choose a subset of appropriate attributes from a larger set of original attributes in terms of a predefined benchmark, such as classification performance or class separability, which plays a significant role in machine learning applications [25–27]. Supervised learning techniques which were extensively used to conduct predictive modeling include k-nearest neighbor (k-NN), Naive Bayes, decision tree, SVM, random forest (RF), and bagging [25–30]. A systematic review of crop yield prediction techniques reports neural networks as a deep learning algorithm used for crop yield prediction besides the application of CNN, LSTM, and DNN algorithms [30].

3 Methods

The following sub-sections describe two (2) methods of the study:

1. Situational Analysis of as-is rural community digital culture: Using two Malaysian state case study findings to understand the preferences of farmers

Table 1 Case study selection

Case study	Region	Research output
1	West Malaysia (9 villages)	Seeking of agriculture information through mobile phone among paddy farmers in Selangor [31]
2	East Malaysia (11 villages)	Digital inclusion and mobile media in remote Sarawak [32]

and the new culture of technology adoption within the rural community (as-is digital culture stage).

2. Modeling Crop Prediction. Using supervised learning techniques such as decision tree, Naive Bayes, and k-nearest neighbors to propose a predictive model.

3.1 *Situational Analysis*

Situational analysis unveils knowledge of as-is community digital culture, strengths, weaknesses, opportunities, and threats. Information for the situational analysis was selected from secondary information from published papers studying rural communities in two Malaysian state (Selangor and Sarawak) case studies of villages in rural parts of West and East Malaysia rural community. Case study selection is a convenient and appropriate representation to inform smart village programs in Malaysia to determine the baseline for future research design in the tropical region and to recognize digital culture in decision-making activities. Information-seeking behavior and digital culture from these case studies can be representative of tropical small-scale farmers, as it can be a pattern in developing countries' farming systems and village community culture (Table 1).

3.2 *Modeling Crop Plan*

A dataset is obtained from a public database [33] to develop a supervised learning model which predicts crops for farmers. Feature selection was conducted to reduce attributes into the following list (Table 2).

Twenty-two crops are categorized according to agricultural classification such as cereal, pulse, fruit, and commercial [34] as shown by introducing a class attribute to ease visualization of the data. See Table 3.

Table 2 Dataset attribute selection

Attribute name	Data type	Description
Label	Categorical	22 types of crops
N	Numerical	Ratio of nitrogen content in soil
P	Numerical	Ratio of phosphorous content in soil
K	Numerical	Ratio of potassium content in soil
Temperature	Numerical	Temperature in degree Celsius
Humidity	Numerical	Relative humidity in %
pH	Numerical	pH value of soil
Rainfall	Numerical	Rainfall in mm

Table 3 Crop classification

Class	Cereal	Pulse	Fruit	Commercial
Crop label	1. Rice 2. Maize	1. Black gram 2. Chickpea 3. Pigeon peas 4. Kidney bean 5. Lentil 6. Moth bean 7. Mung bean	1. Apple 2. Banana 3. Coconut 4. Grapes 5. Mango 6. Orange 7. Papaya 8. Pomegranate 9. Watermelon 10. Muskmelon	1. Coffee 2. Jute 3. Cotton

4 Results and Discussion

The result is explained from two aspects: the situational analysis of case studies and the crop prediction model. Discussion proposes a strategic crop planning application model to enhance farmers' decision-making culture.

4.1 As-is Digital Culture Stage

The situational analysis pointed out strengths, weaknesses, opportunities, and threats of the community's digital culture.

The digital vision promoted and continuously supported by all levels of government provides strength to the rural community in terms of resources, digital infrastructure, and change management programs. As a result, usage of hand phones among the rural community is overwhelmingly leading to improved digital fluency and skill. However, the reliability of Internet connection is still a problem hindering a more active information-seeking behavior using the Net. Instead, farmers rely on voice

STRENGTHS 1. Presence of Digital Vision programs 2. Digital literacy enhanced 3. Wide use of smart phone and Internet 4. Use of social media & social networking tool-WhatsApp 5. Strongly community engagement 6. Seeking information behavior change	WEAKNESSES 1. Little knowledge of agriculture digital innovation 2. High dependency on government agencies' experts 3. Digital infrastructure readiness not optimal, limited connectivity persist 4. Biophysical and environmental data collection devices not installed 5. Farmers age range does not reflect young talents
OPPORTUNITIES 1. Can realize better crop planning 2. Can improve digital skill 3. Can collect data from crop planning 4. Can enhance community engagement with portal and collaborative tools 5. Can supply in advance product information to consumers	THREATS 1. Rural-urban migration still attractive 2. Consumers limited knowledge of farmers' products 3. Fast changing technology 4. Emergence of urban farming 5. Digital culture issues

Fig. 1 SWOT analysis as-is digital culture

communication to get agriculture information and social network tool, WhatsApp to seek information. Social media participation emerges as a new culture for social and business activities.

Localized biophysical and environment data collection and processing are still challenging issues faced by the rural community. Digital innovation is limited by a lack of knowledge. The age range did not show young talents in agriculture. Problem solving is highly dependent on information from agriculture-related government officers as experts. There is a wide opportunity for crop planning solutions and product information dissemination as the digital culture permeates into the rural community. Besides rural–urban migration persists, the rural community now must be competitive as urban farming programs kick-off. Digital social issues are a potential threat to the health of rural community digital culture. Figure 1 shows the SWOT analysis output.

4.2 Crop Predictor Modeling

Figure 2 shows the distribution of crop class in the dataset where fruit class is the larger crop class in the dataset while cereal class is the smallest. The average of N, P, and K ratio is presented based on each class where the average of N ratio in commercial class is highest while the average of K ratio is the lowest for this class.

A correlation matrix is constructed using RapidMiner as tabulated below in Table 4. From the table, K has the highest correlation value with P. Therefore, before removing the correlated attribute K, a comparison of before and after is done.

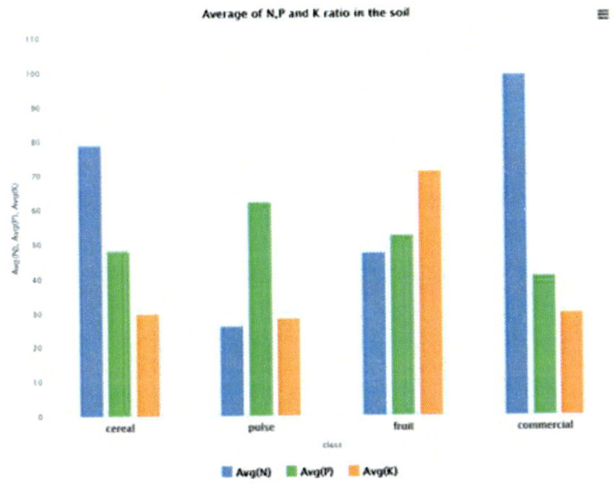

Fig. 2 Ratio of nutrient and crop

Table 4 Correlation matrix

Attributes	N	P	K	Temperature	Humidity	pH	Rainfall
N	**1**	−0.231	−0.141	0.027	0.191	0.097	0.059
P	−0.231	**1**	*0.736*	−0.128	−0.119	−0.138	−0.054
K	−0.141	*0.736*	**1**	−0.16	0.191	−0.17	−0.053
Temperature	0.027	−0.128	−0.16	**1**	0.205	−0.018	−0.03
Humidity	0.191	−0.119	0.191	0.205	**1**	−0.008	0.094
pH	0.097	−0.138	−0.17	−0.018	−0.008	**1**	−0.109
Rainfall	0.059	−0.064	−0.053	−0.03	0.094	−0.109	**1**

The classification algorithm was used due to the polynomial target variable. Specifically, models such as decision tree, Naive Bayes, and k-nearest neighbors were used for comparison. In RapidMiner, the cross-validation operator used will automatically split the train and test datasets. Results of the comparison are tabulated in Table 5.

The highest accuracy of 99.45% can be achieved using the Naive Bayes technique with the presence of the K attribute. Although the K attribute is highly correlated, the attribute is proved to be important in the models as the accuracy is decreased after the K attribute is removed. The existing crop list is exemplary, with the new crop list, and the Naïve Bayes model needs to be re-trained.

Table 5 Comparison of models

Model	Before K removed (%)	After K removed (%)
Decision tree	95.68	93.05
Naïve Bayes	99.45	98.73
K-NN	98.23	95.45

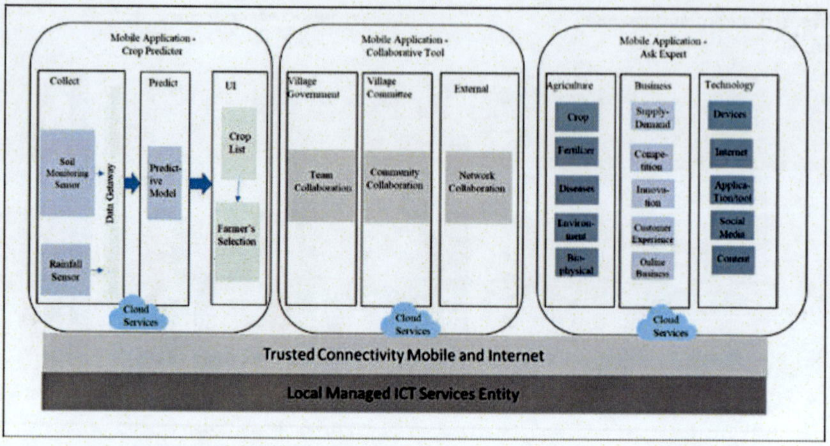

Fig. 3 Smart village crop planning application model

4.3 Smart Village Crop Planning Application Model

The proposal smart village crop planning application model focuses on providing data-driven crop planning predictive model with situational awareness of decision-makers' information-seeking behavior. Two aspects of decision-making are included: (a) situational awareness, where farmers access information to build up knowledge of the scenario from the Ask Expert application, and (b) option selection of crop listing provided by the crop predictor component.

The model as illustrated in Fig. 3 has three components.

1. Crop Predictor

   The crop predictor composition includes the collector which requires the installation of soil monitoring and rainfall sensor with data getaway, managed by local ICT provider, and the data collected will be input to the trained crop predicting model to produce the crop list for selection. The crop predictor model is classifiers in a family of elegant probabilistic categorization methods in machine learning is the Naive Bayes. Every class label is predicted by the likelihood of a given instance. Naive Bayes classifiers conclude that, provided the class attribute, the value of a particular quality is independent of the value of any other quality. Accessibility to the crop predictor is by mobile application.
2. Collaborative Tool

   The smart village crop planning application model's second component is the collaborative tool available, and team collaboration is to be conducted and managed by a team leader, community collaboration is for farmers to participate in discussion and sharing sessions, and network collaboration tool is for farmers to collaborate with external parties by networking.

3. Ask Expert application

   The use of mobile phones to seek information and advice is still a popular community culture when dealing with decisions that require input from external experts. The application which encourages communication with experts to empower farmers with effective decision-making for crop planning must be easy to use with mobile phones, and since small screens are challenging to the older generation of farmers, voice communication is still preferred. The features of the Ask Expert application are divided into three components of situational awareness information. Figure 3 depicts the agriculture, business, and technology components.

5 Conclusion

The research emphasizes sociotechnical aspects when developing a data-driven predictive model to enhance farmers' decision-making pragmatic solution for crop planning. A situational analysis identifies as-is strengths, weaknesses, opportunities, and threats of the rural community digital culture maturity stage to provide crucial non-functional elements of the smart village crop planning application model. The work presented in this paper is preliminary exploration, with a drawback of using secondary data, and with assumption, farmers decide on crop plan with the basis of experience and risk perceptions within a naturalistic decision-making framework.

Future work of situation awareness is recommended to study the access of information using digital technology, to understand the decision phase, how more experienced farmers use a solution developed using the smart village crop planning application model.

Acknowledgments The authors would like to thank Business Innovation & Technology Commercialization Centre (BITCOM), Universiti Teknologi MARA (UiTM) for support, trust, and appreciation of the research effort and the Faculty of Computer and Mathematical Sciences, UiTM, Shah Alam.

References

1. Adesipo A, Fadeyi O, Kuca K, Krejcar O, Maresova P, Selamat A, Adenola M (2020) Smart and Climate-Smart Agricultural Trends as Core Aspects of Smart Village Functions. Sensors 20(21):5977
2. Singh R, Singh GS (2017) Traditional agriculture: a climate-smart approach for sustainable food production. Energy, Ecology, Environ 2(5):296–316
3. Limnirankul B (2007) Collective action and technology development: up-scaling of innovation in rice farming communities in Northern Thailand
4. Jarvis DI, Hodgkin T, Sthapit BR, Fadda C, Lopez-Noriega I (2011) A heuristic framework for identifying multiple ways of supporting the conservation and use of traditional crop varieties within the agricultural production system. Crit Rev Plant Sci 30(1–2):125–176

5. Rajak RK, Pawar A, Pendke M, Shinde P, Rathod S, Devare A (2017) Crop recommendation system to maximize crop yield using machine learning technique. Int Res J Engineering Technology. 4(12):950–953
6. Maheswari R, Azath H, Sharmila P, Gnanamalar SS (2019) Smart village: Solar-based smart agriculture with IoT enabled for climatic change and fertilization of the soil. In: 2019 IEEE 5th International Conference on Mechatronics System and Robots (ICMSR), May 3, pp 102–105. IEEE
7. Muangprathub J, Boonnam N, Kajornkasirat S, Lekbangpong N, Wanichsombat A, Nillaor P (2019) IoT and agriculture data analysis for the smart farm. Comput Electron Agric 1(156):467–474
8. Ramlan SZ, Mohd Deni S (2021) Rainfall prediction in flood prone area using deep learning approach. In: International Conference on Soft Computing in Data Science 2021 Nov 2. Springer, Singapore, pp 71–88
9. Aziiza AA, Susanto TD (2011) The smart village model for rural area (Case Study: Banyuwangi Regency). In: IOP Conference Series: Materials Science and Engineering 2020, vol 722, No 1. IOP Publishing, p. 012011
10. Adamowicz M, Zwolińska-Ligaj M (2020) The, "Smart Village" as a way to achieve sustainable development in rural areas of Poland. Sustainability 12(16):6503
11. Dury J, Schaller N, Garcia F, Reynaud A, Bergez JE (2012) Models to support cropping plan and crop rotation decisions. A review. Agronomy Sustainable Develop. 32(2):567–580
12. Brooker RW, George TS, Homulle Z, Karley AJ, Newton AC, Pakeman RJ, Schöb C (2021) Facilitation and biodiversity–ecosystem function relationships in crop production systems and their role in sustainable farming. J Ecol 109(5):2054–2067
13. Boyabatlı O, Nasiry J, Zhou Y (2019) Crop planning in sustainable agriculture: Dynamic farmland allocation in the presence of crop rotation benefits. Manage Sci 65(5):2060–2076
14. Waldman KB, Todd PM, Omar S, Blekking JP, Giroux SA, Attari SZ, Baylis K, Evans TP (2020) Agricultural decision making and climate uncertainty in developing countries. Environ Res Lett 15(11):113004
15. Klein GA (1993) A recognition-primed decision (RPD) model of rapid decision making. Decision making in action: Models and methods 5(4):138–147
16. Dury J, Garcia F, Reynaud A, Therond O, Bergez JE. Modelling the complexity of the cropping plan decision-making
17. Dury J, Garcia F, Reynaud A, Bergez JE (2013) Cropping-plan decision-making on irrigated crop farms: A spatio-temporal analysis. Eur J Agron 1(50):1
18. Bradford Lori EA, A complicated chain of circumstances: Decision making in the New Zealand wool supply chains (Doctoral dissertation, Lincoln University)
19. Von Ketteler L. Factors influencing farmer's decision-making and resilience: The case of banana production in Amubri, Costa Rica
20. Slee B (2019) Delivering on the concept of smart villages–In search of an enabling theory. European Countryside. 11(4):634–650
21. Abdul Razak N, Abdul Malik J, Saeed M, A development of smart village implementation plan for agriculture: A pioneer project in Malaysia.
22. Uzelac A (2008) How to understand digital culture: Digital culture-a resource for a knowledge society. Digital Culture: The Changing Dynamics. Institute for International Relations, Zagreb, pp 7–21
23. Uzelac A (2010) Digital culture as a converging paradigm for technology and culture: Challenges for the culture sector. Digithum 27(12)
24. Churi AJ, Mlozi MR, Tumbo SD, Casmir R, Mahoo MR, A decision support system for enhancing crop productivity of smallholder farmers in semi-arid agriculture
25. Suruliandi A, Mariammal G, Raja SP (2021) Crop prediction based on soil and environmental characteristics using feature selection techniques. Math Comput Model Dyn Syst 27(1):117–140
26. Maya Gopal PS, Bhargavi R (2019) Selection of important features for optimizing crop yield prediction. Int J Agricultural Environmental Information Systems (IJAEIS) 10:54–71. https://doi.org/10.4018/IJAEIS.2019070104

27. Zala DH, Chaudhri MB (2018) Review on use of BAGGING technique in agriculture crop yield prediction. Int J Scientific Research Development 6:675–677
28. Dela Cruz GB, Gerardo BD, Tanguilig BT (2014) Agricultural crops classification models based on PCA-GA implementation in data mining. Int J Modeling Optimization 4(III):375–382. https://doi.org/10.7763/IJMO.2014.V4.404
29. Rosli N, Forecasting paddy production at Muda Agricultural Development Authority (MADA)/Norziela Binti Rosli (Doctoral dissertation, Universiti Teknologi MARA Cawangan Kelantan).
30. Van Klompenburg T, Kassahun A, Catal C (2020) Crop yield prediction using machine learning: A systematic literature review. Comput Electron Agric 1(177):105709
31. Ramli NS, Hassan MS, Man N, Samah BA, Omar SZ, Rahman NA, Yusuf S, Shamsul M (2019) Seeking of agriculture information through mobile phone among paddy farmers in Selangor. Int J Academic Research Business Social Sciences 9(6):527–538
32. Horn C, Rennie E, Gifford S, Riman R, Hoo G (2018) Digital inclusion and mobile media in remote Sarawak. RMIT University
33. Atharva Ingle ttps://www.kaggle.com/atharvaingle/crop-recommendation-dataset,2020, December 19, last accessed 2021/08/21
34. Cobley LS (1976) An introduction to the botany of tropical crops. Longman

Closed-Domain Multiple-Choice Question Answering System for Science Questions

Kedar P. Vaidya, Sanya A. Chetwani, and Mansi A. Radke

Abstract There is a huge amount of textual information in digital form growing exponentially over the years. With this data explosion, retrieving relevant information through information retrieval (IR) systems has become crucial. Due to the growing popularity of various tools such as voice assistants and chatbots, that rely on human–machine interaction, it is important that such systems are able to answer any query directly, rather than leading the user to a set of references. Recently developed question answering (QA) systems aim at providing direct answers to the user's queries, hence, becoming of great interest to a large community. In this work, we propose an end-to-end pipeline that performs this task by creating a knowledge graph from the corpus. It uses embeddings to predict the missing links in the knowledge graph and a unique answer selection module in order to reach the correct answer. We test the proposed methodology on the SciQ dataset and obtain an accuracy of 62%. We also test this methodology on a curated knowledge base, Aristo, and present a comparative study highlighting the aspects on which information extraction models can improve, thus opening possibilities for future work in this field.

Keywords Question answering · Textual entailment · Information extraction · Knowledge graph · Vector embeddings · Multiple-choice QA

1 Introduction

Question answering is a fundamental sub-domain in information retrieval. It heavily relies on natural language processing (NLP) and artificial intelligence (AI) techniques in order to return concise answers to queries in natural language. The rising popularity

K. P. Vaidya (✉) · S. A. Chetwani · M. A. Radke
Visvesvaraya National Institute of Technology, Nagpur, India
e-mail: kedarvaidya0504@gmail.com

S. A. Chetwani
e-mail: ssanyachetwani@gmail.com

M. A. Radke
e-mail: mansi.radke@cse.vnit.ac.in

X.-S. Yang et al. (eds.), *Proceedings of Seventh International Congress on Information and Communication Technology*, Lecture Notes in Networks and Systems 447,
https://doi.org/10.1007/978-981-19-1607-6_65

of QA systems is due to the increasing abundance of information in the digital age. While a multitude of researches have used different techniques for QA, a growth in the usage of one particular technique can be observed in recent times, knowledge graphs [1]. They are a simplified representation of a large amount of interconnected data in the form of a graph, with the nodes representing entities, i.e. people, places, objects, etc. and their connecting edges representing the relationship between two (neighbouring) entities (refer Fig. 1). The applications of knowledge graphs are wide-ranging, the most significant of which are recommendation systems, chatbots, smart search and question answering. Hence, in this work, we leverage the use of knowledge graphs in building a pipeline to answer multiple-choice questions.

We evaluate the proposed methodology over the SciQ dataset [2] which is a popular crowdsourced QA dataset, that consists of questions from a number of science domains, along with accompanying multiple choices with one correct answer for each question (refer Fig. 2). The prerequisite for solving over this dataset is that the model should have relevant domain information. Therefore, we collect a set of corpuses helpful in answering the questions (refer Fig. 3).

As most of the information in modern days is in the form of complex natural language, the corpuses for QA systems pose many challenges. The questions require

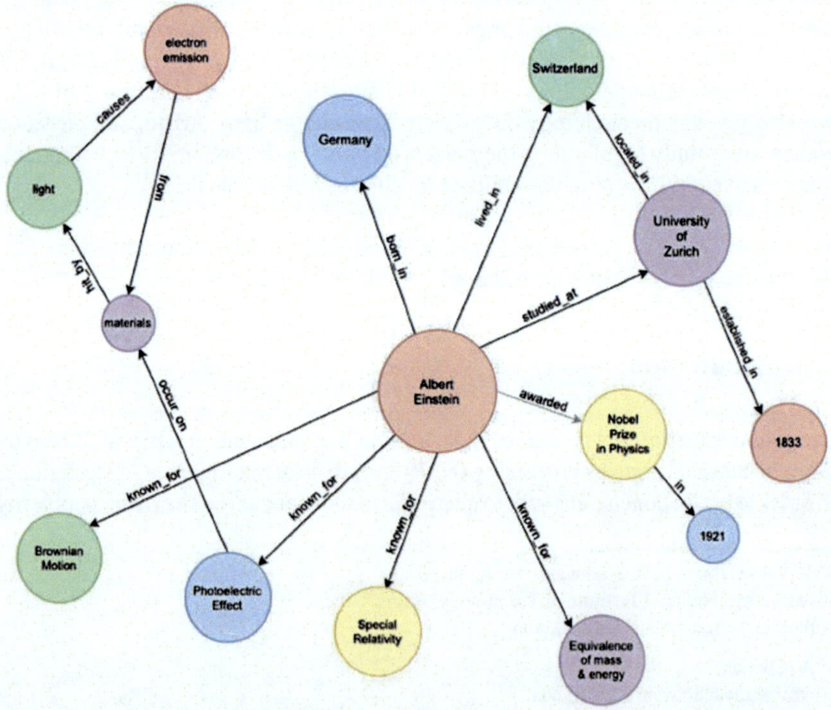

Fig. 1 An example of knowledge graph

Fig. 2 An example from the SciQ dataset

Q. *What is the food stored in plant seed called ?*
distractor 1 : Larval
distractor 2 : Pollin
distractor 3 : Membrane
correct answer : Endosperm

Reference Paragraph
One sperm and the egg combine, forming a diploid nucleus—the future embryo. The other sperm fuses with it in the center of the embryo sac, forming a triploid cell that will develop into the endosperm. It is a tissue that serves as a food reserve.The zygote develops into an embryo with a radicle, or small root, and one or two leaf-like organs called cotyledons. Seed food reserves are stored outside embryo, and cotyledons serve as conduits to transmit the broken-down food reserves to the developing embryo. .

Fig. 3 Reference material required to answer the question mentioned in Fig. 2

the model to infer over multiple sentences spread out in a paragraph to find the information entailed by it and thus reach to the correct answer. This necessitates textual entailment approaches that go beyond lexical-level matching and reason over multiple edges of the knowledge graph in order to arrive at the correct answer, also termed as "multi-hop" knowledge graph question answering (KGQA) [3].

In this study, we also compare the two paths that can be taken while developing a QA model, namely the end-to-end proposed pipeline and the partial pipeline using ground truth for extracted tuples. Both these paths produce different outcomes and vary significantly when compared in terms of accuracy. These differences are reasoned by noting key aspects on which each model operates. Through the course of experiments in this work, it is observed that there is a necessity of information extraction systems that work on a particular schema. This work also highlights the importance of rule-based knowledge bases in the field of question answering. The study concludes by emphasising the aspects that current techniques can be improved upon, pointing out some directions for future research in the domain of natural language processing.

In Sect. 2, we present the literature review of the previous research works carried out in the field of question answering and knowledge graphs. In Sect. 3, we describe the proposed methodology. The Sect. 4 presents a detailed description of the datasets used and experiments conducted, along with Sect. 5 describing the evaluation metrics used. Following this, we present the results and discussion in Sect. 6. Section 7 concludes the paper by pointing out some directions for future work.

2 Background and Related Work

Over the years, researchers have experimented with numerous techniques in order to build an efficient question answering system and several of them have been able to achieve remarkable results on numerous open-domain and closed-domain datasets such as SQuAD [4] and emrQA [5], respectively.

Holistic pipelines for question answering tasks prove to be important tools as they can process huge amounts of information. They are able to give specific answers to the users' queries instead of directing them to lengthy and complicated resources to deduce the answer, in which case they have to sift through the returned documents and dig out the answers themselves. Basic structure of a generic QA pipeline involves information retrieval, query processing and answering module. Research works have shown remarkable performances on each of the aforementioned topics individually, but development of an end-to-end QA system remains less explored.

One of the important pre-processing tasks in a QA system is to identify the pronouns, referring to an entity in the text, mentioned previously. This is termed as coreference resolution. For example, consider the following text "John likes ice cream. His favourite flavour is chocolate". The word "His" refers to "John" which occurs in the previous sentence. In order to find correct answers to a given question, the computer needs to understand such references and infer the information accordingly. To achieve this task of coreference resolution, numerous approaches have been proposed in the literature which are based on graphical models, logical models, rule-based models and word embeddings. In addition to these, a number of methods make use of span representations. SpanBERT [6] proposes a pre-training method that is used to predict spans of the text. The end-to-end neural coreference resolution [7] approach combines context-dependent boundary representations with a head-finding attention mechanism. Another important sub-task of the QA challenge is the information extraction. It refers to the extraction of unseen relations in an unstructured text and separating them from their arguments in an unsupervised manner. Traditionally, methods involved were statistical and rule-based, but recent works have revolved around using neural networks for extracting these relations. Some of the information extraction techniques include OLLIE [8] which employs a context-analysis step to add attribution and clausal modifiers to the output triples. DeepEx [9] proposes a zero-shot learning framework that formalises the task of triple extraction as a translation between task-specific input text and output triples. Stanford OpenIE [10] uses a small set of predefined patterns for canonically structured sentence and then trains a classifier to extract self-contained clauses from longer sentences. Another tool, Graphene [11], transforms the input sentences into core facts and contexts. It also identifies the rhetorical relation between them, thus preserving the context of every extracted tuple. In addition to such approaches, some highly curated domain-specific knowledge bases are available which can be used directly [12]. The major difference between such knowledge bases and the unsupervised approach is the lack of a fixed set of relations that can define the structure of the tuples.

An equally important aspect of QA systems is their ability to translate extracted triples into a form that can facilitate answer search. Popular approaches for this task generally involve the use of knowledge graph embeddings. Various models have been developed for this purpose. TransE [13] interprets relations as translations operating on the low-dimensional embeddings of the entities. TransR [14] embeds the entities and relations in separate spaces and then projects them into the relation space where embeddings are learned by translations between projected entities. ComplEx [15] maps the entities and relations in a complex space, and RotatE [16] defines relations as a rotation from source to target entity. RESCAL [17] proposes a tensor factorisation-based relational learning approach.

With the growing use of knowledge graphs, a graphical representation of knowledge tuples, some pieces of work have also resulted in models that directly utilise a KG in its original form, without using methods to change its form and reason over it. Answering complex questions using OpenIE [18] selects most relevant tuples from KB with respect to question, to answer the query. Techniques like SimpleIR have also been proposed for NY Regents science exam data and involve reasoning over semi-automatically prepared KB [19]. Various methods have been adopted in KGQA, ranging from graph-pattern isomorphism to standard neural networks. Xu et al. [20] demonstrate neural network-based relation extraction to answer over knowledge base.

3 Methodology

Through this section, we aim to introduce the methodology followed in building the end-to-end pipeline. In the initial step of the pipeline, it targets every occurrence of pronouns in the corpus using the coreference resolution technique. By replacing all pronouns with the entity they refer to, we make each sentence in the corpus independent. Following this, the task is to construct or build a knowledge base in the form of knowledge graph tuples. Using information extraction technique, each sentence is converted into one or more possible tuples of the form <entity1, relation, entity2> . The pipeline makes use of embeddings to reach the correct answer and also counters the problem of knowledge graph sparsity. For this purpose, the extracted tuples, questions and answer choice data are converted into their respective embeddings. Once the complete dataset has been embedded, direct links are predicted between the topic entity of questions and all entities present in the knowledge graph. The model predicts an answer based on the distance between entities, and the predicted answer is thus obtained. The model compares the predicted answer with given answer choices choosing the option that is the closest through an answer selection module (refer Fig. 4A).

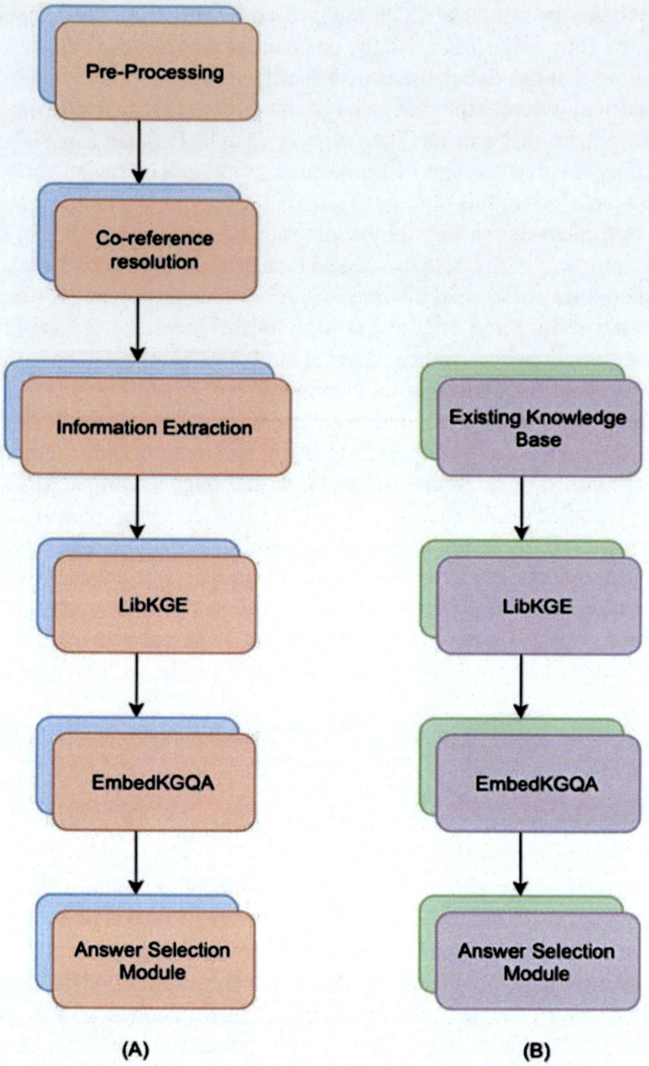

Fig. 4 **a** Structure of the proposed pipeline and **b** modified pipeline for testing robustness against knowledge base variation

3.1 Pipeline Components

Each task of the pipeline is performed by different components that are either tailor-made to suit the application or make use of existing open-source models with state-of-the-art results on specific datasets.

Data Pre-Processing: The SciQ dataset contains .json files that are processed for obtaining question and answer in a required format. Later stages of the pipeline require that each query file must contain questions along with their respective topic entities and correct answers. This topic entity must also be present in the list of entities generated while extracting the KB tuples from corpus. Thus, SpaCy library is used for noun parsing to obtain noun entities in each question, and then, using fuzzy match and WordNet, we select those entities that are present in the KB entity list.

SpanBERT for coreference resolution: The model is supplied with corpuses that contain relevant information for solving the SciQ question dataset. The major challenge with using these corpuses is that they contain numerous occurrences of pronouns referring to objects and subjects mentioned previously in the passage. This indicates that not only is the model required to find the answer spread over two or three sentences but it is also expected to understand the links connecting these individual sentences, i.e. the pronouns. Therefore, coreference resolution is applied as an indispensable step of the pipeline. By employing SpanBERT [6], an enhanced version of Bidirectional Encoder Representations from Transformers (BERT) [21], which provides improved representation and predictions of spans of textual data, the system component is able to replace all pronouns with their respective noun (refer Fig. 5).

IMoJIE for Information Extraction: Various models for information extraction have benchmarked their results on domain-specific datasets as well as open-domain sentences collected from Wikipedia and other sources. Each model reports its shortcomings like redundancy, wrong extractions, non-uniformity, etc. We compare the scores achieved by each model and choose the model based on its correctness of deriving one or more tuple(s) from a given sentence. IMoJIE [22] is chosen for this

Coreferenced Paragraph
S1 : One sperm and the egg combine, forming a diploid nucleus—the future embryo.
S2 : The other sperm fuses with *diploid nucleus* in the center of the embryo sac, forming a triploid cell that will develop into the endosperm.
S3 : *Endosperm* is a tissue that serves as a food reserve.
S4 : The zygote develops into an embryo with a radicle, or small root, and one or two leaf-like organs called cotyledons.
S5 : Seed food reserves are stored outside embryo, and cotyledons serve as conduits to transmit the broken-down food reserves to the developing embryo.

Fig. 5 Coreferenced sentences obtained from SpanBERT

Tuples Obtained		
entity1	***relation***	***entity2***
One sperm and the egg	forming	a diploid nucleus
The other sperm	fuses	with the diploid nucleus in the center of the embryo sac
a triploid cell	will develop	into the endosperm
Endosperm	is	a tissue
a tissue	serves	a food reserve
The zygote	develops	into an embryo with a radicle, or small root, and one or two leaf-like organs
one or two leaf-like organs	called	cotyledons
Seed food reserves	are stored	outside the embryo
the cotyledons	serve	as conduits to transmit the broken-down food reserves to the developing embryo

Fig. 6 Tuples obtained after information extraction

task and fed with the coreferenced paragraphs to extract three tuples of the form <entity1, relation, entity2> (refer Fig. 6).

LibKGE for Knowledge Graph Embedding: The next step requires converting the obtained knowledge graph into embeddings for which an open-source PyTorch library LibKGE [23] is used. Word or vector embeddings are common sub-tasks performed in NLP and IR. It involves conversion of words into low-dimensional forms, i.e. into vectors. Hence, the words which are similar in meaning are located closer to each other in the vector space. The LibKGE library includes common KGE algorithms like RESCAL [17], TransE [13], ComplEx [15] and RotatE [16].

In this study, the ComplEx model is employed owing to its extremely high compatibility with many datasets including SciQ. It can handle a large variety of symmetric and antisymmetric relations and outperforms many KGE models with a fairly simple approach towards the task [15]. Since this method of using embeddings is not only able to overcome several restrictions posed by multi-hop QA, but also eliminates the problem of knowledge graph sparsity, we translate the obtained knowledge graph into its embeddings.

EmbedKGQA for Answer Prediction: After converting KG into embeddings, the EmbedKGQA [24] model is used to convert the question and its topic entity into embeddings. It uses RoBERTa [25] as the question embedding module. After obtaining the question, topic entity and answer, it uses the ComplEx scoring function to train the model and predicts direct links between the question and answer entity. In this way, the restriction posed by multi-hop questions is resolved and the model gets trained in such a way that, provided a question and its topic entity, it gives the predicted answer directly in the form of an embedding.

Answer Selection Module: The last step of the pipeline is based on selecting an answer from a list of given choices. For this, an answer selection module is introduced. After predicting the answer embedding by EmbedKGQA, all given answer choices are also converted into embedding form by employing the RoBERTa model [25]. Euclidean distance is calculated for each set of predicted answer and corresponding options. The lesser the distance, the closer is the option, hence indicating their similarity. Therefore, depending upon their closeness, the option with least Euclidean distance is chosen as the final answer by the model. Since grammar and vocabulary play a major role in QA tasks, there are instances where predicted answer is close to the correct answer but differs lexically. Therefore, using WordNet, four synonyms are assigned to each answer choice. Once all the original choices and their respective synonyms are converted into embeddings, their closeness to the predicted answer embedding helps in choosing the option to which the closest embedding belongs. This ensures the model can detect the correct answer despite the lexical differences.

4 Experiments

In order to evaluate the proposed methodology, an experiment is conducted on the SciQ dataset, wherein only one option is correct among the given answer choices. In an attempt to understand and trace back the errors, it is found that one step in particular required more improvements, and to confirm this, we modify our pipeline, consequently making use of a curated tuple dataset (refer Fig. 4b).

Datasets: The SciQ [2] dataset contains 13k crowdsourced questions from science exams, based on physics, chemistry, biology and earth science. This dataset has been designed for two tasks: direct QA, for which, the model can make use of the question and supporting fact to detect a span of answer; multiple-choice QA, where the model can use only the question and four option choices and mark the correct answer based on whatever background information it has on the topic. For the experiments in this work, the multiple-choice version is used. To provide the model with domain information, following knowledge sources are considered:

1. A set of corpuses (M) on SciQ-targeted domains are collected from open-source science textbooks on OpenStax[1] which shares this material under the Creative Commons licence. The corpuses generally have scientific grammar, and the information is distributed over a set of sentences, therefore requiring complex inference to extract the entailed information. The set M is initially processed with SpanBERT to encounter every instance of pronoun and then fed into IMoJIE to extract KB tuples from the corpus.

[1] https://openstax.org/.

2. Additionally, the curated knowledge base tuples of the Aristo Tuple KB dataset [12] are used in this work, during the experimentation phase, to gauge the correctness of the pipeline. Aristo knowledge base is a set of 294,000 highly precise tuples, that spread over a range of domains in the field of science, and reports an accuracy of 77.4% in answering questions of the SciQ dataset. The tuples are in the form of <subject:relation:object> and make use of certain standard NLP relations that prove to be useful during the course of this study.

Testing Robustness against Knowledge Base Variation: Carrying out all the steps sequentially on the collected reference material M, the pipeline is able to obtain an accuracy of 23.6%, but interestingly, when the curated tuple KB is used, the accuracy significantly rises to about 62%. Hence, we can infer that pipeline can detect correct answers, provided an accurate underlying KB is fed to the model. We also observe that while building the knowledge base from reference set M, the information extractor generally tends to break the sentence into three major parts. So the entity1 and entity2 contain more than one word, and this diverts the attention of the system to other words of the phrase, that are less important. As mentioned earlier, information extraction models also do not extract rule-based relations between entity1 and entity2. They are highly dependent on the grammar and vocabulary of the sentence, which creates a set of varied relations.

5 Evaluation Metrics

In this study, potential of the pipeline can be gauged solely on its ability to mark the correct choice from a pool of four choices. Thus, accuracy is the sole metric which indicates the number of questions the model is able to answer correctly against the total number of questions.

$$\text{Accuracy} = \frac{\text{Number of Correctly Answered Questions}}{\text{Total Number of Questions}} \times 100 \tag{1}$$

6 Results and Discussion

Through the course of experiments, it is observed that each component of the pipeline proves to play a vital role, and its individual accuracy can dramatically affect the overall result of the pipeline. One essential observation made is that the accuracy and precision of the KB tuples can affect the results significantly (refer Table 1). Tuples obtained by information extraction (IE) methods prove to be inefficient especially when compared to curated, high-precision and domain-specific KB, in this case, Aristo Tuple KB. The dependency of IE techniques on lexical aspects of a sentence

Table 1 Results from KB variations

Pipeline	Accuracy over SciQ (%)
Pipeline with information extraction step	23.6
Pipeline with curated KB tuples	62

proves to have a great impact on the results. With the tuple entities, namely entity1 and entity2, being a span of the sentence, and the relations not following a certain schema, the uniformity of KB is greatly compromised. These issues are resolved when the Aristo KB is introduced since it has been built on rule-based relations and contains one- or two-word entities. Hence, there is a strong need for the knowledge base to incorporate these features, as the variation in relations and entities with each sentence ultimately affects the training of subsequent components.

7 Conclusion

In this work, an end-to-end QA pipeline is presented for multiple-choice questions that can be answered using pre-existing domain knowledge. Further, this pipeline is modified to test against variations in knowledge base. This reveals how the accuracy and precision of KB can be crucial for the overall accuracy of the pipeline. These facts are demonstrated by the results achieved through the original and modified pipelines and the significant difference in the accuracy points towards the IE step. We found how the current generation IE techniques are rendered inadequate for the purpose of question answering and highlight the areas where they can be improved in the future works. In conclusion, the pipeline is based on the concept of fusing coreference resolution, information extraction and vector embeddings in order to find the correct answer. It also takes into account the lexical differences which can possibly occur when the option choices are synonyms of the predicted answer. This approach makes our work unique and delivers a pipeline that can be applied to any QA task, given that it is supplied with supporting knowledge required to answer the question.

References

1. Hogan A, Blomqvist E, Cochez M, D'amato C, De Melo G, Gutierrez C, Kirrane S, Gayo JEL, Navigli R, Neumaier S et al (2021) Knowledge graphs. ACM Comput Surv 54(4):1–37
2. Welbl J, Liu NF, Gardner M (2017) Crowdsourcing multiple choice science questions
3. Saxena A, Tripathi A, Talukdar P (2020) Improving multi-hop question answering over knowledge graphs using knowledge base embeddings. In: Proceedings of the 58th Annual Meeting of the Association for Computational Linguistics, pp 4498–4507, Online, July 2020. Association for Computational Linguistics

4. Rajpurkar P, Zhang J, Lopyrev K, Liang P (2016) Squad: 100,000+ questions for machine comprehension of text
5. Pampari A, Raghavan P, Liang J, Peng J (2018) emrQA: A large corpus for question answering on electronic medical records. In: Proceedings of the 2018 Conference on Empirical Methods in Natural Language Processing, pages 2357–2368, Brussels, Belgium, October-November 2018. Association for Computational Linguistics
6. Joshi M, Levy O, Zettlemoyer L, Weld D (2019) Bert for coreference resolution: Baselines and analysis. In: Proceedings of the 2019 Conference on Empirical Methods in Natural Language Processing and the 9th International Joint Conference on Natural Language Processing (EMNLP-IJCNLP)
7. Lee K, He L, Lewis M, Zettlemoyer L (2017) End-to-end neural coreference resolution. In: Proceedings of the 2017 Conference on Empirical Methods in Natural Language Processing, pp 188–197, Copenhagen, Denmark, September. Association for Computational Linguistics
8. Mausam MS, Bart R, Soderland S, Etzioni O (2012) Open language learning for information extraction. In: Proceedings of Conference on Empirical Methods in Natural Language Processing and Computational Natural Language Learning (EMNLP-CONLL)
9. Wang C, Liu X, Chen Z, Hong H, Tang J, Song D (2021) Zero-shot information extraction as a unified text-to-triple translation. arXiv preprint arXiv:2109.11171
10. Angeli G, Johnson Premkumar MJ, Manning CD (2015) Leveraging linguistic structure for open domain information extraction. In: Proceedings of the 53rd Annual Meeting of the Association for Computational Linguistics and the 7th International Joint Conference on Natural Language Processing (Volume 1: Long Papers), pp 344–354
11. Cetto M, Niklaus C, Freitas A, Handschuh S (2018) Graphene: a context-preserving open information extraction system. arXiv preprint arXiv:1808.09463
12. Dalvi Mishra B, Tandon N, Clark P (2017) Domain-targeted, high precision knowledge extraction. Trans Assoc Computational Linguistics 5:233–246
13. Wang Z, Zhang J, Feng J, Chen Z (2014) Knowledge graph embedding by translating on hyperplanes. In: Proceedings of the AAAI Conference on Artificial Intelligence, vol 28
14. Lin Y, Liu Z, Sun M, Liu Y, Zhu X (2015) Learning entity and relation embeddings for knowledge graph completion. In: Twenty-ninth AAAI conference on artificial intelligence
15. Trouillon T, Welbl J, Riedel S, Gaussier E, Bouchard G (2016) Complex embeddings for simple link prediction. In International conference on machine learning, pp 2071–2080. PMLR
16. Sun Z, Deng Z-H, Nie J-Y, Tang J (2019) Rotate: Knowledge graph embedding by relational rotation in complex space. arXiv preprint arXiv:1902.10197
17. Kong X, Chen X, Hovy E (2019) Decompressing knowledge graph representations for link prediction. arXiv preprint arXiv:1911.04053
18. Khot T, Sabharwal A, Clark P (2017) Answering complex questions using open information extraction
19. Clark P, Etzioni O, Khot T, Sabharwal A, Tafjord O, Turney P, Khashabi D (2016) Combining retrieval, statistics, and inference to answer elementary science questions. In: Proceedings of the AAAI Conference on Artificial Intelligence 30(1)
20. Xu K, Reddy S, Feng Y, Huang S, Zhao D (2016) Question answering on freebase via relation extraction and textual evidence. arXiv preprint arXiv:1603.00957
21. Devlin J, Chang M-W, Lee K, Toutanova K (2019) Bert: Pretraining of deep bidirectional transformers for language understanding
22. Kolluru K, Aggarwal S, Rathore V, Chakrabarti S, et al (2020) Imojie: Iterative memory-based joint open information extraction. arXiv preprint arXiv:2005.08178
23. Broscheit S, Ruffinelli D, Kochsiek A, Betz P, Gemulla R (2020) Libkge-a knowledge graph embedding library for reproducible research. In: Proceedings of the 2020 Conference on Empirical Methods in Natural Language Processing: System Demonstrations, pp 165–174
24. Saxena A, Tripathi A, Talukdar P (2020) Improving multi-hop question answering over knowledge graphs using knowledge base embeddings. In: Proceedings of the 58th Annual Meeting of the Association for Computational Linguistics, pp 4498–4507

25. Liu Y, Ott M, Goyal N, Du J, Joshi M, Chen D, Levy O, Lewis M, Zettlemoyer L, Stoyanov V (2019) Roberta: A robustly optimized bert pretraining approach. arXiv preprint arXiv:1907.11692

Quality Management Within and Visiting e-cultural Tourist Destinations: Case Study Rural Parish of San Miguelito

Alicia Porras-Angulo, Alba Hernández-Freire, Johana Porras-Quispe, and Adriana Cuesta-Chiriboga

Abstract Cultural tourist sites in search of an intercultural approach, participatory in the face of quality management processes and focused on the interest of the visitor with a cultural visit profile, through continuous improvement, ISO standards and indicators that contribute to organizational effectiveness that provide responses to the changes generated in cultural sites and allow the consolidation of the administration, operation and structure of cultural institutions, to which, the study analyzed the quality levels of the representative cultural centers of the rural parish of San Miguelito. The objective of this study is to analyze quality management in the satisfaction of tourists in visiting cultural tourist destinations. This research is correlational with the use of Pearson's correlation and simple descriptive, due to the detail of the phenomena influenced in visiting cultural tourist destinations, it is a determining factor in quality, through the perception of quality, through satisfaction it is possible to define different categories, in addition to taking the character of non-experimental design research, with a cross-sectional design because the level and state of the variables are analyzed at a single point in time; to verify the relational trend of the variables, using a validated data collection instrument focused on the adapted SERVPERF model. Furthermore, high positive correlation results were achieved between quality management in cultural sites, the results at the time of making the relationship through the t statistic, showed a correlation with a level of significance < 1%, with which the hypothesis was validated, it was supported with a level of 1% (0.610).

A. Porras-Angulo (✉) · A. Hernández-Freire · A. Cuesta-Chiriboga
Facultad de Ciencias Humanas Y de La Educación, Universidad Técnica de Ambato, Ambato, Ecuador
e-mail: am.porras@uta.edu.ec

A. Hernández-Freire
e-mail: albaphernandezf@uta.edu.ec

A. Cuesta-Chiriboga
e-mail: ccuesta4606@uta.edu.ec

J. Porras-Quispe
Universidad Técnica Particular de Loja, Loja, Ecuador
e-mail: Jbporras@utpl.edu.ec

X.-S. Yang et al. (eds.), *Proceedings of Seventh International Congress on Information and Communication Technology*, Lecture Notes in Networks and Systems 447,
https://doi.org/10.1007/978-981-19-1607-6_66

Keywords E-tourism · Culture · Cultural tourism · Management · Quality standards · SERVPERF model

1 Introduction

The quality management of visiting cultural sites constitutes a process that in the future will contribute to integral sustainability; through the application of the correlational methodology, the quality factor and the sustainability of the cultural tourist destination are articulated [1] consider that tourist destinations are linked to their natural and cultural environment, together with the integration of the community. [2] argues that the organization of spaces ... is a public responsibility, and the cultural tourism organization. From the point of relation with quality factors, problems can be described as: cultural products without quality management approaches, progressive loss of attractions, and cultural manifestations.

The current situation in the cultural visit sites in the rural parish, San Miguelito, involves factors such as: the destruction of historical sites and the limited development of tourist activities due to the low influx to cultural sites. The components that directly affect the progressive loss of cultural identity, represented in cultural sites, which narrate the beginnings of the Cosanga Pillaro culture, losing a cultural site, involves losing a portion of the history, culture, and tradition of the peoples. [3] states that, it is vital to control, plan, and organize the pressure that civilization exerts on visiting cultural assets and, above all, anticipate their alteration.

It is prudent to generate strategies based on the needs of visiting cultural sites, through management models that consider the purposes of quality and tourist satisfaction, through quality programs, analysis of the management applied in this type of attractions, balancing the cultural sustainability and social development of residents and tourists.

The links with tourism and culture each time frame sustainability, [4] points out that, in his determined research; the tourist management of territorial heritage in rural areas, in an analysis of the context of the city of Madrid (Spain), defines that the possibilities of developing tourism in a cultural environment are achieved through the construction of collaboration models. And consider, prior to developing strategies, address:

- Local cultural factors
- Aspects of valuation and respect for heritage sites and cultural expressions

[5] proposes that tourism management is a dynamic field that involves analyzing both the demand and supply of tourism goods and services, as well as the impact caused by tourists, establishments and services in the spaces, economies and in the cultural and natural heritage of the places of destination. [6] identifies that cultural tourism allows the coexistence of artistic and historical interest, specifying as a result that culture contributes to the cohesion of the community, considering it vital

to generate projects with strategies marked in the development and conservation of cultural heritage assets.

[7] proposes planning and management strategies in cultural tourism, which present results that are favorable to the process of conservation of cultural heritage, based on a management model. The approach that considers the research of [1] on: the notes for the construction of cultural tourism from the analysis of the heritage cultural offer and its demand by the tourism sector, carried out in Quindío, (Colombia) identifies the importance of the conservation and preservation of cultural resources with sustainable management, through the participation of local populations, in front of local opportunities.

In the local social context, the research carried out by [4] on the cultural tourist: typologies and analysis of the valuations of the destination from the case study of Cuenca-Ecuador can give a sense of conservation and preservation through the visit of tourists to visiting cultural centers, and determines in its results that the tourist can promote education and respect, conservation, and preservation of cultural sites.

Tourist satisfaction plays a fundamental role in defining the competitiveness of the tourist center. Regarding satisfaction, it can define that, it is considered as the level of a person's state of mind that results from comparing the perceived performance of a product or service, with their expectations. Tourist expectations are related to satisfaction, thus [8] determine it, focused on the quality of service as the strategic axis to achieve satisfaction.

2 State-of-the-Art

Quality is a fundamental piece of services, goods, or products, and it takes as a stage of perfection the Japanese managers in 1950, [9], who perfected the theories of quality, based on the improvement of the products, goods, and services. Quality focuses on several theories, which go from Deming (1986) cited in [10], determines the 14 principles of quality: that allow to achieve perfection through the application of instruments that allow to determine the potentiality based on the quality.

The intangibility, characteristic of tourist services, allows them to be heterogeneous, since the level of variation depends on one product to another, the visiting cultural sites, has its own theme, illustrating the tourist about the history and antecedents of the site, [10], in the publication: total quality in service management. (Spain).

The construction of a management system or model must consider establishing the appropriate results for the operation and administration, that is why from the point of view of Dr. Walter Shewart, cited in (Beltrami, 2017), when he coined the methodology of Plan, Do, Verify and Act, known by its acronym PHVA from the dissemination of Edwards Deming's work, planning was considered, the same as applied to quality management. A model is a complex structure which allows determining aspects, such as the planning, control, systematization, and operation of tourist

Table 1 Quality management parameters

Quality management	
Applied to	All the processes that make up a management system
Take as central objective	Achieve the goals proposed by the planned objectives
The participation	The participation of the staff, who make up the organization, is essential
Participation objective	Achieve continuous improvement in an effective process control system

processes, in an establishment, [2] focuses the study of the process management in a cultural environment in models.

The service performance SERVPERF model, according to [11], has been used to evaluate the perception of quality and satisfaction, research on similar aspects has used it, allowing defining qualitative research as a quality dimension in services, in museums, taking peculiarities with cultural visit sites, this type of model presents indicators based on quality, satisfaction, intention to recommend, and effective learning.

Quality management, through the application of concepts, delimited by ISO standards, contemplates certain parameters that are essential to determine (Table 1):

The ISO 9001 family of norms does not determine the standards or criteria for its applicability to the service, on the other hand, the Q certificate includes all the standards and specifies the means to be able to apply it. The Q certification based its standard on UNE 182,001, [12], considers it essential to consolidate standards on the basis of standards that adhere to the reality of the environment tourist.

Attached to the historical value that the cultural visiting site presents, it is essential to emphasize the level of preservation and maintenance, in the ICOMOS publication in the Letter of Cultural Tourism (1976) cited in [13], it defines cultural tourist attractions as "that form of tourism whose object is, among other purposes, the knowledge of monuments and historical-artistic sites.

The visiting cultural sites, according to [5], in their book structure of the tourist market, determines the visiting cultural sites, within the classification of cultural tourism, identified by, including, museums, monuments, historical sites, and the purpose is to present the history, art, and culture of the destination. Framed in the definition proposed by the authors, it can be said that they are places where tourists can interact with the history and folkloric essence of the community.

3 Methodology

The present research is of a correlational and descriptive type, due to the detail of the phenomena influenced in visiting cultural tourist destinations, and it is a determining factor in quality, through the perception of quality, through satisfaction, different categories can be defined, in addition to take the character of non-experimental design research, with a cross-sectional design because the level and state of the variables are analyzed at a single point in time.

The methodology used focuses on the use of the interval scale, based on the Likert scale, which allows the quantitative definition of the difference between variables that involve quality management in visiting cultural centers, in addition, it measures real values and not relative values, and therefore, the Pearson correlation coefficient will be elaborated because the variables are distributed approximately one from the other, and it avoids taking outliers. To finally define if the correlation is significant or if the correlation is significant, in relation to the verification of variables and hypotheses.

The service performance SERVPERF model was used in the collection of information because according to [13], they affirm that the SERVPERF has been found empirically superior to the SERVQUAL scale to be able to explain a greater variance of the total quality of the service. In addition to having greater theoretical support, it uses 50% of the SERVQUAL questions, avoiding doubting the attributes, allowing a more effective factor measurement.

For the purposes of the study, the population to apply the questionnaire based on the SERVPERF model was considered as the number of tourists, visiting cultural sites (museums, monuments with registration, historical sites), that is, according to data from the Decentralized Autonomous Government. Santiago de Píllaro and the San Miguelito Parish Autonomous Decentralized Government are a total of 873 visitors.

4 Results

The instrument applied by generalizing the items within the two variables (quality management and tourist satisfaction) provides a result, through Cronbach's alpha of 0.824, which indicates a good level of reliability, considering that it is > 8, in an acceptable scale (Tables 2 and 3).

Table 2 Case *processing*

		N	%
Cases	Valid	101	100.0
	Excluded[a]	0	0.0
	Total	101	100.0

Source Data output from IBM SPSS Statistics 2.1

Table 3 Reliability statistics (Cronbach's alpha)

Reliability statistics	
Cronbach's alpha	Number of elements
0.824	20

Table 4 Kaiser Meyer test, and Bartlett

KMO and Bartlett test		
Kaiser–Meyer–Olkin measure of sampling		0.706
Bartlett's Test of Sphericity	Approximate chi-square	148.048
	Gl	171
	Sig	0.000

4.1 Nonparametric Correlation

The results of the parametric test have been used through the KMO test or Bartlett's test, in which a measure of 0.706 is obtained, demonstrating that there is positive correlation in the data, which determines the level of coercion of the management of quality in visiting sites (Table 4).

4.2 Construct Analysis

The internal consistency of the constructs was estimated through the application of Cronbach's alpha, by evaluating the external loads, determining at the output of the SPSS software, as follows, by means of the composite reliability analysis, which measures all composite constructs (Tables 5 and 6).

The significance value of 0.000 which indicates that it is under 0.05 (and is marked with two asterisks for the same reason). It indicates that there is a correlation of variables, then the Pearson coefficient $p = 0.610$, shows that there is a moderate positive linear correlation between quality management and satisfaction. This means that if quality management increases, the satisfaction of tourists will also increase.

Table 5 Analysis of the cultural sites construct

Co Construct	Number of indicators	Reliability	Cronbach's alpha
APREN	5	0.90	0.871
REC	4	0.97	0.967
SAT	5	0.93	0.911
CONF	3	0.89	0.860
EMP	3	0.90	0.873

Table 6 Pearson's *correlation*

Correlations

		Management_quality	Sites_visit
Management quality	Pearson's correlation	1	0.610**
	Sig. (bilateral)		0.000
	N	101	101
Satisfaction_cultural sites	Pearson's correlation	0.610**	1
	Sig. (bilateral)	0.000	
	N	101	101

Fig. 1 Constructs diagram

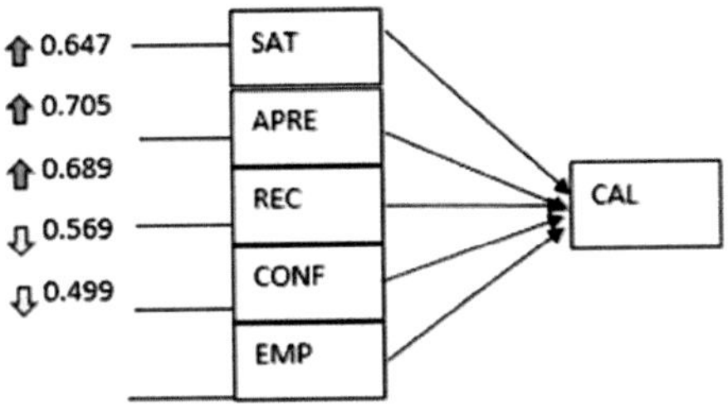

And on the contrary, if quality management decreases the satisfaction variable, it would tend to negative.

With the results, the correlation can be defined in a range from 0.4 to 0.69, which determines a moderate positive correlation. The Pearson correlation coefficient = 0.610, indicates a moderate positive linear correlation, indicates that there is a strong correlation between the variables and that the correlation found by Pearson's statistical method is true. Therefore, H1 is accepted = quality management influences the satisfaction of tourists, in the cultural sites of visit of the rural San Miguelito parish, that is, quality management if it influences the satisfaction of tourists in the parish rural of San Miguelito.

The correlation of the quality dimension with the dimensions of tourist satisfaction in the visiting cultural sites, determined the following correlations (Fig. 1).

4.3 *Comparative Analysis of Correlation Results*

The results contrast with the research of (Gosling, Silva, & Coelho, 2016), which determine a latent correlation with the level of satisfaction with the tourist, who is the key character to define the level and development of quality management, with a level of relationship through GoF of 63.1%, in addition, (Chiriboga, Guaman, Perez, &

Hidalgo, 2018), in the research of the quality of service, and its impacts on cultural tourism, using the SERVPERF method, adapted to the investigative reality, presents its correlation result at 0.607, weighted in the dimensions of the SERVPERF.

5 Conclusions

The development of the study, allowed in the first instance to determine the quality management processes applied in the cultural tourist centers of visit in the rural San Miguelito Parish, which are scarce, despite the operating history of the four cultural sites of visit implanted in the study area, which ranges between 7 and 15 years; however, as it does not have a quality management system, it is not included in the projects of competition of the GAD Santiago de Píllaro with respect to tourism promotion.

The quality levels, once analyzed and subjected to the Pearson correlation, show a linear correlation with a moderate positive trend with 0.610, and support the relationship between quality management and tourist satisfaction in the cultural sites visited, the analysis focused on four dimensions of quality, with their respective constructs that statistically through communality and factorial load differ the importance of quality management as a skillful process for the construction of a continuous improvement system, which is considered a fundamental axis for development of processes.

The quality levels, once analyzed and subjected to the Pearson correlation, show a linear correlation with a moderate positive trend with 0.610, and support the relationship between quality management and tourist satisfaction in the cultural sites visited, the analysis focused on four dimensions of quality, with their respective constructs that statistically through communality and factorial load differ the importance of quality management as a skillful process for the construction of a continuous improvement system, which is considered a fundamental axis for development of processes.

The proposal frames strategies that are oriented in the construction and monitoring of processes, for the implementation and control of a quality management system, through tools such as: sufficiency matrix of ISO standards, IDEFO box, process control, application of the PHVA cycle, documentary pyramid, considering aspects framed in the ISO 9001: 2015 and ISO 9004: 2018 regulations. Responsible for granting confidence to the products and services offered and focused on the company's capacity for sustained success.

References

1. Ponsignon F, Derbaix M (2020) The impact of interactive technologies on the social experience: An empirical study in a cultural tourism context. Tourism Management Perspectives, 100723

2. Wei C, Dai S, Xu H, Wang H (2020) Cultural worldview and cultural experience in natural tourism sites. J Hosp Tourism Manag, 241–249
3. Vergori A, Arima S (2020) Cultural and non-cultural tourism: Evidence from Italian experience. Tourism Manag, 104058
4. Haigh M (2020) Cultural tourism policy in developing regions: The case of Sarawak. Tourism Manag, Malaysia, p 104166
5. Liu L-T (2020) Comparing the perspectives of municipal tourism departments and cultural departments on urban cultural-tourism development. J Destination Marketing & Manag, 100432
6. Zhang T, Yin P (2020) Testing the structural relationships of tourism authenticities. J Destination Marketing & Manag, 100485
7. Seo K, Jordan E, Woosnam K-M, Lee C, Lee E-J (2021) Effects of emotional solidarity and tourism-related stress on residents' quality of life. Tourism Manag Perspec, 100874
8. Nguyen C, Dinh-Su T (2021) Tourism, institutional quality, and environmental sustainability, Sustain Production Consump, 786–801
9. Xiaobin M, Xiaobin S, Guolin H, Xing Z, Li L (2021) Evaluation and spatial effects of tourism ecological security in the Yangtze River Delta. Ecol Indicators, 108190
10. Pertusa-Ortega E, Tarí J, Pereira-Moliner J, Molina-Azorín J-F, López-Gamero M (2021) Developing ambidexterity through quality management and their effects on performance. Int J Hospitality Manag, 102720
11. Dzisi E, Atuah-Obeng D, Tuffour Y (2021) Modifying the SERVPERF to assess paratransit minibus taxis trotro in Ghana and the relevance of mobility-as-a-service features to the service. Heliyon, e07071
12. Leong L-Y, Hew T-s, Lee V-H, Ooi K (2015) An SEM–artificial-neural-network analysis of the relationships between SERVPERF, customer satisfaction and loyalty among low-cost and full-service airline. Expert Systems with Appl, 6620–6634
13. Duque Oliva J, Baquero JAC (2017) Validación del modelo SERVPERF en el ámbito de internet: un caso colombianoValidation of the SERVPERF model in the internet environment: a Colombian case. Suma de Negocios, 115–123

Use of GeoGebra in Learning to Solve the Problem of Calculating the Root of a Nonlinear Equation

Judith Keren Jiménez-Vilcherrez, Felicita Marcela Velásquez-Fernández, Araceli Margarita Acevedo-Ruiz, Ricardo Velezmoro-León, and Robert Ipanaqué-Chero

Abstract Generally, when starting a first undergraduate numerical methods course, the first method taught to calculate the root of a root of a nonlinear equation in a single variable is the bisection method, in which the initial interval is divided into two subintervals taking the midpoint of the segment as a reference, the subinterval containing the root is bisected again, and so on until the desired root is approximated. The question that naturally arises from students is why would the interval necessarily have to be bisected? What if instead of bisecting the initial interval, we divide it according to a given ratio? This chapter describes the interval method divided by a given reason to approximate the root of a nonlinear equation in a single variable as a generalization of the bisection method. Proposing a new method for teaching the calculation of roots of a nonlinear equation.

Keywords Bisection · Reason given · Roots · Nonlinear equation · GeoGebra

1 Introduction

The COVID-19 pandemic has brought about an acceleration in the digitization of education, and this transformation of the educational process was the only way to make possible the continuity of academic activities during isolation [1], forcing university teachers to incorporate the use of digital resources in the teaching process

J. K. Jiménez-Vilcherrez (✉) · A. M. Acevedo-Ruiz · R. Velezmoro-León
Universidad Tecnológica del Perú, Av. Vice Cdra 1, Piura, Peru
e-mail: C19863@utp.edu.pe

R. Velezmoro-León
e-mail: rvelezmorol@unp.edu.pe

F. M. Velásquez-Fernández · A. M. Acevedo-Ruiz · R. Velezmoro-León
Universidad César Vallejo, Avenida Chulucanas s/n-Distrito 26 de Octubre, Piura, Peru
e-mail: fmvelasquezf@ucvvirtual.edu.pe

A. M. Acevedo-Ruiz · R. Velezmoro-León · R. Ipanaqué-Chero
Universidad Nacional de Piura, Urb. Miraflores s/n Castilla, Piura, Peru
e-mail: ripanaquec@unp.edu.pe

X.-S. Yang et al. (eds.), *Proceedings of Seventh International Congress on Information and Communication Technology*, Lecture Notes in Networks and Systems 447,
https://doi.org/10.1007/978-981-19-1607-6_67

[2]; one of the most complicated subjects perhaps for students is mathematics, since mathematics is considered the basis of complex knowledge processes, where it is necessary for people to have critical thinking, reflective, and analytical [3], requiring the use of digital tools in order to improve the academic performance of students; GeoGebra is a digital tool that facilitates the learning of mathematics and problem-solving, constituting an ideal tool for use as a strategy in the teaching of exact sciences [4, 5]. In this paper, he addresses a new methodological strategy to address a classic problem in the course of numerical methods, the calculation of the roots of a nonlinear equation of one variable, and to solve this problem, there are many methods: one of the basic methods is the bisection method that consists of dividing the initial interval into two subintervals taking the midpoint of the segment as a reference, the subinterval containing the root is bisected again, and so on until the desired root is approached, but during the teaching process appearing naturally among students the question of why should the interval necessarily have to be divided in two? What if instead of bisecting the initial interval, we divide it according to a given ratio? Thus, emerging a new way of approaching this problem through a new method in which the initial interval is going to be divided by a given ratio and the subinterval containing the root is divided again and so on until the desired root is approximated. The objective of this paper is to offer the teacher and the student a new method to approach the problem of calculating the root of a nonlinear equation of a single variable as a generalization of the bisection method. The paper is organized as follows: Sect. 2 presents the theoretical basis for the construction of the segment division method using a given ratio to calculate the root of a nonlinear equation of one variable. In Sect. 3, he presents the construction of the method in GeoGebra. Finally, Sect. 4 with the main conclusions of this work.

2 The Interval Method Divided by a Given Ratio

Lemma 1 *Let $f \in C^0([a,b] \subset \mathbb{R})$ with $f(a) \cdot f(b) < 0$. Given the succession $\{x_n\}_{n\in\mathbb{N}}$, such that*

$$x_n(a,b) = \begin{cases} \frac{\nu a+\mu b}{\mu+\nu} & n = 1 \vee f(x_1(a,b)) = 0 \\ \begin{cases} x_{n-1}(a, x_1(a,b))\ f(a) \cdot f(x_1(a,b)) < 0, \\ x_{n-1}(x_1(a,b), b)\ f(x_1(a,b)) \cdot f(b) < 0. \end{cases} & n > 1. \end{cases}$$

with $\mu, \nu > 0$, It is true that $|x_m - x_n| \le \frac{\max\{\mu,\nu\}}{\mu+\nu}|b_n - a_n|$, for any $m \ge n$.

Theorem 1 *Let $f \in C^0([a,b] \subset \mathbb{R})$, with $f(a) \cdot f(b) < 0$. The succession $\{x_n\}_{n\in\mathbb{N}}$, such that*

$$x_n(a,b) = \begin{cases} \frac{\nu a+\mu b}{\mu+\nu} & n = 1 \vee f(x_1(a,b)) = 0 \\ \begin{cases} x_{n-1}(x_1(a,b), b)\ f(a) \cdot f(x_1(a,b)) > 0, \\ x_{n-1}(a, x_1(a,b)) \quad \textit{in another case}. \end{cases} & n > 1. \end{cases}$$

2.1 Speed of Convergence

Since the given ratio is $\mu : \nu$ it is clear that μ, ν, they are positive integers. Further,

$$\max\{\mu, \nu\} = \begin{cases} \mu \textit{ if } \mu \geq \nu \\ \nu \textit{ if } \mu < \nu \end{cases}$$

So that,

$$\frac{\max\{\mu, \nu\}}{\mu + \nu} = \begin{cases} \frac{\mu}{\mu+\nu} \textit{ if } \mu > \nu, \\ \frac{1}{2} \textit{ if } \mu = \nu, \\ \frac{\nu}{\mu+\nu} \textit{ if } \mu < \nu \end{cases}$$

For the first case

$$\nu < \mu < \mu + 2\nu \Leftrightarrow \mu + \nu < 2\mu < 2\mu + 2\nu$$

$$\Leftrightarrow \frac{\mu + \nu}{2} < \mu < \mu + \nu$$

$$\Leftrightarrow \frac{1}{2} < \frac{\mu}{\mu + \nu} < 1$$

For the third case

$$\mu < \nu < 2\mu + \nu \Leftrightarrow \mu + \nu < 2\nu < 2\mu + 2\nu$$

$$\Leftrightarrow \frac{\mu + \nu}{2} < \nu < \mu + \nu$$

$$\Leftrightarrow \frac{1}{2} < \frac{\nu}{\mu + \nu} < 1$$

Thus

$$\frac{1}{2} \leq \frac{\max\{\mu, \nu\}}{\mu + \nu} < 1.$$

This result reaffirms the convergence of the divided interval method in a given ratio. Furthermore, it allows us to conclude with complete certainty that the bisection method converges faster than any other method of division of an interval that considers a ratio other than 1 : 1.

2.2 *Algorithm*

It presents the algorithm of the interval method divided by a given ratio to approximate the root of a nonlinear equation in a variable with GeoGebra.

Algorithm 1 Method of the Interval Divided by Given ratio

Require: extremes: a , b; reason: μ, ν; tolerance: TOL; máx núm of iter: N
Ensure: approximation value q, zero, or error message

$i = 1$
$FA = f(a)$
while $i \leq N$ **do**
  $q = (\nu a + \mu b) / (\mu + \nu)$
  $FQ = f(q)$
  if $FQ = 0$ o $\max\{\mu, \nu\}(b - a) / (\mu + \nu) < TOL$ **then**
    Do return q ▷ Satisfying ending
    stop
  end if
  $i = i + 1$
  if $FA \cdot FQ > 0$ **then**
    Do $a = q$
    $FA = FQ$
  else
    $b = q$
  end if
end while
return The method failed after N iterations
Stop

3 Construction of the Method in GeoGebra

Next we will show the steps to follow in the construction of the method of the interval divided by a given ratio to approximate the root of a nonlinear equation in a variable with GeoGebra. Steps to follow:
Step 1: Input elements To start, we enter the function that we are going to use on the input line. See Fig 1

$$f(x) = x^3 - 2x + 1$$

The graph of the function $f(x)$ is shown in Fig. 6. We also enter in the GeoGebra spreadsheet the headings where we are going to locate the values a_n, b_n, c_n, $f(c_n)$ and *error* and also the values a, b, c, $f(c)$ and *error* in cells B2, C2, D2, E2, respectively; see Fig. 2, where c is calculated with the following formula:

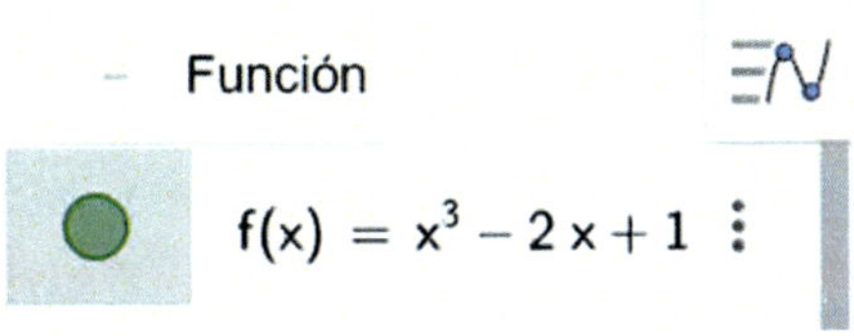

Fig. 1 We enter the function that we are going to use

A	B	C	D	E	F
n	a_n	b_n	c_n	$f(c_n)$	error
1	-3	0	-1.7647058824	-0.9662120904	

Fig. 2 We enter the spreadsheet where we are going to locate the values a_n, b_n, c_n, $f\ (c_n)$ and *error*

Fig. 3 We enter the value of *c* in box D2 of the spreadsheet

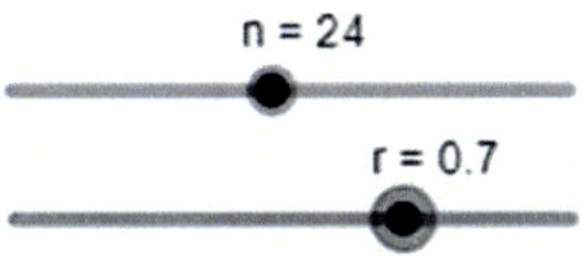

Fig. 4 We enter the constants *n*, *r* as sliders

$$c = \frac{a + rb}{1 + r}$$

to do this we enter the following in box D2; see Fig. 3.

Next we define the constants *n*, *r* (where $r = \frac{u}{v}$) and show the objects as sliders. See Fig. 4. Through an input box, we enter a and b associated with boxes B2 and C2 of the spreadsheet; see Fig. 5.

Fig. 5 We enter *a* and *b* associated with cells B2 and C2 of the spreadsheet

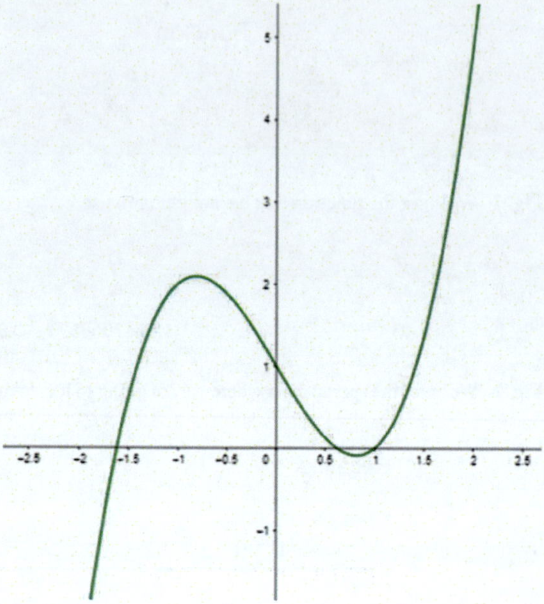

Fig. 6 Graph of the function $f(x)$

Step 2: Interval sequence using the spreadsheet

In the next row of the spreadsheet, we must obtain the values of the new interval and the new value of c. Taking into account that the new interval values depend on the signs of $f(a)$, $f(b)$ and $f(c)$ so you must make use of the instruction conditional (Fig. 6)

```
If [<Condition>, <Then>, <If not>]
```

in cells B3, C3, D3. In cell B3, we must consider the condition if the product $f(a)f(b) > 0$, then the cell takes the value of D2; otherwise, it takes the value of B2; similarly in cell C3, if the product $f(a)f(b) > 0$, then the cell takes the value of C2; otherwise, it takes the value of D2, and the new value of c is obtained by copying (Ctrl + C) cell D2 in cell D3; in the same way, $f(c)$ is obtained and the error is obtained by entering in cell F3 the absolute value of the difference of E2 and E3. Once row 3 is obtained, it can be copied in the following rows until completing the number of iterations n set when defining the slider. See Figs. 7 and 8.

4 Conclusions

This chapter shows the use of GeoGebra to solve the problem of calculating the roots of a nonlinear equation of one variable using the interval division method with

	A	B	C	D	E	F
1	n	a_n	b_n	c_n	$f(c_n)$	error
2	1	-3	0	-1.7647058824	-0.9662120904	
3	2	-1.7647058824	0	-1.0380622837	1.9575363617	2.923748452
4	3	-1.7647058824	-1.0380622837	-1.4654996947	0.7835612951	1.1739750665
5	4	-1.7647058824	-1.4654996947	-1.6415033345	-0.140078559	0.9236398542
6	5	-1.6415033345	-1.4654996947	-1.5690312475	0.2753287097	0.4154072687
7	6	-1.6415033345	-1.5690312475	-1.6116618869	0.0371060982	0.2382226115
8	7	-1.6415033345	-1.6116618869	-1.6292156796	-0.0660670659	0.1031731641
9	8	-1.6292156796	-1.6116618869	-1.6219876473	-0.0232210587	0.0428460072
10	9	-1.6219876473	-1.6116618869	-1.6177358636	0.0017448235	0.0249658822
11	10	-1.6219876473	-1.6177358636	-1.6202369129	-0.0129197094	0.0146645329
12	11	-1.6202369129	-1.6177358636	-1.6192070691	-0.0068740131	0.0060456962
13	12	-1.6192070691	-1.6177358636	-1.6186012786	-0.0033225348	0.0035514783
14	13	-1.6186012786	-1.6177358636	-1.6182449312	-0.0012350949	0.00208744
15	14	-1.6182449312	-1.6177358636	-1.6180353152	-0.000007765	0.0012273299
16	15	-1.6182449312	-1.6180353152	-1.6181586187	-0.0007296721	0.0007219071
17	16	-1.6181586187	-1.6180353152	-1.6181078467	-0.0004323983	0.0002972737
18	17	-1.6181078467	-1.6180353152	-1.6180779808	-0.0002575431	0.0001748552
19	18	-1.6180779808	-1.6180353152	-1.6180604126	-0.0001546912	0.000102852
20	19	-1.6180604126	-1.6180353152	-1.6180500783	-0.0000941914	0.0000604998
21	20	-1.6180500783	-1.6180353152	-1.6180439994	-0.0000586038	0.0000355876
22	21	-1.6180500783	-1.6180439994	-1.6180475752	-0.0000795376	0.0000209338
23	22	-1.6180500783	-1.6180475752	-1.6180490477	-0.0000881575	0.0000086199
24	23	-1.6180500783	-1.6180490477	-1.6180496539	-0.0000917069	0.0000035494
25	24	-1.6180500783	-1.6180496539	-1.6180499036	-0.0000931684	0.0000014615
26						

Fig. 7 Values of the 24 iterations in the GeoGebra spreadsheet

a given ratio, as an alternative to the bisection method. It was possible to check the importance of ICT in teaching numerical methods, and GeoGebra is a tool for teaching and solving this type of problem since it provides us a graphical view of the problem and in a very simple way allows us to perform the numerical calculations necessary for a solution. It was also found that the bisection method converges faster than any other interval division method that considers a ratio other than 1 : 1. This paper seeks to propose a new method to calculate the root of a nonlinear equation in a simple and practical variable. It would be convenient for a later study to modify the algorithm proposing that the given reason be random in each iteration and check if in this way it converges faster compared to bisection method.

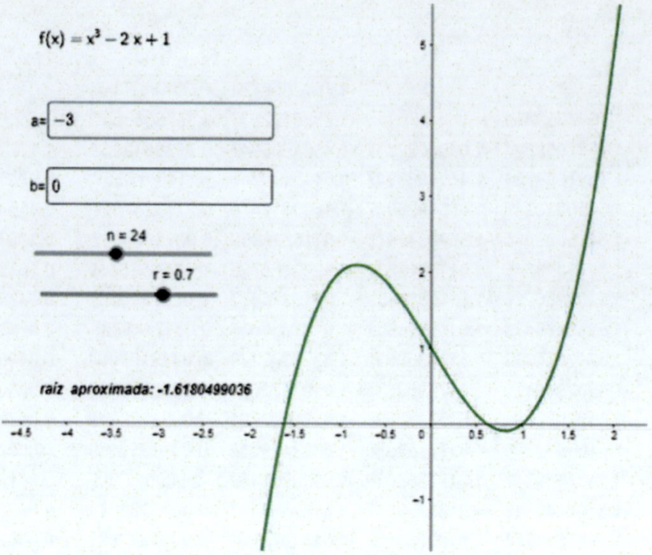

Fig. 8 Values of the 24 iterations in the GeoGebra spreadsheet

References

1. Delgado T (2020) Influencia de la pandemia COVID-19 en la aceleración de la transformación digital. Revista Cubana de transformación digital. 1(3):01–05
2. Rambay M, De La Cruz J (2021) Desarrollo de las competencias digitales en los docentes universitarios en tiempo pandemia: Una revisión sistemática. In Crescendo 11(4):511–527
3. Rambay M, De La Cruz J (2015) GeoGebra para la enseñanza de la matemática y su incidencia en el rendimiento académico estudiantil. Revista Tecnológica - ESPOL 28(5):121–132
4. Jiménez J, Jiménez S (2017) GeoGebra, una propuesta para innovar el proceso enseñanza-aprendizaje en matemáticas. Revista Electrónica sobre Tecnología, Educación y Sociedad 4(7):131–142
5. Del Río L (2020) Recursos para la enseñanza del Cálculo basados en GeoGebra. Revista Do Instituto GeoGebra Internacional de São Paulo 9(1):120–131. https://doi.org/10.23925/2237-9657.2020.v9i1p120-131

Usability Evaluation Using Unmoderated Remote Usability Testing on Angkasa LMS Website Case Study

Veronikha Effendy, Dana Sulistiyo Kusumo, Nungki Selviandro, and Kusuma Ayu Laksitowening

Abstract The pandemic has made digital transformation faster, and Indonesia is no exception, especially in education. Learning management system (LMS) is a learning media that is widely used in educational institutions. However, not all educational institutions have sufficient resources to build an LMS from scratch. "Angkasa LMS" Web is a Web that allows educational institutions to order a ready-to-use LMS easily. The target users of this service are pretty varied because they consist of education management and educational management foundations, especially in areas of Indonesia that have Internet access. To be used easily by these varied target users, this Website must have good usability in its user interface. For this reason, usability testing needs to be carried out in order to get feedback for improving the user interface design before the application is delivered to the public. However, the developer has obstacles related to the COVID-19 pandemic policy to carry out usability testing, which limits direct interaction with people. Moreover, the project time is quite narrow, and the schedule is quite tight for the developer team. Based on these limitations, this study conducted usability testing of the "Angkasa LMS" Web case study using the unmoderated remote usability testing method. The experimental results show that this method can be used to obtain insightful feedback from the participants, with additional treatment such as the use of convenience sampling, periodic reminders for participants, and the increasing number of participants exceeding the target.

Keywords LMS · Unmoderated remote usability testing · Developer limitation · Convenience sampling

1 Introduction

Since entering the COVID-19 pandemic, the Indonesian government has encouraged educational institutions to run online learning. Based on data released by the Spada Indonesia Website, it is recorded that the use of LMS in universities is increasing

V. Effendy (✉) · D. S. Kusumo · N. Selviandro · K. A. Laksitowening
Telkom University, Bandung, Indonesia
e-mail: veffendy@telkomuniversity.ac.id

X.-S. Yang et al. (eds.), *Proceedings of Seventh International Congress on Information and Communication Technology*, Lecture Notes in Networks and Systems 447,
https://doi.org/10.1007/978-981-19-1607-6_68

significantly [1]. This shows the acceleration of LMS use in Indonesia [1]. Unfortunately, not all educational institutions have sufficient resources to implement LMS in their institutions. This has prompted Angkasa to create a ready-to-use LMS ordering service Web so that an institution that wants to use an LMS does not have to bother thinking about implementing its LMS from scratch. With this Web application, users will only need to order LMS in LMS packages according to their needs, just like buying an Internet quota package. The target users of this service are educational institutions and educational management foundations, especially those in Indonesia. Considering that the target user is quite broad, in developing the Web application, it is necessary to consider usability testing so that the developer can produce the usable product [2].

Usability testing generally involves a number of participants and requires sufficient time to gain insight in order to improve the user interface design. On the other hand, the obstacles faced by software developers, in general, are limited time, tight schedules, the number of developers, and costs. Plus, due to the COVID-19 pandemic, offline usability testing with face-to-face contact with participants is also indirectly limited.

In this study, usability testing was carried out using unmoderated remote usability testing to accommodate these limitations. The selection of remote usability testing is to accommodate the limitations of direct meetings with respondents, while unmoderated is chosen due to the limited number of developers and also the determination of meeting schedules is difficult due to the developer team's schedule and the respondent's schedule being unmatched [2, 3].

2 Literature Review

2.1 Remote Usability Testing

Remote usability testing allows the involvement of participants from different locations without high costs. Almost anyone who has a device such as a smartphone and an Internet connection can participate [2]. This convenience is the most significance advantage of remote testing. Some other advantages such as no extra time required to travel to meet participants, no travel expenses required, does not require special lab, the use of hardware that is familiar to the participants because they use their own devices, and a more natural participant environment [2]. The drawback of this remote testing is that it takes extra time and effort to ensure that the test equipment can run well on the participant's device [2]. With the condition of software developers who have limited time and cost, it is clear that the selection of remote testing is more feasible to do.

There are two types of remote user testing, namely (1) moderated remote testing and (2) unmoderated remote testing. The main difference between the two methods is in the direct interaction of the researcher during the evaluation [3]. Moderated remote testing requires researchers to directly guide the testing process carried out by users using online communication tools. Unmoderated remote testing does not require researchers to be directly involved but requires tools that provide clear instructions to the user. The tools should have the ability to record user activities and display follow-up questions to be answered by the user [3].

2.2 *Unmoderated Remote Testing*

In unmoderated remote usability testing, participants participate when they feel comfortable doing so and without moderation from the UX researcher. This allows participants to complete the evaluation tasks at the time and place they find most natural [2].

Since this method does not require a direct meeting between the researcher and the participant, the problem of setting the schedule that often occurs due to the unmatched time between the researcher and the participant can be eliminated [2]. According to Nielsen, there are six things that need to be prepared to perform unmoderated remote usability testing [3]: define study goals, select testing software, write tasks and questions, pilot test, recruit participants, and analyze results.

2.3 *Samples*

The number of samples used in this study was above 5 (five) people. According to Nielsen, conducting usability testing with the aim of gaining insight from respondents by involving five participants is close to the maximum benefit–cost ratio for user testing [3]. However, in this study, we added more samples of respondents because by using the unmoderated remote testing method, it is possible that respondents did not complete the test until the end due to various things such as lack of motivation or other technical obstacles when running the prototype [3].

2.4 *Convenience Sampling Technique*

The sampling technique used in this study is a convenience sampling technique with the aim of reducing the risk of incomplete testing and improving the quality of the insights obtained. This technique is done by selecting the appropriate respondents and agreeing to provide input so that it is expected to provide sufficient insight [4].

2.5 System Usability Scale (SUS)

SUS is a commonly used questionnaire to evaluate the usability of a product [5]. This SUS was chosen because it is easy and fast to get usability from a product, so it can indirectly reduce costs. This study uses SUS, which has been adapted in Indonesian [6] because it will make it easier for respondents who use Indonesian in their daily activities.

3 Research Methodology

Figure 1 shows the research methodology, and we used in this study.

3.1 Define Study Goals and Selecting Testing Tools

This usability testing aims to understand user behavior, obtain information about usability problems in user interface design, and gain insight from participants through feedback from open questions.

To accommodate those goals, we use Maze tools as software testing tools to deliver information about the testing purpose and display what tasks the participants must do, record their activities on the user interface when performing those tasks, and gain insight from participants through open feedback questions.

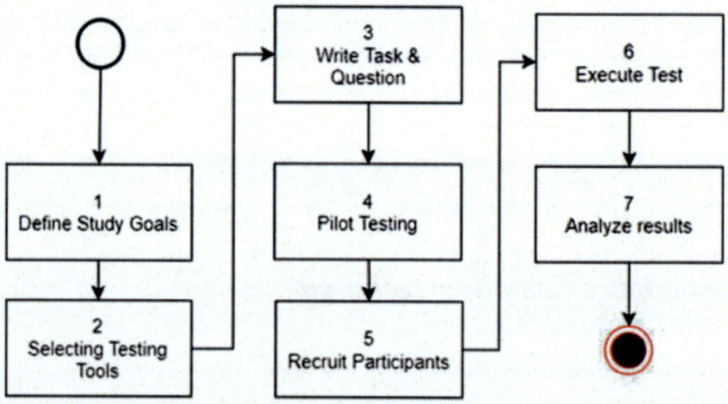

Fig. 1 Research methodology

Table 1 Examples of task instruction in Bahasa

No	Task's purpose	Task instruction (scenario task)
1	Melihat informasi tentang Angkasa LMS	Semisal Anda berminat untuk menggunakan LMS pada institusi Anda. Kemudian, Anda mendapatkan informasi tentang layanan Angkasa dari teman Anda. Karena tertarik lebih lanjut, Anda kemudian mengakses web angkasa, dan mulai mencari tahu tentang Angkasa LMS. Tunjukkan bagaimana Anda mencari informasi terkait Angkasa LMS pada web ini
2	Melihat informasi Produk & layanan	Anda juga ingin melihat produk dan layanan apa saja yang ditawarkan Angkasa. Tunjukkan bagaimana caranya Anda mendapatkan informasi tersebut

3.2 Write Tasks and Questions

In unmoderated remote usability testing, this stage is the most challenging [2]. The absence of a moderator can cause the instructions to be misunderstood by the participants. One of the factors that need to be considered in designing task instructions is the length of the instructions [7]. Instructions should not be so long as to cause confusion for participants. Writing tasks must be made as realistic as possible so that participants can understand in what environment they will perform the task [3]. However, task writing should not describe the steps that participants must take on the prototype so that we can get better insight [3].

To create a realistic task, a scenario task model is used. Scenario tasks are arranged in order and are related to each other to increase engagement and make it easier for participants to condition themselves when performing these tasks. The following is an example of writing task instructions that have been created (Table 1).

3.3 Pilot Testing

Before conducting the actual test involving real participants, we conducted a pilot test run on the testing instrument that had been set on the software testing tools. This pilot testing aims to validate its feasibility for use in the actual testing later. Pilot testing is very crucial because the execution of the actual test will not involve the researcher as a moderator [8].

The pilot test involved five respondents outside the research team in this study. This test aims to ensure the task instructions, and questions are understandable by the user and to reveal obstacles that might arise in the execution of usability testing.

There were some typos in some parts of the instructions, the use of different terms between the task description and user interface design, and ambiguous questions that caused the respondents' answers to be inaccurate. We then refined the instruction and questions based on these findings before being used for real testing.

Another finding from this pilot testing is that only 3 out of 5 participants completed the usability test to the end. The other 2 participants had Internet connection issues after being confirmed.

3.4 *Recruit Participants and Execute Test*

Participant recruitment is carried out in a closed manner using convenience sampling, with the target users: (1) educational institution management foundations and (2) educational institution management apparatus from high school and college level. We also determine the minimum criteria for participants, which is that they must have good literacy related to the Internet and Web applications because Internet connection is a factor that we cannot predict. Based on the success ratio of getting a complete response from pilot testing, we recruit a minimum of 10 participants to achieve the target of at least five complete responses.

In carrying out usability testing, we distribute URL links containing testing tools to target participants and provide a testing period of 14 (fourteen) days. We also monitor the responses that come in every week and provide reminders to participants until the specified testing period ends. The problem we encountered was that there was an unstable network constraint on the part of the participants, and this affected the motivation of the participants during the testing period. In addition, after experiencing problems, some participants forgot to repeat the test again. For this reason, the provision of reminders was carried out more intensively.

3.5 *Analyze Result and Discussion*

Considering the implementation of the test that we cannot monitor (unmoderated), we need to analyze the several types of data we get, namely statistical data on test results, participant activity records, feedback on each task, SUS questionnaire responses, and open question at the end of SUS questionnaire.

Table 2 shows statistical data from the test results. From the statistical data in Table 2, we explored the activity records and found that some participants stopped in the

Table 2 Statistical data result

Item	Value
Number of response	22
Number of participants	16
Number of participants who completed all tasks	7
Misclick average rate	19%

middle of test execution. Some of them tried to repeat the test again and succeeded, but some failed. After confirmation, we found that they were experienced in Internet connection trouble and ran out of time for repeating the test again.

Internet connection problems, which are still a major problem in several regions in Indonesia, are undeniably the biggest obstacle to this method. Therefore, participants must have high motivation and commitment to provide feedback. In this unmoderated remote evaluation, participants also have flexible time to do usability testing. Because of this flexibility, the provision of periodical reminders is quite important to get responses from participants according to the time frame specified.

We also calculate the SUS score from the results of the SUS questionnaire responses. The SUS score was 67.5. This score is in the lower margin of the acceptability range of "acceptable", so it still needs a lot of improvement. This is in line with the overall feedback provided by participants after filling out the SUS questionnaire. The feedback is about the unfamiliar menu and hidden menu, which makes it difficult for users to complete their tasks.

Table 3 shows the tasks ID, misclick rate, and feedback assigned to each task. From the feedback assigned to each task, we get quite a lot of insightful feedback from participants. This input is then taken into consideration to improve the user interface design of the Angkasa LMS Website.

We also looked for the relationship between the misclick calculations of each task with the feedback given by the participant from each task that has been done. From those data, it can be seen that the presence of a misclick does not necessarily indicate a problem, as seen by the feedback response that is quite good. Likewise, the low number of misclicks does not necessarily indicate that there are any usability issues. The data show that the misclick rate can indicate the severity of usability problems if only there is any feedback from the participant that shows usability problem existence. This can be seen in task 4 and task 7 that require extra attention to be redesigned and become a critical priority to do first. Even so, further exploration is still needed to conclude the relationship between misclick rate and usability problems.

4 Conclusion

In this study, we relied not only on the test results of unmoderated remote usability but also used the results of the SUS questionnaire and the limited follow-up open questions. The test results are then combined and analyzed to gain insight. This is done to avoid bias from certain test results. In future research, exploring other combinations of testing methods might be interesting in order to get more insightful feedback by focusing on specific user experience goals, such as the aspect of help, utility, and other aspects related to Websites providing certain public services.

Remote unmoderated usability testing can be done by combining it with a convenience sampling technique to increase participant engagement so that they are willing to complete all tasks in the test. Due to several factors such as Internet connection problems or internal participant problems that we cannot control, we need to recruit

Table 3 Misclick rate and feedback responses of each task

Task ID	Misclick rate (%)	Feedback
1	26	Short information about space is better placed on the main Web page
2	0	Good
3	0	Information is easy to find, but I get lost then The server resource information obtained is still lacking It should be made to fit only one page, no need to scroll
4	**30.50**	I finally found it, but the symbol is not familiar Ambiguous with the "account registration" task I thought I could do "event registration" after login
5	15.70	Good
6	22	Good
7	**46.50**	Lack of information on payment instructions
8	31	Good
9	25	I think it will be better to use a button than a link? What if I need to downgrade?
10	29	It is necessary to add information related to the statistics of enrolled users, the number of subjects, the number of activities, etc. It would be better if there was an automatic offer for upgrades if needed
11	13	Maybe using a tooltip/popup when hovering over the button is also more helpful/informative? There need to be some improvements to (1) what if changing the package is a downgrade? (2) What if the package replacement is not done at the end of the package's active period?
12	0	I do not understand the "tools" menu. Apparently, it is for export Export data should not only be the database file. How about exporting the LMS content?
13	9.70	Need double confirmation when deleting (maybe entering the password again?) The menu name is a little confusing at the beginning It is difficult. It is better if you just have a non-active button on the main page Was the site really deleted? Or is it just deactivated? It seems that deleting the site does not need to be a mandatory tool provided

participants with a number that exceeds the targeted number. In this study, a periodical reminder for the participants plays an important role in order to get the responses.

Follow-up questions that are asked every time the task is done by participants are quite effective in getting user feedback focused on a specific task. Open questions asked at the end of the SUS questionnaire can also provide an opportunity for participants to provide their feedback from their experience in completing the overall tasks.

References

1. LMS Perguruan Tinggi dan LLDIKTI, Kementrian Pendidikan dan Kebudayaan, 2021 [Online]. Available: https://spada.kemdikbud.go.id/course/lldikti/lldikti-iv. Accessed 2 Nopember 2021
2. Bleecker ID, Okoroji R (2018) Remote usability testing: Actionable insights in user behavior across geographies and time zones. Packt Publishing Ltd
3. Nielsen J (2012) Nielsen Norman Group, NN/g, 3 June [Online]. Available: https://www.nngroup.com. Accessed 5 November 2021
4. Wohlin C, Runeson P, Host M, Ohlsson MC, Regnell B, Wesslen A (2012) Experimentation in software engineering. Springer, New York
5. Lewis JR (2006) Usability testing. IBM Software Group
6. Sharfina Z, Santoso HB (2016) An Indonesian adaptation of the system usability. In: International Conference on Advanced Computer Science and Information Systems (ICACSIS), Malang
7. Alexander KP (2013) The usability of print and online video instructions. Technical Communication Quarterly, pp 237–259, 15 April
8. He S (2021) Applying stepped task in remote unmoderated test: A case report. In: International Conference on Human-Computer Interaction

Distributed Deep Reinforcement Learning for Resource Allocation in Digital Twin Networks

Jie Luo, Jie Zeng, Ying Han, and Xin Su

Abstract With the rapid growth of the wireless network scale and the aggressive development of communication technology, the communication network connection is required to drift to digits in order to ameliorate the network efficiency. Digital twin (DT) is one of the most promising techniques, which promotes the digital transition of communication networks by establishing mappings between virtual models and physical objects. Nevertheless, due to the limitation and heterogeneity of equipment resources, it is a great challenge to provide efficient network resource allocation. To solve this problem, the authors propose a novel network paradigm based on digital twin to build the topology and model of the communication system. Then a distributed deep reinforcement learning (DRL) method is designed to dispose the problem of resource allocation in cellular networks, and an online–offline learning framework is proposed. Firstly, the offline training is carried out in the simulation environment, and the DRL algorithm is applied to train the deep neural network (DNN). Secondly, in the process of online learning, the real data are further utilized to fine-tune the DNN. Numerical results illustrate the superiority of the proposed method in terms of average system capacity. In the case of different user densities, the performance of the proposed algorithm has more advantages than that of benchmark algorithms and has better generalization ability.

Keywords Digital twin (DT) · Communication networks · Resource allocation · Deep reinforcement learning (DRL)

This work was supported by National Key R&D Program of China (2020YFB1806702) and DATANG MOBILE COMMUNICATIONS EQUIPMENT CO., LTD Funded Project (20202000659).

J. Luo (✉)
China Academy of Telecommunications Technology, Beijing, China
e-mail: nsluojie1016@163.com

J. Zeng · X. Su
Tsinghua University, Beijing, China

Y. Han
Chongqing University of Posts and Telecommunications, Chongqing, China

X.-S. Yang et al. (eds.), *Proceedings of Seventh International Congress on Information and Communication Technology*, Lecture Notes in Networks and Systems 447,
https://doi.org/10.1007/978-981-19-1607-6_69

1 Introduction

Wireless communication is going through rapid technological development in the late years, and the network scale is growing exponentially. According to the report of IDC [1], the number of devices connected to the communication network will reach 41.6 billion, and it is expected that these devices will generate nearly 8×10^6 TB of data by 2025. The huge amount of data and computation require the wireless communication network to speed up the digital transformation.

Digital twin (DT) is a great potential technology that enables digital conversion by virtualizing physical objects into digital models, as shown in Fig. 1. The digital models are able to comprehend the condition of the physical entity through data sensing, which provides a channel of prediction and estimation of the dynamic changes. The conception of DT was first proposed in [2] and then applied in air force vehicles to diagnose abnormality. Recently, DT has expanded to smart cities, manufacturing, industrial Internet of things, and so on. With the help of DT, the network topology and the other physical components in the communication network are able to be reflected excellently, and so that the network can be managed systematically. However, many technical challenges need to be conquered in applying DT to communication networks. Firstly, large amounts of data collected from various devices require real-time processing while local servers are hard to support fast signal processing and DT modeling due to the limit computational resources [3]. Secondly, huge channel resources are required owing to the frequent communication between virtual models and physical objects. In addition, the randomness of the

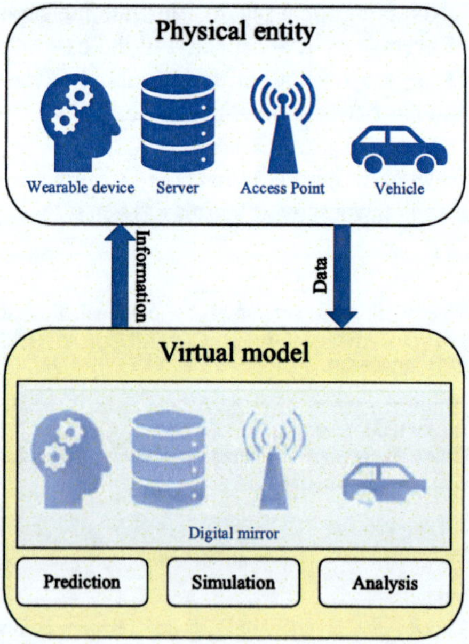

Fig. 1 Concept of DT

wireless channel may lead to unfavorable transmission links and high service delays, correspondingly.

The existing works have proposed a variety of model-driven algorithms for resource allocation in wireless communication networks, such as fractional programming (FP) [4], weighted minimum mean square error (WMMSE) [5], and so on. It is obvious that they perform excellent performance in terms of numerical simulation in theory, but are also faced with serious obstacles in practical deployment. First, these methods have high dependence on accurate models of mathematics that are easy to deal with. Second, these schemes are usually centralized and require instantaneous global channel state information (CSI). In addition, the solutions are hard to extend to a large number of cellular devices because of its computational complexity. Therefore, under the actual varying channel conditions, the implementation of these algorithms is quite challenging.

Deep reinforcement learning (DRL) is a new technology to tackle time-variant problems. State-of-the-art research has used DRL to optimize computing offload in wireless communications. In [6], a task offloading scheme based on a deep Q network (DQN) is proposed to choose the optimum edge server and the corresponding transmission mode, in order to optimize the effectiveness of offloading in vehicle networks. The authors in [7] propose a hybrid data offload scheme based on double DQN backscatter to decrease the consumption of information transmission. In [8], a computational offload and resource allocation scheme based on depth deterministic policy gradient is proposed to minimize average energy expenditure for wireless vehicle networks. However, these works mainly concentrate in static centralized optimization problems. None of them can be directly exploited to complex and dynamic wireless communication networks.

In this paper, we propose a fresh network paradigm which is employed with DT to establish an effective mapping between the physical communication network and the digital system and establish a virtual model of the network entity by monitoring the real-time status of devices. Then, an optimization problem is described for the constrained resource allocation. Based on the virtual model and monitoring data of the network, an algorithm based on DRL is designed to deal with the highly complex problem. For the sake of clarity, we conclude the main contributions in following three parts.

- An architecture of integrated DT and communication network is proposed to model the network topology and physical devices.
- The problem of resource allocation is discussed, and a distributed DRL method based on the online–offline learning framework is proposed. Firstly, the DRL algorithm is used to train DQN offline in the simulation environment. Then, with the help of transfer learning, the trained DQN can be further adjusted and updated in real scenarios.
- The total rate of the current system is designed as the reward function. An input super-parameter set which can effectively approach the optimal solution is designed. The average system capacity of the proposed algorithm is higher

than those in model-driven way and shows good generalization ability in a series of benchmark simulation tests.

2 System Model

In this paper, we suppose a system with M cells, which serves N users and shares K subchannels. $M = \{1, 2, \ldots, m\}$, $N = \{1, 2, \ldots, n\}$, $K = \{1, 2, \ldots, k\}$ represents the collection of their indices, respectively. The base station composed of transmitter and transmitter is settled in the center of the corresponding cell, and the fully synchronous time slot system with fixed slot duration is considered. Due to the relative scarcity of available spectrum, we assume that M is much greater than K. Each user is made to select a subchannel at the beginning of each time slot. Similar to [9], the channel model of the system consists of the large-scale fading and the small-scale one. For simplicity, it is assumed that large-scale fading is the same on all subchannels, while small-scale fading is frequency selective; that is, it is different on all subchannels [10]. In each subchannel, assuming that the small-scale fading is flat, the channel coefficient of the kth independent subchannel between the time slot t, the m base station and the nth user can be expressed as follows,

$$g_{m,n}^{k}(t) = \beta_{m,n}^{k}\left|h_{m,n}^{k}(t)\right|^{2}, t = 1, 2, \ldots, \tag{1}$$

where h is small-scale flatness fading, and β is large-scale fading considering path loss.

Let $A = \left\{a_n^k(t), n \in N, k \in K\right\}$ denote the indicator for the channel selection, and the binary variable $a_n^k(t)$ represents the subchannel selection of the time slot t. If user n selects subchannel k, there is $a_n^k(t) = 1$, and vice versa $a_n^k(t) = 0$. Presuming that the transmission power of the time slot t is $p_n(t)$, the signal to interference plus noise ratio (SINR) of the time slot t can be described by the following formula,

$$\gamma_n^k(t) = \frac{\alpha_n^k(t) g_n^k(t) p_n(t)}{\sum_{l \neq n} \alpha_l^k(t) g_l^k(t) p_l(t) + \sigma^2}, \tag{2}$$

where σ^2 represents the normalized noise at the receiver. Assuming that the bandwidth is normalized, the downlink spectral efficiency of the link in subchannel k in time slot t is calculated as,

$$C_n^k(t) = \log\left(1 + \gamma_n^k(t)\right). \tag{3}$$

Let the vector $\alpha(t)$ and the vector $p(t)$ represent the selection state and power of the subchannels in the time slot t, respectively, and the optimization problem of maximizing the system rate [11] can be formulated as follows.

$$
\begin{aligned}
&(P1)\ \max_{p(t),u(t)} \sum_{n=1}^{N} C_n(t) \\
s.t.\quad & C1: 0 \le p_n(t) \le p_{\max}, \forall n \in \mathcal{N} \\
& C2: \alpha_n^k(t) \in \{0,1\}, \forall n \in \mathcal{N}, \forall k \in \mathcal{K} \\
& C3: \sum_{k \in \mathcal{K}} \alpha_n^k(t) = 1, \forall n \in \mathcal{N}
\end{aligned} \tag{4}
$$

This problem is non-convex and NP-hard, so we propose a distributed deep reinforcement learning algorithm to tackle it in the next section.

3 Deep Reinforcement Learning for Resource Allocation

Q-learning is one of the most popular methods based on reinforcement learning to solve Markov decision process (MDP) problems. The agent examines the state $s(t) \in S$ and takes the action $a(t) \in A$ at time t and next communicates with the environment to get the reward and the next state $s(t+1)$. A and S are action sets and state sets, respectively. Since state S can be continued, a DQN which combines Q-learning with flexible DNN is proposed to solve the infinite state space. The cumulative discount reward function is shown below.

$$R^t = \sum_{\tau=0}^{\infty} \gamma^{\tau} r^{t+\tau+1}, \tag{5}$$

where r is the reward and $\gamma \in [0, 1]$ is the discount factor to compromise the influence of current and approaching rewards. Under a certain strategy π, the Q function of the agent with action a and state s is expressed by

$$Q_{\pi}(s, a; \theta) = \mathbb{E}_{\pi}\left[\mathrm{R^t | s^t = s, a^t = a}\right], \tag{6}$$

where θ denotes the DQN parameter and $\mathbb{E}[\cdot]$ represents the expectation operator. The method focuses on the communication between agents and unknown environments in order to maximize Q functions. The maximization of the above formula is equivalent to the Bellman optimality equation [12], which is described as follows

$$y^t = r^t + \gamma \max_{a'} Q\left(s^{t+1}, a'; \theta^t\right), \tag{7}$$

where y^t is the best Q value. DQN is trained as an approximate Q function, and the standard Q-learning update of parameter θ is calculated according to the following formula

$$\theta^{t+1} = \theta^t + \eta\left(y^t - Q\left(s^t, a^t; \theta^t\right)\right) \nabla Q\left(s^t, a^t; \theta^t\right) \tag{8}$$

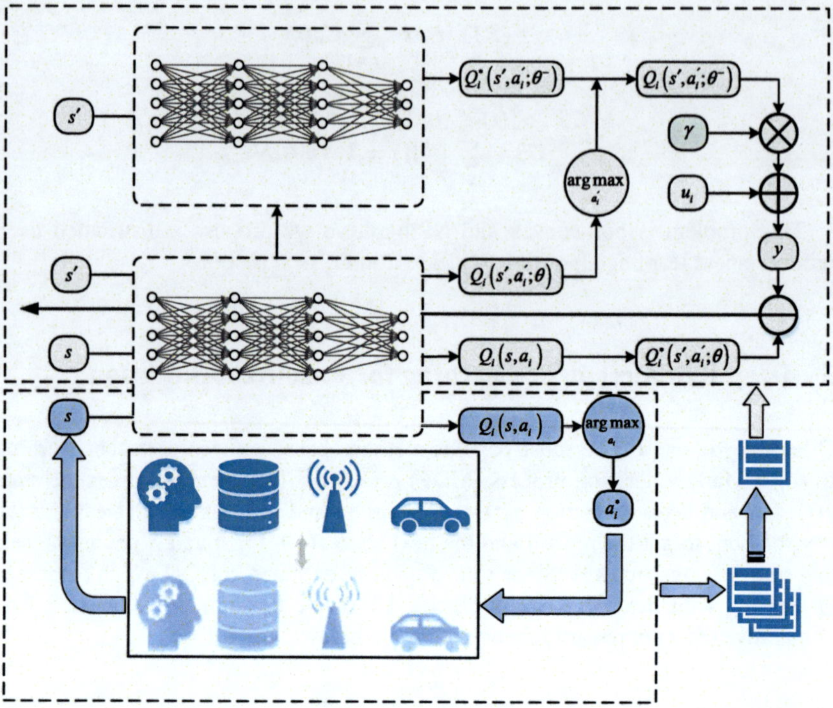

Fig. 2 Proposed distributed DRL algorithm

where η is the learning rate. The calculation is similar to a random gradient descent, gradually updating the current value $Q\left(s^t, a^t; \theta^t\right)$ toward y^t. The experienced data of the agent is formalized as a quaternion of s^t, a^t, r^t, s^{t+1}. The DQN is trained periodically with record batch data which is sampled from an experience replay buffer randomly.

In order to deal with the NP-hard problem in Section II, this paper proposes a distributed DRL algorithm. In traditional DQN, maximizing the inaccurate Q value will lead to a large approximation error, which will lead to overestimation. The use of a deep double Q network (DDQN) can decouple selection and evaluation to prevent this from happening [13]. In addition, the competitive DQN is used to speed up the training process [14], and the Q value is divided into a state value flow and an action dominance flow. Therefore, the algorithm proposed in this paper combines the above two network architectures and introduces a distributed mechanism, which is distributed double dueling DQN (D4QN).

Compared with the centralized learning algorithm, this distributed learning algorithm is easy to extend and has outstanding adaptive ability. Firstly, an agent is trained offline, which is able to be develop to larger scales. Secondly, the proposed method can well apply to the dynamic wireless communication channel scenarios.

Algorithm 1 and Fig. 2 describe the flow of the proposed distributed DRL algorithm.

Algorithm 1: D4QN-based DRL algorithm

Initialization:
Initialize experience replay buffer with fixed memory
Initialize online network DQN and target network DQN
Initialize environment according to the initial state
Training:
for each episode and each step **do**
  Examine the environment and get the state
  Select an action:
  Randomly choose an action from the set with probability ε
  Otherwise, choose the action that maximizes the Q function
  Drop the probability ε
  Save the transition tuple:
  Take actions and get the reward as well as the next state
  Save experience tuples in experience replay memory
  Update the online network and the target network:
  Sample the batch in memory, approximate the objective function
  Calculate loss function and use gradient descent training, minimize the loss function
  Update the target network DQN
  Update the learning rate and drop the probability ε
end for

4 Performance Evaluation

4.1 Simulation Settings

In this paper, we make a hypothesis that a base station is deployed in the center of each cell in a large-scale communication network, which serves the users in the cell at the same time. The small-scale fading is simulated as a Rayleigh distribution. According to the LTE standard, the model of large-scale fading is $\beta = -120.9 - 37.6\log_{10}(d) + 10\log_{10}(z)$, where z is a lognormal random variable with a standard deviation of 8 dB and d is the transceiver distance. The rest of the communication simulation parameters are summarized in Table 1.

Table 1 Simulation parameters

Parameter	Value
The number of cells	25
The number of users in each cell	4
Channels	20
Transmission power threshold	40 dBm
Cell radius	{0.01 km, 1 km}
Noise power	−114 dBm/Hz

Table 2 Training hyperparameters

Parameter	Value	Parameter	Value
Observation episode	100	Initial learning rate	0.001
Exploration episode	9900	Final learning rate	0.0001
Steps	50	maximum ε	0.2
Training interval	10	ε greedy increment	0.0001
Replay memory	50,000	hide layer 1	128
Mini-batch	256	hide layer 2	64

In the offline training phase, the D4QN is initialized randomly, and then the D4QN is trained set by set. In the first 100 episodes, the agent only acts randomly and then enters the next exploration period according to the adaptive greedy learning strategy [12]. In each episode, the number of training sets must be set higher to surmount the generalization problem due to the constantness of large-scale fading. Each set has 50 time slots, and D4QN randomly selects 256 samples from the experience playback memory for training every 10 time slots. In this paper, the ADAM algorithm [15] is utilized, and the learning rate attenuates from 0.001 to 0.0001. A four-layer neural network is selected to apply to D4QN, where the number of the hide layers is 128 and 64, respectively. The activation function is set to be linear in the output layer, and the activation function of the two hidden layers is ReLU.

For better illustration, all training hyperparameters are listed in Table 2.

4.2 Simulation Settings

In this section, we first investigate the performance of different discount factors γ. $\gamma \in \{0.0, 0.1, 0.3, 0.7, 0.9\}$ is set, and the average rate during training is shown in Fig. 3a. In the same time period, rates with the higher γ (0.7 and 0.9) were significantly lower than those with lower γ values. Then, the trained D4QN is tested in several cellular networks with different cell numbers. As shown in Fig. 3b, the D4QN with $\gamma = 0$ gets the highest rate value, while the D4QN with the highest γ value gets the lowest rate value. The simulation results show that nonzero γ has an

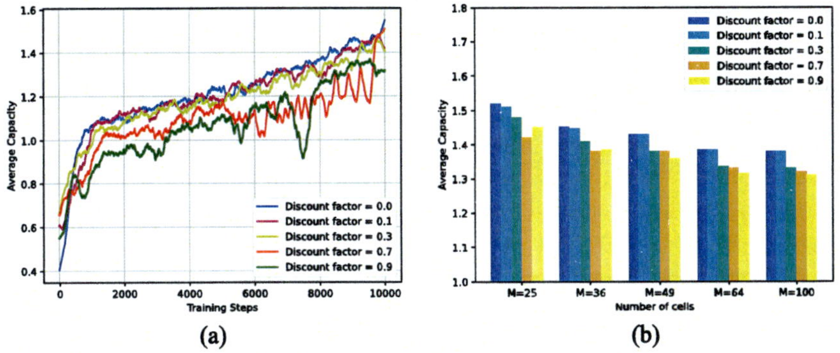

Fig. 3 **a** Training curve of the proposed algorithm with various discount factors. **b** the average system capacity and network scalability of the proposed algorithm

adverse effect on the property of D4QN. Therefore, for the proposed algorithm, we use the zero discount factor.

In Fig. 4a, D4QN trained with $\gamma = 0$ is tested with several benchmark algorithms. In the actual network, the user density changes with time. In order to solve this problem, D4QN must have good generalization ability. Suppose the number of users in each network varies in a triple set $\{2, 4, 6\}$. The average simulation results are obtained after 500 repeats. D4QN reached the highest rate in all test scenarios. With the number of users grows up, the gap between random scheme, maximum scheme, and other optimization algorithms increases. This is principally due to the fact that the interference in the cell increases with the increase of user density, indicating that in the cellular network with higher user density, the optimization of resource allocation is more significant.

Figure 4b shows an example of a test set (with 4 users). It can be seen that the performance of D4QN, FP, WMMSE, and water filling (WF) algorithms is unstable,

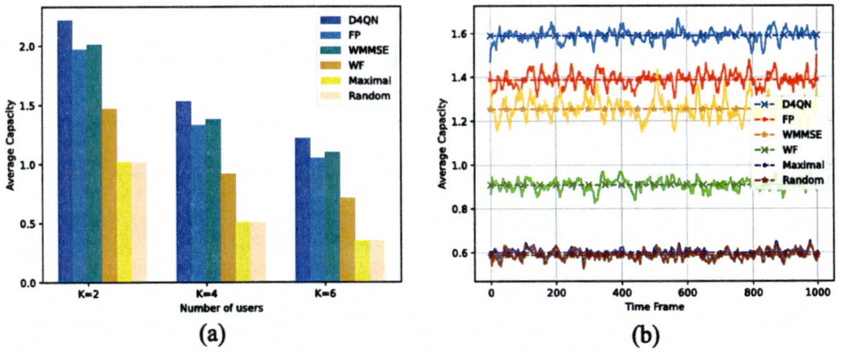

Fig. 4 Performance comparison between D4QN and five benchmark algorithms. **a** user density versus average system capacity. **b** average rate curve in 1000 time frames

especially dependent on specific large-scale fading effects. In addition, in some scenarios, D4QN does not outperform other algorithms over time, which means that the performance of D4QN still has the potential to improve. In terms of computational complexity, FP, WF, and WMMSE are all iterative algorithms, so the time cost is not constant and depends upon initialization as well as CSI. The time cost of D4QN is linearly related to the number of neurons in the DNN output layer, so the proposed algorithm has a good advantage in terms of computational complexity.

5 Conclusion and Future Work

Under the background of the digital transformation of wireless communication and the rapid expansion of network scale, this work studies the resource allocation problem in DTN. Firstly, a new DTN paradigm is proposed to establish the network topology and model of the communication system. Then, a distributed DRL algorithm with online–offline separation is proposed to cope with the resource allocation problem. DRL algorithm is used to train DQN offline in the simulation scene. Then, the trained DQN can be further updated in the real communication scenario dynamically, and the DNN can be fine-tuned using real data. Secondly, this paper designs a reasonable reward function for the proposed algorithm and designs an input super-parameter set which can effectively approach the optimal solution. Simulation results show that, in comparison with classic optimization algorithms and existing standard DRL training, the proposed algorithm framework achieves the highest average system speed. In the case of different user densities, the proposed algorithm shows better performance than that of the benchmark algorithm and has better generalization ability. In our future research, we will further testify the online learning phase of the algorithm to adapt to the specific communication environment and the real situation of different model distributions.

References

1. Shirer M, MacGillivray C (2019) The growth in connected IoT devices is expected to generate 79.4 zb of data in 2025, according to a new idc forecast. IDC
2. Glaessgen E, Stargel D (2012) The digital twin paradigm for future NASA and US Air Force vehicles. In Proc. 53rd AIAA/ASME/ASCE/AHS/ASC Structures, 1818
3. Lu Y, Huang X, Zhang K, Maharjan S, Zhang Y (2020) Communication-efficient federated learning for digital twin edge networks in industrial iot. IEEE Trans Ind Informat 17(8):5709–5718
4. Shen K, Yu W (2018) Fractional programming for communication systems—Part I: Power control and beamforming. IEEE Trans on Signal Process 66(10):2616–2630
5. Shi Q, Razaviyayn M, Luo ZQ, He C (2011) An iteratively weighted MMSE approach to distributed sum-utility maximization for a MIMO interfering broadcast channel. IEEE Trans Signal Process 59(9):4331–4340

6. Zhang K, Zhu Y, Leng S, He Y, Maharjan S, Zhang Y (2019) Deep learning empowered task offloading for mobile edge computing in urban informatics. IEEE Internet Things J 6(5):7635–7647
7. Xie Y, Xu Z, Xu J, Gong S, Wang Y (2019) Backscatter-aided hybrid data offloading for mobile edge computing via deep reinforcement learning. In: Proc. Int Conf Mach Learn Intell Commun, pp 525–537
8. Dai Y, Zhang K, Maharjan S, Zhang Y (2020) Edge intelligence for energy-efficient computation offloading and resource allocation in 5G beyond. IEEE Trans Veh Technol 69(10):12175–12186
9. Liang L, Kim J, Jha SC, Sivanesan K, Li GY (2017) Spectrum and power allocation for vehicular communications with delayed CSI feedback. IEEE Wireless Commun. Lett. 6(4):458–461
10. Tan J, Liang YC, Zhang L, Feng G (2020) Deep reinforcement learning for joint channel selection and power control in D2D networks. IEEE Trans Wireless Commun 20(2):1363–1378
11. Luo ZQ, Zhang S (2008) Dynamic spectrum management: Complexity and duality. IEEE J. Sel. Top. Signal Process. 2(1):57–73
12. Sutton RS, Barto AG (2018) Reinforcement learning: An introduction. MIT press
13. Hasselt H, Guez A, Silver D (2016) Deep reinforcement learning with double Q-learning. In: Proc. 30th AAAI Conf. Arti. Inte
14. Wang Z, Freitas N, Lanctot M (2015) Dueling network architectures for deep reinforcement learning. arXiv:1511.06581
15. Kingma DP, Ba J (2014) Adam: A method for stochastic optimization. arXiv:1412.6980

The COVID-Enforced Adoption of Technology for Reluctant Entrepreneurial Businesses: A Systematic Literature Review

Gareth Mclean and Adriana A. Steyn

Abstract This article presents a systematic literature review aiming to understand how the COVID-19 pandemic enforced reluctant entrepreneurial businesses to adopt technology into their business. A total of 32 academic literature articles published after 2004 in English were identified and analysed using the Preferred Reporting Items for Systematic Reviews and Meta-Analyses (PRISMA) principles. This article focuses on five main discussion points, namely COVID-19 effect on business, cause for technology adoption, hesitancy towards technology adoption, reasons for technology adoption, the government's role in technology adoption, and the outcome of forced technology adoption. COVID-19 is a relatively recent and developing topic; however, based on the outcome of the discussion, it was found that from a business continuation and survival perspective for entrepreneurial businesses, the COVID-19 pandemic has in many ways enforced the adoption of technology for reluctant entrepreneurial businesses. There is a need for further studies at a later stage to understand the effect the COVID-19 pandemic had on entrepreneurial businesses in conjunction with the adoption of technology. These studies should aid in understanding the effectiveness of technology adoption in response to business disruption.

Keywords COVID-19 · Entrepreneurs · Small- and medium-sized enterprises (SME) · Technology adoption · Systematic review

1 Introduction

On 11 March 2020, the World Health Organization declared the outbreak of the coronavirus (COVID-19) a pandemic [1]. This was just the beginning of many challenges and changes ahead for people all around the world, including entrepreneurs.

G. Mclean (✉) · A. A. Steyn
University of Pretoria, Pretoria, South Africa
e-mail: garethcmclean@gmail.com

A. A. Steyn
e-mail: riana.steyn@up.ac.za

X.-S. Yang et al. (eds.), *Proceedings of Seventh International Congress on Information and Communication Technology*, Lecture Notes in Networks and Systems 447,
https://doi.org/10.1007/978-981-19-1607-6_70

The COVID-19 pandemic has caused extensive business disruption, with start-ups, small businesses, and entrepreneurs being the most vulnerable and greatly impacted [2]. Considering this disruption, many of these businesses were faced with several unprecedented challenges, which presented the need for crucial decision-making on how to maintain and progress their businesses under the challenging new conditions brought on by COVID-19. Some solutions to these challenges include the adoption of technology into the business; however, there are entrepreneurs who are reluctant or otherwise hesitant towards the adoption of technology in their business [3].

COVID-19 has halted many of the daily operations for numerous businesses, limiting them to operate under strict protocol, and in some instances allowing no operations at all [2, 4]. In such unprecedented times like the COVID-19 pandemic, businesses may try find comfort in defaulting to their normal processes, if permitted. However, it is times like these where new approaches such as the adoption of technology into the business may prove to be most valuable and, in some cases, the only solution for the business to continue operating [3].

Despite the potential COVID-19-enforced adoption of technology into the business, if an entrepreneur or entrepreneurial business has found their business to have been affected either directly or indirectly by COVID-19, the question arises for the business whether they should capitalise, or at least consider capitalising on the opportunity to adopt technology into their business if the technology may provide potential benefit, not only for the present time but also the foreseeable future of the business [4].

In order to gain knowledge on how COVID-19 forced reluctant entrepreneurs and entrepreneurial businesses to adopt technology into the businesses, we need to understand the factors causing them to be reluctant, and how they overcame these reluctant factors resulting in the forced adoption of technology into their businesses.

For the purposes of this article, "entrepreneur" and "entrepreneurial business" refers to and includes entrepreneurs, small businesses, and SMEs. There are various definitions of a small- and medium-sized enterprise (SME) depending on the country or organisation, which generally consider various quantitative measures such as the number of employees, annual revenue, and gross assets. There is a vast outcome in the definition requirements of an SME. However, for the purposes of this article, SMEs are enterprises or businesses with up to 250 employees and a total annual revenue of up to US$15 million. This is similar to the World Bank Group guideline of less than 300 employees and less than US$15 million in annual revenue, which is used when no local definition of an SME is available [5].

This systematic literature review proceeds as follows: the next section, Section 2 explains the research method. Section 3 contains the results and quality evaluation of findings. Section 4 presents the discussion, data, and findings. Following on, Sect. 5 concludes the results and discussion.

2 Research Method

This systematic literature review is conducted using the Preferred Reporting Items for Systematic Reviews and Meta-Analyses (PRISMA) principles which were used as a bias for conducting systematic reviews and contains guidelines to ensure transparency and clarity are adhered to.

To construct this systematic literature review, the following high-level question was formulated:

How did COVID force reluctant entrepreneurs to adopt technology in their business?

To gather research material focussed on, and related to the research question at hand, various formulated search terms were required. Various formulated search terms were constructed using words or phrases that relate to entrepreneurial businesses and how COVID-19 forced the adoption of technology into these businesses.

To better refine the research material, there was a need to further adjust the academic literature findings with selection criteria including inclusion criteria and exclusion criteria, to filter through the academic literature findings. English academic literature such as journal articles, research articles, and conference proceedings focussing on the search terms were included as well as literature that directly aided in answering the research question. Academic literature published prior to 2005, where the full text was not available, and duplicate and unclear academic literature were excluded.

The following includes the source types that were used to obtain relevant academic literature used in this literature review:

(i) IEEE Xplore, (ii) ScienceDirect, (iii) Google Scholar, (iv) SSRN, and (v) EBSCO.

An initial number of 1290 articles were identified by searching the relevant databases using the search terms. After the execution of the PRISMA process, a final of 32 studies were included.

All these papers were further assessed by the answering the quality assessment questions:

(i) Does the academic literature provide evident implications of technology adoption by entrepreneurial businesses or SMEs? (ii) Does the academic literature provide reasoning regarding the reluctance of technology adoption by entrepreneurial businesses or SMEs? (iii) Is there research regarding technology adoption by entrepreneurial businesses or SMEs during unexpected events (such as COVID-19)? (iv) Is the research of an empirical nature or make use of empirical studies? (v) Does the academic literature refer to similar studies?

The possible answers to the above-mentioned questions consist of the following:

- Yes (Y) = 1
- Partial (P) = 0.5
- No (N) = 0

The total assessment score for each included academic literature was calculated based on the combined total scores of the above-mentioned assessment questions and the associated answers per academic literature assessed. The closer the total assessment score of the assessed academic literature is to 5, the higher the quality and relevance of the assessed academic literature is to answering the research question.

3 Results

The academic literature referenced in this systematic literature review was sourced from five electronic databases, namely Google Scholar (11), ScienceDirect (10), SSRN (5), EBSCO (4), and IEEE Xplore (2). The literature is made up of 28 (88%) journal articles, 3 (9%) conference proceedings, and 1 (3%) book.

Most of the academic literature included in this systematic literature review was published in the year 2020 given that the coronavirus (COVID-19) outbreak was declared a pandemic by the World Health Organization in early 2020.

4 Discussion

The discussion contains five main discussion points identified through the academic literature screening process.

4.1 COVID-19 Effect on Business

Entrepreneurs are crucial drivers of economic growth and development as well as a source of employment opportunities in both developing and developed countries, with entrepreneurial businesses constituting more than 95% of all businesses globally and more than 50% employment around the world [6].

The International Labour Organization has estimated that around 300 million full-time employees have either lost their job or have had a reduction in pay or work hours in 2020 due to COVID-19 [6]. The implementation of social distancing, lockdowns and various other COVID-19 protocols has in most cases caused many businesses to restrict interactions and many business processes, or even completely shut down [7]. This lead to significant economic downturn, even in countries with sound economies [2].

Bartik et al. [8] presented evidence from a survey conducted between 28th March and 4th April 2020 by Alignable, a network of 4.6 million entrepreneurial businesses, revealing that the pandemic had already caused major disruption with 43% of businesses temporarily closed, of which nearly all the closures due to COVID-19. Further investigation showed that the COVID-19 closures were caused mostly due to businesses experiencing reductions in business demand from customers, and employee health concerns [8]. Additional reasoning for the closures included supply chain disruptions and delays in processing business [2].

According to Bartik et al. [8], the impact of COVID-19 varies across industries with a 50% employment decline in the retail, personal services, arts and entertainment, hospitality, and food services. However, the professional services, finance, and real estate-related businesses were able to transition to a remote working style more efficiently, experiencing less major disruption [8].

Despite the major impact the COVID-19 pandemic brought to business, it has already greatly altered the business environment on a global scale. The post-COVID-19 pandemic world is likely to provide important opportunities for entrepreneurial businesses, which may lead to a great transformation of the global business environment [9].

Entrepreneurs are therefore faced with questions regarding many uncertainties such as how long the COVID-19 pandemic will last, the extent of business disruption it will cause, and how to prevent or otherwise accommodate for the disruptions caused thereof. This calls for the need for entrepreneurial businesses to adapt to the changing environment, looking for solutions aiding in business continuations with minimal disruption, not only in present times but also for potential future disruptions. According to many articles, it was found that one of the most common effective solutions to the COVID-19 pandemic uncertainties is the adoption of technology in various approaches [1, 3, 10, 11].

4.2 *Cause for Technology Adoption*

Despite the COVID-19 pandemic, it was evident, according to Cant & Wild [12] that entrepreneurial businesses were already facing increased pressure due to increasing market expansion and globalisation due to the adoption of new technologies and innovation [12]. The opportunity for entrepreneurs and entrepreneurial businesses to adopt technology provides them with the possibility to broaden their marketplace and furthermore allows them to communicate and provide their products and services globally [13].

According to Akpan et al. [11], technologies that enable new communication channels, social business channels, customer relation management systems, Internet of Things integration, and virtual reality technologies allowing for remote working are all crucial in lowering business costs and allowing business continuation. Furthermore, critical enablers such as visual, predictive, and big data analytics are crucial in aiding these technologies when making complex business decisions during current and potential future challenging business conditions [11].

It was discovered by Bruce & Moyano [14] in as early as 2007 that business growth requires businesses to adopt new and more powerful technological solutions [14]. However, according to Dubihlela & Kupangwa [15], various studies have confirmed that entrepreneurs and entrepreneurial businesses are slow with the adoption of technology into the business [15].

With the arrival of the COVID-19 pandemic and all its major disruptions, technologies that were once viewed as "nice to have" technologies immediately became "critical to have" technologies in order to aid in business continuation and survival [11]. This created a sense of urgency for entrepreneurial businesses to readdress their view on what technologies are no longer just "nice to have" but rather crucial at the very least for business continuation. However, Lorente-Martínez et al. [16] says it is crucial that entrepreneurial businesses owners and top level management have the knowledge and skills to correctly assess what technology adoption is needed as well as the potential of the technology for the business [16].

4.3 *Hesitancy of Technology Adoption*

Entrepreneurs and entrepreneurial businesses are confronted with different challenges when looking at the adoption of technology. The most common constraint, according to many articles typically being limited resources, which include budget, time, information and knowledge constraints, and potential legal constraints [17–20]. Additional barriers on the technical side majorly are related to technical issues, security risks, and inadequate facilities with little technical infrastructure [17, 21]. Various constraints make it difficult for entrepreneurial businesses to capitalise on the adoption of latest technologies, or technology in general, as well as restrict potential first-mover advantages [11]. Further concerns include drastic changes to business models, and organisational structure and culture [10].

Dubihlela & Kupangwa [15] found that entrepreneurial business leaders and top level management who do not possess traits such as flexibility, proactivity, creative orientation, openness to risk, and are enthusiastic in nature are more hesitant towards the adopting technology into the business [15].

Chen [18] discovered that budget constraints are generally the root cause of many of the barriers of technology adoption by entrepreneurs and entrepreneurial businesses. The lack of funding limits businesses from adopting technology, which in turn prevents employees from exposure to, and education on said technologies, bringing on greater information and knowledge constraints [18]. In addition, Dubihlela & Kupangwa [15] found that employee technological knowledge and skills are crucial for employees to be committed to using technology and supporting top level management in achieving the goals of the business. Employees that possess technology skills and knowledge have often been found to demonstrate a desire for business growth and innovation [15].

Given various barriers and issues that may arise leading to the hesitancy of technology adoption as given in Table 1 above, entrepreneurs and entrepreneurial businesses should take into consideration and compare all potential benefits against all potential downside that may arise from technology adoption to understand whether

Table 1 Reasons for hesitancy of technology adoption

Hesitancy reasons	Source
Limited resources: • Financials • Time • Knowledge and information	[3, 17–20, 22]
Technical issues: • Security risks • Inadequate facilities • Little technical infrastructure	[17, 21]
Business issues: • Business model • Organisational structure • Culture	[10, 15]

there is a need for technology adoption, especially during a disruptive period such as the COVID-19 pandemic.

4.4 Reasons to Adopt Technology

The adoption of certain technologies is regarded as critical enablers of new or improved business models with the potential to disrupt current processes, operations, and strategic techniques. Akpan [23] further discovered that rapid advancements in technology can lead to sudden changes in business models, which often re-engineer or improve the business process, potentially paving the way for market leadership and long-term growth [23]. According to Matarazzo et al. [24], the adoption of technology can assist in creating a new or improved digitally integrated business model which has shown to create and appropriate additional value for the business [24].

Kergroach [25] found that the adoption of technology by entrepreneurial businesses can aid in the reduction of costs and becoming more cost effective, as well as save time and resources. This was especially true for entrepreneurial businesses that handle lower volumes of production and have less internal capacity to handle complex business environments [25]. Kurnia et al. [26] further discovered that the adoption of certain technologies by entrepreneurial businesses enables them to have access to larger markets without the need to expand their physical presence [26].

According to Chen et al. [18], by adopting technology into the business, entrepreneurial businesses can potentially experience an increase in competitive advantage, achieve business growth, and improve business performance [18]. The adoption of technology is reshaping the traditional interaction between entrepreneurial businesses and their clients, with rapid increases in the number of touchpoints realised on their customer journey [24]. In addition, Nugroho et al. [27] considers technology to be a great enhancer of the relationship between the seller and prospective buyer as it creates potential for better communication and interaction [27].

The adoption of Internet technologies allows entrepreneurs and entrepreneurial businesses to contend on a more equal playing field, not only in local markets but globally too [12]. Konstantinou [28] further found that with the right technology adoption by entrepreneurial businesses, they were able to assess the environment and identify potential opportunities to access new markets [28]. Moghavvemi et al. [29] considers technology adoption a crucial competitive weapon, allowing for the development of a sustainable competitive advantage, enhancing not only the competitiveness of the business but also flexibility and productivity [29].

Matarazzo et al. [24] discovered that the adoption of certain technologies also allow for entrepreneurial businesses operational tasks such as accounting, marketing, human resource management, and many other operational and repetitive tasks to be better controlled and managed with more ease [24]. In addition, Moghavvemi et al. [29] found that by enchaining operational efficiency and effectiveness with the use of technology, entrepreneurial businesses can reimagine and improve the way the business competes to create better strategic opportunity and reassess competitive boundaries [29].

Much of the technology used in business during the current era is cloud based, and according to Kergroach [25], in 2018 large businesses were twice as likely to implement cloud computing services as small businesses [25]. However, smaller businesses should take the opportunity to review the adoption of technology via cloud computing given that it is now readily available, easier to use, and cheaper than ever before [23]. In addition, with the exponential growth and advancements of technology, Bai et al. [10] found that in many cases, entrepreneurial businesses that have integrated technology into their business models and processes have been found to have a more positive environmental impact than businesses that have not adopted technology [10].

Riom & Valero [1] discovered that businesses which had adopted technology or technological capabilities 3 years or more prior to the COVID-19 pandemic began were significantly more likely to continue business during the COVID-19 pandemic with little to no disruption [1]. Steyn [19] found that technology adoption capabilities can enable quick responses to opportunities and changing environments such like the COVID-19 pandemic and further observed that entrepreneurial businesses have the ability to adapt to change at a more rapid rate than larger businesses, if conditions allow for it [19].

The adoption of technology into the business may play a crucial role in business continuation and survival, not just during business disruptions but even in the general business environment. Thus, leaving entrepreneurs and entrepreneurial businesses with an important decision in preparation for any potential business disruption, with the COVID-19 pandemic disruption presenting as a potential opportunity for technology adoption.

Given various reason to adopt technology as given in Table 2 above, entrepreneurs and entrepreneurial businesses must consider the need for technology adoption in

Table 2 Reasons for technology adoption

Adoption reasons	Source
Business improvements of: • Business model • Business processes • Business performance • Operational tasks	[18, 23, 24]
Improved utilisation of: • Budget • Time • Resources	[25]
Improved business environments: • Competitive advantage • New markets • Virtual expansion	[18, 26, 28, 29]
Cheap and easy-to-use technology options available	[23]
Improved environmental impact	[10]
Effective response to potential business disruptions	[1, 19]

their business and take into consideration their business orientation as well as business strategies before adopting the technology to ensure technology adoption readiness.

4.5 Government's Role in Technology Adoption

Given entrepreneurs and entrepreneurial businesses large contribution to the economy, governments are concerned with entrepreneurial business growth, competitiveness, and performance. Therefore, a push and positive acceptance for technology adoption by governments may be necessary [18]. According to Chen et al. [18], much research shows the importance of government's roles with aiding entrepreneurial businesses in the technology adoption process. The government should play its part by creating policies, rules, and processes that aid entrepreneurs and entrepreneurial businesses with their potentially inadequate capabilities and limited resources, as it is extremely challenging and in some cases not possible for entrepreneurial businesses to overcome technology adoption barriers using only their own capabilities and resources [18].

In addition to the need for government intervention with technology adoption during disruptive periods such like the COVID-19 pandemic, Chen et al. [18] provides four government roles which could support in the adoption of technology by entrepreneurs and entrepreneurial businesses. The four roles include the following: providing a digital platform for small businesses, providing digital training, promoting digital payments, and building a digital ecosystem for small business collaboration. In doing so, governments can then use findings from the above-mentioned roles to determine policies, roles, and programmes to support technology adoption for entrepreneurial businesses [18].

However, given the heterogeneity of entrepreneurial businesses and the diversity of their business models and environments, Kergroach [25] highlights the need for differentiated, multilevel government policy solutions in order for governments and other stakeholders to understand how they can support entrepreneurial businesses with the adoption of technology and accurately accommodate the challenges thereof [25]. Turkyilmaz et al. [21] found that the lack of clarity and implementation of technology adoption measures results in additional challenges, leaving no clear assessment of the technology adoption needs of entrepreneurial businesses, potentially hindering development and innovation [21].

Evidence from De Vera et al. [30] shows that government initiatives for entrepreneurial and technological capabilities are one of the most crucial influences for entrepreneurs and entrepreneurial businesses with regard to technology adoption and innovation [30]. A government's acceptance towards and aid in technology adoption for entrepreneurs and entrepreneurial businesses may in turn impact both the local and global economy by enhancing the business sustainability and advancements of entrepreneurial businesses.

Despite current efforts by governments to encourage and aid entrepreneurial businesses with technology adoption, many studies have shown that it is still the larger businesses that reap the benefits of technology adoption [18, 27, 31]. Therefore, the extent to which governments can help with the adoption of technology by entrepreneurial businesses, especially in a time of disruption or need, is still a work in progress and requires further development and understanding.

4.6 Outcome of Forced Technology Adoption

It is almost as if many, if not all businesses were effectively forced to innovate and adopt technology in some way, shape or form for business continuation. Even if involuntarily, evidence has shown that technology adoption may not only offer a competitive advantage, but more importantly provide a means for survival [11]. It was further found by Chen et al. [18] that technology adoption increases business survival in a time of crisis, such like the COVID-19 pandemic, with many businesses even experiencing a rapid increase in online business [18].

According to Bai et al. [10], the actual adoption of technology by entrepreneurial businesses is less of a technical issue and more a managerial one. In most cases, businesses experience issues such as but not limited to the redesign of business processes, investment in training and human resources, and organisational capabilities. However, businesses that do overcome the above-mentioned issues may experience improved efficiency, lowered costs, and better innovation [10]. It was further discovered by Konstantinou [28] that technology adoption was more of a matter of strategy and culture rather than funding and familiarity of the technology [28]. However, when it comes to the actual adoption of technology, Bai et al. [10] found that the choice of technology is generally related to the businesses' existing equipment and current basic digital competencies to make use of these technologies [10].

As much as technology adoption may aid entrepreneurial businesses, there are some circumstances where the adoption may hinder certain businesses. Businesses that are unable to adapt technology may see the adoption of technology caused by COVID-19 as a threat. This threat is caused due to businesses that do manage to adopt technology, as they may see a competitive advantage over businesses that do not adopt technology [10].

Despite the potential benefits entrepreneurs and entrepreneurial businesses may realise from the adoption of technology, Kurnia [26] recommends that if the business is not organisationally prepared for the adoption of technology, it may be better for the business to delay the adoption of technology until they have improved their organisational readiness or are under extreme pressure to adopt technology [26]. According to Naushad & Sulphey [32], technology self-efficiency, which is the belief to learn and apply technology in a given environment, is one of the most essential predictors of technology adoption readiness [32].

Although technology adoption may present itself as a challenging process in some cases, it has become a competitive necessity in various aspects for many entrepreneurial businesses to ensure business continuation through business disruptions such like the COVID-19 pandemic [18]. In the light of this, it was found by Riom & Valero [1] that over 90% of businesses that had adopted technology reported that they intend to continue with the use of technology and further innovate [1].

5 Conclusion

The COVID-19 pandemic has to some extent accelerated, projected, and magnified the impact of technology adoption not just for entrepreneurial businesses, but all businesses. It has created many opportunities for entrepreneurs and entrepreneurial businesses to adapt and improve on their business models by utilising new techniques and processes brought by the adoption of technology.

It is important to note that in almost all situations, only technology can assist in the continuation of business operations during the absence of physical interaction, and as though technology adoption by businesses is unavoidable. With the exponential growth rate of technology, it is essential for businesses to increase their awareness of the technological opportunities that are available to them. In doing so, businesses should take into consideration and compare all potential benefits against all potential downside that may arise from the adoption of technology into the business. If entrepreneurial businesses are to adopt technology, they must ensure technology adoption readiness.

Another important factor discussed is the extent to which governments can help with the adoption of technology by entrepreneurial businesses, especially in a time of disruption or need. Government initiatives for entrepreneurial and technological capabilities are a crucial influence for entrepreneurial businesses with regard to technology adoption and innovation and is therefore a critical matter which requires further development and understanding.

It is important to remember that the COVID-19 pandemic is still a relatively recent and developing topic at the time of writing this article. However, from the findings, it is evident that from a business continuation and survival perspective for entrepreneurial businesses, the COVID-19 pandemic has in many ways enforced the adoption of technology for reluctant entrepreneurial businesses.

References

1. Riom C, Valero A (2020) The business response to COVID-19: the CEP-CBI survey on technology adoption. Centre for Economic Performance, London School of Economics and Political Science
2. Meahjohn I, Persad P (2020) The impact of COVID-19 on entrepreneurship globally. J Econ Bus 3:1165–1173
3. Kumar A, Ayedee N (2020) Technology adoption: A solution for SMEs to overcome problems during COVID-19. Forthcoming, Academy of Marketing Studies J 25(1)
4. Masood T, Sonntag P (2020) Industry 4.0: Adoption challenges and benefits for SMEs. Computers in Industry 121:103261
5. Ardic OP, Mylenko N, Saltane V (2011) Small and medium enterprises: A cross-country analysis with a new data set. World Bank Policy Research Working Paper, 2011(5538)
6. Beglaryan M, Shakhmuradyan G (2020) The impact of COVID-19 on small and medium-sized enterprises in Armenia: Evidence from a labor force survey. Small Business Int Rev 4(2):e298
7. Papadopoulos T, Baltas KN, Balta ME (2020) The use of digital technologies by small and medium enterprises during COVID-19: Implications for theory and practice. Int J Inf Manage 55:102192
8. Bartik AW et al (2020) The impact of COVID-19 on small business outcomes and expectations. Proc Natl Acad Sci 117(30):17656–17666
9. Zahra SA (2021) International entrepreneurship in the post Covid world. J World Bus 56(1):101143
10. Bai C, Quayson M, Sarkis J (2021) COVID-19 pandemic digitization lessons for sustainable development of micro-and small-enterprises. Sustainable Production and Consumption
11. Akpan IJ, Soopramanien D, Kwak D-H (2020) Cutting-edge technologies for small business and innovation in the era of COVID-19 global health pandemic. J Small Business & Entrepreneurship, pp 1–11
12. Cant MC, Wiid JA (2016) Internet-based ICT usage by South African SMEs: The barriers faced by SMEs. Journal of Applied Business Research (JABR) 32(6):1877–1888
13. Jere JN, Ngidi N (2020) A technology, organisation and environment framework analysis of information and communication technology adoption by small and medium enterprises in Pietermaritzburg. South African J Information Management 22(1):1–9
14. Bruque S, Moyano J (2007) Organisational determinants of information technology adoption and implementation in SMEs: The case of family and cooperative firms. Technovation 27(5):241–253
15. Dubihlela J, Kupangwa W (2016) Employee perspectives of factors influencing e-business technology adoption and use by small and medium retail enterprises. Int J Business Manag Stu 8(1):1–19

16. Lorente-Martínez J, Navío J, Rodrigo-Moya B (2020) Analysis of the adoption of customer facing InStore technologies in retail SMEs. J Retail Consum Serv 57:102225
17. MacGregor R, Vrazalic L (2008) A profile of Australian regional SME non-adopters of e-commerce. Small Enterp Res 16(1):27–46
18. Chen C-L et al (2021) Role of Government to Enhance Digital Transformation in Small Service Business. Sustainability 13(3):1028
19. Steyn RA (2018) Changing thoughts towards digital literacy interventions for South African entrepreneurs. Reading & Writing-Journal of the Reading Association of South Africa 9(1):1–9
20. Kumar M, Syed AA, Pandey D (2020) Impact of online resources/technology adoption on SMEs performance. PIMT J Res
21. Turkyilmaz A et al (2021) Industry 4.0: Challenges and opportunities for Kazakhstan SMEs. Procedia CIRP 96:213–218
22. Giotopoulos I et al (2017) What drives ICT adoption by SMEs? Evidence from a large-scale survey in Greece. J Bus Res 81:60–69
23. Akpan IJ, Udoh EAP, Adebisi B (2020) Small business awareness and adoption of state-of-the-art technologies in emerging and developing markets, and lessons from the COVID-19 pandemic. J Small Business & Entrepreneurship, pp 1–18
24. Matarazzo M et al (2021) Digital transformation and customer value creation in Made in Italy SMEs: A dynamic capabilities perspective. J Bus Res 123:642–656
25. Kergroach S (2020) Giving momentum to SME digitalization. J Int Council Small Business 1(1):28–31
26. Kurnia S et al (2015) E-commerce technology adoption: A Malaysian grocery SME retail sector study. J Bus Res 68(9):1906–1918
27. Nugroho MA et al (2017) Exploratory study of SMEs technology adoption readiness factors. Procedia Computer Sci 124:329–336
28. Konstantinou JK (2016) Digitization of European SMEs in tourism and hospitality: The case of Greek hoteliers. Int J Social Business Sci 10(5):1558–1562
29. Moghavvemi S, Hakimian F, Tengk Feissal TMF (2012) Competitive advantages through IT innovation adoption by SMEs. Socialinės technologijos [Social Technologies] 2(1):24–39
30. de Vera IJM, et al (2018) Key drivers and critical success factors in the technology adoption and use by Asia-Pacific SMEs. In: 2018 5th Asia-Pacific World Congress on Computer Science and Engineering (APWC on CSE). IEEE
31. Susanty A, Sari DP, Anastasia D (2016) Critical success factors for the internet technology adoption by SMEs and its impact for the performance. In: 2016 2nd International Conference on Science in Information Technology (ICSITech). IEEE
32. Naushad M, Sulphey M (2020) Prioritizing technology adoption dynamics among SMEs. TEM J 9(3):983

Integrated Remote Primary Care Infrastructure: A Framework for Adoption and Scaling of Remote Patient Management Tools and Systems

Barimwotubiri Ruyobeza, Sara S. Grobbelaar, and Adele Botha

Abstract Digital health technologies have for number years now been expected to reduce the skyrocketing health related costs as well as the care burden on traditional healthcare systems. However, their adoption and scaling have consistently been unsatisfactory and sometimes, outright disappointing. Scholars have offered several valuable, insightful, and pertinent contributions to address the above challenge. However, these contributions are in most cases atomistic, transitory, non-spatial, and dispersed. The above state of affairs has left practitioners in limbo as to where and when to apply which insights to what type of digital health intervention and in which context. In this article, a new holistic and integrated theoretical framework, specifically focusing on remote patients' health management tools and systems (RPMTSs) used to engage patients and potential patients at distance or away from healthcare facilities is proposed and introduced to address the above existing fragmentation and gaps. The new framework demonstrates how a clear and holistic understanding of "adoption" and "scaling" processes in the context of a given type and nature of digital health intervention along with an adaptive complex, processual and systems thinking approach can help confront the complexity of the healthcare apparatus while at the same time responding to its constantly evolving, dynamic nature with "agents" who may sometimes act irrationally or behave in unpredictable ways. In the process, a new framework is added to the knowledge base to guide and support the adoption and scaling of RPMTSs.

Keywords RPMTS · Adoption · Scaling · Process · Theory · Preconception · Uptake · Adaptation

B. Ruyobeza (✉) · S. S. Grobbelaar
Stellenbosch University, Stellenbosch 7600, South Africa
e-mail: ruyobeza@sun.ac.za

S. S. Grobbelaar
e-mail: ssgrobbelaar@sun.ac.za

A. Botha
Council for Scientific and Industrial Research, Pretoria, South Africa
e-mail: abotha@csir.co.za

X.-S. Yang et al. (eds.), *Proceedings of Seventh International Congress on Information and Communication Technology*, Lecture Notes in Networks and Systems 447,
https://doi.org/10.1007/978-981-19-1607-6_71

1 Background

In confronting the problem of low adoption and limited scope for scaling of digital health interventions in general and remote patients' health management tools and systems (RPMTSs) in particular; scholars have made a number of excellent and helpful suggestions. However, these suggestions are scattered and not necessarily streamlined to focus on the adoption and scaling of given, specific types of digital health interventions such as RPMTSs. Furthermore, these proposals and theories are not necessarily positioned at specific lifecycle phases or stages of digital health interventions making it difficult for practitioners to establish when and where to apply them.

In the context of RPMTSs, we argue that "adoption" ought to be understood to mean not just the initial, once-off or occasional use of an RPMTS but a behavioural change process which induces target adopters to permanently change their previous habits (prior to the introduction of an RPMTS intervention) and consistently make use of these tools and systems whenever they need to access primary healthcare services [1]. The mere download or access to a mobile application as part of an RPMTS or the initial use of such an application perhaps followed by abandonment thereafter, should not qualify as "adoption" in this context. RPMTS's use has to be embedded into its adopter's routines [2]. Only under these circumstances should an RPMTS be understood to have been successfully adopted by its target users, at least in a given specific setting. Understood in this way, three key observations can be made regarding the "adoption" of RPMTS interventions:

Firstly, the adoption of an RPMTS intervention ought to be conceptualised as "a process" with distinct phases and related stages and activities rather than a once-off event. It is a series of stages in which designers ought to deliberate seek to influence the perceptions and attitudes of potential adopters, not only to start using their new RPMTS intervention when accessing primary healthcare services but to also embed the use of such tools and systems into their habits and routines [3–5].

Secondly, RPMTS's designers ought to conceive them as a consumer service package rather than a consumer technology product and intentionally or deliberately seek to design them for desirability, uptake, acceptance, and routinised use right from their inception [6–8]. If stakeholders such as developers or designers conceive the adoption of an RPMTS intervention to mean the mere acquisition of a related piece of equipment or software, then the focus of the design will solely be on factors which may trigger initial acquisition of the relevant equipment or application rather than on sustained, routinised use of such equipment or software [9].

Finally, given that "adoption" is here considered to be a change process, designers need to consider temporal and spatial dimensions of the theoretical tools [10], they operationalise to achieve specific goals across the different lifecycle phases of the adoption process [11]. For example, applying the famous technology acceptance model (TAM) or the unified theory of acceptance and use of technology (UTAUT) to predict or explain mechanisms of technology adoption, without due regard to the temporal dimensions of Roger's diffusion of innovation theory, might be tantamount to taking a once-off picture of a moving object in an attempt to understand its entire journey [12], and this might partially explain why researchers have often obtained conflicting results while testing the validity of one of these models or frameworks [13, 14].

The above short discussions demonstrates the need for stakeholders such as owners, designers, and developers to adopt a holistic view of the adoption and scaling processes by firstly understanding the specifics and particularities of a given type of digital health intervention such as RPMTSs, and then using a phased process framework to guide the adoption and scaling of that type of intervention in specific healthcare contexts with a sustained focus on user-centred experience across the entire intervention's lifecycle [15, 16]. This is what this article does for RPMTSs.

In the next section, a description of the methodology followed to develop the proposed holistic, integrated process framework to guide the adoption and scaling of RPMTSs is presented. The section thereafter discusses the nature of RPMTSs as a specific type of digital health intervention and the ideal adoption and scaling processes thereof. Next, the new framework is introduced and its phases, stages and stage-gates are discussed. Finally, the article concludes with the limitations, recommendations and future research directions relating to the proposed framework.

2 Framework Development Methodology

The development of the proposed framework is taking place within the broader context of a design science research (DSR) methodology which draws both from the task environment as well as from the existing knowledge base [17]. We are following the six steps proposed by Peffers et al. which include: (1) identifying the problem, (2) defining the solution's objectives, (3) designing and developing or creating the artefact, (4) demonstrating the use of the artefact, (5) evaluating it, and (6) communicating its development, utility, and novelty [18]. The present article is situated at step 3 of the above process and presents the initial development or creation of the proposed, process framework which is positioned at level 2 of Gregor and Hevner's "Design Science Research Contribution types" as a *nascent design theory* or *knowledge as operational principles/architecture* [19, p. 342].

The above type of theory that the researchers herein seek to contribute to existing knowledge is known as "theory for design and action" and generally refers to the theory or set of theories of the middle range that are positioned between working hypotheses and a unified theory or theories which may explain or predict social behaviour, organisation, or change [19]. As eloquently articulated by Gregor (2006), the most basic of all theories, the analytic theory is required for developing all of the other types of theory. Using Gregor's proposed "interrelationships among theory types" [20], Fig. 1 below demonstrates how the researchers moved from a general topology for analysing the adoption process to the present, initial design of the proposed framework. Green shaded theories represent potential contributions to the existing knowledge base while light-purple-shaded theories are types of existing

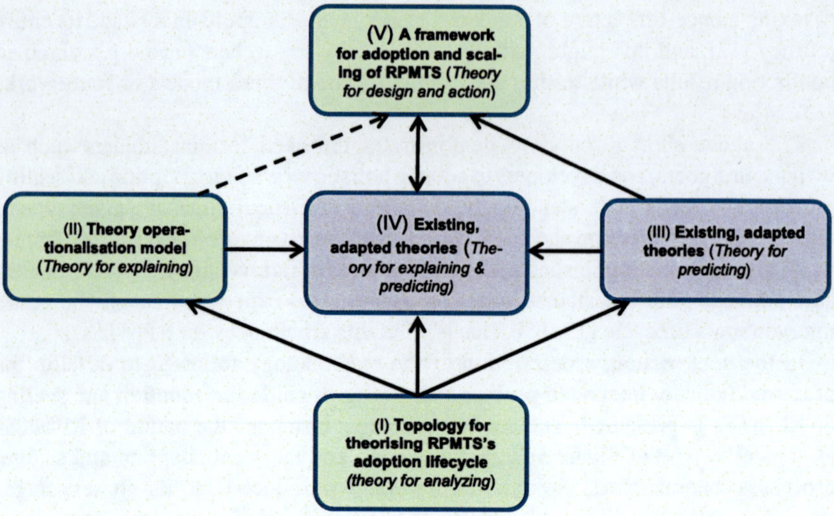

Fig. 1 Interrelationships among theory types for framework's building, adapted from [20]

theories and frameworks that we relied on. Given the call to increase reliance on theoretical tools across the design and implementation phases of digital health interventions in general [21, 22], the topology we introduce here can itself be viewed as another contribution of the analytic type, as a different group of researchers may populate it with different theories than we have done herein.

To come up with the above foundational topology for analysing RPMTS's adoption and scali

ng lifecycle as a starting point for the development of the proposed framework, a critical review was undertaken. We preferred a critical review because of the opportunity that it offers researchers, not only to analytically review and evaluate the existing bodies of research works such as existing technology adoption and scaling theories, but also a real chance for conceptual innovation and development based on the reviewed pieces of literature, which could potentially and did lead to the early, tentative design of the proposed framework [23, 24] presented herein. A critical review of this nature was especially favoured because it is consistent with the recommended approach for sociotechnical IS design science research in which a review of extant theories, knowledge and data allows researchers to propose and refine design theory and knowledge [25] as displayed in Fig. 2 below:

To confine the scope of the above critical review to relevant adoption and scaling theories, the adoption process ideally suitable for RPMTS interventions, in general, was firstly modelled (as will be discussed in the next section) and then used to propose a related topology for identifying, selecting and analysing relevant, existing constructs, frameworks, models, instruments, concepts and theories, thought to promote and enhance the adoption and scaling of novel technologies and then assess their suitability for use at various stages of RPMTS's adoption process [21].

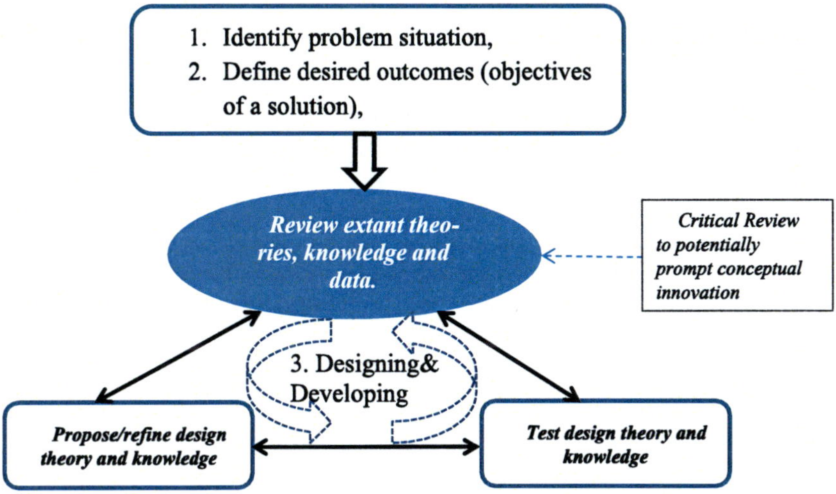

Fig. 2 Sociotechnical IS design theory development, adapted from [25]

The aim here was to critically analyse relevant frameworks and constructs' alignment with various tasks and activities involved in the conceptualisation, design, deployment, and sustainability of RPMTS interventions. Despite the fact that critical reviews are generally criticised for not demonstrating features such as replicability and transparency which are evident in other more structured, systematic approaches to literature reviews, the researchers followed the SALSA (Search, Appraisal, Synthesis and Analysis) analytical framework discussed below [24, 26].

In the search phase, the researchers started with Sovacool and Hess's list of 96 theories of adoption as provided in [27] and augmented it with additional relevant theories mainly from the health, behavioural economics, sociology, cognitive and ecological psychology, human–computer interaction (HCI), and science and technology studies domains or fields. We ended up with more than 150 theories. A two-step process was then undertaken in the appraisal phase; the first one of which initially considered four factors relating to the type and nature of RPMTS adoption to reduce the number of listed theories from 150 to potential 35 RPMTS' adoption related theories. Appendix 1 provides a list of 35 selected theories. In the second step, we summarised the selected 35 theories and then considered the purpose of each phases' stage and stage-gates, at which they had been proposed to be operationalised. The theory most aligned to the purpose was selected.

The analysis then involved a careful reading of the seminal works on each of the selected theories and then critically evaluating and analysing their assumptions, strengths, weaknesses, purpose, and responsiveness to the temporal and spatial requirements for adoption and scaling phenomena at that given phase' stage or stage-gate. Within each framework or theory, we sought to identify constructs and propositions to potentially drive the tasks or activities associated with the stage or stage-gates of RPMTS' adoption process at which these theories were proposed to be applied.

Appendix 2 explains why each theory or framework was selected at each phase's stage or stage-gate as well as gauged based on the credentials of its author or authors and the quality of major journals they published in. Finally, the synthesis took the form of establishing whether or not the purpose of each phase's stage and stage-gates could be fully supported by the single theory selected to support it, and where a clear deficit existed; identifying additional constructs to complement the one previously selected in order to reach the overall objective of that stage or stage-gate. In the next section, we reinforce the above methodology by explaining how the adoption and scaling processes were modelled as a starting point for the development of the proposed framework.

3 Nature and Ideal RPMTS Adoption Process

As has already been proposed, "adoption" in the context of RPMTS and indeed of many other types of digital health interventions ought to be viewed as being more than the mere decision or action of choosing to take up the use of a given RPMTS, to encompass embedding its consistent use into the adopters' care-seeking routines, habits, and behaviours. As such, it becomes a process which begins when there is recognition that a need exists for such interventions [28, 29]. In fact, Nechully, Pokhriyal and Thomas [30] who summarised many theories and models of adoption of innovations consistently demonstrated that public awareness of a particular innovation precedes and often determines subsequent adoption or non-adoption thereof. For example, they refer to Wilkening (1953) who posited that awareness is the first of the four stages of adoption [31, p. 17]. They also quote the famous Roger's diffusion of innovation theory (1961), wherein "awareness" is the first of the five-step process of diffusion in which customers should be given relevant information regarding the product [31, p. 18]. They further point to the "multistage innovation diffusion process model of Dodson and Muller (1978) who proposed that companies must convert an "ignorant" to a "prospect" and a "prospect" to a "customer" [31, p. 21]. Finally, they refer to Triandis' theory of interpersonal behaviour which suggests that companies with innovative products should "match" the expectations of their customers with their product's functionality and features "to create a favourable attitude towards the innovation" [31, p. 14]. Furthermore, Rice and Rogers agreed with the tool of Eveland et al. (1977) for the analysis of technology transfer decisions, for which they conceptualised "innovation adoption" as a sequence of sub-processes, the first one of which ought to be the agenda-setting stage during which problems are defined [31]. Therefore, RPMTS design companies ought to develop methods or ways of learning customers' expectations around their proposed, future innovations, ahead of the design process [32].

Relatively more recently, Lundvall patently demonstrated the need for information exchange between users and producers of innovations prior to the launch of new innovative products or services into the market place. He observed that in the current innovation dispensation, producers have little or no information about potential, future users of their innovations while users have no knowledge about the "use value" characteristics of new or future products to be anticipated. He argues that information needed by users should not only involve product' "awareness" but also "specific information about how new "use value" characteristics relate to the users' specific needs." According to him when new knowledge (not a new product or service) is produced as part of an innovation process, there is a need for feedback from users of the knowledge and resultant innovations to those producing it [33]. We have therefore conceptualised a "preconception" phase to accommodate this necessary exchange of information between future users and producers prior to the design and development of an RPMTS intervention [34].

In addition to the above "preconception" phase, users' decision to acquire and try out a new innovation such as a newly introduced RPMTS intervention, remains central to the adoption process and cannot be overemphasised [35]. However, it is now herein referred to as "acquisition or uptake" to reflect the fact that such initial decision to start using or try out a particular technology is not always permanent as users may subsequently abandon its use. Finally, in the context of RPMTSs, where successful diffusion and scalability of any new RPMTS intervention may significantly depend on potential users' willingness to continuously use their own devices for access to healthcare services; "adaptation" to customise online applications related to RPMTSs or to configure them in ways that allows their owners to continue to carry out their usual tasks on their own devices without hindrance, is a critical step in routinising and sustaining an RPMTS intervention's use and embedding it into the habits of its users [12, 36]. The process of RPMTS adoption was, therefore, conceptualised as depicted in Fig. 3 below:

We argue that viewing adoption and scaling from this perspective (as being a related social-behavioural process but parallel to formal, traditional design and implementation processes often associated with project or programme management) would help innovators, designers, developers, and other interested parties to begin with "the end in mind," when conceptualising new RPMTS interventions by clarifying what "successful adoption and scaling" of their contemplated interventions should ultimately mean or will mean to them and all of their stakeholders. It would help them to preview "the adoption and scaling road ahead" along with its contours and corners that they have to cross and turn in order to successfully arrive at their destination [9, 37]. While most current research efforts on RPMTS interventions are rightly focused on

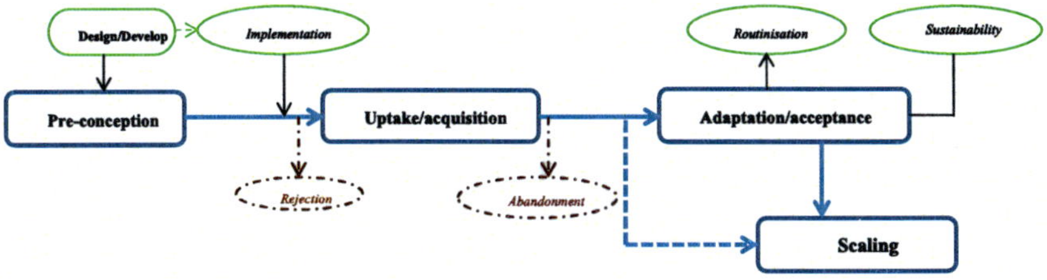

Fig. 3 Processes of adoption and scaling for RPMTS

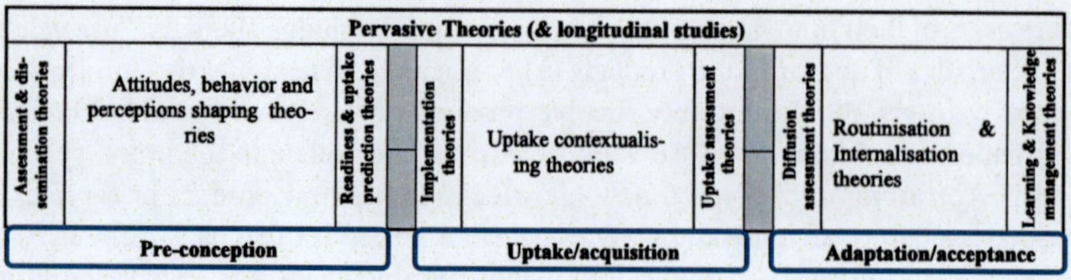

Fig. 4 Topology for theorising RPMTS's development

design and implementation sciences; pre-conceptualisation, routinisation, scalability, and sustainability are equally important aspects of successful adoption and scaling which ought to be considered right from the inception of new RPMTS interventions, especially in the healthcare context [38, 39].

With the above brief discussion, the researchers relied on the topology in Fig. 4 below which may be used to identify and select perhaps a different set of relevant, existing theories whenever a new RPMTS intervention is being contemplated or conceptualised:

Since the suggestion is to continuously theorise as the intervention is being designed, developed, implemented, scaled, and maintained [22, 40, 41]; apart from pervasive theories which ought to be identified prior to the conceptualisation of a new healthcare intervention, theories selected for the different stages and tasks ought to be justified based on prevailing contextual realities and the aims of those tasks and activities carried out at the relevant stage of the adoption process. We contend that theorising costs should always be part and parcel of all RPMTS interventions' development budgets.

We take the position that RPMTS designers and developers ought to focus on selecting and relying on the most appropriate framework(s) or theories for use at "appropriate moments" in the adoption process and for the appropriate purpose. Designers ought to also be capable of justifying and demonstrating the relevance of chosen theories at the intersection of digital health technologies and their context of use as well as showcasing how such chosen theories improve the adoption of the targeted digital health intervention such as an RPMTS, at salient points in time and space of the adoption process [42]. Figure 5 below depicts the updated populated topology with theories relied on for developing the proposed framework:

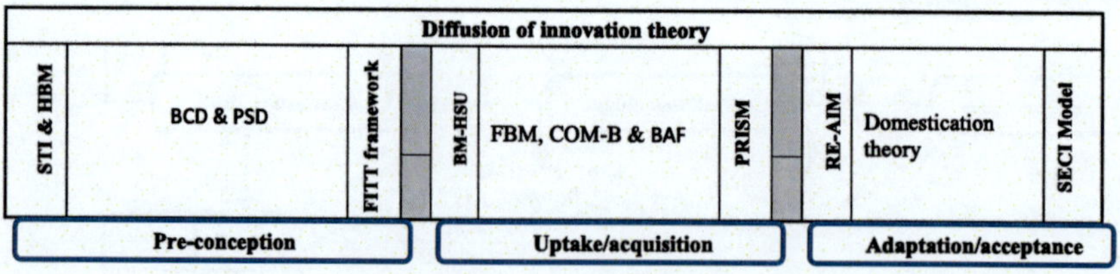

Fig. 5 Updated topology for theorising RPMTS's planning

4 A New Integrated Framework for the Adoption and Scaling of Remote Patient Management Tools and Systems (RPMTSs)

Following the selection of above theories, the researchers integrated emergent theoretical concepts and working mechanisms with existing and well established theoretical models, frameworks and constructs, drawn from previously selected theories with the aim of supporting each adoption phase's stage and stage-gates [43]. Existing theoretical frameworks are operationalised in new ways while emergent constructs such as "sociotechnical imaginaries" (STI) are assessed and positioned to meet the purpose of a given stage or stage-gate. For example, the researchers recognised the "temporal dimensions" of the FITT framework and therefore, within the proposed framework, recommended that the degree of "task-user fit" be assessed prior to the RPMTS design process. The researchers also recognised that the widely known concept of "perceived ease of use" is largely anchored in the users' current technology skills and past experience with similar or related technologies. Similarly, we proposed that the existing construct of "perceived usefulness" of a given technology be understood with the lenses of existing contextual constraints or the significance of problems or limitations that the technology is meant to address. Generally, the model is structured to allow practitioners, to use the "assess, plan, do, evaluate, and report (APDER) cycle" at each phase of the adoption process [44] with ample opportunities to reposition an RPMTS as and when needed, to improve its chances for adoption and scaling. As has already been indicated, to the extent possible, the integration adopted a systems thinking approach in an attempt to confront the complexity of the healthcare apparatus [45, 46], while at the same time recognising its constantly evolving, dynamic nature with "agents" who may sometimes act irrationally or behave in unpredictable ways [37]. The resultant, integrated, process framework is provided in Appendix 3.

5 Conclusions, Limitations Recommendations

We posited that one of the main reason contributing to low adoption and scaling is the lack of a holistic, integrated view of "adoption" as a process from the inception of new RPMTS interventions, which leads to some critical stages or stage-gates being skipped or neglected. Further, we demonstrated that if adoption and scaling are to succeed, focused attention across all RPMTS lifecycle phases and reliance on theoretical tools to guide adoption and scaling ought to be mandatory, especially in the healthcare context. The preliminary framework presented in Appendix 3 starts by acknowledging that RPMTS adoption is a behaviour change process which, consist of three distinct phases: preconception, uptake, and adaptation. The researchers here have, therefore, integrated multiple constructs and adapted a number of existing frameworks to conceptually achieve the aims of each stage and stage-gate.

Further, this discussion sought to clarify that each phase ought to begin with an assessment stage-gate to establish whether the timing is ripe for it to start or not, and be concluded by an evaluative stage-gate to establish the extent to which the objective of the phase was achieved. While the plan and report tasks are generally well established within the traditional technology development practice, the "assess, do and evaluate" tasks are less highlighted and are, therefore, strongly emphasised within the proposed framework. The "assess and evaluate" gates offer stakeholders important opportunities for continuous reflection, careful targeting and tailoring of their RPMTS interventions, and ongoing attention to the nature of dynamic stakeholder relationships at each process' phase. Overall, we have proposed a framework consisting of three stages and six stage-gates in total. Given that some included constructs have no empirical basis yet, research efforts should now focus on testing these concepts before validating the model with experts in the field.

Appendix 1: Topology for Theorizing RPMTS's Development Populated with 35 Potential Theories

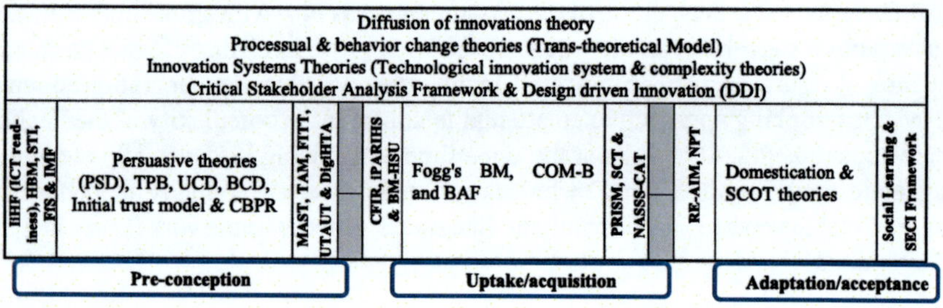

Appendix 2: Theory Selection Details and Suitability Assessment for CR

	Selected theory/framework	Explaining the match/link to the stage or stage-gate	Main author (and outcome)	*h*-index (publications)	5 year impact factor of major journal
	Pervasive theory				
1	***Diffusion of Innovations***	The entire adoption process essentially aims at understanding how the new RPMTS intervention will spread over time. Designers and developers, therefore, seek to predict and respond to the needs of the present adopter segment on the diffusion curve. Rogers' diffusion of innovation theory was thus included as a pervasive theory not only to help designers and developers to take a long term view of any new RPMTS intervention but to also deliberately track its diffusion over time at critical stage-gates. This ability to track an intervention's diffusion over time, would allow developers and implementers to take appropriate action, at the right time to evolve the intervention to meet the needs of successive segments identified by the theory and overcome barriers to the adoption and scaling of RPMTSs as and when they arise	Everett M. Rogers (included)	40 (172)	4.498 (2019-Communication Research)
Assessment/dissemination theory					

(continued)

(continued)

	Selected theory/framework	Explaining the match/link to the stage or stage-gate	Main author (and outcome)	*h*-index (publications)	5 year impact factor of major journal
2	***Sociotechnical Imaginaries***	Assessment and dissemination theories are aimed at both evaluating collectively held views around the proposed RPMTS intervention as well as the current state and infrastructure of primary healthcare services and at the same time, informing target potential adopters of a new intervention in the pipeline. The main activities and tasks here include the assessment of mobile device penetrations, level of technology literacy and skills (mobile device use), identifying opinion leaders and innovators as well as attitudes and opinions about current access to and quality of primary healthcare services and the proposed new intervention, its benefits, main features and perceived risks. Of the four theories listed, "sociotechnical imaginaries" was selected because it is future oriented and seeks to assess collectively held views rather than individual opinions to be aggregated	Sheila Jasanoff (included)	23 (61)	2.366 (2019-Minerva); 1.705 (2019-Science as Culture)
Attitudes, behaviour and perceptions shaping theory					
3	***Community-Based Participatory Research***	After evaluating collectively held views; attitudes, behaviour and perceptions shaping theories seek to influence potential adopters towards a favourable position in relation to the proposed intervention. While the technology itself may be designed in a such way that it accomplishes this objective, the researchers primarily believe that meaningful involvement of concerned users in solving their specific challenges is the most appropriate approach and thus selected "community-based participatory research (CBPR)" from among the listed theories as the most suitable framework for systematically involving users in solving their own problems	Barbara A. Israel (excluded and replaced)	1 (1)	2.818 (2019-Journal of Urban health)
Readiness and uptake prediction framework					

(continued)

(continued)

	Selected theory/framework	Explaining the match/link to the stage or stage-gate	Main author (and outcome)	*h*-index (publications)	5 year impact factor of major journal
4	***Fit between Individual, Task and Technology***	To conclude the preconception stage, we proposed the use of readiness and uptake prediction theories to assess the level of preparedness of the adopting context to take up the use of the new intervention. Activities here include measuring attitudinal changes around the proposed new intervention, its benefits, main features, and perceived risks; soliciting inputs from innovators and opinion leaders within the target adopter segment; projecting the new RPMTS intervention's uptake and spread based on the diffusion of innovation theory and assessing improvements in technology literacy and skills (assuming sufficient time has passed since the initial assessment [e.g., 2 years]). Among the listed theories here, it was thought that the FITT framework was the most appropriate because of its focus on how the individual uses the technology to accomplish a specific task or collection of tasks	D. Goodhue (Included)	33 (79)	2.410 (2020-Decision Sciences); 8.180 (2020-MIS quarterly)
Implementation model					

(continued)

(continued)

	Selected theory/framework	Explaining the match/link to the stage or stage-gate	Main author (and outcome)	*h*-index (publications)	5 year impact factor of major journal
5	***Behavioural Model of Health Service Utilisation***	The uptake/acquisition stage begins with implementation theories aimed primarily at planning for implementation and initial uptake or use of a given, new RPMTS intervention. Activities here go beyond mere project management tasks aimed at rolling out the new RMPTS intervention (which like design, ought to be informed by contextual realities) but also include understanding what would move target adopters to start using the new intervention and on what occasions. Therefore, to be able to subsequently measure improvements in access to and quality of healthcare services as a result of the new RPMTS intervention, it was felt that "the behavioural model of health service utilisation (BM-HSU)" might offer insights beyond project management activities related to implementation and chosen	Ronald Max Andersen (included)	88 (256)	3.675 (2019-Journal of Health and Social Behavior)
Uptake contextualising model					
6	***Fogg's Behaviour Model***	Next, we propose the use of uptake contextualising theories to focus on the actual setting in which the initial decision to try out the new intervention is anticipated to take place and seek to bring together all the elements necessary to kick start the initial use of the new intervention from the target adopters' perspective. Fogg behaviour model (FBM) was specifically selected for its simplicity and insights on role of "triggers" in inducing desired behaviour (uptake) in contexts where motivation and ability are pre-existing	B. J. Fogg (Included)	33 (63)	2.449 (2019-Research Technology Management Journal)
Uptake assessment model					

(continued)

(continued)

	Selected theory/framework	Explaining the match/link to the stage or stage-gate	Main author (and outcome)	*h*-index (publications)	5 year impact factor of major journal
7	***P ractical, Robust Implementation and sustainability model***	To end the uptake stage, we make use of uptake assessment theories to evaluate how the deployment of the new intervention in its context of use, may be impacting its uptake and sustainability thereafter. Now that the intervention is already being used at least by some in the target adopting population, we focus on evaluating the perspectives of adopters and non-adopters alike, on the deployed intervention as well as characteristics or attributes of each of these groups to understand how either the intervention or its implementation may be improved to subsequently increase uptake. For these tasks, "the practical, robust implementation and sustainability model (PRISM)" was thought to be the most fitting framework when compared to other listed theories at this stage-gate	Adrianne C. Feldstein (included)	29 (66)	6.084 (2019-BMJ Quality & Safety)
Diffusion assessment framework					
8	***Reach, Effectiveness, Adoption, Implementation and Maintenance***	We started "the adaptation/acceptance" stage with diffusion assessment theories to primarily measure the achieved level of adoption so far, to determine whether the intervention has reached the critical mass (reach) required to now begin embedding its use in the habits and routines of its target users. At this stage-gate, we found that RE-AIM framework not only combines the concepts of reach and adoption that we are seeking to evaluate while at the same time being well integrated with the PRISM framework but also enables us to double up on the evaluation of implementation and effectiveness of the newly deployed intervention along with the evaluation of maintenance. RE-AIM was, therefore, thought to be the best framework for the job at this stage-gate	Russell E. Glasgow (included)	108 (482)	4.210 (2020-American Journal of Public Health)
Routinisation and internalisation theory					

(continued)

(continued)

	Selected theory/framework	Explaining the match/link to the stage or stage-gate	Main author (and outcome)	*h*-index (publications)	5 year impact factor of major journal
9	***Domestication theory***	If it is established that the critical (diffusion) mass for the newly deployed intervention has been reached based on the RE-AIM framework, we suggest the use of routinisation and internalisation theories to embed the seamless use of the new intervention in the habits and routines of its users as well as to ensure that it is part of their lives and lifestyles. For the above purpose, domestication theory proved to be the most appropriate because of its clarity, simplicity, and seamless integration with adaptation stage of the adoption process	Roger Silverstone (included)	16 (46)	1.929 (2019-Media, Society and Culture)
Learning and knowledge management theory					
10	***Socialisation, Externalisation, Combination and Internalisation model***	Finally, we proposed that the adoption process as a whole be completed with learning and knowledge management theories to draw and document lessons from the adoption process and to potentially create new knowledge about the process for scaling the intervention. While knowledge management theories and practices such as "case-based reasoning" could have done the job, the socialisation, externalisation, combination, and internalisation (SECI) model was found to be most appropriate framework for offering formal mechanisms for creating new knowledge as opposed to merely learning from experience	Ikujiro Nonaka (included)	30 (65)	13.21 (2019-Harvard Business Review)

Appendix 3: A Framework for Adoption and Scaling of Remote Patient Management Tools and Systems

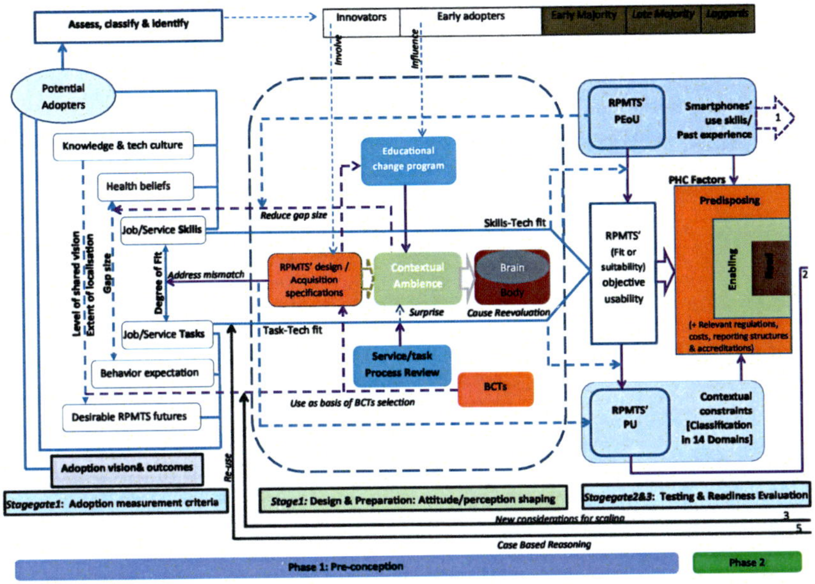

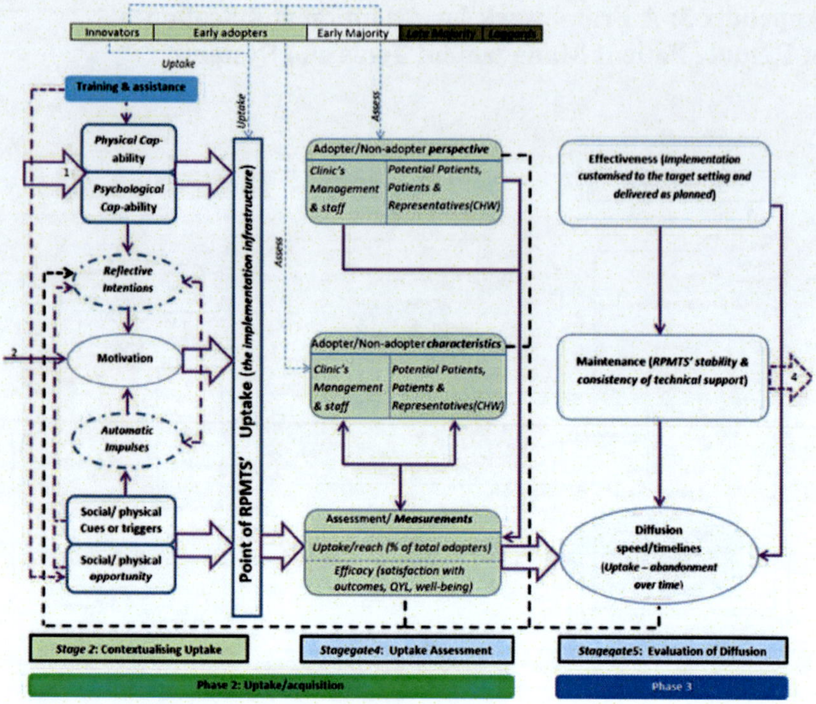

Innovators
Early adopters
Early Majority
Uptake
Training & assistance
Physical Cap-ability
Psychological Cap-ability
1
Uptake
Assess
Assess
Point of RPMTS' Uptake (the implementation infrastructure)
Adopter/Non-adopter perspective
Clinic's Management & staff
Potential Patients, Patients & Representatives(CHW)
Effectiveness (implementation customised to the target setting and delivered as planned)
Reflective intentions
2
Motivation
Automatic impulses
Adopter/Non-adopter characteristics
Clinic's Management & staff
Potential Patients, Patients & Representatives(CHW)
Maintenance (RPMTS' stability & consistency of technical support)
4
Social/ physical Cues or triggers
Social/ physical opportunity
Assessment/ Measurements
Uptake/reach (% of total adopters)
Efficacy (satisfaction with outcomes, QYL, well-being)
Diffusion speed/timelines (Uptake – abandonment over time)
Stage 2: Contextualising Uptake
Stagegate4: Uptake Assessment
Stagegate5: Evaluation of Diffusion
Phase 2: Uptake/acquisition
Phase 3

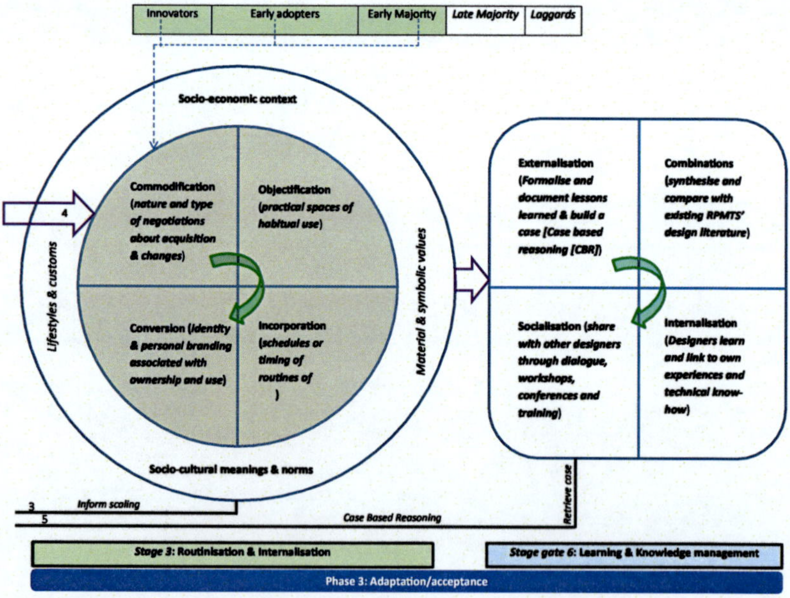

Innovators
Early adopters
Early Majority
Late Majority
Laggards
Socio-economic context
4
Lifestyles & customs
Commodification (nature and type of negotiations about acquisition & changes)
Objectification (practical spaces of habitual use)
Conversion (identity & personal branding associated with ownership and use)
Incorporation (schedules or timing of routines of)
Material & symbolic values
Socio-cultural meanings & norms
Externalisation (Formalise and document lessons learned & build a case [Case based reasoning [CBR])
Combinations (synthesise and compare with existing RPMTS' design literature)
Socialisation (share with other designers through dialogue, workshops, conferences and training)
Internalisation (Designers learn and link to own experiences and technical know-how)
3
Inform scaling
5
Case Based Reasoning
Retrieve case
Stage 3: Routinisation & Internalisation
Stage gate 6: Learning & Knowledge management
Phase 3: Adaptation/acceptance

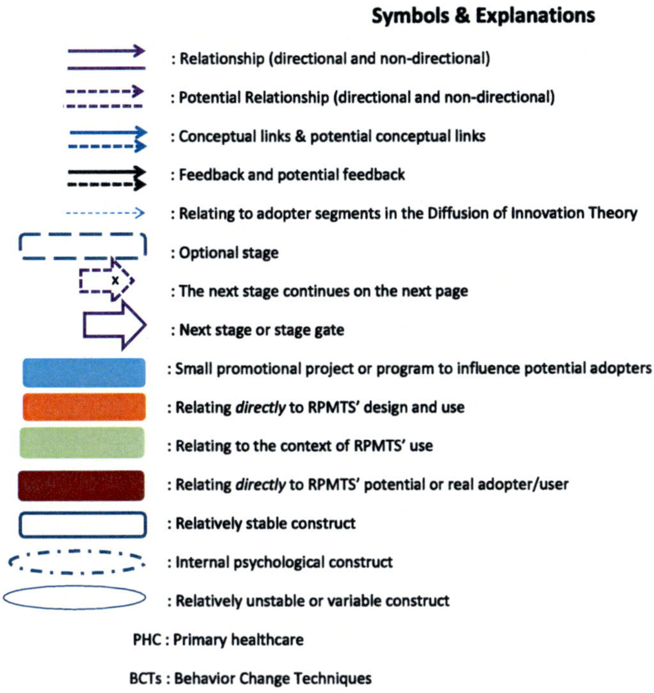

References

1. Kho J, Gillespie N, Martin-Khan M (2020) A systematic scoping review of change management practices used for telemedicine service implementations. BMC Health Serv Res 20(1):1–16. https://doi.org/10.1186/s12913-020-05657-w
2. Ravneberg B (2012) Usability and abandonment of assistive technology. J Assist Technol 6(4):259–269. https://doi.org/10.1108/17549451211285753
3. Materia FT, Faasse K, Smyth JM (2020) Understanding and preventing health concerns about emerging mobile health technologies. JMIR mHealth uHealth 8(5), https://doi.org/10.2196/14375
4. Peters D, Calvo RA, Ryan RM (2018) Designing for motivation, engagement and wellbeing in digital experience. Front Psychol 9(MAY):1–15. https://doi.org/10.3389/fpsyg.2018.00797
5. Picazo-Vela S, Fernandez-Haddad M, Luna-Reyes LF (2013) IT's alive!! Social media to promote public health. ACM Int. Conf. Proceeding Ser, 111–119, https://doi.org/10.1145/2479724.2479743
6. Lockton D (2012) POSIWID and determinism in design for behaviour change. SSRN Electron J, April. https://doi.org/10.2139/ssrn.2033231
7. Shaw J et al (2018) Beyond 'implementation': digital health innovation and service design. NPJ Digit Med 1(1). https://doi.org/10.1038/s41746-018-0059-8
8. Serrano-Santoyo A, Rojas-Mendizabal V (2017) Exploring a complexity framework for digital inclusion interventions. Procedia Comput Sci 121:212–217. https://doi.org/10.1016/j.procs.2017.11.029

9. Dearing JW, Smith DK, Larson RS, Estabrooks CA (2013) Designing for diffusion of a biomedical intervention. Am J Prev Med 44(1 SUPPL. 2):S70–S76. https://doi.org/10.1016/j.amepre.2012.09.038
10. Sovacool BK, Hess DJ (2017) Ordering theories: Typologies and conceptual frameworks for sociotechnical change. Soc Stud Sci 47(5):703–750. https://doi.org/10.1177/0306312717709363
11. Kowatsch T, Otto L, Harperink S, Cotti A, Schlieter H (2019) A design and evaluation framework for digital health interventions. IT—Inf Technol 61(5–6):253–263. https://doi.org/10.1515/itit-2019-0019
12. Kerr D, Talaei-Khoei A, Ghapanchi AH (2018) A paradigm shift for bring your own device (BYOD). Am Conf Inf Syst 2018 Digit. Disruption AMCIS 2018(2014):1–10
13. Dhiman N, Arora N, Dogra N, Gupta A (2019) Consumer adoption of smartphone fitness apps: an extended UTAUT2 perspective. J Indian Bus Res 12(3):363–388. https://doi.org/10.1108/JIBR-05-2018-0158
14. Abbas RM, Carroll N, Richardson L (2019) Assessing the need of decision-making frameworks to guide the adoption of health information systems in healthcare. Heal. 2019—12th Int Conf Heal Informatics, Proceedings; Part 12th Int. Jt. Conf. Biomed. Eng. Syst. Technol. BIOSTEC 2019, no. Biostec, pp 239–247. https://doi.org/10.5220/0007363202390247
15. Wickramasinghe N, Bodendorf F (2020) Delivering superior health and wellness management with IoT and analytics. Springer, Richmond
16. Xu W (2012) User experience design: Beyond user interface design and usability. Ergon—A Syst Approach. https://doi.org/10.5772/35041
17. Hevner AR, March ST, Park J, Ram S (2004) Design science in information systems research. MIS Q Manag Inf Syst. 28(1):75–105. https://doi.org/10.2307/25148625
18. Peffers K, Tuunanen T, Rothenberger MA, Chatterjee S (2007) A design science research methodology for information systems research. J Manag Inf Syst 24(3):45–77. https://doi.org/10.2753/MIS0742-1222240302
19. Gregor S, Hevner AR (2013) Positioning and presenting design science research for maximum impact types of knowledge in design science research. MIS Q 37(2):337–355
20. Gregor S (2006) The nature of theory in information systems. MIS Q Manag Inf Syst 30(3):611–642. https://doi.org/10.2307/25148742
21. Davidoff F (2019) Understanding contexts: How explanatory theories can help. Implement Sci 14(1):1–9
22. Kuru H (2018) Behavior change techniques used in mobile applications targeting physical activity: A systematic review chapter. In: Current and emerging mHealth technologies: adoption, implementation, and use Sezgin E et.al (Ed). Springer International Publishing AG, Ankara, pp 1–311
23. Grant MJ, Booth A (2009) A typology of reviews: An analysis of 14 review types and associated methodologies. Health Info Libr J 26(2):91–108. https://doi.org/10.1111/j.1471-1842.2009.00848.x
24. Samnani SS, Vaska M, Ahmed S, Turin TC (2017) Review typology: The basic types of reviews for synthesizing evidence for the purpose of knowledge translation. J Coll Physicians Surg Pakistan 27(10):635–641
25. Carlsson SA, Henningsson S, Hrastinski S, Keller C (2011) Socio-technical IS design science research: Developing design theory for IS integration management. Inf Syst E-bus Manag 9(1):109–131. https://doi.org/10.1007/s10257-010-0140-6
26. Turin TC (2016) Conducting a literature review in health research: basics of the approach, typology and methodology. JNHFB 5:44–51. Available: http://www.equator-network.org/reporting-guidelines/
27. Dietz T, Stern PC, Wells P, Brown M, Anable J, Steg L (2016) Sovacool & hess supplementary online material for 'ordering theories' 1:1–19, 2016.
28. Wisdom JP, Chor KHB, Hoagwood KE, Horwitz SM (2014) Innovation adoption: a review of theories and constructs. Adm Policy Ment Heal 41(4):480–502. https://doi.org/10.1007/s10488-013-0486-4.Innovation

29. Hamine S, Gerth-Guyette E, Faulx D, Green BB, Ginsburg AS (2015) Impact of mHealth chronic disease management on treatment adherence and patient outcomes: A systematic review. J Med Internet Res 17(2):1–15. https://doi.org/10.2196/jmir.3951
30. Nechully S, Pokhriyal SK, Thomas SE (2018) A journey through the evolution of theories and models of adoption of innovations (years: 1798–1980). Int J Mech Eng Technol 9(10):1–36
31. Rice RE, Rogers EM (1980) Reinvention in the innovation process. Sci Commun 1(4):499–514. https://doi.org/10.1177/107554708000100402
32. Cocosila M, Archer N, Yuan Y (2009) Early investigation of new information technology acceptance: A perceived risk-Motivation model. Commun Assoc Inf Syst 25(1):339–358. https://doi.org/10.17705/1cais.02530
33. Lundvall B-A (2016) Innovation as an interactive process: from user–producer interaction to the national systems of innovation. In: The learning economy and the economics of hope (Sampath PG, Narula R Eds). Anthem Press, London, pp 61–81
34. Kreuter MW, Casey CM, Bernhardt JM (2012) Enhancing dissemination through marketing and distribution systems: a vision for public health. In: Kreuter MW, Casey CM, Bernhardt JM (eds) Dissemination and implementation research in health: translating science to practice, 1st edn. Oxford Univesity Press, New York, pp 213–222
35. Sitorus HM, Govindaraju R, Wiratmadja II, Sudirman I (2016) Technology adoption: an interaction perspective. IOP Conf Ser Mater Sci Eng 114(1). https://doi.org/10.1088/1757-899X/114/1/012080
36. Furlong E et al (2019) Adaptation and implementation of a mobile phone-based remote symptom monitoring system for people with cancer in Europe. J Med Internet Res 5(1):1–19. https://doi.org/10.2196/10813:10.2196/10813
37. Crabtree BF et al (2011) Primary care practice transformation is hard work: Insights from a 15-year developmental program of research. Med Care 49(12 SUPPL. 1):28–35. https://doi.org/10.1097/MLR.0b013e3181cad65c
38. Van Velthoven MH, Cordon C (2019) Sustainable adoption of digital health innovations: Perspectives from a stakeholder workshop. J Med Internet Res 21(3). https://doi.org/10.2196/11922
39. Barker PM, Reid A, Schall MW (2016) A framework for scaling up health interventions: Lessons from large-scale improvement initiatives in Africa. Implement Sci 11(1):1–12. https://doi.org/10.1186/s13012-016-0374-x
40. Vandenberk T et al (2019) Vendor-independent mobile health monitoring platform for digital health studies: Development and usability study. JMIR mHealth uHealth 7(10):1–10. https://doi.org/10.2196/12586
41. Wang Y, Fadhil A, Lange JP, Reiterer H (2017) Towards a holistic approach to designing theory-based mobile health interventions. arXiv
42. Damschroder LJ (2019) Clarity out of chaos: Use of theory in implementation research. Psychiatry Res 283(June):2020. https://doi.org/10.1016/j.psychres.2019.06.036
43. Lockton D (2018) Design, behaviour change and the Design with Intent toolkit. In: Design for Behaviour Change: Theories and practices of designing for change, Lockton D (Ed). Routledge, London, pp 58–73
44. Harden SM, Balis LE, Strayer T, Wilson ML (2021) Assess, plan, do, evaluate, and report: iterative cycle to remove academic control of a community-based physical activity program. Prev Chronic Dis 18:E32. https://doi.org/10.5888/pcd18.200513
45. van der Bijl-Brouwer M, Malcolm B (2020) Systemic design principles in social innovation: A study of expert practices and design rationales. She Ji 6(3):386–407. https://doi.org/10.1016/j.sheji.2020.06.001
46. van Dyk L (2014) A review of telehealth service implementation frameworks. Int J Environ Res Public Health 11(2):1279–1298. https://doi.org/10.3390/ijerph110201279

A Comprehensive Virtual Classroom Dashboard

Amber Kimberling and Sampson Akwafuo

Abstract Learning Management Systems are of utmost importance in our technological world for educating young minds in the courses they are passionate about. The recent pandemic has only shown us how utilizing an online platform for learning and integrating it within the classroom has become vital. Education is crucial because it provides us with the tools and knowledge to move through our careers and life. Moving forward from this arduous time, there will be a need for various learning tools that we can provide to students and use as educators. This project provides an understanding of learning management systems and provides an LMS web application. In developing this web platform, there were various steps involved, such as research, design, implementation, testing, and integration. With this application, students, teachers, and administration will create a dynamic and cohesive learning environment.

Keywords Learning management system (LMS) · Virtual classroom dashboard (VCD) · Accessibility · Modifiability

1 Introduction

Education is at the forefront of developing the tools individuals need to succeed in their careers, whether in a classroom or online. For most people, schooling is in a traditional setting, or it used to be. "By 24 April 2020, education institutions in approximately 180 countries were closed, affecting 85% of the world's student population and causing nearly 1.5 billion students to stay out of the classroom" [1]. COVID-19 has caused a strain on the educational industry and has forced many institutions to turn to online learning. E-learning is not a new concept and has been around for many years. E-learning (Electronic learning) is education through electronic devices

A. Kimberling (✉) · S. Akwafuo
California State University, Fullerton, CA 92831, USA
e-mail: Akimberling3@csu.fullerton.edu

X.-S. Yang et al. (eds.), *Proceedings of Seventh International Congress on Information and Communication Technology*, Lecture Notes in Networks and Systems 447,
https://doi.org/10.1007/978-981-19-1607-6_72

and digital media. One of the leading tools in online learning is Learning Management Systems (LMS). An LMS is a software platform for tracking online courses or programs.

1.1 Background

In 1924, the testing machine was created as an automatic teacher. The automatic teacher was used for rote-and-drill learning. This machine was also considered a big failure. In 1954, the teaching machine was developed by B.F. Skinner and was used for students to learn at their own pace [2]. In the 1960s, training on the computer was developed and was called Programmed Logic for Automated Teaching Operations (PLATO). PLATO implemented drills and questions. From these inventions and developments emerged Digital Native in the 1990s. This period is where the concept of electronic learning began to be recognized as a tool for learning. In the 2000s, businesses began to implement the concept of E-learning by training their workers electronically. From 2010 and up, social media began to inspire more tools for E-learning, such as YouTube and Twitter.

Today, there are over 1000 LMS to choose from, each one containing a lot of useful learning features [3]. The educational system utilizes E-learning to teach its students. Corporate companies or lower-level businesses use online learning for training their workers. There are several different types of Learning Management Systems, including cloud-based and open-source. Cloud-based systems can be accessed anywhere since they do not require hardware specifications or software to be installed. Open-source systems provide free code allowing organizations to customize the environment they will be using. These includes the Moodle (Modular Object-Oriented Dynamic Learning) is a popular open-source online learning platform. It has many enticing features such as fast grading, gamification, online testing, assignment submission, analytics, and group options [4]. Another LMS is Canvas. Canvas is web-based and is used for all education levels. It includes various features such as speed grading and provides learning outcomes, integrated calendars, online testing, originality checks for submissions, and discussion boards [5]. Google Classroom is another free web-based learning service. This LMS was created with the primary purpose of streamlining file sharing. It also has several great features like class management, announcement posting, question-driven discussions, work tracking, and a student summary that can be shared with a parent or guardian. Our virtual classroom however provides personalized approach to teaching and learning, with added security and flexible features.

2 Literature Review

2.1 *Understanding Learning Management Systems*

Learning Management Systems can aid in the assistance of educators not only in the physical classroom but also virtually with quite a few activities such as quizzes, group works, and turning in assignments [6]. The core of online or hybrid learning is utilizing the web-based tools given within an LMS. "Learning management systems can offer a great diversity of pathway and for its users it provides a framework so that they can promote knowledge distribution and exchanging information between learners in a course, allow teachers to spread information to learners, generate learning material, make tests and assignments, take on discussions, operate distance classes and permit cooperative learning with forums, chats, file storage areas, news services, etc." [7]. Bloom's Taxonomy is how we can define and distinguish the levels of educational learning through the human intellectual brain with models. This ideology is used for guidance learning that begins with educators prepping work for students and ends with homework-driven tasks for students to complete [8]. Tools to help LMS users include tracking the students' learning, providing areas for discussion, and, most importantly, providing a space that has high accessibility and is user friendly.

2.2 *User Impact*

A study was conducted by several students in Australia looking for the connection between Learning Management Systems and positive outcomes. "The more the students use and interact with the LMS platform, the more they can be engaged with the materials and be more satisfied" [9]. This concept of using an online platform for learning produced actual positive results for the students in the study. Another study was conducted in the Philippines but was focused on the Unified Theory of Acceptance and Use of Technology (UTAUT) within Canvas. UTAUT explains user intentions and behaviors with the use of technology. The results from the study showed that it was highly recommended for admins, teachers, and students. The results also suggested that "to improve implementation of canvas in teaching and learning process, encouragement and support from peers is imperative" [10].

2.3 *Performance Benefits*

How do we improve performance benefits for students and understand their needs? "By using easy and efficient analysis tools for teachers and researchers, L.A. (Learning Analytics) allows an improvement in the quality of student performance"

[11]. There are many ways to improve students' conduct. Analyzing tools are one of the most efficient ways to get more accurate and direct results. With analytics, we can improve how the students learn and how the teachers present the information for different types of learners. The first step to understanding the analytics is comprehending what e-learning is theoretical and how analyzing performance benefits work. E-learning is the interconnection between training education, teaching on digital platforms, and the class format (in person or online) [12]. Online learning targets the students who need exceptionally engaging education by learning through gaming or puzzles. The in-person classes target the students who need better clarification or attention from a teacher.

2.4 Improvement and Enhancement

Over the years, LMSs have been enhanced with features such as authorization, authentication, monitoring materials, and so much more. Moodle is one of the top LMSs due to its flexibility, modifiability, and open-source format. With these improvements in LMSs, there are significant advantages for the students and the teachers. Some of the benefits from an LMS do not help everyone because of disadvantages such as "not all users like or are familiar with this online learning approach due to lack of computing skills or are computer illiterate to access the learning material" [13]. Ultimately, this hybrid learning style or online learning with LMS can outweigh the disadvantages because a person can learn computer skills. Unfortunately, we are not considering the students who do not have access to the technology itself.

3 Project and Significance

3.1 Problem Statement

Learning Manage Systems have been around for several years. Each type of system contains some unique and similar sets of features. Each year new technology arises as well as new tools for learning. With these developments come changes to the educational system more technologically. In 2020, the world was hit with the COVID-19 pandemic leaving institutions with no choice but to turn to online learning. Having multiple tools and platforms can prove difficult, especially for lower levels of education. Why cannot these tools be all in one? There is a need for a uniform, all-in-one, easy-to-use platform where teachers can video chat, communicate, and provide material, submissions, and assessments. "The best online learning combines elements where students go at their own pace, on their own time, and are set up to think deeply and critically about subject matter combined with elements where students go online at the same time and interact with other students, their teacher, and content" [14].

3.2 Virtual Classroom Dashboard

This project is called Virtual Classroom Dashboard (VCD). It is a free web-based application that contains three types of users: students, educators, and administrators. VCD contains grading, discussion boards, content management, and online testing features. Educators need one application to provide for their students, and students need a single place to find all their materials for a given class. Creating an application that benefits the teachers and students is about bridging the gap between communication and defining a diverse learning environment. "From 2021 to 2024, the learning management system market is expected to expand to $25.4 billion at a CAGR (Compound annual growth rate) of 23.8%" [15]. Learning Management Systems are becoming very important and continue to be essential in the growth of the educational industry. Schools are not the only ones who are invested in LMS; companies are as well.

3.3 Methodology

Virtual Classroom Dashboard was developed by following the software development life cycle phases of planning, analysis, design, building, testing, deploying, and maintaining. The initial process started with the planning phase that includes defining the problem, the objectives, activities, environment, reports and products, and the project milestones. The first phase included analysis, research, and design. Within the analysis portion, the project's requirements were defined and established the users' needs through a product backlog. In the research portion of the process, a series of LinkedIn Learning videos were watched to understand how to build an application with the defined requirements. The design side included charts, diagrams, flowcharts, and any graphic necessary to understand how the software will function on the frontend and backend and how the web application will look. The next phase included build and testing. The software went into physical development in this phase, meaning the coding process started. The third phase entailed deploying and maintaining the application. From this point, perfective, adaptive, preventative, or corrective maintenance were documented and published through Azure. The last part of this project included documentation and demonstrations of the application. The documentation included a paper on the technical details developed from the previous steps, user manuals, and a PowerPoint describing the high-level details of the application and the process for creating it.

3.4 Development Environment

The Full-Stack Application is a C# ASP.Net Internet application that is hosted using a web server. It is cross-browser compatible and responsive for all device types. This application works on any operating system because it is a web-based application. The main languages used for software development are HTML5, CSS3, JavaScript, embedded SQL statements, and C#. It also uses an MSSQL relational database using constraints and keys. The tools used to develop this application will be Visual Studio 2019, Microsoft SQL Server (for development), GitHub, Azure Database (for deployed application), and a web server hosted in a cloud provider such as Azure.

3.5 Virtual Classroom Dashboard

In Fig. 1, number 1 depicts a user to move to the login page and have the option to decide whether they have an account already or not. If a user has an account, they may log in, and if they do not and are an administrator of the school, they may go to the registration page. Number 2 represents the student flow diagram where they have similar capabilities as other users, such as viewing their courses or updating their personal information. Students can also respond to discussion postings, take tests, or view their grades. Number 3 shows how the capabilities of the role of the administration. The role of the admin is allowed to manage the other types of users (students and teachers). They can also update their personal information in the account/profile section. Number 4 displays the complex flow of a teacher/educator. In the teacher role, the user begins at their dashboard and can view their profile or courses. They can also add discussions, assignments, and assessments. The assessments can either be a quiz or an examination. From this decision, they can add questions and answers that can be added to the database. The educators can also submit grades for the students.

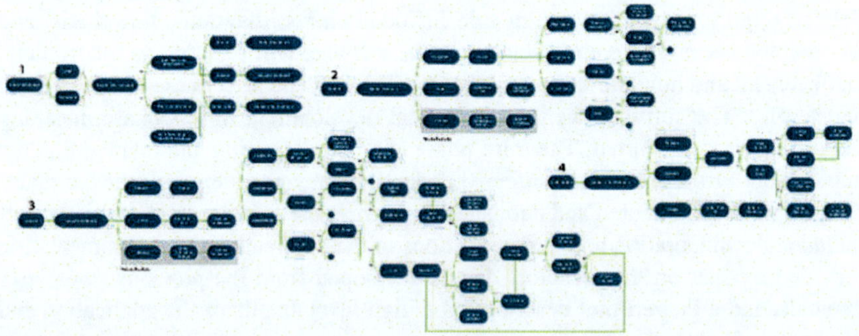

Fig. 1 User flow diagrams

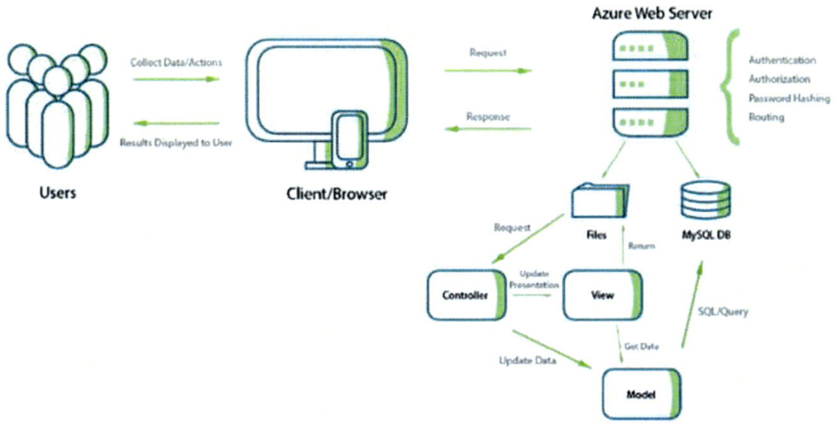

Fig. 2 System security and architecture diagram

Figure 2 displays the architecture of the web application and the security procedures. Users will enter the data, and the website will collect the data entered by the users or respond to the actions made by users. Then, the application sends a request to the Azure server, and the server responds with either a session id or a different response depending on the request. The website has the MVC pattern, which separates concerns for higher modifiability. It also has security measures of authorization, authentication (Future Feature), and encryption (Hashing algorithms).

4 Results and Discussions

4.1 Results

Virtual Classroom Dashboard has a long way to go before being used in educational institutions. There are more features needed to help improve the application, including functionality to help improve the accessibility of the website. More tests need to be conducted to help improve the sustainability of the application, including load testing and hacking. With more time, development, and research, this web application can be implemented into all levels of education.

4.2 Learning Management Systems and Accessibility

LMSs are the future of education to manage and support students' and teachers' academics. These systems have the capabilities of helping students with disabilities by providing accessibility tools such as text to speech, speech to text, and making the text more prominent. COVID-19 was a big test against how well academics could adapt to online learning. Utilizing learning management systems can help simplify switching between online and traditional in-person simple. If an LMS is equipped with the correct tools, any professor or student should succeed within the classroom no matter how academics are presented.

Implementing an online learning institution places higher importance on self-reliability between students and educators. The future of the Virtual Classroom Dashboard is limitless, and it is intended to be continually developed with more research and tools to help aid users with their learning or teaching needs. User experience and accessibility are top concerns because people need engaging and inclusive experiences.

5 Conclusion

Learning Management Systems are the future of education, and having software or application that is flexible and heavily geared toward their users is eminent. User experience is the future of technology, meaning the way we interact with the applications we use needs to be our focus. COVID-19 has shown us how providing flexibility in our education can help connect each other no matter where we are in the world and provide more diverse learning environments. LMSs can help extend the learning process beyond the physical classroom. This project shows how this is possible, provides understanding into an LMS, provides critical features in a user-driven product, and, most importantly, provides a system that can be easily manipulated to be upgraded with ease. It also shows how accessibility can be taken into account when developing an online platform.

References

1. Five facts on e-learning that can be applied to COVID-19—Institute for Environment and Human Security", United Nations University, 2021
2. Gogos B (2021) A brief history of elearning (infographic)—eFront Blog. eFront Blog, 2021 [Online]. Available: https://www.efrontlearning.com/blog/2013/08/a-brief-history-of-elearning-infographic.html. Accessed 25 April 2021
3. Pappas C (2021) The best learning management systems (2020 Update), eLearning Industry, 2021 [Online]. Available: https://elearningindustry.com/the-best-learning-management-systems-top-list. Accessed 31 August 2021
4. Raga R, Rodavia M (2018) Perceptions and utilization of a learning management system: An analysis from two perspectives. 2018 International Symposium on Educational Technology (ISET), pp 33–36. Available: https://doi.org/10.1109/iset.2018.00017. Accessed 31 August 2021
5. Key features: Canvas Course Design. Canvas.uw.edu, 2021 [Online]. Available: https://canvas.uw.edu/courses/866251/pages/key-features. Accessed 24 April 2021

6. Huilan S, Wang W, Zhongua D, Xiaolonge Q (2020) Educational management in Critical Thinking Training Based on Bloom's Taxonomy and SOLO Taxonomy. 2020 International Conference on Information Science and Education (ICISE-IE), pp 518–520
7. Saroha K, Mehta P (2016) Analysis and evaluation of learning management system using data mining techniques. 2016 International Conference on Recent Trends in Information Technology (ICRTIT), pp 1–4. Available: https://doi.org/10.1109/icrtit.2016.7569542. Accessed 27 August 2021
8. Shanavaz SF (2021) 8 reasons why moodle is still awesome. eLearning Industry [Online]. Available: https://elearningindustry.com/why-moodle-still-awesome-8-reasons. Accessed 19 April 2021
9. Ghapanchi A, Purarjomandlangrudi A, Miao Y (2020) Uncovering the impact of university students' adoption of learning management systems on positive learning outcomes. 2020 24th International Conference Information Visualisation (IV), pp 684–687
10. Endozo A, Oluyinka S, Daenos R (2019) Teachers' experiences towards usage of learning management system. Proceedings of the 2019 11th International Conference on Education Technology and Computers, pp 91–94
11. Costa L, Souza M, Salvador L, Amorim R (2019) Monitoring students performance in e-learning based on learning analytics and learning educational objectives. 2019 IEEE 19th International Conference on Advanced Learning Technologies (ICALT), pp 192–193. Available: https://doi.org/10.1109/icalt.2019.00067. Accessed 30 August 2021
12. Rahim Y, Mohd O, Sahari M, Safie N, Rahim Z (2018) A study on the effects of learning material handling procedures towards information integrity in moodle learning management system (LMS). 2nd International Conference on Electrical Engineering and Informatics (ICon EEI), pp. 81–85
13. Meyliana H, Widjaja S, Santoso S, Fernando E, Condrobimo A (2020) Improving the quality of learning management system (LMS) based on student perspectives using UTAUT2 and trust model. 2020 4th International Conference on Informatics and Computational Sciences (ICICoS), pp 1–5, 2020
14. Greenhow C (2021) Ask the expert: Online learning vs. classroom learning. MSUToday|Michigan State University [Online]. Available: https://msutoday.msu.edu/news/2020/ask-the-expert-online-learning-vs-classroom-learning. Accessed 31 August 2021
15. Dykes A (2021) Predictions for the future of LMS|technology advice. TechnologyAdvice [Online]. Available: https://technologyadvice.com/blog/human-resources/predictions-for-the-future-of-lms/. Accessed 25 April 2021

Freddy Render: A Horizontally Scaled Blender-Based Solution for 3D Graphics Rendering

Mike Peralta and Sampson Akwafuo

Abstract Rendering animations into 2D or 3D involves sequential proceeding of inputs. This causes operational bottleneck, resulting in expensive and lengthy inefficient processes. In this paper, an efficient Blender-based software for creating 3D animations from related sources is proposed. It introduces a horizontally scaled and concurrent rendering of multiple Blender-based projects. It runs in three modes that combine to accomplish its task: *Master*, *Client*, and *Slave*. A single *Master* instance exposes a web GUI to the user, maintains the master list of render job states, and controls *Slave* instances. *Client* instances are launched on demand to provide users with a GUI to submit render jobs to the *Master* instance. *Slave* instances run on rendering machines and will, in turn, launch subprocesses of Blender to render individual frames when instructed to by the *Master* instance. Test implementation of our solutions indicate improvements over vertical scaling (increasing the power of a single rendering machine) and greatly reduces the overall time taken to render a complex animation project.

Keywords Blender · Rendering · 3D animations · Render farm · Horizontal scaling · EmberJs

1 Introduction

Blender is a 2D and 3D motion graphics application with a massive community of independent and commercial artists. However, most artists struggle to render their animations within a reasonable amount of time. Single frames of even moderately complex animations can take hours or days to render. If we supposed a frame rate of 24 frames per second and a very modest 10 min to render a single animation frame, a 15-min animation would cost an artist 15 * 60 * 24 * 10 = 216,000 min or 150 days to render their animation just once. Spending several months just to render an animation is prohibitive and unacceptable. The problem is compounded further

M. Peralta (✉) · S. Akwafuo
California State University, Fullerton, CA 92831, USA
e-mail: mikeperalta@csu.fullerton.edu

X.-S. Yang et al. (eds.), *Proceedings of Seventh International Congress on Information and Communication Technology*, Lecture Notes in Networks and Systems 447,
https://doi.org/10.1007/978-981-19-1607-6_73

that artists sometimes need to render animations multiple times to fix mistakes and adjust animations. Vertical scaling to improve performance works to a certain extent but is generally expensive and insufficient for problems "at scale." Instead, modern computing problems "at scale" require horizontal scaling.

Freddy Render functions as Render Farm software for Blender. The term "Render Farm" refers to a horizontally scaled render workflow to reduce long render times. Multiple computers work together on the same render job (or multiple render jobs), sharing the work of rendering all frames. Each computer still renders a single frame at one time. However, multiple frames are rendered simultaneously on different computers, leading to an overall reduction in the time it takes to render all frames. With horizontal scaling, rendering time can be reduced simply by adding more hardware to the Render Farm. The typical workflow for rendering in Blender goes something like the following:

1. Create an animation in Blender
2. Begin rendering the animation by instructing Blender to render each frame of the animation one at a time.
3. Wait for all frames to render
4. Potentially periodically recover from system crashes, required system downtime, and other incidents that halt a render and continue from the last successful frame.
5. When all frames finish rendering, convert them into a video file using a third-party tool such as ffmpeg, a standalone video conversion program invoked via the command line ("ffmpeg" stands for "Fast Forward MPEG"; MPEG was a popular video codec)

Step 4 of the sequence above means it is incredibly unwise to instruct Blender to render directly to a video file, rather than rendering individual frames. Any crash during direct video rendering would mean a video had to be re-rendered in its entirety. Rendering whole movies at a time would also prevent horizontal scaling. The workflow for rendering with Blender using a Render Farm software package would look a little different:

1. Create an animation in Blender.
2. Upload the Blender project to the Render Farm software and configure the render job.
3. Begin rendering the animation by instructing the Render Farm software to begin the render job.
4. Wait while the Render Farm software manages the rendering of the animation as needed.
5. Download the final rendered frames from the Render Farm software.
6. Convert all rendered frames into a video file using a third-party tool such as ffmpeg (a standalone video conversion program invoked via the command line; "ffmpeg" stands for "Fast Forward MPEG"; MPEG was a popular video codec).

There is, therefore, a need for a less-expensive, closed-source, and secure render farm. Some available solutions are usually not secure, are difficult to set up, and may not work with the current version of Blender. Our approach aims to fill this gap and unmet needs.

2 Literature Review

While some Render Farm solutions for Blender presently exist, there is not an abundance of selection of robust, high-quality solutions. Many desirable characteristics and features appear in some solutions but not others, making it difficult to find a solution which covers them all. For example, some render farms are expensive (read as: "not free"), while others are insecure; some are free, but lack essential features such robustness, security, and user-friendliness.

2.1 Horizontal Scaling

Rendering 3D images takes an immense amount of time. According to studies by Vallejo et al., rendering a single frame costs a great deal of time. The rendering stage is often considered as a bottleneck due to the huge amount of time needed just to generate one image, which could take hours or days [1]. Vertical scaling is one potential way to reduce rendering times when computational needs are reasonably low by increasing the power of the rendering machine. Vertical scalability (improving a single machine) is an option. However, this is not desirable. It quickly becomes prohibitively expensive and even infeasible beyond a certain point. Additionally, vertical scalability is very expensive. The second biggest issue with vertical scalability is that it actually has hard limits. No matter how much money you may be willing to spend, it is not possible to continually add memory. "Similar limits apply to CPU speed, number of cores per server, and hard drive speed" [2]. Glez-Morcillo et al. noted that the use of Render Farms is a common way to decrease rendering time for many frames and used in most professional production studios. They suggested reducing rendering time by distributing the processing. Generating X images that take Y time units can be done in only about Y time units using horizontal scaling [3]. Use of horizontally scaled Render Farms at the highest levels of 3D movie production is mentioned in the paper *A Distributed Render Farm System for Animation Production as well*: "The first full length animation film, Toy Story, used 117 Sun workstations and the Pixar Render-Man system. The film, Shrek 3, is rendered by more than 4000 HP workstations, where every second requires 3000 h CPU time. Render farm is also widely used in architecture design, advertising, and the visual effects industry" [4]. One article summarized "the best" paid commercial Render Farm services on iRendering.net. None of the render farms listed came anywhere close to the raw price of electricity (which is the minimum cost of using a Render Farms on personal hardware). According to this website, a service called Rebus Farm cost roughly $1.00 per day (for a 3Ghz CPU) with additional costs related to the power of the GPU you chose, while another (iRender) cost up to $22/hr for one GPU [5].

2.2 *Insecure Rendering Solutions*

It is sometimes tempting to participate in a free community Render Farm service. However, these services tend to expose private Blender projects to the public. Note the following quote from the FAQ page from one of the most popular Blender render farm solutions, which essentially shows that community machines receive your unprotected project files: "When the owner of a project adds their scene to the jobs to queue, the service splits the animation into single frames to render, sends each frame to a connected computer and aims to optimize its choice based on the available memory, as well as the CPU/GPU power" [6]. Another popular community Render Farm service essentially says the same thing. Each user renders projects belonging to other participants, which means no user's project is really kept private [7]. Also of note, based on the above quote, community-based Render Farm services generally only render as many frames as one has already rendered in the first place for others. Users do not get "free" frames rendered. Instead, users may render community frames when they have no other active jobs of their own to sort of "bank frames." Then, when needed, a user can have the community render their frames, but only up to the amount already credited to their account. This makes it possible to leverage the render farm to render a large number of frames in a short amount of time, but the requirement of pre-"banking" of frames means the overall throughput (averaged over all renders and all projects one artist or studio would perform) would be no better than simply rendering with one machine.

3 Solution Description and Methodology

3.1 *Installation and Compatibility*

Installing *Freddy Render* is a fairly simple process. It is written primarily in pure Java, with some extra code for the front end (EmberJs) and control of Blender (python scripts). The entirety of *Freddy Render* is packaged as a single JAR file and only needs the appropriate version of Blender and Java to run. Thus, "installation" consists of installing Blender, Java, and copy/pasting the *Freddy Render* JAR file anywhere the user deems convenient. It is also OS-agnostic and can run on any operating system where the correct version of Java is installed. There are three different modes of operation (*Master, Slave, and Client*). Each can be launched with different but related commands.

3.2 User Interaction with Each Mode

Exactly one instance of *Master* mode should run continuously. One or more instances are then configured to run in *Slave* mode. *Slave* instances may be launched and stopped as desired, as the *Master* instance will automatically detect the status of the farm and assign frames appropriately. Each instance would run on a dedicated machine, and be left running indefinitely while Blender projects render, or the system waits for new projects to render. The first step toward rendering a Blender project is launching an instance in *Client* mode. A *Client* instance allows the user to select a Blender project file for rendering, then sends that file and all dependencies over the network to the Master instance. Once the project is sent to the Master instance, the user closes the Client instance until the next time it is needed (Fig. 1).

Once Freddy Render receives one or more Blender projects from a *Client* instance, the user can control and manage Freddy Render and its render jobs by interacting with the Master instance. User directs their browser to the hostname and port of the Master instance and is presented with a web front-end interface, written in EmberJs. Freddy Render's homepage presents a summary of all active render jobs, along with web GUI controls that allow the user to resume/pause jobs, change job priority/weight, and see the status of Slave instances. The user can click an individual job to see its details on a dedicated page. A render job's dedicated page also contains controls to resume/pause the job and adjust its weight and priority. It shows an animated preview of the most recent frames rendered at the top. It also has buttons to delete the job or to purge its rendered frames. This job detail view also allows the user to view all participating *Slave* instances, blacklist individual *Slave* instances, or de-blacklist *Slave* instances. The user can click yet another link to see a page that shows all rendered frames in a gallery view. The thumbnail gallery view is a simple page, which shows each rendered frame as an individual image in the gallery. Each image in the gallery displays which *Slave* instance rendered the image and can be enlarged

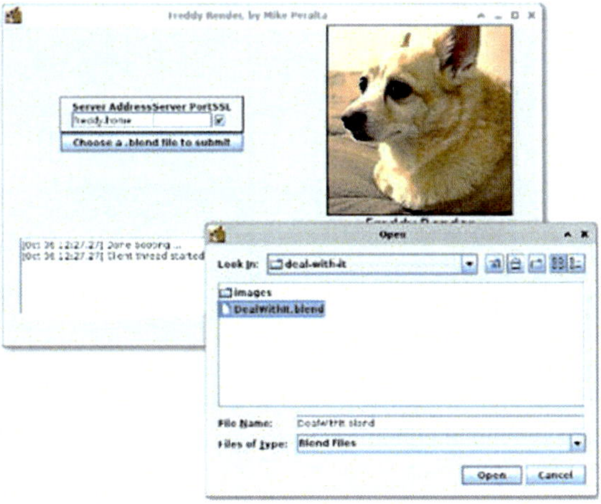

Fig. 1 Client instance—Submitting render job

by clicking it. Once enlarged, an individual frame also has a button that allows the user to force the frame to render again (reset).

After a project/job had finished rendering, the user can go back to the job details page and click a button to generate a zip file containing all rendered frames, preview images, and logs. The web GUI will update to display its progress as it generates the zip file. Eventually, when the zip file has finished generating, a button will appear to let the user download the entire zip file directly from the web GUI. Once the user has downloaded the final zip file, they would typically purge the render job from the *Master* instance using a button on the job details page. The user typically does not interact with *Slave* instances after initial setup, so their details are omitted from this perspective (Figs. 2, 3, 4, and 5).

Fig. 2 Master GUI—Rendered thumbnails

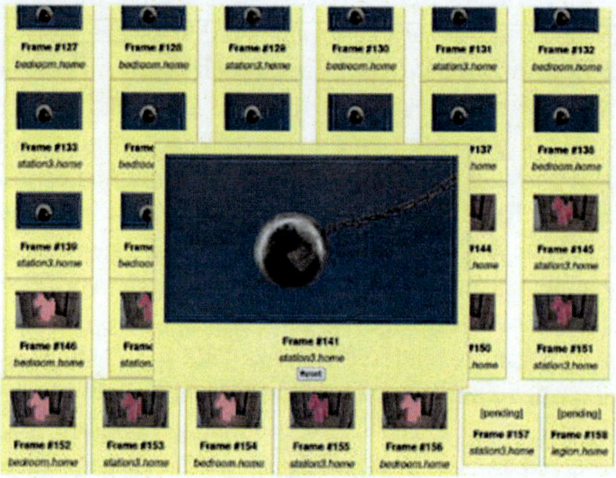

Fig. 3 Master GUI—Job details

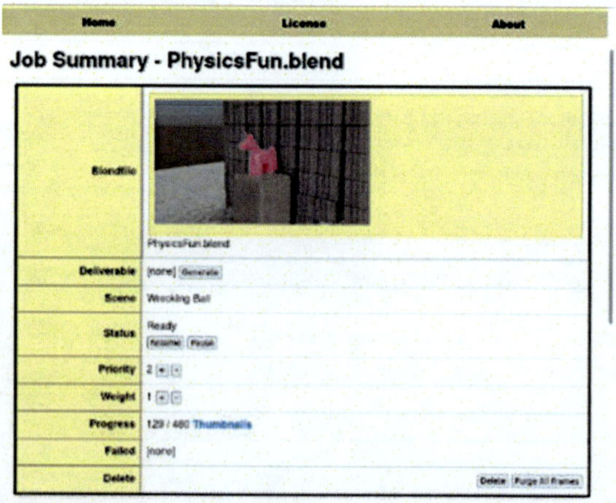

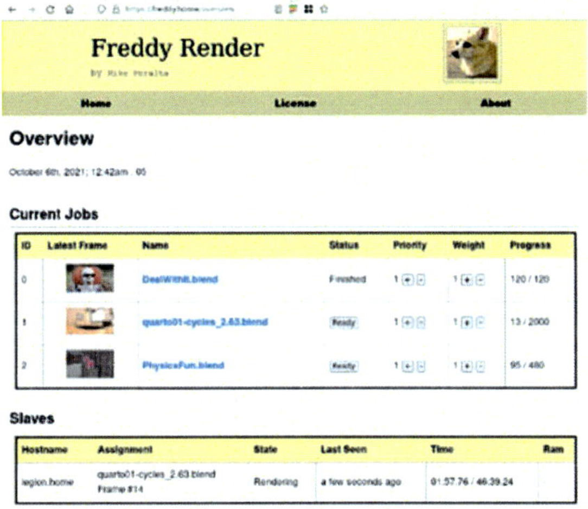

Fig. 4 Master GUI—Overview

Fig. 5 Master GUI—generating zip delivery

4 Case Study and Results

Four Blender projects were used in a case study to verify the functionality of *Freddy Render*. Each project was a slightly different animation of some sort, involving a slow camera movement over a basic scene. One project ("spinner") utilized Blender's ability to compute physics simulations. Each project was modified to render very small frames with low quality, to ensure rendering time for a complete project could be completed in less than an hour (as opposed to days). It is worth noting that such reduced quality settings cause Blender to spend a disproportionately high amount of time initializing and finalizing each frame, compared to actual render time. Additionally, reduced quality frames also result in more time disproportionally spent distributing dependency files to each *Slave* instance. This means the results grossly underestimate the time savings that would be achieved under typical usage. As discussed previously, the workflow for rendering a project with *Freddy Render* is simple. The user makes sure the *Master* instance is running somewhere. The user then launches a *Client* instance and uses it to upload a Blender project to *Freddy Render*. Once uploaded, the Blender project (referred to as a "render job") shows up in the GUI, and the *Master* instance begins attempting to assign frames for rendering to individual *Slave* instances. The user then makes sure as many *Slave* instances are running (one per machine) as possible and watches the *Master's* GUI for progress.

The first measurement needed for this case study was a baseline for each project's rendering time without horizontal scaling. Each project was rendered with the traditional method (directly inside Blender), one-by-one, on one of two average desktop machines from the render farm. The project "Tron Set" rendered 1440 frames in 47 min (1.95 s per frame), "Turn Table" rendered 72 frames in 19 min (15.8 s per frame), "Spinner" rendered 500 frames in 34 min, and "Dungeon" rendered 300 frames in 44 min (8.8 s per frame).

For the next measurement, each project was rendered one-by-one within a small *Freddy Render* farm using four modest machines: The two average desktop machines used for the baseline test, plus a somewhat fast laptop and another average performance headless machine. The project "Tron Set" rendered in 36 min (1.5 s per frame), "Turn Table" rendered in 25 min (20.8 s per frame), "Spinner" rendered in 17 min (2.04 s per frame), and "Dungeon" rendered in 11 min (2.2 s per frame). The resulting improvements in total rendering times were 30.56% ("Tron Set"), −24% ("Turn Table"), 100% ("Spinner"), and 300% ("Dungeon"), for an average improvement of 101.64%. Again, it is worth noting that these improvements are much lower than would be expected when using full quality render settings for each project (e.g., "Turn Table" performed worse due to relatively high setup/finalizing time costs per frame) (Fig. 6).

Finally, *Freddy Render* was sent all four projects at once to test its ability to coordinate multiple projects and multiple *Slave* instances at the same time. As expected, *Freddy Render* was able to handle this situation without any issues, and without significant strain on the *Master* instance. Based on the low CPU and network utilization shown in the screenshot of the *Master* instance, *Freddy Render* will most likely be able to handle a much larger number of *Slave* instances and simultaneous projects. After 32 min, all jobs were paused, and progress was inspected. The render farm running under Freddy Render was able to render 486 frames in 32 min, for an average of 3.95 s per frame. This was also an improvement over the baseline of 7.66 s per frame, by nearly double.

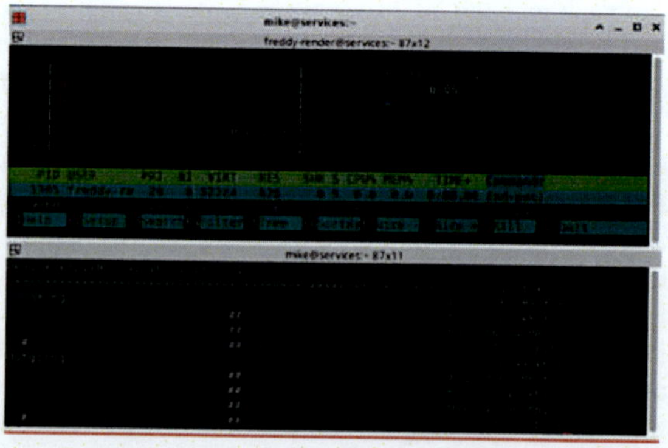

Fig. 6 Master—low utilization

5 Conclusion

Freddy Render presents a fairly more robust and intuitive solution to rendering problems. It is also a Free Open Source Software (FOSS). Unlike some solutions, it can handle Blender projects with external dependencies and nearly unlimited size per file. Freddy Render's architecture allows administration and management of render jobs from any browser on the local network rather than only one workstation. In the future, Freddy Render should be stressed to its maximum extent to determine just how large of a farm it can support. Based on the developer's own small network and experience, a rough guess is that Freddy Render may support a hundred Slave instances on a modest network, with a modest server running the Master instance. However, empirical measurements with many more machines would be much better proof. Additionally, Freddy Render does not currently support 1-Frame jobs. It is hoped that Freddy Render will become easier for all artists to install and use. It may turn out that Windows users (for example) only have easy access to a version of Java that is too old for Freddy Render. If this obstacle presents itself, Freddy Render might need upgraded documentation or perhaps be compiled against a more appropriate version.

In the future, Freddy Render will implement SSL listening for the Master instance and automated SSL certificate generation, to improve security. Authentication over an SSL connection, combined with brute force protection may make Freddy Render much more secure over untrusted networks. Additionally, Freddy Render currently recovers its state across reboots of the master by analyzing the file system. In the future, Freddy Render will support database persistence with SQLite, so recovery across reboots happens much more quickly. Additionally, Freddy Render will implement blackbox testing against the API and front-end web GUI.

References

1. Vallejo D, Glez-Morcillo C, Angulo J, Albusac J (2011) Distributed rendering of images through intelligent task distribution. Proc. -2011 Int. Conf. P2P, Parallel, Grid, Cloud Internet Comput. 3PGCIC 2011, pp 242–247. https://doi.org/10.1109/3PGCIC.2011.44
2. Ejsmont A (2015) Web scalability for startup engineers. McGraw-Hill Education
3. Glez-Morcillo C, Vallejo D, Albusac J, Jimenez L, Castro-Schez JJ (2011) A new approach to grid computing for distributed rendering. Proc. -2011 Int. Conf. P2P, Parallel, Grid, Cloud Internet Comput. 3PGCIC 2011, pp 9–16. https://doi.org/10.1109/3PGCIC.2011.12
4. Yao J, Pan Z, Zhang H (2009) A distributed render farm system for animation production. In: Lecture Notes in Computer Science (including subseries Lecture Notes in Artificial Intelligence and Lecture Notes in Bioinformatics), vol. 5709 LNCS, pp 264–269. https://doi.org/10.1007/978-3-642-04052-8_31
5. Best Render Farms 2021 ComparisonliRender GPURental. https://irendering.net/best-render-farms-2021-comparison/ (accessed Sep. 07, 2021)
6. SheepIt Render Farm. https://www.sheepit-renderfarm.com/faq.php (accessed Sep. 07, 2021)
7. Free Blender Render Farm—GradedBlue Renderfarm. https://render.gradedblue.com/free (accessed Sep. 07, 2021)

Medical X-Ray Image Classification Employing DCGAN and CNN Transfer Learning Techniques

Md. Asif Talukdar, Ayesha Siddika, Ahasanul Haque Abir, Mohammed Ziad Hassan, and Muhammad Iqbal Hossain

Abstract Over the decades, a typical imaging test that has been used is an X-ray. It allows doctors to see into the body without an incision. As a result, an X-ray can aid in diagnosing, monitoring, and treating a variety of medical disorders by detecting diseases beforehand. Among the diseases, pneumonia got major heed because of its intensity. As the lungs are the most vulnerable part of the body when it comes to pneumonia, doctors rely on the chest X-ray to diagnose the disease. In this research, we have worked on the X-ray images to discern pneumonia using our custom CNN model and different types of transfer learning models and manifested a comparison of those methods in terms of their ability to detect the disease. Furthermore, we performed generative adversarial networks (GAN) with deep convolutional layers to generate and merge a new training dataset using existing image data. Then, we executed the models anew after acquiring a new artificial dataset. Before using GAN, we got accuracy of 94%, 94%, 73%, 73%, 96%, 97%, and 94% in Custom CNN, InceptionV3, ResNet50, EfficientNetB0, VGG16, DenseNet201, and Xception, respectively. However, we observed improved accuracy from all models applying GAN except for DenseNet201. Moreover, VGG16, DenseNet201, and custom CNN acquired the higher accuracy overall.

Keywords DCGAN · Transfer learning · X-ray · Pneumonia · Synthetic image · Convolutional neural network

Md. A. Talukdar (✉) · A. Siddika · A. H. Abir · M. Z. Hassan · M. I. Hossain
BRAC University, Dhaka, Bangladesh
e-mail: md.asif.talukdar@g.bracu.ac.bd

A. Siddika
e-mail: ayesha.siddika2@g.bracu.ac.bd

A. H. Abir
e-mail: ahasanul.haque.abir@g.bracu.ac.bd

M. Z. Hassan
e-mail: mohammed.ziad.hassan@g.bracu.ac.bd

M. I. Hossain
e-mail: iqbal.hossain@bracu.ac.bd

X.-S. Yang et al. (eds.), *Proceedings of Seventh International Congress on Information and Communication Technology*, Lecture Notes in Networks and Systems 447,
https://doi.org/10.1007/978-981-19-1607-6_74

1 Introduction

Pneumonia is a respiratory infection in which the air sacs in the lungs become inflamed. Coughing, fever, chills, and trouble breathing can occur when the air sacs become blocked with fluid or pus. A complication such as an empyema or the growth of an abscess could result in a lack of response. Doctors usually perform a physical exam and a chest X-ray to examine the lungs, heart, and blood vessels to diagnose pneumonia. When reading the X-ray, the radiologist will look for white areas in the lungs that indicate an infection. Other pneumonia-related disorders, such as abscesses or pleural effusions, will be visible on this X-ray image [1].

Medical imaging findings can be processed more quickly with computer-aided detection and diagnosis using machine learning techniques. Convolutional neural networks (CNN), one of the most well-known machine learning algorithms for image categorization, will be used to achieve our goal. This technique enables machines to predict and successfully label new images [2]. Furthermore, this method has been shown to extract beneficial characteristics from images in image classification applications [3]. There have been numerous CNN architectures produced to date. In this study, we attempt to demonstrate and compare several of these CNN architectures so that clinicians may make better decisions when diagnosing pneumonia. We utilized DCGAN to pit two neural networks against one other to generate new, synthetic data samples that may pass for accurate data to achieve improved outcomes. Our ultimate goal is to develop a process that would allow doctors to diagnose a condition quickly.

2 Literature Review

In the last few years, the application of deep learning (DL) has grown exponentially in the healthcare sector, and various researches prove that DL models can be used for the detection of different diseases using image classification. Some of those are: In comparing two CNN networks Xception and VGG16, Ayan et al. [4] showed that vgg16 performs better for detecting the typical case, and on the other hand, Xception performs better for the detection of pneumonia, and the combination of both will be more accurate. However, on a profound scale, deep learning algorithms were utilized by Srivastav et al. to categorize chest X-ray images to diagnose pneumonia [5]. First, deep convolutional generative adversarial networks (DCGAN) were trained to supplement synthetic images and oversample the dataset to improve the model's performance. Then, using VGG16 as the foundation model for image classification, transfer learning was applied with convolutional neural networks (CNN). On the validation set, the model had a 94.5% accuracy rate. Furthermore, the accuracy of the proposed model was found to be significant compared to the naive models. Lastly, Rodŕiguez et al. used a neuroevolution algorithm based on particle swarm optimization for the construction and training of GANs to develop biomedical chest X-ray (CXR) pictures of pneumonia caused by COVID-19 [6]. The suggested method allows for creating a swarm of GAN topologies, each of which grows incrementally while being trained at the same time. For the synthesis of CXR pictures, the suggested approach achieves better FID outcomes than handcrafted GANs.

3 Methodology

Our study explores the usability and application of deep convolutional generative adversarial network (DCGAN) in detecting pneumonia from X-ray images. Generative adversarial networks are unsupervised learning using the deep learning method, which looks for patterns in input data to generate new synthetic output [7]. We used DCGAN to generate new X-ray data using the existing dataset. Firstly, we used our original dataset in our models and later used the original dataset with DCGAN generated data to improve the final result (Fig. 1).

3.1 Data Acquisition and Preprocessing

We used an X-ray dataset for our study from a subsidiary of Google called Kaggle [8]. Our dataset contains chest X-ray images, which include normal chest X-rays

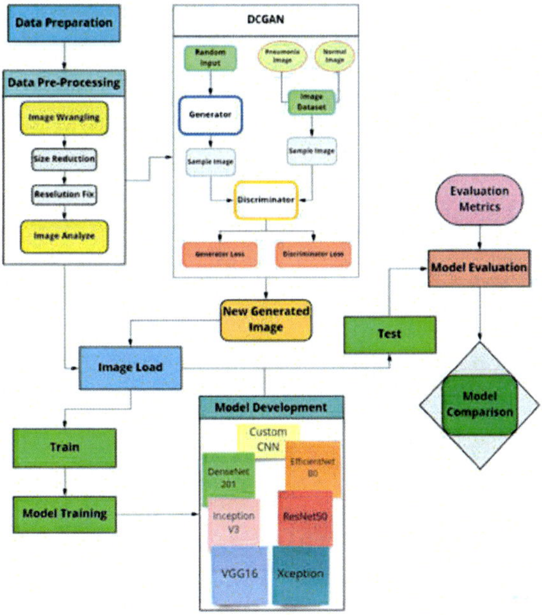

Fig. 1 Workflow diagram with DCGAN architecture

and pneumonia chest X-rays. Dataset was divided and stratified into a training set containing 80% of data and a test set containing 20% of data. Our dataset contains a total of 5856 images (73% pneumonia and 27% normal). All the images were resized to 128 × 128 pixels. For the DCGAN model, we used training dataset of X-ray images to generate synthetic image dataset of 64 × 64 pixels which were upscaled to 128 × 128 pixels for optimal result. We applied 150-epoch approach to generate 1500 synthetic pneumonia and 100-epoch approach to generate 1000 synthetic normal image data using DCGAN model. For both of the cases, we selected approximately the last 7% of the newly generated synthetic images.

3.2 *Model Selection*

Deep Convolutional Generative Adversarial Network (DCGAN) DCGAN is a process with deep convolutional layers that can create synthetic images referencing authentic images that look like an actual image. DCGAN consists of two models, generator, also known as artist, and discriminator, also known as the critic [9]. In DCGAN, the artist generator creates images using random data. Then, the discriminator gets the synthetic image created by generator and takes sample images from the actual images dataset for reference. During the process, it produces generator loss and discriminator loss which will determine whether the generated synthetic image is acceptable or not. For standard loss function, if generator loss is decreasing and discriminator loss is increasing, that means the generator is moving toward making a synthetic image that looks like an actual image [10]. The generator will keep generating images using random data, and after each approach, it will try to improve its creation from the previous one based on the output of the loss function.

Custom Convolutional Neural Network Convolutional neural networks (CNN) are commonly used to analyze visual imagery. CNN uses a unique convolution technique rather than traditional matrix multiplication. The setup of our custom convolutional neural network is given in Table 1.

CNN Transfer Learning Techniques Our study includes transfer learning techniques such as InceptionV3, ResNet50, EfficientNetB0, VGG16, DenseNet201, and Xception.

4 Result and Analysis

4.1 *Experimental Setup*

The deep learning CNN models, including DCGAN and learning models, were trained on a laptop using a Google Colab environment using GPU and high ram configuration. Libraries used within this work include Pandas, Numpy, Seaborn, Matplotlib, Scikit-learn, Keras, PIL, and TensorFlow.

Table 1 DCGAN and DCNN model setup

Techniques	DCGAN		Deep Convolutional Neural Network	
Models	**Generator**	**Discriminator**	**Custom CNN**	**Transfer Learning CNN**
Model Structure	Input (100)	Input (64,64,3)	Conv2D filter = 32 kernel = 3	
	Dense_1 (8192)	Conv2D filter = 64,	Conv2D filter = 64 kernel = 3	
	Reshape (8,8,128)	kernel = 4, stride = 2	MaxPool2D = 2	
	Conv2DT filter = 128,	Leaky Relu (alpha = 0.2)	Conv2D filter = 128 kernel = 3	
	kernel = 4, stride = 2	Conv2D filter = 128,	MaxPool2D = 2	Base Model output
	Leaky Relu (alpha = 0.2)	kernel = 4, stride = 2	Conv2D filter = 256 kernel = 3	layer trainable = False layers
	Conv2DT filter = 256,	LeakyRelu (alpha = 0.2)	MaxPool2D = 3	Flatten Dense_1024
	kernel = 4, stride = 2	Conv2D filter = 128,	Flatten	Dense_512
	Leaky Relu (alpha = 0.2)	kernel = 4, stride = 2	Dense_120	Dropout = 0.2
	Conv2DT filter = 512,	Leaky Relu (alpha = 0.2)	Dense_120	
	kernel = 4, stride = 2	Flatten	Dense_60	
	Leaky Relu (alpha = 0.2)	Dropout (0.2)	Dropout (0.2)	
	Conv2D filter = 3, kernel = 5	Dense_1	Dense_1	
Padding	same			valid
Number of epochs	100-150		20	20
Activities	Leaky Relu, sigmoid		Relu, sigmoid	Relu, sigmoid
Parameters (Trainable)	3,750,275	404,801	1,177,801	

4.2 Model Description

Model setups for our DCGAN and deep convolutional neural network (DCNN) are given in Table 1. All the models are trained with loss function binary cross-entropy, and learning rRate was set to 0.0001. All the models in our work are optimized by Adam optimizer except for InceptionV3 and ResNet50 using RMSprop and SGD as optimizer, respectively.

4.3 Comparison of Custom CNN

Tables 2 and 3 illustrate the accuracy, precision, recall, and F1-score of our custom CNN and other six models before and after applying DCGAN, respectively. Before using DCGAN, our custom CNN obtained 0.92 Precision, 0.93 Recall, and an F1-score of 0.92. Our custom CNN performed better when we utilized DCGAN, with 0.95 Precision, 0.95 Recall, and 0.95 F1-score. Before utilizing DCGAN, we had a test accuracy of 94.03%, and after using DCGAN, we had a test accuracy of 96.09%. For both cases, the custom CNN model was trained for 20 epochs (Fig. 2).

Table 2 Comparison of CNN models (without DCGAN)

Model	Precision	Recall	F1 Score	Train accuracy(%)	Test accuracy(%)	Loss	Train time (s)
Custom CNN	0.92	0.93	0.92	99.15	94.03	0.2923	160
Inception V3	0.92	0.92	0.92	99.22	93.69	**0.9448**	129
Resnet50	0.36	0.50	0.42	73.90	72.95	0.5270	133
EfficientNetB0	0.36	0.50	0.42	73.03	72.95	0.5847	129
VGG16	0.96	0.94	0.95	**1.00**	96.16	0.2279	93
DenseNet201	**0.97**	**0.95**	**0.96**	1.00	**96.84**	0.1961	**258**
Xception	0.93	0.92	0.92	99.74	94.03	0.2504	132

Table 3 Comparison of CNN models (DCGAN applied)

Model	Precision	Recall	F1 Score	Train accuracy(%)	Test accuracy(%)	Loss	Train time (s)
Custom CNN	0.95	0.95	0.95	99.23	96.09	0.1865	179
Inception V3	0.95	0.94	0.94	99.63	95.53	**0.7313**	163
Resnet50	0.36	0.50	0.42	75.05	72.97	0.5249	182
EfficientNetB0	0.36	0.50	0.42	74.21	72.97	0.5845	144
VGG16	**0.96**	**0.96**	**0.96**	**1.00**	**96.89**	0.1838	152
DenseNet201	0.96	0.96	0.96	1.00	96.81	0.1563	**328**
Xception	0.92	0.94	0.93	98.88	94.26	0.2593	209

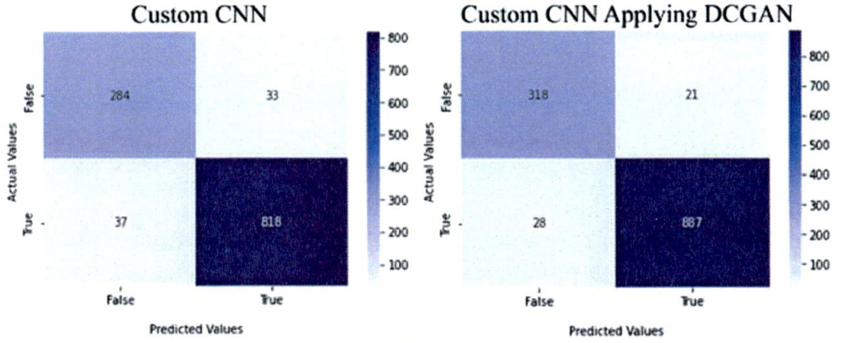

Fig. 2 Custom CNN confusion matrix

Furthermore, as shown in the heatmap diagram, our custom CNN correctly recognized 818 cases as pneumonia before applying DCGAN and 887 cases as pneumonia after applying DCGAN. For normal cases, the model correctly recognized 284 cases before DCGAN was applied and accurately recognized 318 cases after applying DCGAN on the dataset. With regard to the classification of pneumonia and normal X-ray images, it can be observed that our custom CNN provided significantly better performance when DCGAN was employed.

4.4 Comparison of Transfer Learning Models

Using transfer learning techniques for convolutional neural networks, we examined six different networks: Inception v3, RestNet50, EfficientNetB0, VGG16, DenseNet201, and Xception before and after using DCGAN. All the models were trained for 20 epochs.

Before applying DCGAN, InceptionV3 scored 0.92 on Precision, Recall, and F1-score. It also successfully detected 822 pneumonia cases and 276 normal cases without DCGAN. RestNet50 and EfficientNetB0 performed 0.36 Precision, 0.50 Recall, and a F1-score of 0.42. Moreover, both the models detected 855 pneumonia cases successfully. Furthermore, VGG16 performed 0.96 Precision, 0.94 Recall, and a F1-score of 0.95, and it detected correctly 839 pneumonia and 288 normal cases. Xception performed 0.93 Precision, 0.92 Recall, and a F1-score of 0.92, and it precisely detected 821 pneumonia cases and 281 cases of normal instances. However, DenseNet201 achieved the highest performance with 0.97 Precision, 0.95 Recall, 0.96 F1-score and a accuracy of 96.84 % (Fig. 3).

Furthermore, after applying DCGAN, we got better performance for almost every transfer learning model. The result shows InceptionV3 scored 0.95 Precision, 0.94 Recall, and an F1-score of 0.94. It also successfully detected 891 pneumonia cases and 307 normal cases after applying DCGAN. However, slight improvement was

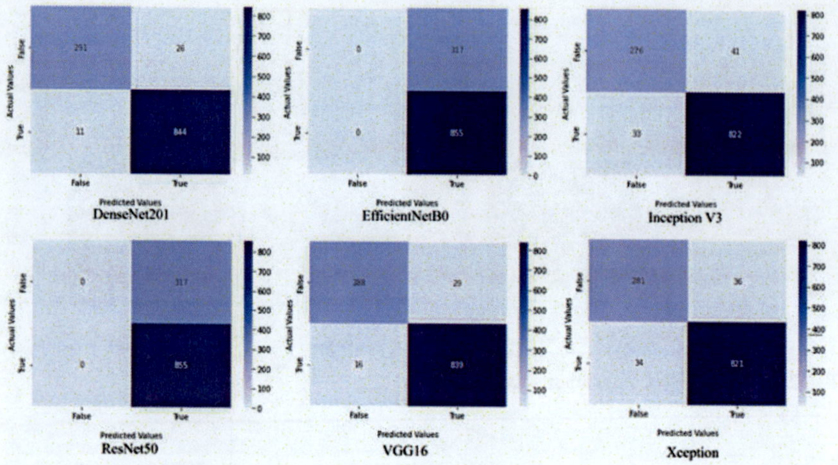

Fig. 3 Transfer learning confusion matrix before DCGAN applied

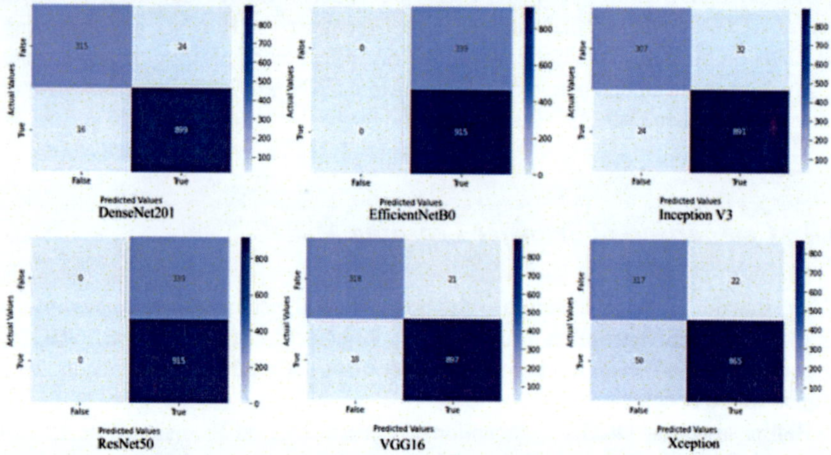

Fig. 4 Transfer learning confusion matrix after DCGAN applied

demonstrated by RestNet50 and EfficientNetB0 in terms of test accuracy and detected 915 pneumonia cases correctly. Moreover, DenseNet201 and VGG16 performed 0.96 on Precision, Recall, and F1-score. VGG16 detected 897 pneumonia cases and 318 normal cases accurately, whereas DenseNet201 detected 899 pneumonia cases and 315 normal cases correctly. Then lastly, Xception performed 0.92 Precision, 0.94 Recall, and an F1-score of 0.93, and it precisely detected 865 pneumonia cases and 317 normal cases after applying DCGAN (Fig. 4).

While DenseNet201 had the best accuracy of all the models before employing DCGAN, VGG16 achieved the highest accuracy after employing DCGAN. Except for DenseNet201, we observed that using DCGAN improved accuracy for other models. InceptionV3 and VGG16 models accuracy was increased 1.84% and 0.73% after DCGAN applied while our custom CNN showed a promising accuracy increase of 2.06%. Moreover, from the confusion matrix heatmaps, we can conclude that all CNN learning models performed better at detecting real situations applying DCGAN.

5 Conclusion and Future Works

Our purpose in this study was to produce analytical results that would allow us to compare CNN models after employing DCGAN. According to the results of the studies, custom CNN, VGG16, and DenseNet201 are the most accurate of all methods. Not only that, but we also acquired a remarkable accuracy increase in the VGG16 and custom CNN approach after fabricating our dataset with DCGAN. Therefore, this method can successfully achieve the ultimate goal of obtaining more accurate results in the X-ray image classification.

However, due to the lack of computational capacity, we might not have achieved the peak results produced by DCGAN. Therefore, in the future, we would like to collect a more extensive dataset with the most computational power possible to produce the most remarkable results. In addition, in the future work, we will experiment with various medical data images that are not limited to chest X-rays. As a result, the overall classification of medical images will be more effective.

References

1. (ACR) R (2021) Pneumonia. In: Radiologyinfo.org. https://www.radiologyinfo.org/en/info/pneumonia. Accessed 15 December 2021
2. Erickson B, Korfiatis P, Akkus Z, Kline T (2017) Machine learning for medical imaging. RadioGraphics 37:505–515. https://doi.org/10.1148/rg.2017160130
3. CNN For Image Classification | Image Classification Using CNN (2021). In: Analytics Vidhya. https://www.analyticsvidhya.com/blog/2021/01/image-classification-using-convolutional-neural-networks-a-step-by-step-guide/ . Accessed 15 Dec 2021
4. Ayan E, Unver H (2019) Diagnosis of pneumonia from chest X-ray images using Deep Learning. In: 2019 Scientific Meeting on Electrical-Electronics & Biomedical Engineering and Computer Science (EBBT). https://doi.org/10.1109/ebbt.2019.8741582
5. Srivastav D, Bajpai A, Srivastava P (2021) Improved classification for pneumonia detection using transfer learning with GAN based synthetic image augmentation. In: 2021 11th international conference on Cloud Computing, Data Science & Engineering (Confluence). https://doi.org/10.1109/confluence51648.2021.9377062
6. Rodriguez-de-la-Cruz J, Acosta-Mesa H, Mezura-Montes E (2021) Evolution of generative adversarial networks using PSO for synthesis of COVID-19 chest X-ray images. In: 2021 IEEE Congress on Evolutionary Computation (CEC). https://doi.org/10.1109/cec45853.2021.9504743
7. Brownlee J (2021) A gentle introduction to Generative Adversarial Networks (GANs). In: Machine learning mastery. https://machinelearningmastery.com/what-are-generative-adversarial-networks-gans/. Accessed 15 Dec 2021

8. Kaggle.com (2021) Chest X-ray images (pneumonia) (online). Available at: https://www.kaggle.com/paultimothymooney/chest-xray-pneumonia/. Accessed 15 Dec 2021
9. Deep Convolutional Generative Adversarial Network|TensorFlow Core (2021). In: TensorFlow. https://www.tensorflow.org/tutorials/generative/dcgan. Accessed 15 Dec 2021
10. Brownlee J (2021) A gentle introduction to generative adversarial network loss functions (online). In: Machine learning mastery. Available at: https://machinelearningmastery.com/generative-adversarial-network-loss-functions. Accessed 15 Dec 2021

A Survey on Counterfeits in the Information and Communications Technology (ICT) Supply Chain

Samar Saleh, Rong Lei, Weihong Guo, and Elsayed A. Elsayed

Abstract One of the major threats to the information and communications technology (ICT) supply chain is the introduction of counterfeit parts and components. Global efforts have been intensified to defend against counterfeiters and counterfeit products due to their detrimental impact on the economy, safety, and security. Among the extensive literature of papers, reviews, books, and articles, this review attempts to include a detailed selection of most significant research work done in the intersection of ICT, supply chains, and counterfeits to provide a reference source for researchers. Citation network and global citation scores have been used to extract and analyze papers and discuss them in different types of clusters (electronic, medical, food, and anti-counterfeiting technologies and approaches). Our review approaches the clustered papers by focusing on (1) their contribution in documenting and modeling the intrusion of counterfeit electronic parts in the ICT supply chain, (2) the proposed counterfeits' detection and avoidance techniques in the ICT supply chain, and (3) the contribution of ICT in thwarting counterfeits in medical, pharmaceutical, and food supply chains. This review provides a better understanding of the global efforts to address counterfeits in the ICT supply chain, as well as the role of ICT in thwarting counterfeits in other supply chains, which can guide future research to minimize the impact of counterfeits on supply chains.

Keyword Supply chain · Counterfeit · Countermeasures · Citation network

---

S. Saleh · R. Lei · W. Guo (✉) · E. A. Elsayed
Rutgers University-New Brunswick, Piscataway, NJ 08854, USA
e-mail: wg152@soe.rutgers.edu

S. Saleh
e-mail: shs164@scarletmail.rutgers.edu

R. Lei
e-mail: rl839@scarletmail.rutgers.edu

E. A. Elsayed
e-mail: elsayed@soe.rutgers.edu

X.-S. Yang et al. (eds.), *Proceedings of Seventh International Congress on Information and Communication Technology*, Lecture Notes in Networks and Systems 447,
https://doi.org/10.1007/978-981-19-1607-6_75

1 Introduction

The supply chain has gained a great amount of interest since it deals with the management of the entire process from the making of a product to its actual delivery to the customer. As supply chains expand globally and become more complex, managing the flow of materials, information, and data becomes more susceptible to threats that interrupt and/or impact its flow. One of the major threats to the information and communications technology (ICT) supply chain is the introduction of counterfeit parts and components, as electronics are an indispensable part of our lives [1].

A *counterfeit* is defined by the Society of Automotive Engineers (SAE International) as "a fraudulent part that has been confirmed to be a copy, imitation, or substitute that has been represented, identified, or marked as genuine, and/or altered by a source without legal right with intent to mislead, deceive, or defraud" [2]. Counterfeits can exist in non-deceptive forms where people intentionally buy counterfeit or fake items and in deceptive forms where people unknowingly buy counterfeit items believing they are genuine. Non-deceptive counterfeits lead to loss of sales and an increase in costs to control these fake products. Deceptive counterfeits that enter the supply chain during the production process could lead to low-quality products and consequently, high product recalls, lost sales, bad reputation, and even legal proceedings.

Whether deceptive or non-deceptive, counterfeit items are a real threat to the global economy, people's safety, overall security, and innovation. Threatening people's lives is the most detrimental impact of counterfeits. Hundreds of cases of deaths caused by counterfeit food or medicine have been reported. Incorporating counterfeit parts in life-supporting equipment, vehicles, aviation, etc., bears ominous consequences. The global economy is affected by the fact that some businesses lose sales or even shut down because of a bad reputation due to counterfeits; this loss was estimated to be around $323 billion in 2018. Besides, the number of jobs lost due to counterfeiting was 2.6 million in 2013 and is estimated to reach around five million by the year 2022 [3]. Moreover, counterfeiting hinders investments in countries notorious for importing or producing counterfeits. On the other side, detecting and mitigating counterfeits have become the interest of a significant proportion of investments and innovations globally. A significant amount of funds and intellectual efforts are dedicated to counterfeits instead of being dedicated to new developments.

Counterfeiting targets any item that can be produced at a lower price, and it accounts for 3.3% of world trade with no hint of declining. A study done in 2019 estimated the percentage of counterfeit parts in industries [4]. As illustrated in the pie chart in Fig. 1, footwear, clothing, and leather goods are the most hit by counterfeits. Electrical equipment goes next (12%). Amidst the other industries with lower percentages, there are critical industries like medical equipment and pharmaceuticals; counterfeited medical items directly threaten people's lives. The counterfeit industry is taking advantage of all circumstances. For instance, the COVID-19 pandemic favors the proliferation of counterfeit products such as substandard sanitizers or masks. Closed borders force authentic industries to rely on untrusted suppliers and

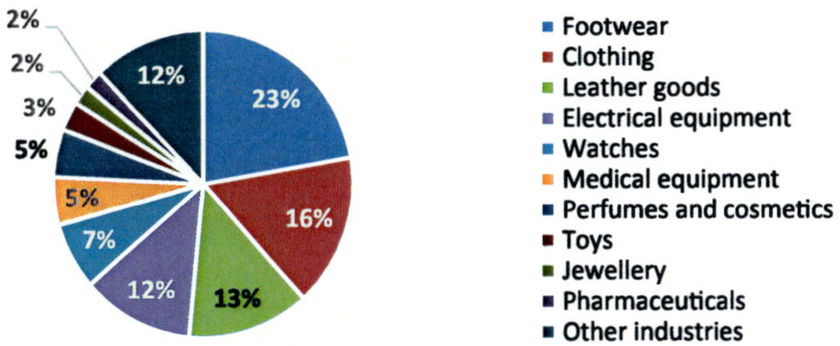

Fig. 1 Distribution of counterfeits on industries [4]

closed stores and social distancing increased online shopping, which is the most favorable venue for counterfeiting.

In addressing this problem, a significant number of detection techniques, mitigation strategies, policies, and campaigns to make people aware of the counterfeits' criticality have been made. The literature focusing on counterfeits in supply chains is vast. In our effort, we gather, connect, and analyze the work done in the intersection area of ICT, supply chains, and counterfeits to set forth an inclusive reference for future studies and enhancements in this field. This review provides a better understanding of the global efforts to address counterfeits in the ICT supply chain, as well as the role of ICT in thwarting counterfeits in other supply chains, which can guide future research to minimize the impact of counterfeits on supply chains.

The method for analyzing publications, including constructing, and visualizing bibliometric networks is described in Sect. 2. In Sect. 3, we analyze the literature in counterfeits in ICT supply chains, including how counterfeit parts are documented and modeled, as well as the related anti-counterfeiting approaches. In Sect. 4, we analyze the literature in ICT-related anti-counterfeiting technologies and approaches (countermeasures) in thwarting counterfeits in supply chains broader than just electronics. Section 5 concludes the review with a discussion of future research directions.

2 Methodology

Searching through Elsevier's Scopus citation database covering over 42,000 titles from approximately 11,680 publishers, we obtain over 900 articles using the keywords "supply chain" and "counterfeit." Scopus is chosen because it covers a big number of articles, and we can obtain bibliographic database files that can be used to construct and visualize bibliometric networks. We analyze this big data by creating a map based on the author keywords (i.e., keywords given by the authors). The co-occurrence of keywords in the network reveals a specific theme or trend in research that we need to identify [5]. Figure 2 shows the keywords co-occurrence network developed in VOSviewer [6], where a bigger circle indicates that the keyword appears

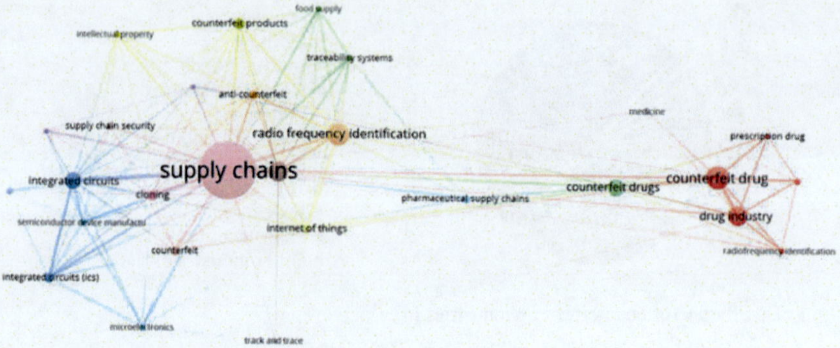

Fig. 2 Keyword co-occurrence network

more frequently in the publication set, the different colors represent the different clusters, these keywords are assigned to, and the lines indicate co-occurrence of the keywords.

According to the size of the circles, their colors, and their connections in Fig. 2, we identify four interconnected clusters of research areas in the literature tackling counterfeits in supply chains:

(1) Area One: Counterfeits in electronic parts in ICT supply chains
(2) Area Two: Counterfeits in medical and pharmaceutical supply chains
(3) Area Three: Counterfeits in food supply chains
(4) Area Four: Anti-counterfeiting technologies and approaches (countermeasures)

In addition to the keyword co-occurrence network, we perform citation analysis by developing a citation network in VOSviewer. The citation network reveals "interconnected" literature and "isolated" literature. Interconnected literature is identified as papers that add value to the studied topic, and it shows the trend in the research done. Figure 3 shows the citation network obtained using VOSviewer, where the size of the nodes illustrates the number of citations of the paper. The nodes in gray represent the isolated papers while nodes in the center represent the interconnected papers. Figure 4 shows a closer view of the interconnected citation network containing 131 references. The network shows the connection between literature and its clustering into four main clusters or areas. These areas coincide with the clusters identified from the author keyword network. Additionally, we notice that some gray nodes in Fig. 3 are relatively large, indicating high citations, but they are not directly linked to the four interconnected clusters. This suggests that although these isolated papers do not share common citations with the papers in the interconnected citation network, they still have significant impact in addressing counterfeits in supply chains. Therefore, the ten most cited isolated papers according to their global citation scores are matched to the preceding four areas and included in our survey. Furthermore, additional papers, beyond Scopus, selected by experts with elaborate research work on counterfeiting are incorporated into our survey.

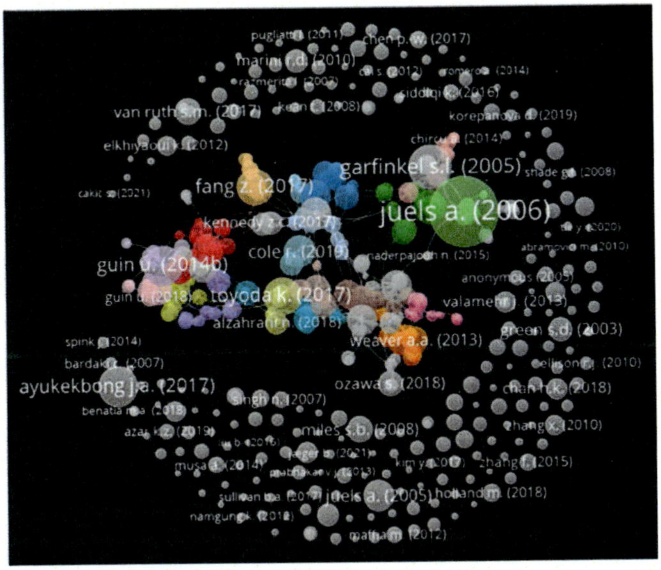

Fig. 3 Citation network for all papers in the publication set

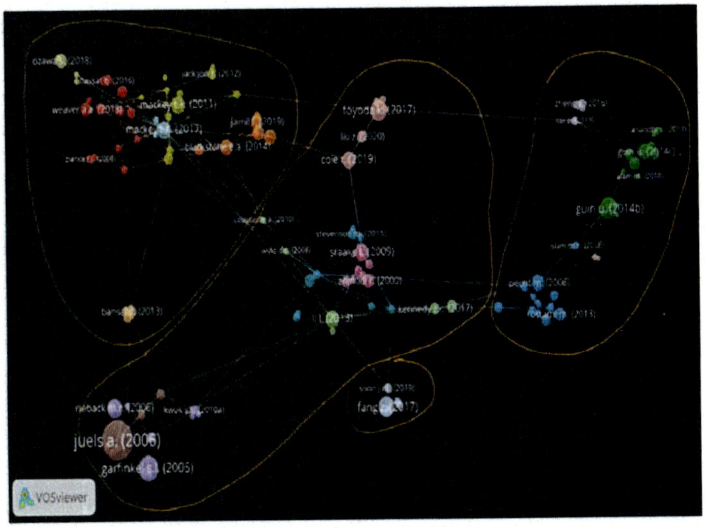

Fig. 4 Citation network for the interconnected papers in the publication set

In this paper, literature identified in the four areas is organized according to their relationship to ICT. Literature in area (1) is reviewed in Sect. 3, including how counterfeit electronic parts in the ICT supply chain are documented and modeled, as well as the detection and avoidance techniques in the ICT supply chain. Literature in areas (2), (3), and (4) is reviewed in Sect. 4, with a focus on how ICT is used in thwarting counterfeits in supply chains beyond electronics. For areas (2) and (3), we focus on the role of ICT in thwarting counterfeits in medical and pharmaceutical supply chains and food supply chains, rather than how the counterfeit drugs or food are introduced.

3 Counterfeits in Information and Communications Technology (ICT) Supply Chains

The increase of counterfeiting in electrical and electronic systems and components has been on the rise recently due to the production shift to less law-enforcing countries, online shopping, and more sophisticated counterfeiting techniques [7]. Counterfeit electronics in the defense and security supply chains represent one of the most critical threats to security and safety [8].

3.1 Documenting and Modeling Counterfeits

The increasing concern about counterfeit parts, especially in the ICT supply chain, has given rise to a robust literature that includes reports from components of the Department of Defense (DoD), NASA, and other government agencies, as well as from Lockheed Martin, IBM, the Aerospace Industries Association, and other companies/organizations in the private sector. Indeed, the presence of counterfeit and other unacceptable products have led to the formation of Government-Industry Data Exchange Program (GIDEP) [9] as a cooperative activity between government and industry participants seeking to reduce or eliminate expenditures of resources by sharing technical information essential during research, design, development, production, and operational phases of the life cycle of systems, facilities, and equipment and report any counterfeit products. Likewise, Electronic Resellers Association International (ERAI) was founded in 1985 as a major resource for checking if a component is counterfeit [10]. It is the world's largest database of suspect counterfeit and nonconforming electronic parts. This tool, alone, has fully changed the way in which sectors of the supply chain research, track, identify, purchase, and sell material. It allows members to better mitigate risks in their procurement process, especially when dealing with end-of-life or obsolete parts. These two sources in addition to the DoD trusted suppliers list constitute the first "line of defense" to check for counterfeits parts or components.

The literature also describes best practices for government and industry. For instance, the Navy's Counterfeit Materiel Process Guidebook [11] aims to "equip DON [Department of the Navy] activities with a practical tool for implementing a risk-based counterfeit materiel prevention program." Wix [12] recommends clearly defining company-wide counterfeit risk mitigation strategies. Among those strategies are to prohibit sourcing electronic components from independent distributors, reducing number of vendors, requiring distributors to use third party test houses or conform to inspection standards, and to develop measurable quality criteria for distributors. Szakal and Pearsall [13] discuss the challenges of increasing reliance on commercial-off-the-shelf ICT components and describe a framework to mitigate the risk.

There is not much literature that looks at counterfeiting theoretically. Bodner [14] lays out a modeling framework to understand the counterfeiting problem, based on "the exogenous environment, policy, enterprise actors, supply chain flows, and system/constituent behavior," and presents a prototype agent-based simulation that implements the framework. Stevenson and Busby [15] describe a theoretical account of counterfeiting based on signaling theory that what counterfeiters' strategies aim to achieve often involves the generation, suppression, or exploitation of signals.

3.2 *Anti-counterfeiting Approaches*

The Society of Automotive Engineers developed a standardized process for detecting counterfeits. The semi-quantitative risk assessment method categorizes components into levels where each level has appropriate types of laboratory testing [16]. Clearly, there is a consensus on the need for a comprehensive approach utilizing technological solutions and decision-making for counterfeits detection and avoidance in the ICT supply chain [17]. Rostami et al. [18] summarize that the state-of-the-art defenses proposed for counterfeits in ICT supply chains are testing, using aging sensors that detect recycled products, and adopting proactive techniques like hardware metering, fingerprinting, and watermarking. Chatterjee and Das [19] state that a major contribution toward combatting counterfeits is done by the parts manufacturers in building trust in the supply chain, and they provide a set of measures for manufacturers to increase trust.

One focus in the ICT supply chain is counterfeit integrated circuits (ICs). Guin et al. [20, 21] provide a comprehensive background of the types of counterfeit ICs available, the testing effectiveness for detecting each type of counterfeit, and the available counterfeit avoidance techniques. More detailed information on the aforementioned topics can be found in the "counterfeit integrated circuits: detection and avoidance" book [22]. Guin et al. [23] also provide a thorough overview of the types of counterfeits in general and the available detection and avoidance methods. They also present a selecting algorithm that enables choosing the optimum method considering test time, cost, and application risks. Alam et al. [24] propose a low-cost and highly-accurate method to detect counterfeit ICs using a ring oscillator (RO) and a nonvolatile memory where the digital signature and the corresponding RO frequency and conditions are stored. A similar attempt is used to detect counterfeit field-programmable gate arrays (FPGA) and ICs, in which the detection is accompanied with an aging analysis to detect recycled or out-of-specification ICs by their performance degradation [25, 26]. Ghosh and Chakraborty [27] propose an enhanced image texture analysis to detect counterfeit ICs. Their method shows high accuracy even when images are taken using ordinary digital cameras. Frazier et al. [28] present a new (near real-time) counterfeit detection process, which is based upon infrared thermal imaging, intensive statistical analysis, and machine learning, to differentiate between authentic and inauthentic electronic parts.

Lowering the cost of counterfeit detection can help widen the adoption of the detection techniques. Huang et al. [29] introduce a low-cost support vector machine to classify the authenticity of parts. This technique is useful in cases of recycled counterfeits. Similarly, Kumari et al. [30] provide a low-cost detection method for recycled memory chips which are incorporated in many electronic systems. The detection method depends on studying the chips' timing characteristics, which are sensitive to high usage, program, and read time [30, 31].

A lot of the existing counterfeit detection approaches depend on human decisions and consequently are vulnerable to error. To improve the accuracy in counterfeit detection, there is ongoing interest in automating the detection by leveraging the strength of data analytics, machine learning, and artificial intelligence. Ahmadi et al. [32] attempt to explore an automatic detection method using image processing techniques and machine learning algorithms. Zheng et al. [33] propose a method utilizing clock phase sweep, with low design and hardware requirements, to identify counterfeit FPGA. Experimental results show 99% accuracy of the proposed method. Zheng et al. [34, 35] claim that detection methods based on aging sensors require more design verification efforts and do not apply to legacy chips. They attempt to detect recycled chips using a more comprehensive approach. This approach combines exploiting differential aging to isolate aged chips under large die process variations and comparing transient current testing results between adjacent similar circuit structures in a chip. An added value to the methods proposed in Refs. [24–26, 34, 35] is that they do not require perfectly functioning chips as a reference that accounts for high-test complexity and cost.

The most common counterfeit avoidance techniques are secure split-test (SST) and physically unclonable function (PUF). Contreras et al. [36] emphasize the SST's effectiveness in securing ICs. Ben Dodo et al. [37] explain how spin transfer torque-magnetic random-access memory technology can make the existing PUF more secure and less vulnerable to tampering. A simulation model proves that this technique detects all tampered parts. Counterfeit avoidance techniques develop continuously. Chakraborty et al. [38] propose logic locking as a solution to thwart piracy and counterfeiting using a "keying" mechanism. Basak et al. [39] propose a similar mechanism by locking ICs using antifuse devices in input/output circuitry.

Traceability, via blockchain, and compliance verification using suitable testing methods and/or embedded PUF in conjunction with risk management are proposed as the major tools to identify the authenticity of products by tracking, tracing, and analyzing parts during the entire life cycle [40–46]. Livingston [47] suggests that test methods in government and industry designed to verify integrity and performance of authentic parts are not as important as traceability programs to help detect counterfeits. He also suggests that documentations of certifications of conformance or test reports accompanying parts be investigated since they are sometimes not authentic.

Traceability can be enabled using RFID [48–50]. The effectiveness of RFID in protecting the supply chain from counterfeiters is modeled by Yang et al. [48] in an IoT supply chain using a printed circuit board prototype [49, 51]. Although experimental results obtained are promising and the cost of the technique proposed in Anandhi et al. [50] is low because the majority of the components needed already exist in modern IoT supply chains, experimental analysis shows that the authentication protocol relying on an asymmetric key cryptosystem, and one-way hash function is computationally expensive and cannot handle big data efficiently. Dimase et al. [52] note that an appropriate level of traceability must be implemented based on risk prioritization because the cost of implementing traceability at a wide range might outweigh its benefits.

4 ICT-Related Anti-counterfeiting Technologies and Approaches in Other Supply Chains

4.1 ICT-Related Anti-counterfeiting Technologies and Approaches (Countermeasures)

In this section, we take a closer look at the research focusing on the ICT-related anti-counterfeiting approaches themselves regardless of the type of the product and the supply chain. Li [53] categorizes and briefly describes all available technologies to combat counterfeits in supply chains, until 2013, into two types: 1) product authentication verification and 2) product tracking and tracing.

The Defense Logistics Agency [54] suggests different kinds of tests, including electrical testing, x-ray testing, and microscopic exam. Gansler et al. [55], in addressing the DOD supply chain, suggest stronger quality assurance standards tied to a risk-based approach to counterfeit mitigation. They also recommend stronger preventive measures. Among preventive measures suggested are tamper-proof packaging and x-ray inspection, debarring suppliers who repeatedly provide components with counterfeit parts, and providing penalties for suppliers not reporting suspect counterfeits. Rogers and O'Donnell [56] point out that the defense industry "routinely failed to report cases of suspect counterfeit parts," and Livingston [47] makes a similar observation. For reporting counterfeit parts, Gansler et al. [55] recommend using GIDEP as does Livingston. The Aerospace Industries Association [57] recommends developing a program of limited liability for those accurately reporting counterfeit parts using GIDEP.

Lockheed Martin [58] describes what makes a good counterfeit prevention plan. One of the ideas described is to allow customers to review what suppliers' processes are without disclosing proprietary information.

Digital watermarking is a growing approach to defend against counterfeiting. For example, DARPA is working on countermeasures that involve marking in an electronic component that can enable the authentication of genuine devices [59]. In applications other than ICT, Lingle [60] discusses digital watermarks on packaging and products for cosmetic and personal care products. The invisible digital watermarks can be viewed with a special app that will then authenticate the product. The Digital Watermarking Alliance [61] discusses how the watermark on packaging or an object can be used to uniquely identify specific items and also carry other information such as lot number and intended destination, and points out that watermarks can be encrypted so only authorized devices can access the data. In digital manufacturing, Chan et al. [62] propose the use of watermarking technology to protect the intellectual property of 3D printers' content. In pharmaceutical applications, watermarking is explored as a tool in producing tamper-resistant prescription forms in an attempt to prevent fraudulent prescriptions for controlled substances [63]. However, watermarking of physical components is also used in the pharmaceutical industry to make it possible to follow a pill bottle (and perhaps eventually an individual pill) from its inception in a plant to its final destination.

RFIDs are suggested as one of the most suitable solutions to track items in a supply chain especially with the decline in their cost in the past few years [64–66] and their capability to integrate with mobile technologies to obtain a self-validated location-based authentication system [67] and data processing and synchronization algorithm [68]. Besides, the development of RFID drives authorities to encourage the use of RFID. For instance, federal agencies and retailers are influencing the adoption of RFID in the pharmaceutical supply chain [69]. Azuara et al. [66] show the efficiency of RFID-enabled systems in counterfeit detection applied only at the manufacturing stage of the supply chain. RFIDs can also be used on unstable and non-uniform surfaces like textiles [70]. Security and privacy concerns are discussed, and proposed solutions are found in Refs. [71–75]. Cai et al. [76] initiate a secure and flexible protocol that enables each supply chain party to securely update tag keys and consequently ensure a safe transfer of RFID tags in the supply chain. However, this protocol is not financially justified for all types of products. Kumar et al. [77] suggest that the use of RFID in the drug supply chain has an important advantage of managing the reverse logistic process besides reducing the risk of counterfeits.

Electronic product code (EPC) tags are a form of RFID but highly vulnerable to cloning and counterfeit attacks. Juels [78] attempts to strengthen the EPC using personal identification number (PIN)-based access-control and privacy enhancement mechanisms. Miles et al. [79] provide comprehensive coverage of information related to RFIDs including their application in anti-counterfeiting.

Singh and Li [80] highlight the importance of trust in an RFID-enabled supply chain and provide an RFID trust framework. Ting and Tsang [81] develop a tool that can identify counterfeit sources in a supply chain by analyzing the relationship between people. They depend on social network analysis in characterizing certain features that identify the relationship.

Quick-response (QR) code linked to blockchain databases have been used in avoiding counterfeit medications [82]. Using blockchain is extensively promoted to avoid counterfeit infiltration into a supply chain. Indeed, blockchains also bring improved quality, enhanced inventory, reduced cost of supply chain transaction, etc. [83]. Pun et al. [84] explicitly describe the effectiveness of adopting blockchain to combat counterfeits in a supply chain and encourage governments to provide subsidies for this technology. Liu and Li [85] propose a blockchain-based framework for e-commerce, which shows to be effective in protecting the supply chain against clone attacks, counterfeit tag attacks, and counterfeit product attacks. Kennedy et al. [86] propose the use of lanthanide nanomaterial in 3D printed parts and link the obtained unique chemical signature into a blockchain database. Smith and Skrabalak [87] review the use of metal nanomaterials in developing optical anti-counterfeit labels. Toyoda et al. [88] attempt to secure RFID-attached products from any tampering in the post-supply chain. They use Bitcoin's blockchain idea to allow customers to identify the genuineness of the item if the seller has the ownership. The proof-of-concept and cost performance are evaluated experimentally. Hepp et al. [89] identify the limitations of adopting blockchain technology for tracking physical assets. They suggest using PUF instead of RFID, which is susceptible to cloning, and using OriginStamp system that can aggregate events, reducing by that the number of transactions instead of logging each event on the blockchain individually.

In their attempt to understand how IoT technologies enable and constrain the actors' control capabilities, Boos et al. [90] use IoT applications for counterfeit detection in supply chains. They discuss how accountability (visibility, responsibility, and liability) and control (transparency, predictability, and influence) are affected by the IoT technologies' range to inform, automate, and transform.

4.2 The Role of ICT in Thwarting Counterfeits in Medical and Pharmaceutical Supply Chains

Counterfeits in the medical and pharmaceutical supply chain pose a huge threat with significant consequences for global health and patient safety. Although only 7% of the counterfeiting actions (as shown in Fig. 1) target medical and pharmaceutical products, counterfeit medication adds another 15% of the medical and pharmaceutical supply chain and is a threat to human lives. Moreover, it costs the legitimate pharmaceutical industry between $37.6 billion and $162.1 billion with 57,500–247,800 lost jobs in the U.S alone [91]. Refs. [92–97] investigate how counterfeit drugs are introduced and their impact on health and economics. Law enforcement, strict surveillance, and awareness are the most common solutions for counterfeit medicines [98–105].

An overview of the projects done by key organizations to halt the proliferation of counterfeit medicines is provided in Nayyar et al. [106], in which they also provide recommendations regarding technologies, communication, and laws to increase pharmaceutical governance. Chaudhry and Stumpf [107] emphasize the importance of the counterfeit detection device #3 or CD3 [108] developed by the U.S. Food and Drug Administration (FDA) that is inexpensive and enables inspectors to identify counterfeit drugs nationally and at border entry. By emitting light in ten different wavelengths, the device scans drug samples and compares them against authentic drugs in its memory. It is also capable of checking tampered packaging. Mackey and Liang [109, 110] propose a global policy framework to enable cooperation and coordination to combat counterfeit drugs. Davison [111] provides a thorough review on combatting counterfeit medicines. It discusses regulations, authentication strategies (packaging, analytical techniques), product tracking, and case studies from around the world.

Hamilton et al. [112] propose a combination of countermeasures at the different levels of the pharmaceutical supply chain; the main suggested countermeasures are global monitoring, pharmacovigilance, pharmacists training, customer awareness, and the adoption of testing technologies convenient to low-resource settings in addition to the emerging consumer verification techniques such as mobile authentication services (MAS). They suggest simple, inexpensive MAS, such as having a hidden barcode that when scratched and texted to a secure hotline a confirmation of the medication genuineness is received. Fadlallah et al. [113] express concern that authentication systems, such as RFID, are effective in the dispensing phase only and are challenging for less developed countries because they require an infrastructure connecting all pharmacies which requires time, effort, and commitment. Cohn et al. [114] investigate a successful response to falsified medications in Nairobi using a transparent quality assurance system where procurement parties, manufacturers, and other stakeholders are instantly notified about falsified medications, followed by immediate testing and recall processes. Cuomo and Mackey [115] and Mackey et al. [116] propose a surveillance mechanism using statistical analysis and geospatial modeling to identify the distribution of counterfeit cancer medication in the USA. This model can play an important role in predicting future counterfeit medicine incidents.

It is clear that there is a tendency of increased counterfeit drug incidents in low- and middle-income countries because of low production costs and weak law governance. Regulations for drug donation and safe drug disposals are especially important in poor countries which have stockpiles of donated drugs [113, 117, 118]. Countermeasures that reduce the impact of counterfeits include ensuring the authenticity of the drug, stock control, and awareness by caregivers. In some instances, these measures are shown to reduce the economic impact and number of deaths by about 40% when simulated using an agent-based model [119]. Low-cost tests to identify falsified drugs are also used to ensure the authentication of medications [120, 121]. Marini et al. [122] build a low-cost detection method for counterfeit medicines which is equipped with a deep ultraviolet radiation detector. To study the accuracy of the prototype, they perform a full validation and method comparison study with other conventional detection methods, and the obtained results are promising.

Other research asserts the importance of monitoring technologies in applying the laws and regulations [123]. Lybecker [124] presents a theoretical model to characterize the implications of these technologies on counterfeiters. Mackey and Nayyar [125] present a review of all digital technologies that protect the supply chain from counterfeit medications and the technologies evolving in preventing the sale of counterfeit medications. Taylor [126] describes the emergence of radio-frequency identification (RFID) into the pharmaceutical barriers and stresses the fact that what RFID brings to patients and brands justifies its technological and investment requirements. Chen et al. [127] describe the use of the quantitative radio-frequency spectroscopic technique for a safer pharmaceutical supply chain, while Kwok et al. [128] construct a prototype to prove RFIDs' effectiveness. Trenfield et al. [82] propose a novel anti-counterfeit method to track 3D-printed medicines by adding a combination of material inks for detection using Raman spectroscopy. Similarly, Cozzella et al. [129] explain the use of white-light speckle theory, which is a speckle visible under ultraviolet fluorescence light, as a fingerprint for drug packages.

Adding to the above-mentioned descriptive papers, Raj et al. [130] explain the use of blockchain technology to increase visibility and traceability and control counterfeit medications, while Kumar et al. [131] suggest that the use of smart contracts along with blockchains for the drug supply chain increases the trust between stakeholders, automatic payments, and quality control. Meyliana et al. [132] propose the use of blockchain with smart contract in supply chain management to comply with the good manufacturing practices (GMP) regulation set by the Indonesian government. Alzahrani and Bulusu [133] propose a decentralized anti-counterfeiting supply chain using blockchain and near field communication technologies (NFC) to detect any modification attack and track products, introducing a new consensus protocol as well utilizing a small number of validators while maintaining a high level of security. Internet of things (IoT) and blockchains are envisioned in managing supply information across healthcare supply chain processes which allow better management for recalls, expiration, shortages, and counterfeits [134, 135]. Sylim et al. [136] develop a pharmaco-surveillance blockchain system prototype running on smart contracts. The use of a connector module connecting supply chain echelons with blockchain is simulated; the results prove increased collaboration, trust, and system performance measured by fill rate [137]. Jamil et al. [138] propose a system for secure drug supply chain records usage which is handled and conducted by Hyperledger Fabric based on blockchain. Moreover, a limited-access to patients' drug and health records is maintained by a smart contract. Experimental analysis validates the usability and efficiency of the proposed system. Kumar and Tripathi [139] propose a blockchain-based framework to enhance drug security and authenticity of manufacturers. Their methodology is based on digital signatures which are provided by the certificate authority following the public key infrastructure protocol. This is claimed to prevent replay and man-in-middle attacks. Global Governance Blockchain is suggested to address the counterfeiting problem and allows surveillance by every participant involved in the supply chain [140].

4.3 *The Role of ICT in Thwarting Counterfeits in Food Supply Chains*

All types of food products are susceptible to counterfeiting. Traceability is considered to be the best anti-counterfeiting technique for ensuring food quality. Shahbazi and Byun [141] propose a blockchain machine learning traceability system addressing the shelf life, weight, evaporation, warehouse transactions, and shipping time of perishable food which are sensitive due to discrepancy and deterioration. Tsang et al. [142] propose a traceability system by integrating blockchain, IoT technology, and fuzzy logic. Blockchain and IoT ensure products traceability and avoidance of counterfeits whereas fuzzy logic is used to evaluate quality decay.

Soon and Manning [143] use Scotch whisky as a case study for counterfeit in the supply chain and highlight the effectiveness of smart packaging in detecting counterfeits. They propose overt smart packaging technologies such as barcodes, RFID, or watermarks and covert ones such as intaglio printing, security threads, and fluorescence artifacts. Besides, they stress the effectiveness of collaboration between all members of the value chain in detecting counterfeiters. Smart packaging can also be combined with antimicrobial and antioxidant material, time–temperature indicators, freshness indicators, gas concentration indicators, etc., to promote microbial safety and longer shelf life [144, 145].

It is clear that the progress in combatting counterfeiting in the food industry is weak as compared to the other sectors since the cost of implementing critical countermeasures is significantly high compared to the product price.

5 Conclusion

In this study, we review a detailed selection of significant research work done in the intersection of ICT, supply chains, and counterfeits. The aim of the review is to provide a comprehensive reference for counterfeit detection and avoidance techniques, methods, and approaches proposed until now in the context of ICT supply chains and beyond. Using clustering and citation network analysis, we identified four main clusters of relevant work done in this topic: (1) counterfeits in electronic parts in ICT supply chains, (2) counterfeits in medical and pharmaceutical supply chains, (3) counterfeits in food supply chains, and (4) anti-counterfeiting technologies and approaches. We analyze the clustered research based on how ICT is incorporated to defeat counterfeits in ICT supply chains and other threatened supply chains. This review provides a better understanding of the global efforts to address counterfeits in the ICT supply chain, as well as the role of ICT in thwarting counterfeits in other supply chains, which can guide future research to minimize the impact of counterfeits on supply chains.

This review reveals the trend of using RFID and blockchains in avoiding counterfeits in all types of supply chains. It is clear that among all the proposed solutions to thwart counterfeits, no single countermeasure can be generalized for all clusters. This is due to the continuous development of counterfeiters' techniques and also the vulnerabilities in the solutions proposed. The best solution for a supply chain should be customized from the set of available solutions and strengthening methods suggested based on the supply chain structure and expansion, cost, and local and international regulations.

Acknowledgements This material is based upon work supported by the U.S. Department of Homeland Security under Grant Award Number 17STQAC00001-05-00.

Disclaimer The views and conclusions contained in this document are those of the authors and should not be interpreted as necessarily representing the official policies, either expressed or implied, of the U.S. Department of Homeland Security. The authors thank Fred S. Roberts for his helpful input in revising this paper.

References

1. CISA Working Group 2 (2021) Information and Communications Technology Supply Chain Risk Management Task Force Threat Evaluation Working Group: Threat Scenarios Version 2.0: https://www.cisa.gov/sites/default/files/publications/ict-scrm-task-force-threat-scenarios-report-v2.pdf
2. SAE International (2019) SAE AS5553 Counterfeit Electrical, Electronic, and Electromechanical (EEE) Parts; Avoidance, Detection, Mitigation, and Disposition Standards
3. Frontier Economics Ltd (2017) The Economic Costs of Counterfeiting and Piracy
4. OECD, European Union Intellectual Property Office (2019) Trends in trade in counterfeit and pirated goods. Illicit Trade, OECD Publishing, Paris
5. Falagas ME, Pitsouni EI, Malietzis GA, Pappas G (2008) Comparison of PubMed, Scopus, Web of Science, and Google Scholar: strengths and weaknesses. FASEB J 22(2):338–342
6. Van Eck NJ, Waltman L (2010) VOSviewer: visualizing scientific landscapes [software]. https://www.vosviewer.com
7. Pecht M, Tiku S (2006) Bogus: electronic manufacturing and consumers confront a rising tide of counterfeit electronics. IEEE Spectr 43(5):37–46
8. Stradley J, Karraker D (2006) The electronic part supply chain and risks of counterfeit parts in defense applications. IEEE Trans Compon Packag Technol 29(3):703–705
9. GIDEP. https://www.gidep.org/about/about.htm
10. ERAI. https://www.erai.com/aboutus_profile
11. Office of the Assistant Secretary of the Navy (2017) Counterfeit Material Process Guidebook: Guidelines for Mitigating the Risk of Counterfeit Materiel in the Supply Chain. NAVSO P-7000
12. Wix SD (2017) Suspect/Counterfeit Electronics Overview. Sandia National Lab.(SNL-NM), Albuquerque, NM (United States)
13. Szakal A, Pearsall K (2014) Open industry standards for mitigating risks to global supply chains. IBM J Res Dev 58(1):1–13
14. Bodner DA (2014) Enterprise modeling framework for counterfeit parts in defense systems. Procedia Computer Science 36:425–431
15. Stevenson M, Busby J (2015) An exploratory analysis of counterfeiting strategies. Int J Oper Prod Manag 35(1):110–144
16. Collier ZA, Linkov I, Keisler JM, Walters S, DiMase D (2014) A semi-quantitative risk assessment standard for counterfeit electronics detection. SAE Int J Aerosp 7(1):171–181

17. Lambert JH, Keisler JM, Wheeler WE, Collier ZA, Linkov I (2013) Multiscale approach to the security of hardware supply chains for energy systems. Environ Syst Decisions 33(3):326–334
18. Rostami M, Koushanfar F, Rajendran J, Karri R (2013) Hardware security: threat models and metrics. in 2013 IEEE/ACM International Conference on Computer-Aided Design (ICCAD). IEEE. pp 819–823
19. Chatterjee K, Das D (2007) Semiconductor manufacturers' efforts to improve trust in the electronic part supply chain. IEEE Trans Compon Packag Technol 30(3):547–549
20. Guin U, Huang K, Dimase D, Carulli JM, Tehranipoor M, Makris Y (2014) Counterfeit integrated circuits: a rising threat in the global semiconductor supply chain. Proc IEEE 102(8):1207–1228
21. Guin U, Dimase D, Tehranipoor M (2014) Counterfeit integrated circuits: detection, avoidance, and the challenges ahead. J Electron Test 30(1):9–23
22. Tehranipoor M, Guin U, Forte D (2015) Counterfeit integrated circuits: detection and avoidance
23. Guin U, Dimase D, Tehranipoor M (2014) A comprehensive framework for counterfeit defect coverage analysis and detection assessment. J Electron Test 30(1):25–40
24. Alam M, Chowdhury S, Tehranipoor MM, Guin U (2018) Robust, low-cost, and accurate detection of recycled ICs using digital signatures. in 2018 IEEE International Symposium on Hardware Oriented Security and Trust (HOST). pp 209–214
25. Dogan H, Forte D, Tehranipoor MM (2014) Aging analysis for recycled FPGA detection. in 2014 IEEE International Symposium on Defect and Fault Tolerance in VLSI and Nanotechnology Systems (DFT). pp 171–176
26. Guo Z, Xu X, Rahman MT, Tehranipoor MM, Forte D (2018) SCARe: an SRAM-based countermeasure against IC recycling. IEEE Transactions on Very Large Scale Integration (VLSI) Systems, 26(4): 744–755
27. Ghosh P, Chakraborty RS (2017) Counterfeit IC detection by image texture analysis. in 2017 Euromicro Conference on Digital System Design (DSD). pp 283–286
28. Frazier PD, Gilmore ET, Collins IJ, Samotshozo WE, Chouikha MF (2018) A novel counterfeit detection approach for integrated circuit supply chain assurance. Journal of Hardware and Systems Security 2(3):240–250
29. Huang K, Carulli JM, Makris Y (2013) Counterfeit electronics: a rising threat in the semiconductor manufacturing industry. in 2013 IEEE International Test Conference (ITC). pp 1–4
30. Kumari P, Talukder BMSB, Sakib S, Ray B, Rahman MT (2018) Independent detection of recycled flash memory: challenges and solutions. in 2018 IEEE International Symposium on Hardware Oriented Security and Trust (HOST). pp 89–95
31. Sakib S, Kumari P, Talukder B, Rahman M, Ray B (2018) Non-invasive detection method for recycled flash memory using timing characteristics. Cryptography 2(3):17
32. Ahmadi B, Javidi B, Shahbazmohamadi S (2018) Automated detection of counterfeit ICs using machine learning. Microelectron Reliab 88–90:371–377
33. Zheng Y, Wang X, Bhunia S (2015) SACCI: scan-based characterization through clock phase sweep for counterfeit chip detection. IEEE Transactions on Very Large Scale Integration (VLSI) Systems, 23(5): 831–841
34. Zheng Y, Basak A, Bhunia S (2014) CACI: dynamic current analysis towards robust recycled chip identification. in Proceedings of the 51st Annual Design Automation Conference. San Francisco, CA, USA: Association for Computing Machinery. pp 1–6
35. Zheng Y, Yang S, Bhunia S (2016) SeMIA: self-similarity-based IC integrity analysis. IEEE Trans Comput Aided Des Integr Circuits Syst 35(1):37–48
36. Contreras GK, Rahman MT, Tehranipoor M (2013) Secure Split-Test for preventing IC piracy by untrusted foundry and assembly. in 2013 IEEE International Symposium on Defect and Fault Tolerance in VLSI and Nanotechnology Systems (DFTS). pp 196–203
37. Ben Dodo S, Bishnoi R, Mohanachandran Nair S, Tahoori MB (2019) A spintronics memory PUF for resilience against cloning counterfeit. IEEE Transactions on Very Large Scale Integration (VLSI) Systems, 27(11): 2511–2522

38. Chakraborty A, Jayasankaran NG, Liu Y, Rajendran J, Sinanoglu O, Srivastava A, Xie Y, Yasin M, Zuzak M (2020) Keynote: a disquisition on logic locking. IEEE Trans Comput Aided Des Integr Circuits Syst 39(10):1952–1972
39. Basak A, Zheng Y, Bhunia S (2014) Active defense against counterfeiting attacks through robust antifuse-based on-chip locks. in 2014 IEEE 32nd VLSI Test Symposium (VTS). pp 1–6
40. Livingston H (2007) Avoiding counterfeit electronic components. IEEE Trans Compon Packag Technol 30(1):187–189
41. Islam MN, Patii VC, Kundu S (2018) On IC traceability via blockchain. in 2018 International Symposium on VLSI Design, Automation and Test (VLSI-DAT). IEEE
42. Skudlarek JP, Katsioulas T, Chen M (2016) A platform solution for secure supply-chain and chip life-cycle management. Computer 49(8):28–34
43. Islam MN, Kundu S (2019) Enabling IC traceability via blockchain pegged to embedded PUF. ACM Transactions on Design Automation of Electronic Systems 24(3):1–23
44. Guin U, Cui P, Skjellum A (2018) Ensuring proof-of-authenticity of IoT edge devices using blockchain technology. in 2018 IEEE International Conference on Internet of Things (iThings) and IEEE Green Computing and Communications (GreenCom) and IEEE Cyber, Physical and Social Computing (CPSCom) and IEEE Smart Data (SmartData). IEEE. pp 1042–1049
45. Cui P, Dixon J, Guin U, Dimase D (2019) A blockchain-based framework for supply chain provenance. IEEE Access 7:157113–157125
46. Negka L, Gketsios G, Anagnostopoulos NA, Spathoulas G, Kakarountas A, Katzenbeisser S (2019) Employing blockchain and physical unclonable functions for counterfeit IoT devices detection. in Proceedings of the International Conference on Omni-Layer Intelligent Systems. Crete, Greece: Association for Computing Machinery. pp 172–178
47. Livingston H (2010) Securing the DOD supply chain from the risks of counterfeit electronic components. BAE Systems
48. Yang K, Forte D, Tehranipoor M (2018) ReSC. ACM Transactions on Design Automation of Electronic Systems, 23(3): 1–27
49. Yang K, Forte D, Tehranipoor M (2015) An RFID-based technology for electronic component and system counterfeit detection and traceability. in 2015 IEEE International Symposium on Technologies for Homeland Security (HST). IEEE. pp 1–6
50. Anandhi S, Anitha R, Sureshkumar V (2019) IoT enabled RFID authentication and secure object tracking system for smart logistics. Wireless Pers Commun 104(2):543–560
51. Yang K, Forte D, Tehranipoor MM (2017) CDTA: a comprehensive solution for counterfeit detection, traceability, and authentication in the IoT supply chain. ACM Transactions on Design Automation of Electronic Systems (TODAES), 22(3): Article 42
52. Dimase D, Collier ZA, Carlson J, Gray RB, Linkov I (2016) Traceability and risk analysis strategies for addressing counterfeit electronics in supply chains for complex systems. Risk Anal 36(10):1834–1843
53. Li L (2013) Technology designed to combat fakes in the global supply chain. Bus Horiz 56(2):167–177
54. Metz C (2012) Defense Logistics Agency, America's Combat Logistics Support Agency Counterfeit Items Detection and Prevention, DLA J-334
55. Gansler JS, Lucyshyn W, Rigilano J (2014) Addressing counterfeit parts in the DOD supply chain, Center for Public Policy and Private Enterprise, School of Public Policy, University of Maryland, UMD-LM-14–012
56. Rogers RSM, O'Donnell J (2017) Supply chain security: DFARS – Detection & Avoidance of Counterfeit Electronic Parts, https://smtnet.com/library/files/upload/supply-chain-security.pdf
57. Aerospace Industries Association (2011) Counterfeit Parts: Increasing Awareness and Developing Countermeasures, https://www.aia-aerospace.org/report/counterfeit-parts-increasing-awareness-and-developing-countermeasures/
58. Lockheed Martin Counterfeit Prevention: What Makes a Good Control Plan?, https://slidetodoc.com/counterfeit-prevention-what-makes-a-good-control-plan/

59. DARPA A DARPA Approach to Trusted Microelectronics, https://www.darpa.mil/attachments/Obscurationandmarking_Summary.pdf
60. Lingle R (2014) In-mold labels use digital watermarking for authentication, https://www.packagingdigest.com/trends-issues/mold-labels-use-digital-watermarking-authentication. Packaging Digest
61. Digital Watermarking Alliance Authentication of content and objects (includes government IDs), https://digitalwatermarkingalliance.org/digital-watermarking-applications/authentication-of-content-and-objects/
62. Chan HK, Griffin J, Lim JJ, Zeng F, Chiu ASF (2018) The impact of 3D Printing Technology on the supply chain: Manufacturing and legal perspectives. Int J Prod Econ 205:156–162
63. CDC Tamper-resistant prescription form requirements, https://www.cdc.gov/phlp/docs/menu-prescriptionform.pdf
64. Staake T, Michahelles F, Fleisch E, Williams JR, Min H, Cole PH, Lee S-G, McFarlane D, Murai J (2008) Anti-counterfeiting and supply chain security. Springer, Berlin Heidelberg, pp 33–43
65. Chen C-l, Chen Y-Y, Huang Y-C, Liu C-S, Lin C-I, Shih T-F (2008) Anti-counterfeit ownership transfer protocol for low cost RFID system. WSEAS Transactions on Computers archive 7:1149–1158
66. Azuara G, Luis Tornos J, Luis Salazar J (2012) Improving RFID traceability systems with verifiable quality. Ind Manag Data Syst 112(3):340–359
67. Kwok SK, Ting JSL, Tsang AHC, Lee WB, Cheung BCF (2010) Design and development of a mobile EPC-RFID-based self-validation system (MESS) for product authentication. Comput Ind 61(7):624–635
68. Choi SH, Yang B, Cheung HH, Yang YX (2015) RFID tag data processing in manufacturing for track-and-trace anti-counterfeiting. Comput Ind 68:148–161
69. Wyld D, Jones M (2007) RFID is no fake: the adoption of radio frequency identification technology in the pharmaceutical supply chain. International Journal of Integrated Supply Management - Int J Integrated Supply Manag, 3
70. Agrawal TK, Koehl L, Campagne C (2018) A secured tag for implementation of traceability in textile and clothing supply chain. Int J Adv Manuf Tech 99(9):2563–2577
71. Juels A (2006) RFID security and privacy: a research survey. IEEE J Sel Areas Commun 24(2):381–394
72. Rieback MR, Crispo B, Tanenbaum AS (2006) The evolution of RFID security. IEEE Pervasive Comput 5(1):62–69
73. Garfinkel SL, Juels A, Pappu R (2005) RFID privacy: an overview of problems and proposed solutions. IEEE Secur Priv 3(3):34–43
74. Lee YK, Batina L, Singelee D, Preneel B, Verbauwhede I (2010) Anti-counterfeiting, untraceability and other security challenges for RFID systems: public-key-based protocols and hardware. Springer, Berlin Heidelberg, pp 237–257
75. Santos BLD, Smith LS (2008) RFID in the supply chain: panacea or Pandora's box? Commun ACM 51(10):127–131
76. Cai S, Li T, Ma C, Li Y, Deng RH (2009) Enabling secure secret updating for unidirectional key distribution in RFID-enabled supply chains. Springer, Berlin Heidelberg, pp 150–164
77. Kumar S, Dieveney E, Dieveney A (2009) Reverse logistic process control measures for the pharmaceutical industry supply chain. Int J Product Perform Manag 58(2):188–204
78. Juels A (2005) Strengthening EPC tags against cloning. in WiSe - 2005 ACM Workshop on Wireless Security. Cologne: Association for Computing Machinery (ACM). pp 67–75
79. Miles SB, Sarma S, Williams JR (2008) RFID technology and applications. RFID Technology and Applications. Vol. 9780521880930. Cambridge University Press. 1–218
80. Singh MKM, Li X (2010) Trust in RFID-enabled supply-chain management. Int. J. Secur. Networks 5:96–105
81. Ting SL, Tsang AHC (2014) Using social network analysis to combat counterfeiting. Int J Prod Res 52(15):4456–4468

82. Trenfield SJ, Xian Tan H, Awad A, Buanz A, Gaisford S, Basit AW, Goyanes A (2019) Track-and-trace: Novel anti-counterfeit measures for 3D printed personalized drug products using smart material inks. Int J Pharmaceutics 567:118443
83. Cole R, Stevenson M, Aitken J (2019) Blockchain technology: implications for operations and supply chain management. Supply Chain Management: An Int J 24(4):469–483
84. Pun H, Swaminathan JM, Hou P (2021) Blockchain adoption for combating deceptive counterfeits. Prod Oper Manag 30(4):864–882
85. Liu Z, Li Z (2020) A blockchain-based framework of cross-border e-commerce supply chain. International J Information Manag 52:102059
86. Kennedy ZC, Stephenson DE, Christ JF, Pope TR, Arey BW, Barrett CA, Warner MG (2017) Enhanced anti-counterfeiting measures for additive manufacturing: coupling lanthanide nanomaterial chemical signatures with blockchain technology. J Materials Chemistry C 5(37):9570–9578
87. Smith AF, Skrabalak SE (2017) Metal nanomaterials for optical anti-counterfeit labels. Journal of Materials Chemistry C 5(13):3207–3215
88. Toyoda K, Mathiopoulos PT, Sasase I, Ohtsuki T (2017) A novel blockchain-based product ownership management system (POMS) for anti-counterfeits in the post supply chain. IEEE Access 5:17465–17477
89. Hepp T, Wortner P, Schönhals A, Gipp B (2018) Securing physical assets on the blockchain: linking a novel object identification concept with distributed ledgers. in Proceedings of the 1st Workshop on Cryptocurrencies and Blockchains for Distributed Systems. Munich, Germany: Association for Computing Machinery. pp 60–65
90. Boos D, Guenter H, Grote G, Kinder K (2013) Controllable accountabilities: The Internet of Things and its challenges for organisations. Behaviour and Info Tech 32(5):449–467
91. Acri KML, Lybecker n (2018) Pharmaceutical counterfeiting: contributing factors. Fraser Institute. pp 6–23
92. Blackstone EA, Fuhr JPJ, Pociask S (2014) The health and economic effects of counterfeit drugs. American health & drug benefits 7(4):216–224
93. Hall A, Koenraadt R, Antonopoulos GA (2017) Illicit pharmaceutical networks in Europe: organising the illicit medicine market in the United Kingdom and the Netherlands. Trends in Organized Crime 20(3–4):296–315
94. Tremblay M (2013) Medicines counterfeiting is a complex problem: a review of key challenges across the supply chain. Curr Drug Saf 8(1):43–55
95. Khan MH, Akazawa M, Dararath E, Kiet HB, Sovannarith T, Nivanna N, Yoshida N, Kimura K (2011) Perceptions and practices of pharmaceutical wholesalers surrounding counterfeit medicines in a developing country: a baseline survey. BMC Health Serv Res 11(1):306
96. Rosen LS, Jacobs IA, Burkes RL (2017) Bevacizumab in Colorectal Cancer: Current Role in Treatment and the Potential of Biosimilars. Target Oncol 12(5):599–610
97. Mackey TK, Liang BA, York P, Kubic T (2015) Counterfeit drug penetration into global legitimate medicine supply chains: a global assessment. Am J Tropical Medicine and Hygiene 92(6_Suppl):59–67
98. Gautam CS, Utreja A, Singal GL (2009) Spurious and counterfeit drugs: a growing industry in the developing world. Postgrad Med J 85(1003):251–256
99. Stewart MW, Narayanan R, Gupta V, Rosenfeld PJ, Martin DF, Chakravarthy U (2016) Counterfeit Avastin in India: punish the criminals, not the patients. Am J Ophthalmol 170:228–231
100. Ozawa S, Evans DR, Bessias S, Haynie DG, Yemeke TT, Laing SK, Herrington JE (2018) Prevalence and estimated economic burden of substandard and falsified medicines in low- and middle-income countries: a systematic review and meta-analysis. JAMA Netw Open 1(4):e181662–e181662
101. Medina E, Bel E, Suñé JM (2016) Counterfeit medicines in Peru: a retrospective review (1997–2014). BMJ Open 6(4):e010387
102. Venhuis BJ, Oostlander AE, Giorgio DD, Mosimann R, du Plessis I (2018) Oncology drugs in the crosshairs of pharmaceutical crime. Lancet Oncol 19(4):e209–e217

103. Jackson G, Patel S, Khan S (2012) Assessing the problem of counterfeit medications in the United Kingdom. Int J Clin Pract 66(3):241–250
104. Chambliss WG, Carroll WA, Kennedy D, Levine D, Moné MA, Douglas Ried L, Shepherd M, Yelvigi M (2012) Role of the pharmacist in preventing distribution of counterfeit medications. J Am Pharm Assoc 52(2):195–199
105. Ziance RJ (2008) Roles for pharmacy in combatting counterfeit drugs. J Am Pharm Assoc 48(4):e71–e91
106. Nayyar GML, Breman JG, Mackey TK, Clark JP, Hajjou M, Littrell M, Herrington JE (2019) Falsified and substandard drugs: stopping the pandemic. Am J Trop Med Hyg 100(5):1058–1065
107. Chaudhry PE, Stumpf SA (2013) The challenge of curbing counterfeit prescription drug growth: Preventing the perfect storm. Bus Horiz 56(2):189–197
108. Ranieri N, Tabernero P, Green MD, Verbois L, Herrington J, Sampson E, Satzger RD, Phonlavong C, Thao K, Newton PN (2014) Evaluation of a new handheld instrument for the detection of counterfeit artesunate by visual fluorescence comparison. Am J Trop Med Hyg 91(5):920
109. Mackey T, Liang B (2011) The global counterfeit drug trade: patient safety and public health risks. J Pharm Sci 100:4571–4579
110. Mackey TK, Liang BA (2013) Improving global health governance to combat counterfeit medicines: a proposal for a UNODC-WHO-Interpol trilateral mechanism. BMC Med 11(1):233
111. Davison M (2011) Pharmaceutical anti-counterfeiting: combating the real danger from fake drugs. John Wiley & Sons
112. Hamilton WL, Doyle C, Halliwell-Ewen M, Lambert G (2016) Public health interventions to protect against falsified medicines: a systematic review of international, national and local policies. Health Policy Plan 31(10):1448–1466
113. Fadlallah R, El-Jardali F, Annan F, Azzam H, Akl EA (2016) Strategies and systems-level interventions to combat or prevent drug counterfeiting: a systematic review of evidence beyond effectiveness. Pharmaceutical Medicine 30:263–276
114. Cohn JE, von Schoen-Angerer T, Jambert E, Arreghini G, Childs ML (2013) When falsified medicines enter the supply chain: description of an incident in Kenya and lessons learned for rapid response. J Public Health Policy 34:22–30
115. Cuomo RE, Mackey TK (2014) An exploration of counterfeit medicine surveillance strategies guided by geospatial analysis: lessons learned from counterfeit Avastin detection in the US drug supply chain. BMJ Open 4(12):e006657
116. Mackey TK, Cuomo R, Guerra C, Liang BA (2015) After counterfeit Avastin®—what have we learned and what can be done? Nat Rev Clin Oncol 12(5):302–308
117. Kamba PF, Ireeta ME, Balikuna S, Kaggwa B (2017) Threats posed by stockpiles of expired pharmaceuticals in low- and middle-income countries: a Ugandan perspective. Bull World Health Organ 95:594–598
118. Reynolds L, McKee M (2010) Organised crime and the efforts to combat it: a concern for public health. Glob Health 6(1):21
119. Ozawa S, Haynie DG, Bessias S, Laing SK, Ngamasana EL, Yemeke TT, Evans DR (2019) Modeling the economic impact of substandard and falsified antimalarials in the Democratic Republic of the Congo. Am J Trop Med Hyg 100(5):1149–1157
120. Weaver AA, Reiser H, Barstis T, Benvenuti M, Ghosh D, Hunckler M, Joy B, Koenig L, Raddell K, Lieberman M (2013) Paper analytical devices for fast field screening of beta lactam antibiotics and antituberculosis pharmaceuticals. Anal Chem 85(13):6453–6460
121. Weaver AA, Lieberman M (2015) Paper test cards for presumptive testing of very low quality antimalarial medications. The American Society of Tropical Medicine and Hygiene 92(6_Suppl):17–23
122. Marini RD, Rozet E, Montes MLA, Rohrbasser C, Roht S, Rhème D, Bonnabry P, Schappler J, Veuthey JL, Hubert P, Rudaz S (2010) Reliable low-cost capillary electrophoresis device for drug quality control and counterfeit medicines. J Pharm Biomed Anal 53(5):1278–1287

123. Bansal D, Malla S, Gudala K, Tiwari P (2013) Anti-counterfeit technologies: a pharmaceutical industry perspective. Sci Pharm 81(1):1–14
124. Lybecker KM (2008) Keeping it real: anticounterfeiting strategies in the pharmaceutical industry. Manag Decis Econ 29(5):389–405
125. Mackey TK, Nayyar GML (2017) A review of existing and emerging digital technologies to combat the global trade in fake medicines. Expert Opin Drug Saf 16:587–602
126. Taylor D (2014) RFID in the pharmaceutical industry: addressing counterfeits with technology. J Med Syst 38:1–5
127. Chen C, Zhang F, Barras J, Althoefer K, Bhunia S, Mandal S (2016) Authentication of medicines using nuclear quadrupole resonance spectroscopy. IEEE/ACM Trans Comput Biol Bioinf 13(3):417–430
128. Kwok SK, Ting SL, Tsang AHC, Cheung CF (2010) A counterfeit network analyzer based on RFID and EPC. Ind Manag Data Syst 110(7):1018–1037
129. Cozzella L, Simonetti C, Schirripa Spagnolo G (2012) Drug packaging security by means of white-light speckle. Opt Lasers Eng 50(10):1359–1371
130. Raj R, Rai N, Agarwal S (2019) Anticounterfeiting in pharmaceutical supply chain by establishing proof of ownership. in TENCON 2019 - 2019 IEEE Region 10 Conference (TENCON). IEEE. pp 1572–1577
131. Kumar A, Choudhary D, Raju MS, Chaudhary DK, Sagar RK (2019) Combating counterfeit drugs: a quantitative analysis on cracking down the fake drug industry by using blockchain technology. in 2019 9th International Conference on Cloud Computing, Data Science & Engineering (Confluence). IEEE. pp 174–178
132. Meyliana, Surjandy, Fernando E, Cassandra C, Marjuki (2021) Propose Model Blockchain Technology Based Good Manufacturing Practice Model of Pharmacy Industry in Indonesia. in 2021 2nd International Conference on Innovative and Creative Information Technology (ICITech). pp 190–194
133. Alzahrani N, Bulusu N (2020) A new product anti-counterfeiting blockchain using a truly decentralized dynamic consensus protocol. Concurrency and Computation: Practice Exp 32(12):e5232
134. Raja J, Khaled S, Nelson K (2019) Improving opportunities in healthcare supply chain processes via the internet of things and blockchain technology. International Journal of Healthcare Information Systems and Informatics (IJHISI) 14(2):49–65
135. Singh R, Dwivedi AD, Srivastava G (2020) Internet of Things based blockchain for temperature monitoring and counterfeit pharmaceutical prevention. Sensors 20(14):3951
136. Sylim PG, Liu F, Marcelo AB, Fontelo PA (2018) Blockchain technology for detecting falsified and substandard drugs in distribution: pharmaceutical supply chain intervention. JMIR Research Protocols, 7(9): e10163
137. Longo F, Nicoletti L, Padovano A, d'Atri G, Forte M (2019) Blockchain-enabled supply chain: an experimental study. Comput Ind Eng 136:57–69
138. Jamil F, Hang L, Kim K, Kim D (2019) A novel medical blockchain model for drug supply chain integrity management in a smart hospital. Electronics (Switzerland) 8(5)
139. Kumar R, Tripathi R (2019) Traceability of counterfeit medicine supply chain through Blockchain. in 11th International Conference on Communication Systems and Networks, COMSNETS 2019. Institute of Electrical and Electronics Engineers Inc. pp 568–570
140. Tseng J-H, Liao Y-C, Chong B, Liao S-W (2018) Governance on the drug supply chain via Gcoin blockchain. Int J Environ Res Public Health 15(6):1055
141. Shahbazi Z, Byun Y-C (2020) A procedure for tracing supply chains for perishable food based on blockchain, machine learning and fuzzy logic. Electronics 10(1):41
142. Tsang YP, Choy KL, Wu CH, Ho GTS, Lam HY (2019) Blockchain-Driven IoT for Food Traceability with an Integrated Consensus Mechanism. IEEE Access 7:129000–129017
143. Soon JM, Manning L (2019) Developing anti-counterfeiting measures: the role of smart packaging. Food Res Int 123:135–143

144. Fang Z, Zhao Y, Warner RD, Johnson SK (2017) Active and intelligent packaging in meat industry. Trends Food Sci Technol 61:60–71
145. Sohail M, Sun D-W, Zhu Z (2018) Recent developments in intelligent packaging for enhancing food quality and safety. Crit Rev Food Sci Nutr 58(15):2650–2662

Banking Credit Risk Analysis using Artificial Neural Network

Charles Maruma, Chunling Tu, and Claude Nawej

Abstract Banking credit risk analysis is a form of evaluation conducted by financial institutions to determine applicants' ability to repay their debt obligation. Financial institutions, such as banks, set objectives to offer credit to creditworthy customers, after spending time trying to evaluate their repaying capacity. In this paper, we propose a credit risk analysis system based on an artificial neural network (ANN) to identify customers who will default. A feedforward propagation algorithm is used to train the model consisting of three layers. Data pre-processing is performed to clean the datasets and check for missing variables. The datasets were normalized using min–max normalization to get the correlation among the variables. The datasets are applied to the proposed model and logistic regression models, and the comparison shows the proposed model which has a better performance.

Keywords Credit risk analysis · Artificial neural network · Logistic regression · Default · Credit · And algorithm

1 Introduction

The banking industry focuses mainly on offering credit/loans [1]. Although the banking industry does offer other services such as policies and insurances. Since banks offer credit as their primary service offerings, they might suffer a great financial loss if customers fail to pay their credit back [2]. For banks to avoid financial loss as a result of customers defaulting on credit, banks must perform a credit risk analysis to predict and minimize the risk by discriminating between customers with a high risk of default and customers with a low risk of default. Credit risk analysis

C. Maruma (✉) · C. Tu · C. Nawej
Tshwane University of Technology, Pretoria, South Africa
e-mail: Charlesmaruma1@gmail.com

C. Tu
e-mail: Duc@tut.ac.za

C. Nawej
e-mail: NawejMC@tut.ac.za

X.-S. Yang et al. (eds.), *Proceedings of Seventh International Congress on Information and Communication Technology*, Lecture Notes in Networks and Systems 447,
https://doi.org/10.1007/978-981-19-1607-6_76

is a process used to predict whether or not a customer will default on a loan or credit [3]. Credit applications for consumer or commercial loans are being processed and evaluated using credit risk analysis before approving or denying them. Credit risk analysis is usually performed based on historical data such as past transactions, credit history, and other relevant information. The decision to approve or deny credit is based on customer's risk of default; if the customer has a high risk of default, then the credit will be denied, however, if a customer has a low risk of default, the credit will be approved [4]. Credit scoring is a method used to predict and analyze credit risk. Credit scoring is categorized into two different scoring types, such as behavioral scoring and application scoring. Behavioral scoring is usually performed after the credit has been approved, it monitors how the customer repays the credit [5]. Application scoring is performed before approving or denying credit, based on the customer's credit history. When a customer applies for a loan, his/ her credit information is obtained from the credit bureau. The credit information includes features such as employment status, past bank transactions, salary, and credit history [1].

Credit scoring is a technique used to analyze and predict credit risk. When a customer applies for credit, a credit risk analysis is performed before approving or denying the credit based on the customer's credit history and other relevant information [6]. This process is performed to minimize and manage the risk of customers faulting on loans. Bank manages the flow of money between investors and borrowers. Investors put money in the bank, then the bank borrows that money to customers/borrowers in the form of credit/loans, then the bank manages the risk on how the borrower pays back the loan with interest [7]. When the borrower pays back the loan, the money is put back into the investor's account with a small portion of the interest, and the other portion of the interest goes to the bank for managing the risk. The study of credit risk analysis is the most researched area in the banking industry. The credit risk analysis technique plays an important part in the banking industry, particularly for big banks with large data that is hard to work with and process. Credit risk analysis is a form of evaluation process conducted by financial analysts to determine applicants' ability to repay their debt obligation. After a person or a company applies for credit at the bank, the bank processes and evaluates the costs and profit related to the credit [8]. The credit risk analysis model is utilized to predict the costs and risks related to the credit. The objective of this study is to address credit risk analysis issues by applying an ANN to credit risk models [9]. ANN algorithm is applied to credit risk/loan application datasets of public data to predict loan risk. The credit risk model produces the output of "1" or "0" to predict whether or not the customer will default.

2 Literature Review

Credit risk analysis using historical data predicts how different customer characteristics determine whether or not the customer will be able to repay the loan. This method uses a score to categorize customers according to their risk of default on credit. The customer with a high score has a low risk of defaulting on the loan and he/she will be able to repay the loan, while the customer with a low score has a high risk of defaulting on the loan and will not be able to repay the loan. Normally, logistic regression and discriminant analysis were the most used machine learning algorithms in credit risk analysis models. The first machine learning algorithm applied to the credit scoring model was discriminant analysis. The use or application of linear discriminant analysis has regularly been disapproved because of how it deals with categorical data of datasets, and the classification of creditworthy and not creditworthy classes is not likely to be accurate. Another machine learning algorithm used in credit risk analysis is logistic regression. Logistic regression is used as a machine learning algorithm of choice in predicting customers who will or will not default on their credit.

Artificial neural network (ANN) algorithm in artificial intelligence is a data modeling technique that is similar to the brain of the human and nervous system, and it works similarly to how the human brain works [10]. ANN has a network of inter-connected neurons to find the functionality of the model. A few tests were performed on different machine learning algorithms to measure the accuracy of the individual models. During the performance testing process, it was discovered that artificial neural networks performed better and produced more accurate results when compared to logistic regression [11]. Normally for the artificial neural network to perform results prediction, it entails being trained on the input and target variables of the given datasets. There are many successful real-world applications of artificial neural networks such as edited file detectors (checks if a file has been modified), unusual banking transaction detectors, and other predicting technologies. The application of machine learning algorithms to credit risk has improved the performance of credit risk analysis [12].

Discriminant analysis technique was considered the foremost common technique for developing customer credit risk analysis models [10]. Even though the discriminant analysis technique has been criticized by the analysts because of the way, it processes and handles datasets of categorical variables to predict customers with a high risk of default and customers with a low risk of default. In machine learning, the neural network algorithm is a nonlinear technique that provides a new alternative to linear methods, especially in situations where the dataset has more composite relationships between the independence of the nonlinear variables [13]. An artificial neural network (ANN) is a machine learning algorithm that develops a relationship between the independent variables and dependent variables, in consideration that there is a correlation among the variables. ANNs are artificial intelligence algorithms that mimic the structure of the human brain and work similarly to the nervous system [14]. An artificial neural network is consists of a network of neurons organized in a matrix form. Neurons are associated by joins with related weights which decide how data are being processed [15].

The credit risk analysis model uses a feedforward neural network [16]. Feedforward neural network is an ANN technique. The inter-connected links between nodes in the feedforward neural network do not a form cycle [17]. In feedforward neural networks, input variables enter the model via the input layer, and the variables are multiplied by the weights. The values of each variable are added to get the total sum of the input variables. We get an output of 1 if the sum of the values is over a given threshold, however, if the sum of the values is below a specific threshold, then less than 1 is a product at the output.

3 Methodology

In this paper, ANN is used to build a credit risk analysis system. Publicly accessible datasets obtained from the Internet are applied to train the model. The dataset consists of about 1000 customer records and 10 categorical and numerical variables.

The independent variables are

- Age: Age of the borrower
- Sex: Gender of the borrower (male or female).
- Job: Employment status of the borrower (employed or unemployed).
- Housing: Checks if the borrower owns or rents a property.
- Saving accounts: Banking history of savings account.
- Checking account: Banking history of a cheque account.
- Credit amount: Loan amount given to the borrower.
- Duration: The term in months the borrower will settle the loan amount.
- Purpose: Reason for taking a loan.

The dependent variable is the risk represented with 1 or 0. The risk predicts whether the customer will be able to repay the full loan amount on time. If the customer's risk prediction is "1", it means that the customer will default on the loan and will not be able to pay back the loan otherwise, it is a creditworthy customer.

The first step is variable-processing. Usually, variables in the dataset do not come in a way that can be directly used. Pre-processing is therefore needed. Categorical variables have labels instead of numbers. For example, the gender has "female" or "male" labels. Numerical variables have numerical values. In datasets, categorical variables could have a lot of null values. For this situation, it is a significant loss if the concerned data are discarded. So we can replace those null values with random labels. The other important pre-processing to improve the performance of the model is data normalization. Normalization is carried out to get the correlation of the data. In this paper, min–max normalization techniques are used to normalize input/independent variables between 0 and 1 as shown in Eq. (1).

$$Z_i = \frac{x_i - \min(x)}{\max(x) - \min(x)} \tag{1}$$

where $x = (x_1, x_2, \ldots, x_n)$ and Z_i are the i-th normalized variable data.

Target/independent variable is normalized between 0 and 1, as shown in Eq. (2).

$$y2 = \log(1 + y) \tag{2}$$

The ANN-based credit evaluation system architecture is set up as follows: the feedforward propagation network is used as shown in Fig. 1. There are three layers in the network: 1 input layer with 9 independent variables, 1 hidden layer with 10 neurons, and 1 output layer with 1 dependent variable representing if the customer is creditworthy or not. The random weight/bias rule is used as the training function to train the neural network. The feedforward propagation algorithm makes the neurons perform better by reducing the error between the actual and the desired results to the least possible amount.

The root mean squared shown in Eq. (3) is used as the training error of the ANN.

$$\mathrm{RMSE} = \sqrt{\frac{1}{n}\sum_{i=1}^{n}(S_i - O_i)^2} \tag{3}$$

where n is the number of observations, O_i is the observations, and S_i is the predicted values.

To reduce the error of neural networks, weights must be adjusted by a small amount. Choosing the correct parameters is crucial, especially for the learning coefficient and the number of hidden neurons. Based on the proposed neuron network structure as depicted in Fig. 1, neurons are represented by nodes in each layer and the lines between them weight. The RSME is reduced by retraining the model or it

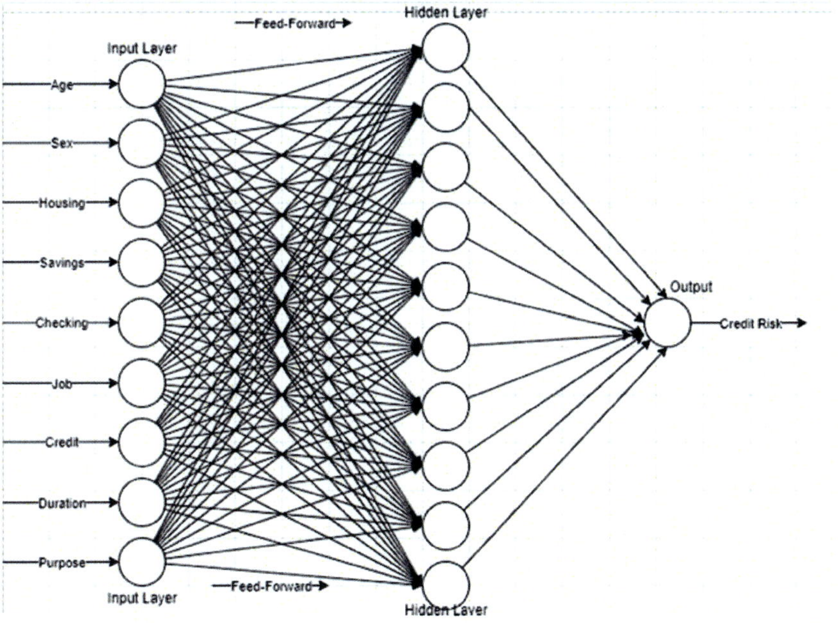

Fig. 1 Feedforward ANN

can be reduced by adjusting the settings on ANN such as reducing the number of neurons.

In this paper, artificial neuron network has been applied to the datasets. Feedforward networks consist of 3 layers, namely input, hidden, and output layer. The dataset is divided into 3 categories; training, test, and validation sets. During the training stage, all variables are tested 1 by 1, to check if they improved the performance of the model. If variables do not improve the performance of the model, they are discarded. The most important variables remain in the dataset. The model was trained until it produced better results.

4 Results

The dataset with 1000 records was fed into ANN with the following parameters and properties:

- Network type: feedforward back prop as shown in Fig. 1
- Datasets division method: random
- Training function: random weight/bias rule
- Adoption learning function: lean GDM
- Performance function: root mean squared error (RMSE)
- Number of hidden layer neurons: 10
- Transfer function: hyperbolic tangent sigmoid

The output error of the model is 0.02 for the trained ANN. The accuracy of the neural network model is compared to the logistic regression model. As per the comparison, neural network performed better than logistic regression, as the error is shown in Table 1. The output values for logistic regression range between 0 and 1.

Table 1 shows that the RMSE for the proposed model is smaller than logistic regression, which means the proposed system is more accurate to predict the risk levels.

The proposed model has a nonlinear activation layer which makes input variables have a nonlinear impact on the credit risk status as the weights have a generalized weight of more than 1. The 9 input variables have been normalized to get the correlation between them before they were added to the neural network. The output of the neural network model is classified as 0 or 1. Data cleaning and pre-processing were performed on the datasets. The datasets were pre-processed to ensure that there were no missing values.

Table 1 Error comparison for proposed system and logistic regression

Proposed model	Logistic regression
0.021005	0.032532

Fig. 2 Total customer credit risk status

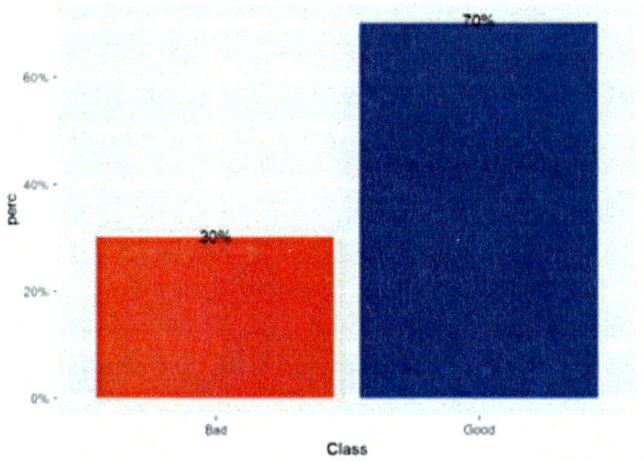

Fig. 3 Age groups by credit risk status

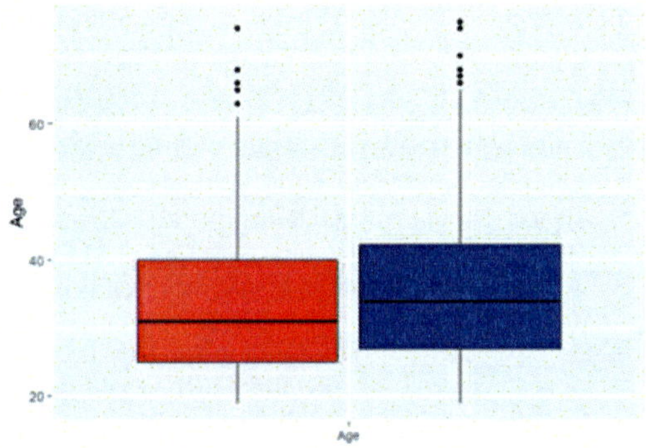

For the total customers in the dataset used in this paper, there are approximately 70% of customers are creditworthy and have a low risk of default, while approximately 30% are at high risk, as shown in Fig. 2, where blue color represents low risk while red color represents a high risk.

In Fig. 3, the age groups show different risk levels, red color shows that customers below 40 years old have a higher risk of default, while the blue color shows the group above 40 years old age has a lower risk of default.

5 Conclusion

In this paper, we studied banking credit risk analysis using ANN and logistic regression algorithms to detect whether customers are likely to default on their credit, based on the given customer information on the dataset. The data were firstly cleaned by a pre-processing stage, to fill in missing values and handle exceptions. The correlations among the independent/input variables are detected. The independent/input variables have been normalized using min–max normalization.

The proposed ANN-based system was compared with the logistic regression model. The experiment results show that the proposed method performed better than logistic regression.

References

1. Lavrushin O, Sokolinskaya N (2020) Confidence level and credit risk analysis in Russian banks. Banks and Bank Syst 15(2):38–46
2. Shan Y (2017) Systemic risk and credit risk in bank loan portfolios. SSRN Electronic J
3. Livshits I (2015) Recent developments in consumer credit and default literature. J Econ Surv 29(4):594–613
4. Singh M, Dixit G (2018) Modeling customers credit worthiness using enhanced ensemble model. Int J Comp Sci Eng 6(7):1466–1470
5. Giannopoulos V (2018) The effectiveness of artificial credit scoring models in predicting NPLs using micro accounting data. J Accounting & Marketing 7(04)
6. Miroshnychenko I, Ivliieva K (2019) Assessing credit risk using machine learning methods. Efektyvna Ekonomika (12)
7. Aslam M, Kumar S, Sorooshian S (2019) Predicting likelihood for loan default among bank borrowers. Int J Financial Res 11(1):318
8. Amat O, Manini R, AntónRenart M (2017) Credit concession through credit scoring: Analysis and application proposal Intangible Cap 13(1):51
9. Demma C (2017) Credit scoring and the quality of business credit during the crisis. Econ Notes 46(2):269–306
10. Ettensperger F (2019) Comparing supervised learning algorithms and artificial neural networks for conflict prediction: performance and applicability of deep learning in the field. Qual Quant 54(2):567–601
11. Maler L (2020) Neural networks: how a multi-layer network learns to disentangle exogenous from self-generated signals. Curr Biol 30(5):R224–R226
12. Abkowitz M, Camp J (2017) Structuring an enterprise risk assessment protocol: traditional practice and new methods. Risk Manag Insurance Rev 20(1):79–97
13. Shah S, Ng J (2020) Hands-on artificial intelligence for banking. Packt Publishing, Limited, Birmingham
14. Kang E, Baek S (2019) Humanistic brain that artificial intelligence can't mimic and artificial intelligence challenging human ambivalence (creativity and limitation). J Contemp Psychoanalysis 21(2):143–154
15. Bondarenko A, Borisov A, Alekseeva L (2015) Neurons vs weights pruning in artificial neural networks. Environment. Technology. Resources. Proceedings of the International Scientific and Practical Conference, 3, p.22.
16. Amardeep R (2017) Training feed forward neural network with backpropogation algorithm. Int J Eng Comp Sci
17. Fuangkhon P (2021) Normalized data barrier amplifier for feed-forward neural network. Neural Net World 31(2):125–157

Implementation, Analysis, and Emulation of Electric Vehicle Powertrain System with Sensorless Field Controlled PMSM Drive

Monika Verma, Mini Sreejeth, and Madhusudan Singh

Abstract The high torque and power density of permanent magnet synchronous motor (PMSM) make it potentially eligible for playing the crucial role of transmitting energy from battery pack to wheel system of electric vehicle (EV) in electric powertrain (PT). This paper presents a sensorless field controlled EVPT test bench emulating the EV propulsion system. The complete EVPT model developed in MATLAB/Simulink is analyzed for different modes of operations. The PMSM is fed by a voltage source inverter (VSI). Speed regulation is performed via TWR-KV46F150 microcontroller in EVPT emulator. The simulation results and experimental results are presented and verified. This work helps to understand the real-time operational environment of EV and is beneficial in expansion of PT system control.

Keywords Powertrain · MATLAB · Microcontroller · Field control · PMSM · VSI

1 Introduction

During last few decades, EV industry has excessively developed at global level. India, one of the largest automobile manufacturers globally, produces various types of EV. Various national electric mobility plans are launched and budgeted by Indian government to consign several issues such as vehicular pollution, promote automobile manufacturers' proficiencies, and ensure nation's energy sustainability [1, 2]. The use of EV helps to curb various environmental issues like emission of greenhouse gases, automobile associated pollution, usage of fossil fuel-based economy. An EV uses electric motor and controllers rather than internal combustion engine (ICEs) in propulsion system of the vehicle. This paper shows the EV emulation system

M. Verma (✉) · M. Sreejeth · M. Singh
Delhi Technological University, Shahbad Extension, New Delhi 110042, India
e-mail: monikaverma_phd2k17@dtu.ac.in

M. Sreejeth
e-mail: minisreejeth@dce.ac.in

M. Singh
e-mail: madhusudan@dce.ac.in

X.-S. Yang et al. (eds.), *Proceedings of Seventh International Congress on Information and Communication Technology*, Lecture Notes in Networks and Systems 447,
https://doi.org/10.1007/978-981-19-1607-6_77

developed to study the EVPT system under real-time environment. It employs PMSM as its traction device to explore its capabilities in powertrain engineering [3].

2 Development of EVPT Test Bench Emulator

2.1 EVPT System with Belt-Pulley Transmission System in the Emulator

In the developed emulator system shown in Fig. 1a, a 3 Hp, 380 V Compage PMSM is coupled with pulley, wheels and 1.68 Hp, 380 V Compage PM synchronous generator where PMSM acts as traction device. The extended motor shafts and pulley system are shown in Fig. 1b. Table 1 presents the EVPT system specifications.

For the belt-pulley system, the vehicle (V_v) or wheel (V_w) linear speed in km/h in terms of motor speed N_m in rpm, pulley ratio p, and wheel radius r_w in m is computed using Eq. (1).

$$V_v = V_w = \frac{3.6 N_m \pi r_w}{30 p^2} \tag{1}$$

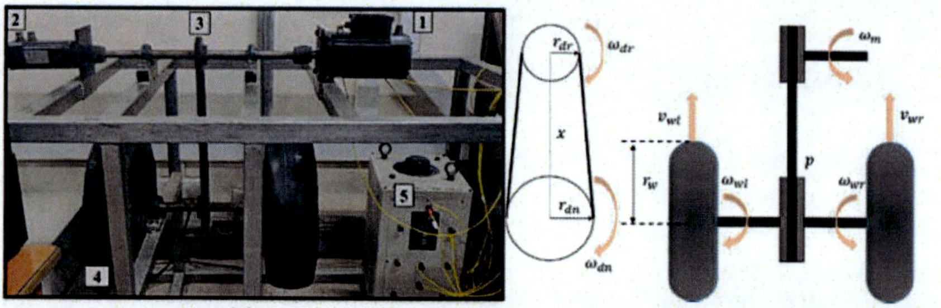

(a) (b)

Fig. 1 **a** EVPT test bench emulator system consisting of 1: PMSM, 2: PMSG, 3: belt-pulley system, 4: wheel, 5: three phase autotransformer, **b** schematic diagram of transmission system; r_{dr}: driver pulley radius in m, r_{dn}: driven pulley radius in m, x: vertical distance between pulleys in m, ω_{dr}: driver pulley rotational speed in rad/s, ω_{dn}: driven pulley rotational speed in rad/s, ω_m: motor rotational speed in rad/s, ω_{wl}: left wheel rotational speed in rad/s, ω_{wr}: right wheel rotational speed in rad/s, v_{wl}: left wheel linear speed in m/s, v_{wr}: right wheel linear speed in m/s, p: pulley ratio, r_w: wheel radius in m

Table 1 Emulator system specifications [4]

Motor specification			
Power	2.28 kW	Field intensity	3.035×10^{-3} Nm-sec
Voltage	380 V	Moment of inertia	1.5×10^{-3} kg-m^2
Torque	24 Nm	Magnetic flux	0.1194 V-sec
Poles	6	Line resistance	2.8 Ω
Phases	3	Line inductance	26 mH
Vehicle specification			
Wheel mass	1100 kg	Wheel velocity threshold	0.05 m/s
Frontal area	3 m^2	Wheel radius	0.25 m
Drag coefficient	0.4	Rolling radius	0.3 m
Load capacity	1600 kg	Driven to drive pulley ratio	1.5

2.2 *Test Bench Control Setup and Sensorless Field-Oriented Control of EVPT System*

Three phase voltage supply is applied to the inverter system in which IGBT modules, SKM75GB12T4, are used for fast switching using the PWM generated using IDE tool called kinetis design studio (KDS), the software for Freescale microcontroller unit (MCU). The specifications of MCU are given in Table 2. The schematic diagram of control circuitry and the power electronics support test bench is shown in Fig. 2a and b, respectively. For simplifying calculation of motor quantities, dynamic model of motor is developed in synchronously rotating frame [5]. In FOC technique, the torque and flux producing components of the stator currents are decoupled to control

Table 2 TWR-KV46F150M microcontroller specifications

Heading level	Example
KV46F256VLL15 processor	150 MHz ARM Cortex M4 + core
Memory	256 KB flash, 32 KB SRAM
Timing and control	18 channels flex timer modules (FTM), four 32 bit periodic interrupt timer (PTT)
Mixed signal channels	Two 12 bit ADC
Connectivity	Two UARTs (RS232), SPI and I^2C module
System power	5 V from USB or power jack J516
Debug interface	2 × 10 pin ARM Cortex JTAG connector
Analog inputs	Four thermistors RT1-RT4 (10kΩ), operating temperature: – 20 to 90 °C

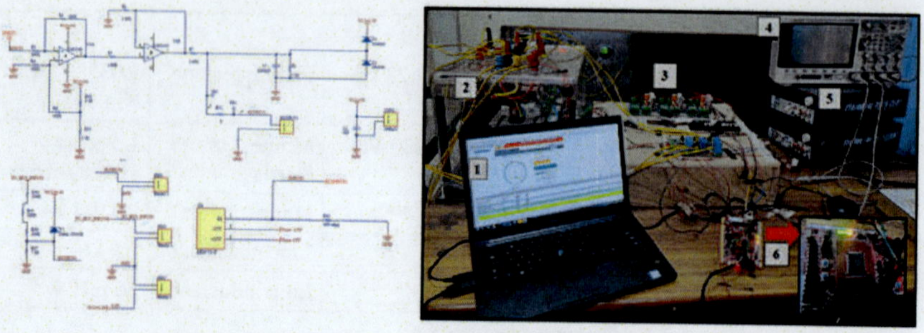

(a) (b)

Fig. 2 **a** Schematic diagram of control circuitry, **b** the test bench control circuitry setup consisting of 1: PC system, 2: three phase inverter, 3: control circuitry, 4: DSO, 5: DC regulated supply, 6: MCU

them independently using conventional PI controllers [6]. In sensorless control, the perspective is shifted from stator to rotor. Figure 3a shows the detailed block diagram for sensorless control of PMSM. It is based upon the back EMF estimation using per phase electrical model of PMSM in simplified form, shown in Fig. 3b. The mathematical relationship is given by Eq. (2). This equation is simplified to obtain digital interpretation of phase current (eqs. 3–5).

$$v_s = Ri_s + L\frac{d}{dt}i_s + e_s \tag{2}$$

$$\frac{d}{dt}i_s = -\frac{R}{L}i_s + \frac{1}{L}(v_s - e_s) \tag{3}$$

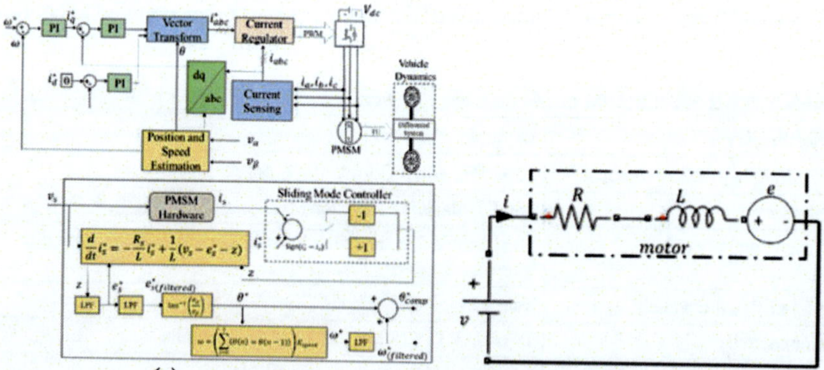

(a) (b)

Fig. 3 **a** Schematic diagram of sensorless field control, **b** simplified per phase electrical model of PMSM,

$$\frac{i_s(n+1) - i_s(n)}{T_s} = -\frac{R}{L} i_s(n) + \frac{1}{L}(v_s(n) - e_s(n)) \tag{4}$$

$$i_s(n+1) = \left(1 - T_s\frac{R}{L}\right) i_s(n) + \frac{T_s}{L}(v_s(n) - e_s(n)) \tag{5}$$

3 Analysis Results and Discussion

3.1 Simulink Model Analysis

The designed MATLAB model, presented in Fig. 4, is analyzed under different operating conditions. Gain constants of the controller are chosen as $K_p = 1000$ and $K_I = 5$ respectively. For a fixed speed of 100 rad/s, by applying the load disturbance of 15 Nm at $t = 1.5$ s, it is observed that initially vehicle runs @ 70 km/h steadily, and the vehicle accelerates (@ 5.5 m/s^2) at the instant of load variation. Stator reference currents, torque, motor, and vehicle speed results are shown in Fig. 5a. Along with the change of speed command from 0 to 150 rad/sec, given at $t = 1$ s, by applying a load disturbance of 10 Nm is given at $t = 2$ s, the vehicle accelerates @ 33.3 m/s^2 at speed command and @ 6.9 m/s^2 at load command. Simulation results for motor and vehicle speed, torque, and current are shown in Fig. 5b. By applying an increasing speed with acceleration of 1 rad/sec^2, the vehicle starts at 40 km/h and then runs at steady speed until a load disturbance of 10 Nm at $t = 1$ s. Thereafter, vehicle accelerates @ 4.16 m/s^2. The analysis results are shown in Fig. 6a. This summarizes traction mode of the vehicle. For the braking mode, the load is varied at $t = 1$ s from 10 to 0 Nm, and the speed is varied from 150 rad/sec to 0 rad/sec at $t = 2$ s. The vehicle runs initially @ 80 km/h and is observed to be decelerating @ 16.6 m/s^2 at the instant of braking and reaches to 0 km/h at 2.29 s (Fig. 6b).

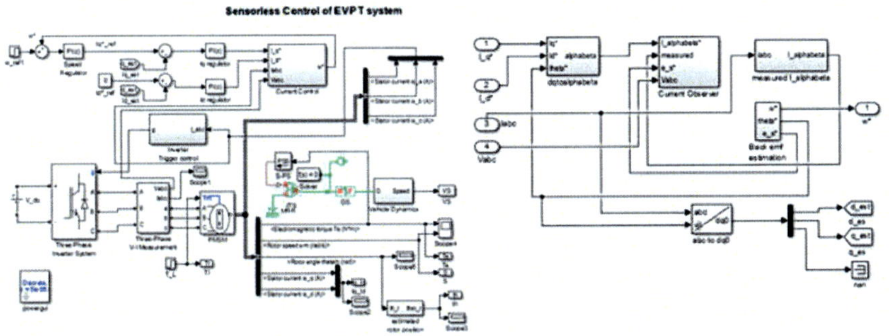

Fig. 4 MATLAB/Simulink model of controlled EVPT system

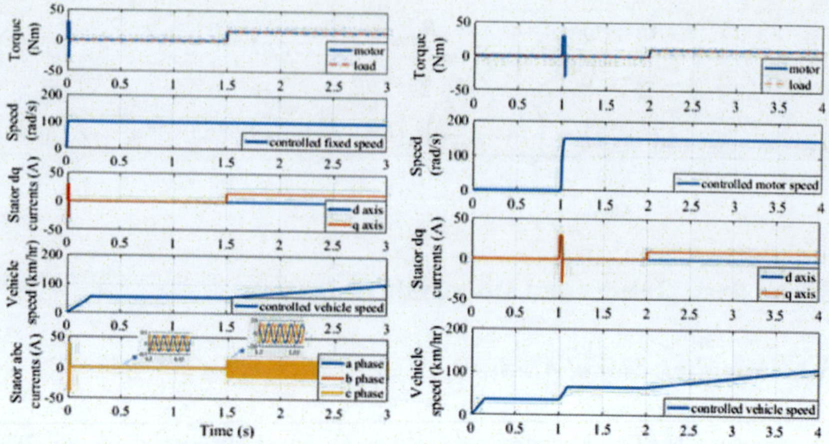

(a) (b)

Fig. 5 Response for **a** fixed speed, **b** step variable speed along with variable load

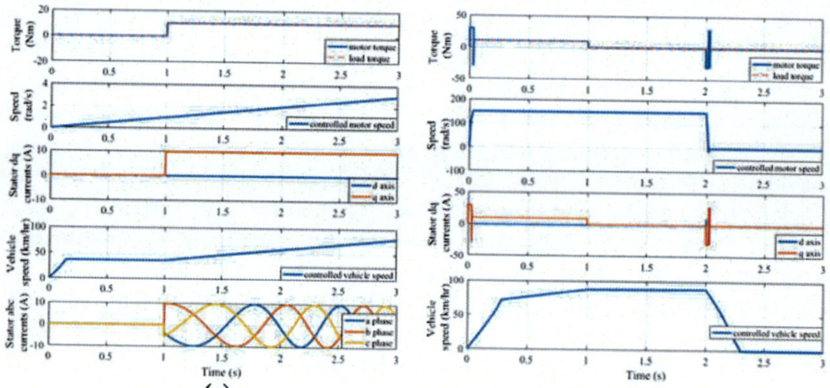

(a) (b)

Fig. 6 Response for **a** increasing speed, **b** decreasing speed along with variable load

3.2 *EVPT Emulator Analysis*

A 290 V, 3φ voltage is applied using 18.28 kVA, 415 V, 3 φ variable autotransformer. The rectified DC voltage is given as input to the MCU unit. At first, FreeMASTER executes the start-up subroutine to provide initial jerk to motor by energizing stator windings with estimated rotor position angle. Initially, 50 V DC is supplied to perform start-up process. The ramps are provided at the interval of 1.4 s to establish the alignment of the rotor with the rotating flux linkage. The wheels attain maximum of 69.9 rpm during start-up mode, shown in Fig. 7a and c. In freewheel or rolling mode,

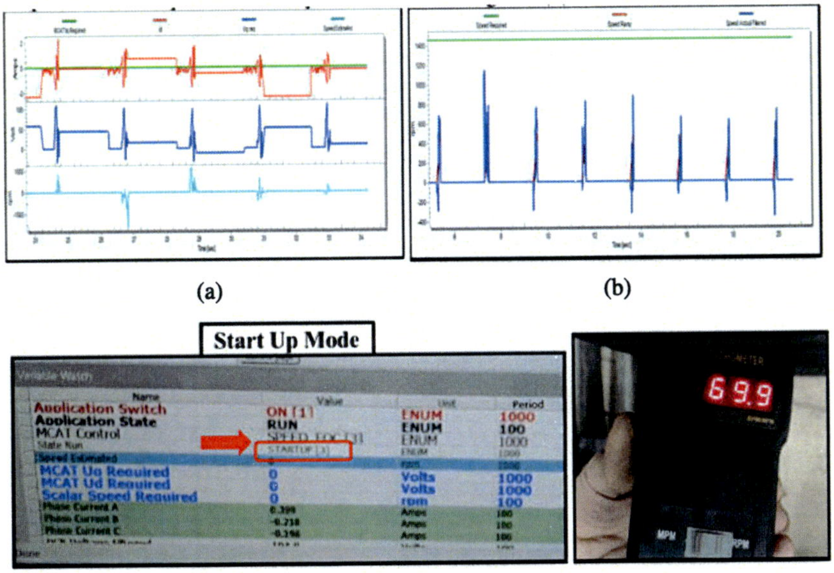

(a) (b)

(c)

Fig. 7 Response during start-up mode **a** speed, Uq, d, and q axes plots, **b** speed response, **c** FreeMASTER 2.0 variable watch window and tachometer reading

vehicle performs the apparent check of currently effective driving speed, shown in Fig. 8a and b. At the end of start-up and freewheel mode, the logic is switched to sensorless closed loop control. For different speeds of the motor, the estimated and measured values of vehicle speeds are recorded in Table 3. The error between estimated and measured speeds is ~0.05% which verifies the feasibility of belt-pulley transmission system used in the emulator system. When speed of the motor is changed from 0 to 700 rpm, the motor reaches to required speed with acceleration of 104.7 rad/s^2 (Fig. 9a). The phase currents in stator winding are shown in Fig. 9b. For braking mode test, the speed trace of the motor during braking from 780 to 0 rpm is shown in Fig. 10a (deceleration @ 81.68 rad/s^2). The comparison of measured

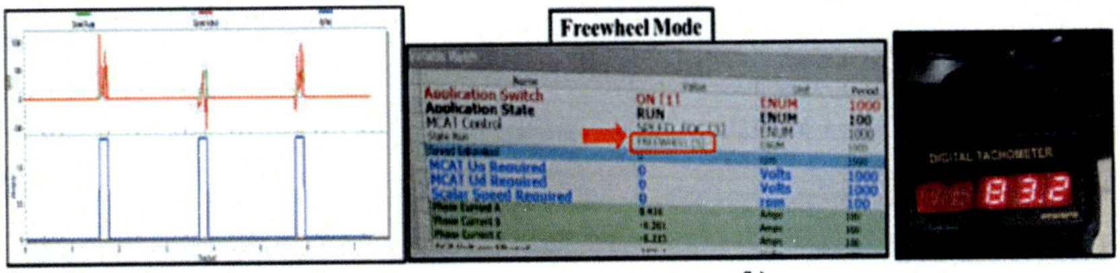

(a) (b)

Fig. 8 Response during freewheel mode **a** current and speed response showing initial speed ramps followed up by the motor, **b** FreeMASTER 2.0 variable watch window and tachometer reading

Table 3 Spin mode test along with belt-pulley system at different speeds of motor

Speed command to motor (rpm)	Estimated angular speed of wheel (rpm)	Measured angular speed of wheel using Tachometer (rpm)	Linear speed of vehicle using Eq. (1) (km/h)
93	41.3	39.1	3.89
814	361.7	360.2	34.09
1870	831.1	829.5	78.33

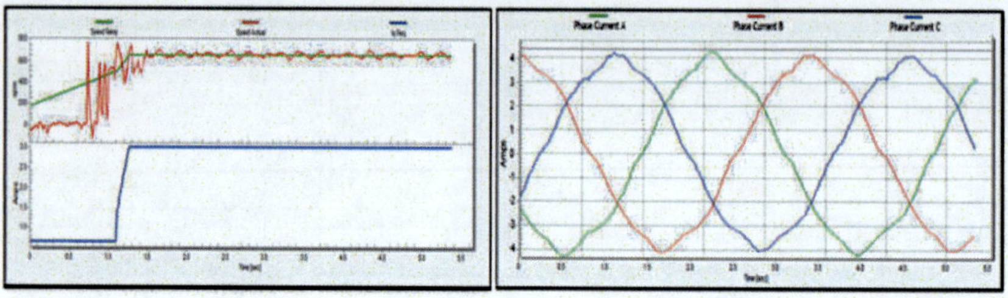

(a) (b)

Fig. 9 **a** Response during acceleration mode (0 rpm–700 rpm), **b** steady-state phase current waveforms in motor

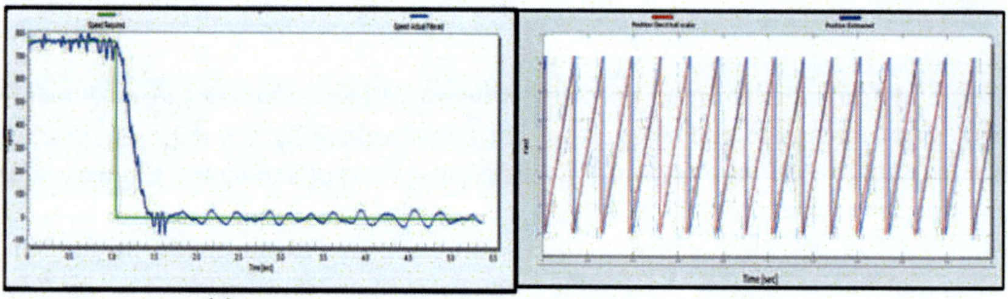

(a) (b)

Fig. 10 **a** Response during braking mode (780 rpm–0 rpm), **b** comparison of measured and estimated rotor position

and estimated rotor position is shown in Fig. 10b. To predict the driving patterns and reducing the road test expense, the drive cycle test is applied to the emulator system. For this, NEDC is designed and fed as input command to the system shown in Fig. 11a. The system speed response with respect to drive cycle is shown in Fig. 11b.

4 Conclusion

The following realizations are encountered in this paper successfully:

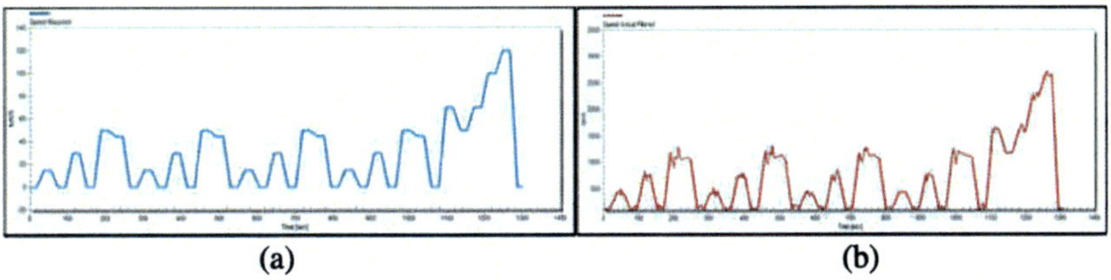

Fig. 11 a Drive cycle for implying vehicle test, b test bench speed response

- The sensorless FOC is implemented and analyzed in MATLAB/Simulink environment. It is thus tested via TWR-KV46F150 MCU in emulator structure which enables the user/tester to analyze the EVPT system in real time for different modes of operation, such as start-up, freewheel, spin/run, or braking modes. The FreeMASTER 2.0 software communicates via USB interfacing and IDE tool KDS.
- The analysis results show the robustness and dynamically fast response of the traction motor with respect to the variation in torque, speed, or power requirements.
- The simulated and tested controller performs the prompt action of providing the accurate control over the EVPT system under different operating conditions. The vehicle speed calculated via belt-pulley arrangement is established in agreement with the analytically estimated positions.

References

1. Mounir Z, Mohamed B, Demba D (2006) Electric motor drive selection issues for HEV propulsion systems: A comparative study. IEEE Trans Veh Technol 55(6):1756–1764. https://doi.org/10.1109/TVT.2006.878719
2. Hossein D (2015) Global role and collaboration of OEM and suppliers in making of successful electric vehicles. IEEE Transportation Electrification Conference and Expo. https://doi.org/10.1109/ITEC.2015.7165722
3. Berman B, Gleb G (1974) Propulsion Systems for electric cars. IEEE Trans Veh Technol 23(3):61–72. https://doi.org/10.1109/T-VT.1974.23575
4. Emma A, Torbjorn T (2016) Performance Analysis of current BEVs based on a comprehensive review of specifications. IEEE Transactions on Transportation Electrification. 2(3):270–289. https://doi.org/10.1109/TTE.2016.2571783
5. Bilin A, Erkin D (2011) A control strategy for parallel hybrid electric vehicles based on extremum seeking. International Journal of vehicle mechanics and mobility 50(2):199–227. https://doi.org/10.1080/00423114.2011.577224
6. Dong Q, Nga T-T, Han HC, Jin-Woo J (2014) Speed control system design and experimentation for interior PMSM drives. Int J Electron 102(5):864–885. https://doi.org/10.1080/00207217.2014.942888

Verification of the Effectiveness of Learning Materials that Support Self-regulation for Learning Considering Differences in Career Resilience: Acquiring Knowledge of Level 3 Automated Driving Vehicles

Maki Arame, Junko Handa, Yoshiko Goda, Masashi Toda, Ryuichi Matsuba, Huiping Zhou, Makoto Itoh, and Satoshi Kitazaki

Abstract This study focused on resilience and examined the effects of individual differences such as self-regulation and learning style on learning. Resilience and self-regulation have a strong relation. Thus, in this study, the value of self-regulation was used the value of resilience instead. The instructional design method was applied to develop text, interactive, and video teaching materials. The teaching materials were designed to support Japanese driver's license holders to recognize learning goals and to create opportunities for reflection when learning about automated driving level 3. If the teaching material development method is effective, it is assumed that the low resilience and self-regulation can be compensated. As a result, developing teaching materials that support goal setting and reflection was effective, and this study found that the influence of learning style differs depending on the media used.

Keywords Conditional driving automation · Resilience · Self-regulation · Learning style · Decision tree analysis

1 Introduction

With the development of automated driving vehicles progressing, automated driving vehicles of level 3 and 4 will be around in the town soon. Drivers need to have

M. Arame (✉) · J. Handa
Polytechnic University of JAPAN, Tokyo, Japan
e-mail: arame@uitec.ac.jp

M. Arame · J. Handa · Y. Goda · M. Toda · R. Matsuba
Kumamoto University, Kumamoto-city, Japan

H. Zhou · M. Itoh
University of Tsukuba, Tsukuba-city, Japan

S. Kitazaki
Automotive Human Factors Research Center, Tsukuba-city, Japan

X.-S. Yang et al. (eds.), *Proceedings of Seventh International Congress on Information and Communication Technology*, Lecture Notes in Networks and Systems 447,
https://doi.org/10.1007/978-981-19-1607-6_78

knowledge about the new types of cars that need a different driving operation [1]. The previous studies indicate that learner has individual differences, including driving experience and learning experience regarding safe driving when they learn about automated driving [2–4]. One of the studies also suggests that the video teaching material has the effect of absorbing career resilience as individual attributes. Another study suggests that motivational teaching material based on narrative has the effect of absorbing individual differences compared to the text-based teaching material.

This study focused on self-regulation as a way to assist learners. Zimmerman (1990) suggests that self-regulation is an important factor influencing academic performance [5]. In recent studies on self-regulation, there are many studies that relate self-regulation to resilience. These previous studies suggest that working on goal setting and providing an opportunity to learn from mistakes are useful ways to improve resilience through self-regulation [6]. This study examined the different types of learning effects of teaching materials that support self-regulation.

2 Theoretical Framework

2.1 *Self-regulation*

Self-regulation has a process including three phases of behavioral management [7]. (A) The forethought phase includes task analysis, motivational beliefs, self-efficacy, and intrinsic motivation. (B) The performance phase includes performance and time management, and assistance requests at the stage of actually conducting tasks. (C) In the self-reflection phase, evaluate what you have done or can improve, manage the emotions, and use self-reflection to restart the cycle.

Studies on teaching methods to promote self-regulation are aimed at individual tasks and teaching materials and examine how to implement interventions that promote self-regulated learning. Self-regulated learning implements scene-limited and context-dependent, and there is no learner who uses self-regulation in all situations.

The study of learning style and self-regulated learning indicates that various learning styles have a positive effect on self-regulation. Learning styles such as *reflective and sensing* based on the Felder-Silverman model tend to show higher value on (b) time management of the circular model. Furthermore, adopting a learning style of *sensing* may promote (c) recognizing the value of his/her tasks and reflect what he/her has done [8].

Based on the self-regulation circulation model, a learner who has basic knowledge and experience about driving a car connects the systematic knowledge about automatic driving level 3 with the learner's own experience and knowledge and reflects on it. If it can be conceptualized, it can be expected to promote learning.

2.2 Teaching Material Development using Instructional Design

As a research field that summarizes methods for enhancing the effectiveness, efficiency, and attractiveness of educational activities, there is instructional design (ID) that utilizes the knowledge taxonomy of educational objective. ID is used to design a whole process of learning with tasks and activities.

In the analysis phase, a needs assessment should be conducted in order to identify problems. Needs indicate a gap between the current situation and what it should be.

In this phase, the current state of learning activities, learning environment, and learning contents is actively investigated and reflects the results in teaching materials and lessons.

Goal analysis is to describe what it should be done as a clear learning goal. Goal analysis has six steps (1) identify an aim, (2) set goals for each aim, (3) refine goals to delete duplicates and to combine similar ones, (4) rank goals by order of importance, (5) refine goals again to find discrepancies between goals and existing tasks, and (6) make a final ranking to determine how important the goal is.

In the design phase, it is important to set the learning goals of each learning activity or task and determine in what order they are treated in a course or lesson.

Determining the order of learning is designing how the learning subjects, as revealed by the subject analysis, reach their learning goals. Furthermore, learning topics and learning procedures should be well examined for the target learners to achieve the main goals. Evaluating learning outcomes should be observable or measurable. Determining the learning goals help understand the scale of the education and identify the elements that make up the clarified learning goals [9, 10].

There are two functions for feedback during learning from an information processing perspective. One is to provide the learner with the correctness of his response or performance. Second one is to provide corrective information to the learner that can be useful to modify or improve his performance [11].

3 Purpose of This Study

Self-regulation has been regarded as important factor to influence resilience [6]. According to research, self-regulation model indicates that planning and goal setting phase are important in the self-regulating process for one's behavior. Furthermore, another study examined that goals were essential for enhancing resilience [7]. In addition, another study pointed out that it is very important to learn from mistakes to enhance resilience through self-regulation [12].

The Felder-Silverman model indicates that each learner prefers different learning styles, but there are few studies that examined the effects of learning styles and resilience during learning [13].

It is estimated that each lesson time at a public training facility for traffic safety education in Japan is about 5 min. If a teaching material that encompasses setting goals and supporting reflection using ID method is developed, it is assumed that it is effective for people with low resilience. In this study, three types of learning materials were developed. They were texts, interactive, and video materials. A learner needs to go through the text material by himself to learn. While, the interactive material includes asking some questions to encourage reflections and giving feedback. The video material provides learning content with explanation and animation. In short, the materials have the same learning contents, yet they have different delivery methods. This study aims to find effective teaching materials for learners with low resilience.

4 Method

4.1 Participants and Procedures

In this study, driver's license holders learned about automated driving systems and driving operation using one of the teaching materials among the text material, the interactive material, and the video materials. They are about five-minute-long learning materials which were developed using ID. Each teaching material was developed to include the same learning goals, which were that the learners were able to explain as follows; (1) the differences of levels of automated driving vehicles and driving operation, (2) the situations or conditions that take over driving, and (3) roles of drivers.

The data were collected using a survey on the Internet in October 2020. The data of the survey were used three different studies in the past [2–4]. This study used the 930 respondents who used only teaching materials without watching motivational materials. The dataset was the first time to be used. The data were the first time to be used. The survey asked questions about demographic data such as gender, age, learning style, and career resilience. Learning effects were examined using the scores of pre-post-tests. The test questions were about the basic knowledge of level 3 automated driving system. Based on the scores of the pre-test and the post-test, the group with a high score is classified the up group and non-up group in which they did not rise or fell. In addition, useful teaching materials for people with low self-regulation and resilience were validated using logistic regression analysis and decision tree analysis. Since Zimmerman pointed out self-regulation as a factor that influences academic performance, in this study, the proxy variable for self-regulation was set from the score of the pre-test.

Table 1 Overview of the four learning styles

Category	pair	
Active-reflective	Active	Active learners prefer learning by doing or actively getting involved in tasks and prefer social interaction
	Reflective	Reflective learners prefer thinking quietly rather than interactively get involved in learning tasks
Sensing-intuitive	Sensing	Sensing learners prefer to deal with facts and concepts
	Intuitive	Intuitive learners prefer managing their own learning. They like finding discovery, innovation, and abstractions
Visual-verbal	Visual	Visual learners prefer learning with photos, diagrams, flowcharts, timelines, videos, demos, etc.
	Verbal	Verbal learners prefer words to visually received information
Sequential-global	Sequential	Sequential learners prefer sequentially organized content such as college courses
	Global	Global learners prefer having a big picture of learning

4.2 Data Analyzes

The learning style used was a simplified version of the model proposed by Felder [7]. The simplified version is used in the survey in the form of a question to the learner, asking three questions each about the four conflicting characteristics. Table 1 outlines the four items.

Four measurements of career resilience based on the previous studies show that factor 1 refers to problem-solving and adaptability, and factor 2 is novelty and diversity of interests. Factor 3 is a positive future-oriented factor, including optimism. Factor 4 is one of the social skills related to helping behavior. It consisted of 14 questions, and the answers were calculated item by item in five stages, from "very (5 points)" to "not at all (1 point)" [14].

5 Results of Basic Analyzes

5.1 Results of Test Scores by Each Teaching Material

The basic attributes of respondents were divided into 6 categories, up to 39 years old, 40–49 years old, 50–59 years old, 60–69 years old, and 70 years old and over, and 62 people in each age group were half men and women each material. In short, 310 people responded for each material, and the total number of respondents was 930 people. Figure 1 indicates test scores results for each material. Post-learning test results were significantly higher for all materials. The score increase rate of the pre-post-test for each teaching material type was 1.08 times for pamphlets, 1.17 times for

Fig. 1 Test score results

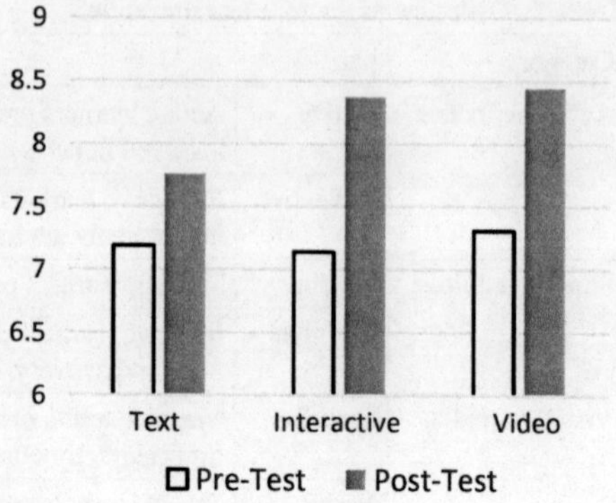

interactives, and 1.15 times for videos. Interactive teaching material has the highest score increase rate and the highest standard deviation decrease rate.

5.2 *Results of Learning Style and Career Resilience for Each Material*

Table 2 shows the mean and *standard deviation of* learning style scores, and there was no significant difference between the interactive and video materials.

In terms of resilience, there was no problem with internal consistency, i.e., factor 1 of α is 0.76, factor 2 of α is 0.80, factor 3 of α is 0.90, factor 4 of α is 0.81, total of all items of α is 0.80.

Table 2 Learning style scores for each teaching material

		Text $N = 310$	Interactive $N = 310$	Video $N = 310$
Active-reflective	Mean	0.38	0.39	0.35
	Standard deviation	0.49	0.49	0.48
Sensing-intuitive	Mean	0.75	0.70	0.67
	Standard deviation	0.43	0.46	0.47
Visual-verbal	Mean	0.65	0.64	0.65
	Standard deviation	0.48	0.48	0.48
Sequential-global	Mean	0.69	0.65	0.64
	Standard deviation	0.46	0.48	0.48

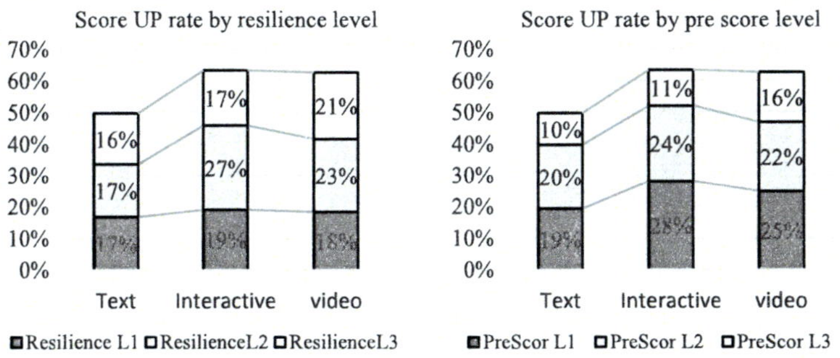

Fig. 2 Comparison of test scores by 3 levels of resilience and 3 levels of pre-score

5.3 *Score Up Group Rate by Teaching Material and Resilience*

Figure 2-left shows the percentage of people in the group whose ex-post-test increased by leveling the total resilience score into three stages, with the low group as level 1, the middle group as level 2, and the high group as level 3. Figure 2-right shows the percentage of people in the group whose ex-post-test increased by leveling the pre-test score into three stages, with the low group as level 1, the middle group as level 2, and the high group as level 3.

As a result of the analysis, it was the interactive teaching material and video teaching material that had the highest percentage of increase in scores by resilience level and pre-scoring. Furthermore, in all the teaching materials, the percentage of the number of people who scored higher at resilience levels 1 and 2 was higher than that at level 3. From this, it is possible that the development of teaching materials using ID is effective for people with low resilience. Similarly, the percentage of people who scored higher in advance score levels 1 and 2 is higher than in level 3. From these, there is a possibility that the development of teaching materials using ID is effective for people with low self-regulation.

6 Factors and Priority Analysis

6.1 *Logistic Regression Analysis*

Logistic regression was used to investigate how much affect one's learning regarding learning material type, learning style, resilience level, and pre-test score level. Table 3 indicates the results of all respondents. In the learning style, (active) had an influence. This result was supported by the previous studies. Tables 4, 5, and 6 show the results by teaching material. It was confirmed that the learning styles that affect each material type differ.

Table 3 Logistic regression analysis of all respondents

	B	Significance	Exp(B)	Lower limit	Upper limit
Pre-test level (3 levels)	– 0.59	0.00**	0.56	0.47	0.66
Resilience level (3 levels)	0.34	0.00**	1.40	1.17	1.68
Teaching material type	0.30	0.00**	1.35	1.14	1.60
Active-reflective	– 0.18	0.02*	0.83	0.71	0.97
Sensing-intuitive	0.06	0.48	1.06	0.90	1.25
Visual-verbal	0.08	0.30	1.08	0.93	1.25
Sequential-global	0.14	0.07	1.15	0.99	1.35

$^{*}p < 0.05$ $^{**}p < 0.01$ Nagelkerke R^2 0.10 $N = 930$

Table 4 Logistic regression analysis of text teaching material

	B	Significance	Exp(B)	Lower limit	Upper limit
Pre-test level (3 levels)	– 0.44	0.00**	0.65	0.48	0.86
Resilience level (3 levels)	0.32	0.04*	1.37	1.02	1.85
Active-reflective	– 0.24	0.07	0.78	0.60	1.02
Sensing-intuitive	0.06	0.67	1.06	0.80	1.42
Visual-verbal	0.11	0.42	1.11	0.86	1.43
Sequential-global	0.22	0.10	1.24	0.96	1.62

$^{*}p < 0.05$ $^{**}p < 0.01$ Nagelkerke R^2 0.07 $N = 310$

Table 5 Logistic regression analysis of interactive teaching material

	B	Significance	Exp(B)	Lower limit	Upper limit
Pre-test level (3 levels)	– 0.75	0.00	0.47	0.34	0.64
Resilience level (3 levels)	0.36	0.03*	1.43	1.03	1.99
Active-reflective	– 0.28	0.04*	0.75	0.58	0.99
Sensing-intuitive	– 0.10	0.49	0.90	0.67	1.21
Visual-verbal	0.15	0.28	1.16	0.89	1.50
Sequential-global	0.17	0.23	1.18	0.90	1.56

$^{*}p < 0.05$ $^{**}p < 0.01$ Nagelkerke R^2 0.15 $N = 310$

Table 6 Logistic regression analysis of video teaching material

	B	Significance	Exp(B)	Lower limit	Upper limit
Pre-test level (3 levels)	−0.59	0.00	0.56	0.41	0.75
Resilience level (3 levels)	0.35	0.03	1.42	1.04	1.93
Active-reflective	−0.05	0.73	0.95	0.73	1.25
Sensing-intuitive	0.21	0.13	1.24	0.94	1.63
Visual-verbal	0.00	1.00	1.00	0.77	1.30
Sequential-global	0.06	0.69	1.06	0.79	1.42

$^{*}p < 0.05$ $^{**}p < 0.01$ Nagelkerke R^2 0.08 $N = 310$

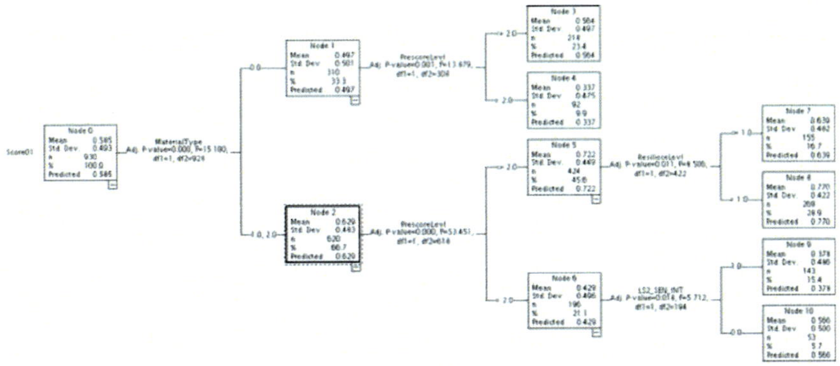

Fig. 3 Results of decision tree analysis

6.2 Decision Tree Analysis

SPSS statistics v27 was used for the decision tree analysis, and exhaustive chi-squared automatic interaction detector (CHAID) was used for the growing method. As a result of the decision tree analysis using two groups as the dependent variables, the influence of pre-scoring was the largest, and the middle and low groups (1 and 2) accounted for 69% (642 people). The top group (3) accounted for 31% (288 people). Decision tree analysis branches in descending order of influence on dependent variables.

The group with a high-pre-test score was influenced by the learning style (sensing-intuitive), and the group with a high-intuitive score in the learning style became the up group. The groups with medium and low pre-scores were strongly influenced by the type of teaching materials, the interactive teaching materials and video teaching materials became the up group, and the group with high resilience became the up group. The misclassification rate was 21.9% when classification and prediction were performed based on the conditions obtained from the results of the decision tree analysis results (Fig. 3).

The result of specifying the teaching material type as the first branching condition was that both interactive and video materials were influenced by the pre-score level. But the video materials were influenced by the learning style (sensing-intuitive), and the group with high sensing became the up group.

7 Discussions

This study considered how to develop effective learning materials that take into accounts individual characteristics such as career resilience and learning style using three patterns of online teaching material and analyzed the effectiveness of individual characteristics. As a result of the verification, it was confirmed that developing a teaching material design that supports self-regulation may be effective for those who are low resilience and low self-regulation. On the other hand, when using teaching materials with different media characteristics, the learning style has an effect. This study found that there were effects of active-reflective and sensing-intuitive. Furthermore, the interactive teaching materials absorb the difference in learning styles most. In short, this study found three points:

1. Developing teaching materials that promote reflection using ID has the effect of supporting self-regulation for learning.
2. Interactive teaching materials may be able to absorb the influence of sensing-intuitive more than video teaching material.
3. As the previous studies show, there is a correlation between resilience and self-regulation.

8 Conclusion

In this study, three types of teaching materials of automated driving level 3 were developed using ID and were examined whether they could support self-regulation for learning.

- Text teaching material that requires a learner to go through the learning contents by himself, i.e., a learner should learn actively to perceive the text information.
- Interactive teaching material that encourages learners' reflection by asking questions and giving feedback, i.e., it promotes a learner to reflect what he learns.
- Video teaching material that learning content with explanation and animations, i.e., a learner can receive clear information of learning contents by just watching it.

The study found that materials that automatically guide the learning process, such as interactive and video materials, are more effective for people with low resilience.

Moreover, as a result of verifying the influence on learning by logistic regression analysis and decision tree analysis, it was confirmed that the learning style was influenced by the teaching material type.

Since developing automated driving vehicles were remarkable and soon to be around in our life, finding effective educational methods for drivers and pedestrians are an urgent issue.

9 Limitations and Recommendations

There are limitations to this research design. The interactive teaching material used in this study did not include the responses differently depending on the learner's understandings. For the future studies, simulation-type interactive teaching materials should be developed.

It may be interesting to examine whether or not adding more various feedback messages depending on the learner's reaction or understanding, the different learning styles could be absorbed using simulation-type interactive materials.

Moreover, the level of self-regulation was estimated from the results of the pre-test and examined its relationship with resilience in this study. There was a correlation between pre-test and resilience levels, but it has a low correlation. This study used pre-test score as the alternative variance for self-regulation abilities and skills, however, it should be considered to use more appropriate or accurate values.

More detailed examination is needed whether supporting self-regulation works for enhancing resilience in the same way.

Acknowledgments This work was supported by Council for Science, Technology and Innovation (CSTI), Cross-ministerial Strategic Innovation Promotion Program (SIP), entitled "Human Factors and HMI Research for Automated Driving" (funding agency: NEDO).

The authors would like to thank Mr. Miki Akamatsu in Tokyo Business Service Co., Ltd. to support this project.

References

1. Zhou H, Itoh M, Kitazaki S (2019) Long-term effect of experiencing system malfunction on driver take-over control in conditional driving automation. In: Proceedings of The 2019 IEEE International Conference on Systems, Man, and Cybernetics (SMC2019), pp 1950–1955. Bari, Italy
2. Arame M, Handa J, Goda Y, Toda M, Matsuba R, Zhou H, Itoh M, Kitazaki S (2020) Learning effects of different learning materials about automated driving level 3: evidence from a propensity score matching estimator. In: Proceedings of Fifth International Congress on Information and Communication Technology, ICICT 2020, London, vol 2, pp 387–394
3. Arame M, Handa J, Goda Y, Toda M, Matsuba R, Zhou H, Itoh M, Kitazaki S (2021) Using narrative based video on gaining safety driving: focusing on career resilience and learning style in automated driving level 3. In: Proceedings of Sixth International Congress on Information and Communication Technology, ICICT 2021, London, vol 2, pp 787–799
4. Arame M, Handa J, Goda Y, Toda M, Matsuba R, Zhou H, Itoh M, Kitazaki S (2020) Effects of learning materials about automated driving level 3 focusing on frequency of driving: verification by propensity score matching. In: 2020 Fourth World Conference on Smart Trends in Systems, Security and Sustainability (WorldS4), pp 454–460. https://doi.org/10.1109/WorldS450073.2020.9210310
5. Zimmerman BJ (2008) Investigating self-regulation and motivation: historical background, methodological developments, and future prospects. Am Educ Res J 45:166–183
6. Nota L, Soresi S, Zimmerman BJ (2004) Self-regulation and academic achievement and resilience: a longitudinal study. Int J Educ Res 41(3):198–215
7. Zimmerman BJ, Labuhn AS (2012) Self-regulation of learning: process approaches to personal development. In: APA Educational Psychology Handbook, Vol. 1: Theories, Constructs, and Critical Issues, Zimmerman BJ, Labuhn BJ (Eds). American Psychological Association, Washington, DC, pp 399–425
8. Çakiroğlu Ü, Er B, Uğur N, Aydoğdu E (2012) Exploring the use of self-regulation strategies in programming with regard to learning styles. Int J Comp Sci Educ Schools 2(2). https://doi.org/10.21585/ijcses.v2i2.29

9. Morrison GR, Ross SM, Kemp JE (2001) Designing effective instruction, 3rd edn. John Wiley & Sons, Inc.
10. Goda Y, Yamada M, Matsuda T, Saito Y, Miyagawa H (2014) Plan and reflection of self-regulated learning: perspectives of outside classroom learning hours and English proficiency. J Japan Society Educational Technol 38(3):269–286
11. Reiser RA, Dempsey JV (Eds) (2018) Trends and issues in instructional design and technology, 4rd edn. Pearson, New York
12. Artuch-Garde R, González-Torres MC, de la Fuente J, Vera MM, Fernández-Cabezas M, López-García M (2017) Relationship between resilience and self-regulation: a study of Spanish youth at risk of social exclusion. *Frontiers Psychol* 8, Article 612
13. Felder RM, Silverman LK (1988) Learning styles and teaching styles in engineering education. Engr. Education 78(7):674–681
14. Kodama M (2015) Examination of construct of career resilience and development of measurement scale. Psychology Research 86(2):150–159

ICT-Enabled Vehicle Theft Detection and Recovery System

Kamlesh Kumawat and Vijay Singh Rathore

Abstract In this research paper, a new automatic system is introduced to detect vehicle and recover the same at toll plaza. This aims to propose and implement a new security system based on RFID, ANPR, GSM, OTP, and car lift up technology. The system helps the police department to detect theft vehicle and to recover that vehicle and also to improve the vehicle theft recovery rate in India. According to the articles published in newspapers, vehicle theft detection and recovery are the cases which are least solving in India. After a vehicle is theft, there are so many situations where police department and toll plazas are not able to detect and recover the same. Currently, the ETC system helps to detect theft vehicle at toll gates, but there are some limitations with this system also. To solve these limitations, a new system is proposed in this research paper. This new automatic vehicle theft detection and recovery system (AVTDRS) based on ICT which will be worked at toll plazas. This paper also approaches currently working technologies like GSM, GPS, OTP, RFID, smart phone applications, QR codes, and fingerprint identification system. These all technologies are currently using for detection and recovery of vehicles in India. The prime motive of written paper is to observe boundaries of present working system and to develop an updated automatic system to solve these problems.

Keywords ETC · RFID FASTag · Vehicle theft detection system · Regional transport office (RTO) · GPS · One time password (OTP) · QR code · Vehicle lift up technology · GSM

1 Introduction

The number of vehicle are continuously increasing as well as theft vehicle cases also rapidly increasing, and it is getting more challenging to detect and recover theft

K. Kumawat (✉) · V. S. Rathore
Department of CS & IT, IIS (Deemed to be University), Jaipur, India
e-mail: kamleshdal.rkd@gmail.com

V. S. Rathore
e-mail: vijaysingh.rathore@iisuniv.ac.in

X.-S. Yang et al. (eds.), *Proceedings of Seventh International Congress on Information and Communication Technology*, Lecture Notes in Networks and Systems 447,
https://doi.org/10.1007/978-981-19-1607-6_79

vehicles. The reports published in some newspapers show that the realization rate of stolen vehicles detection is very inferior. These theft vehicle cases are the minimal solving cases also. There are so many technologies currently working to detect these theft vehicles. These technologies are GSM, GPS, RFID, OTP, QR code, smart phone applications, and fingerprint identification system.

The Indian government mandates radio frequency identification (RFID)-enabled tags called FASTag in India. These tags are mounted on the windshield of vehicle and help to deduct the amount of toll without stopping vehicle at toll plazas. These FASTag is very helpful in detection of theft vehicle while crossing the toll gates. After any vehicle stolen by thief, the registered vehicle owner reports FIR to the police department and the police department shares detail information of vehicle to regional transport office (RTO). Then, RTO blacklisted the RFID FASTag of that particular vehicle. Whenever the blacklisted tag mounted vehicle reaches at toll tax plaza, RFID reader detects it as blacklisted tag by police and detects the vehicle as theft vehicle and also the authorities do not allow that vehicle to cross the gates. Simultaneously, the system automatically sends information to nearest police stations, and the vehicles registered owner through SMS or mail. Still in so many cases, the existing system not works properly.

The following are the detailed analysis of the problem:

1.1 *Statistics of Theft Vehicles in India*

- According to article published on Jan 20, 2018 by Times of India shows that at Mohali, India, the realization rate of theft vehicles was 13% only. Also the continuously enhancing theft cases show the stealers not so scares of cops [1].
- Article published in another newspaper dated Jan 10, 2019 shows that over 5 vehicles/hour reported stolen at Delhi. According to the report, the theft vehicle cases were 44,158 in 2018, which were 39,084 in 2017. Among the stolen vehicles, 8,036 (18.20%) were only cars and 4,619 (10.46%) stolen vehicles were found and only 6,751 thieves arrested [2].
- "The Hindu" published that according to the data provided by police departments shows that the vehicle theft cases are the least solved crimes in India [3].

1.2 *Present Theft Vehicle FIR Lodge Process*

Whenever a vehicle stolen, the first process takes place is to lodge FIR of the same. Here, the vehicles owner can lodge FIR through two processes. Either owner can go through manual FIR process at police station or can submit the information through online application for lodging FIR. In some states, state governments have been launched Web application to lodge online FIR for theft vehicle from anywhere and anytime. These online applications bring transparency. The uploaded data can be

Fig. 1 FIR lodge Web application [4]

accessible by both theft vehicle owner and police officials. Now, these applications are for limited FIR's includes vehicle theft and vehicle lifting cases. If someone submit forge information and lodge fake FIR, then the required action can be taken by police department.

The Web app initiated by New Delhi Govt. of India contains following links:

1. FIR of vehicle register against the theft vehicle
2. Retrieve FIR of lodged FIR
3. Retrieve final report of lodged FIR
4. Status of lodged FIR
5. FAQs

With the help of above links, the vehicle owner can easily lodged FIR by entering information about vehicle, owner of the vehicle, and place of vehicle theft. The person also can get FIR copy in print out format [4].

The whole process makes the system hassle free, but the recovery rate of the theft cases is still less as per database (Fig. 1).

1.3 Present Theft Vehicle Detection and Process of Recovery

In present theft vehicle detection and recovery process, when a vehicle theft and FIR lodged against vehicle, the police authorities submit information of stolen vehicle to RTO. Then, RTO blacklists the FASTag which is RFID tag applied on the theft vehicle. Whenever the vehicle reaches at toll plazas, the RF reader reads the tag as refused to deduct money, and the whole information automatically sent to the nearest police departments with vehicles registered owner through email and SMS. Still in so many situations, toll plazas cannot detect vehicles as theft.

Insurance Information Bureau (IIB) introduced V-Seva service app helps in detection of the theft vehicle in year 2014. This application accommodates whole information about theft vehicle. The facts have been collecting from public, department of police, and insurers of the vehicles. Police authorities can detect vehicles by using this, but they requires sufficient database of the vehicle like chassis number and engine number [5].

Another application is "Vahan Saman- vaya" launched in 2016, works for checking and tracing the current status of stolen vehicle [6].

1.4 Toll Plaza's Role in Theft Vehicle Detection and Recovery

Figure 2 shows automated toll collection system; according to this, the vehicle owner apply FASTag on the front windscreen. Whenever the car arrives at plaza, the RF reader scans the tag by sending radio waves. This activates the vehicles FASTag. This tag sends information of the car to the reader. Then, the reader conveys tag details to the toll plaza lane controller. This lane controller transfers the vehicle information to

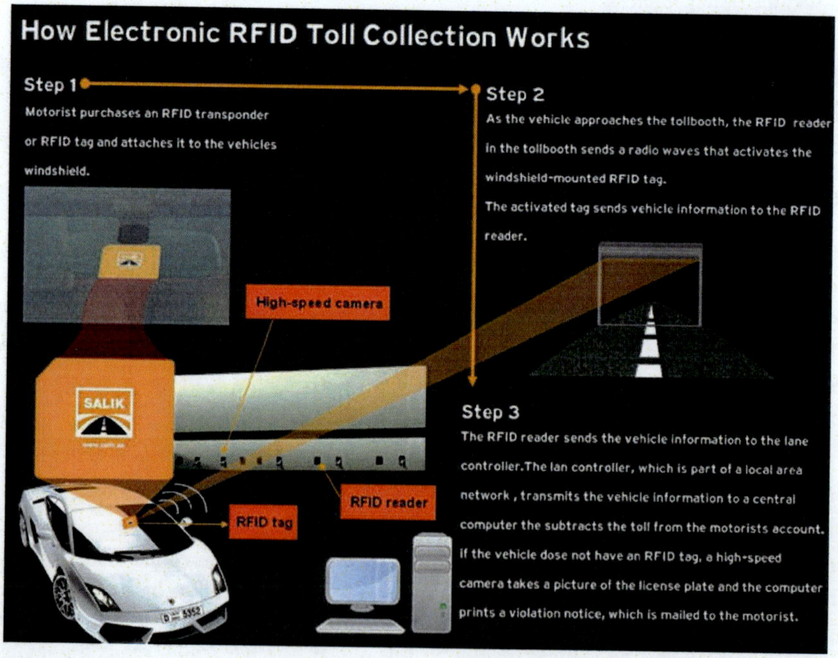

Fig. 2 Roll of ATCS [7]

the central server and server withdraws the toll amount from the car owner's available balance.

2 Literature Review

A friendly system proposed in year 2016. In this system, the theft vehicle owner can inform to toll gate authorities about the stolen vehicle directly and also owner can send a secret number. The system is, whenever the motor vehicle crosses the toll gate, the software exhibits information of that on the screen, also detects it as theft. Then, toll authorities ask a secret number to vehicles driver. If the given number is valid, toll gate opens otherwise the gate remains close and a secret message automatically sent to the registered contact number of vehicle owner along with plaza details [8].

A system introduced in year 2018, exercised communication protocol, API, and RFID. The motive of that paper was to develop fresh approach for apprehending thief using RFID technology. Introduced system also narrates the uses of current RFID technology and ETC, which automatically deducts the toll from vehicle owner's account during cross the toll gates. The system also allows the owner to change RFID tags like passwords. The paper also describes RCTDAS application from which vehicle owner can forge FIR for missing car anytime from anywhere [9].

A new system proposed by Velantina and Viniatha in year 2019 [10]. This toll collection system is based on IoT. The toll plazas can deduct the toll amount of the vehicle by presenting a RFID card. The reader reads it and gets details of the card. The LCD screens also use to show cards balance message. Also amount of toll reduces from cards available amount. The information of amount deduction will be sent to registered contact number after toll deduction. Red color LED lights indicates toll operation as default, yellow color light defines toll collection in process, and green color light shows toll deducted successfully. After completing whole process, the vehicle will be able to pass toll gates [10].

A fingerprint and driver identification also introduced in year 2019. In this, an application developed for improvement of car security and identification. A new identification model and driver profiling also introduced in this research, which is based on various data collections. The car's driver can login to the application also can get his detailed profile from there. In any case of sold vehicle, owner can log out and can reset the application to its initial settings. This paper was very helpful for vehicles driver identification [11].

Another article used vehicle number plate and color identification system to identify whether it is modified or not. The system used microcontrollers with some modules. The police authorities also uploaded theft vehicles every detail for the more accurate detection. There was a digital signature built within the system, helped in vehicle tracking. This activated breakdown mode and immobilizes the vehicle, few seconds later it crosses the toll plaza [12].

New number plate recognition system proposed by Akhtar and Ali in year 2020 [13]. The system firstly captures vehicles number plate before collecting toll amount at plaza. According to this system, major four steps have to be followed: (1) reprocessing, (2) number plate localization, (3) character segmentation, (4) character recognition. According to the experimental results, this methods accuracy is 90.9%. Thus, the system uses character recognition, automatic number plate recognition system (ANPR) and edge detection technologies [13].

2.1 Constraints of the Existing Theft Vehicle Detection and Recovery System

1. Need to propose model to detect and recover vehicles passing through toll plaza in below situations:
   - Theft vehicle approaches at plaza with original FASTag and number plate.
   - Theft vehicle approaches plaza with altered number plate and fake or no FASTag.
   - Theft vehicle approaches plaza with altered number plate and fake or no FASTag.
2. Need to propose an application from which the owner and police department can get the overall status of the vehicle like recovery, place of theft, place of recovery, etc.
3. No secure system available to detect weapon carried by thief in theft vehicle, and it is extreme risky to stop vehicle after detecting as theft vehicle at toll plaza.
4. There is a speed limit in existing system of the speed which is more than the limited speed, and the detectors cannot detect the same.

3 Proposed System: Automatic Vehicle Theft Detection and Recovery System at Toll Plazas in India

3.1 Working Mechanism

According to current process, whenever the stolen vehicle reaches at plaza, RF reader scans the tag applied on the windscreen. In any situation if the FASTag is showing blacklisted, silent alarm activates and plazas gate do not open. Also the authorities do not allow vehicle to cross the gate. Then, the electronic toll collection (ETC) system sends a message or mail automatically to vehicle owners registered detail and nearest police station.

In this system, whenever a theft vehicle will be reached at any plaza, RF reader will scan FASTag, and after detecting vehicle as theft vehicle, it will not be permitted to pass the gates. Proposed system will automatically inform police station nearby the plaza and owner of the vehicle by sending a SMS. If there is no RFID FASTag available on the vehicle, the vehicle information will be getting by using ANPR technology by just capturing image of theft vehicles number plate, and an OTP will be sent to the owner's registered contact details. If OTP is valid, the vehicle will be permitted to cross the gate; but if there will be forge number plate on vehicle, then the driver will not be able to provide the OTP and the silent buzzer buzz, and SMS will be sent to nearest police authorities and registered owner.

During this detection and recovery process, the vehicle will be lifted up to avoid long waiting queue and any violence by vehicle driver.

The overall system will be very helpful to detect the theft vehicle as well as recover the same.

3.2 Elements used in Proposed System

- **FASTag (RFID Technology)**: This RFID technology contains two parts: (1) RF reader and (2) RF tag which is FASTag. The FASTag contains frequency waves, which helps to identify objects as well as people. This is a no wire technology. It is long distance technology works without requiring a sightline between RFID tag and RF reader. For buying RFID FASTag, owner of the vehicle have to register himself inserting following details. Then, fix tag on vehicles windshield and reader reads the tag to deduct toll amount at plaza [14].
- **CENTRAL HUB/SERVER**: It is big database storage. CS handles various documentations at a time and has structured searching algorithm and also provides fast response to the scanned inputs. Central server helps to manage all necessary documentations and maintain toll transactions with balance amount status [15].
- **AUTOMATIC NUMBER PLATE RECOGNITION SYSTEM (ANPR)**: This is system vision technology which helps to recognize number plate of vehicle without any person intervention. The cameras automatically capture vehicle number plate image and extract the characters from the same. These extract characters will be searched in the collected data to identify the owner [13].
- **GLOBAL SYSTEM FOR MOBILE COMMUNICATION (GSM)**: This helps the system to communicate by sending and receiving messages [16].
- **BUZZER**: This is a alarming device uses at plazas. It needs oscillator circuit and speaker for beeping noise with DC voltage [16].
- **VEHICLE LIFT UP TECHNOLOGY (VLT)**: This technology is useful to lift up the vehicle for parking at upper floors. In our proposed system, this technology will be used to lift up the theft detected or suspect vehicles to the upper floor avoid traffic jam and make the traffic smoother. On the upper floor, specific team will be available to identify, detect, and recover the theft vehicle. This technology includes following components: pallet, turn table mechanism, elevator rails, geared machine, lift cart, geared machine, control system, car buffer, and counter weight buffer [17].

4 Methodology

4.1 **Data Collection:** In this step, the data will be collected of theft and non-theft vehicle. This data will be contained following information: (1) vehicle registration number, (2) owner detail, (3) RFID number of the vehicle, (4) vehicle type and number of wheels of the same.

4.2 **Designing a proposed model:** After data collection, a model will be designed for detection and recovery of theft vehicle. The model will be an ICT-enabled automatic vehicle theft detection and recovery system. This will include different technologies which are RFID, ANPR, GSM, OTP system, and central server.

4.3 **Implementation:** The proposed model will be implemented at toll plazas where the RFID-enabled vehicle will be passed through the toll gates.

4.4 **Analyze and validate the proposed model:** In this stage after implementing the proposed model, it will be analyzed, the results will be compared, and the system will be validated.

5 Experimental Setup

There will be three steps to design the proposed system. (1) Implement the system to detect the vehicle, (2) design a mechanism to inform the nearby police department and vehicles owner, (3) recover the theft vehicle. This whole system will be included different technologies: central server (CS), camera, GSM module, RFID reader, RFID FASTag, sensors, and vehicle lift up technology.

Below flowchart shows the framework of the proposed system. In this, the camera will take pictures of number plate of vehicle, and data will be shared to central server by USB channel. The database of vehicle stored in central server. Detection process will be processed after image processing within the CS. The GSM module will be used for communication between toll plaza server and the user, and through this technology, the OTP and SMS will be sent u authorized person. To make the process smooth and hassle free, vehicle lift up technology will be used. In this, the vehicle will be lift up by authorities in doubt of theft vehicle.

5.1 Data Flow Diagram

The following diagram shows the working flow of the proposed system. The system starts working after theft vehicle FIR lodged by owner at police department and vehicle reaches at any toll plaza:

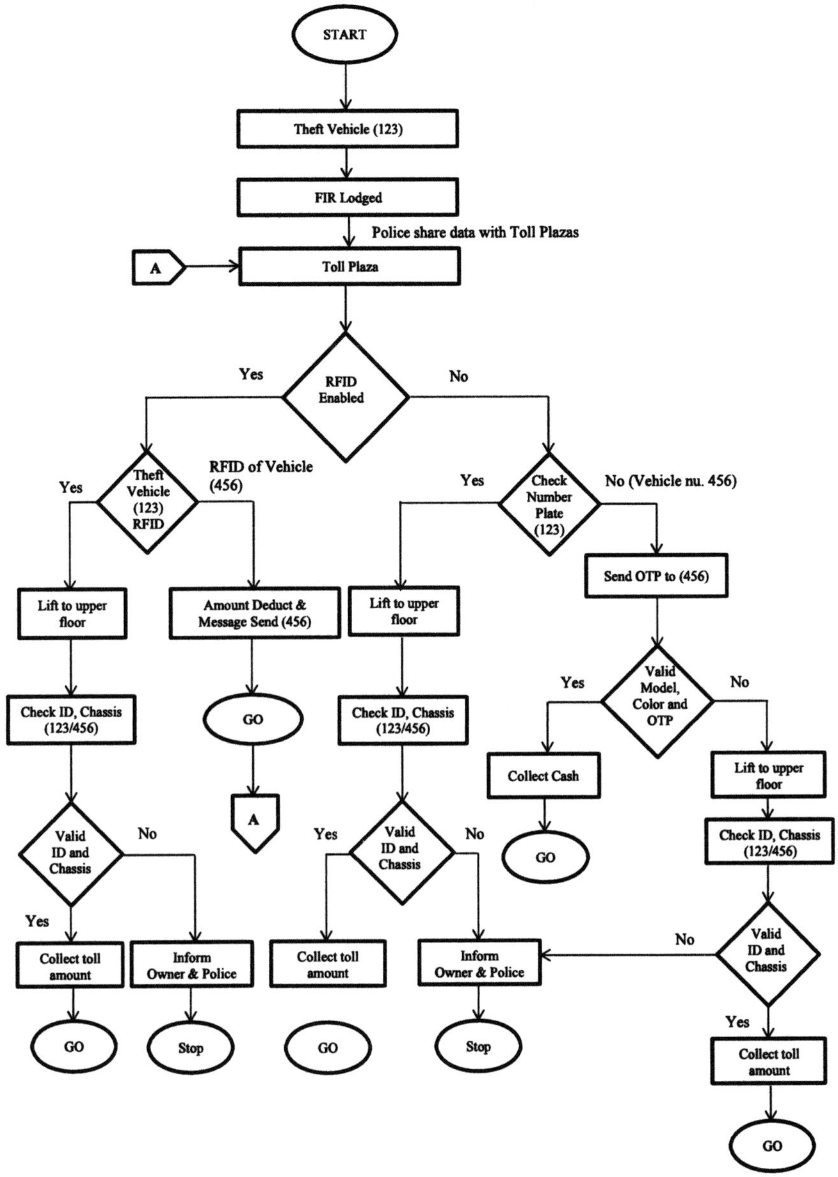
START
Theft Vehicle (123)
FIR Lodged
Police share data with Toll Plazas
A
Toll Plaza
RFID Enabled
Yes
No
Theft Vehicle (123) RFID
RFID of Vehicle (456)
Yes
Check Number Plate (123)
Yes
No (Vehicle nu. 456)
Send OTP to (456)
Lift to upper floor
Amount Deduct & Message Send (456)
Lift to upper floor
Valid Model, Color and OTP
Yes
No
Check ID, Chassis (123/456)
GO
Check ID, Chassis (123/456)
Collect Cash
Lift to upper floor
A
GO
Check ID, Chassis (123/456)
Valid ID and Chassis
No
Yes
Valid ID and Chassis
No
Yes
Valid ID and Chassis
No
Yes
Collect toll amount
Inform Owner & Police
Collect toll amount
Inform Owner & Police
GO
Stop
GO
Stop
Collect toll amount
GO

6 Results and Discussion

- This system will be built with central server connected with RFID technology, camera, sensors, lift up technology, and GSM module.
- Training database of theft and non-theft vehicles will be stored in the server at first.
- Then, system will be processed to read details RF FASTag from the vehicle and automatically extract information about the vehicle.
- There could be four type of databases entered for system testing and results checking: (1) information of theft vehicle with RFID FASTag, (2) information of theft vehicle without RFID and with original number plate, (3) information of theft vehicle without RFID and with forged number plate, (4) information of non-theft vehicle.
- After checking all the situations according to the proposed system, the system will be detected the vehicle as theft or not. If the vehicle is detected as theft vehicle, GSM module will be sent a message to the registered mobile number of vehicles owner and nearby police station with details of toll plaza.

7 Conclusion and Future Scope

- The proposed method will be very helpful for automatic toll collection, theft vehicle detection, and recovery.
- Proposed system will be money efficient because RFID is the very thrifty, rapid growing technology.
- This system might be very effective to decrease the vehicle theft cases in India and will be increased the recovery rate of theft vehicle.
- The RFID technology with ANPR and vehicle lift up at toll plazas will make the entire traffic smoother and hassle free.
- The hurdles of this research are to gather actual theft vehicles documentations from police departments, RTO, and National Highways.
- Proposed system could not work correctly in low network or no network areas.
- For upcoming scope, weapon detection system can also be implemented at the toll plaza like airports which will be helpful to reduce the accidental cases.

References

1. Times of India, https://timesofindia.indiatimes.com/city/chandigarh/only-13-stolen-vehicles-recovered-in-2017/articleshow/62575258.cms. last accessed 2021/10/21
2. Jakhar A, Over 5 vehicles were reported stolen every hour in Delhi during 2018: Police. https://www.news18.com/news/auto/over-5-vehicles-were-reported-stolen-every-hour-in-delhi-during-2018-police-1997933.html. Last accessed 2021/10/21

3. Bhandari H, Police data show motor vehicle theft the least solved crime. https://www.thehindu.com/news/cities/Delhi/police-data-show-motor-vehicle-theft-the-least-solved-crime/article25954331.ece. Last accessed 2021/10/21
4. http://mvt.delhipolice.gov.in/. Last accessed 2021/10/21
5. Times of India, Tracing stolen vehicle just a click away. https://timesofindia.indiatimes.com/city/hyderabad/Tracing-stolen-vehicle-just-a-click-away/articleshow/39319578.cms. Last accessed 2021/10/21
6. Gadgetsnow, 'Vahan Samanvaya' app with data on stolen vehicles across the country launched. https://www.gadgetsnow.com/apps/Vahan-Samanvaya-app-with-data-on-stolen-vehicles-across-the-country-launched/articleshow/51394172.cms. Last accessed 2021/10/21
7. Parimi J, How will electronic toll collection work in India. https://www.quora.com/How-will-electronic-toll-collection-work-in-India. Last accessed 2021/10/21
8. Mahesh B, Prabu SF, Kumar M, Balamurugan P (2016) Theft vehicle identification system in toll gate by using RFID,GSM and visual basic front end. International J Scientific Engineering Appl Science (IJSEAS) 2:2395–3470
9. Murugan K, Gobu R, Zabiyullah GS, Gunasekaran R, Santhosh V (2018) An automation of vehicle theft detection in the toll plaza by using the RF technology. IJEEE 1(3):10–17
10. Vinitha V, Velantina V (2019) Advanced automatic toll collection and vehicle detection system using internet of things. SSRG-IJEEE 6(8):5–10
11. Mekki AE, Bouhoute A, Berrada I (2019) Improving driver identification for the next-generation of in-vehicle software systems. IEEE Transactions on Vehicular Tech 68(8)
12. Mallikalava V, Vengatesan K, Kumar A, Punjabi S, Sadara SSA (2020) Theft vehicle detection using image processing integrated digital signature based ECU. 978-1-7281-5821-1/20, IEEE (2020)
13. Akhtar Z, Ali R (2020) Automatic number plate recognition using random forest classifier. Springer. Department of Computer Engineering, Aligarh Muslim University, Aligarh, India. SN Computer Science 1:120
14. Prathiba S, Viji A, Mary A (2017) Online payment of tolls and tracking of theft vehicles using number plate image. Global J Pure Applied Mathematics 13(7):3005–3012. ISSN 0973–1768
15. Raj U, Nidhi N, Nath V (2019) Automated toll plaza using barcode-laser scanning technology. Springer, Department of Electronics and Communication Engineering, Birla Institute of Technology Mesra, Ranchi 835215, India. © Springer Nature Singapore Pte Ltd. Nath V, Mandal JK (eds) (2019) Nanoelectronics, Circuits and Communication Systems, Lecture Notes in Electrical Engineering 511, https://doi.org/10.1007/978-981-13-0776-8_44
16. Mohanasundaram S, Krishnan V, Madhubala V (2019) Vehicle theft tracking, detecting and locking system using open CV. International Conference on Advanced Computing & Communication Systems (ICACCS), 978–1–5386–9533–3/19/$31.00 ©2019 IEEE, pp 1075–1078
17. Rahul JK, Gawade SS (2014) Design and development of lift for an automatic car parking system. Int J Theoretical Appl Res Mech Eng (IJTARME) 2(2):2319–3182

Determination of Antibiotic Resistance Level in *Klebsiella* using Machine Learning Models

Snehal Gupta, Sreemoyee Chatterjee, Amita Sharma, Marina Popolizio, Vincenzo Di Lecce, Mariantonietta Succi, Patrizio Tremonte, Rita Dario, and Vijay Singh Rathore

Abstract Antimicrobial drug resistance (AMR) in bacteria is a public health hazard and is growing alarmingly. There is a development of multidrug-resistant organisms due to the selective pressure exerted on organisms by drugs. Due to delay in antibiotic susceptibility testing results, artificial intelligence (AI) is employed to control the organism's resistance against the last resort drugs and speeding up the AMR detection process. Therefore, machine learning (ML), a mathematical tool for AI, is used. For this study, 6 classification ML models were used to train and forecast the resistance of β-lactam drugs in *Klebsiella pneumoniae* and were carried out on orange tool. Out of the 6 ML classifier models, KNN and random forest outperformed the remaining 4 classifiers. The purpose of this research was to develop an AI-based model to classify strains based on specific features.

Keywords Antimicrobial drug resistance · *Klebsiella pneumoniae* · Artificial intelligence · Machine learning · β-lactam drugs

1 Introduction

Antibiotics are among one of the most successful forms of therapy for fighting deadly bacteria. Sir Alexander Fleming's discovery of penicillin in 1928 provided a

S. Gupta · S. Chatterjee · A. Sharma (✉) · V. S. Rathore
IIS (deemed to be University), Jaipur, India
e-mail: amita.1983@iisuniv.ac.in

M. Popolizio · V. Di Lecce
Department of Electrical and Information Engineering, Polytechnic of Bari, Bari, Italy

M. Succi
Department of Food Microbiology, University of Molise, Campobasso, Italy

P. Tremonte
Department of Agricultural, Environmental and Food Science, University of Molise, Campobasso, Italy

R. Dario
Medical Management, Hospital Medical Management, Polytechnic of Bari, Bari, Italy

X.-S. Yang et al. (eds.), *Proceedings of Seventh International Congress on Information and Communication Technology*, Lecture Notes in Networks and Systems 447,
https://doi.org/10.1007/978-981-19-1607-6_80

new dimension to the world of antibiotics. Penicillin was successful in controlling bacterial infections among World War II soldiers. Antibiotics not only save patient's lives, but it is also a boon for medicine and surgery. It helped in preventing and treating infections in patients with comorbidity like diabetes, end-stage renal disease, rheumatoid arthritis, or patients undergoing chemotherapy treatments, or the patients who have undergone organ transplants, joint replacements, or cardiac surgery. In countries where sanitation is still poor, antibiotics help in decreasing morbidity and mortality. Although antibiotics have such a great positive impact but efficiency of antibiotics is compromised by a growing number of antibiotic-resistant pathogens. Antimicrobial resistance (AMR) is a natural feature of microbial ecosystems. AMR is the ability of a microorganism to stop the medication from working against it [1]. AMR in bacteria is the major threat in controlling bacterial infections [2]. Antibiotic resistance, a reason for elevated morbidity and mortality rates as well as in the increased treatment costs, has become a one of the major global public health hazards so much so that, first ever list of antibiotic-resistant "priority pathogens" has been recently published by WHO. It is a catalog of 12 families of bacteria which pose the greatest danger to human health (*antimicrobial resistance*, no date).

The main causes of AMR include overuse of antibiotics, inappropriate prescribing, and extensive agricultural use. There is a global antibiotic drug pressure that is due to unnecessary prescriptions in medical settings and extensive agricultural use [3].

In response to this resistance, new β-lactam antibiotics were discovered and used. Unfortunately, during that same decade, the first case of methicillin-resistant *Staphylococcus aureus* (MRSA) was identified in the UK in 1962 and the US in 1968 [3]. Over the years, due to selective pressure of drugs, organisms have developed different kinds of resistance mechanisms that have led to the development of multidrug-resistant organisms (MDROs). The problematic MDROs include *Pseudomonas aeruginosa, Acinetobacter baumannii, Escherichia coli*, and *Klebsiella pneumonia* possessing extended-spectrum beta-lactamases (ESBL), vancomycin-resistant *enterococci* (VRE), methicillin-resistant *Staphylococcus aureus* (MRSA) etc. [4].

Among these MDROs, *Klebsiella pneumoniae*, a member of the Enterobacteriaceae family, which is the natural inhabitant of the GI tract microbiota of healthy humans and animals, is one of the common opportunistic hospital-associated pathogen and accounts for about one-third of all of the gram-negative bacterial infections in general. Since these pathogens are part our microbiome, they appear as a major concern. These infections result in high mortality rate, increased duration of hospitalization and therefore the higher cost [5]. In the era of antibiotic resistance, *Klebsiella pneumoniae* is one of the significant pathogens implicated in bacterial resistance to antibiotics, and it is categorized as an ESKAPE bacterium, along with other key multidrug-resistant pathogens [6]. Due to the limited therapeutic treatment options as a result of antibiotic resistance, it has become a troublesome process to control diseases caused by Enterobacteriaceae.

Members of Enterobacteriaceae family like *K. pneumoniae* and *E. coli* produce extended-spectrum β-lactamases (ESBLs). ESBLs are resistance mechanisms that have been introduced into gram-negative bacilli [7]. ESBLs are a group of enzymes that cause resistance to a variety of lactam antibiotics, including Aztreonam, Ceftazidime, Cefotaxime, related Oxyimino—lactams, cephalosporins, and penicillins but are inhibited by Clavulanic acid. Patients with extended hospital stays in intensive care units in Europe demonstrated a high level of resistance and were the first cases of resistance to be observed. As the isolates were discovered in Africa, Asia, the Middle East, and South and North Americas, ESBL GNB quickly became a global problem, causing fear [8]. ESBL-producing bacteria effectively hydrolyze lactam antibiotics, which include broad-spectrum lactam drugs and monobactams, excluding cefamycins and lactam inhibitors.

The β-lactamases are the hydrolytic enzymes that extend bacterial resistance to β-lactam antibiotics, such as the penicillin, cephalosporin, and carbapenem families which are the common antimicrobial drugs that are used all around the world. This particular class of antibiotics makes up to 65% of the total antibiotics in the market. β-lactam antibiotics potrays the most common drug class of antimicrobial drugs which have broad clinical indications [9].

Common ESBL genes responsible for β-lactam resistance for isolates of *K. pneumoniae* were chosen for this study viz., blaKPC-2 (*K. pneumoniae* carbapenemases), blaSHV-11, and blaSHV-12 (sulphydryl variable reagent), blaCTX-M-65 (cefotaximase that preferentially hydrolyzes cefotaxime).

For the AMR treatment, the organism's susceptibility or resistance against the available drugs is checked. Traditionally, it is predicted through disk diffusion or minimum inhibitory concentration (MIC) of the antibiotics [10]. PCR and microarray help in verifying phenotypic results along with identifying multiple genes each of which is different and can be responsible for AMR. But, the results of these antibiotic susceptibility testing (AST) took at least 2 days to arrive, which steer to the administration of broad-spectrum antibiotics. To control the organism's resistance against the last resort drugs and speeding up the AMR detection process, artificial intelligence (AI) is employed.

Artificial intelligence (AI) is the developing field of computer science where the computer system performs the tasks that require human intelligence like decision-making, recognition, and processing of speech and language and visual perception. AI helps in establishing a decision by studying the patterns in the training data and recognizing these same patterns in the new data (test data). AI data mining can help in detecting the AMR infection which otherwise is a tedious and time demanding task for the epidemiological investigations. For AMR detection, next-generation sequencing is used [11]. These include identification of resistant determinants like single nucleotide polymorphism (SNPs) or from whole-genome sequencing (WGS) for a better understanding of the AMRs biological mechanism.

WGS employs the entire genome of the organism to identify the entire set of known resistance factors in an organism, as well as target-mediated resistance. It can be used to identify new resistance gene variants, which aids in the identification of novel resistance gene families. In WGS, neither the DNA preparation nor the resistance gene identification steps contribute to an increase in time or cost as the number of antibiotics under evaluation expands. Pathogen WGS has the potential for a single test to be employed for numerous reasons like for species identification and assessment of strain relatedness in clinical diagnostics. WGS has many potential benefits but apart from that, it has many limitations that need to be trimmed before it can be even used as a possible alternative to in vitro phenotypic tests. WGS has several drawbacks, including a slower speed and higher cost than traditional phenotypic susceptibility testing, as well as the possibility of inaccuracies because antibiotic resistance genes are selected directly via the technique, which speeds up the analysis time and thus increases the risk of error. Simple gene identification only gives us an idea about the antibiotic that may be resistant to an infection; it does not explain the treatment that will be needed. Transferable resistance to last resort drugs like colistin and carbapenems in existing multidrug-resistant bacteria has resulted in the emergence of practically pan-resistant isolates. Therefore, it is very important to identify precisely any antibiotic susceptibility the isolate may still have. For all these challenges machine learning is employed to make the task easier, faster and can make accurate predictions from the complex dataset [12].

Machine learning (ML) is a mathematical tool for AI, and its main focus is to mimic the intelligence property of the human brain. ML can easily work in the presence of large data by learning new rules without human interference which makes it less fragile and more independent of human expertise [13]. Since the clinical data available are vast, ML helps in classifying the data by knocking out the uninformative features and retaining the suitable one that helps in our prediction [14]. In data mining, classification of data is a popular tool of machine learning technique. Classification is the learning method that aids in classifying the data into class labels which are already defined in the provided dataset. Classification is a process constituting two-step viz., learning and classification. In first step, i.e., the learning step, a classification model is constructed from the given dataset, and such dataset is referred as the training set that is helpful in learning of the classification model. In the classification step, the class labels in a separate unknown given data are tested via the models. The dataset which helps in testing the classification models is regarded as the training set. For classification different algorithms are available viz., random forest, Naïve Bayes, support vector machine (SVM), K-nearest neighbors (KNN), neural network, logistic regression, etc. [15].

For data mining, orange a graphical user interface (GUI)-based tool is used. This tool helps in visualizing patterns and understanding data without learning to code. Orange is a data mining platform that works on GUI-based workflow. Various tasks can be executed in orange starting from basic visuals to data manipulation, transformation, and data mining. A single workflow covers all the functions for an entire process. Apart from its flexibility and ease to use, there are various data exploration widgets. It helps in selection of experiment, predictive modeling, and recommendation systems. It is used in the field of genomic research, biomedicine, bioinformatics, and teaching [16].

2 Methodology

The research methodology in this research has five phases.

Stage 1: Data collection

The major steps followed in this stage are as follows:

1. Meta-data for *Klebsiella pneumoniae* were collected from the National Center for Biotechnology Information (NCBI) by applying *Klebsiella* as the keyword filter and further filtering the data via AMR genotype.
2. Total 1890 matched clusters were retrieved after applying the desired filters.
3. Out of these matched clusters, the csv file of the first 2 clusters was downloaded having SNP cluster ID as PDS000045296.40 and PDS000048063.38.
4. The dataset consists of organism group, SNP cluster, species TaxID, AMR genotype, isolation source, isolation type, assembly, and other factors.
5. Drug data of beta-lactam drugs were collected from Resfinder. Many antimicrobial resistant (AMR) genes were listed under the β- lactam drug. 4 genes viz., blaKPC-2, blaCTX-M-65, blaSHV-11, and blaSHV-12 were selected from this list of AMR genes.

Stage 2: Data Pre-processing

Further, the raw data are refined to make it suitable for machine learning. Two clusters of *Klebsiella pneumoniae* were used to prepare the test dataset and the training dataset. There are a total of 1233 samples in the dataset prepared using 2 clusters. The dataset table has the attributes –organism, strain, assembly, BlaKPC-2, BlaSHV-11, BlaSHV-12, and BlaCTX-M-65. The dataset was prepared by converting all the information under the gene's classes in binary form. Here, 1 indicates the presence of the gene, and 0 indicates the absence of the gene in a strain. Resistance level is defined as:1 in the resistance level means all 4 genes are present in that strain, 2 means any 3 genes are present, 3 means any 2 genes are present, and 1 means only 1 gene is present.

Stage 3: Data Splitting

1. The prepared dataset was then divided in two subsets: training dataset and testing dataset. Training dataset that is used to fit the model for ML and testing dataset which is used to evaluate the fitted model.
2. Out of 1233 samples, we prepared a training dataset containing 200 samples.
3. The testing dataset was prepared via two ways viz., balanced and unbalanced dataset.
4. The balanced dataset contained 40 samples, and the unbalanced dataset contained 30 samples.

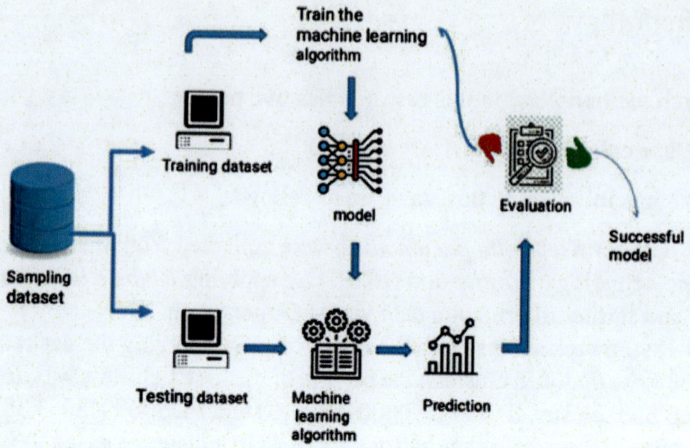

Fig. 1 Flowchart depicting stages of sampling dataset to get final model using machine learning

Stage 4: Model Framing

Selection of model and tool

1. Orange data mining tool was used for data analysis.
2. 6 classification models were chosen viz., SVM model, KNN model, Naive Bayes model, random forest model, neural network model, and logistic regression model.

Implementation of model and experiments

1. On the orange workflow stage, we loaded the class labeled dataset, and resistance level was set as the target variable.
2. 6 supervised learning models were loaded on the stage and scored in cross validation.
3. The central widget for training dataset was test and score widget and for testing dataset was predictions.
4. ROC curve and confusion matrix widget were also loaded and linked to the central widget for additional analysis of cross validation results.

Figure 1 shows the flowchart of machine learning model framing.

Stage 5: Result Analysis—Results obtained from the models are analyzed, and comparative study is conducted.

3 Algorithms for Model Framing

The algorithms considered for study are K-nearest neighbor (KNN), Naïve Bayes, logistic regression (LR), support vector machine (SVM), random forest (RF), and artificial neural networks. The simplest classification algorithm is K-nearest neighbor, which is a type of case-based learning, also known as lazy learning. It considers local approximation and suspends all computations until classification. Naive Bayes is commonly used algorithm based on the probability. This classifying algorithm is based on a fundamental probability concept that predicts the likelihood of class membership. The attributes on each class in Naive Bayes are independent of each other, which is known as class conditional independence. Like Naïve Bayes, logistic regression (LR) is also a probability-based classification algorithm. It uses a logistic function also known as sigmoid function for calculating probability. Its function is limited between 0 and 1. For speculating analytics, support vector machine (SVM) classification algorithm is used. It mainly works as the binary classifier, i.e., to infer the information by their class either 0 or done via finding a hyperplane that aids in separating the d-dimensional data into two classes. Linearly inseparable data can be separated linearly by mapping them into a higher dimensional space [17]. Random forest (RF) is another classification algorithm, where several learning algorithms are used together for result prediction. RF forms a group of decision trees that showcase controlled variation by selecting a combination of bootstrap aggregation (bagging) and random features. Artificial neural networks (ANNs) are a computational model-based algorithms that work on the basic principle of neural networks similar to the one found in the human brain. Backpropagation is some of the famous neural network algorithm that learns on a multilayer feed-forward neural network. These algorithms are versatile in nature and frequently used in medical datasets.

4 Results and Discussion

4.1 Training Dataset

Figure 2 depicts a flowchart of orange tool for designing ML models. As already mentioned, the target variable is resistance level, and independent variables are BlaKPC-2, BlaSHV-11, BlaSHV-12, and BlaCTX-M-65. Our aim was to determine highly resistant genes with respect to beta-lactam drugs (Fig. 3).

The 6 ML models were built using tenfold cross validation with 66% training dataset. The performance of each model is compared using area under curve (AUC), cumulative accuracy (CA), F1 score, which is the weighted harmonic mean of precision and recall. Figure 4 represents the learning performance of KNN, SVM, random forest, ANN, Naive Bayes, and logistic regression for gene dataset. According to the results, SVM dominates with AUC score of 100% and 95% score each in CA, F1, precision, and recall. It also states that the dataset size is sufficient enough for training.

Figure 5 provides the confusion matrix generated by each model. According to the results of the confusion matrix, SVM, neural network, and logistic regression are able to predict the antimicrobial resistance more accurately. In all the four categories, 5 out of 6 ML models find difficulty in classifying category 2, whereas 3 out of 6 finds

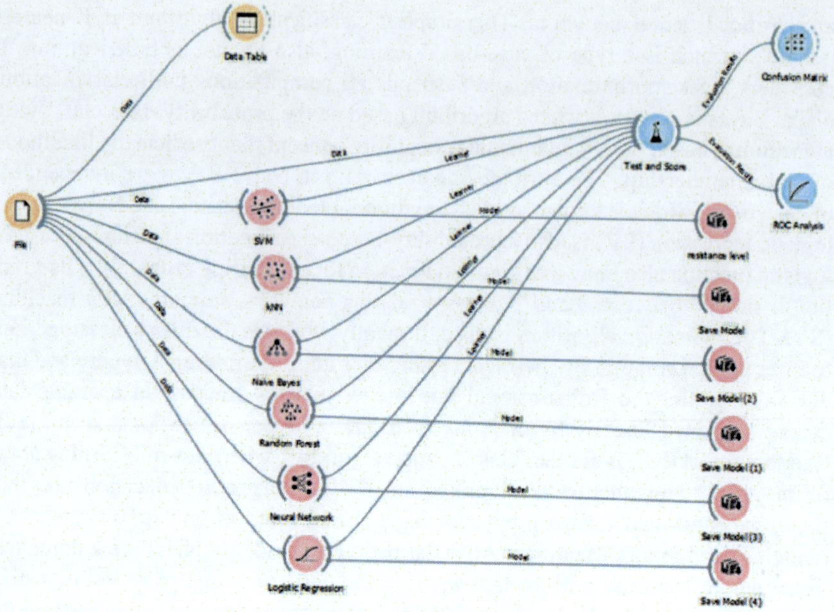

Fig. 2 Flowchart of workflow in orange tool for training dataset

Fig. 3 Performance comparison of 6 ML models

Evaluation Results

Model	AUC	CA	F1	Precision	Recall
kNN	0.997	0.975	0.975	0.977	0.975
SVM	1.000	0.995	0.995	0.995	0.995
Random Forest	0.997	0.990	0.990	0.990	0.990
Neural Network	0.997	0.995	0.995	0.995	0.995
Naive Bayes	0.991	0.930	0.929	0.936	0.930
Logistic Regression	0.995	0.995	0.995	0.995	0.995

difficulty in classifying category 4. Naive Bayesian classification model has difficulty in classifying category 3 and 4. This misclassification has reduced its performance. In case of training datasets, SVM, neural network, and logistic regression are the best performers.

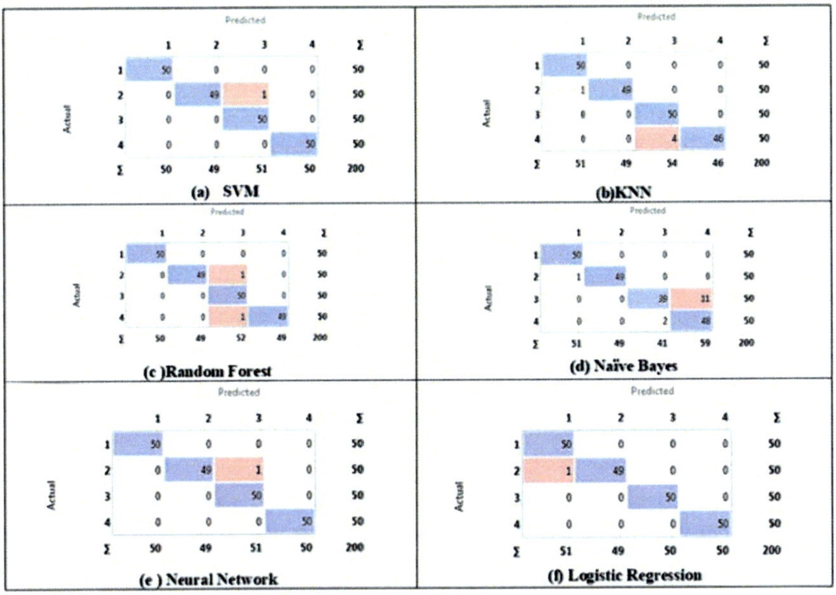

Fig. 4 Confusion matrix for all the ML models

Fig. 5 Comparison of prediction of different algorithms chosen

Model	AUC	CA	F1	Precision	Recall
SVM	1.000	1.000	1.000	1.000	1.000
kNN	1.000	1.000	1.000	1.000	1.000
Random Forest	1.000	1.000	1.000	1.000	1.000
Naive Bayes	1.000	0.950	0.949	0.958	0.950
Neural Network	1.000	1.000	1.000	1.000	1.000
Logistic Regression	1.000	1.000	1.000	1.000	1.000

4.2 Testing Dataset

A balanced datasets were provided to previously created 6 ML models for testing. In this case, we have provided 40 test dataset to 6 trained ML models. Figure 7 depicts the score of all the 6 ML models.

All the 6 models showed 100% accuracy. According to the prediction results, all the algorithms except Naive Bayes showed 100% score. According to the confusion matrix generated for each ML model depicted by Fig. 6, manifests that out of 6 ML models only Naive Bayes found difficulty in classifying category 3. Figure 7 shows values for AUC from the ROC curve that are plotted between false positive and false negative. It is summarized as follows-AUC for SVM: 0.958, AUC for KNN:

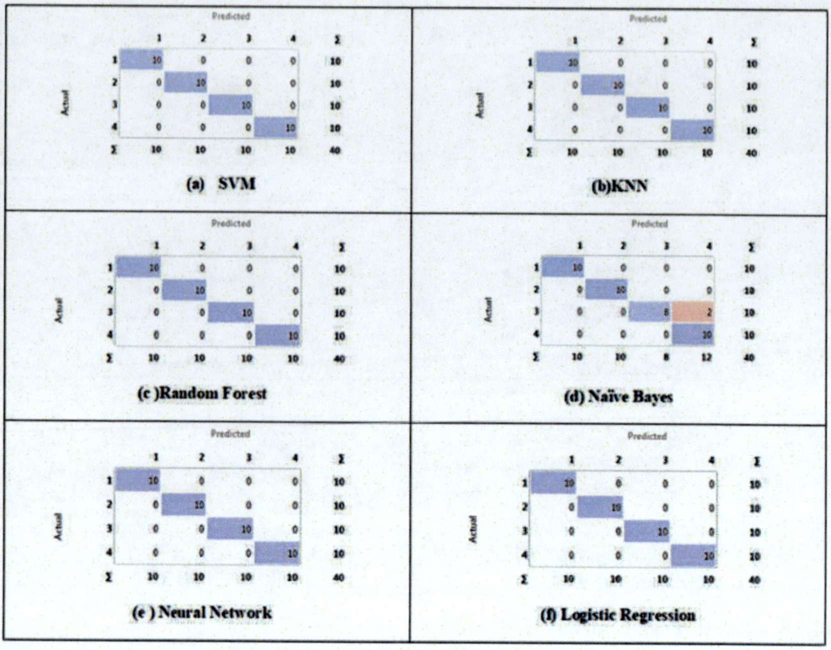

Fig. 6 Confusion matrix for 6 ML models

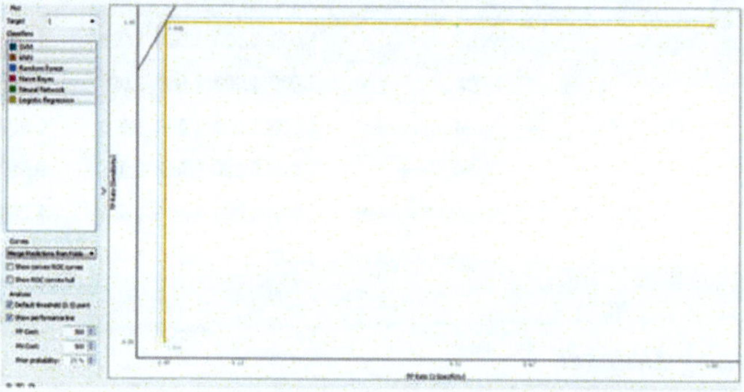

Fig. 7 Receiver operating characteristics curve (ROC) of the antimicrobial resistance dataset in *Klebsiella*

1.000, AUC for random forest: 1.000, AUC for Naive Bayes: 0.778, AUC for neural network: 0.971, and AUC for logistic regression: 0.882.

Cumulative Summary of Test Case Study: Algorithms like KNN and random forest gave highly accurate results in comparison with others. These experiments helped to finalize the best classification model for *Kp* strain resistant study.

5 Conclusions and Future Work

In this research, we have designed an ML model to classify highly resistant *Kp* strains to β-lactam antibiotics. In the dataset, four proteins are considered: i.e., blaKPC-2, blaSHV-11, blaSHV-12, and blaCTX-M-65, and these genes are responsible for the resistance against β-lactam antibiotics. For testing the model, balanced dataset is used. 6 ML algorithms are used to construct the resistance model for antimicrobial drug resistance in *Kp*. Out of these 6 algorithms; KNN and random forest are better and highly accurate in predictions than the remaining 4 algorithms. Both the outperformed algorithms have AUC-ROC score as 1.000 and precision of 100% for balanced test dataset.

The purpose of this research is to develop an AI-based model to classify strains based on features like presence of certain resistant protein source of sample and strain assembly. Further enhancement of the predictive model can help in identifying high-risk patients and implementing more targeted treatment for the AMR before it becomes life threatening.

References

1. Shi J et al (2019) Antimicrobial resistance genetic factor identification from whole-genome sequence data using deep feature selection. BMC Bioinformatics 20(15):535. https://doi.org/10.1186/s12859-019-3054-4
2. Liu Z et al (2020) Evaluation of machine learning models for predicting antimicrobial resistance of Actinobacillus pleuropneumoniae from whole genome sequences. Front Microbiol 11:48. https://doi.org/10.3389/fmicb.2020.00048
3. Ventola CL (2015) The antibiotic resistance crisis: part 1: causes and threats. P & T: A Peer-Reviewed J Formulary Management 40(4):277–283
4. Alekshun MN, Levy SB (2007) Molecular mechanisms of antibacterial multidrug resistance. Cell 128(6):1037–1050. https://doi.org/10.1016/j.cell.2007.03.004
5. Navon-Venezia S, Kondratyeva K, Carattoli A (2017) Klebsiella pneumoniae: a major worldwide source and shuttle for antibiotic resistance FEMS Microbiol Rev 41(3)252–275 https://doi.org/10.1093/femsre/fux013
6. Boucher HW et al (2009) Bad bugs, no drugs: no ESKAPE! An update from the Infectious Diseases Society of America. Clinical Infectious Diseases: An Official Publication of the Infectious Diseases Society of America 48(1):1–12. https://doi.org/10.1086/595011
7. Ojdana D et al (2014) The Occurrence of blaCTX-M, blaSHV, and blaTEM Genes in Extended-Spectrum β-Lactamase-Positive Strains of Klebsiella pneumoniae Escherichia coli, and Proteus mirabilis in Poland. Int J Antibiotics 2014:e935842. https://doi.org/10.1155/2014/935842
8. Pishtiwan AH, Khadija KM (2019) Prevalence of blaTEM, blaSHV, and blaCTX-M Genes among ESBL-Producing Klebsiella pneumoniae and Escherichia coli Isolated from Thalassemia Patients in Erbil, Iraq Mediterranean. J Hematology Infecti Diseases 11(1):e2019041. https://doi.org/10.4084/MJHID.2019.041
9. Avershina E et al (2021) AMR-daig: neural network based genotype-to-phenotype prediction of resistance towards β-lactams in E.coli and K. pneumoniae. Computational Struct Biotech J 19:1896–1906. https://doi.org/10.1016/j.csbj.2021.03.027
10. Camp PJ, David BH, Porollo A (2020) Prediction of antimicrobial resistance in gram-negative bacteria from whole-genome sequencing data. Front Microbiol 11(1013). https://doi.org/10.3389/fmicb.2020.01013

11. Fitzpatrick F, Doherty A, Lacey G (2020) Using artificial intelligence in infection prevention. Curr Treat Options Infect Dis 12:135–144. https://doi.org/10.1007/s40506-020-00216-7
12. Pesesky MW et al (2016) Evaluation of machine learning and rules-based approaches for predicting antimicrobial resistance profiles in gram-negative Bacilli from whole genome sequence data. Front Microbiol 7:1887. https://doi.org/10.3389/fmicb.2016.01887
13. Fanelli U et al (2020) Role of artificial intelligence in fighting antimicrobial resistance in pediatrics, 9(11):767. https://doi.org/10.3390/antibiotics9110767
14. Bhargava H, Sharma A, Valadi JK (2021) Machine learning for bioinformatics. In: Suravajhala PN (eds) Your passport to a career in bioinformatics. Springer, Singapore. https://doi.org/10.1007/978-981-15-9544-8_11
15. Sarker I, Kayes ASM, Watters P (2019) Effectiveness analysis of machine learning classification models for predicting personalized context-aware smartphone usage. J Big Data 6. https://doi.org/10.1186/s40537-019-0219-y
16. Sharma A, Jain S, Chatterjee S (2021) Applications of machine learning algorithms in cancer diagnosis. In: Saxena A., Chandra S. (eds) Artificial intelligence and machine learning in healthcare. Springer, Singapore. https://doi.org/10.1007/978-981-16-0811-7_8
17. Hosseinzadeh H, Nassiri-Asl M (2013) Avicenna's (Ibn Sina) the Canon of Medicine and saffron (Crocus sativus): a review. Phytotherapy Res: PTR 27(4):475–483. https://doi.org/10.1002/ptr.4784

Author Index

A
Abdulkader, Omar Ahmed, 75
Abdurrahman, Usman Alhaji, 645
Abid, Dm. Mehedi Hasan, 145
Abir, Ahasanul Haque, 839
Abouelatta, Mohamed, 677
Abu Bakar Sajak, Aznida, 431
Acevedo-Ruiz, Araceli Margarita, 753
Afroz, Atia, 707
Ahmad, Nur Zaimah, 431
Ahmed, B. M., 607
Ahmed, Suhaib, 493
Ajayi, Makinde Oluwafemi, 389
Akhloufi, Moulay, 415
Akwafuo, Sampson, 819, 829
Aleksandrova, Marina, 93
Aleksandrova, Svetlana, 93
AL Ghadmi, Bandar Ali Alrami, 75
Alikina, Elena, 657
Alim Al Islam, A. B. M., 493
Ali, Md. Yeasin, 493
Alwarhi, Ahmad Abdulaziz, 75
Aman, Reem, 213
Andrews, Edward G., 191
Annuar, Nursyamilah, 19
Arame, Maki, 889
Ayala-Inca, Carlos, 349

B
Babichenko, Dmitriy, 191
Benaddy, Mohamed, 415
Blankinship, Zachary, 507
Botha, Adele, 797
Boukdir, Abdelbasset, 415
Burkard, Simon, 311
Burlaka, S. A., 181
Bustamante-Gonzales, Jenifer Diana, 423

C
Cabanillas-Carbonell, Michael, 339, 349
Canbaloğlu, Gülay, 223
Canton, Stephen P., 191
Casillo, Mario, 461
Chan, Eang Teng, 359
Chatzopoulos, Avraam, 633
Chen, Yongfei, 597
Chipangura, Baldreck, 517
Choi, HeeSeok, 557
Choi, Seiwoong, 557
Choi, Sing, 369
Chondrogiannis, Elefterios, 633
Christoph, Uran, 379
Chumpitaz-Caycho, Hugo Eladio, 423
Chung, Kwang Sik, 557
Colace, Francesco, 471
Cordova-Buiza, Franklin, 423
Cuesta-Chiriboga, Adriana, 743

D
Dahal, Binay, 369
Dario, Rita, 913
Daud, Dazmin, 19
De Belie, Frederik, 155
Delport, Petrus M. J., 257
De Santo, Massimo, 461
Didych, V. M., 181
Diouf, Pape Made, 57
Do, Lam Thanh, 297
Doll, Konrad, 687

X.-S. Yang et al. (eds.), *Proceedings of Seventh International Congress on Information and Communication Technology*, Lecture Notes in Networks and Systems 447,
https://doi.org/10.1007/978-981-19-1607-6

Dzator, Janet, 213

E
Effendy, Veronikha, 761
Egger, Valentin, 379
Ekogha, Ely Ondo, 537
Elsayed, Elsayed A., 849

F
Faizar, Faizah Ahmad, 717
Fall, Ibrahima, 57
Fuada, Syifaul, 275
Fuchs-Kittowski, Frank, 311
Fujiyanti, Vina, 275
Fukuhara, Yoshihisa, 401

G
Georgiev, Georgi, 203
Goda, Yoshiko, 889
Gris, Olga, 547
Grobbelaar, Sara S., 797
Guida, Caterina Gabriella, 471
Gulova, Shirin, 251
Guo, Weihong, 849
Gupta, Brij, 471
Gupta, Snehal, 913

H
Halima, Donia Ben, 699
Handa, Junko, 889
Han, Ying, 771
Haruna, Usman, 645
Hassan, Mohammed Ziad, 839
Hassan, Shah, 443
Hernández-Freire, Alba, 743
Horvath, Kurt, 379
Hossain, Mohammad-Sahadet, 707
Hossain, Muhammad Iqbal, 493, 839
Hossain, Musannan, 707
Hossain, Ramim, 145
Hsieh, Pi-Jung, 615
Hubert, Andreas, 687
Hussein, Nebras, 699

I
Innes, Mia, 137
Ipanaqué-Chero, Robert, 753
Islam, Tariqul, 145
Itoh, Makoto, 889

J
Jagadeesh Anmala, 525
Javed, Khalid, 155
Jiban, Md Jibanul Haque, 443
Jiménez-Vilcherrez, Judith Keren, 753
Joo, Tang Mui, 285
Jung, Janis, 687

K
Kalogiannakis, Michail, 633
Kalovrektis, Konstantinos, 633
Kardouchi, Mustapha, 415
Kedar P. Vaidya, 729
Khudaiberdina, Marina, 657
Kimberling, Amber, 819
Kisliy, Nikolay, 93, 251
Kitazaki, Satoshi, 889
Kochemasova, Tatiana, 93
Kopyltsov, Alexander, 85
Kristaly, Dominic Mircea, 323
Ksantini, Mohamed, 699
Kumawat, Kamlesh, 901
Kurnikova, Irina, 93, 251
Kushnina, Lyudmila, 657
Kusumo, Dana Sulistiyo, 761

L
Labaze, Dukens, 191
Lahyani, Rahma, 99
Lai, Hui-Min, 615
Laksitowening, Kusuma Ayu, 761
Lampert, Pascal, 687
Laseinde, Opeyeolu Timothy, 389
Lecce Di, Vincenzo, 913
Lei, Rong, 849
Lê, Ngoc C., 297
Lennon, Ruth G., 567
Littleton, Eliza Beth, 191
Lombardi, Marco, 461
López, William Manuel Montalvo, 451
Lorusso, Angelo, 471
Louati, Ali, 99
Lou, Jingjing, 597
Lukyanov, Gennadi, 85
Luo, Jie, 771

M
Mahalanobis, Abhijit, 443
Mahmood, Sohaib A., 699
Malemath, V. S., 127
Manolescu, Vasile Denis, 39

Mansi A. Radke, 729
Marongiu, Francesco, 471
Maruma, Charles, 871
Masmoudi, Fatma, 99
Matsuba, Ryuichi, 889
Mayayise, Thembekile, 665
Mbodji, Senghane, 57
Mclean, Gareth, 783
Meenakshi Sundaram, K., 127
Meslouhi, Othmane El, 415
Miah, Shah J., 213
Mia, Kausar, 145
Mills, Andrew, 191
Mohd Nor, Antashah, 19
Moraru, Sorin-Aurel, 323
Mosca, Rosalba, 461
Mousa, Mohamed, 677
Mushayi, Precious, 665

N
Nawej, Claude, 871
Nedyalkov, Ivan, 203
Ngo, Giang Nam, 297
Nguyen, Tung Dinh, 297
Nikiforov, Igor, 171
Noor, Jannatun, 493
Nordin, Ariza, 717
Nosseir, Ann, 1

O
Om Prakash, 241
Ovchynnykov, K. V., 181
Owolawi, Pius A., 537

P
Palacios-Loayza, Jesus, 349
Papoutsidakis, Michail, 633
Patel, Ravi, 191
Peralta, Mike, 829
Perlova, Irina, 657
Permiakova, Kristina, 657
Phung, Hieu Trong, 297
Pitogo, Vicente A., 587
Ponce, Darío Fernando Yépez, 451
Ponce, Héctor Mauricio Yépez, 451
Popolizio, Marina, 913
Porras-Angulo, Alicia, 743
Porras-Quispe, Johana, 743
Pshehotskaya, Ekaterina, 483
Psycharis, Sarantos, 633

R
Rahman, Tanvir, 145
Razzaq, Ali Abdul, 699
Ruyobeza, Barimwotubiri, 797

S
Sadio, Amy, 57
Saeed, Ahmed, 607, 677
Said, Ghada Refaat El, 107
Salah, Mostafa M., 677
Saleh, Samar, 849
Salih, Nazar, 699
Salman, Ahmad, 507
Sanjana, G. B., 29
Santaniello, Domenico, 461, 471
Sanya A. Chetwani, 729
Sargar, Ramchandra, 251
Sari, Nadia Tiara Antik, 275
Sayed, M., 607
Secco, Emanuele Lindo, 39, 137
Selviandro, Nungki, 761
Serizawa, Hinata, 401
Serov, Igor, 85
Sevastianov, V. M., 181
Shaker, Ahmed, 677
Sharma, Amita, 913
Shemyakinskaya, Anastasia, 171
Shikha Yadav, 241
Shivali Agarwal, 29
Shoukralla, E. S., 607
Siddika, Ayesha, 839
Singh, Madhusudan, 879
Sosnovskaya, Anna, 547
Sow, Papa Lat Tabara, 57
Sreejeth, Mini, 879
Sreemoyee, Chatterjee, 913
Steyn, Adriana A., 783
Sufola Das Chagas Silva E Araujo, 127
Sundar Guntnur, 29
Succi, Mariantonietta, 913
Sun, Zhubing, 597
Surco-Anacleto, Juan, 339
Su, Xin, 771

T
Taghva, Kazem, 369
Talip, Bazilah A., 431
Talukdar, Md. Asif, 839
Tang, Mui Joo, 359
Teng, Chan Eang, 285
Timosan, Jesterlyn Q., 587
Toda, Masashi, 889

Tran, Trung Thanh, 297
Tremonte, Patrizio, 913
Treur, Jan, 223
Tu, Chunling, 871
Turuganti, Venkateswarlu, 525

V
Valentino, Carmine, 461
Vandevelde, Lieven, 155
Vasilevskyi, O. M., 181
Velásquez-Fernández, Felicita Marcela, 753
Velezmoro-León, Ricardo, 753
Verma, Monika, 879
Vijay Singh Rathore, 901, 913
Vinnikov, Vladimir, 483
Von Solms, Rossouw, 257
Vu, Anh Tuan, 297

W
Wasek, Mashrur, 707
Wiewiora, Anna, 223
Wöllik, Helmut, 379

X
Xenakis, Apostolis, 633

Y
Yu, Xujiang, 597

Z
Zaifri, Lutfil Hadi, 431
Zaman, Zahura, 145
Zavalina, Maria, 93
Zekry, A., 677
Zeng, Jie, 771
Zheng, Lirong, 645
Zheng, Pengfei, 597
Zhou, Huiping, 889